CW00709311

J-Contractive Matrix Valued Functions and Related Topics

J-Contractive Matrix Valued Functions and Related Topics

Damir Z. Arov
Department of Mathematical Analysis
South-Ukranianian Pedagogical University
65029 Odessa, Ukraine
email arov_damir@mail.ru

and

Harry Dym
Department of Mathematics
The Weizmann Institute of Science
Rehovot 76100, Israel
email harry.dym@weizmann.ac.il

CAMBRIDGE UNIVERSITY PRESS
Cambridge, New York, Melbourne, Madrid, Cape Town, Singapore, São Paulo, Delhi

Cambridge University Press
The Edinburgh Building, Cambridge CB2 8RU, UK

Published in the United States of America by Cambridge University Press, New York

www.cambridge.org
Information on this title: www.cambridge.org/9780521883009

First published 2008

Printed in the United Kingdom at the University Press, Cambridge

A catalogue record for this publication is available from the British Library

ISBN 978-0-521-88300-9 hardback

Dedicated to our wives Natasha and Irene, for their continued support and encouragement, and for being ideal companions on the path of life.

Contents

Preface

This book is one of the products of the joint work of the authors over the last 15 years. We first met at the IWOTA and MTNS conferences in Japan in 1991. At the time one of us knew very little English and the other's Russian was limited to *da* and *net*. Fortunately, our mutual friend and colleague Israel Gohberg was able and ready to act as an interpreter, and tentative arrangements were made for the first author to visit the second at the Weizmann Institute. These visits were repeated each year for three or more months, and a research program that focused on direct and inverse problems for canonical systems of integral and differential equations was initiated. This program made extensive use of the existing theory of J-contractive and J-inner mvf's (matrix valued functions) as well as requiring new developments. This monograph is a comprehensive introduction to that theory, and a number of its applications. Much of it is either new, or not conveniently accessible. A second volume on canonical systems of integral and differential equations is planned.

The authors gratefully acknowledge and thank: Victor Katsnelson for supplying helpful remarks and a number of references; Ruby Musrie, Diana Mandelik, Linda Alman and Terry Debesh for typing assorted sections; Clare Lendrem for her careful copy editing of a not so final draft; the administration of South Ukranian Pedagogical University for authorizing extended leaves of absence to enable the first author to visit the second and finally, and most importantly, the Minerva Foundation, the Israel Science Foundation, the Arthur and Rochelle Belfer Institute of Mathematics and Computer Science, and the Visiting Professorship program at the Weizmann Institute for the financial support that made these visits possible and enabled the authors to work together under ideal conditions.

May 26, 2008
Rehovot. Israel

1
Introduction

This book is devoted to the theory of J-contractive and J-inner **mvf's** (matrix valued functions) and a number of its applications, where J is an $m \times m$ signature matrix, i.e., J is both unitary and self adjoint with respect to the standard inner product in $\mathbb{C}^m$. This theory plays a significant role in a number of diverse problems in mathematical systems and networks, control theory, stochastic processes, operator theory and classical analysis. In particular, it is an essential ingredient in the study of direct and inverse problems for canonical systems of integral and differential equations, since the matrizant (fundamental solution) $U_x(\lambda) = U(x, \lambda)$ of the canonical integral equation

$$u(x, \lambda) = u(0, \lambda) + i\lambda \int_0^x u(s, \lambda) dM(s) J, \quad 0 \leq x < d, \tag{1.1}$$

based on a nondecreasing $m \times m$ mvf $M(x)$ on the interval $0 \leq x < d$ is an entire mvf in the variable λ that is J-inner in the open upper half plane $\mathbb{C}_+$ for each point $x \in [0, d)$:

(1) $U_x(\lambda)$ is J-contractive in $\mathbb{C}_+$:

$$U_x(\lambda)^* J U_x(\lambda) \leq J \quad \text{for} \quad \lambda \in \mathbb{C}_+$$

and

(2) $U_x(\lambda)$ is J-unitary on the real axis $\mathbb{R}$:

$$U_x(\lambda)^* J U_x(\lambda) = J \quad \text{for} \quad \lambda \in \mathbb{R}.$$

Moreover, $U_x(\lambda)$ is monotone in the variable x in the sense that

$$U_{x_2}(\lambda)^* J U_{x_2}(\lambda) \leq U_{x_1}(\lambda)^* J U_{x_1}(\lambda) \quad \text{if} \quad 0 \leq x_1 \leq x_2 < d$$

and $\lambda \in \mathbb{C}_+$. These properties follow from the fact that the matrizant $U_x(\lambda) = U(x, \lambda)$ is a solution of the system (1.1) with $U_0(\lambda) = I_m$, i.e.,

$$U(x, \lambda) = I_m + i\lambda \int_0^x U(s, \lambda) dM(s) J, \quad 0 \leq x < d,$$

and hence satisfies the identity

$$U_{x_2}(\lambda) J U_{x_2}(\omega)^* - U_{x_1}(\lambda) J U_{x_1}(\omega)^* = -i(\lambda - \overline{\omega}) \int_{x_1}^{x_2} U_x(\lambda) dM(x) U_x(\omega)^*.$$

The family $U_x(\lambda)$ is also continuous in the variable x and normalized by the condition

$$U_x(0) = I_m \quad \text{for} \quad 0 \leq x < d.$$

The most commonly occuring signature matrices (except for $J = \pm I_m$) are the matrices

$$j_{pq} = \begin{bmatrix} I_p & 0 \\ 0 & -I_q \end{bmatrix}, \; J_p = \begin{bmatrix} 0 & -I_p \\ -I_p & 0 \end{bmatrix} \text{ and } \mathcal{J}_p = \begin{bmatrix} 0 & -iI_p \\ iI_p & 0 \end{bmatrix},$$

$-j_{pq}$, $-J_p$ and $-\mathcal{J}_p$. The equivalences

$$\begin{bmatrix} \varepsilon^* & I_q \end{bmatrix} j_{pq} \begin{bmatrix} \varepsilon \\ I_q \end{bmatrix} \leq 0 \Longleftrightarrow \varepsilon^* \varepsilon \leq I_q;$$

$$\begin{bmatrix} \varepsilon^* & I_p \end{bmatrix} J_p \begin{bmatrix} \varepsilon \\ I_p \end{bmatrix} \leq 0 \Longleftrightarrow \varepsilon + \varepsilon^* \geq 0$$

and

$$\begin{bmatrix} \varepsilon^* & I_p \end{bmatrix} \mathcal{J}_p \begin{bmatrix} \varepsilon \\ I_p \end{bmatrix} \leq 0 \Longleftrightarrow \frac{\varepsilon - \varepsilon^*}{i} \geq 0$$

indicate a connection between the signature matrices j_{pq}, J_p and $\mathcal{J}_p$ and the classes

$$\begin{aligned} \mathcal{S}_{const}^{p \times q} &= \{\varepsilon \in \mathbb{C}^{p \times q} : \varepsilon^* \varepsilon \leq I_q\} && \text{of contractive } p \times q \text{ matrices;} \\ \mathcal{C}_{const}^{p \times p} &= \{\varepsilon \in \mathbb{C}^{p \times p} : \varepsilon + \varepsilon^* \geq 0\} && \text{of positive real } p \times p \text{ matrices;} \\ i\mathcal{C}_{const}^{p \times p} &= \{\varepsilon \in \mathbb{C}^{p \times p} : (\varepsilon - \varepsilon^*)/i \geq 0\} && \text{of positive imaginary} \\ & && p \times p \text{ matrices.} \end{aligned}$$

Moreover, if an $m \times m$ matrix U is J-contractive, i.e., if

$$U^* J U \leq J, \tag{1.2}$$

then the inequality

$$[x^* \quad I] J \begin{bmatrix} x \\ I \end{bmatrix} \leq 0 \tag{1.3}$$

implies that

$$[x^* \quad I] U^* J U \begin{bmatrix} x \\ I \end{bmatrix} \leq 0 \tag{1.4}$$

and hence, the linear fractional transformation

$$T_U[x] = (u_{11}x + u_{12})(u_{21}x + u_{22})^{-1}, \tag{1.5}$$

based on the appropriate four block decomposition of U, maps a matrix x in the class $\mathcal{F}_{const}(J)$ of matrices that satisfy the condition (1.3) into $\mathcal{F}_{const}(J)$, if x is admissible, i.e., if $\det(u_{21}x + u_{22}) \neq 0$.

Conversely, if $J \neq \pm I_m$ and U is an $m \times m$ matrix with $\det U \neq 0$ such that T_U maps admissible matrices $x \in \mathcal{F}_{const}(J)$ into $\mathcal{F}_{const}(J)$, then

$$\rho U \text{ is a } J\text{-contractive matrix for some } \rho \in C \setminus \{0\}. \tag{1.6}$$

Moreover, if T_U also maps (admissible) matrices x that satisfy (1.3) with equality into matrices with the same property, then the matrices ρU, considered in (1.6) are automatically J-unitary, i.e., $(\rho U)^* J (\rho U) = J$. These characterizations of the classes of J-contractive and J-unitary matrices are established in Chapter 2. The proofs are based on a number of results in the geometry of the space $\mathbb{C}^m$ with indefinite inner product

$$[\xi, \eta] = \eta^* J \xi$$

defined by an $m \times m$ signature matrix J, which are also presented in Chapter 2.

Analogous characterizations of the classes $\mathcal{P}(J)$ and $\mathcal{U}(J)$ of meromorphic J-contractive and J-inner mvf's in $\mathbb{C}_+$ are established in Chapter 4. These characterizations are due to L. A. Simakova. They are not simple corollaries of the corresponding algebraic results in Chapter 2: if the given $m \times m$ mvf $U(\lambda)$ is meromorphic in $\mathbb{C}_+$ with $\det U(\lambda) \not\equiv 0$ in $\mathbb{C}_+$ and $\rho(\lambda)U(\lambda) \in \mathcal{P}(J)$, then $\rho(\lambda)$ must be a meromorphic function in $\mathbb{C}_+$. To obtain such characterizations of mvf's in the classes $\mathcal{P}(J)$ and $\mathcal{U}(J)$ requires a number of results on inner-outer factorizations of scalar holomorphic functions in the Smirnov class $\mathcal{N}_+$ in $\mathbb{C}_+$ and inner denominators of scalar meromorphic functions in the Nevanlinna class $\mathcal{N}$ of functions with bounded characteristic in $\mathbb{C}_+$, and the Smirnov maximum principle in the class $\mathcal{N}_+$. This material and generalizations to $p \times q$ mvf's in the classes $\mathcal{N}_+^{p\times q}$ and $\mathcal{N}^{p\times q}$ with entries in the classes $\mathcal{N}_+$ and $\mathcal{N}$, respectively, is presented in Chapter 3. In particular, the Smirnov maximum principle, inner-outer factorization and a number of denominators for mvf's $f \in \mathcal{N}^{p\times q}$ are discussed in this chapter. Thus, Chapters 2 and 3 are devoted to topics in linear algebra and function theory for scalar and matrix valued functions that are needed to study J-contractive and J-inner mvf's as well as the other problems considered in the remaining chapters.

The sets $\mathcal{P}(J)$ and $\mathcal{U}(J)$ are multiplicative semigroups. In his fundamental paper [Po60] V. P. Potapov obtained a multiplicative representation for mvf's $U \in \mathcal{P}(J)$ with $\det U(\lambda) \not\equiv 0$ that is a far reaching generalization of the Blaschke-Riesz-Herglotz representation

$$u(\lambda) = b(\lambda)\exp\{i\alpha + i\beta\lambda\}\exp\left\{-\frac{1}{\pi i}\int_{-\infty}^{\infty}\frac{1+\mu\lambda}{\mu-\lambda}d\sigma(\mu)\right\} \tag{1.7}$$

of scalar holomorphic functions $u(\lambda)$ in $\mathbb{C}_+$ with $|u(\lambda)| \leq 1$. In formula (1.7) $b(\lambda)$ is a Blaschke product, $\alpha \in \mathbb{R}$, $\beta \geq 0$ and $\sigma(\mu)$ is a bounded nondecreasing function on $\mathbb{R}$. To obtain his multiplicative representation, Potapov used the factors that are now known as elementary Blaschke-Potapov factors. If $J \neq \pm I_m$, there are four kinds of such factors according to whether the pole is in the open lower half plane $\mathbb{C}_-$, in $\mathbb{C}_+$, in $\mathbb{R}$, or at ∞. He obtained criteria for the convergence of infinite products of normalized elementary factors that generalizes the Blaschke condition, using his theory of the J modulus. The Potapov **multiplicative representation** of mvf's $U \in \mathcal{P}(J)$ with $\det U(\lambda) \not\equiv 0$, leads to factorizations of U of the form

$$U(\lambda) = B(\lambda)U_1(\lambda)U_2(\lambda)U_3(\lambda),$$

where $B(\lambda)$ is a BP (Blaschke-Potapov) product of elementary factors, $U_1(\lambda)$ and $U_3(\lambda)$ are entire J-inner mvf's that admit a representation as a multiplicative integral that is a generalization of the second factor in (1.7), and $U_2(\lambda)$ is a holomorphic, J-contractive invertible mvf in $\mathbb{C}_+$ that admits a representation as a multiplicative integral that is a generalization of the third factor in (1.7).

In view of Potapov's theorem, every entire J-inner mvf $U(\lambda)$ with $U(0) = I_m$ admits a multiplicative integral representation

$$U(\lambda) = \int_0^{\curvearrowright d} \exp\{i\lambda dM(x)J\}, \tag{1.8}$$

where $M(x)$ is a nondecreasing $m \times m$ mvf on $[0, d]$. Moreover, $M(x)$ may be chosen so that $M(x)$ is absolutely continuous on $[0, d]$ with derivative $H(x) = M'(x) \geq 0$ normalized by the condition trace $H(x) = 1$ a.e. on $[0, d]$. But even under these last conditions, $H(x)$ is not uniquely defined by $U(\lambda)$, in general.

Multiplicative integrals were introduced in the theory of integral and differential equations by Volterra. In particular, the matrizant $U_x(\lambda)$ of the integral equation (1.1) may be written in the form of a multiplicative integral.

$$U_x(\lambda) = \int_0^{\curvearrowright x} \exp\{i\lambda dM(s)J\}, \quad 0 \leq x < d, \tag{1.9}$$

and if $d < \infty$ and $M(x)$ is bounded on $[0, d]$, then formula (1.8) coincides with formula (1.9) with $x = d$, and $U(\lambda) = U_d(\lambda)$ is the monodromy matrix of the system (1.1). Thus, in view of Potapov's theorem, every entire mvf $U \in U(J)$ with $U(0) = I_m$ may by interpreted as the monodromy matrix of a system of the form (1.1) on $[0, d]$.

A number of Potapov's results on finite and infinite BP products and on the multiplicative representation of mvf's in $\mathcal{P}(J)$ are presented in Chapter 4, sometimes without proof.

The problem of describing all normalized $m \times m$ mvf's $H(x) \geq 0$ in a differential system of the form

$$\frac{d}{dx}u(x,\lambda) = i\lambda u(x,\lambda)H(x)J \quad \text{a.e. on } [0, d] \tag{1.10}$$

with a given monodromy matrix $U_d(\lambda) = U(\lambda)$ ($U \in \mathcal{E} \cap \mathcal{U}(J)$ and $U(0) = I_m$) is one of a number of inverse problems for systems of the form (1.10). The system (1.10) arises by applying the Fourier-Laplace transform

$$u(x,\lambda) = \int_0^\infty e^{i\lambda t} v(x,t)dt$$

to the solution $v(x,t)$ of the Cauchy problem

$$\begin{aligned} \frac{\partial v}{\partial x}(x,t) &= -\frac{\partial v}{\partial t}(x,t)H(x)J, \quad 0 \le x \le d,\ 0 \le t < \infty, \\ v(x,0) &= 0. \end{aligned} \tag{1.11}$$

Since $u(x,\lambda) = u(0,\lambda)U_x(\lambda)$, $0 \le x \le d$, the monodromy matrix $U_d(\lambda)$ is the transfer function of the system with distributed parameters on the interval $[0,d]$ specified by $H(x)$; with input $v(0,t)$, output $v(d,t)$ and state $v(x,t)$ at time t. Thus, the inverse monodromy problem is the problem of recovering the distributed parameters $H(x)$, $0 \le x \le d$, described by the evolution equation (1.11) from the transfer function of this system.

Potapov's theorem establishes the existence of a solution of the inverse monodromy problem. The uniqueness of the solution is established only under some extra conditions on $U(\lambda)$ or $H(x)$.

If $J = \pm I_m$ the Brodskii-Kisilevskii condition

$$\text{type } \{U(\lambda)\} = \text{type } \{\det U(\lambda)\}$$

on the exponential type of the entire mvf $U(\lambda)$ is necessary and sufficient for uniqueness. If $J \ne \pm I_m$, then the problem is much more complicated, even for $m = 2$.

A fundamental theorem of L. de Branges states that every entire $\mathcal{J}_1$-inner 2×2 mvf $U(\lambda)$ with $U(0) = I_2$ and the extra symmetry properties

$$\overline{U(-\overline{\lambda})} = U(\lambda) \quad \text{and} \quad \det U(\lambda) = 1$$

is the monodromy matrix of exactly one canonical differential system of the form (1.10) with $J = \mathcal{J}_1$ and with real, normalized Hamiltonian $H(x) \ge 0$ a.e. on $[0,d]$.

The Brodskii-Kisilevskii criteria was obtained in the sixties as a criteria for the unicellularity of a simple dissipative Volterra operator with a given characteristic mvf $U(\lambda)$.

Characteristic functions of nonselfadjoint (and nonunitary) operators were introduced in the 1940's by M. S. Livsic, who showed that these functions define the operator up to unitary equivalence under the assumption of simplicity and that they are J-contractive in $\mathbb{C}_+$ (in the open unit disc $\mathbb{D}$, respectively). Moreover, he discovered that to each invariant subspace of the operator there corresponds a divisor of the characteristic function and, to an ordered chain of invariant subspaces, there corresponds a triangular representation of the operator that generates a multiplicative representation of the characteristic function of the operator. Livsic also proposed a triangular model of the operator based on the multiplicative representation of the characteristic function. This was one of the main motivations for the development of the theory of multiplicative representations of J-contractive mvf's by V. P. Potapov.

L. de Branges obtained his uniqueness theorem and a number of other results in harmonic analysis, by consideration of the reproducing kernel Hilbert spaces of entire vvf's (vector valued functions) with reproducing kernels $K_\omega(\lambda)$ defined by the entire J-inner 2×2 mvf's $U(\lambda)$ by the formula

$$K_\omega(\lambda) = \frac{J - U(\lambda)JU(\omega)^*}{\rho_\omega(\lambda)}, \quad \text{where } \rho_\omega(\lambda) = -2\pi i(\lambda - \overline{\omega}).$$

The theory of RKHS's (reproducing kernel Hilbert spaces) with kernels of this form (and others) was developed by him, partially in collaboration with J. Rovnyak for $m \times m$ mvf's $U \in \mathcal{P}(J)$ for $m \geq 2$ and even for operator valued functions $U(\lambda)$.

A number of results on the spaces $\mathcal{H}(U)$ for $U \in \mathcal{P}(J)$, and for the de Branges spaces $\mathcal{B}(\mathfrak{E})$ are discussed in Chapter 5. In particular it is shown that if $U \in \mathcal{P}(J)$ and $\det U(\lambda) \not\equiv 0$, then the vvf's f in the corresponding RKHS $\mathcal{H}(U)$ are meromorphic in $\mathbb{C} \setminus \mathbb{R}$ with bounded Nevanlinna characteristic in both $\mathbb{C}_+$ and $\mathbb{C}_-$. Thus, every vvf $f \in \mathcal{H}(U)$ has nontangential boundary values

$$f_+(\mu) = \lim_{\nu \downarrow 0} f(\mu + i\nu) \quad \text{and} \quad \lim_{\nu \downarrow 0} f(\mu - i\nu) = f_-(\mu) \quad \text{a.e. on } \mathbb{R}.$$

Moreover,

$$U \in \mathcal{U}(J) \Longleftrightarrow f_+(\mu) = f_-(\mu) \quad \text{a.e. on } \mathbb{R} \quad \text{for every } f \in \mathcal{H}(U).$$

Connsequently, every $f \in \mathcal{H}(U)$ may be be identified with its boundary values if $U \in \mathcal{U}(J)$.

The space $\mathcal{H}(U)$ is R_α invariant with respect to the generalized backwards shift operator R_α that is defined by the formula

$$(R_\alpha f)(\lambda) = \frac{f(\lambda) - f(\alpha)}{\lambda - \alpha}, \quad \lambda \neq \alpha,$$

for points λ and α in the domain of holomorphy of $U(\lambda)$.

The subclasses $\mathcal{U}_S(J)$, $\mathcal{U}_{rR}(J)$, $\mathcal{U}_{rsR}(J)$, $\mathcal{U}_{\ell R}(J)$ and $\mathcal{U}_{\ell sR}(J)$ of singular, right regular, right strongly regular, left regular and left strongly regular

J-inner mvf's are introduced in Chapter 4 and are characterized in terms of the properties of the boundary values of vvf's from $\mathcal{H}(U)$ in Chapter 5: if $U \in \mathcal{U}(J)$, then

$$\begin{aligned} U \in \mathcal{U}_{rsR}(J) &\iff \mathcal{H}(U) \subset L_2^m, \\ U \in \mathcal{U}_{rR}(J) &\iff \mathcal{H}(U) \cap L_2^m \quad \text{is dense in } \mathcal{H}(U), \\ U \in \mathcal{U}_S(J) &\iff \mathcal{H}(U) \cap L_2^m = \{0\}. \end{aligned}$$

Moreover, if $U \in \mathcal{U}(J)$, then the transposed mvf $U^\tau \in \mathcal{U}(J)$ and

$$U \in \mathcal{U}_{\ell R}(J) \Longleftrightarrow U^\tau \in \mathcal{U}_{rR}(J) \quad \text{and} \quad U \in \mathcal{U}_{\ell sR}(J) \Longleftrightarrow U^\tau \in \mathcal{U}_{rR}(J).$$

Furthermore, the following implications hold when $\omega \notin \mathbb{R}$:

$$U \in \mathcal{U}_{rsR}(J) \cup \mathcal{U}_{\ell sR}(J) \Longrightarrow \rho_\omega^{-1} U \in L_2^{m\times m} \Longrightarrow U \in \mathcal{U}_{rR}(J) \cap \mathcal{U}_{\ell R}(J).$$

There are a number of other characterizations of these classes. Thus, for example, in Chapter 4, an mvf $U \in \mathcal{U}(J)$ is said to belong to the class $US(J)$ of singular J-inner mvf's if it is an outer mvf in the Smirnov class $\mathcal{N}_+^{m\times m}$ in $\mathbb{C}_+$, i.e., if $U \in \mathcal{N}_+^{m\times m}$ and $U^{-1} \in \mathcal{N}_+^{m\times m}$. Then an mvf $U \in \mathcal{U}(J)$ is said to be right (resp., left) regular J-inner, if it does not have a nonconstant right (resp., left) divisor in the multiplicative semigroup $\mathcal{U}(J)$ that belongs to $\mathcal{U}_S(J)$. Characterizations of the subclasses $\mathcal{U}_{rsR}(J)$ and $\mathcal{U}_{\ell sR}(J)$ in terms of the Treil-Volberg matricial version of the Muckenhoupt (A_2)-condition are established in Chapter 10.

Every mvf $U \in \mathcal{U}(J)$ admits a pair of essentially unique factorizations:

$$U(\lambda) = U_1(\lambda)U_2(\lambda), \quad \text{where } U_1 \in \mathcal{U}_{rR}(J) \text{ and } \mathcal{U}_2 \in US(J), \tag{1.12}$$

and

$$U(\lambda) = U_3(\lambda)U_4(\lambda), \quad \text{where } U_4 \in \mathcal{U}_{\ell R}(J) \text{ and } U_3 \in \mathcal{U}_S(J).$$

The second factorization follows from the first (applied to the transposed mvf's $U^\tau(\lambda)$). The first factorization formula is established in Chapter 7 by considering the connection between mvf's $W \in \mathcal{U}_{rR}(j_{pq})$ and the GSIP (generalized Schur interpolation problem) in the class

$$\mathcal{S}^{p\times q} = \{s \in H_\infty^{p\times q} : \|s\|_\infty \leq 1\},$$

where $H_\infty^{p\times q}$ is the Hardy space of holomorphic bounded $p \times q$ mvf's in $\mathbb{C}_+$. In this problem, three mvf's are specified: $s^\circ \in \mathcal{S}^{p\times q}$ and two inner mvf's

$b_1 \in \mathcal{S}^{p\times p}$ and $b_2 \in \mathcal{S}^{q\times q}$ and

$$\mathcal{S}(b_1, b_2; s^\circ) = \{s \in \mathcal{S}^{p\times q} : b_2^{-1}(s - s^\circ)b_2^{-1} \in H_\infty^{p\times q}\}$$

is the set of solutions to this problem. The GSIP based on s°, b_1 and b_2 is said to be completely indeterminate if, for every nonzero vector $\xi \in \mathbb{C}^q$, there exists an mvf $s \in \mathcal{S}(b_1, b_2; s^\circ)$ such that $s(\lambda)\xi \not\equiv s^\circ(\lambda)\xi$. An mvf $W \in \mathcal{U}(j_{pq})$ is the resolvent matrix of this GSIP if

$$\mathcal{S}(b_1, b_2; s^\circ) = \{T_W[\varepsilon] : \varepsilon \in \mathcal{S}^{p\times q}\}. \tag{1.13}$$

There are infinitely many resolvent matrices $W \in \mathcal{U}(j_{pq})$ for each completely indeterminate GSIP (a description is furnished in Chapter 7) and every such W automatically belongs to the class $\mathcal{U}_{rR}(j_{pq})$. Conversely, every mvf $W \in \mathcal{U}_{rR}(j_{pq})$ is the resolvent matrix of a completely indeterminate GSIP. The correspondence between the class $\mathcal{U}_{rR}(j_{pq})$ and the completely indeterminate GSIP's is established in Chapter 7. Moreover, $W \in \mathcal{U}_{rsR}(j_{pq})$ if and only if W is the resolvent matrix of a strictly completely indeterminate GSIP; i.e., if and only if there exists at least one $\varepsilon \in \mathcal{S}^{p\times q}$ such that $\|T_W[\varepsilon]\|_\infty < 1$. The correspondence between the subclasses $\mathcal{U}_{rR}(J_p)$ and $\mathcal{U}_{rsR}(J_p)$ and completely indeterminate and strictly completely indeterminate GCIP's (generalized Carathéodory interpolation problems) are discussed in Chapter 7 too. This chapter also contains formulas for resolvent matrices $U(\lambda)$ that are obtained from the formulas in Chapter 5 for $U \in \mathcal{U}_{rsR}(J)$ with $J = j_{pq}$ and $J = J_p$ from the description of the corresponding RKHS's $\mathcal{H}(U)$.

The results on GCIP's that are obtained in Chapter 7 are used in Chapter 8 to study bitangential generalizations of the Krein extension problem of extending a continuous mvf $g(t)$, given on the interval $-a \le t \le a$, with a kernel

$$k(t, s) = g(t + s) - g(t) - g(-s) + g(0)$$

that is positive on $[0, a] \times [0, a]$ to a continuous mvf $\widetilde{g}(t)$ on $\mathbb{R}$ which is subject to analogous constraints on $[0, \infty) \times [0, \infty)$. In particular, the classes of entire mvf's U in $\mathcal{U}_{rR}(J_p)$ and $\mathcal{U}_{rSR}(J_p)$ are identified as the classes of resolvent matrices of completely indeterminate and strictly completely indeterminate bitangential extension problems for mvf's $g(t)$. A bitangential generalization of Krein's extension problem for continuous positive definite mvf's and Krein's extension problem for accelerants and the resolvent matrices for these problems are also considered in this chapter.

In Chapter 11 extremal values of entropy functionals for completely indeterminate generalized interpolation and extension problems are established in a uniform way that is based on the parametrizations of j_{pq} and J_p inner mvf's that was discussed in earlier chapters.

Every mvf $U \in \mathcal{U}(J)$ has a pseudocontinuation from $\mathbb{C}_+$ into $\mathbb{C}_-$ that is a meromorphic mvf of Nevanlinna class in $\mathbb{C}_-$. Consequently, every submatrix $s \in \mathcal{S}^{p\times q}$ of an inner mvf $S \in \mathcal{S}^{m\times m}$ admits such an extension to $\mathbb{C}_-$, as do mvf's of the form $s = T_W[\varepsilon]$ and and $c = T_A[\tau]$, where $W \in \mathcal{U}(j_{pq})$, $A \in \mathcal{U}(J_p)$, ε is a constant $p \times q$ contractive matrix and τ is a constant $p \times p$ matrix with $\tau + \tau^* \geq 0$. Such representations of the mvf's s and c arose in the synthesis of passive linear networks with losses by a lossless system with a scattering matrix S, a chain scattering matrix W or a transmission matrix A, repectively. The representations of s as a block of an $n \times n$ inner mvf S and $s = T_W[\varepsilon]$ and $c = T_A[\tau]$ with constant matrices $\varepsilon \in \mathcal{S}^{p\times q}$ and $\tau \in \mathcal{C}^{p\times p}$, respectively, are called Darlington representations, even though Darlington only worked with scalar rational functions $c \in \mathcal{C}$, and the scattering formalism described above was introduced by Belevich for rational mvf's $s \in \mathcal{S}^{p\times q}$. In the early seventies Darlington representations for mvf's $s \in \mathcal{S}^{p\times q}$ and $c \in \mathcal{C}^{p\times p}$ that admit pseudocontinuations into $\mathbb{C}_-$ were obtained independently by D. Z. Arov [Ar71] and P. Dewilde [De71]; generalizations to operator valued functions were obtained in [Ar71] and [Ar74a] and by R. Douglas and J. W. Helton in [DoH73]. Descriptions of the sets of representations and solutions of other inverse problems for J-inner mvf's are discussed in Chapter 9, which includes more detailed references.

In the study of bitangential interpolation problems and bitangential inverse problems for canonical systems, a significant role is played by a set $ap(W)$ of pairs $\{b_1, b_2\}$ of inner mvf's $b_1 \in \mathcal{S}^{p\times p}$ and $b_2 \in \mathcal{S}^{q\times q}$ that are associated with each mvf $W \in \mathcal{U}(j_{pq})$ and a set $ap_{II}(A)$ of pairs $\{b_3, b_4\}$ of $p \times p$ inner mvf's that is associated with each mvf $A \in \mathcal{U}(J_p)$. The inner mvf's in $\{b_1, b_2\}$ are defined in terms of the blocks w_{11} and w_{22} of W by the inner-outer factorization of $(w_{11}^{\#})^{-1} = (w_{11}(\overline{\lambda})^*)^{-1}$, which belongs to $\mathcal{S}^{p\times p}$ and the outer-inner factorization of w_{22}^{-1}, which belongs to $\mathcal{S}^{q\times q}$:

$$(w_{11}^{\#})^{-1} = b_1\varphi_1 \quad \text{and} \quad w_{22}^{-1} = \varphi_2 b_2.$$

The pair $\{b_3, b_4\} \in ap_{II}(A)$ is defined analogously in terms of the entries in the blocks of the de Branges matrix

$$\mathfrak{E}(\lambda) = \begin{bmatrix} E_-(\lambda) & E_+(\lambda) \end{bmatrix} = \begin{bmatrix} a_{22}(\lambda) - a_{21}(\lambda) & a_{22}(\lambda) + a_{21}(\lambda) \end{bmatrix}$$

that is defined in terms of the bottom blocks of A via the inner-outer and outer-inner factorizations of $(E_-^{\#}(\lambda))^{-1} = (E_-(\overline{\lambda})^*)^{-1}$ and $E_+(\lambda)^{-1}$ in the Smirnov class $\mathcal{N}_+^{p\times p}$:

$$(E_-^{\#})^{-1} = b_3\varphi_3 \quad \text{and} \quad E_+^{-1} = \varphi_4 b_4.$$

If the mvf A is holomorphic at the point $\lambda = 0$, then b_3 and b_4 are also holomorphic at the point $\lambda = 0$ and may be uniquely specified by imposing the normalization conditions $b_3(0) = I_p$ and $b_4(0) = I_p$.

To illustrate the role of associated pairs we first consider a system of the form (1.1) or (1.10) with $J = j_{pq}$. Then the matrizant W_x, $0 \leq x < d$ is a monotonic continuous chain (with respect to the variable x) of entire j_{pq}-inner mvf's that is normalized by the condition $W_x(0) = I_m$. Correspondingly there is a unique chain of associated pairs $\{b_1^x(\lambda), b_2^x(\lambda)\}$ of entire inner mvf's with $b_1^x(0) = I_p$ and $b_2^x(0) = I_q$, and this chain is monotonic and continuous with respect to the variable x.

The class $\mathcal{U}_{rsR}(J)$ plays a significant role in a number of inverse problems for canonical systems of the forms (1.1) and (1.10). In particular, the matrizant $U_x(\lambda), 0 \leq x < d$, of every canonical system that can be reduced to a Dirac system with locally summable potential belongs to the class $\mathcal{U}_{rsR}(J)$ for every $x \in [0, d)$; see e.g., [ArD05c], which includes applications to matrix Schrödinger equations with potentials of the form $q(x) = v^2(x) \pm v'(x)$ (even though the matrizant of the Schrödinger equation belongs to the class $\mathcal{U}_S(J)$).

In the authors' formulation of bitangential inverse problems, the given data is a monotonic continuous chain of pairs $\{b_1^x(\lambda), b_2^x(\lambda)\}$, $0 \leq x < d$, and a spectral characteristic (e.g., a monodromy matrix, an input scattering or impedance matrix, or a spectral function) and the problem is to find a system with the given spectral characteristic that satisfies the two restrictions:

(1) $W_x \in \mathcal{U}_{rR}(j_{pq})$ for every $x \in [0, d)$.

(2) $\{b_1^x, b_2^x\} \in ap(W)$ for every $x \in [0, d)$.

These inverse problems were solved by Krein's method, which is based on identifying the matrizant with a family of resolvent matrices of an appropriately defined completely indeterminate extension problem; see e.g., [ArD05b], [ArD05c] and [ArD07b].

The Krein method works because for each completely indeterminate GSIP with given data b_1, b_2, s°, there is an mvf $W \in \mathcal{U}(j_{pq})$ such that (1.13) holds and $\{b_1, b_2\} \in ap(W)$ that is unique up to a right constant j_{pq} unitary multiplier. Moreover, if b_1 and b_2 are holomorphic at the point $\lambda = 0$, then W is holomorphic at the point $\lambda = 0$ and then may be uniquely specified by imposing the normalization $W(0) = I_m$. Furthermore, $W(\lambda)$ is entire if b_1 and b_2 are entire. These relationships are discussed in Chapters 7 and 8.

Descriptions of the RKHS's $\mathcal{H}(W)$ and $\mathcal{H}(A)$ based on associated pairs are discussed in Chapter 5.

The theory of the RKHS' $\mathcal{H}(U)$ and $\mathcal{B}(\mathfrak{E})$ is developed further and applied to construct functional models for Livsic-Brodskii operator nodes in Chapter 6. In this chapter the mvf's $U \in \mathcal{U}(J)$ that are holomorphic and normalized at the point $\lambda = 0$ (and in the even more general class $\mathcal{LB}(J)$) are identified as characteristic mvf's of Livsic-Brodskii nodes. Connections with conservative and passive linear continuous time invariant systems are also discussed.

Necessary and sufficient conditions for the characteristic mvf of a simple Livsic-Brodskii node to belong to the class $\mathcal{U}_{rsR}(J)$ are furnished in Chapter 10, and functional models of these nodes are given in terms of the associated pairs of the first and second kind of the characteristic function U of the node.

An $m \times m$ mvf $U \in \mathcal{P}(J)$ may be interpreted as the resolvent matrix of a symmetric operator with deficiency indices (m, m) in a Hilbert space. This theory was developed and applied to a number of problems in analysis by M. G. Krein; see e.g., Krein [Kr49] and the monograph [GoGo97]. The latter focuses on entire symmetric operators and, correspondingly, entire resolvent mvf's $U \in \mathcal{U}(J)$. Connections between the Krein theory of resolvent matrices and and characteristic mvf's of Livsic-Brodskii J-nodes with the de Branges theory of RKHS' $\mathcal{H}(U)$ were considered in [AlD84] and [AlD85]. Resolvent matrices of symmetric operators were identified as characteristic mvf's of generalized LB J-nodes by M. G. Krein and S. N. Saakjan [KrS70], A. V. Shtraus [Sht60], E. R. Tsekanovskii and Yu. L. Shmulyan [TsS77] and others.

An $m \times m$ mvf $U \in \mathcal{P}(J)$ may also be interpreted as the resolvent matrix of a completely indeterminate commutant lifting problem; see e.g., [SzNF70] and [FoFr90].

Finally, we remark that although we have chosen to focus on the classes $\mathcal{P}(J)$ and $\mathcal{U}(J)$ for the open upper half plane $\mathbb{C}_+$, most of the considered results have natural analogues for the open unit disc $\mathbb{D}$ with boundary $\mathbb{T}$.

2
Algebraic preliminaries

This chapter introduces a number of the concepts that will play a significant role in the analysis of the mvf's (matrix valued functions) that are considered in this monograph in the special setting of matrices, i.e., constant mvf's. Particular attention is paid to the linear fractional transformations that map the classes $\mathcal{S}_{const}^{p\times q}$ and $\mathcal{C}_{const}^{p\times p}$ into themselves. These transformations are defined in terms of certain subblocks of matrices $U \in \mathbb{C}^{m\times m}$ that are contractive with respect to an appropriately chosen signature matrix J and for the even larger class of J-minus matrices, which is defined in terms of the indefinite inner product

$$[u, v] = v^* J u \quad \text{for} \quad u, v \in \mathbb{C}^m.$$

The geometry of the space $\mathbb{C}^m$ with respect to this indefinite inner product is studied and then applied to obtain properties of J-contractive matrices, J-unitary matrices and minus matrices and the corresponding linear fractional transformations and linear transforms. Some other properties of matrices that will be needed in the sequel are also presented in this chapter.

2.1 The classes $\mathcal{P}_{const}(J)$ and $\mathcal{U}_{const}(J)$

A matrix $J \in \mathbb{C}^{m\times m}$ is said to be a **signature matrix** if it is both selfadjoint and unitary with respect to the standard inner product in $\mathbb{C}^m$, i.e., if $J = J^*$ and $J^*J = I_m$. In view of these assumptions, the matrices

$$P = \frac{I_m + J}{2} \quad \text{and} \quad Q = \frac{I_m - J}{2} \tag{2.1}$$

are complementary orthogonal projectors on $\mathbb{C}^m$ and every $m \times m$ signature matrix $J \neq \pm I_m$ is unitarily equivalent to the matrix

$$j_{pq} = \begin{bmatrix} I_p & 0 \\ 0 & -I_q \end{bmatrix}, \quad p \geq 1, \quad q \geq 1, \quad p+q=m, \tag{2.2}$$

where

$$p = \text{rank } P \quad \text{and} \quad q = \text{rank } Q. \tag{2.3}$$

In addition to j_{pq} itself, the main examples of signature matrices that will be of interest in the sequel are:

$$J_p = \begin{bmatrix} 0 & -I_p \\ -I_p & 0 \end{bmatrix}, \quad \mathcal{J}_p = \begin{bmatrix} 0 & -iI_p \\ iI_p & 0 \end{bmatrix} \quad \text{and} \quad j_p = j_{pp} \quad \text{for } 2p = m. \tag{2.4}$$

The signature matrices J_p and j_p are connected by the formula

$$J_p = \mathfrak{V}^* j_p \mathfrak{V}, \quad \text{where} \quad \mathfrak{V} = \frac{1}{\sqrt{2}} \begin{bmatrix} -I_p & I_p \\ I_p & I_p \end{bmatrix}. \tag{2.5}$$

A matrix $U \in \mathbb{C}^{m \times m}$ is said to be:

(1) J**-contractive** if $U^* J U \leq J$.

(2) J**-expansive** if $U^* J U \geq J$.

(3) J**-unitary** if $U^* J U = J$.

The class of J-contractive matrices will be designated $\mathcal{P}_{const}(J)$ in honor of V. P. Potapov, who systematically investigated their properties; the class of J-unitary matrices will be designated $\mathcal{U}_{const}(J)$.

2.2 The Potapov–Ginzburg transform

Let

$$\mathcal{S}_{const}^{p \times q} = \left\{ A \in \mathbb{C}^{p \times q} : A^* A \leq I_q \right\}$$

and

$$\mathring{\mathcal{S}}_{const}^{p \times q} = \left\{ A \in \mathbb{C}^{p \times q} : A^* A < I_q \right\}.$$

In this subsection we will introduce the Potapov–Ginzburg transform $\text{PG}(U)$, which maps $U \in \mathcal{P}_{const}(J)$ into the set $\mathcal{S}_{const}^{m \times m}$. It is convenient, however, to first establish two preliminary lemmas.

Lemma 2.1 *If $U \in \mathbb{C}^{m\times m}$, then the six matrices*

$$P \pm QUQ, \quad P \pm QU, \quad P \pm UQ$$

are either all invertible or all singular.

Proof To verify the assertion for the first four matrices, it suffices to observe that if $u \in \mathbb{C}^m$, then

$$\begin{aligned}(P \pm QUQ)u = 0 &\iff Pu = 0 \text{ and } QUQu = 0\\ &\iff Pu = 0 \text{ and } QUu = 0\\ &\iff (P \pm QU)u = 0.\end{aligned}$$

This analysis also implies that

$$P \pm QU^*Q \text{ is invertible} \iff P \pm QU^* \text{ is invertible}$$

and hence, as P and Q are orthogonal projectors, that

$$P \pm QUQ \text{ is invertible} \iff P \pm UQ \text{ is invertible},$$

as needed to complete the proof. □

Lemma 2.2 *Let $U \in \mathbb{C}^{m\times m}$. Then $P + QU$ is invertible if and only if $P - UQ$ is invertible. Moreover, if one (and hence both) of these matrices is invertible, then*

$$(PU + Q)(P + QU)^{-1} = (P - UQ)^{-1}(UP - Q).$$

Proof The first assertion was verified in the preceding lemma; the second is a straightforward calculation. □

The mapping from

$$\{U \in \mathbb{C}^{m\times m} : \ P + QU \ \text{ is invertible}\}$$

into $\mathbb{C}^{m\times m}$ that is defined by the formula

$$S = (PU + Q)(P + QU)^{-1} = (P - UQ)^{-1}(UP - Q) \tag{2.6}$$

is called the **Potapov–Ginzburg transform**. We shall refer to it as the **PG transform** of U and shall write $S = PG(U)$.

Lemma 2.3 *Let $U \in \mathcal{P}_{const}(J)$ with $J \neq \pm I_m$. Then the matrices $P \pm QUQ$, $P \pm QU$ and $P \pm UQ$ are all invertible. Moreover, if $S = PG(U)$, then:*

(1) $S \in \mathcal{S}_{const}^{m\times m}$.

(2) $I_m - S^*S = (P+U^*Q)^{-1}(J - U^*JU)(P+QU)^{-1}$.

(3) $I_m - SS^* = (P-UQ)^{-1}(J - UJU^*)(P-QU^*)^{-1}$.

(4) *$P \pm QSQ$, $P \pm QS$ and $P \pm SQ$ are invertible.*

(5) $U = PG(S) = (PS+Q)(P+QS)^{-1} = (P-SQ)^{-1}(SP-Q)$.

(6) $J - U^*JU = (P+S^*Q)^{-1}(I_m - S^*S)(P+QS)^{-1}$.

(7) $J - UJU^* = (P-SQ)^{-1}(I_m - SS^*)(P-QS^*)^{-1}$.

(8) *The matrices $P \pm QUQ$ are both expansive, i.e., $P \pm QUQ$ is invertible and $(P \pm QUQ)^{-1} \in \mathcal{S}_{const}^{m\times m}$.*

(9) $(P+QS)(P+QU) = (P+QU)(P+QS) = I_m$.

(10) $QSQUQ = QUQSQ = Q$.

(11) $(P+QSQ)(P+QUQ) = (P+QUQ)(P+QSQ) = I_m$.

Proof The assumption that $U \in \mathcal{P}_{const}(J)$ is equivalent to the inequality

$$U^*PU + Q \leq P + U^*QU.$$

But this in turn implies that

$$QU^*PUQ + Q \leq QU^*QUQ$$

and hence that

$$QU^*PUQ + I_m \leq P + QU^*QUQ = (P \pm QU^*Q)(P \pm QUQ),$$

which clearly implies that the two matrices $P \pm QUQ$ are expansive and hence invertible. Therefore, (8) holds and all six of the matrices referred to in the statement of the lemma are invertible, by Lemma 2.1. Assertion (1) is immediate from the identities in (2) and (3), which are verified by straightforward calculations.

The next step is to verify (4). In view of Lemma 2.1, it suffices to check that $P + QS$ is invertible, which follows easily from the observation that

$$(P+QS)(P+QU) = P + Q(PU+Q) = P + Q = I_m.$$

This calculation also justifies the formulas in (9). Items (5), (6) and (7) are again routine calculations, (10) is obtained by multiplying (9) on the left and on the right by Q, and (11) follows easily from (10). $\square$

Theorem 2.4 *If J is an $m \times m$ signature matrix and $J \neq \pm I_m$, then*

(1) *The PG transform is well defined on the set $\mathcal{P}_{const}(J)$.*

(2) *The PG transform is a one to one map of $\mathcal{P}_{const}(J)$ onto the set*

$$\{S \in \mathcal{S}_{const}^{m\times m} : P + QSQ \text{ is invertible}\}.$$

(3) *If $U \in \mathcal{P}_{const}(J)$ and $S = PG(U)$, then $U = PG(S)$.*

(4) *$U \in \mathcal{U}_{const}(J)$ if and only if $PG(U) \in \mathcal{U}_{const}(I_m)$.*

Proof (1) holds because $P + QU$ is invertible if $U \in \mathcal{P}_{const}(J)$. (2) and (3) follow from (4) and (5) of Lemma 2.3, and (4) follows from (6) of Lemma 2.3. □

Corollary 2.5 *If J is an $m \times m$ signature matrix and $J \neq \pm I_m$, then*

$$U \in \mathcal{P}_{const}(J) \iff U^* \in \mathcal{P}_{const}(J)$$

and

$$U \in \mathcal{U}_{const}(J) \iff U^* \in \mathcal{U}_{const}(J).$$

Lemma 2.6 *Let $U \in \mathcal{P}_{const}(J)$ with $J \neq \pm I_m$, let $S = PG(U)$ and let*

$$U_P = Q + PUP, \quad U_Q = P + QUQ, \quad S_P = Q + PSP \quad \text{and} \quad S_Q = P + QSQ. \tag{2.7}$$

Then:

(1) *$S_P \in \mathcal{S}_{const}^{m\times m}$ and $S_Q \in \mathcal{S}_{const}^{m\times m}$.*

(2) *$S_Q U_Q = I_m$.*

(3) *The matrix S_Q is invertible, $U_Q = S_Q^{-1}$ and the matrix U_Q is expansive.*

(4) *$\|U\| \geq 1$.*

(5) *$\|U_P\| \leq \|U\|$ and $\|U_Q\| \leq \|U\|$.*

(6) *The formulas*

$$U = (I_m + PSQ)S_P U_Q (I_m - QSP) \tag{2.8}$$

and

$$S = (I_m + PUQ)U_P S_Q (I_m - QUP) \tag{2.9}$$

hold.

(7) $\|U\| \leq 3\|U_Q\|$.

(8) $\det U = \dfrac{\det S_P}{\det S_Q}$ *and* $\det S = \dfrac{\det U_P}{\det U_Q}$.

(9) *U is invertible if and only if S_P is invertible.*

(10) *If U is invertible, then*

$$U^{-1} = (I_m + QSP)S_Q S_P^{-1}(I_m - PSQ) \tag{2.10}$$

and

$$\|S_P^{-1}\| \leq \|U^{-1}\| \leq 3\|S_P^{-1}\|. \tag{2.11}$$

(11) *If $U \in \mathcal{U}_{const}(J)$, then U is invertible,*

$$PSPU^*P = P \quad and \quad S_P U_P^* = I_m$$

and hence S_P is invertible, $U_P^* = S_P^{-1}$ and U_P is expansive.

Proof The first assertion in (1) follows easily from the fact that

$$\begin{aligned} S_P^* S_P &= (Q + PS^*P)(Q + PSP) = Q + PS^*PSP \\ &\leq Q + P = I_m. \end{aligned}$$

The inequality $S_Q^* S_Q \leq I_m$ may be checked in the same way.

Assertion (2) rests on the identity

$$S_Q U_Q = P + QSQUQ = P + Q = I_m,$$

which is immediate from (11) of Lemma 2.3, and (3) is immediate from (2).

Let u be a nonzero vector in the range of Q. Then, since U_Q is expansive,

$$\begin{aligned} \|u\|^2 &\leq \|U_Q u\|^2 = \|QUQu\|^2 \\ &\leq \|U\|^2 \|u\|^2. \end{aligned}$$

This proves (4), which is then used to verify (5):

$$\begin{aligned} \|U_P u\|^2 &= \|Qu\|^2 + \|PUPu\|^2 \\ &\leq \|Qu\|^2 + \|U\|^2 \|Pu\|^2 \\ &\leq \|U\|^2(\|Qu\|^2 + \|Pu\|^2) \\ &= \|U\|^2 \|u\|^2 \end{aligned}$$

for every $u \in \mathbb{C}^m$. Therefore, $\|U_P\| \le \|U\|$ and, by much the same argument, $\|U_Q\| \le \|U\|$.

Next, upon substituting the identities

$$PS + Q = (I_m + PSQ)(Q + PSP) \quad \text{and}$$

$$QS + P = (I_m + QSP)(P + QSQ)$$

into the formula $U = (PS + Q)(QS + P)^{-1}$, it is readily seen that

$$U = (I_m + PSQ)S_P(S_Q)^{-1}(I_m + QSP)^{-1},$$

which leads easily to (2.8), with the help of the formulas in (1) and (2) and the fact that $(I_m + QSP)^{-1} = I_m - QSP$.The verification of formula (2.9) is similar.

Formula (2.8) yields the bound

$$\|U\| \le \|I_m + PSQ\|\|S_P\|\|U_Q\|\|I_m - QSP\|,$$

which leads easily to (7), since $\|S_P\| \le 1$ and $\|(I_m + PSQ)\| \le \sqrt{3}$, as follows from the sequence of inequalities:

$$\begin{aligned} \|(I_m + PSQ)u\|^2 &= \|u\|^2 + 2\Re\langle SQu, Pu\rangle + \|PSQu\|^2 \\ &\le \|u\|^2 + 2\|SQu\|\|Pu\| + \|PSQu\|^2 \\ &\le \|u\|^2 + 2\|Qu\|\|Pu\| + \|Qu\|^2 \\ &\le \|u\|^2 + \|Qu\|^2 + \|Pu\|^2 + \|Qu\|^2 \\ &\le 3\|u\|^2. \end{aligned}$$

The formulas in (8) follow from the formulas in (6) and (2) and the fact that for any matrix $B \in \mathbb{C}^{m\times m}$,

$$\det(I_m \pm PBQ) = \det(I_m \pm BQP) = \det I_m = 1$$

and

$$\det(I_m \pm QBP) = \det(I_m \pm BPQ) = \det I_m = 1.$$

Moreover, the first formula in (8) yields (9).

If U is invertible, then the formulas in (2) and (6) lead easily to (2.10) and the second inequality in (2.11). The first inequality in (2.11) then follows from the observation that

$$PU^{-1}P = PS_P^{-1}P$$

and hence that

$$\|U^{-1}\| \geq \|PU^{-1}P\| = \|PS_P^{-1}P\| = \|S_P^{-1}\|.$$

To verify the equality $\|PS_P^{-1}P\| = \|S_P^{-1}\|$, we first note that

$$S_P^{-1} = (PS_PP + Q)^{-1} = PS_P^{-1}P + Q$$

and hence that $\|PS_P^{-1}P\| \geq 1$ and

$$\begin{aligned} \|S_P^{-1}x\|^2 &= \|PS_P^{-1}Px\|^2 + \|Qx\|^2 \\ &\leq \|PS_P^{-1}P\|^2\|Px\|^2 + \|Qx\|^2 \\ &\leq \|PS_P^{-1}P\|^2(\|Px\|^2 + \|Qx\|^2) \\ &\leq \|PS_P^{-1}P\|^2\|x\|^2 \end{aligned}$$

for every $x \in \mathbb{C}^m$. Therefore,

$$\|S_P^{-1}\| \leq \|PS_P^{-1}P\|$$

and thus, equality holds, since the opposite inequality is obvious.

Finally, if $U \in \mathcal{U}_{const}(J)$, then the formulas

$$(Q + SP)U^* = (P - UQ)^{-1}U^*JU$$

and

$$(P + SQ) = (P - UQ)^{-1}J$$

imply that

$$(Q + SP)U^* = P + SQ$$

and hence that

$$PSPU^*P = P$$

and

$$\begin{aligned} S_PU_P^* &= (Q + PSP)(Q + PU^*P) \\ &= Q + PSPU^*P = Q + P = I_m. \end{aligned}$$

This completes the proof of the displayed formulas in (11). The remaining assertions in (11) follow easily from these formulas. □

In the special case that $J = j_{pq}$ with $p \geq 1$ and $q \geq 1$, the orthogonal projectors defined in (2.1) can be written explicitly as

$$P = \begin{bmatrix} I_p & 0 \\ 0 & 0_{q\times q} \end{bmatrix} \quad \text{and} \quad Q = \begin{bmatrix} 0_{p\times p} & 0 \\ 0 & I_q \end{bmatrix}.$$

Correspondingly, if $W \in \mathcal{P}_{const}(j_{pq})$ and $S = PG(W)$ are written in block form as

$$W = \begin{bmatrix} w_{11} & w_{12} \\ w_{21} & w_{22} \end{bmatrix} \quad \text{and} \quad S = \begin{bmatrix} s_{11} & s_{12} \\ s_{21} & s_{22} \end{bmatrix} \tag{2.12}$$

with blocks w_{11} and s_{11} of size $p \times p$ and w_{22} and s_{22} of size $q \times q$, then the formulas of Lemma 2.3 lead to the following conclusions:

Lemma 2.7 *Let $W \in \mathcal{P}_{const}(j_{pq})$ and $S = PG(W)$ be written in the standard block form (2.12). Then*

(1) *w_{22} and s_{22} are invertible.*

(2) $S = \begin{bmatrix} w_{11} & w_{12} \\ 0 & I_q \end{bmatrix} \begin{bmatrix} I_p & 0 \\ w_{21} & w_{22} \end{bmatrix}^{-1} = \begin{bmatrix} I_p & -w_{12} \\ 0 & -w_{22} \end{bmatrix}^{-1} \begin{bmatrix} w_{11} & 0 \\ w_{21} & -I_q \end{bmatrix}$, *i.e.,*

$$\begin{aligned} s_{11} &= w_{11} - w_{12}w_{22}^{-1}w_{21}, \quad s_{12} = w_{12}w_{22}^{-1}, \\ s_{21} &= -w_{22}^{-1}w_{21} \ \textit{and} \ s_{22} = w_{22}^{-1}. \end{aligned} \tag{2.13}$$

(3) $W = \begin{bmatrix} s_{11} & s_{12} \\ 0 & I_q \end{bmatrix} \begin{bmatrix} I_p & 0 \\ s_{21} & s_{22} \end{bmatrix}^{-1} = \begin{bmatrix} I_p & -s_{12} \\ 0 & -s_{22} \end{bmatrix}^{-1} \begin{bmatrix} s_{11} & 0 \\ s_{21} & -I_q \end{bmatrix}$,

i.e.,

$$\begin{aligned} w_{11} &= s_{11} - s_{12}s_{22}^{-1}s_{21}, \quad w_{12} = s_{12}s_{22}^{-1}, \\ w_{21} &= -s_{22}^{-1}s_{21} \ \textit{and} \ w_{22} = s_{22}^{-1}. \end{aligned} \tag{2.14}$$

(4) $s_{11} \in \mathcal{S}^{p\times p}_{const}$, $s_{12} \in \mathring{\mathcal{S}}^{p\times q}_{const}$, $s_{21} \in \mathring{\mathcal{S}}^{q\times p}_{const}$ *and* $s_{22} \in \mathcal{S}^{q\times q}_{const}$.

(5) $\det W = \dfrac{\det s_{11}}{\det s_{22}}$ *and* $\det S = \dfrac{\det w_{11}}{\det w_{22}}$.

(6) W *is invertible* $\iff$ s_{11} *is invertible.*

(7) *The formulas for* W *and* S *can also be expressed as*

$$W = \begin{bmatrix} I_p & s_{12} \\ 0 & I_q \end{bmatrix} \begin{bmatrix} s_{11} & 0 \\ 0 & w_{22} \end{bmatrix} \begin{bmatrix} I_p & 0 \\ -s_{21} & I_q \end{bmatrix} \tag{2.15}$$

and

$$S = \begin{bmatrix} I_p & w_{12} \\ 0 & I_q \end{bmatrix} \begin{bmatrix} w_{11} & 0 \\ 0 & s_{22} \end{bmatrix} \begin{bmatrix} I_p & 0 \\ -w_{21} & I_q \end{bmatrix}. \tag{2.16}$$

(8) $w_{22} = s_{22}^{-1}$ *and* $\|w_{22}\| \leq \|W\| \leq 3\|w_{22}\|$.

(9) *If* W *is invertible, then*

$$W^{-1} = \begin{bmatrix} I_p & 0 \\ s_{21} & I_q \end{bmatrix} \begin{bmatrix} s_{11}^{-1} & 0 \\ 0 & s_{22} \end{bmatrix} \begin{bmatrix} I_p & -s_{12} \\ 0 & I_q \end{bmatrix}$$

and

$$\|s_{11}^{-1}\| \leq \|W^{-1}\| \leq 3\|s_{11}^{-1}\|.$$

(10) *If* $W \in \mathcal{U}_{const}(j_{pq})$, *then* $w_{11}^* s_{11} = I_p$.

Proof Items (1), (2) and (3) are immediate from Lemma 2.3. Moreover, since

$$\begin{bmatrix} s_{11}^* & s_{21}^* \\ s_{12}^* & s_{22}^* \end{bmatrix} \begin{bmatrix} s_{11} & s_{12} \\ s_{21} & s_{22} \end{bmatrix} \leq \begin{bmatrix} I_p & 0 \\ 0 & I_q \end{bmatrix},$$

it is readily seen that

$$s_{11}^* s_{11} + s_{21}^* s_{21} \leq I_p \quad \text{and} \quad s_{12}^* s_{12} + s_{22}^* s_{22} \leq I_q.$$

Therefore, $s_{11} \in \mathcal{S}_{const}^{p\times p}$, $s_{22} \in \mathcal{S}_{const}^{q\times q}$ and, since s_{22} is invertible, $s_{12} \in \mathring{\mathcal{S}}_{const}^{p\times q}$. The supplementary inequality

$$s_{21} s_{21}^* + s_{22} s_{22}^* \leq I_q,$$

which is obtained from the 22 block of the inequality $SS^* \leq I_m$, serves to guarantee that $s_{21} \in \mathring{\mathcal{S}}_{const}^{q\times p}$, to complete the proof of (4). Item (6) is immediate from (5), which, in turn, follows from (3) and (2). Item (7) is obtained from the Schur complement formula

$$\begin{bmatrix} w_{11} & w_{12} \\ w_{21} & w_{22} \end{bmatrix} = \begin{bmatrix} I_p & w_{12}w_{22}^{-1} \\ 0 & I_q \end{bmatrix} \begin{bmatrix} w_{11} - w_{12}w_{22}^{-1}w_{21} & 0 \\ 0 & w_{22} \end{bmatrix} \begin{bmatrix} I_p & 0 \\ w_{22}^{-1}w_{21} & I_q \end{bmatrix} \tag{2.17}$$

and the formulas in (2.13). Similar arguments serve to justify formula (2.16). The upper bound in (8) follows from (2.15), since the matrices s_{ij} are contractive. The remaining assertions of the lemma follow from the corresponding assertions in Lemma 2.6. □

Lemma 2.8 *If $W \in \mathcal{U}_{const}(j_{pq})$, then there exists a unique choice of parameters*

$$k \in \mathring{\mathcal{S}}^{p\times q}_{const}, \quad u \in \mathcal{U}_{const}(I_p) \quad and \quad v \in \mathcal{U}_{const}(I_q) \tag{2.18}$$

such that

$$W = \begin{bmatrix} (I_p - kk^*)^{-1/2} & k(I_q - k^*k)^{-1/2} \\ k^*(I_p - kk^*)^{-1/2} & (I_q - k^*k)^{-1/2} \end{bmatrix} \begin{bmatrix} u & 0 \\ 0 & v \end{bmatrix}. \tag{2.19}$$

Conversely, if k, u, v is any set of three matrices that meet the condition (2.18), then the matrix W defined by formula (2.19) is j_{pq}-unitary.

Proof Let $W \in \mathcal{U}_{const}(j_{pq})$, let w_{ij} and s_{ij} be the blocks in the four block decomposition of W and $S = PG(W)$ and let $k = s_{12}$. Then, by Lemma 2.7, $k \in \mathring{\mathcal{S}}^{p\times q}_{const}$,

$$s_{11}s_{11}^* = I_p - kk^* \quad s_{22}^*s_{22} = I_q - k^*k \quad \text{and} \quad s_{21}s_{11}^* = -s_{22}k^*,$$

since $S \in \mathcal{U}_{const}(I_m)$, where $m = p + q$. The first two equalities imply that

$$s_{11} = (I_p - kk^*)^{1/2}u \quad \text{and} \quad s_{22} = v^*(I_q - k^*k)^{1/2}, \tag{2.20}$$

where $u \in \mathcal{U}_{const}(I_p)$ and $v \in \mathcal{U}_{const}(I_q)$ are uniquely defined by s_{11} and s_{22}, respectively. Moreover, these formulas lead easily to the asserted parametrization of W with parameters that are uniquely defined by W and satisfy the constraints (2.18), since

$$w_{11} = s_{11}^{-*}, \; w_{22} = s_{22}^{-1}, \; w_{12} = ks_{22}^{-1} \text{ and } w_{21} = -s_{22}^{-1}s_{21} = k^*s_{11}^{-*}. \tag{2.21}$$

The converse is easily checked by direct calculation. □

Lemma 2.9 *If $W \in \mathcal{U}_{const}(j_{pq})$, then $u = I_p$ and $v = I_q$ in the parametrization formula (2.19) if and only if $W > 0$.*

Proof If $W \in \mathcal{U}_{const}(j_{pq})$ and $W > 0$, then parametrization formula (2.19) implies that $u = I_p$ and $v = I_q$, since $w_{11} > 0$ and $w_{22} > 0$. Conversely, if $W \in \mathcal{U}_{const}(j_{pq})$ and $u = I_p$ and $v = I_q$, then, by Schur complements, formula (2.19) can be written as

$$W = \begin{bmatrix} I_p & k \\ 0 & I_q \end{bmatrix} \begin{bmatrix} (I_p - kk^*)^{1/2} & 0 \\ 0 & (I_q - k^*k)^{-1/2} \end{bmatrix} \begin{bmatrix} I_p & 0 \\ k^* & I_q \end{bmatrix},$$

since $w_{11} - w_{12}w_{22}^{-1}w_{21} = (I_p - kk^*)^{1/2}$. But this clearly displays W as a positive definite matrix. □

2.3 Linear fractional transformations

Let

$$U = \begin{bmatrix} u_{11} & u_{12} \\ u_{21} & u_{22} \end{bmatrix} \tag{2.22}$$

be a constant $m \times m$ matrix with blocks u_{11} of size $p \times p$ and u_{22} of size $q \times q$, respectively, and define the **linear fractional transformation**

$$T_U[x] = (u_{11}x + u_{12})(u_{21}x + u_{22})^{-1} \tag{2.23}$$

for every $x \in \mathbb{C}^{p \times q}$ for which the indicated inverse exists, i.e., for x in the set

$$\mathcal{D}(T_U) = \{x \in \mathbb{C}^{p \times q} : \ u_{21}x + u_{22} \ \text{ is invertible}\}.$$

We shall refer to this set as the **domain** of T_U and, for a set $X \subseteq \mathcal{D}(T_U)$, we let

$$T_U[X] = \{T_U[x] : \ x \in X\}.$$

Let $V = [v_{ij}]_{i,j=1}^2$ be a second constant $m \times m$ matrix which is partitioned conformally with U. Then if $x \in \mathcal{D}(T_U)$ and $T_U[x] \in \mathcal{D}(T_V)$, it is readily checked that $x \in \mathcal{D}(T_{VU})$ and

$$T_V[T_U[x]] = T_{VU}[x]. \tag{2.24}$$

Lemma 2.10 *If* $U, V \in \mathbb{C}^{m\times m}$ *and* $UV = I_m$*, then*

(1) $T_U[\mathcal{D}(T_U)] = \mathcal{D}(T_V)$.

(2) $T_V[T_U[x]] = x$ *for every* $x \in \mathcal{D}(T_U)$.

(3) $\mathcal{D}(T_U) = \{x \in \mathbb{C}^{p\times q} : \det(v_{11} - xv_{21}) \neq 0\}$.

(4) *If* $x \in \mathcal{D}(T_U)$*, then*

$$T_U[x] = -(v_{11} - xv_{21})^{-1}(v_{12} - xv_{22}). \tag{2.25}$$

Proof (1) and (2) are readily checked by direct calculation. Suppose next that $u_{21}x + u_{22}$ is invertible, but $\xi^*(v_{11} - xv_{21}) = 0$ for some vector $\xi \in \mathbb{C}^p$. Then the identity

$$0_{p\times q} = [I_p \quad -x]I_m \begin{bmatrix} x \\ I_q \end{bmatrix} = [I_p \quad -x]VU \begin{bmatrix} x \\ I_q \end{bmatrix}$$

yields the formula

$$(v_{11} - xv_{21})(u_{11}x + u_{12}) = -(v_{12} - xv_{22})(u_{21}x + u_{22}), \tag{2.26}$$

which implies that $\xi^*(v_{12} - xv_{22}) = 0$ too. Therefore,

$$[\xi^* \quad -\xi^* x]V = 0$$

and hence, since V is invertible, $\xi = 0$. Thus, $v_{11} - xv_{21}$ is invertible.

Since the argument can be reversed it follows that

$$u_{21}x + u_{22} \text{ is invertible} \iff v_{11} - xv_{21} \text{ is invertible.}$$

This establishes (3); (4) is now immediate from formula (2.26). □

Lemma 2.11 *Let* $U \in \mathbb{C}^{m\times m}$ *be an invertible matrix and let* $V \in \mathbb{C}^{m\times m}$ *be such that*

$$V j_{pq} U = j_{pq}, \quad i.e., \ V = j_{pq}U^{-1}j_{pq}.$$

Then, in terms of the standard four block decompositions $[u_{ij}]_{i,j=1}^2$ *and* $[v_{ij}]_{i,j=1}^2$ *described above,*

$$\mathcal{D}(T_U) = \{x \in \mathbb{C}^{p\times q} : \ v_{11} + xv_{21} \ \text{ is invertible}\}.$$

Moreover, if $x \in \mathcal{D}(T_U)$*, then the linear fractional transformation (2.23) can also be written as*

$$T_U[x] = (v_{11} + xv_{21})^{-1}(v_{12} + xv_{22}). \tag{2.27}$$

Proof The lemma follows directly from Lemma 2.10, since $j_{pq}Vj_{pq} = U^{-1}$. It can also be established independently in much the same way by exploiting the identity

$$[I_p \quad x]Vj_{pq}U \begin{bmatrix} x \\ I_q \end{bmatrix} = [I_p \quad x]j_{pq} \begin{bmatrix} x \\ I_q \end{bmatrix} = 0,$$

which is valid for every $x \in \mathbb{C}^{p\times q}$.

Remark 2.12 *We shall refer to the linear fractional transformation T_U defined above that maps $\mathcal{D}(T_U) \subseteq \mathbb{C}^{p\times q}$ into $\mathbb{C}^{p\times q}$ as a* **right linear fractional transformation** *and shall sometimes denote it by T_U^r. There is also a* **left linear fractional transformation**, *T_U^ℓ with domain*

$$\mathcal{D}(T_U^\ell) = \{y \in \mathbb{C}^{q\times p} : \; yu_{12} + u_{22} \quad \textit{is invertible}\}$$

and range in $\mathbb{C}^{q\times p}$ that is defined on $\mathcal{D}(T_U^\ell)$ by the formula

$$T_U^\ell[y] = (yu_{12} + u_{22})^{-1}(yu_{11} + u_{21}).$$

Theorem 2.13 *Let $U \in \mathbb{C}^{m\times m}$. Then the linear fractional transformation T_U meets the three conditions:*

(1) $\mathcal{S}_{const}^{p\times q} \subseteq \mathcal{D}(T_U)$.

(2) $T_U[\mathcal{S}_{const}^{p\times q}] \subseteq \mathcal{S}_{const}^{p\times q}$.

(3) $T_U[\mathring{\mathcal{S}}_{const}^{p\times q}] \subseteq \mathring{\mathcal{S}}_{const}^{p\times q}$.

if and only if

(4) *$U = cW$ for some choice of $c \in \mathbb{C} \setminus \{0\}$ and $W \in \mathcal{P}_{const}(j_{pq})$.*

Proof The proof that (4) $\Longrightarrow$ (1)–(3) is relatively straightforward. Indeed, if $W \in \mathcal{P}_{const}(j_{pq})$ and $x \in \mathcal{S}_{const}^{p\times q}$, then (1) follows from the observation that w_{22} is invertible, $s_{21} \in \mathring{\mathcal{S}}_{const}^{q\times p}$ and

$$w_{21}x + w_{22} = w_{22}(I_q - s_{21}x).$$

Next, (2) and (3) are easily extracted from the formulas

$$(w_{21}x + w_{22})^*\{T_W[x]^*T_W[x] - I_q\}(w_{21}x + w_{22})$$

$$= [x^* \quad I_q]W^*j_{pq}W \begin{bmatrix} x \\ I_q \end{bmatrix} \leq [x^* \quad I_q]j_{pq} \begin{bmatrix} x \\ I_q \end{bmatrix} = \; x^*x - I_q.$$

The same conclusions prevail if W is replaced by $U = cW$ for any $c \in \mathbb{C}\setminus\{0\}$, because then $T_U[x] = T_W[x]$.

The converse implication lies deeper and requires the development of the theory of finite dimensional spaces with indefinite metric. A proof will be furnished later in Corollary 2.41. □

Corollary 2.14 *If $W \in \mathcal{U}_{const}(j_{pq})$, then*

$$\mathcal{S}^{p\times q}_{const} \subseteq \mathcal{D}(T_W) \quad \textit{and} \quad T_W[\mathcal{S}^{p\times q}_{const}] = \mathcal{S}^{p\times q}_{const}.$$

Proof The inclusions $\mathcal{S}^{p\times q}_{const} \subseteq \mathcal{D}(T_W)$ and $T_W[\mathcal{S}^{p\times q}_{const}] \subseteq \mathcal{S}^{p\times q}_{const}$ follow from Theorem 2.13. However, since $W \in \mathcal{U}_{const}(j_{pq}) \Longleftrightarrow W^{-1} \in \mathcal{U}_{const}(j_{pq})$, the inclusion $T_{W^{-1}}[\mathcal{S}^{p\times q}_{const}] \subseteq \mathcal{S}^{p\times q}_{const}$ is also valid. Therefore,

$$\mathcal{S}^{p\times q}_{const} = T_W[T_{W^{-1}}[\mathcal{S}^{p\times q}_{const}]] \subseteq T_W[\mathcal{S}^{p\times q}_{const}] \subseteq \mathcal{S}^{p\times q}_{const}.$$

□

Lemma 2.15 *Let $U \in \mathbb{C}^{m\times m}$ and let T_U denote the linear fractional transformation defined in terms of the blocks u_{ij} of U by formula (2.23). Then*

(1) $\mathring{\mathcal{S}}^{p\times q}_{const} \subseteq \mathcal{D}(T_U) \Longleftrightarrow u_{22}$ *is invertible and* $u_{22}^{-1}u_{21} \in \mathcal{S}^{q\times p}_{const}$.

(2) $\mathcal{S}^{p\times q}_{const} \subseteq \mathcal{D}(T_U) \Longleftrightarrow u_{22}$ *is invertible and* $u_{22}^{-1}u_{21} \in \mathring{\mathcal{S}}^{q\times p}_{const}$.

Proof If $\mathring{\mathcal{S}}^{p\times q}_{const} \subseteq \mathcal{D}(T_U)$, then $0_{p\times q} \in \mathcal{D}(T_U)$ and hence u_{22} is invertible and the matrix $\chi = u_{22}^{-1}u_{21}$ is well defined. Moreover, if $\|\chi\| > 1$, then the matrix $x = -\|\chi\|^{-2}\chi^*$ belongs to the set $\mathring{\mathcal{S}}^{p\times q}_{const}$. However, this is not viable, since $\det(I_q + \chi x) = 0$, because $\|\chi\|^2$ is an eigenvalue of $\chi\chi^*$. Therefore, $\chi \in \mathcal{S}^{q\times p}_{const}$. Conversely, if u_{22} is invertible and $\|\chi\| \leq 1$, then it is readily checked that

$$\mathcal{D}(T_U) = \{x \in \mathbb{C}^{p\times q} : \det(I_q - \chi x) \neq 0\} \supseteq \mathring{\mathcal{S}}^{p\times q}_{const},$$

since $\|\chi x\| \leq \|\chi\|\|x\| < 1$, for $x \in \mathring{\mathcal{S}}^{p\times q}_{const}$. This completes the proof of (1). The proof of (2) is similar. □

Lemma 2.16 *Let $U \in \mathbb{C}^{m\times m}$ and assume that $\mathring{\mathcal{S}}^{p\times q}_{const} \subseteq \mathcal{D}(T_U)$ and that T_U is injective on $\mathring{\mathcal{S}}^{p\times q}_{const}$. Then U is invertible.*

Proof Lemma 2.15 implies that u_{22} is invertible and that the matrix $\chi = u_{22}^{-1}u_{21} \in \mathcal{S}_{const}^{q\times p}$. Thus, if

$$U\begin{bmatrix}\xi\\ \eta\end{bmatrix} = 0 \quad \text{for some choice of } \xi \in \mathbb{C}^p \text{ and } \eta \in \mathbb{C}^q,$$

then the formulas

$$0 = u_{21}\xi + u_{22}\eta = u_{22}(\eta + \chi\xi)$$

imply that $\eta = -\chi\xi$ and consequently that $\eta^*\eta \leq \xi^*\xi$. Now choose $y \in \mathbb{C}^q$ with $y \neq 0$ such that $I_q + \eta y^*$ is invertible and the matrix $\varepsilon = \xi y^*(I_q+\eta y^*)^{-1}$ is in $\mathring{\mathcal{S}}_{const}^{p\times q}$. Then

$$U\begin{bmatrix}\xi y^*\\ I_q + \eta y^*\end{bmatrix} = U\begin{bmatrix}\xi y^*\\ \eta y^*\end{bmatrix} + U\begin{bmatrix}0\\ I_q\end{bmatrix} = U\begin{bmatrix}0\\ I_q\end{bmatrix}$$

and it is readily checked that

$$T_U[\varepsilon] = T_U[0].$$

Therefore, since T_U is assumed to be injective on $\mathring{\mathcal{S}}_{const}^{p\times q}$, $\varepsilon = 0$. Thus, $\xi = 0$ and $\eta = 0$ and hence U is invertible. □

Lemma 2.17 *Let $U \in \mathbb{C}^{m\times m}$ with a four block decomposition as in (2.22) and suppose that u_{22} is invertible and that $\chi = u_{22}^{-1}u_{21}$ is strictly contractive. Then there exists a matrix $W_0 \in \mathcal{U}_{const}(j_{pq})$ such that the blocks $u_{ij}^{(1)}$ of the matrix $U_1 = UW_0$ satisfy the conditions*

$$u_{11}^{(1)} \geq 0, \quad u_{22}^{(1)} > 0 \quad \textit{and} \quad u_{21}^{(1)} = 0. \tag{2.28}$$

Such a matrix U_1 is uniquely defined by the matrix U. Moreover:

(1) *The blocks of U_1 may be expressed in terms of the matrices*

$$k = -(u_{22}^{-1}u_{21})^* = -\chi^*, \tag{2.29}$$

$$\delta = \{u_{22}(I_q - k^*k)u_{22}^*\}^{1/2} \quad \textit{and} \quad v = (I_q - k^*k)^{1/2}u_{22}^*\delta^{-1} \tag{2.30}$$

by the formulas

$$u_{22}^{(1)} = \delta,\ u_{11}^{(1)} = \{(u_{11}+u_{12}k^*)(I_p - kk^*)^{-1}(u_{11}+u_{12}k^*)^*\}^{1/2} \tag{2.31}$$

and

$$u_{12}^{(1)} = (u_{11}k + u_{12})(I_q - k^*k)^{-1/2}v. \tag{2.32}$$

(2) *The factor W_0 is unique if and only if* $\det U \neq 0$.

(3) $\det U \neq 0 \Longleftrightarrow u_{11}^{(1)} > 0$.

(4) *The matrix W_0 admits a parametrization of the form (2.19) in which k and v are uniquely defined by formulas (2.29) and (2.30) and u is a $p \times p$ unitary matrix such that*

$$(u_{11} + u_{12}k^*)(I_p - kk^*)^{-1/2} = u_{11}^{(1)} u^*. \tag{2.33}$$

(Thus u and, hence W_0, is uniquely defined by U if and only if $\det U \neq 0$.*)*

Proof Let $W_0 \in \mathcal{U}_{const}(j_{pq})$ and let $U_1 = UW_0$. Then, if W_0 is parametrized as in (2.19),

$$u_{11}^{(1)} = (u_{11} + u_{12}k^*)(I_p - kk^*)^{-1/2}u, \tag{2.34}$$

$$u_{22}^{(1)} = (u_{21}k + u_{22})(I_q - k^*k)^{-1/2}v, \tag{2.35}$$

$$u_{12}^{(1)} = (u_{11}k + u_{12})(I_q - k^*k)^{-1/2}v \tag{2.36}$$

and

$$u_{21}^{(1)} = (u_{21} + u_{22}k^*)(I_p - kk^*)^{-1/2}u. \tag{2.37}$$

The last equality implies that $u_{21}^{(1)} = 0$ if and only if $k = -\chi*$, i.e., if and only if (2.29) holds. Moreover, since $u_{22}^{(1)}$ is invertible, formula (2.35) implies $u_{22}^{(1)} > 0$ if and only if the unitary matrix v is defined by formula (2.30), whereas (2.34) implies that $u_{11}^{(1)} \geq 0$ if and only if it is defined as in formula (2.31). Furthermore, the $p \times p$ unitary matrix u may be chosen to satisfy (2.33). This choice of u is unique if and only if $u_{11}^{(1)} > 0$.

Finally, the formula

$$\det U \det W_0 = \det u_{11}^{(1)} \det u_{22}^{(1)}$$

yields (3) and the rest of (4), since $u_{22}^{(1)}$ and W_0 are invertible. □

2.4 Matrix balls

A **matrix ball** $\mathcal{B}$ in $\mathbb{C}^{p\times q}$ is a set of the form

$$\mathcal{B} = \{C + LER : E \in \mathcal{S}_{const}^{p\times q}\} \tag{2.38}$$

where $C \in \mathbb{C}^{p\times q}$, $L \in \mathbb{C}^{p\times p}$, $R \in \mathbb{C}^{q\times q}$, $L \geq 0$ and $R \geq 0$. The matrix C is called the **center** of the ball and the matrices L and R are called **left and right semi radii**, respectively.

Lemma 2.18 *A matrix $X \in \mathbb{C}^{p\times q}$ belongs to the matrix ball $\mathcal{B}$ with center C and left and right semi radii L and R if and only if*

$$|\xi^*(X - C)\eta| \leq \|L\xi\| \, \|R\eta\| \tag{2.39}$$

for every choice of $\xi \in \mathbb{C}^p$ and $\eta \in \mathbb{C}^q$.

Proof Suppose first that (2.39) is in force. Then $(X - C)\eta$ is orthogonal to every vector ξ in the null space of L, i.e., range $(X - C) \subseteq$ range (L). Therefore,

$$LL^\dagger(X - C) = X - C,$$

where $L^\dagger$ denotes the Moore-Penrose inverse of L. Similar considerations imply that

$$(X - C)R^\dagger R = X - C,$$

where $R^\dagger$ denotes the Moore-Penrose inverse of R. Thus,

$$(X - C) = LL^\dagger(X - C)R^\dagger R$$

and $L^\dagger(X - C)R^\dagger \in \mathcal{S}_{const}^{p\times q}$, since

$$|u^*L^\dagger(X - C)R^\dagger v| \leq \|LL^\dagger u\| \, \|RR^\dagger v\| \leq \|u\| \, \|v\|$$

for every choice of $u \in \mathbb{C}^p$ and $v \in \mathbb{C}^q$ by (2.39) and $LL^\dagger$ and $RR^\dagger$ are orthogonal projections. This completes the proof that (2.39) implies that X belongs to the matrix ball defined by C, L and R. The converse implication is self-evident. □

Lemma 2.19 *Let $\mathcal{B}_j$ denote the matrix ball with center $C_j \in \mathbb{C}^{p\times q}$ and left and right semi radii $L_j \in \mathbb{C}^{p\times p}$ and $R_j \in \mathbb{C}^{q\times q}$, respectively, for $j = 1, 2$. Then $\mathcal{B}_1 \subseteq \mathcal{B}_2$ if and only if*

$$|\xi^*(C_1 - C_2)\eta| + \|L_1\xi\| \, \|R_1\eta\| \leq \|L_2\xi\| \, \|R_2\eta\| \tag{2.40}$$

for every choice of $\xi \in \mathbb{C}^p$ and $\eta \in \mathbb{C}^q$.

Proof Suppose first that $\mathcal{B}_1 \subseteq \mathcal{B}_2$ and let $E_1 \in \mathcal{S}_{const}^{p\times q}$. Then there exists a matrix $E_2 \in \mathcal{S}_{const}^{p\times q}$ such that

$$C_1 - C_2 + L_1E_1R_1 = L_2E_2R_2$$

and hence

$$|\xi^*(C_1 - C_2)\eta + \xi^*L_1E_1R_1\eta| = |\xi^*L_2E_2R_2\eta| \le \|L_2\xi\|\,\|R_2\eta\|$$

for every choice of $\xi \in \mathbb{C}^p$ and $\eta \in \mathbb{C}^q$. If $L_1\xi = 0$ or $R_1\eta = 0$, then (2.40) is clear. Now fix $\zeta \in \mathbb{C}$ with $|\zeta| = 1$, set $u = L_1\xi$, $v = R_1\eta$ and suppose that $\|u\|\,\|v\| \neq 0$. Then the matrix

$$E_1 = \zeta\frac{uv^*}{\|u\|\,\|v\|}$$

belongs to $\mathcal{S}_{const}^{p\times q}$, and by appropriate choice of ζ

$$\begin{aligned} |\xi^*(C_1 - C_2)\eta + \xi^*L_1E_1R_1\eta| &= |\xi^*(C_1 - C_2)\eta| + \|u\|\,\|v\| \\ &= |\xi^*(C_1 - C_2)\eta| + \|L_1\xi\|\,\|R_1\eta\|, \end{aligned}$$

which, when substituted into the preceding inequality, yields (2.40).

Suppose next that (2.40) is in force and let $X \in \mathcal{B}_1$. Then, in view of Lemma 2.18,

$$\begin{aligned} |\xi^*(X - C_2)\eta| &= |\xi^*(X - C_1)\eta + \xi^*(C_1 - C_2)\eta| \\ &\le |\xi^*(X - C_1)\eta| + |\xi^*(C_1 - C_2)\eta| \\ &\le \|L_1\xi\|\|R_1\eta\| + |\xi^*(C_1 - C_2)\eta| \\ &\le \|L_2\xi\|\,\|R_2\eta\|, \end{aligned}$$

for every choice of $\xi \in \mathbb{C}^p$ and $\eta \in \mathbb{C}^q$. Therefore, by another application of Lemma 2.18, $X \in \mathcal{B}_2$. □

Corollary 2.20 *If $\mathcal{B}_1 = \mathcal{B}_2$, then $C_1 = C_2$.*

Proof If $\mathcal{B}_1 = \mathcal{B}_2$, then the supplementary inequality

$$|\xi^*(C_1 - C_2)\eta| + \|L_2\xi\|\,\|R_2\eta\| \le \|L_1\xi\|\,\|R_1\eta\| \tag{2.41}$$

is also in force for every choice of $\xi \in \mathbb{C}^p$ and $\eta \in \mathbb{C}^q$. The conclusion now follows easily by adding (2.40) and (2.41). □

Lemma 2.21 *If $\mathcal{B}_1 \subseteq \mathcal{B}_2$ and $\mathcal{B}_1 \neq \{C_1\}$, then there exists a finite constant $\rho > 0$ such that*

$$L_1 \leq \rho L_2 \text{ and } R_1 \leq \frac{1}{\rho} R_2.$$

Proof If $\mathcal{B}_1 \neq \{C_1\}$, then there exist a pair of vectors $\xi^\circ \in \mathbb{C}^p$ and $\eta^\circ \in \mathbb{C}^q$ such that $\|L_1\xi^\circ\| \neq 0$ and $\|R_1\eta^\circ\| \neq 0$. Thus, in view of (2.40), $\|L_2\xi^\circ\| \neq 0$, $\|R_2\eta^\circ\| \neq 0$ and

$$\frac{\|L_1\xi^\circ\|}{\|L_2\xi^\circ\|} \leq \frac{\|R_2\eta^\circ\|}{\|R_1\eta^\circ\|} < \infty.$$

Therefore, the number

$$\kappa = \sup\left\{\frac{\|L_1\xi\|}{\|L_2\xi\|} : L_1\xi \neq 0\right\}$$

is finite and positive and

$$\|L_1\xi\| \leq \kappa\|L_2\xi\|$$

first for $\xi \in \mathbb{C}^p$ with $L_1\xi \neq 0$ and then for all $\xi \in \mathbb{C}^p$, and , similarly

$$\|R_1\eta\| \leq \frac{1}{\kappa}\|R_2\eta\|,$$

first for $\eta \in \mathbb{C}^q$ with $R_1\eta \neq 0$ and then for all $\eta \in \mathbb{C}^q$. Consequently,

$$L_1^2 \leq \kappa L_2^2 \text{ and } R_1^2 \leq \frac{1}{\kappa} R_2^2.$$

The conclusion follows by taking square roots. □

Corollary 2.22 *If $\mathcal{B}_1 = \mathcal{B}_2$ and $\mathcal{B}_1 \neq \{C_1\}$, then there exists a finite constant $\gamma > 0$ such that*

$$L_1 = \gamma L_2 \text{ and } R_1 = \gamma^{-1} R_2.$$

Proof If $\mathcal{B}_1 = \mathcal{B}_2$, then there exist a pair of finite positive constants α and β such that the inequalities

$$L_1^2 \leq \alpha L_2^2,\ R_1^2 \leq \frac{1}{\alpha} R_2^2,\ L_2^2 \leq \beta L_1^2 \quad \text{and} \quad R_2^2 \leq \frac{1}{\beta} R_1^2$$

are all in force. Therefore,

$$L_1^2 \leq \alpha\beta L_1^2 \quad \text{and} \quad R_1^2 \leq \frac{1}{\alpha\beta} R_1^2,$$

which implies that $\alpha\beta = 1$ and hence, since $L_1 > 0$ and $R_1 > 0$, that $L_1^2 = \alpha L_2^2$ and $R_1^2 = \alpha^{-1}R_2^2$. Thus, by the uniqueness of the positive semidefinite square root of a positive semidefinite matrix,

$$L_1 = \gamma L_2 \quad \text{and} \quad R_1 = \gamma^{-1} R_2$$

for some finite positive constant γ. □

In the future we shall usually indicate the center of a matrix ball $\mathcal{B}$ in $\mathbb{C}^{p\times q}$ by m_c and the left and right semiradii by the symbols R_ℓ and R_r, respectively. In terms of this notation,

$$\mathcal{B} = \left\{ m \in \mathbb{C}^{p\times q} : m = m_c + R_\ell \varepsilon R_r \quad \text{and} \quad \varepsilon \in \mathcal{S}_{const}^{p\times q} \right\}. \tag{2.42}$$

If $R_\ell > 0$ and $R_r > 0$, then, for each choice of $\kappa > 0$, there exists a unique choice of ρ such that

$$\frac{\det(\rho R_\ell)}{\det(\rho^{-1} R_r)} = \kappa. \tag{2.43}$$

Lemma 2.23 *Let $U \in \mathbb{C}^{m\times m}$ and suppose that $\mathcal{S}_{const}^{p\times q} \subseteq \mathcal{D}(T_U)$. Then $T_U[\mathcal{S}_{const}^{p\times q}]$ is a matrix ball with positive right semiradius R_r, i.e.,*

$$T_U[\mathcal{S}_{const}^{p\times q}] = \left\{ m \in \mathbb{C}^{p\times q} : m = m_c + R_\ell \varepsilon R_r \quad \text{and} \quad \varepsilon \in \mathcal{S}^{p\times q} \right\}, \tag{2.44}$$

where the center of the ball m_c is uniquely specified by the formula

$$m_c = u_{12}^{(1)} (u_{22}^{(1)})^{-1} \tag{2.45}$$

and the left and right semiradii R_ℓ and R_r may be specified by the formulas

$$R_\ell = u_{11}^{(1)} \quad \text{and} \quad R_r = (u_{22}^{(1)})^{-1}, \tag{2.46}$$

in which the $u_{ij}^{(1)}$ are are given by formulas (2.29)–(2.37). Moreover, for this choice of R_ℓ and R_r,

$$\frac{\det R_\ell}{\det R_r} = |\det U|. \tag{2.47}$$

If $\det U \neq 0$, *then the last equality holds only for this choice of R_ℓ and R_r.*

Proof This is an easy consequence of Lemma 2.17 and the preceding discussion. □

Theorem 2.24 *Let $U \in \mathbb{C}^{m\times m}$. Then*

$$\mathcal{S}_{const}^{p\times q} \subseteq \mathcal{D}(T_U) \quad \text{and} \quad T_U[\mathcal{S}_{const}^{p\times q}] = \mathcal{S}_{const}^{p\times q} \tag{2.48}$$

if and only if

$$U = cW \quad \text{for some} \quad W \in \mathcal{U}_{const}(j_{pq}) \quad \text{and some scalar} \quad c > 0. \qquad (2.49)$$

Proof If (2.48) holds, then the the center of the matrix ball $T_U[\mathcal{S}^{p\times q}_{const}]$ is $m_c = 0_{p\times q}$ and the left and right semiradii may be chosen equal to $R_\ell = I_p$ and $R_r = I_q$, respectively. Thus, by formulas (2.45) and (2.46), the blocks of the matrix U_1 that is defined in Lemma 2.17 are $u^{(1)}_{12} = 0$, $u^{(1)}_{11} = cI_p$ and $u^{(1)}_{22} = cI_q$. Therefore, $U_1 = cI_m$ and hence (2.49) is in force.

Conversely, (2.49) implies (2.48), by Corollary 2.14. □

Remark 2.25 *In formula (2.49),*

$$c = |\det U|^{1/m}, \qquad (2.50)$$

since $|\det W| = 1$ *when* $W \in \mathcal{U}_{const}(j_{pq})$.

Lemma 2.26 *Let* $W \in \mathcal{P}_{const}(j_{pq})$. *Then* $T_W[\mathcal{S}^{p\times q}_{const}]$ *is a matrix ball in* $\mathcal{S}^{p\times q}_{const}$ *with positive right semiradius* R_r, *i.e.,*

$$T_W[\mathcal{S}^{p\times q}_{const}] = \left\{ s^{(1)}_{12} + s^{(1)}_{11}\varepsilon s^{(1)}_{22} : \varepsilon \in \mathcal{S}^{p\times q}_{const} \right\}, \qquad (2.51)$$

where the $s^{(1)}_{ij}$ *are the blocks in the four block decomposition of* $S_1 = PG(W_1)$ *and* $W_1 \in \mathcal{P}_{const}(j_{pq})$ *is identified with the matrix* U_1 *considered in Lemma 2.23, with* W *in place of* U. *Moreover, for this choice of* R_ℓ *and* R_r,

$$\frac{\det R_\ell}{\det R_r} = |\det W|. \qquad (2.52)$$

If $\det W \neq 0$, *then the last equality holds only for this choice of* R_ℓ *and* R_r.

Proof This is essentially a special case of Lemma 2.23, which is applicable, since $\mathcal{S}^{p\times q}_{const} \subseteq \mathcal{D}(T_W)$, when $W \in \mathcal{P}_{const}(j_{pq})$. □

2.5 Redheffer transforms

If $W \in \mathcal{P}_{const}(j_{pq})$, then the linear fractional transformation T_W may be rewritten in terms of the **Redheffer transform**, which is defined in terms of the blocks s_{ij} of the PG transform $S = PG(W)$,

$$R_S[\varepsilon] = s_{12} + s_{11}\varepsilon(I_q - s_{21}\varepsilon)^{-1}s_{22} \qquad (2.53)$$

on the set

$$\mathcal{D}(R_S) = \{\varepsilon \in \mathbb{C}^{p\times q} : \det(I_q - s_{21}\varepsilon) \neq 0\}, \tag{2.54}$$

since $\mathcal{D}(R_S) = \mathcal{D}(T_W)$, and, if $\varepsilon \in \mathcal{D}(R_S)$, then by formula (2.13),

$$\begin{aligned} R_S[\varepsilon] &= w_{12}w_{22}^{-1} + (w_{11} - w_{12}w_{22}^{-1}w_{21})\varepsilon(I_q + w_{22}^{-1}w_{21}\varepsilon)^{-1}w_{22}^{-1} \\ &= T_W[\varepsilon]. \end{aligned}$$

The Redheffer transform R_S may also be defined for $(p + k) \times (r + q)$ matrices

$$S = \begin{bmatrix} s_{11} & s_{12} \\ s_{21} & s_{22} \end{bmatrix}$$

with blocks

$$s_{11} \in \mathbb{C}^{p\times r}, \quad s_{12} \in \mathbb{C}^{p\times q}, \quad s_{21} \in \mathbb{C}^{k\times r} \quad \text{and} \quad s_{22} \in \mathbb{C}^{k\times q}$$

on the set

$$\mathcal{D}(R_S) = \{\varepsilon \in \mathbb{C}^{r\times k} : \det(I_k - s_{21}\varepsilon) \neq 0\} \tag{2.55}$$

by the formula

$$R_S[\varepsilon] = s_{12} + s_{11}\varepsilon(I_k - s_{21}\varepsilon)^{-1}s_{22}, \tag{2.56}$$

and not just for $S \in \mathcal{S}_{const}^{m\times m}$ with s_{22} invertible. If $E \subseteq \mathcal{D}(R_S)$, then

$$R_S[E] = \{R_S[\varepsilon] : \varepsilon \in E\}. \tag{2.57}$$

Theorem 2.27 *Let $S \in \mathcal{S}_{const}^{(p+k)\times(r+q)}$ be decomposed into four blocks of the sizes indicated above. Then:*

(1) $\mathring{\mathcal{S}}_{const}^{r\times k} \subseteq \mathcal{D}(R_S)$.

(2) $R_S[\mathcal{S}_{const}^{r\times k} \cap \mathcal{D}(R_S)] \subseteq \mathcal{S}_{const}^{p\times q}$.

(3) $\mathcal{S}^{r\times k} \subseteq \mathcal{D}(R_S)$ *if and only if* $s_{21} \in \mathring{\mathcal{S}}_{const}^{k\times r}$.

(4) $R_S[\mathring{\mathcal{S}}_{const}^{r\times k}] \subseteq \mathring{\mathcal{S}}_{const}^{p\times q}$ *if and only if* $s_{12} \in \mathring{\mathcal{S}}_{const}^{p\times q}$.

(5) *If S is an isometric matrix and $\varepsilon \in \mathcal{D}(R_S)$, then $R_S[\varepsilon]$ is an isometric matrix if and only if ε is an isometric matrix.*

Proof Under the given assumptions, there exist a positive integer $m \le r+q$ and a pair of matrices $M \in \mathbb{C}^{m\times r}$ and $N \in \mathbb{C}^{m\times q}$ such that

$$\begin{aligned} I_{r+q} &- \begin{bmatrix} s_{11}^* & s_{21}^* \\ s_{12}^* & s_{22}^* \end{bmatrix} \begin{bmatrix} s_{11} & s_{12} \\ s_{21} & s_{22} \end{bmatrix} \\ &= \begin{bmatrix} I_r - (s_{11}^* s_{11} + s_{21}^* s_{21}) & -(s_{11}^* s_{12} + s_{21}^* s_{22}) \\ -(s_{12}^* s_{11} + s_{22}^* s_{21}) & I_q - (s_{12}^* s_{12} + s_{22}^* s_{22}) \end{bmatrix} \\ &= \begin{bmatrix} M^*M & M^*N \\ N^*M & N^*N \end{bmatrix}. \end{aligned}$$

Let $\varepsilon \in \mathcal{S}_{const}^{r\times k} \cap \mathcal{D}(R_S)$ and let $g = (I_k - s_{21}\varepsilon)^{-1} s_{22}$. Then

$$\begin{aligned} I_q - R_S[\varepsilon]^* R_S[\varepsilon] &= I_q - (s_{12}^* + g^*\varepsilon^* s_{11}^*)(s_{12} + s_{11}\varepsilon g) \\ &= N^*N + s_{22}^* s_{22} + g^*\varepsilon^*(M^*N + s_{21}^* s_{22}) \\ &\quad + (s_{22}^* s_{21} + N^*M)\varepsilon g \\ &\quad + g^*\varepsilon^*(M^*M + s_{21}^* s_{21} - I_r)\varepsilon g. \end{aligned}$$

Moreover, since

$$s_{22} + s_{21}\varepsilon g = g,$$

it is readily checked that

$$\begin{aligned} s_{22}^* s_{22} &+ g^*\varepsilon^* s_{21}^* s_{22} + s_{22}^* s_{21}\varepsilon g + g^*\varepsilon^* s_{21}^* s_{21}\varepsilon g - g^*\varepsilon^*\varepsilon g \\ &= s_{22}^*(s_{22} + s_{21}\varepsilon g) + g^*\varepsilon^* s_{21}^*(s_{22} + s_{21}\varepsilon g) - g^*\varepsilon^*\varepsilon g \\ &= s_{22}^* g + g^*\varepsilon^* s_{21}^* g - g^*\varepsilon^*\varepsilon g \\ &= g^*(I_k - \varepsilon^*\varepsilon) g \end{aligned}$$

and hence that

$$I_q - R_S[\varepsilon]^* R_S[\varepsilon] = (N + M\varepsilon g)^*(N + M\varepsilon g) + g^*(I_k - \varepsilon^*\varepsilon) g. \tag{2.58}$$

Consequently, $R_S[\varepsilon] \in \mathcal{S}_{const}^{p\times q}$ if $\varepsilon \in \mathcal{S}^{r\times k} \cap \mathcal{D}(R_S)$.

If $s_{21} \in \mathring{\mathcal{S}}^{k\times r}$ and $\varepsilon \in \mathcal{S}_{const}^{r\times k}$, then $s_{21}\varepsilon \in \mathring{\mathcal{S}}^{k\times k}$ and hence $\varepsilon \in \mathcal{D}(R_S)$.

It remains to check that if $\mathcal{S}_{const}^{r\times k} \subseteq \mathcal{D}(R_S)$, then $s_{21} \in \mathring{\mathcal{S}}^{k\times r}$. But if this is not so, then there exists a vector $\eta \in \mathbb{C}^r$ such that $\|s_{21}\eta\| = \|\eta\| = 1$. Thus, the matrix $\varepsilon = \eta\eta^* s_{21}^* \in \mathcal{S}_{const}^{r\times k}$. However,

$$\det(I_k - s_{21}\eta\eta^* s_{21}^*) = \det(1 - \eta^* s_{21}^* s_{21}\eta) = 0.$$

Therefore, the constructed matrix $\varepsilon \notin \mathcal{D}(R_S)$.

Next, to verify (4), suppose first that $\varepsilon \in \mathring{\mathcal{S}}^{r\times k}$, $s_{21} \in \mathring{\mathcal{S}}^{p\times q}$ and $(I_q - R_S[\varepsilon]^* R_S[\varepsilon])\eta = 0$ for some $\eta \in \mathbb{C}^q$. Then, in view of (2.58),

$$(N + M\varepsilon g)\eta = 0 \quad \text{and} \quad g\eta = 0.$$

Therefore, $N\eta = 0$, $s_{22}\eta = 0$ and hence $(I_q - s_{21}^* s_{21})\eta = 0$, which in turn implies that $\eta = 0$, i.e., $s_{21} \in \mathring{\mathcal{S}}^{p\times q} \Longrightarrow R_S[\mathring{\mathcal{S}}^{p\times q}] \subseteq \mathring{\mathcal{S}}^{p\times q}$. The converse implication is self-evident, since $s_{12} = R_S[0]$.

Assertion (5) is self-evident from formula (2.58), since $M = 0$ and $N = 0$ if S is isometric. This completes the proof, since assertion (1) is self-evident. □

2.6 Indefinite metrics in the space $\mathbb{C}^m$

In this section we shall consider subspaces of $\mathbb{C}^m$ endowed with the indefinite inner product

$$[u, v] = \langle Ju, v\rangle = v^* Ju,$$

where $J \in \mathbb{C}^{m\times m}$ is a signature matrix with $J \neq \pm I_m$ (and hence, $m \geq 2$). Then there exist vectors u, v and $w \neq 0$ in $\mathbb{C}^m$ such that

$$[u, u] > 0, \quad [v, v] < 0 \quad \text{and} \quad [w, w] = 0.$$

Thus, for example, if $J = j_1$, then the vectors

$$u = \begin{bmatrix} 1 \\ 0 \end{bmatrix}, \quad v = \begin{bmatrix} 0 \\ 1 \end{bmatrix} \quad \text{and} \quad w = \begin{bmatrix} 1 \\ 1 \end{bmatrix}$$

meet the stated conditions. In general, if $x \in \mathcal{R}(P)$, the range of $P = (I_m + J)/2$; $y \in \mathcal{R}(Q)$, the range of $Q = (I_m - J)/2$; and if $x \neq 0$ and $y \neq 0$, then

$$[x, x] = \|x\|^2 > 0 \quad \text{and} \quad [y, y] = -\|y\|^2 < 0.$$

Moreover, if also $\|x\| = \|y\|$, then

$$[x + y, x + y] = [x, x] + [y, y] = \|x\|^2 - \|y\|^2 = 0.$$

It is convenient to set

$$\begin{aligned}
\mathcal{M}_{++} &= \{u \in \mathbb{C}^m : [u, u] > 0\} \cup \{0\}, \\
\mathcal{M}_{+} &= \{u \in \mathbb{C}^m : [u, u] \geq 0\}, \\
\mathcal{M}_{0} &= \{u \in \mathbb{C}^m : [u, u] = 0\}, \\
\mathcal{M}_{-} &= \{u \in \mathbb{C}^m : [u, u] \leq 0\}, \\
\mathcal{M}_{--} &= \{u \in \mathbb{C}^m : [u, u] < 0\} \cup \{0\}.
\end{aligned}$$

A subspace $\mathcal{L}$ of $\mathbb{C}^m$ is said to be **positive** with respect to $[\,,\,]$ if $\mathcal{L} \subseteq \mathcal{M}_{++}$, **nonnegative** if $\mathcal{L} \subseteq \mathcal{M}_{+}$, **negative** if $\mathcal{L} \subseteq \mathcal{M}_{--}$, **nonpositive** if $\mathcal{L} \subseteq \mathcal{M}_{-}$ and **neutral** if $\mathcal{L} \subseteq \mathcal{M}_0$.

A subspace $\mathcal{L}$ of $\mathbb{C}^m$ is said to be a **maximal nonpositive subspace** if $\mathcal{L} \subseteq \mathcal{M}_-$ and $\mathcal{L}$ is not a proper subspace of any other nonpositive subspace of $\mathbb{C}^m$. A subspace $\mathcal{L}$ of $\mathbb{C}^m$ is said to be a **maximal nonnegative subspace** if $\mathcal{L} \subseteq \mathcal{M}_+$ and $\mathcal{L}$ is not a proper subspace of any other nonnegative subspace of $\mathbb{C}^m$.

Lemma 2.28 *Let $\mathcal{L}$ be a subspace of $\mathbb{C}^m$ and let $p = \operatorname{rank}(I_m + J)$ and $q = \operatorname{rank}(I_m - J)$. Then:*

(1) $\mathcal{L} \subseteq \mathcal{M}_+ \implies \dim \mathcal{L} \le p$.

(2) $\mathcal{L} \subseteq \mathcal{M}_- \implies \dim \mathcal{L} \le q$.

(3) $\mathcal{L} \subseteq \mathcal{M}_0 \implies \dim \mathcal{L} \le \min\{p, q\}$.

Proof Since $J \ne \pm I_m$, it suffices to restrict attention to the case $J = j_{pq}$. In this setting it is convenient to write $u \in \mathbb{C}^m$ as $u = \operatorname{col}(x, y)$, where $x \in \mathbb{C}^p$ and $y \in \mathbb{C}^q$. Then

$$[u, u] = \langle x, x\rangle - \langle y, y\rangle = x^*x - y^*y$$

and hence

$$\begin{aligned} x^*x \ge y^*y &\iff u \in \mathcal{M}_+ \\ x^*x \le y^*y &\iff u \in \mathcal{M}_- \\ x^*x = y^*y &\iff u \in \mathcal{M}_0. \end{aligned}$$

Let $u_1, \ldots, u_t$ be a basis for $\mathcal{L}$ and write $u_j = \operatorname{col}(x_j, y_j)$ with $x_j \in \mathbb{C}^p$ and $y_j \in \mathbb{C}^q$ for $j = 1, \ldots, t$. Then, if $\mathcal{L} \subseteq \mathcal{M}_+$, it is readily checked that $x_1, \ldots, x_t$ is a linearly independent set of vectors in $\mathbb{C}^p$:

$$\begin{aligned} x = \sum_{j=1}^{t} c_j x_j = 0 &\implies u = \sum_{j=1}^{t} c_j u_j = 0 \quad (2x^*x \ge u^*u \text{ if } u \in \mathcal{M}_+) \\ &\implies c_1 = \cdots = c_t = 0, \end{aligned}$$

since $u_1, \ldots, u_t$ is a linearly independent set of vectors in $\mathbb{C}^m$. Consequently, $t \le p$, as claimed in (1). Next, it follows in much the same way that if

$\mathcal{L} \subseteq \mathcal{M}_-$, then

$$y = \sum_{j=1}^{t} c_j y_j = 0 \implies u = \sum_{j=1}^{t} c_j u_j = 0,$$

since now $2y^*y \geq u^*u$. Thus, $y_1, \ldots, y_t$ is a linearly independent set of vectors in $\mathbb{C}^q$ and consequently, $t \leq q$, as claimed in (2).

Finally, (3) follows from (1) and (2), since $\mathcal{M}_0 \subseteq \mathcal{M}_- \cap \mathcal{M}_+$. □

We remark that the formulas

$$\mathcal{R}(P) \subseteq \mathcal{M}_{++}, \quad \dim \mathcal{R}(P) = p$$

and

$$\mathcal{R}(Q) \subseteq \mathcal{M}_{--}, \quad \dim \mathcal{R}(Q) = q$$

imply that $\mathcal{R}(P)$ is a maximal nonnegative subspace and $\mathcal{R}(Q)$ is a maximal nonpositive subspace.

Lemma 2.29 *Let $\mathcal{L}$ be a k-dimensional subspace of $\mathbb{C}^m$ such that $u^* j_{pq} u \leq 0$ for every $u \in \mathcal{L}$. Then:*

(1) *There exists a matrix $S \in \mathcal{S}_{const}^{p\times q}$ and a k-dimensional subspace $\mathcal{Y}$ of $\mathbb{C}^q$ such that*

$$\mathcal{L} = \left\{ \begin{bmatrix} Sy \\ y \end{bmatrix} : \; y \in \mathcal{Y} \right\}. \tag{2.59}$$

(2) *The matrix S in the representation of $\mathcal{L}$ in (1) is unique if and only if either $k = q$ or $k < q$ and S is coisometric.*

(3) *Conversely, if $\mathcal{L}$ is defined by formula (2.59) for some choice of $S \in \mathcal{S}_{const}^{p\times q}$ and some k-dimensional subspace $\mathcal{Y}$ of $\mathbb{C}^q$, then $\mathcal{L} \subseteq \mathcal{M}_-$ and $\mathcal{L}$ is a k-dimensional subspace of $\mathbb{C}^m$.*

Proof Let $v_1, \ldots, v_k$ be a basis for $\mathcal{L}$; let $v_j = \mathrm{col}(x_j, y_j)$ with $x_j \in \mathbb{C}^p$ and $y_j \in \mathbb{C}^q$ for $j = 1, \ldots, k$; and let

$$V = [v_1 \cdots v_k], \quad X = [x_1 \cdots x_k] \text{ and } Y = [y_1 \cdots y_k].$$

Then, under the given assumptions, $V^* j_{pq} V \leq 0$, and hence

$$X^*X \leq Y^*Y.$$

By the proof of the preceding lemma, the rank of the $q \times k$ matrix Y is equal to k and hence it admits a polar decomposition $Y = UD$, where $U \in \mathbb{C}^{q\times k}$ is isometric and $D \in \mathbb{C}^{k\times k}$ is positive definite. Thus, as

$$Y^*Y = DU^*UD = D^2,$$

it follows that

$$XD^{-1} \in \mathcal{S}_{const}^{p\times k} \quad \text{and} \quad XD^{-1}U^* \in \mathcal{S}_{const}^{p\times q}$$

and hence, upon setting

$$S = XD^{-1}U^*,$$

that

$$SY = XD^{-1}U^*UD = X.$$

Thus, (2.59) holds for this choice of $S \in \mathcal{S}^{p\times q}$ and $\mathcal{Y} = \mathcal{R}(Y)$.

Next, to obtain (2), observe first that if $k = q$, then Y is invertible, $\mathcal{Y} = \mathbb{C}^q$ and the matrix $S = XY^{-1}$ is uniquely defined by $\mathcal{L}$. If $k < q$, then there exists a matrix $V \in \mathbb{C}^{q\times(q-k)}$ such that $\widetilde{U} = [U \quad V]$ is unitary and the $p \times q$ matrix

$$S_1 = \begin{bmatrix} XD^{-1} & E \end{bmatrix} \widetilde{U}^* = XD^{-1}U^* + EV^*$$

meets the condition $S_1Y = X$ for every choice of $E \in \mathbb{C}^{p\times(q-k)}$. However, since

$$S_1S_1^* = XD^{-2}X^* + EE^* = SS^* + EE^*,$$

it follows that S_1 is contractive if and only if

$$EE^* \le I_p - SS^*.$$

But this in turn implies that there exists a nonzero matrix $E \in \mathbb{C}^{p\times(q-k)}$ that meets the last constraint if and only if the contractive matrix S is not coisometric. This serves to jusify (2). Thus, as (3) is self-evident, the proof is complete. □

Lemma 2.30 *Let J be a signature matrix that is unitarily equivalent to j_{pq}. Then:*

(1) *$\mathcal{L} \subseteq \mathcal{M}_-$ is a maximal nonpositive subspace of $\mathbb{C}^m$ with respect to J, if and only if* $\dim \mathcal{L} = q$*; and $\mathcal{L} \subseteq \mathcal{M}_+$ is a maximal nonnegative subspace of $\mathbb{C}^m$ with respect to J, if and only if* $\dim \mathcal{L} = p$.

(2) *If* $\mathcal{L} \subseteq \mathcal{M}_-$ *is a nonpositive subspace of* $\mathbb{C}^m$ *and* $\dim \mathcal{L} < q$, *then* $\mathcal{L}$ *is a proper subspace of a maximal nonpositive subspace* $\mathcal{L}_-$ *of* $\mathbb{C}^m$.

(3) *If* $\mathcal{L} \subseteq \mathcal{M}_+$ *is a nonnegative subspace of* $\mathbb{C}^m$ *and* $\dim \mathcal{L} < p$, *then* $\mathcal{L}$ *is a proper subspace of a maximal nonnegative subspace* $\mathcal{L}_+$ *of* $\mathbb{C}^m$.

Proof Since there is no loss of generality in taking $J = j_{pq}$, the first assertion follows easily from the preceding two lemmas. To verify (2), let $\mathcal{L} \subseteq \mathcal{M}_-$ and suppose that $\dim \mathcal{L} = k$ and $k < q$. Then, by Lemma 2.29, (2.59) holds for some $S \in \mathcal{S}_{const}^{p\times q}$. Let

$$\mathcal{L}_- = \left\{ \begin{bmatrix} Sy \\ y \end{bmatrix} : \ y \in \mathbb{C}^q \right\}.$$

Then $\mathcal{L} \subset \mathcal{L}_- \subseteq \mathcal{M}_-$ and $\mathcal{L} \neq \mathcal{L}_-$, since $\dim \mathcal{L} = k < q = \dim \mathcal{L}_-$.

Assertion (3) follows from (2) by replacing J by $-J$. □

If $\mathcal{L}_1$ and $\mathcal{L}_2$ are subspaces of a subspace $\mathcal{L}$ of $\mathbb{C}^m$, then we will write

$$\mathcal{L} = \mathcal{L}_1[+]\mathcal{L}_2$$

if the following three conditions are met:

(1) $\mathcal{L}_1 \cap \mathcal{L}_2 = \{0\}$.

(2) $[x, y] = 0$ for every $x \in \mathcal{L}_1$ and $y \in \mathcal{L}_2$.

(3) Every vector $u \in \mathcal{L}$ can be expressed in the form $u = x + y$, where $x \in \mathcal{L}_1$ and $y \in \mathcal{L}_2$. (In view of (1), this representation is unique.)

We shall also let

$$\mathcal{L}^{[\perp]} = \{v \in \mathbb{C}^m : \ [u, v] = 0 \text{ for every } u \in \mathcal{L}\}.$$

If $J \neq \pm I_m$, then there exist subspaces $\mathcal{L}$ of $\mathbb{C}^m$ such that $\mathcal{L} \cap \mathcal{L}^{[\perp]} \neq \{0\}$. This will happen only if $\mathcal{L} \cap \mathcal{M}_0 \neq \{0\}$. It is readily checked that

$$\mathbb{C}^m = \mathcal{L}[+]\mathcal{L}^{[\perp]} \iff \mathcal{L} \cap \mathcal{L}^{[\perp]} = \{0\}.$$

In this case, we write

$$\mathcal{L}^{[\perp]} = \mathbb{C}^m[-]\mathcal{L}$$

and say that $\mathcal{L}^{[\perp]}$ is the **orthogonal complement** of $\mathcal{L}$ with respect to the indefinite J-metric in $\mathbb{C}^m$. In a similar vein, if $\mathcal{L} = \mathcal{L}_1[+]\mathcal{L}_2$, then we sometimes write $\mathcal{L}_2 = \mathcal{L}[-]\mathcal{L}_1$ and $\mathcal{L}_1 = \mathcal{L}[-]\mathcal{L}_2$.

Lemma 2.31 *If $\mathcal{L} = \mathcal{L}_1[+]\mathcal{L}_2$, where $\mathcal{L}_1 \subseteq \mathcal{M}_{++}$ and $\mathcal{L}_2 \subseteq \mathcal{M}_{--}$, then $\mathcal{L} \cap \mathcal{L}^{[\perp]} = \{0\}$.*

Proof Let $u \in \mathcal{L} \cap \mathcal{L}^{[\perp]}$. Then, in view of the given assumptions, $u = u_1 + u_2$, where $u_1 \in \mathcal{L}_1$, $u_2 \in \mathcal{L}_2$ and

$$[u_1, u_2] = [u, u_1] = [u, u_2] = 0.$$

Therefore,

$$[u_1, u_1] = [u_2, u_2] = 0$$

and hence $u_1 = u_2 = 0$. Thus, $u = 0$ too. □

Remark 2.32 *If $\mathcal{L} \cap \mathcal{L}^{[\perp]} = \{0\}$, then $\mathcal{L}$ admits the decomposition $\mathcal{L} = \mathcal{L}_1[+]\mathcal{L}_2$, where $\mathcal{L}_1 \subseteq \mathcal{M}_{++}$ and $\mathcal{L}_2 \subseteq \mathcal{M}_{--}$ (as follows from Lemma 2.35, that is established below). Moreover, if $\mathcal{L} = \mathcal{L}_1'[+]\mathcal{L}_2'$ for a second pair of positive and negative subspaces $\mathcal{L}_1'$ and $\mathcal{L}_2'$, then* dim $\mathcal{L}_j =$ dim $\mathcal{L}_j'$ *for $j = 1, 2$; see Remark 2.36.*

Lemma 2.33 *Let $\mathcal{L}$ be a subspace of $\mathbb{C}^m$ such that $\mathcal{L} \cap \mathcal{L}^{[\perp]} = \{0\}$ and let u be a nonzero vector in $\mathcal{L} \cap \mathcal{M}_0$. Then there exists a vector $v \in \mathcal{L} \cap \mathcal{M}_0$ such that $[u, v] = 1$.*

Proof Under the given assumptions, the vector $u \notin \mathcal{L}^{[\perp]}$. Therefore, there exists a vector $y \in \mathcal{L}$ such that $[u, y] = 1$. It is readily checked that the vector

$$v = y - \frac{1}{2}[y, y]u$$

meets the two stated requirements. □

Lemma 2.34 *Let $\mathcal{L}$ be a subspace of $\mathbb{C}^m$ such that $\mathcal{L} \cap \mathcal{L}^{[\perp]} = \{0\}$ and $\mathcal{L} \cap \mathcal{M}_0 \neq \{0\}$. Then there exists a nonzero vector $u \in \mathcal{L} \cap \mathcal{M}_{++}$ and a nonzero vector $v \in \mathcal{L} \cap \mathcal{M}_{--}$ such that $[u, v] = 0$.*

Proof Under the given assumptions, the preceding lemma guarantees the existence of a pair of vectors $x, y \in \mathcal{L} \cap \mathcal{M}_0$ such that $[x, y] = 1$. Then as

$$[x + ty, x + ty] = 2t, \quad t \in \mathbb{R},$$

it follows that the vectors $u = x + y$ and $v = x - y$ meet the stated requirements. □

Lemma 2.35 *Let $\mathcal{L}$ be a nonzero subspace of $\mathbb{C}^m$. Then $\mathcal{L}$ admits a J-orthogonal decomposition of the form*

$$\mathcal{L} = \mathcal{L}_+[+]\mathcal{L}_0[+]\mathcal{L}_-, \tag{2.60}$$

where the subspaces $\mathcal{L}_+$, $\mathcal{L}_0$, $\mathcal{L}_-$ meet the conditions

$$\mathcal{L}_+ \subseteq \mathcal{M}_{++}, \quad \mathcal{L}_0 \subseteq \mathcal{M}_0 \quad \textit{and} \quad \mathcal{L}_- \subseteq \mathcal{M}_{--}.$$

Moreover, in any such decomposition,

(1) $\mathcal{L}_0 = \mathcal{L} \cap \mathcal{L}^{[\perp]}$.

(2) $\dim\ \mathcal{L}_+ + \dim\ \mathcal{L}_0 \leq p$ *and* $\dim\ \mathcal{L}_- + \dim\ \mathcal{L}_0 \leq q$.

(3) $\dim\ \mathcal{L}_0 \leq \min\{p, q\}$.

Proof Let $\mathcal{L}$ be a k-dimensional subspace of $\mathbb{C}^m$ with $1 \leq k \leq m$. If $\mathcal{L} \subseteq \mathcal{M}_0$, then the asserted decomposition of $\mathcal{L}$ holds with $\mathcal{L}_+ = \{0\}$, $\mathcal{L}_0 = \mathcal{L}$ and $\mathcal{L}_- = \{0\}$. If $\mathcal{L}$ is not a subset of $\mathcal{M}_0$, then either $\mathcal{L} \cap \mathcal{M}_{++} \neq \{0\}$ or $\mathcal{L} \cap \mathcal{M}_{--} \neq \{0\}$, or both. Suppose for the sake of definiteness that $\mathcal{L} \cap \mathcal{M}_{++} \neq \{0\}$. Then there exists a vector $u_1 \in \mathcal{L}$ with $[u_1, u_1] = 1$. Let $\mathcal{L}_1 = \{u \in \mathcal{L} : \ [u, u_1] = 0\}$. If $\mathcal{L}_1 \cap \mathcal{M}_{++} \neq \{0\}$, then there exists a vector $u_2 \in \mathcal{L}_1$ such that $[u_1, u_2] = 0$ and $[u_2, u_2] = 1$. Let

$$\mathcal{L}_2 = \{u \in \mathcal{L} : \ [u, u_j] = 0 \ \text{ for } \ j = 1, 2\}.$$

If $\mathcal{L}_2 \cap \mathcal{M}_{++} \neq \{0\}$, then the construction continues to generate vectors $u_3, \ldots, u_r$ with

$$[u_j, u_i] = \begin{cases} 0 & \textit{if} \quad i \neq j, \quad i, j = 1, \ldots, r, \\ 1 & \textit{if} \quad i = j, \quad i = 1, \ldots, r, \end{cases}$$

until

$$\mathcal{L}_r = \{u \in \mathcal{L} : \ [u, u_j] = 0, \quad j = 1, \ldots, r\}$$

is such that

$$\mathcal{L}_r \cap \mathcal{M}_{++} = \{0\}.$$

Then, if $\mathcal{L}_r \cap \mathcal{M}_{--} \neq \{0\}$, in a similar way one finds vectors $v_1, \ldots, v_s$ such that

$$[v_i, v_j] = \begin{cases} 0 & if \quad i \neq j, \quad i,j = 1, \ldots, s, \\ -1 & if \quad i = j, \quad i = 1, \ldots, s, \end{cases}$$

$$[v_i, u_j] = 0, \quad i = 1, \ldots, s, \quad j = 1, \ldots, r,$$

$$\mathcal{L}_{r+s} = \{u \in \mathcal{L} : [u, u_j] = 0, \ j = 1, \ldots, r \quad \text{and} \quad [u, v_i] = 0, \ i = 1, \ldots, s\},$$

and

$$\mathcal{L}_{r+s} \cap \mathcal{M}_{--} = \{0\}.$$

We conclude that $\mathcal{L}_{r+s}$ is a $(k - r - s)$-dimensional subspace of $\mathcal{M}_0$ and set $\mathcal{L}_0 = \mathcal{L}_{r+s}$,

$$\mathcal{L}_+ = \text{span}\,\{u_j : 1 \leq j \leq r\} \quad \text{if} \quad \mathcal{L} \cap \mathcal{M}_{++} \neq \{0\}$$

and

$$\mathcal{L}_- = \text{span}\,\{v_i : 1 \leq i \leq s\} \quad \text{if} \quad \mathcal{L} \cap \mathcal{M}_{--} \neq \{0\}.$$

This completes the verification of the decomposition (2.60).

Next, if $u \in \mathcal{L}_0$, then $u \in \mathcal{L}_\pm^{[\perp]}$ and (as follows by invoking Cauchy's inequality in $\mathcal{L}_0$), $u \in \mathcal{L}_0^{[\perp]}$. Therefore, $u \in \mathcal{L} \cap \mathcal{L}^{[\perp]}$. Conversely, if $u \in \mathcal{L} \cap \mathcal{L}^{[\perp]}$, then, in view of the decomposition (2.60), $u = u_+ + u_0 + u_-$, where $u_\pm \in \mathcal{L}_\pm$ and $u_0 \in \mathcal{L}_0$. Consequently,

$$0 = [u, u_\pm] = [u_\pm, u_\pm] \Longrightarrow u_\pm = 0$$

and hence that $u = u_0$, i.e., $u \in \mathcal{L}_0$. This completes the proof of (1).

The asserted bounds on the dimensions follow from Lemma 2.28, since

$$\mathcal{L}_\pm[+]\mathcal{L}_0 \subseteq \mathcal{M}_\pm \quad \text{and} \quad \dim(\mathcal{L}_\pm[+]\mathcal{L}_0) = \dim \mathcal{L}_\pm + \dim \mathcal{L}_0.$$

□

Remark 2.36 *The decomposition exhibited in the last lemma is not unique. Thus, for example, if* $\mathcal{L} = \text{span}\{u, w\}$ *where*

$$[u, u] = 1, \quad [u, w] = 0, \quad \text{and} \quad [w, w] = 0,$$

then $\mathcal{L}$ *admits the two decompositions*

$$\begin{aligned} \mathcal{L} &= \text{span}\{u\}[+]\text{span}\{w\} \\ &= \text{span}\{u + w\}[+]\text{span}\{w\}. \end{aligned}$$

Nevertheless, the dimensions of the subspaces in the sum (2.60) are unique: If the space $\mathcal{L}$ with decomposition (2.60) admits a second decomposition

$$\mathcal{L} = \widetilde{\mathcal{L}}_+[+]\widetilde{\mathcal{L}}_0[+]\widetilde{\mathcal{L}}_-,$$

in which $\widetilde{\mathcal{L}}_+ \subseteq \mathcal{M}_{++}$, $\widetilde{\mathcal{L}}_- \subseteq \mathcal{M}_{--}$ and $\widetilde{\mathcal{L}}_0 \subseteq \mathcal{M}_0$, then

$$\dim \mathcal{L}_\pm = \dim \widetilde{\mathcal{L}}_\pm \text{ and } \mathcal{L}_0 = \widetilde{\mathcal{L}}_0.$$

To verify Remark 2.36, let $u_1, \ldots, u_k$ and $v_1, \ldots, v_k$ be two bases of $\mathcal{L}$ and let $U = \begin{bmatrix} u_1 & \cdots & u_k \end{bmatrix}$ and $V = \begin{bmatrix} v_1 & \cdots & v_k \end{bmatrix}$. Then $U = VX$ for some invertible matrix $X \in \mathbb{C}^{k \times k}$ and hence, as

$$U^*JU = X^*V^*JVX,$$

Sylvester's Law of Inertia (see e.g., [Dy07] p. 431) implies that the matrix U^*JU has the same number of positive (resp., negative) eigenvalues as the matrix V^*JV.

2.7 Minus matrices

A matrix $U \in \mathbb{C}^{m \times m}$ is said to be a **minus matrix** with respect to the indefinite inner product $[\ ,\ ]$ (or with respect to the signature matrix J) if

$$[u, u] \le 0 \implies [Uu, Uu] \le 0.$$

Theorem 2.37 *Let $U \in \mathbb{C}^{m \times m}$ be a minus matrix with respect to an $m \times m$ signature matrix $J \ne \pm I_m$. Let*

$$\mu_-(U) = \inf \left\{ \frac{[Uv, Uv]}{[v, v]} : v \in \mathbb{C}^m \quad \textit{and} \quad [v, v] < 0 \right\}$$

and

$$\mu_+(U) = \sup \left\{ \frac{[Uu, Uu]}{[u, u]} : u \in \mathbb{C}^m \quad \textit{and} \quad [u, u] > 0 \right\}.$$

Then:

(1) $\mu_+(U) \le \mu_-(U)$.

(2) $\mu_-(U) \ge 0$.

(3) $[Uu, Uu] \le \mu_-(U)[u, u]$ *for every* $u \in \mathbb{C}^m$.

(4) $\mu_+(U) \le 0 \iff [Uu, Uu] \le 0$ *for every* $u \in \mathbb{C}^m$.

(5) $\det U \neq 0 \Longrightarrow \mu_+(U) > 0$.

(6) *If* $\mu_-(U) > 0$ *and* $\rho \in \mathbb{C}\setminus\{0\}$, *then*

$$\rho^{-1}U \in \mathcal{P}_{const}(J) \iff \mu_+(U) \leq |\rho|^2 \leq \mu_-(U).$$

Proof Since $J \neq \pm I_m$, there exist a pair of vectors $u, v \in \mathbb{C}^m$ such that $[u, u] = 1$ and $[v, v] = -1$. Suppose first that $[u, v] = re^{i\theta}$ for $r \geq 0$ and some $\theta \in [0, 2\pi)$ and let $\rho = ie^{i\theta}$. Then

$$[u + \rho v, u + \rho v] = [u, u] + 2\Re\{\rho[v, u]\} + [v, v] = 0$$

and hence, since U is a minus matrix,

$$[U(u + \rho v), U(u + \rho v)] \leq 0.$$

Thus, as the same inequality holds if ρ is replaced by $-\rho$,

$$[Uu, Uu] + 2\Re\{\rho[Uv, Uu]\} + [Uv, Uv] \leq 0$$

and

$$[Uu, Uu] - 2\Re\{\rho[Uv, Uu]\} + [Uv, Uv] \leq 0.$$

Consequently,

$$[Uu, Uu] \leq -[Uv, Uv]$$

for every choice of $u \in \mathbb{C}^m$ with $[u, u] = 1$ and $v \in \mathbb{C}^m$ with $[v, v] = -1$. But this is equivalent to the inequality

$$\frac{[Uu, u]}{[u, u]} \leq \frac{[Uv, Uv]}{[v, v]}$$

for every $u \in \mathbb{C}^m$ with $[u, u] > 0$ and $v \in \mathbb{C}^m$ with $[v, v] < 0$. The last inequality clearly justifies assertion (1).

Assertion (2) is clear from the definition.

Next, to justify (3), take any $u \in \mathbb{C}^m$ and observe that:

(a) $[u, u] > 0 \Longrightarrow [Uu, Uu] \leq \mu_+(U)[u, u] \leq \mu_-(U)[u, u]$, thanks to (1), whereas,

(b) $[u, u] < 0 \Longrightarrow \; -[Uu, Uu] \geq -\mu_-(U)[u, u]$, by the definition of $\mu_-(U)$, and

(c) $[u, u] = 0 \Longrightarrow [Uu, Uu] \leq 0 = \mu_-(U)[u, u]$.

The verification of (4) is similar: if $\mu_+(U) \leq 0$, then the first inequality for $[Uu, Uu]$, in (a) just above, implies that

$$[Uu, Uu] \leq 0$$

when $[u, u] > 0$. However, the same inequality must hold if $[u, u] \leq 0$ too, since U is a minus matrix. The implication $\Longleftarrow$ in (4) follows from the definition of $\mu_+(U)$.

Next, (5) is a consequence of (4), because if $\mu_+(U) \leq 0$, then the space $\mathcal{L} = \{Uu : u \in \mathbb{C}^m\}$ is a subspace of $\mathcal{M}_-$. Therefore, by Lemma 2.28, dim $\mathcal{L} \leq q$. However, this is not possible if U is invertible (and $J \neq -I_m$). Thus, U invertible $\Longrightarrow \mu_+(U) > 0$.

Finally, to verify (6), suppose first that $\rho^{-1}U \in \mathcal{P}_{const}(J)$ for some $\rho \in \mathbb{C}\backslash\{0\}$. Then, clearly,

$$[Uu, Uu] \leq |\rho|^2[u, u]$$

for every $u \in \mathbb{C}^m$. Thus,

$$[u, u] < 0 \Longrightarrow \frac{[Uu, Uu]}{[u, u]} \geq |\rho|^2 \Longrightarrow \mu_-(U) \geq |\rho|^2$$

and

$$[u, u] > 0 \Longrightarrow \frac{[Uu, Uu]}{[u, u]} \leq |\rho^2| \Longrightarrow \mu_+(U) \leq |\rho^2|.$$

Moreover, the argument can be run backwards to show that if $\mu_+(U) \leq |\rho^2| \leq \mu_-(U)$, then

$$[Uu, Uu] \leq |\rho^2|[u, u],$$

and hence that $\rho^{-1}U \in \mathcal{P}_{const}(J)$, as claimed. □

Corollary 2.38 *An matrix $U \in \mathbb{C}^{m\times m}$ is a minus matrix with respect to an indefinite inner product [,] if and only if there exists a constant $c \geq 0$ such that*

$$[Uu, Uu] \leq c[u, u]. \tag{2.61}$$

Proof One direction follows from (2) and (3) in Theorem 2.37. The other direction is self-evident. □

2.8 Strictly minus matrices

A matrix $U \in \mathbb{C}^{m\times m}$ is said to be a **strictly minus matrix** with respect to the indefinite inner product $[\ ,\ ]$ (or with respect to the signature matrix J) if there exists a scalar $c > 0$ such that

$$[Uu, Uu] \le c[u, u] \tag{2.62}$$

for every $u \in \mathbb{C}^m$.

Remark 2.39 *Clearly, U is a strictly minus matrix if and only if*

$$U^*JU \le cJ \quad \textit{for some} \quad c > 0,$$

i.e., if and only if $c^{-\frac{1}{2}}U \in \mathcal{P}_{const}(J)$. By Theorem 2.37, **every invertible minus matrix is a strictly minus matrix** *and the inequality (2.62) holds for every $u \in \mathbb{C}^m$ if and only if $\mu_+(U) \le c \le \mu_-(U)$.*

Theorem 2.40 *Let $U \in \mathbb{C}^{m\times m}$ and let J be an $m \times m$ signature that is unitarily equivalent to j_{pq}. Then the following statements are equivalent:*

(1) *U is a strictly minus matrix with respect to J.*

(2) *U is a minus matrix with respect to J and $\mu_-(U) > 0$.*

(3) *$U\mathcal{M}_- \subseteq \mathcal{M}_-$ and $U\mathcal{M}_{--} \subseteq \mathcal{M}_{--}$.*

(4) *The linear mapping $X \longrightarrow UX$ maps each of the two sets*
$$\{X \in \mathbb{C}^{m\times q} :\ X^*JX \le 0\}, \quad \{X \in \mathbb{C}^{m\times q} : X^*JX < 0\}$$
into itself.

(5) *There exists a nonzero constant c such that $cU \in \mathcal{P}_{const}(J)$.*

Proof The implications (1) $\Longrightarrow$ (2), (1) $\Longrightarrow$ (3), (3) $\Longrightarrow$ (4) and (1) $\Longleftrightarrow$ (5) are clear from the definitions, while (2) $\Longrightarrow$ (1) follows from Theorem 2.37.

Suppose next that (4) is in force. Then U is clearly a minus matrix with respect to J and hence, in view of Theorem 2.37, $\mu_-(U) \ge 0$. We wish to show that $\mu_-(U) > 0$. If this were not the case, i.e., if $\mu_-(U) = 0$, then, by another application of Theorem 2.37,

$$[Uu, Uu] \le 0$$

for every vector $u \in \mathbb{C}^m$. Thus, the subspace

$$\mathcal{L} = \{Uu : u \in \mathbb{C}^m\}$$

is a subspace of $\mathcal{M}_-$. Without loss of generality, we can assume that $J = j_{pq}$ and hence that there exists a matrix $s \in \mathcal{S}_{const}^{p\times q}$ such that

$$\{Uu : u \in \mathbb{C}^m\} \subseteq \left\{ \begin{bmatrix} s \\ I_q \end{bmatrix} y : y \in \mathbb{C}^q \right\}.$$

In particular,

$$\left\{ U \begin{bmatrix} \varepsilon \\ I_q \end{bmatrix} y : y \in \mathbb{C}^q \right\} \subseteq \left\{ \begin{bmatrix} s \\ I_q \end{bmatrix} y : y \in \mathbb{C}^q \right\}$$

for every choice of $\varepsilon \in \mathcal{S}_{const}^{p\times q}$. Let $\varepsilon \in \mathring{\mathcal{S}}_{const}^{p\times q}$ and let

$$X = \begin{bmatrix} \varepsilon \\ I_q \end{bmatrix}.$$

Then $X^* j_{pq} X < 0$ and hence, by assumption (4), $(UX)^* j_{pq} (UX) < 0$, i.e.,

$$(u_{11}\varepsilon + u_{12})^*(u_{11}\varepsilon + u_{12}) < (u_{21}\varepsilon + u_{22})^*(u_{21}\varepsilon + u_{22}).$$

Therefore, $u_{21}\varepsilon + u_{22}$ is invertible and consequently

$$\begin{aligned} \left\{ U \begin{bmatrix} \varepsilon \\ I_q \end{bmatrix} y : y \in \mathbb{C}^q \right\} &= \left\{ \begin{bmatrix} T_U[\varepsilon] \\ I_q \end{bmatrix} (u_{21}\varepsilon + u_{22}) y : y \in \mathbb{C}^q \right\} \\ &= \left\{ \begin{bmatrix} T_U[\varepsilon] \\ I_q \end{bmatrix} y : y \in \mathbb{C}^q \right\} \\ &\subseteq \left\{ \begin{bmatrix} s \\ I_q \end{bmatrix} y : y \in \mathbb{C}^q \right\}. \end{aligned}$$

Therefore

$$s = T_U[\varepsilon] \quad \text{for every} \quad \varepsilon \in \mathring{\mathcal{S}}^{p\times q} \tag{2.63}$$

and $s \in \mathring{\mathcal{S}}_{const}^{p\times q}$. Now let $v_1, v_2, \ldots$ be a sequence of nonzero vectors in $\mathcal{M}_{--}$ such that

$$\lim_{n\uparrow\infty} \frac{[Uv_n, Uv_n]}{[v_n, v_n]} = \mu_-(U).$$

We can assume that $[v_n, v_n] = -1$ and then, upon setting $v_n = \text{col}(x_n, y_n)$ with $x_n \in \mathbb{C}^p$ and $y_n \in \mathbb{C}^q$, it is readily seen that

$$\|y_n\|^2 = \|x_n\|^2 + 1 \geq 1.$$

Consequently, the vectors $w_n = \|y_n\|^{-1} v_n$ are subject to the bounds

$$1 \leq \|w_n\|^2 = \frac{\|x_n\|^2 + \|y_n\|^2}{\|y_n\|^2} = \frac{2\|y_n\|^2 - 1}{\|y_n\|^2} \leq 2.$$

Thus, we can assume that the vectors w_n tend to a limit $w \in \mathcal{M}_-$ as $n \uparrow \infty$. Moreover, if $\mu_-(U) = 0$ and $\delta > 0$, then there exists an integer N such that

$$[Uw_n, Uw_n] \geq \delta[w_n, w_n] \quad \text{when} \quad n \geq N,$$

which in turn implies that

$$[Uw, Uw] \geq \delta[w, w].$$

The assumption $w_n \in \mathcal{M}_{--}$ means that in the representation $w_n = \text{col}(\|y_n\|^{-1} x_n, \|y_n\|^{-1} y_n)$, $x_n = \varepsilon_n y_n$ for some $\varepsilon_n \in \mathring{\mathcal{S}}^{p \times q}_{const}$ and hence, in view of (2.63), that

$$\begin{aligned} Uw_n &= \begin{bmatrix} T_U[\varepsilon_n] \\ I_q \end{bmatrix} (u_{21}\varepsilon_n + u_{22}) \frac{y_n}{\|y_n\|} \\ &= \begin{bmatrix} s \\ I_q \end{bmatrix} (u_{21}\varepsilon_n + u_{22}) \frac{y_n}{\|y_n\|} \\ &\longrightarrow \begin{bmatrix} s \\ I_q \end{bmatrix} (u_{21}\varepsilon + u_{22})\eta \end{aligned}$$

for some choice of $\varepsilon \in \mathcal{S}^{p \times q}_{const}$ and $\eta \in \mathbb{C}^q$ with $\|\eta\| = 1$, as $n \uparrow \infty$ along an appropriately chosen subsequence. But this implies that

$$0 > [Uw, Uw] \geq \delta[w, w] \quad (\text{since} \quad s \in \mathring{\mathcal{S}}^{p \times q}_{const})$$

and hence that $\mu_-(U) > 0$. Thus, as

$$[Uu, Uu] \leq \mu_-(U)[u, u] \quad \text{for every} \quad u \in \mathbb{C}^m,$$

by Theorem 2.37; and $\mu_-(U) > 0$, by the preceding analysis, (5) holds with $c = \mu_-(U)^{-1/2}$. □

We are now able to complete the proof of Theorem 2.13.

Corollary 2.41 *If conditions (1)–(3) in the setting of Theorem 2.13 are in force, then condition (4) is in force also.*

Proof Let $u = \text{col}(x, y)$ be a nonzero vector in $\mathcal{M}_{--}$ with components $x \in \mathbb{C}^p$ and $y \in \mathbb{C}^q$. Then the matrix

$$\varepsilon = \frac{xy^*}{y^*y} \quad \text{belongs to} \quad \mathring{\mathcal{S}}^{p\times q}_{const} \quad \text{and} \quad u = \begin{bmatrix} \varepsilon \\ I_q \end{bmatrix} y.$$

Therefore, by (1) and (3), the vector

$$Uu = U \begin{bmatrix} \varepsilon \\ I_q \end{bmatrix} y = \begin{bmatrix} T_U[\varepsilon] \\ I_q \end{bmatrix} (u_{21}\varepsilon + u_{22})y$$

clearly belongs to $\mathcal{M}_{--}$. Similar considerations show that $u \in \mathcal{M}_- \implies Uu \in \mathcal{M}_-$ and hence that U satisfies condition (3) of Theorem 2.40. Therefore, in view of the equivalences established in Theorem 2.40, U also satisfies condition (4) of Theorem 2.13, as claimed. □

The numbers $\mu_+(U)$ and $\mu_-(U)$ associated with a matrix $U \in \mathbb{C}^{m\times m}$ that is a strictly minus matrix with respect to a signature matrix $J \neq \pm I_m$ are eigenvalues of the matrix

$$G = U^\times U = JU^*JU,$$

where $U^\times$ denotes the adjoint of U with respect to the indefinite inner product [,] based on J, i.e.,

$$[Uu, v] = [u, U^\times v]$$

for every choice of $u, v \in \mathbb{C}^m$. Before turning to the proof of this assertion, it is convenient to establish some preliminary results:

Lemma 2.42 *Let $U \in \mathbb{C}^{m\times m}$ be a strictly minus matrix with respect to a signature matrix $J \neq \pm I_m$ and let $G = JU^*JU$. Then:*

(1) *The eigenvalues of G are nonnegative.*

(2) *If $Gu = \lambda u$ and $u \neq 0$, then:*

 (a) $[u, u] > 0 \implies \lambda \leq \mu_+(U)$.

 (b) $[u, u] < 0 \implies \lambda \geq \mu_-(U)$.

(c) $[u,u] = 0 \implies \lambda = \mu_-(U) = \mu_+(U)$.

(d) $\lambda < \mu_-(U) \implies [u,u] > 0$.

(e) $\lambda > \mu_+(U) \implies [u,u] < 0$.

(f) *If* $\mu_+(U) < \mu_-(U)$, *then* $\lambda \notin (\mu_+(U), \mu_-(U))$.

Proof In view of Remark 2.39, there exists a constant $c > 0$ such that $U^*JU \le cJ$. Let $H = cI_m - G$ and let $Gy = \mu y$ for some nonzero vector $y \in \mathbb{C}^m$. Then

$$JH = cJ - U^*JU \ge 0$$

and

$$y^*JHy = (c - \mu)y^*Jy.$$

Moreover,

$$cJG - G^*JG = U^*J(cJ - U^*JU)JU$$

and

$$y^*(cJG - G^*JG)y = (c - \mu)\overline{\mu}y^*Jy = \overline{\mu}y^*JHy \ge 0.$$

Thus, if $y^*JHy > 0$, then $\mu \ge 0$. On the other hand, if $y^*JHy = 0$, then $JHy = 0$, since $JH \ge 0$. Consequently, $Hy = (c - \mu)y = 0$, which forces $\mu = c > 0$. Therefore (1) holds.

Next, if $[y,y] \ne 0$, then (2a) and (2b) follow easily from the definitions of $\mu_-(U)$ and $\mu_+(U)$ and the auxiliary identity

$$\mu = \frac{[Gy,y]}{[y,y]} = \frac{[Uy,Uy]}{[y,y]}.$$

On the other hand, if $[y,y] = 0$, then $(c - \mu)y = 0$ for every point $c > 0$ that meets the inequality $\mu_+(U) \le c \le \mu_-(U)$. Since $\mu_-(U) > 0$ for a strictly minus matrix, this is only viable if $\mu_+(U) = \mu_-(U)$. Thus (c) is proved; assertions (d), (e) and (f) are easy consequences of (a), (b) and (c) and the definitions of $\mu_-(U)$ and $\mu_+(U)$, respectively. □

Lemma 2.43 *Let $U \in \mathbb{C}^{m\times m}$ be a strictly minus matrix with respect to a signature matrix $J \ne \pm I_m$ and let λ be an eigenvalue of the matrix $G = JU^*JU$. Suppose further that the algebraic multiplicity of λ as an eigenvalue of G exceeds its geometric multiplicity. Then*

$$\lambda = \mu_+(U) = \mu_-(U)$$

and there exists a J-neutral eigenvector of G with eigenvalue λ.

Proof If $(G - \lambda I_m)^k u = 0$ and $(G - \lambda I_m)^{k-1} u \neq 0$ for some integer $k > 1$, then clearly $y = (G - \lambda I_m)^{k-1} u$ is an eigenvector of G corresponding to the eigenvalue λ. Moreover,

$$\begin{aligned} [y, y] &= [(G - \lambda I_m)^{k-1} u, y] \\ &= [u, (G - \lambda I_m)^{k-1} y] = 0. \end{aligned}$$

Therefore, by Lemma 2.42, $\lambda = \mu_-(U) = \mu_+(U)$, as claimed. □

Corollary 2.44 *Let $U \in \mathbb{C}^{m \times m}$ be a strictly minus matrix with respect to a signature matrix $J \neq \pm I_m$ and let $G = JU^*JU$. Then:*

(1) *The algebraic multiplicity of every eigenvalue λ of G with $\lambda < \mu_+(G)$ or $\lambda > \mu_-(G)$ is equal to its geometric multiplicity.*

(2) *If $\mu_+(G) < \mu_-(G)$, then G is diagonalizable.*

Lemma 2.45 *Let $U \in \mathbb{C}^{m \times m}$ be a strictly minus matrix with respect to a signature matrix $J \neq \pm I_m$. Let $u_1, \ldots, u_k$ be a linearly independent set of eigenvectors of the matrix $G = JU^*JU$ with the same eigenvalue λ and let $\mathcal{L} = \operatorname{span}\{u_1, \ldots, u_k\}$.*

(1) *If $\lambda < \mu_-(G)$, then $\mathcal{L} \subseteq \mathcal{M}_{++}$.*

(2) *If $\lambda > \mu_+(G)$, then $\mathcal{L} \subseteq \mathcal{M}_{--}$.*

Proof Clearly, $Gu = \lambda u$ for every vector $u \in \mathcal{L}$. Therefore, if $\lambda < \mu_-(G)$ and $u \neq 0$, then $[u, u] > 0$, by Lemma 2.42. Similarly, if $\lambda > \mu_+(G)$ and $u \neq 0$, then $[u, u] < 0$, by the same lemma. □

Lemma 2.46 *Let $U \in \mathbb{C}^{m \times m}$ be a strictly minus matrix with respect to a signature matrix $J \neq \pm I_m$. Let $u_1, \ldots, u_k$ be a linearly independent set of eigenvectors of the matrix $G = JU^*JU$ corresponding to the eigenvalues $\lambda_1, \ldots, \lambda_k$ and let $\mathcal{L} = \operatorname{span}\{u_1, \ldots, u_k\}$.*

(1) *The eigenvectors of G corresponding to distinct eigenvalues are orthogonal with respect to the indefinite inner product.*

(2) *If $\lambda_j < \mu_-(G)$ for $j = 1, \ldots, k$, then $\mathcal{L} \subseteq \mathcal{M}_{++}$.*

(3) *If $\lambda_j > \mu_+(G)$ for $j = 1, \ldots, k$, then $\mathcal{L} \subseteq \mathcal{M}_{--}$.*

Proof The formula

$$\lambda_i[u_i, u_j] = [Gu_i, u_j] = [u_i, Gu_j] = \lambda_j[u_i, u_j]$$

(which utilizes the fact that the eigenvalues of G are real) clearly implies that

$$[u_i, u_j] = 0 \text{ if } \lambda_i \neq \lambda_j.$$

Thus, in view of the previous lemma, $\mathcal{L} = \text{span}\{v_1, \ldots, v_k\}$, where

$$[v_i, v_j] = \begin{cases} 0 & \text{if} \quad i \neq j \\ 1 & \text{if} \quad i = j \quad \text{and } \lambda_j < \mu_-(G) \\ -1 & \text{if} \quad i = j \quad \text{and } \lambda_j > \mu_+(G). \end{cases}$$

Thus, for example, if $\lambda_j > \mu_+(G)$ and $u \in \mathcal{L}$, then

$$u = \sum_{j=1}^{k} c_j v_j \implies [u, u] = -\sum_{j=1}^{k} |c_j|^2 \implies u \in \mathcal{M}_{--}. \qquad \blacksquare$$

Theorem 2.47 *Let $U \in \mathbb{C}^{m\times m}$ be a strictly minus matrix with respect to an $m \times m$ signature matrix $J \neq \pm I_m$ and let $G = JU^*JU$. Then:*

(1) *The eigenvalues of the matrix G are all nonnegative.*

(2) *If $\lambda_1(G) \leq \cdots \leq \lambda_m(G)$ is a list of the eigenvalues of G repeated according to algebraic multiplicity, then*

$$\mu_+(U) = \lambda_p(G) \quad \textit{and} \quad \mu_-(U) = \lambda_{p+1}(G).$$

(3) *If $\gamma \in \mathbb{C}\backslash\{0\}$, then*

$$\gamma^{-1}U \in \mathcal{P}_{const}(J) \iff \lambda_p(G) \leq |\gamma|^2 \leq \lambda_{p+1}(G).$$

(4) *The matrix G has a J-neutral eigenvector with eigenvalue λ if and only if $\lambda_p(G) = \lambda_{p+1}(G)$. In this case $\lambda = \lambda_p(G) = \lambda_{p+1}(G)$.*

Proof The first assertion is verified in Lemma 2.42. The rest of the proof is broken into steps.

Step 1 *Assertion (2) holds if $\mu_+(U) < \mu_-(U)$.*

By Lemma 2.43, the matrix G is diagonalizable in this setting. Let $u_1, \ldots, u_m$ be a set of eigenvectors corresponding to $\lambda_1 \leq \cdots \leq \lambda_m$ that are orthogonal with respect to [,] and let

$$\mathcal{L}_+ = \begin{cases} 0 & \text{if } \lambda_j \geq \mu_-(U) \text{ for } j = 1, \ldots, m \\ \text{span}\{u_j : \lambda_j < \mu_-(U)\} & \text{otherwise} \end{cases}$$

and

$$\mathcal{L}_- = \begin{cases} 0 & \text{if } \lambda_j \leq \mu_+(U) \text{ for } j = 1, \ldots, m \\ \text{span}\{u_j : \lambda_j > \mu_+(U)\} & \text{otherwise.} \end{cases}$$

Then, in view of Lemmas 2.42 and 2.46, $\mathcal{L}_+ \subseteq \mathcal{M}_{++}$ and $\mathcal{L}_- \subseteq \mathcal{M}_{--}$. Moreover, since G has no eigenvalues in the open interval $(\mu_+(U), \mu_-(U))$ by Lemma 2.42,

$$\mathcal{L}_+[+]\mathcal{L}_- = \mathbb{C}^m.$$

Therefore,

$$\dim \mathcal{L}_+ = p \quad \text{and} \quad \dim \mathcal{L}_- = q,$$

$$\lambda_1 \leq \cdots \leq \lambda_p \leq \mu_+(U) < \mu_-(U) \leq \lambda_{p+1} \leq \cdots \leq \lambda_m$$

and every vector $u \in \mathbb{C}^m$ admits a decomposition

$$u = u_+ + u_-$$

with components $u_\pm \in \mathcal{L}_\pm$ that are orthogonal with respect to [,]. Consequently,

$$[u, u] = [u_+, u_+] + [u_-, u_-]$$

and, since $G\mathcal{L}_\pm \subseteq \mathcal{L}_\pm$, $[u_+, u_+] \geq 0$ and $[u_-, u_-] \leq 0$, it is readily seen that

$$\begin{aligned} [Gu, u] &= [Gu_+, u_+] + [Gu_-, u_-] \\ &\leq \lambda_p[u_+, u_+] + \lambda_{p+1}[u_-, u_-] \\ &\leq \gamma[u, u] \end{aligned}$$

for every choice of γ in the interval $\lambda_p \leq \gamma \leq \lambda_{p+1}$. But this implies that

$$\mu_+(U) \leq \lambda_p \quad \text{and} \quad \mu_-(U) \geq \lambda_{p+1}$$

and hence, as the opposite inequalities are also in force, that (2) holds.

Step 2 *Assertion (2) holds if* $\mu_+(U) = \mu_-(U)$.

Let $\lambda = \mu_+(U) = \mu_-(U)$. Then even though G is not necessarily diagonalizable, Corollary 2.44 guarantees that the algebraic multiplicity of every eigenvalue λ_j of G with $\lambda_j \neq \lambda$ is equal to its geometric multiplicity. Thus, we can choose an orthogonal set of eigenvectors u_j for each such choice of λ_j and can define

$$\mathcal{L}_+ = \begin{cases} 0 \text{ if } \lambda_j \geq \mu_+(U) \text{ for } j = 1, \ldots, m \\ \text{span}\{u_j : \lambda_j < \mu_+(U)\} \text{ otherwise} \end{cases}$$

$$\mathcal{L}_- = \begin{cases} 0 \text{ if } \lambda_j \leq \mu_-(U) \text{ for } j = 1, \ldots, m \\ \text{span}\{u_j : \lambda_j > \mu_-(U)\} \text{ otherwise.} \end{cases}$$

Then, in view of Lemma 2.46, $\mathcal{L}_+ \subseteq \mathcal{M}_{++}$, $\mathcal{L}_- \subseteq \mathcal{M}_{--}$ and, consequently, $\mathcal{L}_+ \cap \mathcal{L}_- = \{0\}$. Moreover, by (1) of Lemma 2.46, $\mathcal{L}_+$ is orthogonal to $\mathcal{L}_-$ with respect to $[\ ,\ ]$. Let $\mathcal{L} = \mathcal{L}_+[+]\mathcal{L}_-$. Then, in view of Lemma 2.31, $\mathcal{L} \cap \mathcal{L}^{[\perp]} = \{0\}$ and hence

$$\mathbb{C}^m = \mathcal{L}[+]\mathcal{L}^{[\perp]}$$

and, by Lemma 2.35,

$$\mathcal{L}^{[\perp]} = \mathcal{K}_+[+]\mathcal{K}_-,$$

where $\mathcal{K}_+ \subseteq \mathcal{M}_{++}$ and $\mathcal{K}_- \subseteq \mathcal{M}_{--}$. Consequently, as

$$(\mathcal{L}_+[+]\mathcal{K}_+)[+](\mathcal{L}_-[+]\mathcal{K}_-) = \mathbb{C}^m$$

and

$$\mathcal{L}_+[+]\mathcal{K}_+ \subseteq \mathcal{M}_{++} \text{ and } \mathcal{L}_-[+]\mathcal{K}_- \subseteq \mathcal{M}_{--},$$

it is readily seen that

$$\dim\ \mathcal{L}_+ + \dim\ \mathcal{K}_+ = p \text{ and } \dim\ \mathcal{L}_- + \dim\ \mathcal{K}_- = q.$$

Moreover, the discussion in Step 1 is easily adapted to show that $\mathcal{K}_+ \neq \{0\}$ and $\mathcal{K}_- \neq \{0\}$. Thus, for example, if $\mathcal{K}_+ = \{0\}$, then $\dim\ \mathcal{L}_+ = p$ and consequently $\lambda_1 \leq \cdots \leq \lambda_p < \mu_+(U)$. Moreover, if $u \in \mathbb{C}^m$, then

$$u = u_+ + u_-,$$

where $u_+ \in \mathcal{L}_+$, $u_- \in \mathcal{K}_-[+]\mathcal{L}_-$, $G\mathcal{L}_+ \subseteq \mathcal{L}_+$ and $G(\mathcal{K}_-[+]\mathcal{L}_-) \subseteq \mathcal{L}_-[+]\mathcal{K}_-$. Therefore,

$$\begin{aligned} [u,u] &= [u_+,u_+] + [u_-,u_-] \\ [Gu,u] &= [Gu_+,u_+] + [Gu_-,u_-] \\ &\leq \lambda_p[u_+,u_+] + \mu_-(U)[u_-,u_-] \\ &\leq \lambda_p[u,u], \end{aligned}$$

since $\lambda_p < \mu_+(U) = \mu_-(U)$ by the definition of $\mathcal{L}_+$. But the last inequality also implies that $\mu_+(U) \leq \lambda_p$ which contradicts the fact that $\lambda_p < \mu_+(U)$. Thus, $\mathcal{K}_+ \neq \{0\}$ and, by a similar argument, $\mathcal{K}_- \neq \{0\}$. Therefore, the numbers $r = \dim\ \mathcal{L}_+$ and $s = \dim\ \mathcal{L}_-$ are subject to the inequalities

$$0 \leq r \leq p-1 \quad \text{and} \quad 0 \leq s \leq q-1.$$

But this means that in the list $\lambda_1 \leq \cdots \leq \lambda_m$,

$$\lambda_j = \lambda \quad \text{for} \quad r+1 \leq j \leq m-s$$

and hence that

$$\lambda_p = \lambda = \mu_+(U) = \mu_-(U) = \lambda_{p+1}.$$

Step 3 *Assertions (3) and (4) hold.*

Assertion (3) follows from (2) and Theorem 2.37. Next, to verify (4), suppose first that $\lambda = \mu_+(U) = \mu_-(U)$. Then, by the analysis in Step 2, λ is an eigenvalue of G and the space

$$\mathcal{K} = \ker((G - \lambda I_m)^m)$$

may be decomposed as

$$\mathcal{K} = \mathcal{K}_+[+]\mathcal{K}_-$$

where $\mathcal{K}_+ \subseteq \mathcal{M}_{++}$, $\mathcal{K}_- \subseteq \mathcal{M}_{--}$, $\dim\ \mathcal{K}_+ \geq 1$ and $\dim\ \mathcal{K}_- \geq 1$. There are two possibilities; either
(a) $\dim\ \mathcal{K} > \dim\ \ker(G - \lambda I_m)$ or (b) $\dim\ \mathcal{K} = \dim\ \ker(G - \lambda I_m)$.
In case (a), Lemma 2.43 guarantees the existence of a neutral eigenvector. In case (b), we may choose $u_+ \in \mathcal{K}_+$ with $[u_+,u_+] = 1$ and $u_- \in \mathcal{K}_-$ with $[u_-,u_-] = -1$. Then $u = u_+ + u_-$ is a neutral eigenvector of G corresponding

to λ. This completes the proof of (4) in one direction. However, the other direction is immediate from (2) and Lemma 2.42. □

Theorem 2.48 *Let $U \in \mathbb{C}^{m\times m}$ be a strictly minus matrix with respect to an $m \times m$ signature matrix $J \neq \pm I_m$. Then the following conditions are equivalent:*

(1) $\{u \in \mathbb{C}^m : u \in \mathcal{M}_0$ *and* $Uu \in \mathcal{M}_0\} \neq \{0\}$.

(2) *There exists a neutral eigenvector of* G.

(3) $\lambda_p(G) = \mu_+(U) = \mu_-(U) = \lambda_{p+1}(G)$.

(4) *There exists exactly one scalar* $\rho > 0$ *such that* $\rho^{-1}U \in \mathcal{P}_{const}(\mathcal{J})$.

Proof The equivalences (2) $\Longleftrightarrow$ (3) $\Longleftrightarrow$ (4) are consequences of the previous theorem. The implication (2) $\Longrightarrow$ (1) is clear. It remains only to prove the implication (1) $\Longrightarrow$ (2).

Let (1) be in force and suppose that $\mu_+(U) < \mu_-(U)$. Then

$$\mathbb{C}^m = \mathcal{L}_+[+]\mathcal{L}_-,$$

where the spaces $\mathcal{L}_\pm$ are defined in Step 1 of the proof of the previous theorem. Thus, if $u \in \mathcal{M}_0$ and $Uu \in \mathcal{M}_0$, $u \neq 0$, then

$$u = u_+ + u_- \quad \text{and} \quad Gu = Gu_+ + Gu_-,$$

where $u_\pm \in \mathcal{L}_\pm$ and $Gu_\pm \in \mathcal{L}_\pm$. Consequently,

$$\begin{aligned} 0 &= [u,u] = [u_+,u_+] + [u_-,u_-], \\ 0 &= [Gu,u] = [Gu_+,u_+] + [Gu_-,u_-] \\ &\leq \mu_+(U)[u_+,u_+] + \mu_-(U)[u_-,u_-] \end{aligned}$$

and hence

$$0 \leq \{\mu_+(U) - \mu_-(U)\}[u_+,u_+].$$

Therefore, since $[u_+,u_+] > 0$, this last inequality implies that $\mu_+(U) \geq \mu_-(U)$, contrary to assumption. Consequently (3) holds, and thus, as (3) is equivalent to (2), so does (2). □

2.9 Linear fractional transformations in $\mathcal{S}^{p\times q}_{const}$

Lemma 2.49 *Let $U \in \mathbb{C}^{m\times m}$ and suppose that*

(1) $\mathring{\mathcal{S}}^{p\times q}_{const} \subseteq \mathcal{D}(T_U)$.

(2) $T_U[\mathring{\mathcal{S}}^{p\times q}_{const}] \subseteq \mathcal{S}^{p\times q}_{const}$.

(3) *The set* $T_U[\mathcal{S}^{p\times q}_{const} \cap \mathcal{D}(T_U)]$ *contains more than one matrix.*

Then:

(4) $U = cW$ *for some choice of* $c \in \mathbb{C} \setminus \{0\}$ *and* $W \in \mathcal{P}_{const}(j_{pq})$.

(5) $\mathcal{S}^{p\times q}_{const} \subseteq \mathcal{D}(T_U)$.

(6) $T_U[\mathcal{S}^{p\times q}_{const}] \subseteq \mathcal{S}^{p\times q}_{const}$.

(7) $T_U[\mathring{\mathcal{S}}^{p\times q}_{const}] \subseteq \mathring{\mathcal{S}}^{p\times q}_{const}$.

Proof Let $u \in \mathbb{C}^m$ be a nonzero vector such that $u^* j_{pq} u \leq 0$. Then by Lemma 2.29, there exists a matrix $\varepsilon \in \mathcal{S}^{p\times q}_{const}$ such that

$$u = \begin{bmatrix} \varepsilon \\ I_q \end{bmatrix} y \quad \text{for some vector} \quad y \in \mathbb{C}^q.$$

In view of assumptions (1) and (2),

$$\begin{bmatrix} r\varepsilon \\ I_q \end{bmatrix}^* U^* j_{pq} U \begin{bmatrix} r\varepsilon \\ I_q \end{bmatrix} \leq 0$$

for every choice of r in the interval $0 \leq r < 1$. Consequently,

$$u^* U^* j_{pq} U u = \lim_{r \uparrow 1} y^* \begin{bmatrix} r\varepsilon \\ I_q \end{bmatrix}^* U^* j_{pq} U \begin{bmatrix} r\varepsilon \\ I_q \end{bmatrix} y \leq 0.$$

Thus, U is a minus matrix with respect to the indefinite inner product based on j_{pq} and hence, in view of Theorem 2.37, $\mu_-(U) \geq 0$. However, if $\mu_-(U) = 0$, then, by another application of Theorem 2.37, the vector space

$$\mathcal{L} = \{Uu : \ u \in \mathbb{C}^m\}$$

is a subspace of $\mathcal{M}_-$ (with respect to the inner product based on j_{pq}) and consequently, by Lemma 2.29, there exists a matrix $s \in \mathcal{S}^{p\times q}_{const}$ such that

$$\mathcal{L} \subseteq \left\{ \begin{bmatrix} s \\ I_q \end{bmatrix} y : \ y \in \mathbb{C}^q \right\}.$$

Therefore,

$$\left\{ \begin{bmatrix} T_U[\varepsilon] \\ I_q \end{bmatrix} y : \ y \in \mathbb{C}^q \right\} \subseteq \left\{ \begin{bmatrix} s \\ I_q \end{bmatrix} y : \ y \in \mathbb{C}^q \right\}$$

for every $\varepsilon \in \mathcal{S}_{const}^{p\times q} \cap \mathcal{D}(T_U)$. But this implies that

$$T_U[\varepsilon] = s$$

for every $\varepsilon \in \mathcal{S}_{const}^{p\times q} \cap \mathcal{D}(T_U)$, which is not viable with assumption (3). Consequently $\mu_-(U) > 0$ and hence by item (6) of Theorem 2.37, the matrix $\mu_-(U)^{-1/2}U$ belongs to the class $\mathcal{P}_{const}(j_{pq})$. This justifies assertion (4). Assertions (5)–(7) then follow from Theorem 2.13. □

Lemma 2.50 *Let W be an invertible minus matrix with respect to j_{pq}. Then*

(1) *w_{22} is an invertible $q \times q$ matrix.*

(2) *$w_{12}w_{22}^{-1} \in \mathring{\mathcal{S}}_{const}^{p\times q}$ and $w_{22}^{-1}w_{21} \in \mathring{\mathcal{S}}_{const}^{q\times p}$.*

(3) *$w_{11} - w_{12}w_{22}^{-1}w_{21}$ is an invertible $p \times p$ matrix.*

(4) *$\mathcal{S}_{const}^{p\times q} \subseteq \mathcal{D}(T_W)$.*

(5) *$T_W[\mathcal{S}_{const}^{p\times q}] \subseteq \mathcal{S}_{const}^{p\times q}$.*

(6) *$T_W[\mathring{\mathcal{S}}_{const}^{p\times q}] \subseteq \mathring{\mathcal{S}}_{const}^{p\times q}$.*

(7) *T_W is injective on $\mathcal{D}(T_W)$.*

Proof Let W be an invertible minus matrix with respect to j_{pq}. Then, in view of items (1), (5) and (6) of Theorem 2.37, there exists a number $\gamma \in \mathbb{C}\setminus\{0\}$ such that the matrix $\widetilde{W} = \gamma^{-1}W$ belongs to the class $\mathcal{P}_{const}(j_{pq})$. Consequently, the first three assertions follow from Lemma 2.7, whereas the next three assertions follow from Theorem 2.13. Finally, (7) follows from Lemma 2.10. □

Theorem 2.51 *Let $U \in \mathbb{C}^{m\times m}$. Then the following implicatations are in force:*

(1) *If $\mathring{\mathcal{S}}_{const}^{p\times q} \subseteq \mathcal{D}(T_U)$, $T_U[\mathring{\mathcal{S}}_{const}^{p\times q}] \subseteq \mathcal{S}_{const}^{p\times q}$ and T_U is injective on $\mathring{\mathcal{S}}_{const}^{p\times q}$, then*

$$cU \in \mathcal{P}_{const}(j_{pq}) \quad \text{for some } c \in \mathbb{C}\setminus\{0\} \text{ and } U \text{ is invertible.}$$

(2) *If $cU \in \mathcal{P}_{const}(j_{pq})$ for some $c \in \mathbb{C}\setminus\{0\}$ and U is invertible, then*

$$\mathcal{S}^{p\times q} \subseteq \mathcal{D}(T_U),\ T_U[\mathcal{S}^{p\times q}] \subseteq \mathcal{S}^{p\times q}$$

and T_U is injective on $\mathcal{D}(T_U)$.

Proof Suppose first that the assumptions in (1) are in force and that

$$U\begin{bmatrix} x \\ y \end{bmatrix} = 0$$

for some choice of $x \in \mathbb{C}^p$ and $y \in \mathbb{C}^q$. Then

$$U\begin{bmatrix} x\eta^* \\ I_q + y\eta^* \end{bmatrix} = U\begin{bmatrix} 0 \\ I_q \end{bmatrix},$$

for every $\eta \in \mathbb{C}^q$. Thus, if $\|\eta\|$ is chosen small enough, then the matrix

$$\varepsilon = (x\eta^*)(I_q + y\eta^*)^{-1}$$

belongs to $\mathcal{S}_{const}^{p\times q}$ and the preceding equality implies that

$$T_U[\varepsilon] = T_U[0].$$

Therefore, in view of the presumed injectiveness, $\varepsilon = 0$ and hence $x = 0$. Consequently, upon writing $U = [u_{ij}]$, $i, j = 1, 2$, we obtain

$$u_{21}x + u_{22}y = u_{22}y = 0.$$

Therefore, since (1) guarantees that u_{22} is invertible, it follows that $y = 0$ also and hence that U is invertible. Moreover, $cU \in \mathcal{P}_{const}(j_{pq})$ for some nonzero constant c by Lemma 2.49.

Next, (2) follows from Lemma 2.50. □

Theorem 2.52 *Let $U \in \mathbb{C}^{m\times m}$. Then the four conditions*

(1) $\mathcal{S}_{const}^{p\times q} \subseteq \mathcal{D}(T_U)$.

(2) $T_U[\mathcal{S}_{const}^{p\times q}] \subseteq \mathcal{S}_{const}^{p\times q}$.

(3) *T_U is injective on $\mathcal{S}_{const}^{p\times q}$.*

(4) *T_U maps the class of $p\times q$ isometric (respectively, coisometric) matrices into itself if $p \geq q$ (respectively, $p \leq q$)*

are in force if and only if:

(5) *$U = cW$ for some $c \in \mathbb{C}\backslash\{0\}$ and $W \in \mathcal{U}_{const}(j_{pq})$.*

Moreover, if (5) holds, then:

(a) *T_U maps $\mathcal{S}_{const}^{p\times q}$ onto itself.*

(b) T_U *maps the class of isometric (resp., coisometric)* $p \times q$ *matrices onto itself if* $p \geq q$ *(resp.,* $p \leq q$*).*

(c) $\mu_-(U) = \mu_+(U)$.

(d) *The modulus* $|c|$ *of the number* c *in (5) is unique:*

$$|c|^2 = \mu_-(U).$$

Proof Suppose first that (1)–(4) are in force. Then, by the preceding theorem, $U = cW$ for some choice of scalar $c \in \mathbb{C}\backslash\{0\}$ and invertible matrix $W \in \mathcal{P}_{const}(j_{pq})$. Therefore, since $T_U[\varepsilon] = T_W[\varepsilon]$, we can assume that $c = 1$ without loss of generality and hence that the $m \times m$ matrix

$$F = j_{pq} - U^* j_{pq} U$$

is positive semidefinite. Moreover, if $p \geq q$, then (4) implies that

$$[\varepsilon^* \quad I_q] F \begin{bmatrix} \varepsilon \\ I_q \end{bmatrix} = 0$$

for every isometric matrix $\varepsilon \in \mathcal{S}^{p\times q}$. Thus upon writing $F = [f_{ij}]$, $i, j = 1, 2,$ in appropriate block form, we see that

$$\varepsilon^* f_{11} \varepsilon + \varepsilon^* f_{12} + f_{21} \varepsilon + f_{22} = 0$$

for all such ε. Consequently, upon replacing ε by $e^{i\theta}\varepsilon$, and viewing the previous formula as a matrix Fourier series with respect to $e^{i\theta}$, it is readily seen that

$$f_{12} = 0, \quad f_{21} = 0 \quad \text{and} \quad \varepsilon^* f_{11} \varepsilon + f_{22} = 0.$$

Therefore, since $f_{11} \geq 0$ and $f_{22} \geq 0$, it follows that $f_{22} = 0$ and $\varepsilon^* f_{11} \varepsilon = 0$ for all isometric ε. In particular, writing

$$f_{11} = U_1 \begin{bmatrix} s_1 & & \\ & \ddots & \\ & & s_p \end{bmatrix} U_1^* \quad \text{and} \quad \varepsilon = U_1 \begin{bmatrix} I_q \\ 0 \end{bmatrix};$$

where U_1 is unitary and $s_1 \geq \cdots \geq s_p$ are the singular values of f_{11}; the formula $\varepsilon^* f_{11} \varepsilon = 0$ implies that $s_1 = \cdots = s_q = 0$, and hence by the monotonicity of the singular values, that $f_{11} = 0$ too. Therefore, $F = 0$, i.e., $U \in \mathcal{U}_{const}(j_{pq})$.

If $q \geq p$, then, it is convenient to use the formula

$$T_U[\varepsilon] = (v_{11} + \varepsilon v_{21})^{-1}(v_{12} + \varepsilon v_{22})$$

based on the block decomposition of the matrix

$$V = j_{pq}U^{-1}j_{pq}$$

that is given in Lemma 2.11 and to now let

$$F = Vj_{pq}V^* - j_{pq}.$$

Then, $F \geq 0$. Moreover, the fact that

$$\varepsilon\varepsilon^* = I_p \Longrightarrow T_U[\varepsilon]T_U[\varepsilon]^* = I_p$$

implies that

$$[I_p \ \varepsilon]F \begin{bmatrix} I_p \\ \varepsilon^* \end{bmatrix} = 0$$

and hence, upon invoking the standard block decomposition $F = [f_{ij}]$, $i, j = 1, 2$, it is readily seen that

$$f_{11} + \varepsilon f_{21} + f_{12}\varepsilon^* + \varepsilon f_{22}\varepsilon^* = 0$$

for every coisometric $p \times q$ matrix ε. Then, since $F \geq 0$, it is readily seen, much as before, that $f_{ij} = 0$ and hence that $V \in \mathcal{U}_{const}(j_{pq})$. This completes the proof that (1)–(4) $\Longrightarrow$ (5). The converse is easy.

It remains only to check that if (5) holds, then (a)–(d) also hold. But, (a) and (b) are easy and (c) follows from Theorem 2.48: $U \in \mathcal{U}_{const}(J) \Longrightarrow U\mathcal{M}_0 = \mathcal{M}_0$. Therefore, (1) and (3) of Theorem 2.48 hold. The latter justifies (c). Then, in view of (3) in Theorem 2.47, (d) holds too. □

2.10 Linear fractional transformations in $\mathcal{C}_{const}^{p\times p}$

Analogues of Theorems 2.51 and 2.52 hold for the linear fractional transformations

$$T_A[\tau] = (a_{11}\tau + a_{12})(a_{21}\tau + a_{22})^{-1}$$

that are defined in terms of matrices $A = [a_{ij}]$, $i, j = 1, 2$, with blocks $a_{ij} \in \mathbb{C}^{p\times p}$ and that act in the class

$$\mathcal{C}_{const}^{p\times p} = \{\tau \in \mathbb{C}^{p\times p} : \ \tau + \tau^* \geq 0\}$$

and map the class

$$\mathring{\mathcal{C}}_{const}^{p\times p} = \{\tau \in \mathbb{C}^{p\times p} : \ \tau + \tau^* > 0\}$$

into $\mathcal{C}_{const}^{p\times p}$. The main conclusions regarding such transforms are summarized in the following theorem:

Theorem 2.53 *Let $A \in \mathbb{C}^{m\times m}$. Then the following implications are in force:*

(1) *If*

$$\gamma^{-1}A \in \mathcal{P}_{const}(J_p) \quad \textit{for some } \gamma \in \mathbb{C}\setminus\{0\}, \tag{2.64}$$

then

$$T_A[\mathcal{C}_{const}^{p\times p} \cap \mathcal{D}(T_A)] \subseteq \mathcal{C}_{const}^{p\times p}, \tag{2.65}$$

$$\mathring{\mathcal{C}}_{const}^{p\times p} \subseteq \mathcal{D}(T_A) \quad \textit{and} \quad T_A[\mathring{\mathcal{C}}_{const}^{p\times p}] \subseteq \mathring{\mathcal{C}}_{const}^{p\times p}. \tag{2.66}$$

(2) *Let $\mathring{\mathcal{C}}_{const}^{p\times p} \subseteq \mathcal{D}(T_A)$. Then T_A is injective on $\mathring{\mathcal{C}}_{const}^{p\times p}$ if and only if A is an invertible matrix. Moreover, if T_A is injective on $\mathring{\mathcal{C}}_{const}^{p\times p}$, then T_A is invertible on $\mathcal{D}(T_A)$.*

(3) *If $\mathring{\mathcal{C}}_{const}^{p\times p} \subseteq \mathcal{D}(T_A)$ and $T_A[\mathring{\mathcal{C}}_{const}^{p\times p}] \subseteq \mathcal{C}_{const}^{p\times p}$, and if $T_A[\mathcal{C}_{const}^{p\times p} \cap \mathcal{D}(A)]$ contains more than one matrix, then condition (2.64) is in force.*

(4) *If $\mathring{\mathcal{C}}_{const}^{p\times p} \subseteq \mathcal{D}(T_A)$ and the linear fractional transformation T_A maps $\mathring{\mathcal{C}}_{const}^{p\times p}$ injectively into $\mathcal{C}_{const}^{p\times p}$ then the condition (2.64) holds and the matrix A is invertible.*

(5) *If the linear fractional transformation T_A meets the constraints in (4) and maps the set*

$$\{\tau \in \mathcal{D}(T_A) : \Re\tau = 0\} \textit{ into } \{c \in \mathbb{C}^{p\times p} : \Re c = 0\}, \tag{2.67}$$

then the modulus $|\gamma|$ of the number γ in condition (2.64) is unique:

$$|\gamma|^2 = \mu_+(A) = \mu_-(A), \quad \textit{and} \quad \gamma^{-1}A \in \mathcal{U}_{const}(J_p). \tag{2.68}$$

(6) *Conversely, if (2.68) holds, then the matrix A is invertible and the transform T_A meets the constraints in (5).*

(7) *If A is a strictly minus matrix with respect to $J = J_p$, then*

$$\mu_+(A) = \lambda_p(A) \quad \textit{and} \quad \mu_-(A) = \lambda_{p+1}(A).$$

Furthermore,

$$\gamma^{-1}A \in \mathcal{P}_{const}(J_p) \Longleftrightarrow \mu_+(A) \leq |\gamma|^2 \leq \mu_-(A).$$

Proof The theorem follows from Theorems 2.13, 2.51 and 2.52, Lemma 2.49 and the properties of linear fractional transformations based on the matrix $\mathfrak{V}$ defined in formula (2.5) that are recorded below:

$$\mathcal{D}(T_{\mathfrak{V}}) = \{s \in \mathbb{C}^{p\times p} : \det(I_p + s) \neq 0\},$$

$$\mathcal{C}_{const}^{p\times p} \subseteq \mathcal{D}(T_{\mathfrak{V}}) \quad \text{and} \quad \mathcal{S}_{const}^{p\times p} \cap \mathcal{D}(T_{\mathfrak{V}}) = T_{\mathfrak{V}}[\mathcal{C}_{const}^{p\times p}],$$

$$\mathcal{C}_{const}^{p\times p} = T_{\mathfrak{V}}[\mathcal{S}_{const}^{p\times p} \cap \mathcal{D}(T_{\mathfrak{V}})],$$

$$\mathring{\mathcal{C}}_{const}^{p\times p} = T_{\mathfrak{V}}[\mathring{\mathcal{S}}_{const}^{p\times p}] \quad \text{and} \quad \mathring{\mathcal{S}}_{const}^{p\times p} = T_{\mathfrak{V}}[\mathring{\mathcal{C}}_{const}^{p\times p}].$$

If $W = \mathfrak{V}A\mathfrak{V}$, then

$$A \in \mathcal{P}_{const}(J_p) \Longleftrightarrow W \in \mathcal{P}_{const}(j_p),$$

$$A \in \mathcal{U}_{const}(J_p) \Longleftrightarrow W \in \mathcal{U}_{const}(j_p),$$

$$J_pA^*J_pA = \mathfrak{V}(j_pW^*j_pW)\mathfrak{V}^*,$$

$$\mu_-(A) = \mu_-(W), \quad \mu_+(A) = \mu_+(W),$$

$$\lambda_j(A) = \lambda_j(W), \quad j = 1, \ldots, m.$$

It remains only to verify assertion (5). The conditions in (4) guarantee that

$$\gamma^{-1}A \in \mathcal{P}_{const}(J_p) \quad \text{for some } \gamma \in \mathbb{C} \setminus \{0\} \text{ and that } A \text{ is invertible.}$$

The next step is to show that if condition (2.67) is in force, then $\gamma^{-1}A \in \mathcal{U}_{const}(J_p)$. Condition (2.67) guarantees that if $W = \mathfrak{V}A\mathfrak{V}$, then T_W maps every $p \times p$ matrix ε that meets the conditions

$$\varepsilon^*\varepsilon = I_p, \quad \det(I_p + \varepsilon) \neq 0 \quad \text{and} \quad \det(I_p + T_W[\varepsilon]) \neq 0 \tag{2.69}$$

into a unitary $p \times p$ matrix. Now let ε be any unitary $p \times p$ matrix. Then there exists a sequence γ_n, $n = 1, 2 \ldots$ with $|\gamma_n| = 1$ such that $\gamma_n\varepsilon$ meets the three conditions in (2.69) and $\gamma_n \to 1$ as $n \uparrow \infty$. Consequently, the matrices $T_W[\gamma_n\varepsilon]$ are unitary and

$$T_W[\varepsilon] = \lim_{n\uparrow\infty} T_W[\gamma_n\varepsilon]$$

is unitary. □

Remark 2.54 *The results presented above can easily be reformulated for linear fractional transformations in the classes*

$$i\mathcal{C}^{p\times p}_{const} = \{v \in \mathbb{C}^{p\times p} : \Im v \geq 0\} \quad \textit{and} \quad i\mathring{\mathcal{C}}^{p\times p}_{const} = \{v \in \mathbb{C}^{p\times p} : \Im v > 0\}.$$

In this representation the signature matrix J_p is replaced by the signature matrix $\mathcal{J}_p$; it is useful to note that

$$\mathcal{J}_p = V_1^* J_p V_1 \quad \textit{where} \quad V_1 = \begin{bmatrix} -iI_p & 0 \\ 0 & I_p \end{bmatrix}. \tag{2.70}$$

2.11 Transformations from $\mathcal{S}^{p\times p}_{const}$ into $\mathcal{C}^{p\times p}_{const}$

Let $\mathcal{P}_{const}(j_p, J_p)$ be the class of (j_p, J_p)-contractive matrices, i.e., matrices B such that

$$B^* J_p B \leq j_p, \tag{2.71}$$

and let $\mathcal{U}_{const}(j_p, J_p)$ be the class of (j_p, J_p)-unitary matrices, i.e., matrices B such that

$$B^* J_p B = j_p. \tag{2.72}$$

If $A = B\mathfrak{V}$, then

$$A \in \mathcal{P}_{const}(J_p) \iff B \in \mathcal{P}_{const}(j_p, J_p)$$

and

$$A \in \mathcal{U}_{const}(J_p) \iff B \in \mathcal{U}_{const}(j_p, J_p).$$

Thus, as

$$A^* J_p A = J_p \iff A J_p A^* = J_p \tag{2.73}$$

and

$$A^* J_p A \leq J_p \iff A J_p A^* \leq J_p \tag{2.74}$$

by Corollary 2.5, it follows that

$$B^* J_p B \leq j_p \iff B j_p B^* \leq J_p \tag{2.75}$$

and

$$B^* J_p B = j_p \iff B j_p B^* = J_p. \tag{2.76}$$

The notation

$$\mathcal{C}_{const}(A) = T_B[\mathcal{S}^{p\times p}_{const} \cap \mathcal{D}(T_B)] \quad \text{for } B = A\mathfrak{V} \tag{2.77}$$

will be useful.

Lemma 2.55 *Let $A \in \mathcal{P}_{const}(J_p)$ and let $B = A\mathfrak{V}$. Then*

$$T_A[\mathcal{C}_{const}^{p\times p} \cap \mathcal{D}(T_A)] \subseteq \mathcal{C}_{const}(A) \subseteq \mathcal{C}_{const}^{p\times p} \tag{2.78}$$

and the first inclusion may be proper.

Proof Let $c \in \mathcal{C}_{const}(A)$, i.e., let $c = T_B[\varepsilon]$, where $\varepsilon \in \mathcal{S}_{const}^{p\times p} \cap \mathcal{D}(T_B)$ and let

$$\begin{bmatrix} u \\ v \end{bmatrix} = B \begin{bmatrix} \varepsilon \\ I_p \end{bmatrix} = \begin{bmatrix} b_{11}\varepsilon + b_{12} \\ b_{21}\varepsilon + b_{22} \end{bmatrix}.$$

Then, since $A^* J_p A \leq J_p \Longrightarrow B^* J_p B \leq j_p$,

$$\begin{bmatrix} \varepsilon \\ I_p \end{bmatrix}^* B^* J_p B \begin{bmatrix} \varepsilon \\ I_p \end{bmatrix} \leq \begin{bmatrix} \varepsilon \\ I_p \end{bmatrix}^* j_p \begin{bmatrix} \varepsilon \\ I_p \end{bmatrix} \leq 0. \tag{2.79}$$

The inequality (2.79) implies that

$$u^*v + v^*u \geq 0 \tag{2.80}$$

and the condition $\varepsilon \in \mathcal{D}(T_B)$ means that the matrix $v = b_{21}\varepsilon + b_{22}$ is invertible. Let $c = uv^{-1}$. Then $c \in \mathcal{C}_{const}^{p\times p}$, since (2.80) holds and $c = T_B[\varepsilon]$. Thus, the second inclusion in (2.78) holds.

Furthermore, if $\tau \in \mathcal{C}_{const}^{p\times p} \cap \mathcal{D}(T_A)$ and $c = T_A[\tau]$, then $\tau = T_{\mathfrak{V}}[\varepsilon]$ for some $\varepsilon \in \mathcal{S}_{const}^{p\times p} \cap \mathcal{D}(T_{\mathfrak{V}})$ and, consequently, $c = T_{A\mathfrak{A}}[\varepsilon] = T_B[\varepsilon]$, i.e., $c \in \mathcal{C}_{const}(A)$. Thus, the first inclusion in (2.78) holds.

The following example shows that this inclusion may be proper: Let $\varepsilon = -\xi\xi^*$, where $\xi \in \mathbb{C}^p$ and $\xi^*\xi = 1$. Then $\varepsilon \in \mathcal{S}_{const}^{p\times p}$ and $\det(I_p + \varepsilon) = 0$, i.e., $\varepsilon \notin \mathcal{D}(T_{\mathfrak{V}})$. Let

$$A = \mathfrak{V} \begin{bmatrix} I_p & 0 \\ 0 & u \end{bmatrix} \mathfrak{V} \quad \text{and} \quad B = A\mathfrak{V} = \frac{1}{\sqrt{2}} \begin{bmatrix} -I_p & u \\ I_p & u \end{bmatrix},$$

where u is a $p \times p$ unitary matrix such that $\det(\varepsilon + u) \neq 0$. Then $A \in \mathcal{U}_{const}(J_p)$, $\varepsilon \in \mathcal{S}_{const}^{p\times p} \cap \mathcal{D}(T_B)$ and

$$c = T_B[\varepsilon] = (-\varepsilon + u)(\varepsilon + u)^{-1} \quad \text{belongs to} \quad \mathcal{C}_{const}(A).$$

However, $c \notin T_A[\mathcal{C}_{const}^{p\times p} \cap \mathcal{D}(T_A)]$ because, if $c = T_A[\tau]$ for some $\tau \in \mathcal{C}_{const}^{p\times p} \cap \mathcal{D}(T_A)$, then $\tau = T_{\mathfrak{V}}[\varepsilon_1]$ for some $\varepsilon_1 \in \mathcal{S}_{const}^{p\times p} \cap \mathcal{D}(T_{\mathfrak{V}})$ and

$$c = T_B[\varepsilon_1] = T_B[\varepsilon].$$

Therefore $\varepsilon_1 = \varepsilon$, since B is invertible and hence T_B is injective by Lemma 2.10. But this contradicts the fact $\varepsilon_1 \in \mathcal{D}(T_{\mathfrak{V}})$ and $\varepsilon \notin \mathcal{D}(T_{\mathfrak{V}})$. Thus, in this example, the first inclusion in (2.78) is proper. □

Lemma 2.56 *Let $A \in \mathcal{P}_{const}(J_p)$, $B = A\mathfrak{V}$ and $\begin{bmatrix} b_{21} & b_{22} \end{bmatrix} = \begin{bmatrix} 0 & I_p \end{bmatrix} B$. Then:*

(1) $b_{22}b_{22}^* - b_{21}b_{21}^* \geq 0$.

(2) *The following three conditions are equivalent:*

(a) $\mathcal{S}_{const}^{p\times p} \subseteq \mathcal{D}(T_B)$.

(b) $b_{22}b_{22}^* - b_{21}b_{21}^* > 0$.

(c) b_{22} *is invertible and*

$$\chi = b_{22}^{-1}b_{21} \quad \text{belongs to the set} \quad \mathring{\mathcal{S}}_{const}^{p\times p}. \tag{2.81}$$

Proof Assertion (1) follows by looking at the 22 blocks in the inequality

$$Bj_{pq}B^* \leq J_p.$$

The equivalence of (a) and (c) in (2) is established in Lemma 2.15; the equivalence of (b) and (c) is easy. □

Lemma 2.57 *Let $A \in \mathcal{P}_{const}(J_p)$ and suppose that the blocks b_{ij} in the four block decomposition of $B = A\mathfrak{V}$ satisfy at least one (and hence all) of the conditions (a)–(c) in Lemma 2.56. Then $\mathcal{C}_{const}(A) = T_B[\mathcal{S}_{const}^{p\times q}]$ and hence $\mathcal{C}_{const}(A)$ is a matrix ball with a positive right semiradius. Its unique center m_c and its left and right semiradii R_ℓ and R_r may be specified by the formulas considered in Lemmas 2.17 and 2.23 with B in place of U and $B_1 = BW_0$ in place of U_1, $B_1 \in \mathcal{P}(j_p, J_p)$.*

Proof This is essentially a special case of Lemma 2.23, which is applicable, since $\mathcal{S}_{const}^{p\times q} \subseteq \mathcal{D}(T_B)$, under the given assumptions. □

2.12 Affine generalizations

The set $\mathcal{C}_{const}(A)$ may also be defined directly in terms of a linear fractional transformation $\widetilde{T}_A$ that is defined on the set

$$\mathcal{D}(\widetilde{T}_A) = \left\{ \begin{bmatrix} u \\ v \end{bmatrix} : u, v \in \mathbb{C}^{p\times p} \text{ and } \det(a_{21}u + a_{22}v) \neq 0 \right\}$$

by the formula

$$\widetilde{T}_A \begin{bmatrix} u \\ v \end{bmatrix} = (a_{11}u + a_{12}v)(a_{21}u + a_{22}v)^{-1}.$$

Let

$$\widetilde{\mathcal{C}}^{p\times p}_{const} = \left\{ \begin{bmatrix} u \\ v \end{bmatrix} : u, v \in \mathbb{C}^{p\times p}, \quad u^*v + v^*u \geq 0 \quad \text{and} \quad u^*u + v^*v > 0 \right\} \tag{2.82}$$

be the affine generalization of the class $\mathcal{C}^{p\times p}_{const}$. Then $r = u + v$ is invertible and

$$\widetilde{\mathcal{C}}^{p\times p}_{const} = \left\{ \mathfrak{V} \begin{bmatrix} \varepsilon \\ I_p \end{bmatrix} r : r \in \mathbb{C}^{p\times p} \text{ is invertible and } \varepsilon \in \mathcal{S}^{p\times p}_{const} \right\}. \tag{2.83}$$

Consequently,

$$\mathcal{C}_{const}(A) = \widetilde{T}_A[\widetilde{\mathcal{C}}^{p\times p}_{const} \cap \mathcal{D}(\widetilde{T}_A)]. \tag{2.84}$$

Let $\mathcal{T}_A$ denote the transform that is defined on the set of block matrices $\left\{ \begin{bmatrix} u \\ v \end{bmatrix} : u, v \in \mathbb{C}^{p\times p} \right\}$ by the formula

$$\mathcal{T}_A \begin{bmatrix} u \\ v \end{bmatrix} = A \begin{bmatrix} u \\ v \end{bmatrix} = \begin{bmatrix} a_{11}u + a_{12}v \\ a_{21}u + a_{22}v \end{bmatrix}. \tag{2.85}$$

If $\gamma^{-1}A \in \mathcal{P}_{const}(J_p)$ for some $\gamma \in \mathbb{C} \setminus \{0\}$, then: $\mathcal{T}_A$ maps the class $\widetilde{\mathcal{C}}^{p\times p}_{const}$ into itself; it is injective if A is invertible. Conversely, if $\mathcal{T}_A$ is an injective map of $\widetilde{\mathcal{C}}^{p\times p}_{const}$ into itself, then A is invertible and $\gamma^{-1}A \in \mathcal{P}_{const}(J_p)$ for some $\gamma \in \mathbb{C} \setminus \{0\}$. This characterization of invertible strictly minus matrices with respect to J_p will be established below in the setting of general signature matrices.

Let J be unitarily equivalent to j_{pq}, let

$$\mathcal{F}_{const}(J) = \{X \in \mathbb{C}^{m\times q} : X^*JX \leq 0 \quad \text{and} \quad \text{rank}\, X = q\} \tag{2.86}$$

and

$$\mathring{\mathcal{F}}_{const}(J) = \{X \in \mathbb{C}^{m\times q} : X^*JX < 0 \quad \text{and} \quad \text{rank}\, X = q\}. \tag{2.87}$$

In terms of this new notation, property (4) in Theorem 2.40 may be reformulated as

$$\mathcal{T}_U(\mathcal{F}(J)) \subseteq \mathcal{F}(J) \quad \text{and} \quad \mathcal{T}_U(\mathring{\mathcal{F}}(J)) \subseteq \mathring{\mathcal{F}}(J).$$

Theorem 2.58 *Let J be a signature matrix that is unitarily equivalent to j_{pq}, let $U \in \mathbb{C}^{m\times m}$ and let $\mathcal{T}_U$ denote the linear map of $\mathbb{C}^{m\times q}$ into itself that is defined by the formula*

$$\mathcal{T}_U X = UX, \ X \in \mathbb{C}^{m\times q}. \tag{2.88}$$

Then:

(1) *$\mathcal{T}_U$ is injective on $\mathring{\mathcal{F}}_{const}(J)$ if and only if the matrix U is invertible.*

(2) *$\mathcal{T}_U$ maps $\mathring{\mathcal{F}}_{const}(J)$ injectively into $\mathcal{F}_{const}(J)$ if and only if U is an invertible minus matrix with respect to J.*

(3) *$\mathcal{T}_U$ maps the set $\mathring{\mathcal{F}}_{const}(J)$ into itself if U is a strictly minus matrix with respect to J.*

Proof If U is an invertible $m \times m$ matrix, then $\mathcal{T}_U$ maps $\mathbb{C}^{m\times q}$ injectively onto itself. If U is an invertible minus matrix, then, by Theorem 2.37, $\gamma^{-1}U \in \mathcal{P}_{const}(J)$ for some $\gamma \in \mathbb{C}\setminus\{0\}$, i.e., U is a strictly minus matrix. Consequently,

$$(\mathcal{T}_U X)^* J \mathcal{T}_U X \le |\gamma|^2 X^* J X \begin{cases} \le 0 \text{ if } X \in \mathcal{F}_{const}(J) \\ < 0 \text{ if } X \in \mathring{\mathcal{F}}_{const}(J), \end{cases}$$

i.e., $\mathcal{T}_U$ maps $\mathcal{F}_{const}(J)$ into itself and $\mathring{\mathcal{F}}_{const}(J)$ into itself. This justifies (3) and one direction of (1) and (2). To complete the proof, it suffices to check the remaining assertions for the case $J = j_{pq}$, i.e., to verify that

(a) If $\mathcal{T}_U$ is injective on $\mathring{\mathcal{F}}_{const}(j_{pq})$, then U is invertible.

(b) If $\mathcal{T}_U$ maps $\mathring{\mathcal{F}}_{const}(J)$ injectively into $\mathcal{F}_{const}(j_{pq})$, then U is an invertible minus matrix with respect to j_{pq}.

To verify (a), let the vector $\text{col}(\xi, \eta)$ with components $\xi \in \mathbb{C}^p$ and $\eta \in \mathbb{C}^q$ belong to the null space of U and let $y \in \mathbb{C}^q$ be a nonzero vector such that $I_q + \eta y^*$ is invertible and the matrices

$$X = \begin{bmatrix} \xi y^*(I_q + \eta y^*)^{-1} \\ I_q \end{bmatrix} \quad \text{and} \quad Y = \begin{bmatrix} 0 \\ (I_q + \eta y^*)^{-1} \end{bmatrix}$$

both belong to the class $\mathring{\mathcal{F}}_{const}(j_{pq})$. It is readily seen that this can be achieved for any vector $y \in \mathbb{C}^q$ with $\|y\|$ small enough. Then

$$X = Y,$$

since $UX = UY$ and $\mathcal{T}_U$ is presumed to be injective on $\mathring{\mathcal{F}}_{const}(j_{pq})$. But this implies that

$$\xi^* y(I_q + \eta y^*)^{-1} = 0 \quad \text{and} \quad I_q = (I_q + \eta y^*)^{-1}$$

and hence that $\xi = 0$ and $\eta = 0$, i.e., U is invertible.

Suppose next that hypothesis (b) is in force and let $\varepsilon \in \mathring{\mathcal{S}}^{p\times q}_{const}$. Then

$$\begin{bmatrix} u \\ v \end{bmatrix} = U \begin{bmatrix} \varepsilon \\ I_q \end{bmatrix} \quad \text{belongs to} \quad \mathcal{F}_{const}(j_{pq}).$$

Therefore, $u^*u \leq v^*v$ and, in view of the rank condition, v is invertible, i.e., $\varepsilon \in \mathcal{D}(T_U)$ and

$$uv^{-1} = T_U[\varepsilon] \in \mathcal{S}^{p\times q}_{const}.$$

Thus,

$$\mathring{\mathcal{S}}^{p\times q}_{const} \subseteq \mathcal{D}(T_U) \quad \text{and} \quad T_U[\mathring{\mathcal{S}}^{p\times q}_{const}] \subseteq \mathcal{S}^{p\times q}_{const}.$$

Therefore, U is an invertible minus matrix with respect to j_{pq}, by Theorem 2.51. □

Theorem 2.59 *Let J be a signature matrix that is unitarily equivalent to j_{pq} with $p \geq q$ and let U be an invertible $m \times m$ matrix such that:*

(1) *$U \in \mathcal{P}_{const}(J)$.*

(2) *$\mathcal{T}_U$ maps the set*

$$\{X \in \mathbb{C}^{m\times q} : X^*JX = 0 \quad \text{and} \quad \operatorname{rank} X = q\}$$

into itself.

Then $U \in \mathcal{U}_{const}(J)$.

Proof Without loss of generality, we can take $J = j_{pq}$. Let ε be a $p \times q$ isometric matrix and let

$$X_k = \begin{bmatrix} i^k \varepsilon \\ I_q \end{bmatrix} \quad \text{for} \quad k = 1, \ldots, 4.$$

Then $X_k^* j_{pq} X_k = 0$ and the condition $X_k^* U^* j_{pq} U X_k = 0$ holds if and only if

$$\varepsilon^*\{u_{11}^* u_{11} - u_{21}^* u_{21}\}\varepsilon + i^k\{u_{12}^* u_{11} - u_{22}^* u_{21}\}\varepsilon$$
$$= u_{22}^* u_{22} - u_{12}^* u_{12} + (-i)^k \varepsilon^*\{u_{21}^* u_{22} - u_{11}^* u_{12}\}.$$

Therefore, since $i + i^2 + i^3 + i^4 = 0$, it is readily checked that

$$\varepsilon^*\{u_{11}^*u_{11} - u_{21}^*u_{21}\}\varepsilon = u_{22}^*u_{22} - u_{12}^*u_{12}$$

and

$$\{u_{12}^*u_{11} - u_{22}^*u_{21}\}\varepsilon = 0$$

for every $p \times q$ isometric matrix ε. The second constraint implies that

$$\{u_{12}^*u_{11} - u_{22}^*u_{21}\} = 0.$$

The assumption $U \in \mathcal{P}_{const}(j_{pq})$ yields the supplementary inequalities

$$u_{11}^*u_{11} - u_{21}^*u_{21} \leq I_p \quad \text{and} \quad u_{22}^*u_{22} - u_{12}^*u_{12} \geq I_q,$$

which, when combined with the first constraint, imply that

$$u_{11}^*u_{11} - u_{21}^*u_{21} = I_p \quad \text{and} \quad u_{22}^*u_{22} - u_{12}^*u_{12} = I_q.$$

Therefore, $U \in \mathcal{U}_{const}(J)$. □

2.13 The J modulus

Lemma 2.60 *If $U \in \mathcal{P}_{const}(J)$ is invertible and $G = JU^*JU$, then*

$$\ker(I_m - G)^m \subseteq \ker(I_m - G)^2. \tag{2.89}$$

Proof Let $U \in \mathcal{P}_{const}(J)$. Then $J - U^*JU \geq 0$. Therefore

$$|\langle (J - U^*JU)x, y\rangle|^2 \leq \langle (J - U^*JU)x, x\rangle \, \langle (J - U^*JU)y, y\rangle$$

for every choice of $x, y \in \mathbb{C}^m$, or, equivalently

$$|[(I_m - G)x, y]|^2 \leq [(I_m - G)x, x] \, [(I_m - G)y, y]$$

for every choice of $x, y \in \mathbb{C}^m$. Assume now that $(I_m - G)^3u = 0$ and let $x = (I_m - G)u$. Then the preceding inequality implies that

$$[(I_m - G)^2u, y] = 0 \quad \text{for every } y \in \mathbb{C}^m.$$

Therefore, $(I_m - G)^{k+3}u = 0 \Longrightarrow (I_m - G)^{k+2}u = 0$ for $k = 0$ and hence for every positive integer k. □

Theorem 2.61 *Let $U \in \mathcal{P}_{const}(J)$ be invertible, let $\lambda_1 \leq \cdots \leq \lambda_m$ denote the eigenvalues of $G = JU^*JU$ and assume that J is unitarily equivalent to j_{pq}. Then*

(1) $0 < \lambda_1 \leq \cdots \leq \lambda_p \leq 1$ *and* $1 \leq \lambda_{p+1} \leq \cdots \leq \lambda_m$.

(2) *If* G *is not diagonalizable, then there exists an invertible matrix* $V \in \mathbb{C}^{m\times m}$ *such that*

$$V^*JV = j_{pq} \text{ and } V^{-1}GV = S, \text{ where } S = \operatorname{diag}\{D_1, E, D_2\},$$

$D_1 = \operatorname{diag}\{\lambda_1, \ldots, \lambda_r\}$, $0 \leq r < p$, $D_2 = \operatorname{diag}\{\lambda_{m-s+1}, \ldots, \lambda_m\}$, $0 \leq s < q$, $E = I_{m-r-s} + N$, $N^2 = 0$ *and* $Sj_{pq} \leq j_{pq}$.

(3) G *is diagonalizable if and only if* $\ker(G - I_m)^2 = \ker(G - I_m)$. *If* G *is diagonalizable, then the formulas in (2) hold with* $N = 0$.

Proof Suppose that G is not diagonalizable. Then, by Theorem 2.47,

$$\mu_+(U) = \lambda_p(G) = 1 = \lambda_{p+1}(G) = \mu_-(G).$$

Assume for the sake of definiteness that $m \geq 3$ and

$$\lambda_r < \lambda_{r+1} = 1 = \cdots = \lambda_{m-s} < \lambda_{m-s+1};$$

and let $D_1 = \operatorname{diag}\{\lambda_1, \ldots, \lambda_r\}$ and $D_2 = \operatorname{diag}\{\lambda_{m-s+1}, \ldots, \lambda_m\}$. Then, in view of Lemma 2.46, there exists an invertible matrix $U = [U_1\ U_2\ U_3]$ with block columns of size $m \times r$, $m \times n$ and $m \times s$, respectively, such that

$$GU_1 = U_1D_1,\ U_1^*JD_1 = I_r,\ GU_3 = U_3D_2,\ U_3^*JU_3 = -I_s,\ U_1^*JU_3 = 0$$

and the columns of U_2 are a basis for $\ker(G - I_m)^2$. Moreover, if $Gu = \lambda u$ and $\xi \in \mathbb{C}^n$, then the formula

$$(\lambda - 1)^2[u, U_2\xi] = [(G - I_m)^2u, U_2\xi] = [u, (G - I_m)^2U_2\xi] = 0$$

implies that $U_2^*JU_1 = 0$ and $U_2^*JU_3 = 0$. Thus,

$$U^*JU = \operatorname{diag}\{I_k, U_2^*JU_2, -I_\ell\}$$

and

$$GU = UC, \quad \text{where} \quad C = \begin{bmatrix} D_1 & 0 & 0 \\ 0 & F & 0 \\ 0 & 0 & D_2 \end{bmatrix} \quad \text{and} \quad (F - I_n)^2 = 0.$$

Since U^*JU is invertible, the $n \times n$ matrix $U_2^*JU_2$ is a Hermitian invertible matrix. Therefore, there exists an invertible matrix $K \in \mathbb{C}^{n\times n}$ such that

$K^*U_2^*JU_2K = j_{k\ell}$ where $k+\ell = n$. Let

$$L = \begin{bmatrix} I_r & 0 & 0 \\ 0 & K & 0 \\ 0 & 0 & I_s \end{bmatrix}, \quad S = L^{-1}CL = \begin{bmatrix} D_1 & 0 & 0 \\ 0 & K^{-1}FK & 0 \\ 0 & 0 & D_2 \end{bmatrix},$$

$K^{-1}FK = I_n + N$ and $V = UL$. Then $N^2 = K^{-1}(F - I_n)^2K = 0$,

$$V^*JV = j_{pq} \quad \text{and} \quad GV = GUL = UCL = VS.$$

Thus, as

$$\begin{aligned} 0 \ &\geq \ J - GJ = J - VSV^{-1}J = V(I_m - S)V^{-1}J \\ &= \ V(I_m - S)j_{pq}j_{pq}V^{-1}J = V(I_m - S)j_{pq}V^*, \end{aligned}$$

it follows that $(I_m - S)j_{pq} \geq 0$. Set $K^{-1}FK = E$ to complete the proof of (2) for the case $r > 0$ and $s > 0$. The cases $r = 0$, $s = 0$ and the proof of (3) are easy consequences of the proof furnished above and are left to the reader. □

Lemma 2.62 *Let $B \in \mathbb{C}^{k\times k}$ be an upper triangular matrix with diagonal elements $b_{ii} = \lambda \neq 0$ for $i = 1, \ldots, k$ and let $A \in \mathbb{C}^{k\times k}$. Then*

$$B^2A = AB^2 \iff BA = AB. \tag{2.90}$$

Proof Under the given assumptions,

$$\sum_{j=0}^{k} \binom{k}{j} B^j(-\lambda I_k)^{k-j} = (B - \lambda I_k)^k = 0.$$

Now assume that $k > 1$ and let

$$P_1(B) = \sum_{even\, j} \binom{k}{j} B^{k-j}(\lambda I_k)^j$$

and

$$BP_2(B) = -\sum_{odd\, j} \binom{k}{j} B^{k-j}(\lambda I_k)^j.$$

Then $P_1(B)$ and $P_2(B)$ are polynomials in B^2 that satisfy the identity

$$P_1(B) = BP_2(B).$$

Therefore,

$$BAP_2(B) = BP_2(B)A = P_1(B)A = AP_1(B) = ABP_2(B).$$

Thus, as $P_2(B)$ is invertible, it follows that $BA = AB$, as needed. □

Theorem 2.63 *Let $A, B \in \mathbb{C}^{n\times n}$ and suppose that the eigenvalues of both A and B are positive. Then*

$$A^2 = B^2 \Longleftrightarrow A = B. \tag{2.91}$$

Proof Let $A = USU^{-1}$ and $B = VTV^{-1}$, where S and T are upper triangular matrices in Jordan form and suppose that $A^2 = B^2$. Then, since $\mu I_n + S$ and $\mu I_n + T$ are invertible for $\mu > 0$, it is readily seen that

$$\begin{aligned} \dim\ker(\mu I_n - S)^k &= \dim\ker(\mu^2 I_n - S^2)^k \\ &= \dim\ker(\mu^2 I_n - T^2)^k \\ &= \dim\ker(\mu I_n - T)^k \quad \text{for} \quad k = 1, 2, \ldots. \end{aligned}$$

Therefore, S and T have the same Jordan block decomposition, up to the order in which the blocks are placed. Thus, there is no loss of generality in assuming that $S = T$, and hence that $US^2U^{-1} = VS^2V^{-1}$, i.e.,

$$QS^2 = S^2Q \quad \text{with} \quad Q = V^{-1}U.$$

Now, if S has ℓ distinct eigenvalues, $\mu_1, \ldots, \mu_\ell$, then $S = \operatorname{diag}\{S_1, \ldots, S_\ell\}$, where S_j is an upper triangular matrix with μ_j on the diagonal. Thus, if Q is written in compatible block form, then

$$QS^2 = S^2Q \Longleftrightarrow Q_{ij}S_j^2 = S_i^2Q_{ij} \quad \text{for} \quad i, j = 1, \ldots, \ell.$$

Therefore, $Q_{ij} = 0$ if $i \neq j$ and $Q_{ii}S_i^2 = S_i^2Q_{ii}$, which, in view of Lemma 2.62 implies that $Q_{ii}S_i = S_iQ_{ii}$, for $i = 1, \ldots, \ell$. Consequently, $QS = SQ$, i.e., $A = B$. This completes the proof of the implication $\Longrightarrow$ in (2.91). The implication $\Longleftarrow$ is self-evident. □

Theorem 2.64 *Let $U \in \mathcal{P}_{const}(J)$, $G = JU^*JU$ and suppose that U is invertible. Then there exists a unique matrix R with positive eigenvalues such that $R^2 = G$. Moreover, this matrix R has the following properties:*

(1) $R = \exp\{-HJ\}$, *where $H \geq 0$ (and hence is selfadjoint).*

(2) $JR^*J = R$.

(3) $R \in \mathcal{P}_{const}(J)$.

(4) $UR^{-1} \in \mathcal{U}_{const}(J)$.

Proof We shall asume that $J \neq \pm I_m$, because if $J = \pm I_m$, then the theorem is obvious. In view of Theorem 2.61, it suffices to focus on the case

$$G = VSV^{-1}, \quad \text{where} \quad S = \text{diag}\,\{D_1, E, D_2\}, \quad V^*JV = j_{pq}$$

and the matrices D_1, E and D_2 are specified in (2) of Theorem 2.61. Then clearly the formula

$$R = VS_1V^{-1} \quad \text{with} \quad S_1 = \text{diag}\,\{D_1^{1/2}, E_1, D_2^{1/2}\},$$

and

$$E_1 = I_n + \frac{1}{2}N = \exp\{\frac{1}{2}N\}$$

exhibits a matrix R with positive eigenvalues such that $R^2 = G$. The asserted uniqueness of such an R follows from Theorem 2.63.

Next, to verify (1), let

$$\begin{aligned} L_1 &= \text{diag}\{\ln\lambda_1, \ldots, \ln\lambda_r\}, \quad L_2 = \text{diag}\{\ln\lambda_{m-s+1}, \ldots, \ln\lambda_m\} \\ S_2 &= \text{diag}\{L_1, N, L_2\}, \quad 2H = -VS_2V^{-1}J, \quad k = p - r \text{ and } \ell = q - s. \end{aligned}$$

Then, as

$$j_{pq} = \text{diag}\{I_r, j_{k\ell}, -I_s\}, \quad L_1 < 0, \quad -j_{k\ell}N \geq 0 \quad \text{and} \quad L_2 > 0,$$

it follows that

$$-S_2j_{pq} = \text{diag}\{-L_1, -N, -L_2\}j_{pq} = \text{diag}\{-L_1, -Nj_{kl}, L_2\} \geq 0.$$

Therefore,

$$R = \exp\{\frac{1}{2}VS_2V^{-1}\} = \exp\{-HJ\}$$

and

$$H = -\frac{1}{2}VS_2j_{pq}j_{pq}V^{-1}J = -\frac{1}{2}VS_2j_{pq}V^* \geq 0.$$

This completes the proof of (1); the implications (1) $\Longrightarrow$ (2) $\Longrightarrow$ (4) $\Longrightarrow$ (3) are easily checked. □

The unique matrix R with positive eigenvalues that meets the condition $R^2 = JU^*JU$ is called the J **modulus** of U. The J modulus was introduced by V. P. Potapov as a basic tool in his study of the convergence of products of mvf's with J-contractive values. Its usefulness stems from the following facts, which were also established by Potapov; see [Po60], [Po88a] and [Po88b].

Theorem 2.65

(1) *If* $R = \exp\{-HJ\}$, *where* $H \geq 0$, *then the eigenvalues of* R *are positive and* $R^2 = JR^*JR$, *i.e.,* R *is the* J *modulus of itself.*

(2) *If* $U_1, U_2 \in \mathcal{P}_{const}(J)$ *are both invertible, and* $U_1JU_1^* \leq U_2JU_2^*$, *then* $R_1J \leq R_2J$ *for the corresponding* J *modulii.*

(3) *If* $R_1J \leq R_2J$ *and* $R_j = \exp\{-H_jJ\}$, *where* $H_j \geq 0$, *then* $H_1 \leq H_2$.

(4) *If* $H \in L_{loc}^{m\times m}([0,d))$ *and* $H(x) \geq 0$ *a.e. on* $[0,d)$, *then the multiplicative integral*

$$U = \overset{\curvearrowright}{\int_0^d} e^{-H(x)Jdx}$$

converges if and only if $H \in L_1^{m\times m}([0,d))$. *(See Section 3.19 for the relevant definitions.)*

Remark 2.66 *In view of assertion (4) of Theorem 2.64 and assertion (1) of Theorem 2.65, an* $m \times m$ *matrix* R *is the* J *modulus of an invertible matrix* $U \in \mathcal{P}_{const}(J)$ *if and only if*

$$R = e^{-HJ}$$

for some $H \in \mathbb{C}^{m\times m}$ *with* $H \geq 0$.

2.14 Infinite products of matrices

Let $B_1, B_2, \ldots, B_n$ be a finite sequence of $m \times m$ matrices and let

$$\overset{\curvearrowright}{\prod_{k=1}^{n}} B_k = B_1B_2\cdots B_n \quad \text{and} \quad \overset{\curvearrowleft}{\prod_{k=1}^{n}} B_k = B_nB_{n-1}\cdots B_1,$$

denote the right and left products respectively. The symbols

$$\overset{\curvearrowright}{\prod_{k=1}^{\infty}} B_k = B_1\,B_2\cdots B_n\cdots \quad \text{and} \quad \overset{\curvearrowleft}{\prod_{k=1}^{\infty}} B_k = \cdots B_n\,B_{n-1}\cdots B_1$$

are called the **right and left products** of the infinite sequence $\{B_k\}_{k=1}^{\infty}$ of $m \times m$ matrices, respectively.

A right (resp., left) infinite product of the matrices $B_1, B_2, \ldots$ is said to be **convergent** if

(1) All the matrices B_k are invertible.

(2) The sequence of right (resp., left) partial products

$$P_n = \prod_{k=1}^{\overset{n}{\curvearrowright}} B_k \qquad \left(\text{resp.,} \quad P_n = \prod_{k=1}^{\overset{n}{\curvearrowleft}} B_k \right)$$

converges to a limit P as $n \uparrow \infty$.

(3) The limit P is invertible.

If one or more of these three conditions fails, then the infinite product is said to be **divergent**.

If these three conditions are met, then P is called the right (resp., left) product of the matrices $B_1, B_2, \ldots$, and is denoted

$$P = \prod_{k=1}^{\overset{\infty}{\curvearrowright}} B_k \qquad \left(\text{resp.,} \quad P = \prod_{k=1}^{\overset{\infty}{\curvearrowleft}} B_k \right).$$

Since

$$\left(\prod_{k=1}^{\overset{n}{\curvearrowright}} B_k \right)^{\tau} = \prod_{k=1}^{\overset{n}{\curvearrowleft}} B_k^{\tau},$$

the right product is convergent if and only if the left product of the matrices B_k^{τ}, $k \geq 1$, is convergent and, if these products are convergent, then

$$\left(\prod_{k=1}^{\overset{\infty}{\curvearrowright}} B_k \right)^{\tau} = \prod_{k=1}^{\overset{\infty}{\curvearrowleft}} B_k^{\tau}.$$

Thus, it is enough to study right products.

If the right product

$$P_n = B_1 \cdots B_n = P_{n-1} B_n,$$

of the matrices B_k, $k \geq 1$, converges, then

$$\lim_{n \to \infty} B_n = I_m. \tag{2.92}$$

Since the matrices P_n and B_n are invertible and tend to invertible limits as $n \uparrow \infty$, the sequences $\|P_n\|$, $\|B_n\|$, $\|P_n^{-1}\|$ and $\|B_n^{-1}\|$ are bounded. Consequently, the identity

$$P_{n+\ell} - P_n = P_n \left(\overset{\curvearrowright}{\prod_{k=n+1}^{n+\ell}} B_k - I_m \right)$$

implies that $\{P_k\}$ is a Cauchy sequence if and only if for every $\varepsilon > 0$ there exists an integer N such that

$$\left\| \overset{\curvearrowright}{\prod_{k=n+1}^{n+\ell}} B_k - I_m \right\| < \varepsilon \text{ for every } n > N \text{ and every } \ell \geq 1. \tag{2.93}$$

Moreover, if this condition holds, then

$$\overset{\curvearrowright}{\prod_{k=n+1}^{\infty}} B_k \quad \text{is invertible for large enough } n$$

and hence, P is invertible, since for such n,

$$P = P_n \overset{\curvearrowright}{\prod_{k=n+1}^{\infty}} B_k$$

is the product of two invertible matrices.

Lemma 2.67 *Let $\{B_k\}_{k=1}^{\infty}$ be a sequence of invertible $m \times m$ matrices such that the series $\sum_{k=1}^{\infty} \|B_k - I_m\|$ converges. Then the right product*

$$\overset{\curvearrowright}{\prod_{k=1}^{\infty}} B_k \quad \textit{converges.}$$

Proof Let $A_k = B_k - I_m$, $k = 1, 2, \ldots$, let B_k be invertible and assume that the series $\sum_{k=1}^{\infty} \|A_k\|$ converges. Then

$$\begin{aligned}\left\| \prod_{k=n}^{\curvearrowright n+\ell} B_k - I_m \right\| &= \left\| \prod_{k=n}^{\curvearrowright n+\ell} (I_m + A_k) - I_m \right\| \le \prod_{k=n}^{n+\ell} (1 + \|A_k\|) - 1 \\ &\le \exp\left\{ \sum_{k=n}^{n+\ell} \|A_k\| \right\} - 1,\end{aligned}$$

since the second inequality is self-evident, and the first inequality is easily verified directly for $\ell = 1$, and then by induction for $\ell > 1$. Thus, the Cauchy condition (2.93) follows from the Cauchy condition for the convergence of the series $\sum_{k=1}^{\infty} \|A_k\|$. □

2.15 Some useful inequalities

In this section we establish some useful inequalities for future use.

Lemma 2.68 *If $V \in \mathbb{C}^{p\times q}$, then*

$$\|V\|^2 \le \operatorname{trace}(V^*V) \le q\|V\|^2 \tag{2.94}$$

and

$$\det(V^*V) \le \|V\|^{2q}. \tag{2.95}$$

If $p = q$ and V is invertible with $\|V^{-1}\| \le 1$, then

$$\|V\| \le |\det V| \le \|V\|^p. \tag{2.96}$$

Proof Let $\mu_1^2 \ge \cdots \ge \mu_q^2$ denote the eigenvalues of the positive semidefinite matrix V^*V. Then (2.94) and (2.95) are immediate from the observation that

$$\operatorname{trace}(V^*V) = \mu_1^2 + \cdots + \mu_q^2, \quad \|V\| = \mu_1$$

and

$$\det(V^*V) = \mu_1^2 \cdots \mu_q^2.$$

Finally, (2.96) is easily obtained from the same set of formulas when $q = p$ and V is invertible with $\|V^{-1}\| \leq 1$, because then $\mu_j^2 \geq 1$ for $j = 1, \ldots, p$. □

Lemma 2.69 *Let $W \in \mathcal{U}_{const}(j_{pq})$ and let $s_{21} = -w_{22}^{-1} w_{21}$. Then the matrix*

$$c = (I_q + s_{21}\varepsilon)(I_q - s_{21}\varepsilon)^{-1} \tag{2.97}$$

is well defined for each choice of $\varepsilon \in \mathcal{S}_{const}^{p\times q}$ and, if $s = T_W[\varepsilon]$, then

$$\Re c \geq \left(1 - \|s\|^2\right) w_{22}^* w_{22}. \tag{2.98}$$

Proof The matrix c is well defined, because w_{22} and $w_{21}\varepsilon + w_{22}$ are both invertible matrices. A straightforward calculation shows that

$$\begin{aligned} \Re c &= (I_q - s_{21}\varepsilon)^{-*}(I_q - \varepsilon^* s_{21}^* s_{21}\varepsilon)(I_q - s_{21}\varepsilon)^{-1} \\ &\geq (I_q - s_{21}\varepsilon)^{-*}(I_q - \varepsilon^*\varepsilon)(I_q - s_{21}\varepsilon)^{-1}. \end{aligned}$$

Therefore, since

$$I_q - s^* s = (w_{21}\varepsilon + w_{22})^{-*}(I_q - \varepsilon^*\varepsilon)(w_{21}\varepsilon + w_{22})^{-1},$$

it is readily seen that

$$\Re c \geq w_{22}^*(I_q - s^* s)w_{22} \geq \left(1 - \|s\|^2\right) w_{22}^* w_{22}.$$

□

2.16 Bibliographical notes

In [Po60] V. P. Potapov considered more general injective linear fractional transformations of the class $\mathcal{P}_{const}(J)$ into $\mathcal{S}_{const}^{m\times m}$ than (2.6). The transformation (2.6) was introduced by Yu. Ginzburg [Gi57]. The transformation defined by formula (2.13) was found independently by Redheffer and used to transform chain scattering (transmission) matrices into scattering matrices in his study of transmission lines with distributed parameters via the scattering formalism.

Matrix and operator balls were extensively studied by [Shm68]. The presented proofs of Lemmas 2.18–2.21 in Section 2.4 are adapted from [DFK92].

An analogue of Theorem 2.51 for the transform (2.53) in a Hilbert space setting was established by Redheffer in [Re60].

Most of the material in Sections 2.7 and 2.8 was adapted from [KrS96a].

Corollary 2.38 is equivalent to a result of P. Finsler [Fi37], which states that *if a pair* A *and* B *of* $n \times n$ *Hermitian matrices are such that* A *is invertible and*

$$\langle Ax, x\rangle \geq 0 \Longrightarrow \langle Bx, x\rangle \geq 0$$

for every $x \in \mathbb{C}^n$*, then there exists a constant* $c \geq 0$ *such that* $B - cA \geq 0$. The recent text [GLK05] is a good introduction to linear algebra in vector spaces with indefinite inner products that includes Finsler's result as Proposition 10.2.1.

The verification of (1) in Lemma 2.42 is adapted from Potapov [Po88a]. The discussion of the J modulus in Section 2.13 is adapted from [Po60], [Po88a] and [Po88b]. These references contain additional useful information on the J modulus; for additional perspective, see also [An04]. Item (4) in Theorem 2.65 is a corollary of a result on a family of multiplicative integrals of J moduli that is established in [Po88b], which is perhaps Potapov's principal result in this direction.

3

The Nevanlinna class of meromorphic mvf's

The first half of this chapter summarizes a number of basic definitions and facts on the Nevanlinna class of meromorphic scalar and mvf's of bounded Nevanlinna type in $\mathbb{C}_+$. Special attention is paid to the subclasses associated with the names of Schur, Carathéodory, Smirnov, and Hardy and a subclass of pseudomeromorphic functions for use in the sequel, mostly without proof. For additional information, the books of de Branges [Br68a], Dym and McKean [DMc76] and Rosenblum and Rovnyak [RR94] are recommended for scalar functions; Helson [He64], Rosenblum and Rovnyak [RR85] and Sz-Nagy and Foias [SzNF70] are good sources for matrix and operator valued functions. The article by Katsnelson and Kirstein [KK95] also contains useful information.

In the second part of this chapter, characterizations of the Nevanlinna class of mvf's and some of its subclasses in terms of the domain and range of the operator of multiplication by a mvf f in the class under consideration acting between two Hardy H_2-spaces of **vvf's** (vector valued functions) will be presented. Inner-outer factorizations and the notions of denominators and scalar denominators will also be developed in this part.

The symbols $\mathbb{C}$, $\mathbb{C}_+$ [resp., $\mathbb{C}_-$] and $\mathbb{R}$ will be used to denote the complex plane, the open upper [resp., lower] half plane and the real line, respectively; $\mathbb{R}_+ = [0, \infty)$ and $\mathbb{R}_- = (-\infty, 0]$. The symbols $\mathfrak{R}A = (A + A^*)/2$ and $\mathfrak{I}A = (A - A^*)/(2i)$ will be used for the real and imaginary parts of A for numbers. matrices and operators.

3.1 Basic classes of functions

A measurable $p \times q$ mvf $f(\mu)$ on $\mathbb{R}$ is said to belong to:

$L_r^{p\times q}$ for $1 \le r < \infty$ if

$$\|f\|_r^r = \int_{-\infty}^{\infty} \text{trace}\{f(\mu)^* f(\mu)\}^{r/2} d\mu \quad \text{is finite.}$$

$\widetilde{L}_r^{p\times q}$ for $1 \le r < \infty$ if

$$\int_{-\infty}^{\infty} (1+\mu^2)^{-1} \text{trace}\{f(\mu)^* f(\mu)\}^{r/2} d\mu \quad \text{is finite.}$$

$L_\infty^{p\times q}$ if

$$\text{ess sup}\{\|f(\mu\| : \mu \in \mathbb{R}\} \quad \text{is finite.}$$

A $p \times q$ mvf $f(\lambda)$ is said to belong to:

$\mathcal{S}^{p\times q}$ (the **Schur class**) if it is holomorphic in $\mathbb{C}_+$ and if $f(\lambda)^* f(\lambda) \le I_q$ for every point $\lambda \in \mathbb{C}_+$;

$H_\infty^{p\times q}$ if it is holomorphic in $\mathbb{C}_+$ and if

$$\|f\|_\infty = \sup\{\|f(\lambda)\| : \lambda \in \mathbb{C}_+\} < \infty;$$

$H_r^{p\times q}$ (the **Hardy class**), for $1 \le r < \infty$, if it is holomorphic in $\mathbb{C}_+$ and if†

$$\|f\|_r^r = \sup_{\nu>0} \int_{-\infty}^{\infty} \text{trace}\{f(\mu+i\nu)^* f(\mu+i\nu)\}^{r/2} d\mu \ < \infty;$$

$\mathcal{C}^{p\times p}$ (the **Carathéodory class**) if $q = p$ and it is holomorphic in $\mathbb{C}_+$ and

$$(\Re f)(\lambda) \ = \ \frac{f(\lambda) + f(\lambda)^*}{2} \ \ge 0$$

for every point $\lambda \in \mathbb{C}_+$;

$\mathcal{W}^{p\times q}(\gamma)$ (the **Wiener class**) for a fixed $\gamma \in \mathbb{C}^{p\times q}$, if it admits a representation of the form

$$f(\lambda) = \gamma + \int_{-\infty}^{\infty} e^{i\lambda t} h(t) dt, \text{ for } \lambda \in \mathbb{R},$$

where $h \in L_1^{p\times q}(\mathbb{R})$;

$\mathcal{W}^{p\times q}$ if it belongs to $\mathcal{W}^{p\times q}(\gamma)$ for some $\gamma \in \mathbb{C}^{p\times q}$;

† If $f \in H_r^{p\times q}$, then its norm coincides with the $L_r^{p\times q}$ norm of its boundary values

$\mathcal{W}_+^{p\times q}(\gamma)$ for a fixed $\gamma \in \mathbb{C}^{p\times q}$, if it admits a representation of the form

$$f(\lambda) = \gamma + \int_0^\infty e^{i\lambda t} h(t)dt, \text{ for } \lambda \in \mathbb{R} \cup \mathbb{C}_+,$$

where $h \in L_1^{p\times q}(\mathbb{R}_+)$;

$\mathcal{W}_-^{p\times q}(\gamma)$ for a fixed $\gamma \in \mathbb{C}^{p\times q}$, if it admits a representation of the form

$$f(\lambda) = \gamma + \int_{-\infty}^0 e^{i\lambda t} h(t)dt, \text{ for } \lambda \in \mathbb{R} \cup \mathbb{C}_-,$$

where $h \in L_1^{p\times q}(\mathbb{R}_-)$;

$\mathcal{N}_+^{p\times q}$ (the **Smirnov class**) and $\mathcal{N}_{out}^{p\times q}$ the **subclass of outer mvf's in** $\mathcal{N}_+^{p\times q}$ will be defined in Section 3.11;

$\mathcal{N}^{p\times q}$ (the **Nevanlinna class** of bounded characteristic) if it can be expressed in the form $f = h^{-1}g$, where $g \in H_\infty^{p\times q}$ and $h \in H_\infty$ $(= H_\infty^{1\times 1})$.

The class $\mathcal{N}^{p\times q}$ is closed under addition, and, when meaningful, multiplication and inversion. Moreover, even though $\mathcal{N}^{p\times q}$ is listed last, it is the largest class in the list of meromorphic $p \times q$ mvf's given above. Analogous classes will be considered for the open lower half plane $\mathbb{C}_-$. In particular, a $p \times q$ mvf f is said to belong to

$K_2^{p\times q}$ if it is holomorphic in $\mathbb{C}_-$ and if

$$\|f\|_2^2 = \sup_{\nu>0} \int_{-\infty}^\infty \operatorname{trace}\{f(\mu - i\nu)^* f(\mu - i\nu)\}d\mu \ < \ \infty.$$

For each such class of functions $\mathcal{X}^{p\times q}$ we shall use the symbol $\mathcal{X}$ instead of $\mathcal{X}^{1\times 1}$ and $\mathcal{X}^p$ instead of $\mathcal{X}^{p\times 1}$.

Theorem 3.1 (Fatou) *Let $f \in H_\infty$. Then there exist nontangential boundary values $f(\mu)$ for alomost all points $\mu \in \mathbb{R}$. In particular*

$$f(\mu) = \lim_{\nu\downarrow 0} f(\mu + i\nu) \quad a.e.\ on \quad \mathbb{R}.$$

Proof This follows from the corresponding theorem for harmonic functions in $\mathbb{C}_+$; see, e.g., Theorem 5.3 on p. 29 in [Ga81]. □

Theorem 3.2 *Let $f \in H_\infty$ and suppose that $f(\lambda) \not\equiv 0$ in $\mathbb{C}_+$, then the boundary value,*

$$f(\mu) \neq 0 \quad \textit{for almost all points} \quad \mu \in \mathbb{R}.$$

Proof See Corollary 2 on p.65 of [Ga81]. □

Corollary 3.3 *Let $f_1, f_2 \in H_\infty$ and suppose that the boundary values $f_1(\mu)$ and $f_2(\mu)$ coincide on a set of positive Lebesgue measure in $\mathbb{R}$. Then $f_1(\lambda) \equiv f_2(\lambda)$ on $\mathbb{C}_+$.*

Theorem 3.4 *Let $f \in \mathcal{N}^{p\times q}$. Then f has nontangential boundary values $f(\mu)$ at almost all points $\mu \in \mathbb{R}$ and $f(\lambda)$ is uniquely defined by its boundary values $f(\mu)$ on a set of positive Lebesgue measure in $\mathbb{R}$.*

Proof Let $f \in \mathcal{N}^{p\times q}$. Then, by definition, $f = h^{-1}g$, where $g \in \mathcal{S}^{p\times q}$, $h \in \mathcal{S}$ and $h \not\equiv 0$. Thus, by Fatou's theorem, g and h have nontangential limits $g(\mu)$ and $h(\mu)$ a.e. on $\mathbb{R}$ and $h(\mu) \neq 0$ a.e. on $\mathbb{R}$, by Theorem 3.2. Consequently, $f = h^{-1}g$ has nontangential boundary values. Moreover, if $f(\mu) = 0$ on a subset of $\mathbb{R}$ with positive Lebesgue measure, then $g(\mu) = 0$ on a subset of $\mathbb{R}$ of positive Lebesgue measure and hence $g(\lambda) \equiv 0$ in $\mathbb{C}_+$, by Theorem 3.2. Thus, the same conclusion holds for $f(\lambda) = h(\lambda)^{-1}g(\lambda)$. □

In view of the preceding discussion, every $f \in H_r^{p\times q}$ has nontangential boundary values, since $H_r^{p\times q} \subset \mathcal{N}^{p\times q}$ for $1 \leq r \leq \infty$. Moreover, the norm in $H_r^{p\times q}$ can be computed in terms of boundary values only, and the corresponding spaces $H_r^{p\times q}$ can be identified as closed subspaces of the Lebesgue spaces $L_r^{p\times q}$ on the line, for $1 \leq r \leq \infty$; see Theorem 3.59. Furthermore, the spaces $H_2^{p\times q}$ and $K_2^{p\times q}$ are mutually orthogonal complementary subspaces of $L_2^{p\times q}$ with respect to the inner product $\langle f, g\rangle = \int_{-\infty}^{\infty} \text{trace}\,\{g(\mu)^* f(\mu)\}d\mu$; i.e.,

$$L_2^{p\times q} = H_2^{p\times q} \oplus K_2^{p\times q}; \quad \text{and} \quad L_2^p = H_2^p \oplus K_2^p \quad \text{if} \quad q = 1.$$

Theorem 3.5 *Let $f \in H_r^{p\times q}$ for $1 \leq r < \infty$. Then the*

$$f(\omega) = \frac{1}{2\pi i}\int_{-\infty}^{\infty} \frac{f(\mu)}{\mu - \omega} d\mu \tag{3.1}$$

and the Poisson formula

$$f(\omega) = \frac{\Im\omega}{\pi}\int_{-\infty}^{\infty} \frac{f(\mu)}{|\mu-\omega|^2}d\mu \tag{3.2}$$

are valid for every point $\omega \in \mathbb{C}_+$. *Formula (3.2) is also valid for* $f \in H_\infty^{p\times q}$.

Proof This follows from Theorems 11.2 and 11.8 in [Du70], since it suffices to verify the asserted formulas for each entry in the mvf f. □

We shall use the symbols

Π_+ to denote the orthogonal projection from $L_2^{p\times q}(\mathbb{R})$ to $H_2^{p\times q}$,

$\Pi_- = I - \Pi_+$ for the complementary projection,

$P_\mathcal{L}$ denotes the orthogonal projection onto a closed subspace $\mathcal{L}$ of a Hilbert space,

$f^\#(\lambda) = f(\overline{\lambda})^*, \quad f^\sim(\lambda) = f(-\overline{\lambda})^*, \quad \rho_\omega(\lambda) = -2\pi i(\lambda - \overline{\omega}),$

$\bigvee_{\alpha\in A}\{\mathcal{L}_\alpha\}$ for the closed linear span of subsets $\mathcal{L}_\alpha$ in a space $\mathcal{X}$,

$\mathcal{E}$ for the class of scalar entire functions,

$$\ln^+|a| = \begin{cases} \ln|a| & \text{if} \quad |a| \geq 1 \\ 0 \quad \text{if} & |a| < 1 \end{cases},$$

$\langle g, h\rangle_{st} = \int_{-\infty}^{\infty} h(\mu)^* g(\mu) du$ for the **standard inner product** in $L_2^k(\mathbb{R})$ and for mvf's $h \in L_2^{k\times p}(\mathbb{R})$ and $g \in L_2^{k\times q}(\mathbb{R})$.

3.2 The Riesz-Herglotz-Nevanlinna representation for mvf's in the Carathéodory class

A $p \times p$ mvf $c(\lambda)$ belongs to the Carathéodory class $\mathcal{C}^{p\times p}$ if and only if it admits an integral representation via the Riesz-Herglotz-Nevanlinna formula

$$c(\lambda) = i\alpha - i\lambda\beta + \frac{1}{\pi i}\int_{-\infty}^{\infty}\left\{\frac{1}{\mu-\lambda} - \frac{\mu}{1+\mu^2}\right\}d\sigma(\mu) \quad \text{for } \lambda \in \mathbb{C}_+, \tag{3.3}$$

where $\alpha = \alpha^* \in \mathbb{C}^{p\times p}, \beta \in \mathbb{C}^{p\times p}, \beta \geq 0$ and $\sigma(\mu)$ is a nondecreasing $p \times p$ mvf on $\mathbb{R}$ such that

$$\int_{-\infty}^{\infty} \frac{d(\text{trace } \sigma(\mu))}{1+\mu^2} < \infty. \tag{3.4}$$

The parameters α and β are uniquely defined by $c(\lambda)$ via the formulas

$$\alpha = \Im c(i), \quad \beta = \lim_{\nu\uparrow\infty} \nu^{-1}\Re c(i\nu). \tag{3.5}$$

The Stieltjes inversion formula

$$\sigma(\mu_2) - \sigma(\mu_1) = \lim_{\epsilon\downarrow 0} \int_{\mu_1}^{\mu_2} \Re(c(\mu + i\epsilon))d\mu, \tag{3.6}$$

which is valid at points of continuity μ_1, μ_2 of $\sigma(\mu)$, serves to define $\sigma(\mu)$ up to normalization. Such a mvf $\sigma(\mu)$ will be called **a spectral function of** $c(\lambda)$. In this monograph we shall always assume that $\sigma(0) = 0$ and that $\sigma(\mu)$ is left continuous on $\mathbb{R}$. Under these normalization conditions, $\sigma(\mu)$ is uniquely defined by $c(\lambda)$. The inclusion

$$\mathcal{C}^{p\times p} \subset \mathcal{N}_+^{p\times p},$$

that will be established in Lemma 3.58, serves to guarantee that every mvf $c \in \mathcal{C}^{p\times p}$ has nontangential boundary limits $c(\mu)$ at almost all points $\mu \in \mathbb{R}$. In particular,

$$c(\mu) = \lim_{\varepsilon\downarrow 0} c(\mu + i\varepsilon) \quad \text{for almost all points } \mu \in \mathbb{R}.$$

The spectral function $\sigma(\mu)$ can be decomposed into the sum

$$\sigma(\mu) = \sigma_s(\mu) + \sigma_a(\mu), \tag{3.7}$$

of two nondecreasing $p \times p$ mvf's $\sigma_s(\mu)$ and $\sigma_a(\mu)$, where $\sigma_s(\mu)$ is the **singular component** of $\sigma(\mu)$, i.e., $\sigma_s'(\mu) = 0$ for almost all points $\mu \in \mathbb{R}$, and $\sigma_a(\mu)$ is the **locally absolutely continuous** part of $\sigma(\mu)$ normalized by the condition $\sigma_a(0) = 0$, i.e.,

$$\sigma_a(\mu) = \int_0^{\mu} f(a)da, \tag{3.8}$$

where

$$f(\mu) \geq 0 \quad \text{a.e. on } \mathbb{R} \quad \text{and} \quad \int_{-\infty}^{\infty} \frac{\text{trace } f(\mu)}{1+\mu^2} d\mu < \infty. \tag{3.9}$$

The convergence of the last integral follows from (3.4). In fact, condition (3.4) is equivalent to the two conditions

$$\int_{-\infty}^{\infty} \frac{d(\text{trace } \sigma_s(\mu))}{1+\mu^2} < \infty \quad \text{and} \quad \int_{-\infty}^{\infty} \frac{\text{trace } f(\mu)}{1+\mu^2} d\mu < \infty.$$

Moreover,

$$f(\mu) = \sigma'(\mu) = \Re c(\mu) \quad \text{a.e. on } \mathbb{R} \tag{3.10}$$

and hence, in view of formula (3.9),

$$\int_{-\infty}^{\infty} \frac{\operatorname{trace}\{c(\mu) + c(\mu)^*\}}{1+\mu^2} d\mu \;<\; \infty. \tag{3.11}$$

If $\sigma(\mu)$ is locally absolutely continuous, i.e., if $\sigma(\mu) = \sigma_a(\mu)$, then the $p \times p$ mvf $f(\mu) = \sigma'(\mu)$ is called the **spectral density** of $c(\lambda)$.

Formula (3.3) implies that

$$\Re c(i) = \beta + \frac{1}{\pi} \int_{-\infty}^{\infty} \frac{d\sigma(\mu)}{1+\mu^2},$$

which leads easily to the following concclusions:

Lemma 3.6 *Let $c \in \mathcal{C}^{p\times p}$. Then in formula (3.3)*

$$\beta = 0 \iff \Re c(i) = \frac{1}{\pi} \int_{-\infty}^{\infty} \frac{d\sigma(\mu)}{1+\mu^2}, \tag{3.12}$$

whereas $\beta = 0$ and $\sigma(\mu)$ is locally absolutely continuous if and only if

$$\Re c(i) = \frac{1}{\pi} \int_{-\infty}^{\infty} \frac{\Re c(\mu)}{1+\mu^2} d\mu. \tag{3.13}$$

The subclass of mvf's $c \in \mathcal{C}^{p\times p}$ with $\beta = 0$ and locally absolutely continuous spectral functions $\sigma(\mu)$ in the representation (3.3) will be denoted $\mathcal{C}_a^{p\times p}$. The subclass of mvf's $c \in \mathcal{C}^{p\times p}$ with singular spectral functions will be denoted $\mathcal{C}_{sing}^{p\times p}$. Every mvf $c \in \mathcal{C}^{p\times p}$ has an additive decomposition

$$c(\lambda) = c_s(\lambda) + c_a(\lambda), \quad \text{where} \quad c_s \in \mathcal{C}_{sing}^{p\times p} \quad \text{and} \quad c_a \in \mathcal{C}_a^{p\times p}. \tag{3.14}$$

This decomposition is unique up to an additive constant purely imaginary $p \times p$ matrix. Thus, in terms of the notation introduced in (3.3), (3.7) and (3.10), we may set

$$c_s(\lambda) = i\alpha - i\beta\lambda + \frac{1}{\pi i} \int_{-\infty}^{\infty} \left\{ \frac{1}{\mu - \lambda} - \frac{\mu}{1+\mu^2} \right\} d\sigma_s(\mu) \tag{3.15}$$

and

$$c_a(\lambda) = \frac{1}{\pi i} \int_{-\infty}^{\infty} \left\{ \frac{1}{\mu - \lambda} - \frac{\mu}{1+\mu^2} \right\} f(\mu) d\mu, \tag{3.16}$$

where $f(\mu) = \Re c(\mu)$ a.e. on $\mathbb{R}$. This decomposition corresponds to the normalization $c_a(i) \geq 0$, and is uniquely determined by this normalization. There are other normalizations that may be imposed on c_a that insure uniqueness of the decomposition (3.14).

Remark 3.7 *A mvf $c \in \mathcal{C}^{p\times p}$ admits a holomorphic extension across a finite interval (a, b) if and only if $\sigma_s(a+) = \sigma_s(b-)$ and the spectral density $f(\mu)$ of $c_a(\lambda)$ has a holomorphic extension to an open set in $\mathbb{C}$ that contains (a, b). This depends essentially upon the observation that if f is holomorphic in the set*

$$\Omega = \{\mu + i\nu : a + \delta \leq \mu \leq b - \delta \quad \textit{and} \quad -\delta \leq \nu \leq \delta\},$$

then

$$\int_{a+\delta}^{b-\delta} \left\{\frac{1}{\mu - \lambda} - \frac{\mu}{1 + \mu^2}\right\} f(\mu)d\mu = \int_{\Gamma} \left\{\frac{1}{\mu - \lambda} - \frac{\mu}{1 + \mu^2}\right\} f(\zeta)d\zeta,$$

where Γ is the intersection of the boundary of Ω with $\mathbb{C}_-$ directed from $a+\delta$ to $b - \delta$.

Remark 3.8 *If $c \in \mathcal{C}^{p\times p}$, $a \in \mathbb{R}$ and $b > 0$, then formula (3.3) implies that*

$$(\Re c)(a + ib) = b\beta + \frac{b}{\pi}\int_{-\infty}^{\infty} \frac{d\sigma(\mu)}{(\mu - a)^2 + b^2}. \tag{3.17}$$

The particular choice $c(\lambda) = I_p$ yields the evaluation

$$1 = \frac{b}{\pi}\int_{-\infty}^{\infty} \frac{1}{(\mu - a)^2 + b^2} d\mu. \tag{3.18}$$

Lemma 3.9 *If $c \in \mathcal{C}^{p\times p}$, $a \in \mathbb{R}$ and $b > 0$, then*

$$\int_{-\infty}^{\infty} \frac{(\Re c)(a + ib)}{a^2 + (b + 1)^2} da \leq 2\pi(\Re c)(i). \tag{3.19}$$

Proof If $b > 0$ and $\nu > 0$, then the evaluation

$$\frac{b}{\pi}\int_{-\infty}^{\infty} \frac{1}{(\mu - a)^2 + b^2} \frac{1}{a + i\nu} da = \frac{1}{\mu + i(b + \nu)}, \tag{3.20}$$

which follows readily from the Poisson formula since the function

$$f_\nu(\lambda) = \frac{1}{\lambda + i\nu}$$

belongs to H_∞ if $\nu > 0$, leads easily to the formula

$$\frac{b}{\pi}\int_{-\infty}^{\infty}\frac{1}{(\mu-a)^2+b^2}\,\frac{1}{a^2+\nu^2}da = \frac{b+\nu}{\nu(\mu^2+(b+\nu)^2)}. \tag{3.21}$$

Thus, in view of formulas (3.17) and (3.21),

$$\begin{aligned}\int_{-\infty}^{\infty}\frac{(\Re c)(a+ib)}{a^2+(b+1)^2}da &= \frac{\pi b}{b+1}\beta + \frac{2b+1}{b+1}\int_{-\infty}^{\infty}\frac{d\sigma(\mu)}{\mu^2+(2b+1)^2}\\ &\le \pi\left\{2\beta+\frac{2}{\pi}\int_{-\infty}^{\infty}\frac{d\sigma(\mu)}{\mu^2+1}\right\}\\ &= 2\pi(\Re c)(i).\end{aligned}$$

□

Lemma 3.10 *Let $c \in \mathcal{C}^{p\times p}$. Then:*

(1) $\Re c(\omega) > 0$ *for at least one point $\omega \in \mathbb{C}_+$ if and only if* $\Re c(\omega) > 0$ *for every point $\omega \in \mathbb{C}_+$.*

(2) *If $\Re c(\mu) > 0$ for almost all points $\mu \in \mathbb{R}$, then $(\Re c)(\omega) > 0$ for every point $\omega \in \mathbb{C}_+$.*

(3) *If $\Re c(\omega) > 0$ for at least one point $\omega \in \mathbb{C}_+$, then $c^{-1} \in \mathcal{C}^{p\times p}$.*

Proof Statements (1) and (2) follow from formula (3.17) and the fact that $\beta \ge 0$ and $\sigma(\mu)$ is a nondecreasing mvf on $\mathbb{R}$. The verification of (2) also uses the inequality

$$\int_{-\infty}^{\infty}\frac{d\sigma(\mu)}{|\mu-\lambda|^2} \ge \int_{-\infty}^{\infty}\frac{\Re c(\mu)}{|\mu-\lambda|^2}d\mu.$$

Statement (3) is immediate from (1). □

3.3 Some subclasses of the Carathéodory class $\mathcal{C}^{p\times p}$

(a) $\mathcal{C}^{p\times p}\cap H_\infty^{p\times p}$.

If $c \in \mathcal{C}^{p\times p}\cap H_\infty^{p\times p}$, then, by a well-known theorem of the brothers Riesz (see e.g., p.74 [RR94]), $\beta = 0$ in formula (3.15), the spectral function $\sigma(\mu)$ of $c(\lambda)$ is locally absolutely continuous and formula (3.15) reduces to

$c_s(\lambda) = i\alpha$. Therefore,

$$c(\lambda) = i\alpha + \frac{1}{\pi i}\int_{-\infty}^{\infty}\left\{\frac{1}{\mu-\lambda} - \frac{\mu}{1+\mu^2}\right\} f(\mu)d\mu, \tag{3.22}$$

where $\alpha^* = \alpha \in \mathbb{C}^{p\times p}$,

$$f(\mu) = (\Re c)(\mu) \geq 0 \quad \text{a.e. on } \mathbb{R} \quad \text{and} \quad f \in L_\infty^{p\times p}(\mathbb{R}). \tag{3.23}$$

It is clear that if condition (3.23) is in force, then the function $c(\lambda)$ defined by formula (3.22) belongs to $\mathcal{C}_a^{p\times p}$ and that

$$\Re c(\lambda) \leq \|f\|_\infty I_p \quad \text{for } \lambda \in \mathbb{C}_+.$$

To the best of our knowledge, necessary and sufficient conditions on $f(\mu)$ which guarantee that $c \in \mathcal{C}^{p\times p} \cap H_\infty^{p\times p}$ are not known.

We turn next to two subclasses of the class $\mathcal{C}^{p\times p} \cap H_\infty^{p\times p}$:

(b) $\mathring{\mathcal{C}}^{p\times p} = \{c \in \mathcal{C}^{p\times p} \cap H_\infty^{p\times p} : (\Re c)(\mu) \geq \delta_c I_p > 0 \text{ a.e. on } \mathbb{R}\}$, where $\delta_c > 0$ depends upon c.

(c) $\mathcal{C}^{p\times p} \cap W_+^{p\times p}(\gamma)$, where $\Re\gamma > 0$.

In case (b) the mvf $f(\mu)$ in the integral representation (3.22) is subject to the bounds

$$\delta_1 I_p \leq f(\mu) \leq \delta_2 I_p \quad \text{for a.e.} \quad \mu \in \mathbb{R},$$

where $0 < \delta_1 \leq \delta_2$.

In case (c) the lower bound may not be in force.

(d) $\mathcal{C}_0^{p\times p} = \{c \in \mathcal{C}^{p\times p} : \sup\{\|\nu c(i\nu)\| : \nu > 0\} < \infty\}$.

It is known that a mvf $c(\lambda)$ belongs to $\mathcal{C}_0^{p\times p}$ if and only if it admits a representation of the form

$$c(\lambda) = \frac{1}{\pi i}\int_{-\infty}^{\infty}\frac{1}{\mu-\lambda}d\sigma(\mu), \quad \text{where trace } \sigma(\mu) < \infty,$$

or, equivalently, if and only if $c \in \mathcal{C}^{p\times p}$ and

$$\sup_{\nu>0}\{\nu \text{ trace } \Re c(i\nu)\} < \infty \quad \text{and} \quad \lim_{\nu\uparrow\infty} c(i\nu) = 0. \tag{3.24}$$

3.4 Inner-outer factorization in the class H_∞

A function $f(\lambda)$ belongs to the class $\mathcal{S}_{in}$ of **scalar inner functions** if $f \in \mathcal{S}$ and $|f(\mu)| = 1$ for almost all points $\mu \in \mathbb{R}$.

Example 3.11 (Blaschke products) *Let*

$$b(\lambda) = \prod_{k=1}^{n} \gamma_k \left(\frac{\lambda - \omega_k}{\lambda - \overline{\omega_k}} \right), \quad \text{for } \lambda \in \mathbb{C}_+ \text{ and } n \leq \infty, \tag{3.25}$$

where the points $\omega_k \in \mathbb{C}_+$ *and are subject to the* **Blaschke condition**

$$\sum_{k=1}^{\infty} \frac{|\Im \omega_k|}{1 + |\omega_k|^2} < \infty \quad \text{if} \quad n = \infty \tag{3.26}$$

(that is necessary for the convergence of the product) and the γ_k *are constants of modulus one that are chosen to insure the convergence of the product if* $n = \infty$*. For example, we can choose*

$$\gamma_k = 1 \text{ if } |\omega_k| \leq 1 \text{ and } \gamma_k = \bar{\omega}_k/\omega_k \text{ if } |\omega_k| > 1.$$

If $|\omega_k| \leq 1$ *for at most finitely many indices* k*, then we can choose* $\gamma_k = \bar{\omega}_k/\omega_k$ *for all* k*. In this case,*

$$b(\lambda) = \prod_{k=1}^{n} \frac{1 - \lambda/\omega_k}{1 - \lambda/\bar{\omega}_k}, \qquad \lambda \in \mathbb{C}_+, \ n \leq \infty. \tag{3.27}$$

Products of the form (3.25) and (3.27) are called **Blaschke products** *and the factors are called* **elementary Blaschke factors.**

Theorem 3.12 *A function* $f \in \mathcal{S}$ *admits the representation*

$$f(\lambda) = b(\lambda) b_s(\lambda), \tag{3.28}$$

where $b(\lambda)$ *is a Blaschke product and* $b_s(\lambda) = \exp(i\alpha + i\beta\lambda)$ *with* $\alpha = \overline{\alpha}$ *and* $\beta \geq 0$ *if and only if*

$$\lim_{\nu \downarrow 0} \int_{-\infty}^{\infty} \frac{\ln |f(\mu + i\nu)|}{1 + \mu^2} d\mu = 0. \tag{3.29}$$

Moreover, if $f \in \mathcal{S}_{in}$ *and*

$$f^{-1} \in \bigvee_{t \geq 0} e_{-t} H_\infty \quad \text{in } L_\infty, \tag{3.30}$$

then $b_s(\lambda) = \exp(i\alpha + i\beta\lambda)$ *as above.*

Proof See Akutowicz [Aku56] for the condition (3.29) and [Dy74] for the sufficiency condition (3.30). □

Theorem 3.13 *Let $f \in \mathcal{S}$ and suppose that $f(\omega) = 0$ for at least one point $\omega \in \mathbb{C}_+$ but that $f(\lambda) \not\equiv 0$. Then the set $\omega_1, \omega_2, \ldots,$ of all the zeros of $f(\lambda)$, repeated according to multiplicity, satisfy the Blaschke condition (3.26) and hence the corresponding Blaschke product $b(\lambda)$ defined by formula (3.25) is convergent if the normalization constants γ_k are chosen appropriately. Moreover,*

$$f(\lambda) = b(\lambda)f_1(\lambda), \text{ where } f_1 \in \mathcal{S} \text{ and } f_1(\lambda) \neq 0 \text{ for every } \lambda \in \mathbb{C}_+.$$

Proof The proof rests on the observation that if $f(\omega) = 0$ for some point $\omega \in \mathbb{C}_+$ and $b_\omega(\lambda) = (\lambda - \omega)/(\lambda - \overline{\omega})$, then $b_\omega^{-1} f \in \mathcal{S}$ by Schwarz's lemma. Clearly a finite number of zeros $\omega_1, \ldots, \omega_n$ can be removed from $f(\lambda)$ by iterating this procedure n times. A more detailed analysis may be found in Section 2.13 of [DMc76]. □

Example 3.14 (Singular inner factors) *Let*

$$\begin{aligned} b_s(\lambda) &= \exp\{-c_s(\lambda)\} \\ &= \exp\left\{-\left[i\alpha - i\beta\lambda + \frac{1}{\pi i}\int_{-\infty}^{\infty} \frac{1+\mu\lambda}{\mu-\lambda}\frac{d\sigma_s(\mu)}{1+\mu^2}\right]\right\}, \end{aligned} \tag{3.31}$$

for $\lambda \in \mathbb{C}_+$, where $c_s \in \mathcal{C}_{sing}^{p\times p}$ is given by formula (3.15) and hence, $\alpha \in \mathbb{R}$, $\beta \geq 0$ and $\sigma_s(\mu)$ is a nondecreasing singular function on $\mathbb{R}$ (i.e., $\sigma_s'(\mu) = 0$ a.e. on $\mathbb{R}$) that is subject to the constraint

$$\int_{-\infty}^{\infty} \frac{d(\text{trace } \sigma_s(\mu))}{1+\mu^2} < \infty.$$

The function $b_s(\lambda)$ is referred to as a **singular inner function**. *It can be characterized as an inner function that has no zeros in $\mathbb{C}_+$.*

The next theorem presents a fundamental factorization formula for functions $f \in \mathcal{S}$ that is due to Riesz and Herglotz.

Theorem 3.15 *Let $f \in \mathcal{S}$ and suppose that $f(\lambda) \not\equiv 0$. Then*

$$f(\lambda) = b(\lambda)b_s(\lambda)\varphi(\lambda), \tag{3.32}$$

where $b(\lambda)$ is a finite or infinite Blaschke product that is defined by the zeros $\{\omega_k\}_1^n$ $(n \leq \infty)$ of $f(\lambda)$ (repeated according to their multiplicity) by formula (3.25), or $b(\lambda) = 1$ if $f(\lambda) \neq 0$ in $\mathbb{C}_+$, $b_s(\lambda)$ is a singular inner function and

$$\varphi(\lambda) = \exp\left\{-\frac{i}{\pi}\int_{-\infty}^{\infty} \frac{1+\mu\lambda}{\mu-\lambda}\frac{\ln|f(\mu)|}{1+\mu^2}d\mu\right\} \quad \text{for } \lambda \in \mathbb{C}_+. \tag{3.33}$$

Conversely, every function $f(\lambda)$ of the form (3.32) in which $b(\lambda)$ is any finite or convergent infinite Blaschke product, $b_s(\lambda)$ is any singular inner function and

$$\varphi(\lambda) = \exp\left\{\frac{i}{\pi}\int_{-\infty}^{\infty} \frac{1+\mu\lambda}{\mu-\lambda}\frac{k(\mu)}{1+\mu^2}d\mu\right\} \quad \text{for } \lambda \in \mathbb{C}_+, \tag{3.34}$$

where

$$\frac{k(\mu)}{1+\mu^2} \quad \text{is summable and} \quad k(\mu) \geq 0 \quad \text{a.e. on } \mathbb{R}, \tag{3.35}$$

belongs to the class $\mathcal{S}$ and $k(\mu) = -\ln|f(\mu)|$ a.e. on $\mathbb{R}$.

Proof Theorem 3.13 guarantees that if $f \in \mathcal{S}$ and $f \not\equiv 0$, then

$$f(\lambda) = b(\lambda)\exp\{-c(\lambda)\} \quad \text{for} \quad \lambda \in \mathbb{C}_+,$$

where $b(\lambda)$ is the Blaschke product that is defined by the zeros of $f(\lambda)$ in $\mathbb{C}_+$ and $c \in \mathcal{C}$. Thus, upon invoking the additive decomposition (3.14) and writing

$$b_s(\lambda) = \exp\{-c_s(\lambda)\},$$

we obtain the formula (3.32) with

$$\varphi(\lambda) = \exp\{-c_a(\lambda)\},$$

which is of the form (3.34), with

$$k(\mu) = (\Re c)(\mu) = -\ln|f(\mu)| \quad \text{a.e. on } \mathbb{R}.$$

The stated conditions (3.35) on $k(\mu)$ follow from this formula and the fact that $(1+\mu^2)^{-1}(\Re c) \in L_1$.

The fact that every function $f(\lambda)$ of the form (3.32) belongs to the class $\mathcal{S}$ when $k(\mu)$ is subject to the constraints (3.35) is self-evident. □

Remark 3.16 *The factors $b(\lambda)$ and $b_s(\lambda)$ in formula (3.32) are unique up to constant factors of modulus one.*

Corollary 3.17 *Every inner function $f(\lambda)$ can be written as a product*

$$f(\lambda) = b(\lambda)b_s(\lambda), \quad \text{for } \lambda \in \mathbb{C}_+, \tag{3.36}$$

where $b(\lambda)$ is a Blaschke product and $b_s(\lambda)$ is a singular inner function. Moreover, the factors $b(\lambda)$ and $b_s(\lambda)$ are uniquely determined by $f(\lambda)$ up to multiplicative constants of modulus one.

Proof This is immediate from the previous theorem, because $k(\mu) = 0$ a.e. on $\mathbb{R}$, i.e., $\varphi(\lambda) \equiv 1$. □

A function $\varphi \in H_\infty$ is called **outer** if

$$\{\varphi f : \ f \in H_2\} \ \text{ is dense in } \ H_2.$$

The class of scalar outer functions that belong to the Schur class $\mathcal{S}$ will be denoted $\mathcal{S}_{out}$. The notation

$$\varphi_k(\lambda) = \left(\frac{\lambda - i}{\lambda + i}\right)^k \quad \text{and} \quad \psi_k(\lambda) = \frac{\varphi_k(\lambda)}{\sqrt{\pi}(\lambda + i)}$$
$$\text{for } k = 0, \pm 1, \ldots \quad \text{and} \quad e_t(\lambda) = e^{i\lambda t} \tag{3.37}$$

will be useful.

Theorem 3.18 *Let $\varphi \in H_\infty$. Then the following conditions are equivalent:*

(1) *φ is outer.*

(2) *$\varphi(\lambda) \neq 0$ for every point $\lambda \in \mathbb{C}_+$ and*

$$\ln |\varphi(\omega)| = \frac{\Im \omega}{\pi} \int_{\infty}^{\infty} \frac{\ln |\varphi(\mu)|}{|\mu - \omega|^2} d\mu \tag{3.38}$$

for at least one point $\omega \in \mathbb{C}_+$.

(3) *$\varphi(\lambda) \neq 0$ for every point $\lambda \in \mathbb{C}_+$ and the identity (3.38) holds for every point $\omega \in \mathbb{C}_+$.*

(4) *φ admits an integral representation of the form*

$$\varphi(\lambda) = \gamma \ \exp\left\{\frac{i}{\pi}\int_{-\infty}^{\infty} \frac{1 + \mu\lambda}{\mu - \lambda} \ \frac{k(\mu)}{1 + \mu^2} d\mu\right\} \quad \text{for } \lambda \in \mathbb{C}_+, \tag{3.39}$$

where γ is a constant of modulus one and $k \in \widetilde{L}_1(\mathbb{R})$.

(5) $\bigvee_{t \geq 0} \{e_t \psi_0 \, \varphi : t \geq 0\} = H_2$.

(6) $\bigvee_{n=0}^{\infty} \{\psi_n \varphi : \; n = 0, 1, \ldots\} = H_2$.

Moreover, if φ is outer, then

$$k(\mu) = -\ln|\varphi(\mu)| \tag{3.40}$$

in the integral representation (3.39).

Proof The equivalence of items (1)–(5) is covered in Sections 2.8–2.10 of [DMc72]. The equivalence of (1) and (6) follows from the definition of an outer function and the fact that

$$\{\psi_n : n = 0, 1, \ldots\} \quad \text{is an orthonormal basis for } H_2.$$

□

Corollary 3.19 *The functions φ in the class $\mathcal{S}_{out}$ can be parametrized by the formula (3.39), where γ is a constant of modulus one and $k(\mu)$ is a function on $\mathbb{R}$ that is subject to the constraints (3.35). Moreover, in formula (3.39),*

$$k(\mu) = -\ln|\varphi(\mu)| \quad \text{a.e. on } \mathbb{R} \quad \text{and} \quad \gamma = \varphi(i)/|\varphi(i)|. \tag{3.41}$$

Theorem 3.20 *If $f \in H_\infty$ and $f \not\equiv 0$, then $f(\lambda)$ admits an inner-outer factorization*

$$f(\lambda) = f_i(\lambda) f_o(\lambda), \tag{3.42}$$

in which the inner factor f_i and the outer factor f_o are unique up to multiplicative constants of modulus one. Moreover:

(1) *$|f(\lambda)|$ is subject to the inequality*

$$\ln|f(a+ib)| \leq \frac{b}{\pi} \int_{-\infty}^{\infty} \frac{\ln|f(\mu)|}{(\mu - a)^2 + b^2} d\mu \quad \text{for} \quad b > 0. \tag{3.43}$$

(2) *The inequality (3.43) is an equality at one point $\omega = a + ib$ in $\mathbb{C}_+$ if and only if equality holds at every point $\omega \in \mathbb{C}_+$.*

(3) *The inequality (3.43) is an equality at one point $\omega = a + ib$ in $\mathbb{C}_+$ if and only if $f(\lambda)$ is an outer function.*

(4) *If $f \in \mathcal{S}$ and $f \not\equiv 0$, then $f_o \in \mathcal{S}_{out}$.*

Proof Let $s(\lambda) = \|f\|_\infty^{-1} f(\lambda)$. Then $s \in \mathcal{S}$ and hence, by Theorems 3.15 and 3.18, $s(\lambda)$ admits an inner-outer factorization

$$s(\lambda) = s_i(\lambda)s_o(\lambda).$$

Thus, the factorization formula (3.42) holds with $f_i(\lambda) = s_i(\lambda)$ and $f_o(\lambda) = \|f\|_\infty s_o(\lambda)$. The inequality (3.43) follows from the inequality

$$|f(\omega)| \leq |f_o(\omega)|$$

for every point $\omega \in \mathbb{C}_+$ and the identity (3.38) applied to the outer factor $f_o(\lambda)$. □

Corollary 3.21 *Let* $f \in H_\infty$. *Then*

$$\sup \{|f(\lambda)| : \lambda \in \mathbb{C}_+\} = \text{ess sup} \{|f(\mu)| : \mu \in \mathbb{R}\}. \tag{3.44}$$

Proof In view of formula (3.18), the inequality (3.43) clearly implies that $|f(\lambda)| \leq \text{ess sup} \{|f(\mu)| : \mu \in \mathbb{R}\}$ for every point $\lambda \in \mathbb{C}_+$ and hence that the left hand side of the asserted identity cannot exceed the right hand side. The identity then follows from the fact that

$$f(\mu) = \lim_{\nu \downarrow 0} f(\mu + i\nu) \quad \text{a.e. on } \mathbb{R}.$$

□

3.5 Factorization in the classes $\mathcal{N}_+$ and $\mathcal{N}$

A function $f \in \mathcal{N}$ is said to belong to the **Smirnov class** $\mathcal{N}_+$ if it admits a representation of the form

$$f = g/h \quad \text{with} \quad g \in \mathcal{S} \quad \text{and} \quad h \in \mathcal{S}_{out}. \tag{3.45}$$

Thus, in view of the preceding analysis, every function $f \in \mathcal{N}_+$ may be expressed as a product of the form

$$f(\lambda) = b(\lambda)b_s(\lambda)\varphi_1(\lambda)/\varphi_2(\lambda), \tag{3.46}$$

where $b(\lambda)$ is a Blaschke product of the form (3.25), $b_s(\lambda)$ is a singular inner function of the form (3.31), $\varphi_j(\lambda) \in \mathcal{S}_{out}, j = 1, 2$, and

$$|\varphi_1(\mu)/\varphi_2(\mu)| = |f(\mu)| \quad \text{for almost all points} \quad \mu \in \mathbb{R}.$$

Formulas (3.39) and (3.40) are still valid for $\varphi(\lambda) = \varphi_1(\lambda)/\varphi_2(\lambda)$. However, the restriction $k(\mu) \geq 0$ a.e. on $\mathbb{R}$ is no longer in force.

Theorem 3.22 (The Smirnov maximum principle) *Let $f \in \mathcal{N}$ and let $1 \le p \le \infty$. Then*

$$f \in H_p \iff f \in \mathcal{N}_+ \cap L_p(\mathbb{R}). \tag{3.47}$$

Moreover, if $f \in H_p$, then

$$\|f\|_{H_p} = \|f\|_{L_p} \tag{3.48}$$

for $1 \le p \le \infty$.

Proof See Theorem A on p. 88 of [RR85]. □

If $f \in \mathcal{N}$ is holomorphic in $\mathbb{C}_+$ and $f(\lambda) \not\equiv 0$, then the number

$$\tau_f^+ = \limsup_{\nu \uparrow \infty} \frac{\ln |f(i\nu)|}{\nu} \tag{3.49}$$

is finite.

Lemma 3.23 *If $f \in \mathcal{N}_+$ and $f(\lambda) \not\equiv 0$ in $\mathbb{C}_+$, then:*

(1) $-\infty < \tau_f^+ \le 0$.

(2) $e^{i\lambda\delta} f(\lambda) \in \mathcal{N}_+ \iff \delta \ge \tau_f^+$.

(3) *In formulas (3.36), (3.28) and (3.32), $\tau_f^+ = \tau_{b_s}^+$ and $\tau_{b_s}^+ = -\beta$.*

(4) *If also $f \in \mathcal{S}$, then $e^{i\lambda\tau_f^+} f(\lambda) \in \mathcal{S}$.*

Proof The first three assertions are immediate from the representation formula (3.46) and the formulas for its factors. In fact $\tau_f^+ = -\beta$, since

$$\lim_{\nu \uparrow \infty} \frac{\ln |\varphi_1(i\nu)|}{\nu} = \lim_{\nu \uparrow \infty} \frac{\ln |\varphi_2(i\nu)|}{\nu} = \limsup_{\nu \uparrow \infty} \frac{\ln |b(i\nu)|}{\nu} = 0 \tag{3.50}$$

and

$$\lim_{\nu \uparrow \infty} \frac{\ln |b_s(i\nu)|}{\nu} = \lim_{\nu \uparrow \infty} \frac{\ln |e_\beta(i\nu)|}{\nu} = -\beta. \tag{3.51}$$

The final assertion then follows from the Smirnov maximum principle, since if $f \in \mathcal{S}$, then

$$|e^{i\mu\tau_f^+} f(\mu)| = |f(\mu)| \le 1$$

for almost all points $\mu \in \mathbb{R}$ and $e^{i\lambda\tau_f^+} f(\lambda) \in \mathcal{N}_+$. □

Remark 3.24 *If, in the setting of the last lemma, $f(\lambda) \neq 0$ for every point $\lambda \in \mathbb{C}_+$, then there is no Blaschke factor $b(\lambda)$ in the factorization formula (3.46) and hence the* lim sup *in (3.49) can be replaced by an ordinary limit.*

Lemma 3.25 *If $f \in \mathcal{S}$ and $f(\lambda) \not\equiv 0$, then $f(\lambda)$ is a Blaschke product if and only if the condition (3.29) is satisfied and $\tau_f^+ = 0$.*

Proof The result follows from Theorem 3.12 and assertion (3) in Lemma 3.23. □

Lemma 3.26 *If $f \in H_\infty$ and if $e_{-t}f \in H_\infty$ for every $t \geq 0$, then $f(\lambda) \equiv 0$.*

Proof If $f \in H_\infty$, then $(1 - i\lambda)^{-1}f(\lambda) \in H_2$. Therefore,

$$\frac{f(\lambda)}{1 - i\lambda} = \int_0^\infty e^{i\lambda u} g(u)du$$

for some function $g(u) \in L_2(\mathbb{R}_+)$. Thus if also $e^{-i\lambda t}f(\lambda) \in H_\infty$ for $t \geq 0$, then

$$e^{-i\lambda t}\int_0^\infty e^{i\lambda u} g(u)du = e^{-i\lambda t}\int_0^t e^{i\lambda u} g(u)du + e^{-i\lambda t}\int_t^\infty e^{i\lambda u} g(u)du$$

also belongs to H_2. But this means that

$$e^{-i\lambda t}\int_0^t e^{i\lambda u} g(u)du \in H_2 \cap H_2^\perp$$

and hence that

$$\int_0^t e^{i\lambda u} g(u)du = 0 \quad \text{for every} \quad t > 0.$$

It is now clear from the formula for $(1 - i\lambda)^{-1}f(\lambda)$ that $f(\lambda) \equiv 0$. □

Lemma 3.27 *If $e_{t_j}f \in H_\infty$ for a sequence of positive numbers $t_1 \geq t_2 \geq \cdots$ which tend to 0 as $j \uparrow \infty$, then $f \in H_\infty$.*

Proof The function $g(\lambda) = (1 - i\lambda)^{-1}f(\lambda)$ belongs to H_2, since $e_{t_j}g \in H_2$ and $\|g - e_{t_j}g\|_{st} \to 0$ as $j \uparrow \infty$. Therefore $f \in L_\infty \cap \mathcal{N}_+ = H_\infty$, by the Smirnov maximum principle. □

A function $\varphi \in \mathcal{N}_+$ is called **outer** if it can be expressed as a ratio $\varphi(\lambda) = \varphi_1(\lambda)/\varphi_2(\lambda)$, where $\varphi_1, \varphi_2 \in \mathcal{S}_{out}$. The class of outer functions in $\mathcal{N}$ will be denoted $\mathcal{N}_{out}$. Thus,

$$\mathcal{N}_{out} \subset \mathcal{N}_+ \subset \mathcal{N}.$$

We remark that

$$f \in \mathcal{N}_{out} \Longleftrightarrow f \in \mathcal{N}_+ \text{ and } f^{-1} \in \mathcal{N}_+. \tag{3.52}$$

The implication $\Longrightarrow$ follows easily from the definitions of the classes $\mathcal{N}_{out}$ and $\mathcal{N}_+$. Conversely, if $f \in \mathcal{N}_+$ and $f^{-1} \in \mathcal{N}_+$, then $f = g/h$ and $f^{-1} = g_1/h_1$, with $g \in \mathcal{S}$, $h \in \mathcal{S}_{out}$, $g_1 \in \mathcal{S}$ and $h_1 \in \mathcal{S}_{out}$. Therefore, since $gg_1 = hh_1$ and $hh_1 \in \mathcal{S}_{out}$, it follows that $gg_1 \in \mathcal{S}_{out}$, and hence that $g \in \mathcal{S}_{out}$, as needed, and $g_1 \in \mathcal{S}_{out}$. This implies that the product of two outer functions in $\mathcal{N}_+$ is outer. Conversely, if $f \in \mathcal{N}_+$, $g \in \mathcal{N}_+$ and fg is outer, then (as follows from the factorization formula (3.46)) both f and g are outer.

Remark 3.28 *If $f \in \mathcal{N}_+$ and $f \not\equiv 0$, then the inequality (3.43) holds; equality prevails for at least one (and hence every) point $\omega = a + ib$ in $\mathbb{C}_+$ if and only if $f \in \mathcal{N}_{out}$. This characterization may be used to show that $(\lambda - \omega)^k \in \mathcal{N}_{out}$ for every integer k and every point $\omega \in \overline{\mathbb{C}_-}$. This last conclusion also follows from Lemma 3.57 (with much less effort), since $\mathcal{N}_{out}$ is multiplicative group and $(\lambda - \omega) \in \mathcal{N}_{out}$ if $\omega \in \overline{\mathbb{C}_-}$.*

Later, in Theorem 3.64, we shall see that

$$\varphi \in \mathcal{N}_{out} \Longleftrightarrow \text{ the set } \{\varphi f : f \in H_2 \text{ and } \varphi f \in H_2\} \text{ is dense in } H_2.$$

If $\varphi \in H_\infty$, then $\varphi f \in H_2$ for every $f \in H_2$. Therefore, this last characterization is consistent with the definition of an outer function in H_∞ that was given earlier.

Every $\varphi \in \mathcal{N}_{out}$ admits a representation of the form (3.32), where $k \in \widetilde{L}_1(\mathbb{R})$. From the representation (3.39) and (3.45), it follows that every $f \in \mathcal{N}$ admits an essentially unique representation of the form

$$f(\lambda) = b_1(\lambda)^{-1} b_2(\lambda)\varphi(\lambda), \tag{3.53}$$

where $b_1(\lambda)$ and $b_2(\lambda)$ are inner functions that have no common nonconstant inner divisors in the class $\mathcal{S}_{in}$ and $\varphi \in \mathcal{N}_{out}$, i.e., φ is an outer function in the class $\mathcal{N}_+$. A function $f \in \mathcal{N}$ belongs to $\mathcal{N}_+$ if and only if the denominator $b_1(\lambda)$ in formula (3.53) is absent.

If $f \in \mathcal{N}_+$, then the set of roots of $f(\lambda)$ (counting multiplicities) coincides with the set of roots in the Blaschke factor $b(\lambda)$ in (3.46) (counting multiplicities). In particular, if $f(\lambda) \neq 0$ for any point $\lambda \in \mathbb{C}_+$, then the Blaschke factor $b(\lambda)$ in (3.46) is absent; if $f(\lambda)$ is continuous in $\overline{\mathbb{C}_+}$, then the singular factor in (3.46) reduces to $b_s(\lambda) = \gamma e_\beta(\lambda)$.

Lemma 3.29 *If $f \in \mathcal{E} \cap \mathcal{N}$ and $f(\lambda) \not\equiv 0$, then τ_f^+ is finite and*

$$f(\lambda) = e^{-i\tau_f^+\lambda} b(\lambda)\varphi(\lambda), \tag{3.54}$$

where $\varphi(\lambda) \in \mathcal{E} \cap \mathcal{N}_{out}$ and $b(\lambda)$ is a Blaschke product of the form (3.27). If $f \in \mathcal{E} \cap \mathcal{N}_+$, then $\tau_f^+ \leq 0$ and the factorization (3.54) yields an inner-outer factorization of f with an inner factor $\exp\{-i\lambda\tau_f^+\}b$ that is meromorphic in $\mathbb{C}$ and an entire outer factor φ.

Proof By a double application of formula (3.32), $f(\lambda)$ admits a representation of the form

$$f(\lambda) = e^{i\beta\lambda} b(\lambda)\varphi(\lambda),$$

where $\beta \geq 0$, $b(\lambda)$ is a Blaschke product and $\varphi \in \mathcal{N}_{out}$. Therefore, as follows with the help of (3.50) and (3.51),

$$\tau_f^+ \;=\; -\beta.$$

The Blaschke product b in formula (3.54) may be written in the form (3.19) because $f(\lambda)$ has at most finitely many roots ω_k with $|\omega_k| \leq 1$. Consequently,

$$\varphi(\lambda) = e^{i\lambda\tau_f^+} \lim_{n\to\infty} f_n(\lambda), \quad \text{where} \quad f_n(\lambda) = \prod_{j=1}^{n} \frac{1-\lambda/\overline{\omega_j}}{1-\lambda/\omega_j} f(\lambda)$$

is entire, since the the functions $f_n(\lambda)$ are entire and converge uniformly on compact subsets of $\mathbb{C}$. □

Theorem 3.30 *Let $g(\mu)$ be a measurable function on $\mathbb{R}$ such that $g(\mu) \geq 0$ a.e. on $\mathbb{R}$. Then:*

(1) *The factorization formula*

$$g(\mu) = |f(\mu)|^2 \quad \text{a.e. on } \mathbb{R} \quad \text{for some } f \in \mathcal{N} \tag{3.55}$$

holds if and only if

$$\int_{-\infty}^{\infty} \frac{|\ln|g(\mu)||}{1+\mu^2} d\mu < \infty. \tag{3.56}$$

(2) *If the constraint (3.56) is met, then there exists a solution* $f(\lambda) = \varphi(\lambda)$ *in the class* $\mathcal{N}_{out}$ *that can be uniquely specified by imposing the normalization condition* $\varphi(\omega) > 0$ *at some fixed point* $\omega \in \mathbb{C}_+$.

(3) *The set of all solutions* $f \in \mathcal{N}$ *to this factorization problem is described by formula (3.53), where* $\varphi \in \mathcal{N}_{out}$ *is uniquely specified as in (2) and* b_1 *and* b_2 *are arbitrary inner functions.*

(4) *The factorization (3.55) holds for some* $f \in H_p$, $1 \leq p \leq \infty$ *if and only if the constraint (3.56) holds and* $g \in L_{p/2}$. *Moreover, if these two conditions are met, then every solution* $f \in H_p$ *is of the form* $f = b\varphi$, *where* $b \in \mathcal{S}_{in}$ *and* $\varphi(\lambda)$ *is an outer function that belongs to* H_p.

Proof See [RR 94]. □

3.6 The rank of meromorphic mvf's

Let $f(\lambda)$ be a mvf that is meromorphic in some open nonempty subset Ω of $\mathbb{C}$. Then $\mathfrak{h}_f$ denotes the set of points $\omega \in \Omega$ at which f is holomorphic,

$$\mathfrak{h}_f^+ = \mathfrak{h}_f \cap \mathbb{C}_+ \quad \text{and} \quad \mathfrak{h}_f^- = \mathfrak{h}_f \cap \mathbb{C}_-.$$

We shall define

$$\text{rank } f = \max\{\text{rank } f(\lambda) : \ \lambda \in \mathfrak{h}_f^+\}$$

for every meromorphic $p \times q$ mvf $f(\lambda)$ in $\mathbb{C}_+$.

Lemma 3.31 *Let* $f(\lambda)$ *be a meromorphic* $p \times q$ *mvf in* $\mathbb{C}_+$, *let* $r = \text{rank } f$ *and assume that* $f(\lambda) \not\equiv 0$. *Then*

$$\text{rank } f(\lambda) = r \quad \textit{for every } \lambda \in \mathfrak{h}_f^+ \tag{3.57}$$

except for at most a countable set of points $\{\omega_j\}$ *that do not have a limit point in* $\mathbb{C}_+$. *Moreover, if* $f \in \mathcal{N}^{p\times q}$, *then the sequence of points* $\{\omega_j\}$ *satisfies the Blaschke condition (3.26) and*

$$\text{rank } f(\mu) = r \quad \textit{for almost all points } \mu \in \mathbb{R}. \tag{3.58}$$

Proof Let $r = \text{rank } f$. Then $r = \text{rank } f(\omega)$ for some point $\omega \in \mathfrak{h}_f^+$ and hence there exists an $r \times r$ submatrix $\Delta(\lambda)$ of $f(\lambda)$ such that its determinant $\delta(\lambda)$ is not equal to zero at the point ω. Consequently, $\delta(\lambda) \not\equiv 0$ in $\mathbb{C}_+$ and hence as $\delta(\lambda)$ is meromorphic in $\mathbb{C}_+$, it has at most countably many zeros $\{\omega_j\}$ in

$\mathbb{C}_+$ and this set does not have a limit point in $\mathbb{C}_+$. Therefore, rank $f(\lambda) = r$, except for the set $\{\omega_j\}$, since rank $f(\lambda) \le r$ for every point $\lambda \in \mathfrak{h}_f^+$.

If $f \in \mathcal{N}^{p\times q}$, then $\delta \in \mathcal{N}$ and consequently, the sequence $\{\omega_j\}$ satisfies the Blaschke condition (3.26). Moreover, nontangential limit values $f(\mu)$ and $\delta(\mu)$ exist at almost all points $\mu \in \mathbb{R}$ and rank $f(\mu) \ge r$ for almost all points $\mu \in \mathbb{R}$, since $\delta(\mu) \ne 0$ a.e. on $\mathbb{R}$. On the other hand, if $\Delta_1(\lambda)$ is a submatrix of $f(\lambda)$ of size $r_1 \times r_1$ such that $\delta_1(\mu) = \det \Delta_1(\mu)$ is not equal to zero on a subset of $\mathbb{R}$ of positive Lebesgue measure, then $\delta_1(\lambda) \not\equiv 0$ in $\mathbb{C}_+$. Consequently, $r_1 \le r$ and hence rank $f(\mu) \le r$ for almost all points $\mu \in \mathbb{R}$. Thus, the proof is complete. □

3.7 Inner and outer mvf's in $H_\infty^{p\times q}$

A mvf $f \in H_\infty^{p\times q}$ is said to be **inner** (resp., **$*$-inner**) if

$$I_q - f(\mu)^* f(\mu) = 0 \qquad (\text{resp.,}\ \ I_p - f(\mu) f(\mu)^* = 0)$$

for almost all points $\mu \in \mathbb{R}$. It is said to be **outer** (resp., **$*$-outer**) if the closure

$$\overline{f H_2^q} = H_2^p \qquad (\text{resp.,}\ \ \overline{f^\sim H_2^p} = H_2^q,\ \text{where}\ f^\sim(\lambda) = f(-\bar{\lambda})^*).$$

It is readily checked that in order for $f \in H_\infty^{p\times q}$ to be inner (resp., outer), it is necessary that $p \ge q$ (resp., $p \le q$). (The inequalities are reversed if inner/outer is replaced by $*$-inner/$*$-outer.) In particular, a square mvf is inner if and only if it is $*$-inner and is outer if and only if it is $*$-outer.

The symbols $\mathcal{S}_{in}^{p\times q}$, $\mathcal{S}_{*in}^{p\times q}$, $\mathcal{S}_{out}^{p\times q}$ and $\mathcal{S}_{*out}^{p\times q}$ will be used to designate the classes of functions $f \in \mathcal{S}^{p\times q}$ which are inner, $*$-inner, outer and $*$-outer, respectively. In the square case, $\mathcal{S}_{in}^{p\times p} = \mathcal{S}_{*in}^{p\times p}$ and $\mathcal{S}_{out}^{p\times p} = \mathcal{S}_{*out}^{p\times p}$. Moreover, for mvf's f in either of these two classes, $\det f(\lambda) \not\equiv 0$ in $\mathbb{C}_+$ if $f \in \mathcal{S}_{in}^{p\times p}$ and $\det f(\lambda) \ne 0$ in $\mathbb{C}_+$ if $f \in \mathcal{S}_{out}^{p\times p}$.

Theorem 3.32 *Let $f \in H_\infty^{p\times q}$. Then f is an outer mvf if and only if*

(1) $\operatorname{rank} f(\omega) = p$ *for at least one point $\omega \in \mathbb{C}_+$.*

(2) *Every mvf $g \in H_\infty^{r\times q}$ that meets the inequality*

$$g(\mu)^* g(\mu) \le f(\mu)^* f(\mu)$$

for almost all points $\mu \in \mathbb{R}$ also meets the inequality

$$g(\lambda)^* g(\lambda) \le f(\lambda)^* f(\lambda)$$

for every point $\lambda \in \mathbb{C}_+$. Moreover, in this case, $\mathrm{rank} f(\lambda) = p$ *for every point $\lambda \in \mathbb{C}_+$,* $\mathrm{rank}\ f(\mu) = p$ *for almost all points $\mu \in \mathbb{R}$ and*

$$g(\lambda) = b(\lambda) f(\lambda)$$

for some $b(\lambda) \in \mathcal{S}^{r \times p}$.

Proof See, e.g., Propositions 4.1 and 4.2 on pp. 200–201 of [SzNF70]. □

Lemma 3.33 *If $b \in \mathcal{S}_{in}^{p\times p}$ and if $\det b(\lambda)$ is constant in $\mathbb{C}_+$, then $b(\lambda) \equiv$ constant.*

Proof If $\det b(\lambda) \equiv \gamma$ in $\mathbb{C}_+$, then $b(\lambda)^{-1} \in H_\infty^{p\times p}$ and $\det\{b(\lambda)^* b(\lambda)\} = |\gamma|^2 = 1$. Therefore, since $b(\lambda)^* b(\lambda) \leq I_p$ in $\mathbb{C}_+$, equality must hold and this in turn implies that $b(\lambda)^* \in H_\infty^{p\times p}$. Therefore, $b(\lambda)$ must be constant. □

3.8 Fourier transforms and Paley-Wiener theorems

The **Fourier transforms**

$$\widehat{f}(\mu) = \int_{-\infty}^{\infty} e^{i\mu t} f(t) dt \quad \text{and} \quad f^\vee(t) = \frac{1}{2\pi} \int_{-\infty}^{\infty} e^{-i\mu t} f(\mu) d\mu \tag{3.59}$$

will be considered mainly for $f \in L_1^{p\times q}(\mathbb{R})$ and $f \in L_2^{p\times q}(\mathbb{R})$.

If $f \in L_2^{p\times q}(\mathbb{R})$, then the integral is understood as the limit of the integrals $\int_{-A}^{A}$ in $L_2^{p\times q}(\mathbb{R})$ as $A \uparrow \infty$. Moreover, the mapping

$$f \to (2\pi)^{-1/2} \widehat{f} \quad \text{is a unitary operator in } L_2^{p\times q}(\mathbb{R}),$$

i.e., it is onto and the **Plancherel formula** holds:

$$\langle \widehat{f}, \widehat{g} \rangle_{st} = 2\pi \langle f, g \rangle_{st} \quad \text{for } f, g \in L_2^{p\times q},$$

and

$$f(t) = (\widehat{f})^\vee(t) \quad \text{a.e. on } \mathbb{R}. \tag{3.60}$$

If $f \in L_2^{p\times q}(\mathbb{R})$, then $\mu \widehat{f}(\mu)$ belongs to $L_2^{p\times q}(\mathbb{R})$ if and only if

$$f \quad \text{is locally absolutely continuous on } \mathbb{R} \quad \text{and} \quad f' \in L_2^{p\times q}(\mathbb{R}). \tag{3.61}$$

Moreover, if these conditions hold, then

$$\mu \widehat{f}(\mu) = i \widehat{f'}(\mu). \tag{3.62}$$

If $f \in L_1^{p\times r}(\mathbb{R})$ and $g \in L_1^{r\times q}(\mathbb{R})$, then $\widehat{f} \in \mathcal{W}^{p\times r}(0)$, $\widehat{g} \in \mathcal{W}^{r\times q}(0)$, $\widehat{f}\widehat{g} \in \mathcal{W}^{p\times q}(0)$ and

$$(\widehat{f}\widehat{g})^\vee(t) = \int_{-\infty}^{\infty} f(t-s)g(s)ds \quad \text{a.e. on } \mathbb{R}. \tag{3.63}$$

Formula (3.63) is also valid if $f \in L_1^{p\times r}(\mathbb{R})$, $g \in L_s^{r\times q}(\mathbb{R})$ and $1 < s < \infty$.

Theorem 3.34 *If $\widehat{f} \in \mathcal{W}^{p\times p}(0)$ and $\gamma \in \mathbb{C}^{p\times p}$, then there exists a matrix $\delta \in \mathbb{C}^{p\times p}$ and a mvf $\widehat{g} \in \mathcal{W}^{p\times p}(0)$ such that*

$$(\gamma + \widehat{f}(\mu))(\delta + \widehat{g}(\mu)) = I_p \tag{3.64}$$

if and only if

$$\det(\gamma + \widehat{f}(\mu)) \neq 0 \quad \textit{for every point } \mu \in \mathbb{R} \textit{ and } \gamma \textit{ is invertible.} \tag{3.65}$$

If $\widehat{f} \in \mathcal{W}_+^{p\times p}(0)$ (resp., $\mathcal{W}_-^{p\times p}(0)$) and $\gamma \in \mathbb{C}^{p\times p}$, then there exists a matrix $\delta \in \mathbb{C}^{p\times p}$ and a mvf $\widehat{g} \in \mathcal{W}_+^{p\times p}(0)$ (resp., $\mathcal{W}_-^{p\times p}(0)$) such that (3.64) holds for all points $\mu \in \mathbb{R}$ if and only if

$$\det(\gamma + \widehat{f}(\lambda)) \neq 0 \quad \textit{for every point } \lambda \in \overline{\mathbb{C}_+} \textit{ (resp., } \overline{\mathbb{C}_-}\textit{) and } \gamma \textit{ is invertible.} \tag{3.66}$$

Proof The stated assertions for mvf's are easily deduced from the scalar versions, the first of which is due to N. Wiener; the second (and third) to Paley and Wiener; see [PaW34] for proofs and, for another approach, [GRS64]. □

If (3.64) holds, then (by the Riemann-Lebesgue lemma) $\gamma\delta = I_p$.

Theorem 3.35 (Paley-Wiener) *Let f be a $p \times q$ mvf that is holomorphic in $\mathbb{C}_+$. Then $f \in H_2^{p\times q}$ if and only if*

$$f(\lambda) = \int_0^\infty e^{i\lambda x} f^\vee(x)dx \quad \textit{for } \lambda \in \mathbb{C}_+$$

and some $f^\vee \in L_2^{p\times q}(\mathbb{R}_+)$. Moreover, if $f \in H_2^{p\times q}$, then its boundary values $f(\mu)$ admit the Fourier-Laplace representation

$$f(\mu) = \int_0^\infty e^{i\mu x} f^\vee(x)dx \quad \textit{a.e. on } \mathbb{R}.$$

Proof This follows easily from the scalar Paley-Wiener theorem; see, e.g., pp. 158–160 of [DMc72]. □

Theorem 3.36 (Paley-Wiener) *A $p \times q$ mvf f admits a representation of the form*

$$f(\lambda) = \int_{-\alpha}^{\beta} e^{i\lambda x} f^{\vee}(x) dx \quad \textit{for } \lambda \in \mathbb{C}$$

and some $f^{\vee} \in L_2^{p\times q}([-\alpha, \beta])$ with $0 \le \alpha, \beta < \infty$ if and only if $f(\lambda)$ is an entire $p \times q$ mvf of exponential type with $\tau_+(f) \le \alpha$ and $\tau_-(f) \le \beta$ and $f \in L_2(\mathbb{R})$.

Proof This follows easily from the scalar Paley-Wiener theorem; see, e.g., pp. 162–164 of [DMc72]. □

3.9 The Beurling-Lax theorem

The functions $\varphi_k(\lambda)$ and $\psi_k(\lambda)$ introduced in (3.37) will be used frequently below.

Lemma 3.37 *If $f \in L_1^p(\mathbb{R})$ and*

$$\int_{-\infty}^{\infty} \varphi_k(\mu) f(\mu) d\mu = 0 \quad \textit{for } k = 0, \pm 1, \ldots,$$

then $f(\mu) = 0$ a.e. on $\mathbb{R}$.

Proof Since

$$\left(\frac{\lambda - i}{\lambda + i}\right)^k = \left(1 - \frac{2i}{\lambda + i}\right)^k \quad \text{and} \quad \left(\frac{\lambda + i}{\lambda - i}\right)^k = \left(1 + \frac{2i}{\lambda - i}\right)^k$$

for $k = 0, 1, \ldots$, it is readily seen that under the given assumptions

$$\int_{-\infty}^{\infty} \frac{f(\mu)}{(\mu \pm i)^k} d\mu = 0 \quad \text{for } k = 0, 1, \ldots.$$

Therefore,

$$\int_{-\infty}^{\infty} \frac{f(\mu)}{\mu - \lambda} d\mu = 0 \quad \text{for } \lambda \in \mathbb{C}_+ \cup \mathbb{C}_-.$$

Thus, as

$$(\mu - \lambda)^{-1} = \begin{cases} -i \int_0^{\infty} \exp(i(\mu - \lambda)t) dt & \text{if } \lambda \in \mathbb{C}_- \\ i \int_{-\infty}^{0} \exp(i(\mu - \lambda)t) dt & \text{if } \lambda \in \mathbb{C}_+, \end{cases}$$

it follows that

$$\int_{-\infty}^{\infty} e^{-iat} e^{-b|t|} \widehat{f}(t)dt = 0 \quad \text{for } a \in \mathbb{R} \quad \text{and } b > 0.$$

Consequently, $\widehat{f}(t) = 0$ on $\mathbb{R}$ and hence $f(\mu) = 0$ a.e. on $\mathbb{R}$, as claimed. □

Theorem 3.38 *Let $\mathcal{L}$ be a proper closed nonzero subspace of H_2^p such that*

$$e_t f \in \mathcal{L} \quad \textit{for every } f \in \mathcal{L} \textit{ and every } \; t \geq 0.$$

Then there exists a positive integer $q \leq p$ and an inner mvf $b \in \mathcal{S}_{in}^{p\times q}$ such that

$$\mathcal{L} = bH_2^q. \tag{3.67}$$

Moreover, this mvf $b(\lambda)$ is uniquely defined by $\mathcal{L}$ up to a unitary constant right multiplier.

Proof The formula

$$\frac{\lambda - \alpha}{\lambda - \overline{\alpha}} = 1 + i(\alpha - \overline{\alpha}) \int_0^{\infty} e^{i\lambda t} e^{-i\overline{\alpha} t} dt \quad \text{for } \alpha \in \mathbb{C}_+ \tag{3.68}$$

guarantees that $\mathcal{L}$ is invariant under multiplication by φ_1. Moreover, $\varphi_1 \mathcal{L}$ is a proper subspace of $\mathcal{L}$, because otherwise $(\varphi_1)^k \mathcal{L} = \mathcal{L}$ for all positive integers k and hence

$$\mathcal{L} = (\varphi_1)^k \mathcal{L} = \cap_{j=0}^k (\varphi_1)^j \mathcal{L} \subset \cap_{j=0}^k (\varphi_1)^j H_2^p \quad \text{for } k = 0, 1, \ldots,$$

which leads to a contradiction, since $\cap_{j=0}^{\infty} (\varphi_1)^j H_2^p = \{0\}$. Thus,

$$\mathfrak{N} = \mathcal{L} \ominus \varphi_1 \mathcal{L}$$

is nonzero, and since $(\varphi_1)^j \mathfrak{N}$ is orthogonal to $(\varphi_1)^k \mathfrak{N}$ for $j \neq k$ and $\varphi_1^k \mathfrak{N} \subset \varphi_1^k H_2^p$, much the same sort of argument leads to the **Wold decomposition**

$$\mathcal{L} = \bigoplus_{j=0}^{\infty} \varphi_j \mathfrak{N}.$$

Let $f_1, \ldots, f_k$ be an orthonormal set of vectors in $\mathfrak{N}$ and let

$$F(\lambda) = [f_1 \quad \cdots \quad f_k] \quad \text{and} \quad G(\lambda) = \sqrt{\pi}(\lambda + i)F(\lambda).$$

Then

$$\int_{-\infty}^{\infty} \varphi_n(\mu) \xi^* F(\mu)^* F(\mu) \eta d\mu = 0 \quad \text{for} \quad n = 1, 2, \ldots$$

and every choice of $\xi, \eta \in \mathbb{C}^k$. The last integral is also equal to 0 for $n = -1, -2, \ldots$, as follows upon taking complex conjugates and interchanging ξ and η. Thus,

$$\frac{1}{\pi}\int_{-\infty}^{\infty} \varphi_n(\mu)\frac{G(\mu)^*G(\mu) - I_k}{1+\mu^2}d\mu = 0 \quad \text{for} \quad n = 0, \pm 1, \pm 2, \ldots,$$

which, in view of the preceding lemma, implies that $G(\mu)^*G(\mu) = I_k$ a.e. on $\mathbb{R}$. Therefore, $k \leq p$, and hence dimension $\mathfrak{N} \leq p$. Let $q =$ dimension $\mathfrak{N}$ and let $k = q$. Thus, as $G \in \mathcal{N}_+^{p\times q} \cap L_\infty^{p\times q}$, the Smirnov maximum principle implies that $G \in H_\infty^{p\times q}$ and hence, as $G^*G = I_q$ a.e. on $\mathbb{R}$ that $G \in \mathcal{S}_{in}^{p\times q}$. Moreover, as $f \in \mathfrak{N}$ implies that

$$f = F\xi = \psi_0 G\xi \quad \text{and} \quad \mathcal{L} = \oplus_{j+0}^{\infty}\varphi_j\mathfrak{N},$$

it follows that $\mathcal{L} = GH_2^q$. This completes the proof of existence.

If there are two inner mvf's b and b_1 such that $\mathcal{L} = bH_2^q = b_1 H_2^{q_1}$, then the identity $\mathfrak{N} = H_2^p \ominus bH_2^q = H_2^p \ominus b_1 H_2^{q_1}$ implies that $q = q_1 =$ dimension $\mathfrak{N}$ and hence that the columns of $\psi_0 b$ form an orthonormal basis for $\mathfrak{N}$ as do the columns of $\psi_0 b_1$. Therefore there exists a constant $q \times q$ unitary matrix V such that $b_1 = b_2 V$. This proves the essential uniqueness of b. □

Lemma 3.39 *Let $b_\alpha(\lambda) = (\lambda - \alpha)/(\lambda - \overline{\alpha})$ for $\alpha \in \mathbb{C}_+$ and let $\mathcal{L}$ be a proper closed subspace of H_2^p. Then the following assertions are equivalent:*

(1) *$e_t\mathcal{L} \subseteq \mathcal{L}$ for every $t \geq 0$.*

(2) *$b_\alpha\mathcal{L} \subseteq \mathcal{L}$ for at least one point $\alpha \in \mathbb{C}_+$*

(3) *$b_\alpha\mathcal{L} \subseteq \mathcal{L}$ for every point $\alpha \in \mathbb{C}_+$.*

Proof If (2) is in force, then the argument in the first part of the proof of Theorem 3.38 is easily adapted to verify the Wold decomposition

$$\mathcal{L} = \bigoplus_{n=0}^{\infty} b_\alpha^n \mathfrak{N}_\alpha \quad \text{with} \quad \mathfrak{N}_\alpha = \mathcal{L} \ominus b_\alpha\mathcal{L}.$$

Let $f \in \mathfrak{N}_\alpha$ and let $h_\alpha(\lambda) = \{\sqrt{2\pi/(i\overline{\alpha} - i\alpha)}(\lambda - \overline{\alpha})\}^{-1}$. Then, since $h_\alpha b_\alpha^k$, $k = 0, 1, \ldots$, is an orthonormal basis for H_2,

$$e_t h_\alpha = \sum_{k=0}^{\infty} c_k(t) h_\alpha b_\alpha^k, \quad \text{where} \quad \sum_{k=0}^{\infty} |c_k(t)|^2 < \infty.$$

Thus, if $f \in \mathfrak{N}_\alpha$, then

$$e_t f = \sum_{k=0}^{\infty} c_k(t) b_\alpha^k f, \quad \text{belongs to } \mathcal{L}.$$

Therefore, (2) $\Longrightarrow$ (1); and, as the implication (1) $\Longrightarrow$ (3) is justified by formula (3.68) and (3) $\Longrightarrow$ (2) is self-evident, the proof is complete. □

The **generalized backwards shift** operator R_α is defined for vvf's and mvf's by the rule

$$(R_\alpha f)(\lambda) = \begin{cases} \dfrac{f(\lambda) - f(\alpha)}{\lambda - \alpha} & \text{if} \quad \lambda \neq \alpha \\ f'(\alpha) & \text{if} \quad \lambda = \alpha \end{cases} \tag{3.69}$$

for every λ, $\alpha \in \mathfrak{h}_f$. In order to keep the typography simple, we shall not indicate the space in which R_α acts in the notation.

Lemma 3.40 *Let $\mathcal{L}$ be a proper closed subspace of H_2^p and let $V(t) = \Pi_+ e_{-t}|_{H_2^p}$ for $t \geq 0$. Then the following assertions are equivalent:*

(1) *$V(t)\mathcal{L} \subseteq \mathcal{L}$ for every $t \geq 0$.*

(2) *$R_\alpha \mathcal{L} \subseteq \mathcal{L}$ for at least one point $\alpha \in \mathbb{C}_+$*

(3) *$R_\alpha \mathcal{L} \subseteq \mathcal{L}$ for every point $\alpha \in \mathbb{C}_+$.*

Proof The proof follows by applying Lemma 3.39 to $\mathcal{L}^\perp$, the orthogonal complement of $\mathcal{L}$ in H_2^p and invoking the identities

$$\langle V(t)f, g\rangle_{st} = \langle f, e_t g\rangle_{st} \quad \text{and} \quad \langle R_\alpha f, g\rangle_{st} = \langle f, b_\alpha g\rangle_{st}$$

for $f \in \mathcal{L}$, $g \in \mathcal{L}^\perp$, $t \geq 0$ and $\alpha \in \mathbb{C}_+$. □

The spaces

$$\mathcal{H}(b) = H_2^p \ominus bH_2^q \quad \text{for } b \in \mathcal{S}_{in}^{p\times q} \tag{3.70}$$

will play a significant role in future developments.

Lemma 3.41 *If $bH_2^r \supseteq b_1 H_2^p$, where $b \in \mathcal{S}_{in}^{p\times r}$ and $b_1 \in \mathcal{S}_{in}^{p\times p}$, then $r = p$ and $b^{-1}b_1 \in \mathcal{S}_{in}^{p\times p}$.*

Proof The stated inclusion implies that

$$\left\| P_{b_1 H_2^p} \frac{\xi}{\rho_\omega} \right\|_{st} \leq \left\| P_{bH_2^r} \frac{\xi}{\rho_\omega} \right\|_{st} \tag{3.71}$$

for every $\xi \in \mathbb{C}^p$ and every $\omega \in \mathbb{C}_+$. Therefore, since the orthogonal projectors in (3.71) can be reexpressed in terms of the isometric multiplication operators M_{b_1} and M_b and the projection Π_+ as

$$P_{b_1 H_2^p} = M_{b_1}\Pi_+ M_{b_1^*} \quad \text{and} \quad P_{bH_2^r} = M_b\Pi_+ M_{b^*},$$

respectively, and

$$\Pi_+ M_{b_1^*}\frac{\xi}{\rho_\omega} = \frac{b_1(\omega)^*\xi}{\rho_\omega} \quad \text{and} \quad \Pi_+ M_{b^*}\frac{\xi}{\rho_\omega} = \frac{b_1(\omega)^*\xi}{\rho_\omega},$$

the inquality (3.71) implies that

$$\frac{\xi^* b_1(\omega)^* b_1(\omega)\xi}{\rho_\omega(\omega)} = \left\|\frac{b_1(\omega)\xi}{\rho_\omega}\right\|_{st}^2 \le \left\|\frac{b(\omega)\xi}{\rho_\omega}\right\|_{st}^2 = \frac{\xi^* b(\omega)^* b(\omega)\xi}{\rho_\omega(\omega)}$$

for every choice of $\xi \in \mathbb{C}^p$ and $\omega \in \mathbb{C}_+$. Therefore,

$$b_1(\omega)b_1(\omega)^* \le b(\omega)b(\omega)^*$$

for every point $\omega \in \mathbb{C}_+$. Thus, as the rank of the matrix on the left is equal to p except for an at most countable set of points $\omega \in \mathbb{C}_+$, it follows that $r \ge p$. At the same time, $r \le p$, since $b \in \mathcal{S}_{in}^{p\times r}$. Therefore, $r = p$ and $b^{-1}b_1 \in \mathcal{S}^{p\times p}$. □

Corollary 3.42 *If $bH_2^r \supseteq \beta H_2^p$, where $b \in \mathcal{S}_{in}^{p\times r}$ and $\beta \in \mathcal{S}_{in}$, then $r = p$ and $\beta b^{-1} \in \mathcal{S}_{in}^{p\times p}$.*

Proof This follows from Lemma 3.41 with $b_1 = \beta I_p$. □

Theorem 3.43 *If $b_j \in \mathcal{S}_{in}^{p\times p}$ for $j = 1, \dots, n$, then there exists a mvf $b \in \mathcal{S}_{in}^{p\times p}$ such that:*

(1) *$b_j^{-1}b \in \mathcal{S}_{in}^{p\times p}$ for $j = 1, \dots, n$.*

(2) *If $\widetilde{b} \in \mathcal{S}_{in}^{p\times p}$ and $b_j^{-1}\widetilde{b} \in \mathcal{S}_{in}^{p\times p}$ for $j = 1, \dots, n$, then $b^{-1}\widetilde{b} \in \mathcal{S}_{in}^{p\times p}$.*

Moreover, a mvf $b \in \mathcal{S}_{in}^{p\times p}$ is uniquely specified by these two constraints up to a constant unitary $p \times p$ multiplier on the right.

Proof Let

$$\mathcal{L} = \bigcap_{j=1}^{n} b_j H_2^p \quad \text{and} \quad \beta = \prod_{j=1}^{n} \det b_j.$$

Then, since $\beta b_j^{-1} \in \mathcal{S}_{in}^{p\times p}$, it follows that $\beta H_2^p \subseteq b_j H_2^p$ for $1, \ldots, n$ and hence that $\beta H_2^p \subseteq \mathcal{L}$. In view of Theorem 3.38, there exists a mvf $b \in \mathcal{S}_{in}^{p\times r}$ such that $\mathcal{L} = bH_2^r$, and hence $\beta H_2^p \subseteq bH_2^r$. Therefore, $r = p$ by Corollary 3.42 and (1) then follows from the inclusion $bH_2^p \subseteq b_j H_2^p$ by Lemma 3.41. Moreover, if $\widetilde{b} \in \mathcal{S}_{in}^{p\times p}$ is such that $b_j^{-1}\widetilde{b} \in \mathcal{S}_{in}^{p\times p}$ for $j = 1, \ldots, n$, then $\widetilde{b}H_2^p \subseteq bH_2^p$ and consequently (2) follows by another application of Lemma 3.41.

Finally, if $\widetilde{b} \in \mathcal{S}_{in}^{p\times p}$ meets the same two conditions as b, then $\widetilde{b}^{-1}b$ and $b^{-1}\widetilde{b}$ both belong to $\mathcal{S}_{in}^{p\times p}$ and therefore $b^{-1}\widetilde{b}$ is a constant unitary $p \times p$ matrix. □

A mvf $b \in \mathcal{S}^{p\times p}$ that meets condition (1) in Theorem 3.43 is called a **common left multiple** of the set $\{b_1, \ldots, b_n\}$; b is called a **minimal common left multiple** of the given set if it meets (1) and (2).

Theorem 3.44 *If $b_\alpha \in \mathcal{S}_{in}^{p\times p}$ for $\alpha \in \mathcal{A}$, then there exists a mvf $b \in \mathcal{S}_{in}^{p\times p}$ such that:*

(1) *$b^{-1}b_\alpha \in \mathcal{S}_{in}^{p\times p}$ for every $\alpha \in \mathcal{A}$.*

(2) *If $\widetilde{b} \in \mathcal{S}_{in}^{p\times p}$ and $\widetilde{b}^{-1}b_\alpha \in \mathcal{S}_{in}^{p\times p}$ for every $\alpha \in \mathcal{A}$, then $b^{-1}\widetilde{b} \in \mathcal{S}_{in}^{p\times p}$.*

Moreover, a mvf $b \in \mathcal{S}_{in}^{p\times p}$ is uniquely specified by these two constraints up to a constant unitary $p \times p$ multiplier on the right.

Proof Let

$$\mathcal{L} = \bigvee_{\alpha\in\mathcal{A}} b_\alpha H_2^p$$

be the minimal closed subspace of H_2^p that contains the subspaces $b_\alpha H_2^p$ for every choice of $\alpha \in \mathcal{A}$. Theorem 3.38 implies that $\mathcal{L} = bH_2^r$ for some $b \in \mathcal{S}_{in}^{p\times r}$. Moreover, since $b_\alpha H_2^p \subseteq bH_2^r$, Lemma 3.41 implies that $r = p$. If $\widetilde{b} \in \mathcal{S}_{in}^{p\times p}$ is such that $\widetilde{b}^{-1}b_\alpha \in \mathcal{S}_{in}^{p\times p}$ for every $\alpha \in \mathcal{A}$, then $b_\alpha H_2^p \subseteq \widetilde{b}H_2^p$ for every $\alpha \in \mathcal{A}$, and hence $bH_2^p \subseteq \widetilde{b}H_2^p$. Therefore, $\widetilde{b}^{-1}b \in \mathcal{S}_{in}^{p\times p}$, by Lemma 3.41. □

A mvf $b \in \mathcal{S}^{p\times p}$ that meets condition (1) in Theorem 3.44 is called a **common left divisor** of the set $\{b_\alpha : \alpha \in \mathcal{A}\}$; b is called a **maximal common left divisor** of the given set if it meets (1) and (2).

There are analogous definitions of common right multiple, minimal common right multiple, common right divisor and maximal common right divisor and corresponding analogues of Theorems 3.43 and 3.44.

Lemma 3.45 *If*

$$\beta(\lambda) = \prod_{j=1}^{n} \beta_j(\lambda), \quad \textit{where} \quad \beta_j(\lambda) = \left(\frac{\lambda - \omega_j}{\lambda - \overline{\omega_j}}\right) \quad \textit{and} \quad \omega_j \in \mathbb{C}_+,$$

is a finite Blaschke product, then $\dim \mathcal{H}(\beta) = n$.

Proof It is readily checked that the functions

$$\varphi_k(\lambda) = \prod_{j=1}^{k} \beta_j(\lambda) \frac{1}{\lambda - \overline{\omega_k}}$$

belong to $\mathcal{H}(\beta)$ for $k = 1, \ldots, n$ and are orthogonal in H_2. Let $f \in H_2 \ominus \beta H_2$ and set

$$c_j = \frac{\langle f, \varphi_j \rangle}{\langle \varphi_j, \varphi_j \rangle} \quad \text{for } j = 1, \ldots, n.$$

Then, since

$$0 = \left\langle f - \sum_{j=1}^{n} c_j \varphi_j, \varphi_k \right\rangle \quad \text{for } k = 1, \ldots, n,$$

it follows that $f(\omega_1) - \sum_{j=1}^{n} c_j \varphi_j(\omega_1) = 0$ and

$$\left(\frac{f - \sum_{j=1}^{n} c_j \varphi_j}{\beta_1 \cdots \beta_{k-1}}\right)(\omega_k) = 0 \quad \text{for} \quad k = 2, \ldots, n,$$

and hence that

$$\left(\frac{f - \sum_{j=1}^{n} c_j \varphi_j}{\beta_1 \cdots \beta_k}\right) \in H_2 \quad \text{for} \quad k = 1, \ldots, n.$$

Therefore,

$$f - \sum_{j=1}^{n} c_j \varphi_j \in \beta H_2 \cap (\beta H_2)^{\perp}.$$

Thus, $f - \sum_{j=1}^{n} c_j \varphi_j = 0$, which serves to prove that $\varphi_1, \ldots, \varphi_n$ is a basis for $H_2 \ominus \beta H_2$ and hence that $\dim H_2 \ominus \beta H_2 = n$. □

Lemma 3.46 *If* $b \in \mathcal{S}_{in}^{p \times p}$ *is a rational mvf, then* $\dim \mathcal{H}(b) < \infty$.

Proof Let b be a rational $p \times p$ inner mvf and let $\beta(\lambda) = \det b(\lambda)$. Then, since $\mathcal{H}(b) \subseteq \mathcal{H}(\beta I_p)$ and $\dim \mathcal{H}(\beta I_p) = p \dim \mathcal{H}(\beta)$, the assertion follows from Lemma 3.45. □

3.10 Inner-outer factorization in $H_\infty^{p\times q}$

Lemma 3.47 *Let $f \in L_\infty^{p\times q}$ and let*

$$fH_2^q \subseteq H_2^p. \tag{3.72}$$

Then $f(\mu)$ is the boundary value of a mvf $f(\lambda)$ that belongs to the space $H_\infty^{p\times q}$.

Proof Fix $\omega \in \mathbb{C}_+$. Then, under the given assumption (3.72), the mvf $g(\mu) = \rho_\omega(\mu)^{-1}f(\mu)$ is the boundary value of a mvf $g(\lambda)$ that belongs to $H_2^{p\times q}$. Consequently, $f(\mu)$ is the boundary value of the holomorphic mvf $f(\lambda) = \rho_\omega(\lambda)g(\lambda)$. Moreover, $f \in \mathcal{N}_+^{p\times q}$, since $\rho_\omega \in \mathcal{N}_+$. Consequently, by the Smirnov maximum principle (or inequality (3.43)), applied to each entry of the mvf $f \in \mathcal{N}_+^{p\times q}$, $f \in H_\infty^{p\times q}$. □

Theorem 3.48 *Let $f \in \mathcal{S}^{p\times q}$ and suppose that* rank $f = r \geq 1$. *Then f admits an inner-outer factorization*

$$f = b\varphi, \quad \textit{where } b \in \mathcal{S}_{in}^{p\times r}, \ \varphi \in \mathcal{S}_{out}^{r\times q}, \tag{3.73}$$

and a $*$-outer-$*$-inner factorization

$$f = \varphi_* b_*, \quad \textit{where } \varphi_* \in \mathcal{S}_{*out}^{p\times r} \quad \textit{and} \quad b_* \in \mathcal{S}_{*in}^{r\times q}. \tag{3.74}$$

Moreover, the factors in (3.73) and (3.74) are defined up to constant unitary $r \times r$ multipliers u and v:

$$b \longrightarrow bu, \quad \varphi \longrightarrow u^*\varphi \ \textit{ and } \ \varphi_* \longrightarrow \varphi_* v\ , \quad b_* \longrightarrow v^* b_*.$$

Proof Let $f \in \mathcal{S}^{p\times q}$, let $r =$ rank f and let $\mathcal{L} = \overline{fH_2^q}$. Then $\mathcal{L}$ is a closed subspace of H_2^p that is invariant under multiplication by e_t for $t \geq 0$. Therefore, by the Beurling-Lax theorem, $\mathcal{L} = bH_2^\ell$ for some $b \in \mathcal{S}_{in}^{p\times \ell}$. Moreover, the mvf φ that is defined by the formula

$$\varphi(\mu) = b(\mu)^* f(\mu) \quad \text{for almost all points } \mu \in \mathbb{R}$$

belongs to $L_\infty^{\ell\times q}(\mathbb{R})$, $\|\varphi\|_\infty \leq 1$ and the closure

$$\overline{\varphi H_2^q} = H_2^\ell.$$

Therefore, by Lemma 3.47 and the definition of the class $\mathcal{S}_{out}^{\ell\times q}$, $\varphi(\mu)$ is the boundary value of a mvf $\varphi(\lambda)$ that belongs to the class $\mathcal{S}_{out}^{\ell\times q}$. The inclusion

$$fH_2^q \subseteq bH_2^\ell$$

implies that for any fixed point $\omega \in \mathbb{C}_+$

$$\frac{f}{\rho_\omega} = bg$$

for some $g \in H_2^{\ell\times q}$. Therefore,

$$\text{range } f(\mu) \subseteq \text{ range } b(\mu)$$

for almost all points $\mu \in \mathbb{R}$. Thus,

$$b(\mu)\varphi(\mu) = b(\mu)b(\mu)^* f(\mu) = f(\mu) \quad \text{for almost all points } \mu \in \mathbb{R},$$

since the matrix $b(\mu)b(\mu)^*$ is the orthogonal projection matrix onto the range of $b(\mu)$. Moreover,

$$\text{rank } f(\mu) = \text{rank } f \quad \text{and} \quad \text{rank } \varphi(\mu) = \text{rank } \varphi \quad \text{a.e. on } \mathbb{R}.$$

Therefore, if rank $\varphi = \ell$, then $\ell = \text{rank } f = r$. This completes the proof of the existence of the asserted inner-outer factorization (3.73).

Suppose next that f admits another inner-outer factorization

$$f = b_1\varphi_1 \quad \text{with} \quad b_1 \in \mathcal{S}_{in}^{p\times r_1} \quad \text{and} \quad \varphi_1 \in \mathcal{S}_{out}^{r_1\times q}.$$

Then the formula

$$bH_2^r = \overline{b\varphi H_2^q} = \overline{b_1\varphi_1 H_2^q} = b_1 H_2^{r_1}$$

implies that $r_1 = r$ and $b_1(\lambda) = b(\lambda)u$ for some unitary matrix $u \in \mathbb{C}^{r\times r}$, by the essential uniqueness of the inner mvf in the Beurling-Lax theorem. Thus,

$$b(\mu)u\varphi_1(\mu) = b(\mu)\varphi(\mu) \quad \text{for almost all points } \mu \in \mathbb{R},$$

and hence, upon multiplying both sides of the last equality by $u^*b(\mu)^*$, we obtain

$$\varphi_1(\mu) = u^*\varphi(\mu) \quad \text{for almost all points } \mu \in \mathbb{R}.$$

The assertions on $*$-outer - $*$-inner factorizations for $f \in \mathcal{S}^{p\times q}$ follow from the inner-outer factorization of $f^\sim \in \mathcal{S}^{q\times p}$. □

Corollary 3.49 *Every nonzero mvf $f \in H_\infty^{p\times q}$ admits an inner-outer factorization of the form*

$$f = b\varphi,$$

where $b \in \mathcal{S}_{in}^{p\times r}$ and φ is an outer function of class $H_\infty^{r\times q}$ for some $r \le \min\{p,q\}$. This factorization is unique up to the replacement of $b(\lambda)$ by $b(\lambda)\chi$ and $\varphi(\lambda)$ by $\chi^*\varphi(\lambda)$, where χ is an $r\times r$ constant unitary matrix. Moreover, $r = \operatorname{rank} f$. Every nonzero mvf $f \in H_\infty^{p\times q}$ also admits a $*$-outer-$*$-inner factorization which can be obtained in a self-evident way from the inner-outer factorization of $f^\sim(\lambda)$.

Proof This follows from Theorem 3.48 applied to the mvf $f/\|f\|_\infty$. □

Every square mvf $f \in H_\infty^{p\times p}$ with $\det f(\lambda) \not\equiv 0$ in $\mathbb{C}_+$ admits both an inner–outer factorization and an outer–inner factorization:

$$f = b_1\varphi_1 = \varphi_2 b_2$$

where $b_j \in \mathcal{S}_{in}^{p\times p}$ and φ_j is outer in $H_\infty^{p\times p}$ for $j = 1, 2$.

Theorem 3.50 *Let $f \in \mathcal{S}^{p\times p}$. Then*

(1) $f \in \mathcal{S}_{out}^{p\times p} \iff \det f \in \mathcal{S}_{out}$.

(2) $f \in \mathcal{S}_{in}^{p\times p} \iff \det f \in \mathcal{S}_{in}$.

Proof Let $f \in \mathcal{S}^{p\times p}$. Then, by a theorem of M. G. Krein (see, e.g., [Roz58]),

$$f \in \mathcal{S}_{out}^{p\times p} \iff \begin{cases} \text{(a)} \quad \det f(\lambda) \not\equiv 0 \quad \text{for} \quad \lambda \in \mathbb{C}_+ \\ \\ \text{(b)} \quad \ln|\det f(\omega)| = \dfrac{\Im\omega}{\pi}\displaystyle\int_{-\infty}^{\infty} \dfrac{\ln|\det f(\mu)|}{1+\mu^2}d\mu \\ \qquad\qquad \text{for every point} \quad \omega \in \mathbb{C}_+. \end{cases}$$

By Theorem 3.18, the equality (b) holds for the scalar function $\det f(\lambda)$ of the class $\mathcal{S}$ if and only if $\det f \in \mathcal{S}_{out}$. This completes the proof of (1).

Suppose next that $f \in \mathcal{S}^{p\times p}$. Then the inequality

$$|\det f(\mu)|^2 I_p \le f(\mu)^* f(\mu) \le I_p,$$

which holds for almost all points $\mu \in \mathbb{R}$, implies that if $\det f \in \mathcal{S}_{in}$, then $f \in \mathcal{S}_{in}^{p\times p}$. The converse is self-evident. □

A scalar function $h \in \mathcal{S}$ is said to be a **scalar multiple** of a mvf $f \in \mathcal{S}^{p\times p}$ with $\det f(\lambda) \not\equiv 0$, if $hf^{-1} \in \mathcal{S}^{p\times p}$.

Lemma 3.51 *Let $s \in \mathcal{S}^{p\times q}$ and let $\xi = s(\omega)\eta$ for some point $\omega \in \mathbb{C}_+$ and some vector $\eta \in \mathbb{C}^q$ such that $\xi^*\xi = \eta^*\eta$. Then $\xi = s(\lambda)\eta$ and $\eta = s(\lambda)^*\xi$ for every point $\lambda \in \mathbb{C}_+$.*

Proof If $\eta = 0$, the asserted conclusion is self-evident. If $\eta \neq 0$, then we may assume that $\eta^*\eta = 1$. Then

$$|\xi^* s(\lambda)\eta| \leq 1 = \xi^*\xi = \xi^* s(\omega)\eta$$

for every point $\lambda \in \mathbb{C}_+$. Therefore, the maximum principle applied to the scalar function $\xi^* s(\lambda)\eta$ implies that $\xi^* s(\lambda)\eta \equiv 1$ in $\mathbb{C}_+$ and hence that

$$\begin{aligned} \|\xi - s(\lambda)\eta\|^2 &= \xi^*\xi - \xi^* s(\lambda)\eta - \eta^* s(\lambda)^*\xi + \eta^* s(\lambda)^* s(\lambda)\eta \\ &= \eta^* s(\lambda)^* s(\lambda)\eta - 1 \leq 0, \end{aligned}$$

which imples that $\xi = s(\lambda)\eta$ in $\mathbb{C}_+$. Similar considerations based on the evaluation of $\|s(\lambda)^*\xi - \eta\|^2$ yield the identity $s(\lambda)^*\xi \equiv \eta$ in $\mathbb{C}_+$. □

Corollary 3.52 *Let $s \in \mathcal{S}^{p\times q}$ and let $s(\omega)^* s(\omega) < I_q$ at a point $\omega \in \mathbb{C}_+$, then $s(\lambda)^* s(\lambda) < I_q$ at every point $\lambda \in \mathbb{C}_+$.*

Corollary 3.53 *Let $s \in \mathcal{S}^{p\times p}$ be such that* $\det(I_p - s(\lambda)) \not\equiv 0$ *in $\mathbb{C}_+$. Then* $\det(I_p - s(\lambda)) \neq 0$ *for every point $\lambda \in \mathbb{C}_+$.*

Proof If $\det(I_p - s(\omega)) = 0$ for some point $\omega \in \mathbb{C}_+$. Then $\xi = s(\omega)\xi$ for some vector $\xi \in \mathbb{C}^p$ with $\xi^*\xi = 1$. Therefore, by Lemma 3.51, $\xi \equiv s(\lambda)\xi$ in $\mathbb{C}_+$ and hence $\det(I_p - s(\lambda)) \equiv 0$ in $\mathbb{C}_+$. □

The next lemma amplifies the corollary.

Lemma 3.54 *If $s \in \mathcal{S}^{p\times p}$ and* $\det\{I_p - s(\lambda)\} \not\equiv 0$ *in $\mathbb{C}_+$, then the mvf $I_p - s$ is outer in $H_\infty^{p\times p}$.*

Proof If $I_p - s$ is not outer in $H_\infty^{p\times p}$, then the closure $\overline{\{I_p - s\}H_2^p} \neq H_2^p$. Therefore there exists a nonzero element $g \in H_2^p$ such that

$$\langle \{I_p - s\}f, g\rangle_{st} = 0$$

for every $f \in H_2^p$. Thus

$$g = \Pi_+ s^* g$$

and hence, since

$$\|g\|_2 = \|\Pi_+ s^* g\|_2 \leq \|s^* g\|_2 \leq \|g\|_2,$$

it follows that $s^*g \in H_2^p$ and thus

$$g(\mu) = s(\mu)^* g(\mu)$$

for almost all points $\mu \in \mathbb{R}$. But this implies that

$$\begin{aligned} \det\{I_p - s(\mu)^*\} &= 0 \quad \text{for almost all points} \quad \mu \in \mathbb{R} \\ \Longrightarrow \det\{I_p - s(\mu)\} &= 0 \quad \text{for almost all points} \quad \mu \in \mathbb{R} \\ \Longrightarrow \det\{I_p - s(\lambda)\} &= 0 \quad \text{for every point} \quad \lambda \in \mathbb{C}_+, \end{aligned}$$

which contradicts the given assumption. $\square$

3.11 The Smirnov maximum principle

A $p \times q$ mvf which is holomorphic in $\mathbb{C}_+$ is said to belong to the **Smirnov class** $\mathcal{N}_+^{p\times q}$ if it can be expressed in the form

$$f = h^{-1} g, \tag{3.75}$$

where $g \in H_\infty^{p\times q}$ and h is a scalar outer function of class H_∞. It is readily checked that $\mathcal{N}_+^{p\times q}$ is closed under addition and also multiplication, whenever the matrix multiplication is meaningful. In particular, if $f \in \mathcal{N}_+^{p\times q}$, then $(1 - i\lambda)^k f(\lambda) \in \mathcal{N}_+^{p\times q}$ and $\lambda^k f(\lambda) \in \mathcal{N}_+^{p\times q}$ for every integer k. If also g is outer in $H_\infty^{p\times q}$, then f is said to belong to the class $\mathcal{N}_{out}^{p\times q}$ of **outer mvf's in** $\mathcal{N}_+^{p\times q}$.

Theorem 3.55 *Let* $f \in \mathcal{N}_+^{p\times p}$. *Then*

$$f \in \mathcal{N}_{out}^{p\times p} \Longleftrightarrow \det f \in \mathcal{N}_{out}. \tag{3.76}$$

Proof Let $f \in \mathcal{N}_{out}^{p\times p}$. Then $f = gh^{-1}$ where $g \in \mathcal{S}_{out}^{p\times p}$ and $h \in \mathcal{S}_{out}$. Therefore, since

$$\det f = \frac{\det g}{(h)^p},$$

it follows from Theorem 3.50 that $\det f \in \mathcal{N}_{out}$. Conversely, let $f \in \mathcal{N}_+^{p\times p}$ and let $\det f \in \mathcal{N}_{out}$. Then, since $f = gh^{-1}$ with $g \in \mathcal{S}^{p\times p}$ and $h \in \mathcal{S}_{out}$, it follows that

$$\det g = h^p \det f$$

belongs to $\mathcal{S}_{out}$. Therefore, by Theorem 3.50 $g \in \mathcal{S}_{out}^{p\times p}$ and consequently $f \in \mathcal{N}_{out}^{p\times p}$. $\square$

Theorem 3.55 may be viewed as a generalization of the first assertion in Theorem 3.50. The example

$$f = \frac{g}{(\det\, g)^{1/p}}, \qquad g \in \mathcal{S}_{out}^{p\times p}$$

indicates that the implication

$$\det f \in \mathcal{S}_{in} \Longrightarrow f \in \mathcal{S}_{in}^{p\times p}$$

does not hold for $f \in \mathcal{N}_+^{p\times p}$.

Lemma 3.56 *If* $f \in \mathcal{N}^{p\times p}$ *and* $\det f(\lambda) \not\equiv 0$ *in* $\mathbb{C}_+$, *then*

(1) $f \in \mathcal{N}_{out}^{p\times p}$ *if and only if both* f *and* f^{-1} *belong to* $\mathcal{N}_+^{p\times p}$.

(2) $f \in \mathcal{N}_{out}^{p\times p}$ *if and only if* $f^{-1} \in \mathcal{N}_{out}^{p\times p}$.

Proof The implication $\Longrightarrow$ is a straightforward consequence of Theorem 3.55. Conversely, if $f \in \mathcal{N}_+^{p\times p}$ and $f^{-1} \in \mathcal{N}_+^{p\times p}$, then

$$f = \frac{g}{h} \quad \text{and} \quad f^{-1} = \frac{g_1}{h_1}$$

where $g \in \mathcal{S}^{p\times p}$, $h \in \mathcal{S}_{out}$, $g_1 \in \mathcal{S}^{p\times p}$ and $h_1 \in \mathcal{S}_{out}$. Therefore, since $gg_1 = hh_1 I_p$, and $hh_1 \in \mathcal{S}_{out}$, it follows that

$$\det g \cdot \det g_1 \in \mathcal{S}_{out}$$

and hence that $\det g \in \mathcal{S}_{out}$. Thus, $g \in \mathcal{S}_{out}^{p\times p}$ by Theorem 3.50. This completes the proof of (1); (2) is immediate from (1). □

Lemma 3.57 *Let* $c \in \mathcal{C}^{p\times p}$. *Then:*

(1) $c \in \mathcal{N}_+^{p\times p}$.

(2) $c^{-1} \in \mathcal{N}_+^{p\times p} \Longleftrightarrow \det c(\lambda) \not\equiv 0$ *in* $\mathbb{C}_+ \Longleftrightarrow c \in \mathcal{N}_{out}^{p\times p}$.

Proof Let $c \in \mathcal{C}^{p\times p}$. Then the mvf $s = (I_p - c)(I_p + c)^{-1}$ belongs to the class $\mathcal{S}^{p\times p}$ and the function $\det\{I_p + s(\lambda)\} \neq 0$ at every point $\lambda \in \mathbb{C}_+$. Therefore, by Lemma 3.54 and Theorem 3.50,

$$\frac{I_p + s}{2} \in \mathcal{S}_{out}^{p\times p} \quad \text{and} \quad \det\left\{\frac{I_p + s}{2}\right\} \in \mathcal{S}_{out}.$$

Thus, $c \in \mathcal{N}_+^{p\times p}$. This proves (1). The second assertion follows from Lemma 3.54, Theorem 3.50 and the observation that

$$\frac{I_p - s}{2} \in \mathcal{S}_{out}^{p\times p} \Longleftrightarrow \det c(\lambda) \neq 0 \text{ in } \mathbb{C}_+.$$

□

Lemma 3.58 *The following assertions hold:*

(1) $\mathcal{C}^{p\times p} = T_{\mathfrak{V}}[\mathcal{S}^{p\times p} \cap \mathcal{D}(T_{\mathfrak{V}})]$.

(2) $\mathcal{C}^{p\times p} \subset \mathcal{N}_+^{p\times p}$.

(3) *If* $c \in \mathcal{C}^{p\times p}$, *then* $c \in \mathcal{N}_{out}^{p\times p}$ *if* $\Re c(\omega) > 0$ *for at least one (and hence every) point* $\omega \in \mathbb{C}_+$.

(4) $\mathring{\mathcal{C}}^{p\times p} = T_{\mathfrak{V}}[\mathring{\mathcal{S}}^{p\times p}]$ *and* $\mathring{\mathcal{C}}^{p\times p} \subset \mathcal{N}_{out}^{p\times p}$.

Proof This follows from Lemma 3.57. □

We turn next to the **Smirnov maximum principle**. In the formulation, $f \in \mathcal{N}_+^{p\times q}$ is identified with its boundary values.

Theorem 3.59 *If* $f \in \mathcal{N}_+^{p\times q}$, *then*

$$\sup_{\nu>0} \int_{-\infty}^{\infty} (\text{trace}\{f(\mu+i\nu)^* f(\mu+i\nu)\})^{r/2} d\mu = \int_{-\infty}^{\infty} (\text{trace}\{f(\mu)^* f(\mu)\})^{r/2} d\mu$$

for $1 \leq r < \infty$,

$$\sup_{\lambda\in\mathbb{C}_+} \text{trace}\{f(\lambda)^* f(\lambda)\} = \text{ess sup}_{\mu\in\mathbb{R}} \text{trace}\{f(\mu)^* f(\mu)\},$$

$$\sup_{\nu>0} \int_{-\infty}^{\infty} \|f(\mu+i\nu)\|^r d\mu = \int_{-\infty}^{\infty} \|f(\mu)\|^r d\mu \quad \textit{for } 1 \leq r < \infty$$

and

$$\sup_{\lambda\in\mathbb{C}_+} \|f(\lambda)\| = \text{ess sup}_{\mu\in\mathbb{R}}\{\|f(\mu)\|\},$$

where in these equalities both sides can be infinite. In particular,

$$\mathcal{N}_+^{p\times q} \cap L_r^{p\times q}(\mathbb{R}) = H_r^{p\times q} \tag{3.77}$$

for $1 \leq r \leq \infty$.

Proof See Theorem A on p. 88 of [RR85]. □

A $p \times q$ mvf which is holomorphic in $\mathbb{C}_-$ is said to belong to the Smirnov class $\mathcal{N}_-^{p\times q}$ if it can be expressed in the form (3.75) but with g and h in the corresponding classes with respect to $\mathbb{C}_-$. In particular, $f \in \mathcal{N}_-^{p\times q}$ if and only if $f^\# \in \mathcal{N}_+^{q\times p}$.

The Smirnov maximum principle will be invoked a number of times in the subsequent developments. The proof of the next lemma, which is useful in its own right, is a good illustration of the application of this principle.

Lemma 3.60 *Let $f(\lambda)$ be a $p \times q$ mvf that is holomorphic in a set $\mathfrak{h}_f \supset \mathbb{C}_+$. Then the following equivalences hold:*

(1) $\rho_i^{-1} f \in H_2^{p\times q}$.

(2) $R_\alpha f \in H_2^{p\times q}$ *for at least one point* $\alpha \in \mathfrak{h}_f$.

(3) $R_\alpha f \in H_2^{p\times q}$ *for every point* $\alpha \in \mathfrak{h}_f$.

Proof Suppose first that (1) is in force and let $\alpha \in \mathfrak{h}_f$. Then $f \in \mathcal{N}_+^{p\times q}$ and $R_\alpha f \in \mathcal{N}_+^{p\times q}$. Moreover, if $\mu \in \mathbb{R}$, then

$$\frac{f(\mu) - f(\alpha)}{\mu - \alpha} = \frac{(\mu + i)}{(\mu - \alpha)} \frac{(f(\mu) - f(\alpha))}{(\mu + i)} \in L_2^{p\times q}(\mathbb{R}).$$

If $\alpha \notin \mathbb{R}$, then the preceding assertion is self-evident, since

$$\frac{\mu + i}{\mu - \alpha} \in L_\infty, \ (\mu + i)^{-1} f \in L_2^{p\times q} \text{ and } (\mu + i)^{-1} f(\alpha) \in L_2^{p\times q}.$$

If $\alpha \in \mathbb{R}$, then the same conclusion holds since $(R_\alpha f)(\mu)$ is bounded in a neighborhood $(\alpha - \delta, \alpha + \delta)$ of α, and $(\mu + i)/(\mu - \alpha)$ is bounded for $\mu \in \mathbb{R}$ with $|\mu - \alpha| \geq \delta$. By the Smirnov maximum principle, this completes the proof of (1) $\Longrightarrow$ (3). The implication (3) $\Longrightarrow$ (2) is self-evident. Finally, if (2) is in force, then

$$\frac{f(\lambda) - f(\alpha)}{\lambda + i} = \frac{(\lambda - \alpha)}{(\lambda + i)} \frac{(f(\lambda) - f(\alpha))}{(\lambda - \alpha)}$$

clearly belongs to $H_2^{p\times q}$, since $(\lambda - \alpha)/(\lambda + i)$ belongs to H_∞. Thus, (2) $\Longrightarrow$ (1). □

3.12 Characterization of the classes $\mathcal{N}^{p\times q}$, $\mathcal{N}_+^{p\times q}$, $\mathcal{N}_{out}^{p\times q}$ and $H_\infty^{p\times q}$ via multiplication operators

Let f be a $p\times q$ mvf that is meromorphic in $\mathbb{C}_+$ and let $\widetilde{M}_f$ be the operator of multiplication by f that acts between the spaces H_2^q and H_2^p with

$$\mathcal{D}(f) = \{h \in H_2^q : \ fh \in H_2^p\} \tag{3.78}$$

and range

$$\Delta(f) = \{g \in H_2^p : \ g = fh \ \text{ for some } \ h \in H_2^q\}. \tag{3.79}$$

Lemma 3.61 *The operator $\widetilde{M}_f$ is a closed operator.*

Proof Let $\{h_n\} \in \mathcal{D}(f)$ and $\{g_n\} = \{\widetilde{M}_f h_n\}$ be two convergent sequences such that

$$h_n \longrightarrow h \ \text{ in } H_2^q \quad \text{and} \quad g_n \longrightarrow g \ \text{ in } H_2^p.$$

Then, since $h_n(\lambda) \longrightarrow h(\lambda)$ and $g_n(\lambda) \longrightarrow g(\lambda)$ at every point $\lambda \in \mathbb{C}_+$,

$$(\widetilde{M}_f h_n)(\lambda) = f(\lambda)h_n(\lambda) \longrightarrow f(\lambda)h(\lambda) = g(\lambda) \ \text{ at every point } \lambda \in \mathfrak{h}_f^+,$$

i.e., $h \in \mathcal{D}(f)$ and $g = \widetilde{M}_f h$. □

The closure $\overline{\mathcal{D}(f)}$ of $\mathcal{D}(f)$ in H_2^q is invariant under multiplication by $e_t(\lambda)$, $t \geq 0$. Therefore, the Beurling-Lax theorem implies that if

$$\mathcal{D}(f) \neq \{0\}, \ \text{ then } \ \overline{\mathcal{D}(f)} = b_f H_2^r \quad \text{where} \quad b_f \in \mathcal{S}_{in}^{q\times r}. \tag{3.80}$$

Moreover, $1 \leq r \leq q$ and b_f is uniquely determined by f up to a right constant unitary multiplier of size $r \times r$.

Theorem 3.62 *Let $f(\lambda)$ be a $p\times q$ mvf that is meromorphic in $\mathbb{C}_+$ such that $\mathcal{D}(f) \neq \{0\}$ and let $b_f \in \mathcal{S}_{in}^{q\times r}$ be defined by the Beurling-Lax theorem as in (3.80). Then:*

$$\overline{\mathcal{D}(f)} = b_f H_2^q \ \text{ with } \ b_f \in \mathcal{S}_{in}^{q\times q} \Longleftrightarrow f \in \mathcal{N}^{p\times q}. \tag{3.81}$$

$$\overline{\mathcal{D}(f)} = H_2^q \Longleftrightarrow f \in \mathcal{N}_+^{p\times q}. \tag{3.82}$$

$$\mathcal{D}(f) = H_2^q \Longleftrightarrow f \in H_\infty^{p\times q}. \tag{3.83}$$

Moreover, if $f \in H_\infty^{p\times q}$, then the operator $\widetilde{M}_f$ of multiplication by the mvf f is a bounded linear operator from H_2^q to H_2^p with norm

$$\|\widetilde{M}_f\| = \|f\|_\infty. \tag{3.84}$$

Proof ($\Longleftarrow$) Suppose first that $f \in \mathcal{N}^{p\times q}$. Then $f = f_1/(b\varphi)$, where $f_1 \in \mathcal{S}^{p\times q}$, $b \in \mathcal{S}_{in}$ and $\varphi \in \mathcal{S}_{out}$. Therefore, $b\varphi H_2^q \subseteq \mathcal{D}(f)$ and thus, as

$$\overline{\varphi H_2^q} = H_2^q \quad \text{and} \quad \overline{bH_2^q} = bH_2^q,$$

we see that

$$bH_2^q \subseteq \overline{\mathcal{D}(f)} = b_f H_2^r.$$

Consequently, by Corollary 3.42, $r = q$.

If $f \in \mathcal{N}_+^{p\times q}$, then $b(\lambda) = 1$ in the preceding argument and hence $b_f(\lambda) = I_q$. If $f \in H_\infty^{p\times q}$, then clearly $\mathcal{D}(f) = H_2^q$.

($\Longrightarrow$) Next, to deal with the opposite implications, suppose first that $\overline{\mathcal{D}(f)} = b_f H_2^q$ and $b_f \in \mathcal{S}_{in}^{q\times q}$. Then for any fixed point $\omega \in \mathbb{C}_+$ there exists a sequence of mvf's $h_n(\lambda)$, $n = 1, 2, \ldots$, in $H_2^{q\times q}$ such that their columns belong to $\mathcal{D}(f)$ and tend to the corresponding columns of the mvf $\rho_\omega(\lambda)^{-1}b_f(\lambda)$ in H_2^q. Thus, the Cauchy formula for vvf's in H_2^q guarantees that $h_n(\lambda)$ tends to $\rho_\omega(\lambda)^{-1}b_f(\lambda)$ at every point $\lambda \in \mathbb{C}_+$ when $n \uparrow \infty$. Therefore, since $\det b_f(\lambda) \not\equiv 0$ in $\mathbb{C}_+$, we may conclude that $\det h_n(\lambda) \not\equiv 0$ in $\mathbb{C}_+$ for large enough n and hence that for such n, $f_1(\lambda) = f(\lambda)h_n(\lambda)$ belongs to $H_2^{p\times q}$ and

$$f = f_1 h_n^{-1}$$

belongs to $\mathcal{N}^{p\times q}$. This completes the proof of (3.81).

Suppose next that $\overline{\mathcal{D}(f)} = H_2^q$. Then, (3.81) guarantees that $f \in \mathcal{N}^{p\times q}$, i.e., $f = f_1/b\varphi$, where $f_1 \in \mathcal{S}^{p\times p}$, $b \in \mathcal{S}_{in}$ and $\varphi \in \mathcal{S}_{out}$. Let

$$f_2 = f_1/b = \varphi f.$$

Then,

$$f_2\mathcal{D}(f) = \varphi f\mathcal{D}(f) = \varphi\Delta(f) \subseteq H_2^p$$

and, since $f_2 \in L_\infty^{p\times q}$ and $\|f_2\|_\infty \leq 1$, we also have

$$f_2 H_2^q = f_2\overline{\mathcal{D}(f)} \subseteq \overline{f_2\mathcal{D}(f)} \subseteq H_2^p.$$

Therefore, by Lemma 3.47, $f_2 \in H_\infty^{p\times q}$ and so $f \in \mathcal{N}_+^{p\times q}$. This completes the proof of (3.82). It remains to verify (3.83) and formula (3.84). The implication $\Longleftarrow$ in (3.83) is self-evident. Conversely, if $\mathcal{D}(f) = H_2^q$, then $f \in \mathcal{N}_+^{p\times q}$ by (3.82), and, as $\widetilde{M}_f$ is a closed operator on the full Hilbert space H_2^q, it must be bounded. Let $\xi \in \mathbb{C}^p$, $\eta \in \mathbb{C}^q$ and $\omega \in \mathbb{C}_+$. Then,

$$\frac{\eta}{\rho_w} \in H_2^q, \quad f\frac{\eta}{\rho_w} \in H_2^p \quad \text{and} \quad \frac{\xi}{\rho_w} \in H_2^p,$$

and it follows that

$$\left| \left\langle \widetilde{M}_f \frac{\eta}{\rho_w}, \frac{\xi}{\rho_w} \right\rangle_{st} \right| \leq \|\widetilde{M}_f\| \left\| \frac{\eta}{\rho_w} \right\|_{st} \left\| \frac{\xi}{\rho_w} \right\|_{st} = \frac{\|\widetilde{M}_f\| \|\eta\|_{st} \|\xi\|_{st}}{\rho_\omega(\omega)}.$$

At the same time, by the Cauchy formula for functions in H_2,

$$\left\langle \widetilde{M}_f \frac{\eta}{\rho_w}, \frac{\xi}{\rho_w} \right\rangle = \frac{\xi^* f(\omega)\eta}{\rho_\omega(\omega)}.$$

Thus,

$$|\xi^* f(\omega)\eta| \leq \|\widetilde{M}_f\| \|\eta\|_{st} \|\xi\|_{st}$$

and hence

$$\|f(\omega)\| \leq \|\widetilde{M}_f\|$$

for every point $\omega \in \mathbb{C}_+$. Consequently, $f \in H_\infty^{p\times q}$ and

$$\|f\|_\infty \leq \|\widetilde{M}_f\|.$$

On the other hand, the bound

$$\|\widetilde{M}_f h\|_2 = \|fh\|_2 \leq \|f\|_\infty \|h\|_2$$

implies that $\|\widetilde{M}_f\| \leq \|f\|_\infty$ and hence that formula (3.84) holds. □

Theorem 3.63 *Let T be a bounded linear operator acting from H_2^q into H_2^p such that*

$$e_t(\lambda)(Th)(\lambda) = (T(e_t h))(\lambda) \quad \textit{for } h \in H_2^q \quad \textit{and every } t \geq 0. \tag{3.85}$$

Then there exists a unique mvf $f \in H_\infty^{p\times q}$ such that $T = \widetilde{M}_f$ is the operator of multiplication by the mvf $f(\lambda)$, i.e.,

$$(Th)(\lambda) = f(\lambda)h(\lambda) \quad \textit{for } h \in H_2^q. \tag{3.86}$$

Moreover,

$$\|T\| = \|f\|_\infty. \tag{3.87}$$

Proof Let ξ_j, $j = 1, \ldots, q$, denote the jth column of the identity matrix I_q. Then the vvf $h_j = \rho_i^{-1}\xi_j$ belongs to H_2^q and, consequently, $g_j = Th_j$ belongs to H_2^p for $j = 1, \ldots, q$ and the $p \times q$ mvf

$$f(\lambda) = \rho_i \left[\; g_1(\lambda) \quad \cdots \quad g_q(\lambda) \; \right]$$

belongs to $\mathcal{N}_+^{p\times q}$. In view of Lemma 3.61 and the equivalence (3.82), the operator $\widetilde{M}_f$ of multiplication by the $p \times q$ mvf f with domain $\mathcal{D}(f) \subseteq H_2^q$ and range $\Delta(f) \subseteq H_2^p$ is a closed linear operator with $\overline{\mathcal{D}(f)} = H_2^q$. Moreover, since (3.86) holds for the vvf's $h = \xi/\rho_i$ for every $\xi \in \mathbb{C}^q$ and

$$\bigvee_{t\geq 0} e_t \rho_i^{-1}\mathbb{C}^q = H_2^q,$$

the relation $Th = \widetilde{M}_f h$ holds for every $h \in H_2^q$, i.e., $\mathcal{D}(f) = H_2^q$ and $T = \widetilde{M}_f$. The equivalence (3.83) guarantees that $f \in H_\infty^{p\times q}$. This completes the proof of existence. suppose next that f_1 and f_2 are two mvf's in $H_\infty^{p\times q}$ such that

$$Th = f_1 h = f_2 h \quad \text{for every} \quad h \in H_2^q.$$

Then, the particular identities

$$(f_1 - f_2)\frac{\xi}{\rho_i} = 0 \quad \text{for every} \quad \xi \in \mathbb{C}^q$$

clearly imply that $f_1 = f_2$.

This completes the proof, since (3.87) was established in Theorem 3.62. □

Theorem 3.64 *Let $f \in \mathcal{N}_+^{p\times q}$. Then*

$$f \in \mathcal{N}_{out}^{p\times q} \iff \overline{\Delta(f)} = H_2^p. \tag{3.88}$$

Moreover, if $q = p$, then

$$f^{-1} \in H_\infty^{p\times p} \iff \Delta(f) = H_2^p \tag{3.89}$$

Proof Suppose first that $f \in \mathcal{N}_{out}^{p\times q}$. Then

$$f = f_1/\varphi,$$

where $f_1 \in \mathcal{S}_{out}^{p\times q}$ and $\varphi \in \mathcal{S}_{out}$. Thus, $\varphi H_2^q \subseteq \mathcal{D}(f)$,

$$f\varphi H_2^q = f_1 H_2^q \subseteq \Delta(f)$$

and

$$\overline{f\varphi H_2^q} = \overline{f_1 H_2^q} \subseteq \overline{\Delta(f)} \subseteq H_2^p.$$

But

$$\overline{f_1 H_2^q} = H_2^p, \quad \text{since} \quad f_1 \in \mathcal{S}_{out}^{p\times q}.$$

Thus, $\overline{\Delta(f)} = H_2^p$.

To obtain the converse, let

$$f \in \mathcal{N}_+^{p\times q} \text{ and } \overline{\Delta(f)} = H_2^p.$$

Then

$$f = f_1/\varphi,$$

where $f_1 \in \mathcal{S}^{p\times q}$ and $\varphi \in \mathcal{S}_{out}$ and, by Theorem 3.62, $\overline{\mathcal{D}(f)} = H_2^q$. The next step is to show that $f_1 \in \mathcal{S}_{out}^{p\times q}$ or, equivalently, that

$$\overline{f_1 H_2^q} = H_2^p.$$

We know that

$$f_1 \mathcal{D}(f) = \varphi f \mathcal{D}(f) = \varphi \Delta(f),$$

and hence that

$$\overline{f_1 \mathcal{D}(f)} = \overline{\varphi \Delta(f)} = \overline{\varphi \overline{\Delta(f)}} = \overline{\varphi H_2^p} = H_2^p.$$

Therefore, since

$$f_1 \mathcal{D}(f) \subseteq f_1 H_2^q \subseteq H_2^p,$$

we must also have

$$\overline{f_1 H_2^q} = H_2^p,$$

as needed. Finally, if $q = p$, then (3.89) follows from the equivalence in (3.83) applied to f^{-1} in place of f. □

3.13 Factorization in $\mathcal{N}^{p\times q}$ and denominators

A pair of mvf's $f_1, f_2 \in \mathcal{N}_+^{p\times q}$ is said to be **right coprime** if

$$b \in \mathcal{S}_{in}^{q\times q} \text{ and } f_j b^{-1} \in \mathcal{N}_+^{p\times q} \text{ for } j = 1,2 \\ \Longrightarrow b(\lambda) \text{ is a constant unitary matrix.} \quad (3.90)$$

Similarly, a pair of mvf's $f_1, f_2 \in \mathcal{N}_+^{p\times q}$ is said to be **left coprime** if

$$b \in \mathcal{S}_{in}^{p\times p} \text{ and } b^{-1}f_j \in \mathcal{N}_+^{p\times q} \text{ for } j = 1,2 \Longrightarrow b(\lambda) \text{ is a constant unitary matrix.} \quad (3.91)$$

The notation

$$(f_1, f_2)_R = I_q \quad \text{and} \quad (f_1, f_2)_L = I_p$$

is used to designate right and left coprime pairs, respectively.

Remark 3.65 *If, for example, $b_1, b_2 \in \mathcal{S}_{in}^{p\times p}$, then, in view of Theorem Theorem 3.44,*

$$(b_1, b_2)_L = I_p \Longleftrightarrow b_1 H_2^p \bigvee b_2 H_2^p = H_2^p \Longleftrightarrow \mathcal{H}(b_1) \cap \mathcal{H}(b_2) = \{0\}.$$

There is a more restrictive definition of left (right) coprime H_∞-mvf's that for $b_1, b_2 \in \mathcal{S}_{in}^{p\times p}$ reduces to

$$[b_1, b_2]_L = I_p \Longleftrightarrow \|P_{\mathcal{H}(b_1)} P_{\mathcal{H}(b_2)}\| < 1;$$

see pp. 268–289 in Fuhrmann [Fu81]. If $b_1, b_2 \in \mathcal{S}_{in}^{p\times p}$ are rational, then the two definitions coincide.

Theorem 3.66 *If $f \in \mathcal{N}^{p\times q}$ and $f \not\equiv 0$, then f admits two representations:*

$$f = f_R d_R^{-1}, \quad \text{where } f_R \in \mathcal{N}_+^{p\times q}, \quad d_R \in \mathcal{S}_{in}^{q\times q}, \quad (d_R, f_R)_R = I_q \quad (3.92)$$

and

$$f = d_L^{-1} f_L, \quad \text{where} \quad f_L \in \mathcal{N}_+^{p\times q},\ d_L \in \mathcal{S}_{in}^{p\times p} \text{ and } (d_L, f_L)_L = I_p. \quad (3.93)$$

The factors f_R and d_R are uniquely defined by f up to a constant unitary multiplier on the right, whereas the factors f_L and d_L are uniquely defined by f up to a constant unitary multiplier on the left.

Proof Let $f \in \mathcal{N}^{p\times q}$. Then, by Theorem 3.62, $\overline{\mathcal{D}(f)} = b_f H_2^q$ where $b_f \in \mathcal{S}_{in}^{q\times q}$. Let $f_1(\lambda) = f(\lambda) b_f(\lambda)$. Then,

$$b_f \mathcal{D}(f_1) = \mathcal{D}(f)$$

and hence

$$b_f\, \overline{\mathcal{D}(f_1)} = \overline{b_f \mathcal{D}(f_1)} = \overline{\mathcal{D}(f)} = b_f H_2^q.$$

But this implies that $\overline{\mathcal{D}(f_1)} = H_2^q$. Therefore, by Theorem 3.62, $f_1 \in \mathcal{N}_+^{p\times q}$.

The next step is to show that $\overline{(f_1, b_f)_R} = I_q$. If $b \in \mathcal{S}_{in}^{q\times q}$ is a common right divisor of f_1 and b_f, then $\overline{\mathcal{D}(fb_fb^{-1})} = H_2^q$ by Theorem 3.62, since $fb_fb^{-1} \in \mathcal{N}_+^{p\times q}$. Moreover, the inclusion

$$(fb_fb^{-1})\mathcal{D}(fb_fb^{-1}) \subseteq H_2^p$$

implies that

$$(b_fb^{-1})\mathcal{D}(fb_fb^{-1}) \subseteq \mathcal{D}(f)$$

and hence that

$$b_fb^{-1}H_2^q = (b_fb^{-1})\overline{\mathcal{D}(fb_fb^{-1})} \subseteq \overline{\mathcal{D}(f)} = b_fH_2^q,$$

and, consequently, that

$$b^{-1}H_2^q \subseteq H_2^q.$$

Therefore, since

$$b^{-1} \in \mathcal{S}_{in}^{q\times q} \quad \text{by Lemma 3.47 and} \quad b \in \mathcal{S}_{in}^{q\times q},$$

$b(\lambda)$ is a constant unitary matrix. Thus, the the first representation formula (3.92) with $f_R(\lambda) = f_1(\lambda)$ and $d_R(\lambda) = b_f(\lambda)$ is justified. Moreover, since

$$d_RH_2^q = d_R\overline{\mathcal{D}(f_R)} = \overline{d_R\mathcal{D}(f_R)} = \overline{\mathcal{D}(f)} = b_fH_2^q,$$

it follows that $d_R = b_fu$ for some $q \times q$ unitary marix u and hence that the mvf's d_R and f_R in (3.92) are essentially unique.

Finally, since

$$f \in \mathcal{N}_+^{p\times q} \iff f^\tau \in \mathcal{N}_+^{q\times p} \quad \text{and} \quad b \in \mathcal{S}_{in}^{q\times q} \iff b^\tau \in \mathcal{S}_{in}^{q\times q},$$

the assertions related to (3.93) follow from the results proved above by passing to transposes. □

Theorem 3.67 *Let $f \in \mathcal{N}^{p\times q}$. Then the factors f_R and d_R in the representation formula (3.92) and the factors f_L and d_L in the representation formula (3.93) are minimal in the sense that:*

(1) *If $f = \widetilde{f}_R\,\widetilde{d}_R^{-1}$ with $\widetilde{d}_R \in \mathcal{S}_{in}^{q\times q}$ and $\widetilde{f}_R \in \mathcal{N}_+^{p\times q}$, then $\widetilde{d}_R = d_Rb$ and $\widetilde{f}_R = f_Rb$ for some mvf $b \in \mathcal{S}_{in}^{q\times q}$.*

(2) *If $f = \widetilde{d}_L^{-1}\,\widetilde{f}_L$ with $\widetilde{d}_L \in \mathcal{S}_{in}^{p\times p}$ and $\widetilde{f}_L \in \mathcal{N}_+^{p\times q}$, then $\widetilde{d}_L = bd_L$ and $\widetilde{f}_L = bf_L$ for some mvf $b \in \mathcal{S}_{in}^{p\times p}$.*

Proof In setting (1), Theorem 3.62 implies that $\overline{\mathcal{D}(\widetilde{f}_R)} = H_2^q$ and $\widetilde{d}_R\,\mathcal{D}(\widetilde{f}_R) \subseteq \mathcal{D}(f)$. Therefore, since $\overline{\mathcal{D}(f)} = b_f H_2^q = d_R H_2^q$,

$$\widetilde{d}_R\, H_2^q \subseteq d_R\, H_2^q$$

and, consequently, the mvf $b = d_R^{-1}\,\widetilde{d}_R$ belongs to $\mathcal{S}_{in}^{q\times q}$ and

$$\widetilde{f}_R = f\widetilde{d}_R = f_R\, d_R^{-1}\,\widetilde{d}_R = f_R b,$$

as claimed. The proof of (2) is similar. □

If $f \in \mathcal{N}^{p\times q}$ and $f \not\equiv 0$, then $\widetilde{d}_R \in \mathcal{S}_{in}^{q\times q}$ is called a **right denominator** of f if $f\widetilde{d}_R \in \mathcal{N}_+^{p\times q}$; and $\widetilde{d}_L \in \mathcal{S}_{in}^{p\times p}$ is called a **left denominator** of f if $\widetilde{d}_L f \in \mathcal{N}_+^{p\times q}$. The mvf's d_R and d_L that intervene in (3.92) and (3.93) are called **minimal right and left denominators** of f, respectively, since d_R is a common left divisor of every right denominator $\widetilde{d}_R$ of f and d_L is a common right divisor of every left denominator $\widetilde{d}_L$ of f, thanks to Theorem 3.67.

Lemma 3.68 *Let f be a rational $p \times q$ mvf. Then its minimal left and right denominators d_L and d_R are rational.*

Proof Let $f(\lambda)$ be a rational $p \times q$ mvf. Then there exists a finite scalar Blaschke product $\beta(\lambda)$ such that $\beta(\lambda)f(\lambda)$ is analytic in $\mathbb{C}_+$ and hence $\beta f \in \mathcal{N}_+^{p\times q}$, i.e., $\widetilde{d}_L = \beta I_p$ is a left denominator of f. Consequently, by Theorem 3.67, $\beta d_L^{-1} \in \mathcal{S}_{in}^{p\times p}$ and thus, as

$$\beta d_L^{\#}\frac{\xi}{\rho_i} \in H_2^p \quad \text{and} \quad d_L^{\#}\frac{\xi}{\rho_{-i}} \in K_2^p \quad \text{for every } \xi \in \mathbb{C}^p,$$

it follows that

$$d_L^{\#}\frac{\xi}{\rho_{-i}} \in K_2^p \ominus \left(\frac{\rho_i}{\rho_{-i}}\right)\beta^{\#} K_2^p.$$

Therefore, since the space on the right is a finite dimensional Hilbert space with basis of rational vvf's, the mvf $d_L^{\#}$ is also rational.

The proof that d_R is a rational mvf is similar. □

An ordered pair $\{d_1, d_2\}$ of inner mvf's $d_1 \in \mathcal{S}_{in}^{p\times p}$ and $d_2 \in \mathcal{S}_{in}^{q\times q}$ is called a **denominator** of the mvf $f \in \mathcal{N}^{p\times q}$ if

$$d_1 f d_2 \in \mathcal{N}_+^{p\times q}.$$

Thus $\widetilde{d}_L$ is a left denominator of f if and only if $\{\widetilde{d}_L, I_q\}$ is a denominator of f, and $\widetilde{d}_R$ is a right denominator of f if and only if $\{I_p, \widetilde{d}_R\}$ is a denominator of f.

A denominator $\{d_1, d_2\}$ of f is called a **divisor** of another denominator $\{\widetilde{d}_1, \widetilde{d}_2\}$ of f if

$$\widetilde{d}_1 d_1^{-1} \in \mathcal{S}_{in}^{p\times p} \quad \text{and} \quad d_2^{-1}\widetilde{d}_2 \in \mathcal{S}_{in}^{q\times q}.$$

It is called a **trivial divisor** if $\widetilde{d}_1 d_1^{-1}$ and $d_2^{-1}\widetilde{d}_2$ are constant unitary matrices. A denominator $\{d_1, d_2\}$ of f is called a **minimal denominator** of f if every denominator of f that is a divisor of $\{d_1, d_2\}$ is a trivial divisor.

Lemma 3.69 *Let $f \in \mathcal{N}^{p\times q}$ and let $\{\widetilde{d}_1, \widetilde{d}_2\}$ be a denominator of f. Then there exists a minimal denominator $\{d_1, d_2\}$ of f that is a divisor of $\{\widetilde{d}_1, \widetilde{d}_2\}$.*

Proof Let $\{\widetilde{d}_1, \widetilde{d}_2\}$ be a denominator of mvf f. Let $f_1 = \widetilde{d}_1 f$. Then $\widetilde{d}_2$ is a right denominator of f_1. By Theorem 3.67, there exists a minimal right denominator d_2 of f_1. Thus, $\widetilde{d}_1 f d_2 \in \mathcal{N}_+^{p\times q}$, $d_2 \in \mathcal{S}_{in}^{q\times q}$ and d_2 is a common left divisor of all right denominators of f_1. Therefore, $d_2^{-1}\widetilde{d}_2 \in \mathcal{S}_{in}^{q\times q}$. Let $f_2 = f d_2$. Then $\widetilde{d}_1$ is a left denominator of f_2. By Theorem 3.67, there exists a minimal left denominator d_1 of f_2, i.e.,

$$d_1 f d_2 \in \mathcal{N}_+^{p\times q} \tag{3.94}$$

and d_1 is a common right divisor of all left denominators of $f d_2$. Therefore, $\widetilde{d}_1 d_1^{-1} \in \mathcal{S}_{in}^{p\times p}$. Thus, $\{d_1, d_2\}$ is a divisor of $\{\widetilde{d}_1, \widetilde{d}_2\}$. In view of (3.94), $\{d_1, d_2\}$ is a denominator of f. It is a minimal denominator of f. Indeed, let $\{b_1, b_2\}$ be a denominator of f that is a divisor of $\{d_1, d_2\}$ i.e.,

$$u = d_1 b_1^{-1} \in \mathcal{S}_{in}^{p\times p} \quad \text{and} \quad v = b_2^{-1} d_2 \in \mathcal{S}_{in}^{q\times q}.$$

Then

$$b_1 f b_2 \in \mathcal{N}_+^{p\times q}, \quad d_1 = u b_1 \quad \text{and} \quad d_2 = b_2 v. \tag{3.95}$$

Consequently, b_1 is a left denominator of f_2 and $b_1 d_1^{-1} \in \mathcal{S}_{in}^{p\times p}$, since d_1 is a minimal left denominator of f_2. Therefore, $u \in \mathcal{S}_{in}^{p\times p}$ and $u^{-1} \in \mathcal{S}_{in}^{p\times p}$ and hence, u is a constant unitary $p\times p$ matrix. Furthermore, since $\widetilde{d}_1 d_1^{-1} \in \mathcal{S}_{in}^{p\times p}$ and (3.95) holds,

$$f_1 b_2 = \widetilde{d}_1 f b_2 = \widetilde{d}_1 d_1^{-1} u b_1 f b_2 \in \mathcal{N}_+^{p\times q}.$$

Consequently, b_2 is a right denominator of f_1. From this it follows that $d_2^{-1}b_2 \in \mathcal{S}_{in}^{q\times q}$. Thus, as $v = (d_2^{-1}b_2)^{-1}$ also belongs to $\mathcal{S}_{in}^{q\times q}$, the mvf v is a constant unitary $q \times q$ matrix. □

A denominator $\{d_1, I_q\}$ of f is minimal if and only if d_1 is a minimal left denominator of f. Similarly, a denominator $\{I_p, d_2\}$ of f is minimal if and only if d_2 is a minimal right denominator of f.

Theorem 3.70 *Let $f \in \mathcal{N}^{p\times q}$ and let $\{d_1, d_2\}$ be a minimal denominator of f. Then:*

(1) *If f is a rational mvf, then d_1 and d_2 are also rational mvf's.*

(2) *If f is an entire mvf, then d_1 and d_2 are also entire mvf's.*

Proof Let f be a rational $p \times q$ mvf. Then clearly $f \in \mathcal{N}^{p\times q}$. Let d_L and d_R be minimal left and right denominators of f, respectively. Then, in view of Lemma 3.68, d_L and d_R are also rational. Let $\{d_1, d_2\}$ be a minimal denominator of f. Then the mvf $b = d_L d_1^{-1}$ belongs to $\mathcal{S}_{in}^{p\times p}$, since d_L is a left denominator of fd_2 and d_1 is a minimal left denominator of fd_2. Thus,

$$\mathcal{H}(b) \subseteq \mathcal{H}(d_L)$$

and, as $\mathcal{H}(d_L)$ is finite dimensional with a basis of rational vvf's, thanks to Lemma 3.46, the vvf

$$\frac{b(\lambda) - b(\alpha)}{\lambda - \alpha}\xi \quad \text{is rational for every} \quad \xi \in \mathbb{C}^p.$$

Therefore, b is rational, as is $d_1 = b^{-1}d_L$. The proof that d_2 is rational is based on similar considerations using d_R.

The proof of (2) will be given in Theorem 3.105. □

Theorem 3.71 *Every mvf $f \in \mathcal{N}_+^{p\times q}$ that is not identically equal to zero admits an inner-outer factorization of the form*

$$f(\lambda) = b_L(\lambda)\varphi_L(\lambda), \quad \textit{where } b_L \in \mathcal{S}_{in}^{p\times r} \textit{ and } \varphi_L \in \mathcal{N}_{out}^{r\times q} \tag{3.96}$$

and a $*$-outer-$*$-inner factorization of the form

$$f(\lambda) = \varphi_R(\lambda)b_R(\lambda), \quad \textit{where } \varphi_R \in \mathcal{N}_{*out}^{p\times r} \textit{ and } b_R \in \mathcal{S}_{*out}^{r\times q}. \tag{3.97}$$

In both of these factorizations,

$$r = \text{rank } f. \tag{3.98}$$

The factors in each of these factorizations are defined uniquely up to replacement of

$$b_L \text{ and } \varphi_L \quad \text{by} \quad b_L u \text{ and } u^* \varphi_L$$

and

$$\varphi_R \text{ and } b_R \quad \text{by} \quad \varphi_R v \text{ and } v^* b_R,$$

where u and v are constant unitary $r \times r$ matrices. Moreover, there exists a $q \times q$ permutation matrix K such that

$$[I_r \quad 0]\, K^\tau \varphi_L(\lambda)^* \varphi_L(\lambda) K \begin{bmatrix} I_r \\ 0 \end{bmatrix} > 0 \quad \text{for every point } \lambda \in \mathbb{C}_+,$$

and, for each such permutation matrix K and each fixed point $\omega \in \mathbb{C}_+$, there exists exactly one factor φ_L in (3.96) and exactly one factor φ_R in (3.97) such that

$$\varphi_L(\omega)\, K \begin{bmatrix} I_r \\ 0 \end{bmatrix} > 0 \quad \text{and} \quad [I_r \quad 0] K^\tau \varphi_R(\omega) > 0. \tag{3.99}$$

Proof Let $f \in \mathcal{N}_+^{p\times q}$. Then $f = \varphi^{-1} f_1$, where $\varphi \in \mathcal{S}_{out}$ and $f_1 \in \mathcal{S}^{p\times q}$. Thus, by Theorem 3.48, f_1 admits an essentially unique inner-outer factorization of the form $f_1 = b_1\varphi_1$, where $b_1 \in \mathcal{S}_{in}^{p\times r}$, $\varphi_1 \in \mathcal{S}_{out}^{r\times q}$ and

$$r = \operatorname{rank} \varphi_1(\lambda) \quad \text{for } \lambda \in \mathbb{C}_+.$$

Therefore, $f = b_1\varphi_1/\varphi$, i.e., (3.96) holds with $b_L = b_1$, $\varphi_L = \varphi_1/\varphi$ and $r = \operatorname{rank} f$. Moreover, $\{\varphi(\lambda)\eta : \eta \in \mathbb{C}^q\}$ is independent of $\lambda \in \mathbb{C}_+$, since $\varphi \in \mathcal{S}_{out}^{p\times q}$, and hence there exists a $q\times q$ permutation matrix K such that the first r columns of the $r \times q$ matrix $\varphi_1(\lambda)K$ are linearly independent at every point $\lambda \in \mathbb{C}_+$. This matrix K meets the first condition in the theorem, and for each choice of $\omega \in \mathbb{C}_+$, there exists a unique unitary matrix $x \in \mathbb{C}^{r\times r}$ such that

$$x\varphi_1(\omega)K \begin{bmatrix} I_r \\ 0 \end{bmatrix} > 0.$$

Thus, fixing ω and x, the matrix $\varphi_L(\lambda) = u\varphi_1(\lambda)$ is outer and satisfies the first condition in (3.99). Moreover, if $\widetilde{\varphi}_L$ is a second $r \times q$ outer factor of f that meets the same normalization condition as φ_L, then $\widetilde{\varphi}_L(\lambda) = y\varphi_L(\lambda)$

for some unitary $r \times r$ constant matrix y,

$$y\varphi_L(\omega)K \begin{bmatrix} I_r \\ 0 \end{bmatrix} = \widetilde{\varphi}_L(\omega)K \begin{bmatrix} I_r \\ 0 \end{bmatrix} > 0 \quad \text{and} \quad \varphi_L(\omega)K \begin{bmatrix} I_r \\ 0 \end{bmatrix} > 0.$$

Therefore, $y = I_r$, i.e., there is only one outer factor of f that meets the normalization condition at ω.

Similar arguments serve to verify the assertions for φ_R, with the same matrix K, since $\varphi_L(\mu)^*\varphi_L(\mu) = \varphi_R(\mu)\varphi_R(\mu)^*$ a.e. on $\mathbb{R}$. □

Theorem 3.72 *Every mvf $f \in \mathcal{N}^{p\times q}$ that is not identically equal to zero admits a representation of the form*

$$\begin{aligned} f = d_L^{-1} b_L \varphi_L, \quad &\textit{where } d_L \in \mathcal{S}_{in}^{p\times p},\ b_L \in \mathcal{S}_{in}^{p\times r}, \\ &\varphi_L \in \mathcal{N}_{out}^{r\times q} \textit{ and } (d_L, b_L\varphi_L)_L = I_p, \end{aligned} \tag{3.100}$$

and a representation of the form

$$\begin{aligned} f = \varphi_R b_R d_R^{-1}, \quad &\textit{where } \varphi_R \in \mathcal{N}_{*out}^{p\times r},\ b_R \in \mathcal{S}_{*in}^{q\times r}, \\ &d_R \in \mathcal{S}_{in}^{q\times q} \textit{ and } (\varphi_R b_R, d_R)_R = I_q. \end{aligned} \tag{3.101}$$

In these two representations of $f(\lambda)$, $r = \operatorname{rank} f$ and the factors are essentially unique, i.e., up to the replacement of d_L by ud_L, φ_L by $v\varphi_L$ and b_L by ub_Lv^* in (3.100), where u and v are constant unitary matrices, with analogous replacements for the factors in (3.101).

Proof This is an immediate consequence of Theorems 3.66 and 3.71. □

3.14 Some properties of outer mvf's

This section is devoted to a number of characteristic properties of outer mvf's. The first of these is an extremal characterization.

Theorem 3.73 *Let $\psi \in \mathcal{N}_+^{p\times q}$ and $\varphi \in \mathcal{N}_{out}^{r\times q}$ be such that*

$$\psi(\mu)^*\psi(\mu) \le \varphi(\mu)^*\varphi(\mu) \ \textit{a.e. on } \mathbb{R}. \tag{3.102}$$

Then:

(1) $\psi(\lambda)^*\psi(\lambda) \le \varphi(\lambda)^*\varphi(\lambda)$ *for every* $\lambda \in \mathbb{C}_+$.

(2) $\psi(\lambda) = s(\lambda)\varphi(\lambda)$ *for exactly one mvf* $s \in \mathcal{S}^{p\times r}$.

(3) *If equality holds in formula (3.102), i.e., if*

$$\psi(\mu)^*\psi(\mu) = \varphi(\mu)^*\varphi(\mu) \ a.e. \ on \ \mathbb{R},$$

then $s \in \mathcal{S}_{in}^{p\times r}$ *in (2), i.e., (2) is an inner-outer factorization of* ψ.

(4) *If equality holds in formula (3.102) and* $\psi \in \mathcal{N}_{out}^{p\times q}$, *then* $p = r$ *and* $\psi = u\varphi$ *for some constant* $p \times p$ *unitary matrix* u.

Proof If $\varphi \in \mathcal{N}_{out}^{r\times q}$, then $\varphi = h/g$ for some choice of $h \in \mathcal{S}_{out}^{r\times q}$ and $g \in \mathcal{S}_{out}$. Thus, as $\psi_1 = g\psi$ belongs to $\mathcal{N}_+^{p\times q}$ and

$$\psi_1(\mu)^*\psi_1(\mu) \leq h(\mu)^*h(\mu) \leq I_q \quad \text{a.e. on } \mathbb{R},$$

the Smirnov maximum principle implies that $\psi_1 \in \mathcal{S}^{p\times q}$. Let

$$Tf = \psi_1 f_1 \quad \text{for } f = hf_1, \ f_1 \in H_2^q.$$

Then, in view of the last inequalities, Tf is well defined by $f, Tf \in H_2^p$ and $\|Tf\| \leq \|f\|$ for every $f \in \Delta(h)$. Thus, Tf is a contractive linear operator from $\Delta(h)$ into H_2^q. Therefore, since $h \in \mathcal{S}_{out}^{r\times q}$, T can be extended to a contractive linear operator acting from H_2^r into H_2^p. Moreover, since

$$e_\tau Tf = Te_\tau f, \quad \text{for every } f \in H_2^r \text{ and } \tau \geq 0,$$

Theorem 3.63 guarantees the existence of a mvf $s \in \mathcal{S}^{p\times r}$ such that

$$(Tf)(\lambda) = s(\lambda)f(\lambda) \quad \text{for every } f \in H_2^r.$$

Thus, if $f = h\frac{\xi}{\rho_i}$, then

$$Tf = \psi_1\frac{\xi}{\rho_i} = sh\frac{\xi}{\rho_i} \quad \text{for every } \xi \in \mathbb{C}^q$$

and hence $\psi_1 = sh$, which verifies (1) and proves the existence of at least one $s \in \mathcal{S}^{p\times q}$ such that (2) holds. To check uniqueness, assume that $\psi = s_1\varphi = s_2\varphi$ for a pair of mvf's s_1, s_2 in $\mathcal{S}^{p\times q}$. Then the equality

$$(s_1(\mu) - s_2(\mu))\varphi(\mu) = 0 \quad \text{a.e. on } \mathbb{R}$$

implies that

$$s_1(\mu) = s_2(\mu) \quad \text{a.e. on } \mathbb{R},$$

since $\operatorname{rank}\varphi(\mu) = r$ a.e. on $\mathbb{R}$, as $\varphi \in \mathcal{N}_{out}^{r\times q}$. Thus, $s_1 = s_2$.

Suppose next that equality holds a.e. on $\mathbb{R}$ in (3.102), then

$$\varphi(\mu)^*\{I_r - s(\mu)^*s(\mu)\}\varphi(\mu) = 0 \quad \text{a.e. on } \mathbb{R}.$$

Thus,

$$\{I_r - s(\mu)^* s(\mu)\}\varphi(\mu) = 0 \quad \text{a.e. on } \mathbb{R}$$

and hence, as $\operatorname{rank}\varphi(\mu) = r$ a.e. on $\mathbb{R}$, it follows that $s \in \mathcal{S}_{in}^{p\times r}$.

Finally, (4) follows from (3) and the uniqueness of the inner-outer factorization of the mvf $\psi \in \mathcal{N}_+^{p\times q}$ that was established in Theorem 3.71. □

Theorem 3.74 *Let $\varphi \in \mathcal{N}_+^{p\times q}$. Then:*

(1) *$\rho_\omega^{-1}\varphi \in H_2^{p\times q}$ for some (and hence for every) $\omega \in \mathbb{C}_+$ if and only if $\rho_i^{-1}\varphi \in L_2^{p\times q}$.*

(2) *If $\rho_i^{-1}\varphi \in H_2^{p\times q}$, then $\varphi \in \mathcal{N}_{out}^{p\times q}$ if and only if*

$$\bigvee_{t\geq 0} R_0 e_t \varphi \mathbb{C}^q = H_2^p. \tag{3.103}$$

Proof The implication $\rho_\omega^{-1}\varphi \in H_2^{p\times q} \implies \rho_i^{-1}(\mu)\varphi(\mu) \in L_2^{p\times q}$ if $\omega \in \mathbb{C}_+$ is obvious. The converse implication follows from the Smirnov maximum principle.

Now let $\rho_\omega^{-1}\varphi \in H_2^{p\times q}$ for some $\omega \in \mathbb{C}_+$. Then $\rho_\omega^{-1}\varphi\xi \in H_2^p$ for every $\xi \in \mathbb{C}^q$ and $\omega \in \mathbb{C}_+$. Consequently

$$\mathcal{L}_t = \{(R_0 e_t)\varphi\xi : \xi \in \mathbb{C}^q\}$$

is a subspace of H_2^p for every $t \geq 0$ and

$$\mathcal{D}_+ = \bigvee_{t\geq 0} \mathcal{L}_t$$

is a closed subspace of H_2^p. Moreover, the formulas

$$e_\tau(\lambda)\,\frac{e_t(\lambda)-1}{\lambda} = \frac{e_{t+\tau}(\lambda)-1}{\lambda} - \frac{e_\tau(\lambda)-1}{\lambda}$$

imply that

$$e_\tau \mathcal{L}_t \subseteq \mathcal{L}_{t+\tau} + \mathcal{L}_\tau \quad \text{if} \quad t \geq 0 \quad \text{and} \quad \tau \geq 0$$

and, consequently, that $e_t\mathcal{D}_+ \subseteq \mathcal{D}_+$ for every $t \geq 0$. Therefore, by the Beurling-Lax theorem, there exists an essentially unique mvf $b \in \mathcal{S}_{in}^{p\times r}$ such that

$$\mathcal{D}_+ = bH_2^r \qquad \text{for some } r \leq p.$$

Consider fixed $\alpha > 0$. Then

$$\frac{e_\alpha(\lambda) - 1}{\lambda}\varphi(\lambda)\xi \in \mathcal{L}_\alpha \quad \text{for every} \quad \xi \in \mathbb{C}^q.$$

But, $\mathcal{L}_\alpha \subset \mathcal{D}_+$. Consequently,

$$\frac{e_\alpha(\lambda) - 1}{\lambda}\,\varphi(\lambda) = b(\lambda)h(\lambda), \quad \text{where } b \in \mathcal{S}_{in}^{p\times r} \text{ and } h \in H_2^{r\times q}.$$

Therefore, since $\lambda/(e_\alpha(\lambda) - 1) \in \mathcal{N}_{out}$, it follows that

$$\varphi(\lambda) = b(\lambda)h_1(\lambda), \quad \text{where } h_1 \in \mathcal{N}_+^{r\times q}.$$

By Theorem 3.71, there exists an essentially unique inner-outer factorization of the mvf h_1, i.e., $h_1 = b_L\varphi_L$, where $b_L \in \mathcal{S}_{in}^{r\times r_1}$, $\varphi_L \in \mathcal{N}_{out}^{r_1\times q}$ and $r_1 =$ rank h_1. But $\varphi(\mu)^*\varphi(\mu) = h_1(\mu)^*h_1(\mu)$ a.e. on $\mathbb{R}$. Consequently, rank h_1 = rank φ. Now let $\varphi \in \mathcal{N}_{out}^{p\times q}$. Then rank $\varphi = p$. Consequently, $r_1 = p$. Thus, $\varphi = bb_L\varphi_L$, where $bb_L \in \mathcal{S}_{in}^{p\times p}$, $\varphi_L \in \mathcal{N}_{out}^{p\times q}$, which implies that bb_L and so too b are constant unitary $p \times p$ matrices. Consequently $r = p$ and $\mathcal{D}_+ = bH_2^p = H_2^p$, i.e., (3.103) holds.

Conversely, if the relation (3.103) is in force, then $\overline{\mathcal{D}(\varphi)} = H_2^q$ and $\overline{\Delta(\varphi)} = H_2^p$ and hence, in view of Theorem 3.64, $\varphi \in \mathcal{N}_{out}^{p\times q}$. □

Theorem 3.75 *Let $\varphi \in H_2^{p\times q}$. Then $\varphi \in \mathcal{N}_{out}^{p\times q}$ if and only if*

$$\bigvee_{t\geq 0} e_t\varphi\mathbb{C}^q = H_2^p$$

Proof The proof is much the same as the proof of (2) in Theorem 3.74. □

3.15 Scalar denominators

A function β is called a **scalar denominator** of the mvf $f \in \mathcal{N}^{p\times q}$, if

$$\beta f \in \mathcal{N}_+^{p\times q} \quad \text{and} \quad \beta \in \mathcal{S}_{in}. \tag{3.104}$$

The set of all scalar denominators of f will be denoted $\mathfrak{D}(f)$.

Lemma 3.76 *Let $f \in \mathcal{N}^{p\times q}$ and let d_L and d_R be minimal left and right denominators of f. Then $\mathfrak{D}(f) \neq \emptyset$ and*

$$\mathfrak{D}(f) = \mathfrak{D}(d_L^{-1}) = \mathfrak{D}(d_R^{-1}). \tag{3.105}$$

Proof Let $\beta = \det d_L$, where d_L is a minimal left denominator of a mvf $f \in \mathcal{N}^{p\times q}$. Then

$$\beta f = (\beta d_L^{-1}) f_L \in \mathcal{N}_+^{p\times q}, \tag{3.106}$$

since $f_L \in \mathcal{N}_+^{p\times q}$ and $\beta d_L^{-1} \in \mathcal{S}_{in}^{p\times p}$. Consequently, $\beta \in \mathfrak{D}(f)$, i.e., $\mathfrak{D}(f) \neq \emptyset$.

Next, since (3.106) holds for every $\beta \in \mathfrak{D}(d_L^{-1})$, the inclusion $\mathfrak{D}(d_L^{-1}) \subseteq \mathfrak{D}(f)$ is in force. Conversely, if $\beta \in \mathfrak{D}(f)$, then $\widetilde{d}_L = \beta I_p$ is a left denominator of f. Therefore, since d_L is a minimal left denominator of f, $\widetilde{d}_L\, d_L^{-1} \in \mathcal{S}_{in}^{p\times p}$, i.e., $\beta d_L^{-1} \in \mathcal{S}_{in}^{p\times p}$ and hence $\beta \in \mathfrak{D}(d_L^{-1})$. Thus $\mathfrak{D}(f) \subseteq \mathfrak{D}(d_L^{-1})$. This proves the first equality in (3.105); the second may be verified in just the same way. □

A scalar denominator β_0 of a mvf $f \in \mathcal{N}^{p\times q}$ is called a **minimal scalar denominator** of f, if it is a common divisor of all scalar denominators of f, i.e., if

$$\beta_0 \in \mathfrak{D}(f) \quad \text{and} \quad \beta/\beta_0 \in \mathcal{S}_{in} \text{ for every } \beta \in \mathfrak{D}(f). \tag{3.107}$$

Lemma 3.77 *Let $f \in \mathcal{N}^{p\times q}$, let d_L and d_R be minimal left and right denominators of f. Then:*

(1) *f has a minimal scalar denominator.*

(2) *A minimal scalar denominator of f is a minimal common multiple of all the entries f_{ij} of f.*

(3) *The sets of minimal scalar denominators of the three mvf's f, d_L^{-1}, and d_R^{-1} coincide.*

Proof Let u_{ij} be a minimal denominator for the entry f_{ij} of f, if $f_{ij} \not\equiv 0$. In view of Theorem 3.43, there exists an essentially unique minimal common multiple u of all the u_{ij}. Since $uf_{ij} \in \mathcal{N}_+$, $uf \in \mathcal{N}_+^{p\times q}$. Thus u is a scalar denominator of f, i.e., $u \in \mathfrak{D}(f)$. Moreover, if $v \in \mathfrak{D}(f)$, then $vf \in \mathcal{N}_+^{p\times q}$ and hence $vf_{ij} \in \mathcal{N}_+$, i.e., v is a common multiple of the $\{u_{ij}\}$, i.e., $vu_{ij}^{-1} \in \mathcal{S}_{in}$. However, since u is a minimal common multiple of $\{u_{ij}\}$, $u^{-1}v \in \mathcal{S}_{in}$. Thus, u is a minimal scalar denominator of f. This completes the proof of (2); (1) follows from (2); and (3) follows from Lemma 3.76. □

Let $u \in \mathcal{S}_{in}$ and let

$$\mathcal{N}^{p\times q}(u) = \{f \in \mathcal{N}^{p\times q} :\ u \in \mathfrak{D}(f)\} = \{f \in \mathcal{N}^{p\times q} : uf \in \mathcal{N}_+^{p\times q}\}.$$

Then it is clear that $\mathcal{N}^{p\times q}(1) = \mathcal{N}_+^{p\times q}$.

A rational $p \times q$ mvf f belongs to $\mathcal{N}^{p\times q}$; it belongs to $\mathcal{N}_+^{p\times q}$ if and only if f has no poles in $\mathbb{C}_+$. If f has poles $\omega_1, \ldots, \omega_n$ in $\mathbb{C}_+$, repeated according to their order, then

$$f \in \mathcal{N}^{p\times q}(u), \quad \text{where} \quad u(\lambda) = \prod_{j=1}^{n} \frac{\lambda - \omega_j}{\lambda - \overline{\omega}_j},$$

and $u(\lambda)$ is a minimal scalar denominator of $f(\lambda)$.

3.16 Factorization of positive semidefinite mvf's

This section focuses on the following factorization problem: given a measurable $p \times p$ mvf $g(\mu)$ on $\mathbb{R}$, find $f \in \mathcal{N}^{q\times p}$ such that

$$f(\mu)^* f(\mu) = g(\mu) \quad \text{a.e. on } \mathbb{R}. \tag{3.108}$$

In view of Lemma 3.31,

$$\operatorname{rank} f(\mu) = r = \operatorname{rank} f \quad \text{a.e. on } \mathbb{R}.$$

Thus, this problem admits a solution only if the given mvf $g(\mu)$ satisfies the conditions

$$g(\mu) \geq 0 \quad \text{and} \quad \operatorname{rank} g(\mu) = r \quad \text{a.e. on} \quad \mathbb{R}, \tag{3.109}$$

where $r \leq p$.

The solutions $f = \varphi_L$ of the factorization problem (3.108) such that $\varphi_L \in \mathcal{N}_{out}^{r\times p}$ are of special interest.

The nondegenerate case $r = p$, i.e., the case $g(\mu) > 0$ a.e. on $\mathbb{R}$ will be considered in this section. A criterion for the existence of at least one solution $f \in \mathcal{N}^{q\times p}$ of the problem (3.108) will be formulated and the set of all solutions to this problem will be described when $r = p$. In Theorem 3.110 the factorization problem (3.108) will be considered under the extra assumption that the mvf $g(\mu)$ is the nontangential limit of a mvf $g \in \mathcal{N}^{p\times p}$ and then the case $r < p$ will not be excluded.

Theorem 3.78 *Let $g \in L_1^{p\times p}$ be such that $g(\mu) > 0$ a.e. on $\mathbb{R}$. Then there exists an outer mvf $\varphi_L \in H_2^{p\times p}$ such that $f = \varphi_L$ is a solution of the factorization problem (3.108) if and only if*

$$\ln \det g \in \widetilde{L}_1.$$

This function f is defined by g up to a constant unitary left factor and it may be normalized by the condition $f(\omega) > 0$ at some fixed point $\omega \in \mathbb{C}_+$.

Proof A complete proof may be found in [Roz67]. The uniqueness also follows from assertion (4) in Theorem 3.73. □

We turn next to a more general theorem on the solution of factorization problem (3.108) in the class $\mathcal{N}^{q\times p}$ with $q \geq p$.

Theorem 3.79 *Let $g(\mu)$ be a measurable $p \times p$ mvf on $\mathbb{R}$ such that $g(\mu) > 0$ a.e. on $\mathbb{R}$. Then:*

(1) *There exists at least one $f \in \mathcal{N}^{p\times p}$ such that (3.108) holds if and only if*

$$\ln \|g^{\pm 1}\| \in \widetilde{L}_1. \tag{3.110}$$

(2) *If condition (3.110) is satisfied, then there exists a mvf $\varphi_L \in \mathcal{N}_{out}^{p\times p}$ such that $f = \varphi_L$ is a solution of problem (3.108). This solution is unique, up to a constant unitary left multiplier and it may be uniquely specified by imposing the normalization condition $\varphi_L(\omega) > 0$ at some fixed point $\omega \in \mathbb{C}_+$.*

(3) *The set of all solutions $f \in \mathcal{N}^{q\times p}$ with $q \geq p$ of the factorization problem (3.108) is described by the formula (3.100) with $\varphi_L \in \mathcal{N}_{out}^{p\times p}$, considered in assertion (2) and arbitrary inner mvf's $d_L \in \mathcal{S}_{in}^{q\times q}$ and $b_L \in \mathcal{S}_{in}^{q\times p}$.*

Proof Let $f \in \mathcal{N}^{q\times p}$ be a solution of the factorization problem (3.108) for a measurable $p \times p$ mvf $g(\mu)$ with $g(\mu) > 0$ on $\mathbb{R}$. Then, by Theorem 3.72,

$$f = d_L^{-1} b_L \varphi_L,$$

where $d_L \in \mathcal{S}_{in}^{q\times q}$, $b_L \in \mathcal{S}_{in}^{q\times r}$, $\varphi_L \in \mathcal{N}_{out}^{r\times p}$ and consequently

$$\varphi_L(\mu)^* \varphi_L(\mu) = g(\mu) \text{ a.e. on } \mathbb{R}.$$

Therefore, since $r \leq p$ and rank $g(\mu) = p$ a.e., we must have $r = p$, i.e., $\varphi_L \in \mathcal{N}_{out}^{p\times p}$. This completes the proof of assertion (3). Moreover, $\varphi_L = \varphi_1/\varphi_2$, where $\varphi_1 \in \mathcal{S}_{out}^{p\times p}$ and $\varphi_2 \in \mathcal{S}_{out}$.

Next, since $\det \varphi_1 \in \mathcal{S}$ and $\det \varphi_1 \not\equiv 0$, it follows that

$$0 \geq \int_{-\infty}^{\infty} \frac{\ln |\det \varphi_1(\mu)|}{1+\mu^2} d\mu > -\infty$$

and

$$|\det \varphi_1(\mu)| \leq \|\varphi_1(\mu)\| \leq 1.$$

Consequently,

$$0 \geq \int_{-\infty}^{\infty} \frac{\ln \|\varphi_1(\mu)\|}{1+\mu^2} d\mu > -\infty.$$

Thus, as

$$\|g(\mu)\| = \frac{\|\varphi_1(\mu)\|^2}{|\varphi_2(\mu)|^2},$$

it follows that

$$0 \geq \int_{-\infty}^{\infty} \frac{\ln \|g(\mu)\|}{1+\mu^2} d\mu + \int_{-\infty}^{\infty} \frac{\ln |\varphi_2(\mu)|^2}{1+\mu^2} d\mu > -\infty.$$

Therefore, $\ln \|g\| \in \widetilde{L}_1$, since the second integral is finite. Similar arguments imply that $\ln \|g^{-1}\| \in \widetilde{L}_1$.

Conversely, let $g(\mu)$ be a measurable $p \times p$ mvf on $\mathbb{R}$ such that $g(\mu) > 0$ a.e. on $\mathbb{R}$ and condition (3.110) is in force. Then the function

$$\varphi_2(\lambda) = \exp\left\{ \frac{1}{2\pi i} \int_{-\infty}^{\infty} \left(\frac{1}{\mu - \lambda} - \frac{1}{1+\mu^2} \right) \ln \|g\| d\mu \right\}$$

belongs to the class $\mathcal{N}_{out}$ and

$$|\varphi_2(\mu)|^2 = \|g(\mu)\| \quad \text{a.e. on } \mathbb{R}.$$

Let

$$g_1(\mu) = (1+\mu^2)^{-1} g(\mu)/|\varphi_2(\mu)|^2.$$

Then, clearly $g_1(\mu) > 0$ a.e. on $\mathbb{R}$ and $g_1 \in L_1^{p\times p}$. Moreover, as

$$\det\, g_1(\mu) = \frac{\det(g(\mu)/|\varphi_2(\mu)|^2)}{(1+\mu^2)^p} \leq \frac{\|g(\mu)\|^p}{(1+\mu^2)^p |\varphi_2(\mu)|^{2p}} = \frac{1}{(1+\mu^2)^p},$$

it is readily seen that

$$\int_{-\infty}^{\infty} \frac{\ln \det\, g_1(\mu)}{1+\mu^2} \leq 0.$$

Furthermore,

$$g_1(\mu)^{-1} = (1+\mu^2)|\varphi_2(\mu)|^2 g(\mu)^{-1}$$

and

$$\begin{aligned} \det g_1(\mu)^{-1} &= (1+\mu^2)^p|\varphi_2(\mu)|^{2p} \det g(\mu)^{-1} \\ &\leq (1+\mu^2)^p|\varphi_2(\mu)|^{2p}\|g(\mu)^{-1}\|^p. \end{aligned}$$

Consequently,

$$\int_{-\infty}^{\infty} \frac{\ln \det g_1(\mu)}{1+\mu^2} d\mu > -\infty,$$

since condition (3.110) holds for $\|g(\mu)^{-1}\|$. Thus,

$$g_1 \in L_1^{p\times p}, \; g_1(\mu) > 0 \quad \text{a.e. on} \quad \mathbb{R} \quad \text{and} \quad \ln\det g_1 \in \widetilde{L}_1.$$

By Theorem 3.78 there exists a mvf $f_1 \in H_2^{p\times p} \cap N_{out}^{p\times p}$ such that

$$f_1(\mu)^* f_1(\mu) = g_1(\mu) \text{ a.e. on } \mathbb{R}.$$

Let

$$\varphi_L(\lambda) = (\lambda + i)\varphi_2(\lambda) f_1(\lambda).$$

Then $\varphi_L \in N_{out}^{p\times p}$ and $f = \varphi_L$ is a solution of the factorization problem (3.108). The essential uniqueness of a solution $f = \varphi_L$ with $\varphi_L \in \mathcal{N}_{out}^{p\times p}$ follows from assertion (4) of Theorem 3.73. This completes the proof of assertions (1) and (2). □

Remark 3.80 *The dual factorization problem to (3.108) is to find a mvf* $f \in \mathcal{N}^{p\times q}$ *such that*

$$f(\mu)f(\mu)^* = g(\mu) \quad \text{a.e. on } \mathbb{R}. \tag{3.111}$$

The equivalences

$$f(\mu)f(\mu)^* = g(\mu) \Longleftrightarrow f^\sim(\mu)^* f^\sim(\mu) = g(-\mu) \quad \text{a.e. on } \mathbb{R}$$

and

$$f \in \mathcal{N}^{p\times q} \Longleftrightarrow f^\sim \in \mathcal{N}^{q\times p}$$

yield dual versions of Theorems 3.78 and 3.79.

3.17 Blaschke-Potapov products in $\mathcal{S}^{p\times p}$

Analogues of Blaschke factors for mvf's $f \in \mathcal{S}^{p\times p}$ with $\det f \not\equiv 0$ were systematically studied by V. P. Potapov. These analogues will be called BP (Blaschke-Potapov) factors and their products will be called BP products. An **elementary BP factor** can be expressed as

$$B_\alpha(\lambda) = I_p - P + b_\alpha(\lambda)P = I_p + (b_\alpha(\lambda) - 1)P \tag{3.112}$$

where $P = P^2 = P^*$ is a $p \times p$ matrix orthoprojector and

$$b_\alpha(\lambda) = \gamma\frac{\lambda - \alpha}{\lambda - \overline{\alpha}} \quad \text{with} \quad \alpha \in \mathbb{C}_+ \quad \text{and} \quad |\gamma| = 1.$$

An elementary BP factor is said to be a **primary BP factor** if $\operatorname{rank} P = 1$ in formula (3.112). If $\operatorname{rank} P = k$ and $v_1, \ldots, v_k$ is a basis for the range of P and

$$V = [v_1 \cdots v_k]$$

is the $p \times k$ matrix with columns $v_1, \ldots, v_k$, then

$$P = V(V^*V)^{-1}V^*.$$

Thus,

$$B_\alpha(\lambda) = I_p + (b_\alpha(\lambda) - 1)V(V^*V)^{-1}V^*.$$

If the basis is chosen to be orthonormal, then V is an isometry, i.e., $V^*V = I_k$, and then $P = VV^*$ and

$$I_p + (b_\alpha(\lambda) - 1)P = \prod_{j=1}^{k}\{I_p + (b_\alpha(\lambda) - 1)P_j\},$$

is the product of k primary factors with orthogonal projections

$$P_j = v_jv_j^*, \quad j = 1, \ldots, k,$$

of rank one such that

$$P_iP_j = P_jP_i = 0 \quad \text{for } i \neq j$$

and, as follows from the formula

$$[v_1 \ \cdots \ v_k] \begin{bmatrix} v_1^* \\ \vdots \\ v_k^* \end{bmatrix} = \sum_{j=1}^{k} v_jv_j^*,$$

$$P_1 + \cdots + P_k = P.$$

It is readily checked that

$$B_\alpha(\lambda) = I_p + (b_a(\lambda) - 1)P \in \mathcal{S}_{in}^{p\times p}$$

and that a finite product of such elementary BP factors is a rational mvf in $\mathcal{S}_{in}^{m\times m}$. The converse is also true:

Theorem 3.81 *Every rational $p \times p$ mvf that belongs to $\mathcal{S}_{in}^{p\times p}$ can be expressed as a finite product of elementary BP factors multiplied by a constant unitary $p \times p$ matrix.*

Proof This is a special case of Theorem 4.7. □

It may also be shown that a rational mvf in $\mathcal{S}_{in}^{p\times p}$ has McMillan degree m if and only if it can be expressed as the product of m primary factors times a constant unitary factor on either the left or the right. (The definition and basic properties of McMillan degree may be found in [Kal63a].)

The infinite BP products

$$\overset{\curvearrowright}{\prod_{j=1}^{\infty}} B_j(\lambda) = B_1(\lambda)B_2(\lambda)\cdots B_n(\lambda)\cdots$$

and

$$\overset{\curvearrowleft}{\prod_{j=1}^{\infty}} B_j(\lambda) = \cdots B_n(\lambda)\cdots B_2(\lambda)B_1(\lambda)$$

are said to be convergent if the sequence of corresponding finite partial products converges to a $p \times p$ mvf $B(\lambda)$ such that $\det B(\lambda) \not\equiv 0$ in $\mathbb{C}_+$.

Theorem 3.82 (V. P. Potapov) *Let*

$$B_j(\lambda) = I_p + (b_{\alpha_j}(\lambda) - 1)P_j, \quad j = 1, 2, \ldots,$$

be a sequence of elementary BP factors that are normalized by the condition

$$b_{\alpha_j}(\omega) > 0$$

at some point $\omega \in \mathbb{C}_+$, $\omega \neq \alpha_j$ for $j = 1, 2, \ldots$. Then the following are equivalent for $\lambda \in \mathbb{C}_+$:

(1) *The infinite BP product* $\overset{\curvearrowright}{\prod_{j=1}^{\infty}} B_j(\lambda)$ *converges.*

(2) *The infinite BP product* $\overset{\curvearrowleft}{\prod_{j=1}^{\infty}} B_j(\lambda)$ *converges.*

(3) $\prod_{j=1}^{\infty} b_{\alpha_j}(\lambda)$ *converges.*

(4) $\sum_{j=1}^{\infty}(1 - |b_{\alpha_j}(\lambda)|) < \infty$.

(5) *The given sequence of points* $\alpha_1, \alpha_2, \ldots$ *satisfies the Blaschke condition given in (3.26).*

Moreover, if (4) holds for at least one point $\lambda \in \mathbb{C}_+$*, then it holds for every point* $\lambda \in \mathbb{C}_+$ *and the products in (1)–(3) converge uniformly in every compact subset* Ω *of* $\mathbb{C}$ *that does not intersect the closure of the points* $\alpha_1, \alpha_2, \ldots$.

Theorem 3.83 (V. P. Potapov) *Let*

$$B(\lambda) = \left(\overset{\curvearrowright}{\prod_{k=1}^{n}} B_k^r(\lambda) \right) U_r \quad (1 \leq n \leq \infty), \tag{3.113}$$

where U_r *is a* $p \times p$ *unitary matrix and*

$$B_k^r(\lambda) = I_p + (b_{\omega_k}(\lambda) - 1)P_k^r, \quad n_k = \text{rank } P_k^r$$

is an elementary BP factor. Then

$$\det B_k^r(\lambda) = b_{\omega_k}(\lambda)^{n_k}, \quad \det B(\lambda) = (\det U_r) \prod_{k=1}^{n} b_{\omega_k}(\lambda)^{n_k}$$

and $B(\lambda)$ *may be factored as a left BP product*

$$B(\lambda) = U_\ell \overset{\curvearrowleft}{\prod_{k=1}^{n}} B_k^\ell(\lambda), \tag{3.114}$$

where U_ℓ *is a* $p \times p$*-unitary matrix and*

$$B_k^\ell(\lambda) = I_p + (b_{\omega_k}(\lambda) - 1)P_k^\ell, \quad \text{rank } P_k^\ell = n_k.$$

Conversely, any left BP product (3.114) may be factored as a right BP product (3.113).

Proof See [Po60] and [Zol03]. □

Corollary 3.84 *If $B(\lambda)$ is a BP product, then* $\det B(\lambda)$ *is a Blaschke product.*

Theorem 3.85 *Let $f \in \mathcal{S}^{p\times p}$ and assume that* $\det f(\lambda) \not\equiv 0$ *in $\mathbb{C}_+$ and that the set E of the zeros of* $\det f(\lambda)$ *in $\mathbb{C}_+$ is nonempty. Then $E = \{\omega_j\}_{j=1}^n$, $n \le \infty$, is at most countable and is subject to the constraint (3.26), counting multiplicities. Moreover, $f(\lambda)$ has both left and right BP factorizations*

$$f(\lambda) = B_\ell(\lambda) g_\ell(\lambda) \quad \textit{and} \quad f(\lambda) = g_r(\lambda) B_r(\lambda),$$

where B_ℓ (resp., B_r) is a left (resp., right) BP product of size $p \times p$; $g_\ell, g_r \in \mathcal{S}^{p\times p}$ and $g_\ell(\lambda)$ and $g_r(\lambda)$ are invertible $p\times p$ matrices for every point $\lambda \in \mathbb{C}_+$.

Theorem 3.86 *Let $f \in \mathcal{S}^{p\times p}$, let $g(\lambda) =$* $\det f(\lambda)$ *and suppose that $g(\lambda) \not\equiv 0$ in $\mathbb{C}_+$. Then the following four conditions are equivalent:*

(1) *$f(\lambda)$ is a left BP product.*

(2) *$f(\lambda)$ is a right BP product.*

(3) *$g(\lambda)$ is a Blaschke product.*

(4) *$\tau_g^+ = 0$ and*

$$\lim_{\nu \downarrow 0} \int_{-\infty}^{\infty} \frac{\ln |g(\mu + i\nu)|}{\mu^2 + 1} d\mu = 0. \tag{3.115}$$

Proof The equivalence of (1), (2) and (3) follows from Theorem 3.85 and Corollary 3.84. The equivalence of (3) and (4) follows from Lemma 3.25. □

Corollary 3.87 *If $f = f_1 f_2$, $f_1, f_2 \in \mathcal{S}^{p\times p}$ and f is a BP product, then both f_1 and f_2 are BP products.*

Proof This follows from the equivalence between statements (1) and (3) in Theorem 3.86. □

Theorem 3.88 *Let $B \in \mathcal{S}_{in}^{p\times p}$ and let v be a minimal scalar denominator of $B^\#$. Then B is a BP product if and only if v is a Blaschke product.*

Proof The function $v^{-1}\det B \in \mathcal{S}_{in}$, since $\det B$ is a scalar denominator of $B^{\#}$ and v is a minimal scalar denominator of $B^{\#}$. Therefore, if B is a BP product, then $\det B$ is a Blaschke product by Corollary 3.84 and hence, v is a Blaschke product by Corollary 3.87. Conversely, if v is a Blaschke product, then vI_m is a BP product, and another application of the preceding corollary to the identity

$$(vB^{\#})B = vI_m$$

guarantees that B is a BP product, since $(vB^{\#}) \in \mathcal{S}_{in}^{p\times p}$. $\square$

Theorem 3.89 *Let* $s_j \in \mathcal{S}_{in}^{p\times p}$, $j = 1, 2, \ldots$, *be such that either* $s_j^{-1}s_{j+1} \in \mathcal{S}_{in}^{p\times p}$ *for* $j = 1, 2, \ldots$ *or* $s_{j+1}s_j^{-1} \in \mathcal{S}_{in}^{p\times p}$ *for* $j = 1, 2, \ldots$ *and assume that* $s(\lambda) = \lim_{j\uparrow\infty} s_j(\lambda)$ *exists at each point* $\lambda \in \mathbb{C}_+$ *and* $\det s(\lambda) \not\equiv 0$. *Then* $s \in \mathcal{S}_{in}^{m\times m}$.

Proof Suppose for the sake of definiteness that $s_j^{-1}s_{j+1} \in S_{in}^{p\times p}$ for $j = 1, 2, \ldots$. Then the inequalities

$$s(\omega)s(\omega)^* \le s_{j+1}(\omega)s_{j+1}(\omega)^* \le s_j(\omega)s_j(\omega)^*,$$

clearly imply that $\det s_j(\omega) \neq 0$ whenever $\omega \in \mathbb{C}_+$ and $\det s(\omega) \neq 0$. Moreover, if $m > n$, then the evaluations

$$\int_{-\infty}^{\infty} (s_m(\mu) - s_n(\mu))^*(s_m(\mu) - s_n(\mu))\frac{d\mu}{|\rho_\omega(\mu)|^2}$$

$$= \int_{-\infty}^{\infty} \{2I_p - s_n(\mu)^*s_m(\mu) - s_m(\mu)^*s_n(\mu)\}\frac{d\mu}{|\rho_\omega(\mu)|^2}$$

$$= 2\Re\int_{-\infty}^{\infty} \{I_p - s_n(\mu)^{-1}s_m(\mu)\}\frac{d\mu}{|\rho_\omega(\mu)|^2}$$

$$= \Re\left\{\frac{I_p - s_n(\omega)^{-1}s_m(\omega)}{2\pi\Im\omega}\right\}$$

imply that $\rho_\omega^{-1}s_n$ tends to a limit $f \in H_2^{p\times p}$ and $\rho_\omega(\lambda)^{-1}s_n(\lambda) \to f(\lambda)$ at every point $\lambda \in \mathbb{C}_+$ as $n \uparrow \infty$. But this in turn implies that $f(\lambda) = \rho_\omega(\lambda)^{-1}s(\lambda)$ and hence that

$$\int_{-\infty}^{\infty} \frac{(I_p - s(\mu)^*s(\mu))}{|\rho_\omega(\mu)|^2}d\mu = \lim_{n\uparrow\infty}\int_{-\infty}^{\infty} \frac{I_p - s_n(\mu)^*s_n(\mu)}{|\rho_\omega(\mu)|^2}d\mu \;=\; 0.$$

Therefore, since

$$I_p - s(\mu)^* s(\mu) \geq 0 \quad \text{for almost all points } \mu \in \mathbb{R},$$

it follows that $s(\mu)^* s(\mu) = I_p$ for almost all $\mu \in \mathbb{R}$, i.e., $s \in \mathcal{S}_{in}^{p\times p}$. □

Corollary 3.90 *If $B \in \mathcal{S}^{p\times p}$ is a BP product, then $B \in \mathcal{S}_{in}^{p\times p}$.*

Proof This is immediate from Theorem 3.89. □

A mvf $g \in \mathcal{S}^{p\times p}$ that is invertible at every point $\lambda \in \mathbb{C}_+$ admits both a multiplicative left integral representation and a multiplicative right integral representation. This representation is a generalization of the multiplicative representation of a scalar function $g \in \mathcal{S}$ with $g(\lambda) \neq 0$ in $\lambda \in \mathbb{C}_+$ that follows from the representation

$$g(\lambda) = \exp\{-c(\lambda)\}$$

with $c \in \mathcal{C}$, and the integral representation (3.3) of c; see Potapov [Po60] and Ginzburg [Gi67]. In the next section, we shall present this representation in the special case that $g(\lambda)$ is an entire inner mvf.

3.18 Entire matrix functions of class $\mathcal{N}^{p\times q}$

A $p \times q$ mvf $f(\lambda) = [f_{jk}(\lambda)]$ is **entire** if each of its entries $f_{jk}(\lambda)$ is an entire function. The class of entire $p \times q$ mvf's $f(\lambda)$ will be denoted $\mathcal{E}^{p\times q}$. If f also belongs to some other class $\mathcal{X}^{p\times q}$, then we shall simply write $f \in \mathcal{E} \cap \mathcal{X}^{p\times q}$.

An entire $p\times q$ mvf is said to be of **exponential type** if there is a constant $\tau \geq 0$ such that

$$\|f(\lambda)\| \leq \gamma \exp\{\tau|\lambda|\}, \quad \text{for all points } \lambda \in \mathbb{C} \tag{3.116}$$

for some $\gamma > 0$. In this case, the exact type $\tau(f)$ of f is the infimum of all such τ. Equivalently, an entire $p\times q$ mvf f, $f \not\equiv 0$, is said to be of exponential type $\tau(f)$, if

$$\tau(f) = \limsup_{r\to\infty} \frac{\ln \| M(r)\|}{r} < \infty, \tag{3.117}$$

where

$$M(r) = \max\{\|f(\lambda)\| : |\lambda| = r\}.$$

The inequalities in (2.94) applied to the matrix $f(\lambda)$ yield the auxiliary formula

$$\tau(f) = \limsup_{r\to\infty} \frac{\ln\ \max\{\text{trace}(f(\lambda)^* f(\lambda)) : |\lambda| = r\}}{2r}. \tag{3.118}$$

Moreover, a mvf $f \in \mathcal{E}^{p\times q}$ is of exponential type if and only if all the entries $f_{ij}(\lambda)$ of the mvf f are entire functions of exponential type. Furthermore, if $f \in \mathcal{E}^{p\times q}$ is of exponential type, then

$$\tau(f) = \max\{\tau(f_{ij}) : f_{ij} \not\equiv 0, 1 \le i \le p, 1 \le j \le q\}. \tag{3.119}$$

To verify formula (3.119), observe first that the inequality

$$|f_{ij}(\lambda)|^2 \le \text{trace}\{f(\lambda)^* f(\lambda)\}$$

implies that $\tau(f_{ij}) \le \tau(f)$, if $f_{ij} \not\equiv 0$. On the other hand, if

$$\tau = \max\{\tau(f_{ij}) :\ f_{ij} \not\equiv 0, 1 \le i \le p, 1 \le j \le q\},$$

then there exists a number $\gamma > 0$ such that

$$\text{trace}\, f(\lambda)^* f(\lambda) = \sum |f_{ij}(\lambda)|^2 \le \gamma \exp\{2(\tau + \varepsilon)|\lambda|\}$$

for every $\varepsilon > 0$. Consequently, $\tau(f) \le \tau$. Also, it is clear that if

$$M_\pm(r) = \max\{|f(\lambda)| : |\lambda| \le r \quad \text{and} \quad \lambda \in \overline{\mathbb{C}_\pm}\},$$

then

$$\tau(f) = \max\{\tau_+(f),\ \tau_-(f)\}, \tag{3.120}$$

where

$$\tau_\pm(f) = \limsup_{r\to\infty} \frac{\ln M_\pm(r)}{r};$$

and that

$$\tau_\pm(f) = \max\{\tau_\pm(f_{ij}) :\ f_{ij} \not\equiv 0,\ 1 \le i \le p,\ 1 \le j \le q\}.$$

Analogues of formula (3.118) are valid for $\tau_\pm(f)$. Moreover, if a mvf $f \in \mathcal{E}^{p\times q}$ has exponential type $\tau(f)$, then the types $\tau_+(f)$ and $\tau_-(f)$ of the mvf f in the closed upper and lower half planes $\overline{\mathbb{C}_+}$ and $\overline{\mathbb{C}_-}$ are not less than the exponential types τ_f^+ and τ_f^- of the mvf f on the upper and lower imaginary half axis respectively, i.e.,

$$\tau_f^\pm \stackrel{\text{def}}{=} \limsup_{\nu\uparrow\infty} \frac{\ln \|\ f(\pm i\nu)\ \|}{\nu} \le \tau_\pm(f). \tag{3.121}$$

Theorem 3.91 (M. G. Krein) *Let $f \in \mathcal{E}^{p\times q}$. Then $f \in \mathcal{N}^{p\times q}$ if and only $\tau_+(f) < \infty$ and f satisfies the Cartwright condition*

$$\int_{-\infty}^{\infty} \frac{\ln^+ \| f(\mu) \|}{1+\mu^2} d\mu < \infty. \tag{3.122}$$

Moreover, if $f \in \mathcal{E} \cap \mathcal{N}^{p\times q}$, then $\tau_+(f) = \tau_f^+$.

Proof See [Kr47a], [Kr51] and Section 6.11 of [RR85]. □

Lemma 3.92 *If $f \in \mathcal{E} \cap \mathcal{N}^{p\times q}$ and $f(\lambda) \not\equiv 0$, then*

$$\tau_f^{\pm} = \limsup_{\nu\uparrow\infty} \frac{\ln \operatorname{trace}\{f(\pm i\nu)^* f(\pm i\nu)\}}{2\nu}. \tag{3.123}$$

Proof This is immediate from Lemma 2.68. □

Theorem 3.93 *If $f(\lambda) = [f_{jk}(\lambda)]$ is a $p \times q$ mvf of class $\mathcal{N}^{p\times q}$ such that f is holomorphic in $\mathbb{C}_+$ and $f(\lambda) \not\equiv 0$, then*

$$\tau_f^+ = \max\{\tau_{f_{jk}}^+ \ : \ j = 1, \ldots, p, \ k = 1, \ldots, q \quad \textit{and} \quad f_{jk}(\lambda) \not\equiv 0\}.$$

Proof The proof is the same as that of (3.119). □

Let

$$\delta_f^{\pm} = \limsup_{\nu\uparrow\infty} \frac{\ln|\det f(\pm i\nu)|}{\nu} \tag{3.124}$$

for $f \in \mathcal{N}^{p\times p}$ with $\det f(\lambda) \not\equiv 0$.

Theorem 3.94 *Let $f \in \mathcal{E}^{p\times p}$ be invertible in $\mathbb{C}_+$ and let $f^{-1} \in \mathcal{N}_+^{p\times p}$. Then:*

(1) $f \in \mathcal{E} \cap \mathcal{N}^{p\times p}$.

(2) $0 \le \tau_f^+ < \infty$.

(3) $\delta_f^+ = \limsup_{\nu\uparrow\infty} \frac{\ln|\det f(i\nu)|}{\nu} = \lim_{\nu\uparrow\infty} \frac{\ln|\det f(i\nu)|}{\nu}$ *exists as a limit and* $\tau_f^+ \le \delta_f^+ \le p\tau_f^+$.

(4) $e^{-i\delta_f^+\lambda} \det f^{-1}(\lambda) \in \mathcal{N}_{out}$.

(5) $\delta_f^+ = p\tau_f^+$ *if and only if* $e^{i\tau_f^+\lambda} f(\lambda) \in \mathcal{N}_{out}^{p\times p}$.

Proof (1) is self-evident (since $f = (f^{-1})^{-1}$). Next, since $f^{-1} = h^{-1}g$ with $h \in \mathcal{S}_{out}$ and $g \in \mathcal{S}^{p\times p}$, formula (3.38) with $\varphi = h$ implies that $\tau_h^+ = 0$. Consequently,

$$\tau_f^+ = \tau_{g^{-1}}^+ \quad \text{and} \quad \delta_f^+ = \delta_{g^{-1}}^+$$

and hence (2) follows from the inequalities

$$1 = \|g(i\nu)g(i\nu)^{-1}\| \leq \|g(i\nu)\|\|g(i\nu)^{-1}\| \leq \|g(i\nu)^{-1}\|,$$

and the inequalities in (3) follow by applying (2.96) to $g(\lambda)^{-1}$. To obtain the rest of (3), let $g = b\varphi$ in $\mathbb{C}_+$ with $b \in \mathcal{S}_{in}^{p\times p}$ and $\varphi \in \mathcal{S}_{out}^{p\times p}$. Since

$$h^p(\lambda) = \det f(\lambda) \times \det b(\lambda) \times \det \varphi(\lambda)$$

for $\lambda \in \mathbb{C}_+$, it follows that $\det b(\lambda)$ has no zeros in $\mathbb{C}_+$. Moreover, since f is entire and g is holomorphic in $\mathbb{C}_+$, $\det b(\lambda)$ is a singular inner function of the form

$$\det b(\lambda) = \gamma e^{i\alpha\lambda}$$

for some $\gamma \in \mathbb{C}$ with $|\gamma| = 1$ and some $\alpha \geq 0$. Thus,

$$\delta_f^+ = \lim_{\nu\uparrow\infty} \frac{\ln|\det f(i\nu)|}{\nu} = \alpha$$

exists. The inequalities in (3) are immediate from Lemma 2.68, and (4) is immediate from the preceding identifications.

Next, in view of (1) and Lemma 3.92,

$$\exp\{i\tau_{f_{jk}}^+\lambda\}f_{jk}(\lambda) \in \mathcal{N}_+$$

for every entry $f_{jk}(\lambda)$ in $f(\lambda)$ which is not identically equal to zero. Therefore, since $\tau_f^+ \geq \tau_{f_{jk}}^+$,

$$e^{i\tau_f^+\lambda}f(\lambda) = e^{i\tau_f^+\lambda}\varphi(\lambda)^{-1}b(\lambda)^{-1}$$

belongs to $\mathcal{N}_+^{p\times p}$ and hence so does

$$h(\lambda) := e^{i\tau_f^+\lambda}b(\lambda)^{-1}.$$

Thus, by the Smirnov maximum principle (see Theorem 3.59), $h \in \mathcal{S}_{in}^{p\times p}$. Moreover,

$$\det h(\lambda) = e^{ip\tau_f^+\lambda}\overline{\gamma}e^{-i\delta_f^+\lambda} = \overline{\gamma}\ \exp\{i(p\tau_f^+ - \delta_f^+)\lambda\},$$

which is consistent with the already established inequality $\delta_f^+ \leq p\tau_f^+$. Thus, if $\delta_f^+ = p\tau_f^+$, then

$$\det h(\lambda) = \overline{\gamma}$$

and hence, by Lemma 3.33, $h(\lambda) \equiv$ constant, i.e.,

$$b(\lambda) = e^{i\tau_f^+ \lambda} I_p$$

(up to a constant $p \times p$ unitary multiplier). Therefore

$$e^{i\tau_f^+ \lambda} f(\lambda) = \varphi(\lambda)^{-1}$$

belongs to $\mathcal{N}_{out}^{p\times p}$, which proves the hard half of (5). The converse statement drops out easily from the observation that if $e_{\tau_f^+} f \in \mathcal{N}_{out}^{p\times p}$, then

$$\limsup_{\nu\uparrow\infty} \frac{\ln\ \det\{e^{-\tau_f^+ \nu}|f(i\nu)|\}}{\nu} = 0. \qquad \square$$

Corollary 3.95 *Let $f \in \mathcal{E}^{p\times p}$ be invertible in $\mathbb{C}_+$ with $f^{-1} \in H_\infty^{p\times p}$. Then the following are equivalent:*

(1) $\lim_{\nu\uparrow\infty} \frac{\ln \|f(i\nu)\|}{\nu} = 0.$

(2) $\lim_{\nu\uparrow\infty} \frac{\ln[\text{trace}\{f(i\nu)^* f(i\nu)\}]}{\nu} = 0.$

(3) $\lim_{\nu\uparrow\infty} \frac{\ln|\det f(i\nu)|}{\nu} = 0.$

(4) $f \in \mathcal{N}_{out}^{p\times p}$.

Proof This formulation takes advantage of the fact that for a sequence of nonnegative numbers $\{x_k\}$,

$$\limsup_{k\to\infty} x_k = 0 \iff \lim_{k\to\infty} x_k = 0.$$

The rest is immediate from the preceding two theorems and the inequalities in Lemma 2.68. $\square$

3.19 Entire inner mvf's

A scalar entire inner function $f(\lambda)$ is automatically of the form $f(\lambda) = f(0)e^{i\lambda d}$, where $|f(0)| = 1$ and $d \geq 0$. The set of entire inner $p \times p$ mvf's is much richer. It includes mvf's of the form

$$f(\lambda) = f(0)e^{i\lambda D},$$

where $f(0)^*f(0) = I_p$ and $D \geq 0$ as well as products of mvf's of this form. More general constructions of entire inner $p \times p$ mvf's may be based on the **right multiplicative integral**

$$\overset{b}{\curvearrowright}\!\!\!\!\!\!\int_a \exp\{\varphi(t)dM(t)\}$$

of a continuous scalar valued function $\varphi(t)$ on the closed interval $[a, b]$ with respect to a nondecreasing $p \times p$ mvf $M(t)$ on $[a, b]$. The integral is defined as the limit of products of the form

$$\exp\{\varphi(\tau_1)[M(t_1) - M(t_0)]\} \exp\{\varphi(\tau_2)[M(t_2) - M(t_1)] \\ \cdots \exp\{\varphi(t_n)[M(t_n) - M(t_{n-1})]\},$$

where $a = t_0 < t_1 < \cdots < t_n = b$, is an arbitrary partition of $[a, b]$, $\tau_j \in [t_{j-1}, t_j]$ for $j = 1, \ldots, n$, and the limit is taken over finer and finer partitions as $n \uparrow \infty$ and

$$\max\{t_j - t_{j-1}, \quad j = 1, \ldots n\} \longrightarrow 0.$$

Under the given assumptions on $\varphi(t)$ and $M(t)$ this limit exists and is called the (right) multiplicative integral of $\varphi(t)$ with respect to $M(t)$. Multiplicative integrals may also be defined under less restrictive assumptions on $\varphi(t)$ and $M(t)$ than were imposed here; see e.g., the appendix of [Po60]. A **left multiplicative integral**

$$\overset{b}{\curvearrowleft}\!\!\!\!\!\!\int_a \exp\{\varphi(t)dM(t)\}$$

may be defined in much the same way as the limit of products of the form

$$\exp\{\varphi(\tau_n)[M(t_n) - M(t_{n-1})]\} \exp\{\varphi(\tau_{n-1})[M(t_{n-1}) - M(t_{n-2})] \\ \cdots \exp\{\varphi(t_1)[M(t_1) - M(t_0)]\}.$$

If $H \in L_1^{m\times m}([a,b])$, then by definition

$$\int_a^{t} \curvearrowright \exp\{\varphi(t)H(t)dt\} = \int_a^{t} \curvearrowright \exp\{\varphi(t)dMt\},$$

$$\text{where } M(t) = M(a) + \int_a^t H(s)ds.$$

Theorem 3.96 (V. P. Potapov) *Let $f \in \mathcal{E} \cap \mathcal{S}_{in}^{p\times p}$. Then*

$$f(\lambda) = U_\ell \int_0^{d} \curvearrowright \exp\{i\lambda dM(t)\}, \tag{3.125}$$

where $U_\ell = f(0)$ is a $p \times p$ unitary matrix and $M(t)$ is a nondecreasing $p \times p$ mvf on the interval $[0, d]$ that may be chosen so that it is absolutely continuous on $[0, d]$.
Conversely, if U_ℓ is a constant $p \times p$ unitary matrix and $M(t)$ is a nondecreasing $p \times p$ mvf on the closed interval $[0, d]$, then the mvf $f(\lambda)$ is well defined by (3.125) and $f \in \mathcal{E} \cap \mathcal{S}_{in}^{p\times p}$.

Proof See [Po60] and [Zol03]. □

Lemma 3.97 *If $b \in \mathcal{S}_{in}^{p\times p}$ and if $\det b(\lambda) = e^{i\delta\lambda}$ for some $\delta \geq 0$, then:*

(1) *$R_\omega b\xi \in H_2^p \ominus e_\delta H_2^p$ for every choice of $\omega \in \mathbb{C}_+$ and $\xi \in \mathbb{C}^p$.*

(2) *$b(\lambda)$ is an entire mvf of exponential type.*

Conversely, if a mvf $b \in \mathcal{S}_{in}^{p\times p}$ is entire, then $\det b(\lambda) = e^{i\lambda\delta} \times \det b(0)$ for some $\delta \geq 0$.

Proof Under the given assumptions,

$$b^{-1}(\lambda) = e^{-i\delta\lambda} h(\lambda)$$

for some choice of $h \in H_\infty^{p\times p}$. Therefore,

$$e^{-i\delta\mu} b(\mu) = h(\mu)^*$$

for almost all points $\mu \in \mathbb{R}$, which in turn implies that

$$\frac{b(\lambda)}{\lambda - \omega}\xi \in e^{i\delta\lambda} K_2^p$$

for $\omega \in \mathbb{C}_+$ and $\xi \in \mathbb{C}^p$. But this leads easily to (1), since $R_\omega b\xi \in H_2^p$; (2) is immediate from (1) and Theorem 3.36.

The converse follows easily from the fact that if a mvf $b \in \mathcal{S}_{in}^{p\times p}$ is entire, then $\det b(\lambda)$ is a scalar inner entire function, and hence it must be of the indicated form. □

Similar considerations lead easily to the following supplementary result.

Lemma 3.98 *If $b \in \mathcal{E} \cap \mathcal{S}_{in}^{p\times p}$ and $b = b_1 b_2$, where $b_1, b_2 \in \mathcal{S}_{in}^{p\times p}$, then $b_1, b_2 \in \mathcal{E}^{p\times p}$.*

Proof By (the converse statement in) Lemma 3.97, $\det b(\lambda) = \gamma e^{i\lambda\delta}$, where $|\gamma| = 1$ and $\delta \geq 0$. Therefore, $\det b_1(\lambda)$ is of the same form and hence $b_1(\lambda)$ is entire by (2) of the same lemma. □

The preceding analysis also leads easily to the following set of conclusions, which will be useful in the sequel.

Theorem 3.99 *Let $b \in \mathcal{E} \cap \mathcal{S}_{in}^{p\times p}$ and let $f = b^{-1}$. Then:*

(1) $\det b(\lambda) = e^{i\lambda\alpha} \times \det b(0)$ *for some* $\alpha \geq 0$.

(2) *The limits (3.121) and (3.124) satisfy the inequalities*

$$0 \leq \tau_f^+ \leq \delta_f^+ \leq p\tau_f^+ < \infty. \tag{3.126}$$

(3) $\delta_f^+ \;=\; -\lim_{\nu\uparrow\infty} \dfrac{\ln|\det b(i\nu)|}{\nu} \;=\; \alpha.$

(4) $p\tau_f^+ \geq \alpha$, *with equality if and only if*

$$b(\lambda) = e^{i\lambda\alpha/p} b(0).$$

(5) $b(\lambda)$ *is an entire mvf of exponential type* τ_f^+.

Proof Item (1) is established in Lemma 3.97. Moreover, in view of (1), $f(\lambda) = b^{-1}(\lambda)$ is entire and satisfies the hypotheses of Theorem 3.94. The latter serves to establish (2); (3) is immediate from (1) and then (4) is immediate from (5) of Theorem 3.94.

Finally, the proof of (5) rests mainly on the observation that

$$\tau_f^+ = \limsup_{\nu\uparrow\infty} \frac{\ln\|f(i\nu)\|}{\nu} = \limsup_{\nu\uparrow\infty} \frac{\ln\|b(-i\nu)\|}{\nu}$$

bounds the growth of $b(\lambda)$ on the negative imaginary axis. Theorem 3.91 applied to entire mvf's that are of the Nevanlinna class in the lower half plane

$\mathbb{C}_-$ implies that $\tau_-(b) = \tau_f^+$. Therefore, $\tau(b) = \max\{\tau_-(b), \tau_+(b)\} = \tau_-(b)$, since $\|b(\lambda)\| \le 1$ in $\mathbb{C}_+$. □

In subsequent developments, the structure of entire inner mvf's will be of central importance. The following three examples illustrate some of the possibilities:

Example 3.100 *If* $b(\lambda) = e^{i\lambda a} I_p$ *with* $a \ge 0$, *then*

$$p\tau_{b^{-1}} = \delta_{b^{-1}} = pa.$$

Example 3.101 *If* $b(\lambda) = e^{i\lambda a} \oplus I_{p-1} = \operatorname{diag}\{e^{i\lambda a}, 1, \ldots, 1\}$ *with* $a \ge 0$, *then*

$$\tau_{b^{-1}} = \delta_{b^{-1}} = a.$$

Example 3.102 *If* $b(\lambda) = \operatorname{diag}\{e^{i\lambda a_1}, e^{i\lambda a_2}, \ldots, e^{i\lambda a_p}\}$ *with* $a_1 \ge a_2 \ge \cdots \ge a_p \ge 0$, *then*

$$\tau_{b^{-1}} = a_1 \quad \text{and} \quad \delta_{b^{-1}} = a_1 + \cdots + a_p.$$

For future use, we record the following observation which is relevant to Example 3.101.

Lemma 3.103 *If* $b \in \mathcal{E} \cap \mathcal{S}_{in}^{p\times p}$ *admits a factorization* $b(\lambda) = b_1(\lambda) b_2(\lambda)$ *with* $b_i \in \mathcal{S}_{in}^{p\times p}$ *for* $i = 1, 2$ *and if* $\delta_{b^{-1}} = \tau_{b^{-1}}$, *then:*

(1) $b_i \in \mathcal{E} \cap \mathcal{S}_{in}^{p\times p}$ *and*

(2) $\delta_{b_i^{-1}} = \tau_{b_i^{-1}}$

for $i = 1, 2$.

Proof Since (1) is available from Lemma 3.98, we can invoke Theorem 3.99 to help obtain the chain of inequalities

$$\delta_{b^{-1}} = \delta_{b_1^{-1}} + \delta_{b_2^{-1}} \ge \tau_{b_1^{-1}} + \tau_{b_2^{-1}} \ge \tau_{b^{-1}}.$$

But now as the upper and lower bounds are presumed to be equal, equality must prevail throughout. This leads easily to (2), since $\delta_{b_i^{-1}} \ge \tau_{b_i^{-1}}$. □

3.20 Minimal denominators of mvf's $f \in \mathcal{E} \cap \mathcal{N}^{p\times q}$

Lemma 3.104 *Let* $f \in \mathcal{E} \cap \mathcal{N}^{p\times q}$. *Then:*

(1) $f \in \mathcal{N}_+^{p\times q} \iff \tau_f^+ \le 0$.

(2) *If* $\tau_f^+ > 0$, *then* $e_{\tau_f^+}$ *is a minimal scalar denominator of* f.

(3) *A minimal scalar denominator of* f *is an entire function.*

Proof Since $f \in \mathcal{N}_+^{p\times q} \iff f_{jk} \in \mathcal{N}_+$ and, by Theorem 3.93, $\tau_f^+ \leq 0 \iff \tau_{f_{jk}^+} \leq 0$ for every entry f_{jk} in the mvf f such that $f_{jk} \not\equiv 0$, it is enough to show that

$$f \in \mathcal{N}_+ \iff \tau_f^+ \leq 0 \quad \text{for scalar functions} \quad f \in \mathcal{E} \cap \mathcal{N},$$

in order to verify (1). But this follows from Lemmas 3.23 and 3.29.

Next, to verify (2), let $\tau_f^+ > 0$. Then $\tau_+(f_{jk}) \leq \tau_f^+$ by Theorem 3.93, and hence, $e_{\tau_f^+} f_{jk} \in \mathcal{N}_+$, by Lemma 3.23. Thus, $e_{\tau_f^+} f \in \mathcal{N}_+^{p\times q}$, i.e., $e_{\tau_f^+}$ is a scalar denominator of f. Thus, if β is a minimal scalar denominator of f, then $\beta \in \mathcal{S}_{in}$ and $\beta^{-1} e_{\tau_f^+} \in \mathcal{S}_{in}$. Therefore, Lemma 3.98 guarantees that $\beta \in \mathcal{E} \cap \mathcal{S}_{in}$ and hence that $\beta = \gamma e_\alpha$ with $|\gamma| = 1$ and $0 \leq \alpha \leq \tau_f^+$. On the other hand,

$$-\alpha + \tau_f^+ = \tau_{e_\alpha f}^+ \leq 0,$$

since $e_\alpha f \in \mathcal{N}_+^{p\times q}$. Thus, $\alpha = \tau_f^+$ and (2) holds.

Finally, if $f \in \mathcal{E} \cap \mathcal{N}_+^{p\times q}$, then $\beta = 1$ is a minmimal scalar denominator of f. If $f \in \mathcal{E} \cap \mathcal{N}^{p\times q}$, but $f \notin \mathcal{N}_+^{p\times q}$, then (1) implies that $\tau_f^+ > 0$ and hence (2) yields (3). □

Theorem 3.105 *Let* $f \in \mathcal{E} \cap \mathcal{N}^{p\times q}$. *Then:*

(1) *The minimal left and right denominators* d_L *and* d_R *of* f *are entire mvf's.*

(2) *The minimal denominators* $\{d_1, d_2\}$ *of* f *are also pairs of entire mvf's.*

Proof If $f \in \mathcal{E} \cap \mathcal{N}_+^{p\times q}$, then d_L and d_R are constant unitary matrices. Therefore, it suffices to consider the case where $f \notin \mathcal{N}_+^{p\times q}$. By the previous lemma, $e_{\tau_f^+}$ is a minimal scalar denominator of f and, therefore, by Lemma 3.77, it is a minimal scalar denominator of the mvf's d_L^{-1} and d_R^{-1}. Consequently,

$$b_R = e_{\tau_f^+} d_R^{-1} \in \mathcal{S}_{in}^{q\times q} \quad \text{and} \quad b_L = e_{\tau_f^+} d_L^{-1} \in \mathcal{S}_{in}^{p\times p}.$$

But this is the same as to say that

$$e_{\tau_f^+} I_q = b_R d_R \quad \text{and} \quad e_{\tau_f^+} I_p = b_L d_L,$$

which implies that $d_R \in \mathcal{E} \cap \mathcal{S}_{in}^{q\times q}$ and $d_L \in \mathcal{E} \cap \mathcal{S}_{in}^{p\times p}$, thanks to Lemma 3.98.

Suppose next that $f \in \mathcal{E} \cap \mathcal{N}^{p\times q}$ and let d_L and d_R be minimal left and right denominators of f, respectively, and let $\{d_1, d_2\}$ be a minimal denominator of f. Then, just as in the proof of Theorem 3.70, it may be shown that $d_L d_1^{-1} \in \mathcal{S}_{in}^{p\times p}$ and $d_2^{-1} d_R \in \mathcal{S}_{in}^{q\times q}$ and hence, since d_L and d_R are entire by part (1), that $b_1 = d_L d_1^{-1}$ is entire and $b_2 = d_2^{-1} d_R$ is entire. Consequently, $b_1 d_1$ and $d_2 b_2$ are entire, as are d_1 and d_2. □

3.21 The class $\Pi^{p\times q}$

A $p\times q$ mvf f_- in $\mathbb{C}_-$ is said to be a **pseudocontinuation** of a mvf $f \in \mathcal{N}^{p\times q}$, if

(1) $f_-^\# \in \mathcal{N}^{p\times q}$, i.e. f_- is a meromorphic $p \times q$ mvf in $\mathbb{C}_-$ with bounded Nevanlinna characteristic in $\mathbb{C}_-$ and

(2) $\lim_{\nu\downarrow 0} f_-(\mu - i\nu) = \lim_{\nu\downarrow 0} f(\mu + i\nu)$ $(= f(\mu))$ a.e. on $\mathbb{R}$.

The subclass of all mvf's $f \in \mathcal{N}^{p\times q}$ that admit pseudocontinuations f_- into C_- will be denoted $\Pi^{p\times q}$. Theorem 3.4 implies that each $f \in \Pi^{p\times q}$ admits only one pseudocontinuation f_-.

Although

$$\Pi^{p\times q} \subset \mathcal{N}^{p\times q},$$

by definition, and $f \in \mathcal{N}^{p\times q}$ is defined only on $\mathbb{C}_+$, we will consider mvf's $f \in \Pi^{p\times q}$ in the full complex plane $\mathbb{C}$ via the formulas

$$f(\lambda) = f_-(\lambda) \quad \text{for } \lambda \in \mathfrak{h}_{f_-} \cap \mathbb{C}_-$$

and

$$f(\mu) = \lim_{\nu\to 0} f(\mu + i\nu) \quad \text{a.e. on } \mathbb{R}.$$

The symbol $\mathfrak{h}_f$ will be used to denote the domain of holomorphy of this extended mvf $f(\lambda)$ in the full complex plane and

$$\mathfrak{h}_f^+ = \mathfrak{h}_f \cap \mathbb{C}_+, \quad \mathfrak{h}_f^- = \mathfrak{h}_f \cap \mathbb{C}_- \quad \text{and} \quad \mathfrak{h}_f^0 = \mathfrak{h}_f \cap \mathbb{R}.$$

We shall also write

$$\Pi^p = \Pi^{p\times 1}, \quad \Pi = \Pi^1 \quad \text{and} \quad \Pi \cap \mathcal{X}^{p\times q} = \Pi^{p\times q} \cap \mathcal{X}^{p\times q}$$

for short.

It is clear that a $p \times q$ mvf f belongs to the class $\Pi^{p\times q}$ if and only if all the entries in the mvf f belong to the class Π. Moreover, $\Pi^{p\times q}$ is a linear manifold and

$$f \in \Pi^{p\times r}, \ g \in \Pi^{r\times q} \Longrightarrow fg \in \Pi^{p\times q};$$

$$f \in \Pi^{p\times p} \Longrightarrow \det f \in \Pi \text{ and } \operatorname{trace} f \in \Pi;$$

$$f \in \Pi^{p\times p} \text{ and } \det f \not\equiv 0 \Longrightarrow f^{-1} \in \Pi^{p\times p};$$

$$f \in \Pi^{p\times q} \Longrightarrow f^\# \in \Pi^{q\times p}, \ f^\sim \in \Pi^{q\times p} \text{ and } f(-\lambda) \in \Pi^{p\times q}.$$

The inclusion

$$\mathcal{S}_{in}^{p\times p} \subset \Pi^{p\times p}$$

is also obvious and, moreover, the pseudocontinuation s_- for a mvf $s \in \mathcal{S}_{in}^{p\times p}$ may be defined by the **symmetry principle**

$$s_-(\lambda) = [s^\#(\lambda)]^{-1}, \quad \lambda \in \mathbb{C}_-,$$

that follows from the equality

$$s(\mu)s(\mu)^* = s(\mu)^* s(\mu) = I_p \quad \text{a.e. on } \mathbb{R}.$$

Thus,

$$s^\#(\lambda)s(\lambda) = s(\lambda)s^\#(\lambda) = I_p \quad \text{for every } \lambda \in \mathfrak{h}_s \cap \mathfrak{h}_{s^\#}$$

if $s \in \mathcal{S}_{in}^{p\times p}$.

The following result is due to Douglas, Shapiro and Shields [DSS70].

Theorem 3.106 *Let $f \in H_2$. Then*

$$f \in \Pi \cap H_2 \Longleftrightarrow f \in H_2 \ominus bH_2 \text{ for some } b \in S_{in}. \tag{3.127}$$

Proof If $f \in H_2 \ominus bH_2$ for some $b \in \mathcal{S}_{in}$, then $h_- = b^{-1}f$ belongs to the space K_2. Let $f_- = h_-/b^\#$. Then $f_-^\# \in \mathcal{N}$ and

$$\lim_{\nu\downarrow 0} f_-(\mu - i\nu) = b(\mu)h_-(\mu) = f(\mu) \quad \text{a.e. on } \mathbb{R}, \tag{3.128}$$

i.e., f_- is a pseudocontinuation of f and, consequently, $f \in \Pi \cap H_2$. Conversely, if $f \in \Pi \cap H_2$, i.e., if $f \in H_2$ and (ii) holds for some function $f_-(\lambda)$ such that $f_-^\# \in \mathcal{N}$, then $f_-^\# = g/b$, where $g \in \mathcal{N}_+$ and $b \in \mathcal{S}_{in}$. Therefore, since

$$|g(\mu)| = |f_-(\mu)| = |f(\mu)| \quad \text{a.e. on } \mathbb{R},$$

$g \in L_2 \cap \mathcal{N}_+$ and the Smirnov maximum principle guarantees that $g \in H_2$. Thus, as

$$f(\mu) = f_-(\mu) = b(\mu)g^\#(\mu) \quad \text{a.e. on } \mathbb{R},$$

it follows that f belongs to $H_2 \cap bK_2 = H_2 \ominus bH_2$. □

Corollary 3.107 *Let $f \in H_2^{p\times q}$. Then $f \in \Pi \cap H_2^{p\times q}$ if and only if $b^{-1}f \in K_2^{p\times q}$ for some $b \in \mathcal{S}_{in}$.*

Proof This follows easily by applying Theorem 3.106 to the entries of the mvf $f(\lambda)$. □

If $f \in \Pi^{p\times q}$ and if the restriction of f to $\mathbb{C}_+$ has a holomorphic extension to $\mathbb{C}_-$, then this extension coincides with the pseudocontinuation f_- of f, as follows from the uniqueness of the holomorphic extension. In particular, the pseudocontinuation f_- of an entire mvf $f \in \Pi^{p\times q}$ is the restriction of f to $\mathbb{C}_-$. Thus, if $f \in \mathcal{E}^{p\times q}$, then

$$\begin{aligned} f \in \Pi^{p\times q} \quad &\Longleftrightarrow \quad f \text{ has bounded Nevanlinna characteristic} \\ &\qquad\quad \text{in both half planes } \mathbb{C}_+ \text{ and } \mathbb{C}_- \\ &\Longleftrightarrow \quad f \in \mathcal{N}^{p\times q} \text{ and } f^\# \in \mathcal{N}^{q\times p}. \end{aligned}$$

Theorem 3.108 (M. G. Krein) *Let $f \in \mathcal{E}^{p\times q}$. Then $f \in \Pi^{p\times q}$ if and only if f is an entire mvf of exponential type and satisfies the Cartwright condition*

$$\int_{-\infty}^{\infty} \frac{\ln_+ \| f(\mu) \|}{1+\mu^2} d\mu < \infty. \tag{3.129}$$

Moreover, if $f \in \mathcal{E} \cap \Pi^{p\times q}$, then

(1) $\tau_+(f) = \tau_f^+, \quad \tau_-(f) = \tau_f^-$.

(2) $\tau_f^- + \tau_f^+ \geq 0$.

(3) $\tau(f) = \max\,\{\tau_f^+, \tau_f^-\}$.

Proof All the assertions except for (2) follow from Theorem 3.91. To verify (2), suppose to the contrary that $\tau_f^+ + \tau_f^- < 0$ and let $g = e_\delta f$, where $\tau_f^+ < \delta < -\tau_f^-$. Then $\tau_g^+ = -\delta + \tau_f^+ < 0$ and $\tau_g^- = \delta + \tau_f^- < 0$, which is impossible, since in this case $g(\lambda) \equiv 0$. Therefore, (2) is also valid. □

Let

$$r_g = \max\{\text{rank } g(\lambda) : \lambda \in \mathfrak{h}_g\}$$

for a mvf $g \in \Pi^{p\times q}$.

Lemma 3.109 *If $g \in \Pi^{p\times q}$, then:*

(1) rank $g(\lambda) = r_g$ *for every point $\lambda \in \mathfrak{h}_g$ except possibly for a set of isolated points.*

(2) rank $g(\mu) = r_g$ *a.e. on $\mathbb{R}$.*

Proof Let

$$r_g^+ = \max\{\text{rank } g(\lambda) : \lambda \in \mathfrak{h}_g^+\}$$

for $g \in \mathcal{N}^{p\times q}$. Then, since $\Pi^{p\times q} \subset \mathcal{N}^{p\times q}$, Lemma 3.31 guarantees that

$$\text{rank } g(\lambda) = r_g^+$$

for every point $\lambda \in \mathbb{C}_+$ except possibly for a set of isolated points and that

$$\text{rank } g(\mu) = r_g^+ \qquad \text{a.e. on } \mathbb{R}.$$

The lemma follows from these facts and their analogues for mvf's $g(\lambda)$ such that $g^\# \in \mathcal{N}^{q\times p}$. □

Let $u, v \in \mathcal{S}_{in}$ and define

$$\Pi^{p\times q}(u,v) = \{f \in \Pi^{p\times q} : f \in \mathcal{N}^{p\times q}(u) \text{ and } f^\# \in \mathcal{N}^{q\times p}(v)\}.$$

If $f \in \Pi^{p\times q}$, then $f \in \Pi^{p\times q}(u,v)$ for some pair $\{u,v\}$ of scalar inner functions u and v. In the set of such pairs there exists a pair $\{u,v\}$ that is minimal in the sense that:

(1) $f \in \Pi^{p\times q}(u,v)$.

(2) If $f \in \Pi^{p\times q}(\widetilde{u},\widetilde{v})$, then $u^{-1}\widetilde{u} \in \mathcal{S}_{in}$ and $v^{-1}\widetilde{v} \in \mathcal{S}_{in}$.

Such a **minimal pair** is uniquely defined by f, up to a pair of constant multipliers with modulus one.

If $f \in \mathcal{N}^{p\times p}$ and $f(\mu) \geq 0$ a.e. on $\mathbb{R}$, then $f \in \Pi^{p\times p}$ and $f^{\#}(\lambda) = f(\lambda)$ and, consequently, $f \in \Pi(v, v)$, where v is a minimal scalar denominator of f.

If $f \in \Pi^{p\times q}$, then all the entries in the factors in formulas (3.92), (3.93), (3.100) and (3.101) belong to Π.

A rational $p \times q$ mvf f belongs to the class $\Pi^{p\times q}$. Moreover, if $f \in \Pi^{p\times q}(u, v)$ and $\{u, v\}$ is a minimal pair for f in the above sense, then u and v are both finite Blaschke products. Indeed, the minimal scalar denominators of the rational mvf's f and $f^{\#}$ are scalar Blaschke products with zeros that coincide with the poles of f in $\mathbb{C}_+$ and the poles of $f^{\#}$ in $\mathbb{C}_+$, respectively.

Theorem 3.110 *Let $g \in \Pi^{p\times p}(v, v)$, let $r = r_g$ and suppose that $g(\mu) \geq 0$ a.e. on $\mathbb{R}$. Then:*

(1) *The factorization problem (3.108) has a solution $f = \varphi_L \in \mathcal{N}_{out}^{r\times p}$ that is uniquely defined up to a left constant $r \times r$ unitary factor.*

(2) *Every solution $\varphi_L \in \mathcal{N}_{out}^{r\times p}$ of the factorization problem (3.108) belongs to the class $\Pi^{r\times p}(1, v)$.*

(3) *Every solution $f \in \mathcal{N}^{q\times p}$ with $q \geq r$ of the factorization problem (3.108) can be described by formula (3.100), where $\varphi_L \in \mathcal{N}_{out}^{r\times p}$ is a solution of the factorization problem (3.108).*

(4) *If the solution $f \in \mathcal{N}^{r\times p}$, i.e., if $q = r$, then*

(a) $f \in \Pi^{r\times p}$

(b) $f^{\#}(\lambda) f(\lambda) = g(\lambda)$ for every point $\lambda \in \mathfrak{H}_f \cap \mathfrak{H}_{f^{\#}}$.

(5) *The following equivalences hold:*

(a) $g \in L_1^{p\times p} \Longleftrightarrow \varphi_L \in H_2^{r\times p}$.

(b) $g \in \widetilde{L}_1^{p\times p} \Longleftrightarrow \varphi_L \in \rho_\omega^{-1} H_2^{r\times p}$ *for some (and hence every) point* $\omega \in \mathbb{C}_+$.

(c) $g \in L_\infty^{p\times p} \Longleftrightarrow \varphi_L \in H_\infty^{r\times p}$.

(d) $g \in \mathcal{R}^{p\times p} \Longleftrightarrow \varphi_L \in \mathcal{R}^{r\times p}$.

(6) *If $g \in \mathcal{R}^{p\times p}$, then the set of solutions $f \in \mathcal{R}^{q\times p}$ of the factorization problem (3.108) is described by formula (3.100) with $d_L \in \mathcal{R} \cap \mathcal{S}_{in}^{q\times r}$ and $b_L \in \mathcal{R} \cap \mathcal{S}_{in}^{q\times q}$.*

Proof Assertions (1) and (2) are due to Rosenblum and Rovnyak; see Section 6.5 in [RR85]. Next, if f admits a representation of the form (3.100), then clearly, f is a solution of the factorization problem (3.108). Conversely, if $f \in \mathcal{N}^{r\times p}$ is a solution of this factorization problem, then (3) follows from the representation (3.100) of f, because the formula

$$g(\mu) = f(\mu)^* f(\mu) = \varphi_L(\mu)^* \varphi_L(\mu) \quad \text{a.e. on } \mathbb{R}$$

implies that $\operatorname{rank} \varphi_L = r_g = r$ and hence that the number r considered in formula (3.100) coincides with the number $r = r_g$ that is considered in this theorem. This completes the proof of (3).

Assertion (4a) is clear from (2) and formula (3.100); (4b) then follows because the equality holds a.e. on $\mathbb{R}$. The implications $\Longleftarrow$ in (5) are self-evident. The opposit implications $\Longrightarrow$ in the first three equivalences follow from the Smirnov maximum principle.

The next step is to verify the implication $\Longrightarrow$ for (5d). Towards this end, let $g \in \mathcal{R}^{p\times p}$ be such that $g(\mu) \geq 0$ a.e. on $\mathbb{R}$. Then $g \in \Pi^{p\times p}(v, v)$ where v is a finite Blaschke product. By (2), $\varphi_L \in \Pi^{r\times p}(1, v)$. Moreover, without loss of generality, it may be assumed that g has no poles on $\mathbb{R}$ or at infinity. This can always be achieved by multiplying $g(\lambda)$ by a scalar outer function of the form $\beta(\lambda)^2/(\lambda^2+1)^k$, where $\beta(\lambda)$ is a polynomial with zeros at the real poles of $g(\lambda)$. Thus we may assume that $g \in \mathcal{R} \cap L_\infty^{p\times p}$ and hence by (5c), $\varphi_L \in H_\infty^{r\times p}$. Furthermore, as $v\varphi_L^\# \in \mathcal{N}_+^{p\times r}$, the Smirnov maximum principle guarantees that $v\varphi_L^\# \in H_\infty^{p\times r}$. Thus,

$$\frac{\varphi_L(\lambda) - \varphi_L(i)}{\lambda - i}\xi \in H_2^r \ominus vH_2^r \text{ for every } \xi \in \mathbb{C}^p.$$

Since $H_2^r \ominus vH_2^r$ is a finite dimensional space with a basis of rational vvf's, $\varphi_L \in \mathcal{R}^{r\times p}$. This completes the proof of (5).

Suppose next that $g \in \mathcal{R}^{p\times p}$ and that $f \in \mathcal{R}^{q\times p}$ is a solution of the factorization problem (3.108). Then in (3.100), $d_L \in \mathcal{R}^{q\times q}$ by Lemma 3.68. Moreover, the mvf $\varphi_L \in \mathcal{R}^{r\times p}$ and can be normalized by the condition

$$\varphi_L(\omega)K \begin{bmatrix} I_r \\ 0_{(p-r)\times r} \end{bmatrix} > 0,$$

where K is a $p \times p$ permutation matrix that is chosen as in Theorem 3.71. Then, in formula (3.100),

$$b_L = d_L f K \begin{bmatrix} I_r \\ 0_{(p-r)\times r} \end{bmatrix} \left(\varphi_L K \begin{bmatrix} I_r \\ 0_{(p-r)\times r} \end{bmatrix} \right)^{-1}$$

belongs to $\mathcal{R}^{q\times r}$. □

This theorem, combined with Theorem 3.108 yields the following generalization of a theorem of Akhiezer:

Theorem 3.111 (M. Rosenblum and J.Rovnyak) *Let g be an entire $p\times p$ mvf of exponential type τ_g, let $r = r_g$ and suppose further that $g(\mu) \geq 0$ on $\mathbb{R}$ and $\int_{-\infty}^{\infty} \frac{\ln^+ \|g(\mu)\|}{1+\mu^2} d\mu < \infty$. Then:*

(1) $g \in \Pi^{p\times p}$.

(2) *The special solution $\varphi_L \in \mathcal{N}_{out}^{r\times p}$ of the factorization problem (3.108) is an entire $r \times p$ mvf of exponential type that satisfies the Cartwright condition,*

$$\int_{-\infty}^{\infty} \frac{\ln^+ \|\varphi_L(\mu)\|}{1+\mu^2} d\mu < \infty,$$

i.e., $\varphi_L \in \mathcal{E} \cap \Pi^{r\times p}$.

(3) *The exponential type of the mvf* $\exp\{-i\tau_g\lambda\}\varphi_L(\lambda)$ *is equal to $\tau_g/2$.*

3.22 Mean types for mvf's in $\mathcal{N}^{p\times q}$ and $\Pi^{p\times q}$

The **mean type** τ_f^+ of a mvf $f \in \mathcal{N}^{p\times q}$ that is holomorphic in $\mathbb{C}_+$ is defined by the formula

$$\tau_f^+ = \limsup_{\nu\to\infty} \frac{\ln \|f(i\nu)\|}{\nu}. \tag{3.130}$$

If $f \not\equiv 0$, then τ_f^+ is finite and may also be evaluated by two auxiliary formulas, just as was the case for the types $\tau_\pm(f)$ of entire mvf's f in Section 3.18:

$$\tau_f^+ = \limsup_{\nu\to\infty} \frac{\ln \operatorname{trace}\{f(i\nu)^* f(i\nu)\}}{2\nu} \tag{3.131}$$

and

$$\tau_f^+ = \max\{\tau_{f_{ij}}^+ : f_{ij} \not\equiv 0,\ 1 \leq i \leq p, 1 \leq j \leq q\}. \tag{3.132}$$

If $f \in \mathcal{N}^{p\times p}$ is holomorphic in $\mathbb{C}_+$ and det $f(\lambda) \not\equiv 0$ in $\mathbb{C}_+$ then

$$\delta_f^+ = \limsup_{\nu\to\infty} \frac{\ln|\det\, f(i\nu)|}{\nu} \tag{3.133}$$

is also finite. Similarly, if f is holomorphic in $\mathbb{C}_-$ and $f^\# \in \mathcal{N}^{q\times p}$, then the mean type τ_f^- in $\mathbb{C}_-$ is defined by the formula

$$\tau_f^- = \limsup_{\nu\to\infty} \frac{\ln\|f(-i\nu)\|}{\nu} \tag{3.134}$$

and, if $q = p$ and det $f(\lambda) \not\equiv 0$,

$$\delta_f^- = \limsup_{\nu\to\infty} \frac{\ln|\det\, f(-i\nu)|}{\nu}. \tag{3.135}$$

If f is holomorphic in $\mathbb{C}\setminus\mathbb{R}$ and $f \in \Pi^{p\times q}$, then

$$\tau_f^\pm = \tau_{f^\#}^\mp \quad \text{and, if } q = p \text{ and } \det\, f(\lambda) \not\equiv 0 \quad \delta_f^\pm = \delta_{f^\#}^\mp. \tag{3.136}$$

Moreover, the representation formula (3.38) for scalar outer functions leads easily to the conclusion

$$f \in \mathcal{N}_{out} \Longrightarrow \tau_f^+ = \lim_{\nu\uparrow\infty} \frac{\ln|f(i\nu)|}{\nu} = 0. \tag{3.137}$$

Lemma 3.112 *Let $f \in \mathcal{N}^{p\times q}$ be holomorphic in $\mathbb{C}_+$ and $f \not\equiv 0$. Then the following implications are in force:*

(1) $f \in \mathcal{N}_+^{p\times q} \Longrightarrow \tau_f^+ \le 0.$

(2) $f \in \mathcal{N}_+^{p\times p}$ *and* det $f(\lambda) \not\equiv 0 \Longrightarrow \delta_f^+ \le 0.$

(3) *If $f \in \mathcal{N}_{out}^{p\times p}$, then*

$$\tau_f^+ = \lim_{\nu\uparrow\infty} \frac{\ln\|f(i\nu)\|}{\nu} = 0 \quad \text{and} \quad \tau_{f^{-1}}^+ = \lim_{\nu\uparrow\infty} \frac{\ln\|f(i\nu)^{-1}\|}{\nu} = 0.$$

(4) *If $q = p$ and $f(\lambda)$ is invertible at every point $\lambda \in \mathbb{C}_+$ and $f^{-1} \in \mathcal{N}_+^{p\times p}$, then*

(a) τ_f^+ *and* δ_f^+ *are subject to the bounds*

$$\tau_f^+ \le \delta_f^+ \le p\tau_f^+. \tag{3.138}$$

(b) *If* $d_L \in \mathcal{S}_{in}^{p\times p}$, $d_R \in \mathcal{S}_{in}^{p\times p}$ *are minimal left and right denominators of* f, *i.e., if*

$$f d_R \in \mathcal{N}_{out}^{p\times p} \quad and \quad d_L f \in \mathcal{N}_{out}^{p\times p}, \tag{3.139}$$

then $d_L(\lambda)$ *and* $d_R(\lambda)$ *are holomorphic in* $\mathbb{C}\setminus\mathbb{R}$,

$$0 \le \tau_f^+ = \tau_{d_L}^- = \tau_{d_R}^- \quad and \quad 0 \le \delta_f^+ = \delta_{d_L}^- = \delta_{d_R}^-. \tag{3.140}$$

Proof If $f \in \mathcal{N}_+^{p\times q}$, then $f = g/\varphi$, where $g \in \mathcal{S}^{p\times q}$ and $\varphi \in \mathcal{S}_{out}$. Therefore, $\|f(i\nu)\| \le 1/|\varphi(i\nu)|$, which implies that $\tau_f^+ \le \tau_{\varphi^{-1}}^+$ and thus, verifies (1), since $\tau_{\varphi^{-1}}^+ = 0$. Then (2) follows from (1), since $f \in \mathcal{N}_+^{p\times p} \Longrightarrow \det f \in \mathcal{N}_+$.

Suppose next that $f \in \mathcal{N}_{out}^{p\times p}$. Then $f^{-1} \in \mathcal{N}_{out}^{p\times p}$ and hence $f = g/\varphi$ and $f^{-1} = h/\psi$, where $g, h \in \mathcal{S}_{out}^{p\times p}$ and $\varphi, \psi \in \mathcal{S}_{out}$. Therefore,

$$|\psi(i\nu)| = \|f(i\nu)h(i\nu)\| \le \|f(i\nu)\| \le |\varphi(i\nu)|^{-1},$$

which, in view of formula (3.137), suffices to verify the first assertion in (3). But the second follows from the first, since $f \in \mathcal{N}_{out}^{p\times p} \Longleftrightarrow f^{-1} \in \mathcal{N}_{out}^{p\times p}$.

To verify (4), let f be holomorphic and invertible at each point $\lambda \in \mathbb{C}_+$ and let $f^{-1} \in \mathcal{N}_+^{p\times p}$. Then

$$f^{-1} = \frac{g}{\varphi} \quad \text{for some} \quad g \in \mathcal{S}^{p\times p} \quad \text{and} \quad \varphi \in \mathcal{S}_{out}.$$

Therefore, $g(\lambda)$ is invertible at each point $\lambda \in \mathbb{C}_+$ and the formula

$$f(\lambda) = \varphi(\lambda)h(\lambda)$$

for $h = g^{-1}$ implies that the left and right denominators of f and h coincide and that

$$\tau_f^+ = \tau_h^+ \quad \text{and} \quad \delta_f^+ = \delta_h^+.$$

Thus, as f can be replaced by h in the proof of (4), we can assume that $f^{-1} \in \mathcal{S}^{p\times p}$. Then (4a) follows immediately from (2.96). Moreover,

$$f^{-1} = \psi_L d_L = d_R \psi_R,$$

where $d_L, d_R \in \mathcal{S}_{in}^{p\times p}$ and $\psi_L, \psi_R \in \mathcal{S}_{out}^{p\times p}$. Then, since f is holomorphic in $\mathbb{C}_+$, d_L^{-1} and d_R^{-1} are also holomorphic in $\mathbb{C}_+$ and

$$\begin{aligned} \|d_L(i\nu)^{-1}\| &= \|f(i\nu)\psi_L(i\nu)\| \le \|f(i\nu)\|\|\psi_L(i\nu)\| \\ &\le \|f(i\nu)\| = \|d_L(i\nu)^{-1}\psi_L(i\nu)^{-1}\| \\ &\le \|d_L(i\nu)^{-1}\|\|\psi_L(i\nu)^{-1}\|. \end{aligned}$$

Moreover,

$$\lim_{\nu\uparrow\infty} \frac{\ln \|\psi_L(i\nu)^{-1}\|}{\nu} = 0,$$

since $\psi_L \in \mathcal{N}_{out}^{p\times p}$. Therefore,

$$\tau_f^+ = \tau_{d_L^{-1}}^+ = \tau_{d_L}^- \quad \text{and} \quad \tau_f^+ = \tau_{d_R^{-1}}^+ = \tau_{d_R}^-,$$

since the estimates are independent of the order of the factors in (3.139).

The same arguments applied to det $f(i\nu)$ justify the second set of equalities in (3.140). □

Remark 3.113 *If $f \in \mathcal{E}^{p\times p}$ is such that* $\det f(\lambda) \neq 0$ *in $\mathbb{C}_+$ and $f^{-1} \in \mathcal{N}_+^{p\times p}$, then $f \in \mathcal{N}^{p\times p}$ and, by Theorem 3.70, minimal left and right denominators d_L and d_R of f are entire mvf's, and, by (4) and (5) of Theorem 3.94,*

$$\det d_L(\lambda) = \gamma_1 e_\alpha(\lambda) \quad \textit{and} \quad \det d_R(\lambda) = \gamma_2 e_\alpha(\lambda)$$

for some scalars γ_i of modulus one and $\alpha = \delta_{d_L}^- = \delta_{d_R}^- = \delta_f^+$. Moreover,

$$\delta_f^+ = p\tau_f^+ \Longleftrightarrow d_L(\lambda) = d_L(0)e_\alpha(\lambda) \quad \textit{and} \quad d_R(\lambda) = d_R(0)e_\alpha(\lambda).$$

3.23 Bibliographical notes

Originally $\mathcal{N}$ was defined as the class of meromorphic scalar functions $f(\lambda)$ in $\mathbb{C}_+$ with bounded Nevanlinna characteristic, i.e., for which

$$\sup_{\nu>0} \int_{-\infty}^{\infty} \frac{\ln^+ |f(\mu + i\nu)|}{1+\mu^2} d\mu < \infty$$

and the poles $\omega_1, \omega_2, \ldots$ of f are subject to the Blaschke condition (3.26). Moreover, the Smirnov class $\mathcal{N}_+$ was originally defined as the class of scalar holomorphic functions in $\mathbb{C}_+$ such that

$$\sup_{\nu>0} \int_{-\infty}^{x} \frac{\ln^+ |f(\mu + i\nu)|}{1+\mu^2} d\mu < \infty$$

is absolutely continuous on $\mathbb{R}$. These definitions are equivalent to the characterizations

$$\mathcal{N} = \{f : \; f = f_1/f_2 \;\text{ where }\; f_1 \in \mathcal{S} \;\text{ and }\; f_2 \in \mathcal{S}\}$$

of the class $\mathcal{N}$ and

$$\mathcal{N}_+ = \{f : f = f_1/f_2 \text{ where } f_1 \in \mathcal{S} \text{ and } f_2 \in \mathcal{S}_{out}\}$$

that we started with.

A proof of the integral representation formula (3.3) for mvf's $c \in \mathcal{C}^{p\times p}$ and in the subclass $\mathcal{C}_0^{p\times p}$ defined in Section 3.3 may be found in pp. 23–28 of [Bro72]. Additional information on subclasses of $\mathcal{C}^{p\times p}$ for the case $p = 1$ may be found in [KaKr74] Theorem S.1.4.1 on p.12; and for $p \geq 1$ and operator valued functions in [Bro72] Theorem 4.8 pp. 25–27.

Analogues of the factorization problems (3.46) and (3.53) for holomorphic functions in $\mathbb{D}$ and of the formulas (3.31) and (3.39) for the factors seem to have been first obtained by V. E. Smirnov. He exploited these *parametric representations* to obtain Theorem 3.22 and to establish a theorem that may be reformulated as follows:

If $f \in H_2(\mathbb{D})$ *and* $\mathcal{L}_f = \bigvee_{n\geq 0} \zeta^n f$ *in* $L_2(\mathbb{T})$, *then* $\mathcal{L}_f = H_2(\mathbb{D}) \iff$

$$f(z) = \gamma \exp\left\{\frac{1}{2\pi}\int_0^{2\pi} \frac{e^{i\theta}+z}{e^{i\theta}-z} \ln|f(e^{i\theta}|d\theta\right\} \; for\ z \in \mathbb{D}\ and\ a\ \gamma \in \mathbb{T};$$

see, e.g., [Smi28] and [Smi32]. Later, functions f that admitted such a representation were termed **outer functions** by Beurling [Be48], who proved that $\mathcal{L}_f = bH_2(\mathbb{D})$, where $b \in \mathcal{S}_{in}$ and $b^{-1}f$ is outer. Theorem 3.38 is due to Lax [La59]. The proof presented here is adapted from the discussion on pp. 8–9 of Helson [He64]. A generalization to Hilbert space valued functions based on wandering subspaces was obtained by Halmos [Ha61].

Theorem 3.50 is obtained in Sz.-Nagy-Foias [SzNF70] as a corollary of a general result (Theorem 6.2 on pp. 217–218) on the inner-outer factorization of contractive holomorphic operator valued functions and their scalar multiples. In particular, their results imply that if a scalar multiple of $f \in \mathcal{S}^{p\times p}$ is inner (respectively, outer), then $f \in \mathcal{S}_{in}^{p\times p}$ (respectively, $f \in \mathcal{S}_{out}^{p\times p}$). This fact is applicable to the proof of the presented theorem, because $\det f$ is a scalar multiple of a mvf $f \in \mathcal{S}^{p\times p}$ with $\det f(\lambda) \not\equiv 0$.

Lemma 3.54 is established by other methods in the the proof of Lemma 3.1 in [Ar73].

The Smirnov maximum principle for matrix and operator valued functions may be found in Rosenblum and Rovnyak [RR85].

The discussion of denominators, left and right denominators, and the corresponding minimal denominators for mvf's in the class $\mathcal{N}^{p\times q}$ is adapted from [Ar 73], where these results were used to study Darlington representations. This circle of ideas also plays a role in the theory of approximation of mvf's in the class $\Pi^{p\times q}$ by rational mvf's; see [Ar78], [Kat93], [Kat94] and [KK95].

Theorem 3.79 is due V. N. Zasukhin and M. G. Krein. Krein's proof may be found in Rozanov [Roz58]; see also [Roz60] and Wiener and Masani [WM57]. Devinatz [Dev61] established an operator theoretic generalization; see, e.g., the notes on pp. 126–127 of [RR85] for more on the history of this circle of ideas and additional references.

Theorem 3.81 was obtained by Potapov [Po60]. Potapov's proof is based on his generalization of the well known Schwarz lemma to the setting of mvf's and a recursive argument based on the Schur-Nevanlinna algorithm.

Theorems 3.82 and 3.96 are due to Potapov [Po60]. A multiplicative integral representation formula for normalized entire inner $m \times m$ mvf's $f(\lambda)$ is equivalent to identifying $f(\lambda)$ as the monodromy matrix of a canonical system of integral equations with mass function $M(t)$ on the interval $[0, d]$. Potapov proved this theorem by approximating a mvf $f \in \mathcal{E} \cap \mathcal{S}_{in}^{p\times p}$ by a sequence of mvf's $f_n \in \mathcal{R} \cap \mathcal{S}_{in}^{p\times p}$ and invoking the representation of these mvf's as BP products. A different proof that is based on a circle of ideas introduced by Yu. P. Ginzburg was presented in Theorem 2.2 of [ArD00a]. Multiplicative representations of matrix and operator valued functions with bounded Nevanlinna characteristic in $\mathbb{C}_+$ were obtained by Yu. P. Ginzburg in [Gi67] by consideration of the minorants of these functions and applying the Beurling-Lax-Halmos theorem; see also [GiZe90] and [GiSh94].

Item (2) in Lemma 3.103 coincides with Lemma 30.1 on p.186 of Brodskii [Bro72], which is proved the same way.

Theorem 3.108 may be found in [Kr51]. It was generalized to entire operator valued functions by Rosenblum and Rovnyak; see Section 6.11 of [RR85], and, for additional discussion and references p. 126 of [RR85]. The recent monograph [RS02] is a good source of supplementary information on the class Π and related issues.

4

J-contractive and J-inner matrix valued functions

In this chapter we shall present a number of facts from the theory of J-contractive and J-inner mvf's that will be needed in the sequel. Most of this information can be found in the papers [Po60], [Ar73], [Ar89], [Ar95a] and the monographs [Dy89b] and [Zol03]. The class of J-inner mvf's is a subclass of the class of J-contractive mvf's, which was investigated by Potapov [Po60].

4.1 The classes $\mathcal{P}(J)$ and $\mathcal{U}(J)$

Let J be an $m \times m$ signature matrix and let P and Q be the orthogonal projection matrices that are defined by formula (2.1) with ranks p and q, respectively, as in (2.3).

An $m \times m$ mvf $U(\lambda)$ which is meromorphic in $\mathbb{C}_+$ is said to belong to the **Potapov class** $\mathcal{P}(J)$ of J-contractive mvf's if

$$U(\lambda)^* J U(\lambda) \leq J \tag{4.1}$$

for every point $\lambda \in \mathfrak{h}_U^+$, the domain of holomorphy of U in $\mathbb{C}_+$. In his paper [Po60], Potapov imposed the extra condition $\det U(\lambda) \not\equiv 0$ on $U \in \mathcal{P}(J)$, in order to develop his multiplicative theory. This subclass will be denoted $\mathcal{P}^\circ(J)$.

If $U \in \mathcal{P}(J)$ and $\lambda \in \mathfrak{h}_U^+$, then, since the PG (Potapov-Ginzburg) transform maps $U(\lambda)$ into the contractive matrix $S(\lambda)$, it is readily seen that it maps U into an $m \times m$ mvf S that is holomorphic and contractive on $\mathfrak{h}_U^+$. This mvf S has a unique extension to a mvf in $\mathcal{S}^{m\times m}$ which will also be designated $S(\lambda)$. Thus, the PG transform

$$S(\lambda) = (PU(\lambda) + Q)(P + QU(\lambda))^{-1} = (P - U(\lambda)Q)^{-1}(U(\lambda)P - Q) \tag{4.2}$$

maps $U \in \mathcal{P}(J)$ into $\mathcal{S}^{m\times m}$. More precisely, it maps the class $\mathcal{P}(J)$ onto the set

$$\begin{aligned}&\{S \in \mathcal{S}^{m\times m} : \det(P \pm QS(\lambda)) \not\equiv 0\}\\ &= \{S \in \mathcal{S}^{m\times m} : \det(P \pm S(\lambda)Q) \not\equiv 0\}\\ &= \{S \in \mathcal{S}^{m\times m} : \det(P \pm QS(\lambda)Q) \not\equiv 0\}.\end{aligned}$$

Moreover, since the PG transform is its own inverse, the formula

$$U(\lambda) = (PS(\lambda) + Q)(P + QS(\lambda))^{-1} = (P - S(\lambda)Q)^{-1}(S(\lambda)P - Q) \quad (4.3)$$

is also valid and serves to justify the inclusion

$$\mathcal{P}(J) \subset \mathcal{N}^{m\times m}.$$

Consequently, every mvf $U \in \mathcal{P}(J)$ has nontangential boundary values.

A mvf $U \in \mathcal{P}(J)$ is said to be ***J*-inner** if (in addition to (4.1))

$$U(\mu)^* JU(\mu) = J \quad \text{for almost all points } \mu \in \mathbb{R}. \quad (4.4)$$

We shall designate the class of J-inner mvf's by $\mathcal{U}(J)$ and shall reserve the symbol $W(\lambda)$ for J-inner mvf's with $J = j_{pq}$ and $A(\lambda)$ for J-inner mvf's with $J = J_p$. It is readily seen that

$$U \in \mathcal{U}(J) \iff PG(U) \in \mathcal{S}_{in}^{m\times m}.$$

Condition (4.4) guarantees that $U(\lambda)$ is invertible in $\mathbb{C}_+$ except for an isolated set of points, i.e.,

$$\mathcal{U}(J) \subset \mathcal{P}^\circ(J).$$

Thus, we can (and shall) define a **pseudocontinuation** of $U(\lambda)$ to $\mathbb{C}_-$ by the **symmetry principle**:

$$U(\lambda) = J\left\{U^\#(\lambda)\right\}^{-1} J$$

for $\lambda \in \mathbb{C}_-$. Consequently,

$$\mathcal{U}(J) \subset \Pi^{m\times m}.$$

Corollary 2.5 guarantees that the inequality (4.1) holds if and only if

$$U(\lambda)JU(\lambda)^* \leq J\ ; \quad (4.5)$$

the position of $U(\mu)$ and $U(\mu)^*$ can of course also be interchanged in formula (4.4).

Lemma 4.1 *Let $U \in \mathcal{P}(J)$ with $J \neq \pm I_m$, let $S = PG(U)$ and let*

$$U_P(\lambda) = Q + PU(\lambda)P \quad \textit{and} \quad U_Q(\lambda) = P + QU(\lambda)Q, \quad (4.6)$$

and

$$S_P(\lambda) = Q + PS(\lambda)P \quad and \quad S_Q(\lambda) = P + QS(\lambda)Q. \tag{4.7}$$

Then

(1) *The identity*

$$S_Q(\lambda)U_Q(\lambda) = I_m \quad \text{holds for every point} \quad \lambda \in \mathfrak{h}_U^+ \tag{4.8}$$

and hence the mvf $U_Q(\lambda)$ is expansive in $\mathbb{C}_+$, i.e.,

$$U_Q(\lambda)^* U_Q(\lambda) \geq I_m \text{ for } \lambda \in \mathfrak{h}_U^+.$$

(2) *If $U \in \mathcal{U}(J)$, the identity*

$$S_P(\lambda)U_P^\#(\lambda) = I_m \quad \text{holds for every point} \quad \lambda \in \mathfrak{h}_{U^\#}^+ \tag{4.9}$$

and hence the mvf $U_P(\lambda)$ is expansive in $\mathbb{C}_-$.

(3) *If $U \in \mathcal{U}(J)$, then the formula*

$$U(\lambda) = (I_m + PS(\lambda)Q)U_P^{-\#}(\lambda)U_Q(\lambda)(I_m - QS(\lambda)P) \tag{4.10}$$

holds for every point $\lambda \in \mathfrak{h}_U^+ \cap \mathfrak{h}_{U^\#}^+$.

Proof This follows from Lemma 2.6. □

In future developments it will be convenient to consider the classes $\mathcal{P}(J_1, J_2)$, $\mathcal{P}^\circ(J_1, J_2)$ and $\mathcal{U}(J_1, J_2)$ of $m \times m$ mvf's that are meromorphic in $\mathbb{C}_+$ and are defined in terms of two unitarily equivalent signature matrices J_1 and $J_2 = V^* J_1 V$ by the following constraints:

$$U \in \mathcal{P}(J_1, J_2) \quad \text{if} \quad U(\lambda)^* J_1 U(\lambda) \leq J_2 \quad \text{for every point} \quad \lambda \in \mathfrak{h}_U^+; \tag{4.11}$$

$$U \in \mathcal{P}^\circ(J_1, J_2) \quad \text{if} \quad U \in \mathcal{P}(J_1, J_2) \quad \text{and} \quad \det U(\lambda) \not\equiv 0 \text{ in } \mathfrak{h}_U^+; \tag{4.12}$$

$$U \in \mathcal{U}(J_1, J_2) \quad \text{if} \quad U \in \mathcal{P}(J_1, J_2) \quad \text{and} \quad U(\mu)^* J_1 U(\mu) = J_2 \quad \text{a.e. on } \mathbb{R}. \tag{4.13}$$

It is readily seen that

$$U \in \mathcal{P}(J_1, J_2) \Longleftrightarrow UV^* \in \mathcal{P}(J_1). \tag{4.14}$$

$$U \in \mathcal{P}^\circ(J_1, J_2) \Longleftrightarrow UV^* \in \mathcal{P}^\circ(J_1). \tag{4.15}$$

$$U \in \mathcal{U}(J_1, J_2) \Longleftrightarrow UV^* \in \mathcal{U}(J_1). \tag{4.16}$$

Moreover, (4.11) is equivalent to the condition

$$U(\lambda)J_2U(\lambda)^* \le J_1 \quad \text{for every point} \quad \lambda \in \mathfrak{h}_U^+ \tag{4.17}$$

and

$$U(\mu)^*J_1U(\mu) = J_2 \quad \text{a.e. on } \mathbb{R} \Longleftrightarrow U(\mu)J_2U(\mu)^* = J_1 \quad \text{a.e. on } \mathbb{R}. \tag{4.18}$$

In view of (4.14) and (4.16), $\mathcal{P}(J_1, J_2) \subset \mathcal{N}^{m\times m}$ and $\mathcal{U}(J_1, J_2) \subset \Pi^{m\times m}$.

The classes $\mathcal{P}(J), \mathcal{P}^\circ(J), \mathcal{U}(J)$ and

$$\mathcal{U}_0(J) = \{U \in \mathcal{U}(J) : 0 \in \mathfrak{h}_U \quad \text{and} \quad U(0) = I_m\}, \tag{4.19}$$

are semigroups with respect to matrix multiplication. Moreover, if $U_1, U_2, \in \mathcal{P}(J)$, then

(1) $U_1U_2 \in \mathcal{P}^\circ(J) \Longrightarrow U_1 \in \mathcal{P}^\circ(J)$ and $U_2 \in \mathcal{P}^\circ(J)$.

(2) $U_1U_2 \in \mathcal{U}(J) \Longrightarrow U_1 \in \mathcal{U}(J)$ and $U_2 \in \mathcal{U}(J)$.

Furthermore, if $U_1, U_2, \ldots \in \mathcal{P}(J)$ and if $\mathbb{C}_+ \setminus \cap_{j=1}^\infty \mathfrak{h}_{U_j}^+$ is a set of isolated points and if

$$U(\lambda) = \lim_{n\uparrow\infty} U_1(\lambda)U_2(\lambda)\cdots U_n(\lambda) \tag{4.20}$$

or

$$U(\lambda) = \lim_{n\uparrow\infty} U_n(\lambda)\cdots U_2(\lambda)U_1(\lambda) \tag{4.21}$$

exists for every point $\lambda \in \cap_{j=1}^\infty \mathfrak{h}_{U_j}^+$, then $U \in \mathcal{P}(J)$. Moreover, if $U \in \mathcal{P}^\circ(J)$, then

$$\mathfrak{h}_U^+ = \cap_{j=1}^\infty \mathfrak{h}_{U_j}^+;$$

see Theorem 4.60, below.

An **infinite right product** $U_1(\lambda)U_2(\lambda)\cdots$ of mvf's $U_j \in \mathcal{P}^\circ(J)$ is said to be **convergent** if the limit $U(\lambda)$ in formula (4.20) exists for every $\lambda \in \mathfrak{h}_U^+$ and $U \in \mathcal{P}^\circ(J)$. In this case we shall write

$$U(\lambda) = \overset{\curvearrowright}{\prod_{j=1}^{\infty}} U_j(\lambda). \tag{4.22}$$

Infinite left products can be defined analogously:

$$U(\lambda) = \overset{\curvearrowleft}{\prod_{j=1}^{\infty}} U_j(\lambda). \tag{4.23}$$

Lemma 4.2 *The following facts will be useful:*

(1) $U \in \mathcal{P}(J) \Longleftrightarrow U^{\sim} \in \mathcal{P}(J)$.

(2) $U \in \mathcal{U}(J) \Longleftrightarrow U^{\sim} \in \mathcal{U}(J)$.

(3) $\mathcal{P}(I_m) = \mathcal{S}^{m\times m}$ *and* $\mathcal{U}(I_m) = \mathcal{S}_{in}^{m\times m}$.

(4) $\mathcal{P}(-I_m) = \{s(\lambda) : s(\lambda)$ *is meromorphic in* $\mathbb{C}_+$ *and* $s(\lambda)^* s(\lambda) \geq I_m$ *for* $\lambda \in \mathfrak{h}_s^+\}$ *and* $\mathcal{U}(-I_m) = \{U : U^{-1} \in \mathcal{S}_{in}^{m\times m}\}$.

Proof The first two depend upon facts established in Chapter 2. □

Example 4.3 *Let* $s_{11} \in \mathcal{S}^{p\times p}$ *and* $s_{22} \in \mathcal{S}^{q\times q}$ *and suppose that* $\det s_{22} \not\equiv 0$ *in* $\mathbb{C}_+$*. Then the mvf*

$$W = \begin{bmatrix} s_{11} & 0 \\ 0 & s_{22}^{-1} \end{bmatrix} \in \mathcal{P}(j_{pq}).$$

Moreover,

$$W \in \mathcal{U}(j_{pq}) \Longleftrightarrow s_{11} \in \mathcal{S}_{in}^{p\times p} \quad \text{and} \quad s_{22} \in \mathcal{S}_{in}^{q\times q}.$$

Example 4.4 *Let* $c \in \mathcal{C}^{p\times p}$*. Then the mvf*

$$A(\lambda) = \begin{bmatrix} I_p & c(\lambda) \\ 0 & I_p \end{bmatrix} \in \mathcal{P}(J_p).$$

Moreover, $A \in \mathcal{U}(J_p) \Longleftrightarrow \Re c(\mu) = 0$ *a.e. on* $\mathbb{R}$*, i.e.,* $c \in \mathcal{C}_{sing}^{p\times p}$.

Lemma 4.5 *If* $U \in \mathcal{U}(J)$*, then*

(1) $|\det U(\mu)| = 1$ *a.e. on* $\mathbb{R}$.

(2) *There exist a pair of scalar inner functions* $b_1, b_2 \in \mathcal{S}_{in}$ *such that*

$$\det U(\lambda) = \frac{b_1(\lambda)}{b_2(\lambda)}.$$

(3) $\operatorname{rank} U(\lambda) = m$ *if* $\lambda \in \mathfrak{h}_U \cap \mathfrak{h}_{U^\#}$.

Proof Assertion (1) is immediate from (4.4); (2) then follows from the fact that $\det U \in \mathcal{N}$ and the representation (3.53) for functions in $\mathcal{N}$; and (3) is immediate from (2). □

4.2 Blaschke-Potapov products

The simplest examples of mvf's $U \in \mathcal{U}(J)$ are rational J-inner mvf's with one simple pole $\overline{\omega}$. They are called **elementary BP (Blaschke-Potapov) factors**. If $J = I_m$, or $J = -I_m$, then there is only one kind of elementary BP factor: with $\omega \in \mathbb{C}_+$ or $\omega \in \mathbb{C}_-$, respectively. If $J \neq \pm I_m$, then there are four kinds and an elementary BP factor is said to be of the first, second, third or fourth kind, according as $\omega \in \mathbb{C}_+$, $\omega \in \mathbb{C}_-$, $\omega \in \mathbb{R}$ or $\omega = \infty$. Let

$$b_\omega(\lambda) \;=\; \gamma_\omega \frac{\lambda - \omega}{\lambda - \overline{\omega}} \text{ with } |\gamma_\omega| = 1, \quad \text{if } \omega \notin \mathbb{R}$$

and

$$c_\omega(\lambda) \;=\; \frac{1}{\pi i(\omega - \lambda)} \quad \text{if } \omega \in \mathbb{R},$$

and let $U \in \mathcal{U}_{const}(J)$.

An elementary BP factor of the first or second kind may be represented in the form

$$U_\omega(\lambda) = U(I_m + (b_\omega(\lambda) - 1)P), \tag{4.24}$$

where

$$P = P^2 \quad \text{and} \quad \begin{cases} PJ \geq 0 & \text{if } \omega \in \mathbb{C}_+ \\ PJ \leq 0 & \text{if } \omega \in \mathbb{C}_- \end{cases}$$

An elementary BP factor of the third or fourth kind may be represented in the form

$$U_\omega(\lambda) = U(I_m - c_\omega(\lambda)E) = U\exp\{-c_\omega(\lambda)E\} \quad \text{if } \omega \in \mathbb{R}$$

and

$$U_\infty(\lambda) = U(I_m + i\lambda E) = U\exp\{i\lambda E\} \quad \text{if } \omega = \infty,$$

respectively, where

$$E^2 = 0, \quad \text{and} \quad EJ \geq 0.$$

An elementary BP factor $U_\omega(\lambda)$ is said to be a **primary** BP factor if $\operatorname{rank} P = 1$ or $\operatorname{rank} E = 1$ in the preceding representations. It is easy to check that

$$\begin{aligned}
P^2 = P,\ PJ &\geq 0 \quad \text{and} \quad \operatorname{rank} P = 1 \\
&\Longleftrightarrow \quad P = vv^*J \quad \text{for some } v \in \mathbb{C}^m \text{ with } v^*Jv = 1. \\
P^2 = P,\ PJ &\leq 0 \quad \text{and} \quad \operatorname{rank} P = 1 \\
&\Longleftrightarrow \quad P = -vv^*J \quad \text{for some } v \in \mathbb{C}^m \text{ with } v^*Jv = -1. \\
E^2 = 0,\ EJ &\geq 0 \quad \text{and} \quad \operatorname{rank} E = 1 \\
&\Longleftrightarrow \quad E = vv^*J \quad \text{for some } v \in \mathbb{C}^m \text{ with } v^*Jv = 0.
\end{aligned}$$

An elementary BP factor $U_\omega(\lambda)$ is said to be **normalized at the point** α if the constant J-unitary factor U is taken equal to I_m in the preceding representations and if, in the factors of the first and second kind, $b_\omega(\alpha) > 0$.

Lemma 4.6 *Let $U_\omega(\lambda)$ be an elementary BP factor of the first or second kind with pole at $\overline{\omega}$ that is normalized by setting $U = I_m$ and $b_\omega(\alpha) > 0$ in formula (4.24). Then:*

(1) $U_\omega(\alpha) = \exp\{-HJ\}$, *where*

$$H = -(\ln b_\omega(\alpha))PJ \geq 0,$$

i.e., in view of Theorem 2.64, $U_\omega(\alpha)$ is a J modulus.

(2) $\exp\{-\|H\|\} \leq \|U_\omega(\alpha)^{\pm 1}\| \leq \exp\{\|H\|\}$.

(3) $\exp\{-\operatorname{trace} H\} \leq \|U_\omega(\alpha)^{\pm 1}\| \leq \exp\{\operatorname{trace} H\}$.

Proof Since $P^2 = P$, it is readily checked that

$$\begin{aligned}
e^{\beta P} &= \sum_{j=0}^{\infty} \frac{\beta^j P^j}{j!} = I_m + \left(\sum_{j=1}^{\infty} \frac{\beta^j}{j!} \right) P \\
&= I_m + (e^\beta - 1)P
\end{aligned}$$

for every $\beta \in \mathbb{C}$. Thus,

$$I_m + (b_\omega(\alpha) - 1)P = \exp\{\ln b_\omega(\alpha) P\},$$

which leads easily to (1). Assertion (2) then follows from the bounds

$$\|U_\omega(\alpha)^{\pm 1}\| = \|\exp\{\mp HJ\}\| \leq \exp\{\|HJ\|\} = \exp\{\|H\|\}$$

and

$$1 = \|U_\omega(\alpha)U_\omega(\alpha)^{-1}\| \leq \|U_\omega(\alpha)\|\|U_\omega(\alpha)^{-1}\|.$$

Finally (3) follows from (2) and the inequality $\|H\| \leq \operatorname{trace} H$, which is in force, since $H \geq 0$. □

Clearly the finite product of elementary BP factors is a rational J-inner mvf. The converse is also true:

Theorem 4.7 *Every rational J-inner mvf $U(\lambda)$ can be represented as a finite product of elementary BP factors. Moreover, if the McMillan degree of $U(\lambda)$ is equal to r, then $U(\lambda)$ may be expressed as the product of r primary BP factors that are each normalized at a point α times a constant J-unitary matrix on either the left or the right.*

Proof This is proved in Section 5.8. □

Infinite BP products Conditions for the convergence of the right infinite product of elementary BP factors $U_k(\lambda)$, $k \geq 1$

$$\prod_{k=1}^{\overset{\curvearrowright}{\infty}} U_k(\lambda) = U_1(\lambda)U_2(\lambda)\cdots U_n(\lambda)\cdots$$

and the left infinite product of elementary BP factors

$$\prod_{k=1}^{\overset{\curvearrowleft}{\infty}} U_k(\lambda) = \cdots U_n(\lambda)\cdots U_2(\lambda)U_1(\lambda)$$

that were established by Potapov [Po60] will be presented below.

Remark 4.8 *Let $\{U_{\omega_k}(\lambda)\}_{k=1}^{\infty}$ be a sequence of elementary BP factors such that the right infinite BP product*

$$B^{(r)}(\lambda) = \prod_{k=1}^{\overset{\curvearrowright}{\infty}} U_{\omega_k}(\lambda)$$

is well defined. Then, as will be shown subsequently in Theorem 4.60 from more general considerations,

$$\mathfrak{h}^+_{B^{(r)}} = \bigcap_{k=1}^{\infty} \mathfrak{h}^+_{U_{\omega_k}} = \mathbb{C}_+ \setminus \{\overline{\omega}_k : \omega_k \in \mathbb{C}_-\}$$

and

$$\mathfrak{h}^+_{B^{(r)-1}} = \bigcap_{k=1}^{\infty} \mathfrak{h}^+_{U^{-1}_{\omega_k}} = \mathbb{C}_+ \backslash \{\omega_k : \omega_k \in \mathbb{C}_+\}.$$

Consequently, both sets $\{\overline{\omega}_k : \omega_k \in \mathbb{C}_-\}$ *and* $\{\omega_k : \omega_k \in \mathbb{C}_+\}$ *have no limit points in* $\mathbb{C}_+$ *under the given assumptions. The same conclusions are valid if the left infinite BP product* $B^{(l)}(\lambda)$ *of these factors is well defined. Consequently, if* $B^{(r)}(\lambda)$ *or* $B^{(l)}(\lambda)$ *is well defined, then the set*

$$\mathbb{C}_+ \setminus (\{\omega_k : \omega_k \in \mathbb{C}_+\} \cup \{\overline{\omega}_k : \omega_k \in \mathbb{C}_-\})$$

is open.

Theorem 4.9 *Let* $\{U_{\omega_k}\}_{k\geq 1}$ *be an infinite sequence of elementary BP factors, that are normalized at a point*

$$\alpha \in \mathbb{C}_+ \setminus (\{\omega_k : \omega_k \in \mathbb{C}_+\} \cup \{\overline{\omega}_k : \omega_k \in \mathbb{C}_-\}) :$$

$$U_{\omega_k}(\lambda) = \begin{cases} I_m + (b_{\omega_k}(\lambda) - 1)P_k & \text{if } \omega_k \in \mathbb{C} \setminus \mathbb{R} \\ I_m - c_{\omega_k}(\lambda)E_k & \text{if } \omega_k \in \mathbb{R} \\ I_m + i\lambda E_k & \text{if } \omega_k = \infty \end{cases},$$

where

$$P_k^2 = P_k,\ b_{\omega_k}(\alpha) > 0,\ H_k = -\{\ln b_{\omega_k}(\alpha)\} P_k J \geq 0 \quad \text{if} \quad \omega_k \in \mathbb{C} \setminus \mathbb{R},$$

and

$$E_k^2 = 0 \text{ and } H_k = E_k J \geq 0 \quad \text{if} \quad \omega_k \in \mathbb{R} \quad \text{or} \quad \omega_k = \infty.$$

Then the following conditions are equivalent:

(1) *The right BP product* $\overset{\curvearrowright}{\prod_{k=1}^{\infty}} U_{\omega_k}(\lambda)$ *converges.*

(2) *The left BP product* $\overset{\curvearrowleft}{\prod_{k=1}^{\infty}} U_{\omega_k}(\lambda)$ *converges.*

(3) *The series* $\sum_{k=1}^{\infty} \operatorname{trace} H_k$ *converges.*

Moreover, if condition (3) is satisfied then the right and left BP products, considered in (1) and (2) converge uniformly on every compact set

$$K \subset \mathbb{C} \setminus \Omega \quad \text{where } \Omega = \text{ the closure of } \{\omega_k, \overline{\omega_k} : k \geq 1\}.$$

Proof Suppose first that condition (3) is in force, let K be a compact subset of $\mathbb{C}\setminus\Omega$ and let

$$\beta_{\omega_k}(\lambda)=\begin{cases}|b_{\omega_k}(\lambda)-1| & \text{if } \omega_k\in\mathbb{C}\setminus\mathbb{R},\\ |c_{\omega_k}(\lambda)| & \text{if } \omega_k\in\mathbb{R},\\ |\lambda| & \text{if } \omega_k=\infty.\end{cases}$$

If $\omega_k\notin\mathbb{R}$, then the identity

$$1-b_{\omega_k}(\lambda)=\frac{1-b_{\omega_k}(\alpha)}{b_{\omega_k}(\alpha)}\left[(1+b_{\omega_k}(\alpha))\frac{(\alpha-\overline{\omega}_k)(\lambda-\overline{\alpha})}{(\alpha-\overline{\alpha})(\lambda-\overline{\omega}_k)}-1\right] \tag{4.25}$$

helps to guarantee the existence of a number $c>0$ such that

$$\beta_{\omega_k}(\lambda)\le\begin{cases}c|1-b_{\omega_k}(\alpha)| & \text{if } \lambda\in K \text{ and } \omega_k\in\mathbb{C}\setminus\mathbb{R}\\ c & \text{if } \lambda\in K \text{ and } \omega_k\in\mathbb{R}\cup\{\infty\}\end{cases}$$

since the functions $(\lambda-\overline{\alpha})/(\lambda-\overline{\omega}_k)$ are uniformly bounded on K for $k\ge1$ and the sequences $\{b_{\omega_k}(\alpha):\, k\ge1\}$ and $\{b_{\omega_k}(\alpha)^{-1}:\, k\ge1\}$ are bounded. Thus, as

$$\frac{1}{t}|1-x|\le\left|\int_x^1\frac{1}{y}dy\right|=|\ln x|\le t|1-x| \quad \text{if } \frac{1}{t}\le x\le t$$

for some $t>1$, it follows that

$$\|U_{\omega_k}(\lambda)-I_m\|\le c_1\,\mathrm{trace}\, H_k \quad \text{for } k\ge1.$$

Therefore, the bound

$$\left\|\overset{\curvearrowright}{\prod_{k=n}^{n+\ell}}U_{\omega_k}(\lambda)-I_m\right\|<\exp\left\{c_1\sum_{k=n}^{n+\ell}\mathrm{trace} H_k\right\}-1 \quad \text{for } \ell\ge1,$$

is uniformly satisfied for $\lambda\in K$; and the same estimate holds for

$$\left(\overset{\curvearrowright}{\prod_{k=n}^{n+\ell}}U_{\omega_k}(\lambda)\right)^{-1}, \quad \text{since } U_{\omega_k}(\lambda)^{-1}=JU^{\#}_{\omega_k}(\lambda)J.$$

Moreover, the sequence of partial right products $P_n(\lambda)$ of the mvf's $U_k(\lambda)$ is uniformly bounded on K:

$$\begin{aligned}\|P_n(\lambda)\| &\leq \prod_{k=1}^{n} \|U_{\omega_k}(\lambda)\| \leq \prod_{k=1}^{n} (\|U_{\omega_k}(\lambda) - I_m\| + 1) \\ &\leq \exp\left\{\sum_{k=1}^{n} \|U_{\omega_k}(\lambda) - I_m\|\right\} \\ &\leq \exp\left\{\sum_{k=1}^{\infty} \|U_{\omega_k}(\lambda) - I_m\|\right\} < M < \infty.\end{aligned}$$

Thus, as

$$\|P_{n+\ell}(\lambda) - P_n(\lambda)\| \leq \|P_n(\lambda)\| \left\| \prod_{k=n+1}^{n+\ell} U_{\omega_k}(\lambda) - I_m \right\|,$$

the sequence $P_n(\lambda)$ converges uniformly in K to an invertible bounded mvf $P(\lambda)$. Moreover, since the partial products $P_n(\lambda)^{-1} = JP_n^{\#}(\lambda)J$ are subject to similar bounds, it follows that $P(\lambda)^{-1}$ is bounded on K and that

$$\|P_n(\lambda)^{-1} - P(\lambda)^{-1}\| \leq \|P_n(\lambda)^{-1}\| \|P(\lambda) - P_n(\lambda)\| \|P(\lambda)^{-1}\|.$$

The implication (3) $\Longrightarrow$ (2) follows from (3) $\Longrightarrow$ (1) applied to

$$\prod_{k=1}^{\curvearrowleft \infty} U_{\omega_k}(\lambda)^{\tau}.$$

The implication (1) $\Longrightarrow$ (3) is more difficult to verify. A proof based on deep arguments that rested on the J modulus was given by V. P. Potapov in [Po60], where he introduced this notion, studied its properties and, in particular, applied them to prove the implication (1) $\Longrightarrow$ (3). Here the implication (1) $\Longrightarrow$ (3) will be presented under the assumption that all the factors in the product are elementary BP factors of the first or second kind. Then, by Lemma 4.6,

$$U_{\omega_k}(\alpha) = \exp\{-H_k J\}, \quad \text{where } H_k = -(\ln b_{\omega_k}(\alpha)) P_k J \geq 0,$$

and hence, in view of (1) in Theorem 2.65 the matrices $U_{\omega_k}(\alpha)$ are J moduli. Consequently, the implication (1) $\Longrightarrow$ (3) follows from assertion (4) in Theorem 2.65. □

A finite or infinite convergent right (resp., left) product of elementary BP factors

$$B^{(r)}(\lambda) = \overset{n}{\overset{\curvearrowright}{\prod_{k=1}}} U_{\omega_k}(\lambda) \quad (B^{(l)}(\lambda) = \overset{n}{\overset{\curvearrowleft}{\prod_{k=1}}} U_{\omega_k}(\lambda)), \quad n \leq \infty,$$

is called a **right BP product** (respect., **left BP product**).

Theorem 4.10 *Let U be a right or left BP product. Then:*

(1) $U \in \mathcal{U}(J)$.

(2) *If U is a right BP product and $\alpha \in \mathfrak{h}_U^+ \cap \mathfrak{h}_{U^{-1}}^+$, then*

$$U(\lambda) = B^{(r)}(\lambda)\, U^{(r)},$$

where $B^{(r)}(\lambda)$ is a right BP product of elementary factors that are normalized at the point α and $U^{(r)} \in \mathcal{U}_{const}(J)$.

(3) *If U is a left BP product and $\alpha \in \mathfrak{h}_U^+ \cap \mathfrak{h}_{U^{-1}}^+$, then*

$$U(\lambda) = U^{(l)}\, B^{(l)}(\lambda),$$

where $B^{(l)}(\lambda)$ is a left BP product of elementary factors that are normalized at the point α and $U^{(l)} \in \mathcal{U}_{const}(J)$.

Proof Assertion (1) coincides with Corollary 4.63, which is presented later. Assertions (2) and (3) are due to Potapov; see [Po60]. □

Theorem 4.11 (V. P. Potapov) *Let $U \in \mathcal{P}^{\circ}(J)$. Then there exist two multiplicative representations:*

$$U(\lambda) = B^{(r)}(\lambda) U^{(r)}(\lambda) \text{ and } U(\lambda) = U^{(l)}(\lambda) B^{(l)}(\lambda), \tag{4.26}$$

where $B^{(r)}(\lambda)$ and $B^{(l)}(\lambda)$ are right and left BP products of elementary factors of the first and second kind, and the two mvf's $U^{(r)}(\lambda)$ and $U^{(\ell)}(\lambda)$ are holomorphic and invertible in $\mathbb{C}_+$. Moreover, both factorizations in (4.26) are unique, up to constant J-unitary factors V_1 and V_2:

$$B^{(r)}(\lambda) \longmapsto B^{(r)}(\lambda)V_1 \quad \text{and} \quad U^{(r)}(\lambda) \longmapsto V_1^{-1} U^{(r)}(\lambda),$$
$$B^{(l)}(\lambda) \longmapsto V_2 B^{(l)}(\lambda) \quad \text{and} \quad U^{(l)}(\lambda) \longmapsto U^{(l)}(\lambda) V_2^{-1}.$$

The order of the elementary factors in the BP product $U(\lambda)$ may be chosen arbitrarily. This choice influences the values of the matrices P_k, but not the values of the poles ω_k.

Proof This theorem is due to Potapov [Po60]; see also [Zol03]. □

4.3 Entire J-inner mvf's

A finite product of normalized elementary BP factors of the fourth kind may be expressed as a multiplicative integral:

$$\begin{aligned} U(\lambda) &= \overset{n}{\underset{k=1}{\overset{\curvearrowright}{\prod}}}(I_m + i\lambda H_k J) = \overset{n}{\underset{k=1}{\overset{\curvearrowright}{\prod}}} \exp\{i\lambda H_k J\} \\ &= \overset{n}{\underset{0}{\overset{\curvearrowright}{\int}}} \exp\{i\lambda dM(t)\}, \end{aligned}$$

where $H_k \geq 0, \quad H_k J H_k = 0, \quad I \leq k \leq n$, and

$$M(t) = \begin{cases} H_1 t & \text{for } 0 \leq t \leq 1 \\ \sum_{k=1}^{m-1} H_k + H_m(t - m + 1) & \text{for } m-1 \leq t \leq m,\ 2 \leq m \leq n, \end{cases}$$

is a continuous nondecreasing $m \times m$ mvf on the interval $[0, n]$.

Here trace $M(n) = \sum_{k=1}^{n}$ trace H_k and $U \in \mathcal{U}(J)$ is a polynomial of degree $\leq n$ with $U(0) = I_m$.

Remark 4.12 *Every J-inner mvf $U(\lambda)$ that is a polynomial of degree n, normalized by the condition $U(0) = I_m$ has such a multiplicative representation.*

A convergent infinite product of normalized elementary BP factors of the fourth kind may be rewritten also in the form of a multiplicative integral. Indeed, let

$$U(\lambda) = \overset{\infty}{\underset{k=1}{\overset{\curvearrowright}{\prod}}}(I_m + i\lambda H_k J) = \overset{\infty}{\underset{k=1}{\overset{\curvearrowright}{\prod}}} \exp\{i\lambda H_k J\},$$

where

$$H_k \geq 0, \quad H_k J H_k = 0,\ k \geq 1,\ \text{and} \sum_{k=1}^{\infty} \text{trace } H_k < \infty.$$

Let

$$M(t) = \begin{cases} 2H_1 t & \text{for } 0 \le t \le \frac{1}{2} \\ \sum_{k=1}^{n-1} H_k + 2^n H_n(t - 1 + \frac{1}{2^{n-1}}) \text{ for } 1 - \frac{1}{2^{n-1}} \le t \le 1 - \frac{1}{2^n} \end{cases} \tag{4.27}$$

and $M(1) = \sum_{k=1}^{\infty} H_k$. Then $M(t)$ is a continuous nondecreasing $m \times m$ mvf on the interval $[0, 1]$ such that $M(0) = 0$ and trace$M(1) = \sum_{k=1}^{\infty}$ trace $H_k < \infty$ and

$$U(\lambda) = \overset{\curvearrowright}{\int_0^1} \exp\{i\lambda dM(t)J\}. \tag{4.28}$$

The mvf U belongs to $\mathcal{E} \cap \mathcal{U}(J)$ because it is the uniform limit of mvf's in this class on every compact subset of $\mathbb{C}$. Moreover, $U(0) = I_m$ and U has exponential type $\tau(U) = 0$. Convergent infinite products of the normalized elementary BP factors of the fourth kind with $J = J_p$ arise as the class of resolvent matrices $A(\lambda)$ that serve to describe the set of solutions of matrix generalizations of the Hamburger moment problem in the completely indeterminate case.

The set of finite and convergent infinite products $U(\lambda)$ of normalized elementary BP factors of the fourth kind is a proper subset of the class of mvf's $U \in \mathcal{E} \cap \mathcal{U}(J)$ with $U(0) = I_m$ and $\tau(U) = 0$. To the best of our knowledge, the problem of characterizing this subset is still open, even for the case $m = 2$. Nevertheless, in view of the following theorem, every mvf $U \in \mathcal{E} \cap \mathcal{U}(J)$ with $U(0) = I_m$ admits a multiplicative integral representation of the form (4.28) for some continuous nondecreasing $m \times m$ mvf $M(t)$ on the interval $[0, 1]$.

Theorem 4.13 (V. P. Potapov) *Let $U \in \mathcal{E} \cap \mathcal{U}(J)$. Then*

$$U(\lambda) = U_0 \overset{\curvearrowright}{\int_0^d} \exp\{i\lambda dM(t)J\}, \tag{4.29}$$

where $U_0 \in \mathcal{U}_{const}(J)$, $M(t)$ is a nondecreasing $m \times m$ mvf in the interval $[0, d]$ that may be chosen so that it is absolutely continuous on $[0, d]$ with $M(0) = 0$ and trace $M(t) = t,\ 0 \le t \le d$.

Conversely, if $U_0 \in \mathcal{U}_{const}(J)$ and $M(t)$ is a nondecreasing $m \times m$ mvf on the closed interval $[0, d]$, then the mvf $U(\lambda)$ is well defined by formula (4.29). Moreover, $U \in \mathcal{E} \cap \mathcal{U}(J)$ and $U(0) = U_0$.

Remark 4.14 *If $M(t)$ is a continuous nondecreasing $m \times m$ mvf on the interval $[0, d]$, then the mvf*

$$U_t(\lambda) = \overset{\curvearrowright}{\int_0^t} \exp\{i\lambda dM(s)J\}, \quad 0 \le t \le d, \tag{4.30}$$

is the unique continuous solution of the integral equation

$$y(t,\lambda) = y(0,\lambda) + i\lambda \int_0^t y(s,\lambda)dM(s)J, \quad 0 \le t \le d, \tag{4.31}$$

that meets the initial condition

$$y(0,\lambda) = I_m.$$

This particular solution $U_t(\lambda)$ is called the **matrizant** *(fundamental solution) of the integral system (4.31) and its value $U_d(\lambda)$ at the right end point of the interval $[0, d]$ is called the* **monodromy matrix** *of the system. Thus, Potapov's theorem guarantees that every mvf $U \in \mathcal{E} \cap \mathcal{U}(J)$ with $U(0) = I_m$ can be identified as the monodromy matrix of an integral system of the form (4.31) based on a continuous nondecreasing $m \times m$ mvf $M(t)$ with $M(0) = 0$.*

Remark 4.15 *If $U_0 = I_m$ and*

$$M(t) = \int_0^t H(s)ds \quad \text{for } 0 \le t \le d, \tag{4.32}$$

where

$$H \in L_1^{m\times m}([0,d]) \quad \text{and} \quad H(t) \ge 0 \text{ a.e. on } [0,d], \tag{4.33}$$

then the matrizant $U_t(\lambda)$ of the integral equation (4.31) is absolutely continuous on the interval $[0, d]$ for every $\lambda \in \mathbb{C}$ and is also the matrizant of the differential system

$$\frac{d}{dt}y(t,\lambda) = i\lambda y(t,\lambda)H(t)J, \quad \text{a.e. on } [0,d], \tag{4.34}$$

i.e., $U_t(\lambda)$ *is the unique absolutely continuous solution of this differential system that satisfies the initial condition*

$$y(0, \lambda) = I_m.$$

Thus, in view of Potapov's theorem, every mvf $U \in \mathcal{E} \cap \mathcal{U}(J)$ *with* $U(0) = I_m$ *is the monodromy matrix of a differential system of the form (4.34) with a mvf* $H(t)$ *that satisfies the conditions (4.33). Conversely, every monodromy matrix* $U(\lambda)$ *of such a differential system belongs to the class* $\mathcal{E} \cap \mathcal{U}(J)$ *with* $U(0) = I_m$ *and then may be written as the multiplicative integral*

$$U(\lambda) = \overset{d}{\underset{0}{\curvearrowright\!\!\!\!\!\!\int}} \exp\{i\lambda H(t)dtJ\} \stackrel{def}{=} \overset{d}{\underset{0}{\curvearrowright\!\!\!\!\!\!\int}} \exp\{i\lambda dM(t)J\},$$

where $M(t)$ *is defined in formula (4.32). The mvf* $H(t)$ *in this representation is not uniquely defined by* $U(\lambda)$*, even if it is subject to the usual normalization condition* trace $H(t) = 1$ *a.e. on* $[0, d]$.

In view of the inclusion

$$\mathcal{E} \cap \mathcal{U}(J) \subset \mathcal{E} \cap \Pi^{m \times m},$$

Theorem 3.108 guarantees that every entire J-inner mvf $U(\lambda)$ is of exponential type and satisfies the Cartwright condition and

$$\tau_+(U) = \tau_U^+ \quad \text{and} \quad \tau_-(U) = \tau_U^-.$$

Therefore,

$$\tau(U) = \max\{\tau_U^+, \tau_U^-\}.$$

The multiplicative integral representation (4.29) yields the bound

$$\tau(U) \leq \text{trace}(M(d)) \tag{4.35}$$

and the formula

$$\det U(\lambda) = \det U(0) \exp\{i\lambda \text{trace}(M(d)J)\}.$$

Thus,

$$\det U(\lambda) = \gamma e_a(\lambda)$$

where $|\gamma| = 1$ and $a \in \mathbb{R}$.

Lemma 4.16 *If $U \in \mathcal{U}(J)$ with $J \neq \pm I_m$ and U_P and U_Q are defined by formula (4.6), then*

$$\tau_U^+ = \tau_{U_Q}^+ \geq 0 \text{ and } \tau_U^- = \tau_{U_P}^- \geq 0$$

Proof The first assertion follows from (4), (5) and (7) of Lemma 2.6; the second follows from (10) and (11) of the same lemma. □

4.4 Multiplicative representations in the class $\mathcal{P}^\circ(J)$

Let $U \in \mathcal{U}(J)$ be a rational mvf with all its poles on $\mathbb{R}$ that is normalized by the condition $U(\infty) = I_m$. Then $U(\lambda)$ can be expressed as a finite product of normalized elementary BP factors of the third kind

$$U(\lambda) = \overset{n}{\overset{\curvearrowright}{\prod_{k=1}}} (I_m - c_{\omega_k}(\lambda) H_k J) = \overset{n}{\overset{\curvearrowright}{\prod_{k=1}}} \exp\{-c_{\omega_k}(\lambda) H_k J\}, \tag{4.36}$$

where

$$\omega_1 \leq \cdots \leq \omega_n, \quad H_k \geq 0 \quad \text{and} \quad H_k J H_k = 0 \text{ for } k = 1, \ldots, n.$$

The mvf $U(\lambda)$ may also be expressed as a multiplicative integral

$$U(\lambda) = \overset{\curvearrowright}{\int_0^d} \exp\left\{\frac{1}{-2\pi i(\alpha(t) - \lambda)} dM(t) J\right\}, \tag{4.37}$$

where $d = 1$,

$$M(t) = \begin{cases} nH_1 t & \text{for } 0 \leq t \leq 1 \\ \sum_{k=1}^{m-1} H_k + H_m(nt - m + 1) & \text{for } \frac{m-1}{n} \leq t \leq \frac{m}{n},\ 2 \leq m \leq n \end{cases}$$

and

$$\alpha(t) = \omega_{[nt]+1} \quad \text{for } 0 \leq t < 1 \text{ and } \alpha(1) = \omega_n.$$

In the representation (4.37) the $m \times m$ mvf $M(t)$ is nondecreasing absolutely continuous on the interval $[0, 1]$ and $\alpha(t)$ is a scalar nondecreasing continuous function on the interval $[0, 1]$.

Theorem 4.17 (V. P. Potapov) *Let $U \in \mathcal{P}^\circ(J)$ and assume that U satisfies the following conditions:*

(1) *The mvf's* $U(\lambda)$ *and* $U(\lambda)^{-1}$ *are holomorphic in* $\mathbb{C}_+$.

(2) $U(\lambda)$ *is holomorphic at the point* $\lambda = \infty$.

(3) $U(\mu)^* JU(\mu) = J$ *for every* $\mu \in \mathbb{R}$ *with* $|\mu| \geq \gamma$ *for some* $\gamma > 0$.

(4) $U(\infty) = I_m$.

Then the mvf $U(\lambda)$ *admits a multiplicative integral representation of the form (4.37), where* $M(t)$ *is a nondecreasing absolutely continuous* $m \times m$ *mvf on the interval* $[0, d]$ *with* $M(0) = 0$ *and* $\alpha(t)$ *is a nondecreasing continuous scalar function on the interval* $[0, d]$.

Conversely, a mvf $U(\lambda)$, *that is defined in* $\mathbb{C}_+$ *by formula (4.37), where* $M(t)$ *is a nondecreasing* $m \times m$ *mvf on the interval* $[0, d]$ *and* $\alpha(t)$ *is a nondecreasing continuous scalar function on the interval* $[0, d]$, *belongs to the class* $\mathcal{P}^\circ(J)$ *and satisfies the conditions (1)–(4).*

Proof The converse part of the theorem is easily checked:

$$U(\lambda) = \lim_{n \uparrow \infty} U_n(\lambda)$$

where

$$U_n(\lambda) = \overset{n}{\curvearrowright}\!\!\!\!\!\prod_{k=1} \exp\{-c\alpha(t_k)(\lambda)[M(t_k) - M(t_{k-1})]J\}$$

and

$$t_k = \frac{kd}{n} \quad \text{for } k = 0, 1, \ldots, n.$$

The mvf's $U_n \in \mathcal{P}^\circ(J)$ and satisfy the conditions (1)–(4). Moreover, $U_n(\lambda)$ converges uniformly to $U(\lambda)$ on every compact subset of $\mathbb{C}_+$ and in a neighborhood of the point $\lambda = \infty$.

The proof of the direct part of the theorem is much more difficult. see Potapov [Po60] and Zolotarev [Zol03], or Brodskii [Bro72] for details. □

Remark 4.18 *Let* $U \in \mathcal{P}^\circ(J)$ *and assume that* U *satisfies conditions (2)-(3) of the previous theorem. Then the factors* $U^{(r)}(\lambda)$ *and* $U^{(\ell)}(\lambda)$ *in (4.26) satisfy the conditions (1)–(3) and can be normalized to meet (4) also.*

Theorem 4.19 *Let* $U \in \mathcal{P}^\circ(J)$ *and assume that* $U(\lambda)$ *and* $U(\lambda)^{-1}$ *are holomorphic in* $\mathbb{C}_+$. *Then*

$$U(\lambda) = U_0 U_1(\lambda) U_2(\lambda) U_3(\lambda),$$

where $U_0 \in \mathcal{U}_{const}(J)$, $U_1, U_3 \in \mathcal{E} \cap \mathcal{U}(J)$, $U_1(0) = U_3(0) = I_m$ *and* $U_2(\lambda)$ *admits a multiplicative integral representation*

$$U_2(\lambda) = \overset{d}{\underset{0}{\curvearrowright\!\!\!\!\!\int}} \exp\left\{\frac{1 + \alpha(t)\lambda}{-2\pi i(\alpha(t) - \lambda)} dM(t)J\right\} \tag{4.38}$$

where $M(t)$ *is a nondecreasing absolutely continuous* $m \times m$ *mvf on* $[0, d]$ *with* trace $M(t) = t$ *and* $\alpha(t)$ *is a scalar nondecreasing function on the open interval* $(0, d)$.

Proof This follows from Theorem 5.3 of Zolotarev [Zol03]. □

Theorems 4.11, 4.19 and Potapov's theorem yield a multiplicative representation for every mvf $U \in \mathcal{P}^\circ(J)$.

4.5 The classes $\mathcal{P}(j_{pq})$ and $\mathcal{U}(j_{pq})$

Our next objective is to reexpress a number of the preceding formulas in more explicit block form for $J = j_{pq}$.

Theorem 4.20 *Let*

$$W(\lambda) = \begin{bmatrix} w_{11}(\lambda) & w_{12}(\lambda) \\ w_{21}(\lambda) & w_{22}(\lambda) \end{bmatrix} \quad \textit{and} \quad S(\lambda) = \begin{bmatrix} s_{11}(\lambda) & s_{12}(\lambda) \\ s_{21}(\lambda) & s_{22}(\lambda) \end{bmatrix}$$

be the standard four block decompositions of a mvf $W \in \mathcal{P}(j_{pq})$ *and its PG transform* $S = PG(W)$ *(i.e., the* 11 *blocks are* $p \times p$ *and the* 22 *blocks are* $q \times q$*). Then:*

(1) $S \in \mathcal{S}^{m \times m}$, *where* $m = p + q$.

(2) det $w_{22}(\lambda) \not\equiv 0$.

(3) *The block entries in the mvf* $S(\lambda)$ *can be calculated explicitly in terms of the block entries of* $W(\lambda)$:

$$\begin{aligned} S(\lambda) &= \begin{bmatrix} w_{11}(\lambda) & w_{12}(\lambda) \\ 0 & I_q \end{bmatrix} \begin{bmatrix} I_p & 0 \\ w_{21}(\lambda) & w_{22}(\lambda) \end{bmatrix}^{-1} \\ &= \begin{bmatrix} I_p & w_{12}(\lambda) \\ 0 & I_q \end{bmatrix} \begin{bmatrix} w_{11}(\lambda) & 0 \\ 0 & w_{22}(\lambda)^{-1} \end{bmatrix} \begin{bmatrix} I_p & 0 \\ -w_{21}(\lambda) & I_q \end{bmatrix}, \end{aligned}$$

i.e.,

$$S(\lambda) = \begin{bmatrix} w_{11}(\lambda) - w_{12}(\lambda)w_{22}(\lambda)^{-1}w_{21}(\lambda) & w_{12}(\lambda)w_{22}(\lambda)^{-1} \\ -w_{22}(\lambda)^{-1}w_{21}(\lambda) & w_{22}(\lambda)^{-1} \end{bmatrix}. \tag{4.39}$$

(4) $\det s_{22}(\lambda) \not\equiv 0$ *and* $s_{22}(\lambda) = w_{22}(\lambda)^{-1}$.

(5) *Every mvf* $S \in \mathcal{S}^{m\times m}$ *with blocks* $s_{11}(\lambda)$ *and* $s_{22}(\lambda)$ *of sizes* $p \times p$ *and* $q \times q$, *respectively, such that* $\det s_{22}(\lambda) \not\equiv 0$ *is the PG transform of a mvf* $W \in \mathcal{P}(j_{pq})$.

(6) *The mvf* $W(\lambda)$ *can be recovered from its PG transform* $S = PG(W)$ *by the formula* $W = PG(S)$:

$$\begin{aligned} W(\lambda) &= \begin{bmatrix} s_{11}(\lambda) & s_{12}(\lambda) \\ 0 & I_q \end{bmatrix} \begin{bmatrix} I_p & 0 \\ s_{21}(\lambda) & s_{22}(\lambda) \end{bmatrix}^{-1} \\ &= \begin{bmatrix} I_p & s_{12}(\lambda) \\ 0 & I_q \end{bmatrix} \begin{bmatrix} s_{11}(\lambda) & 0 \\ 0 & s_{22}(\lambda)^{-1} \end{bmatrix} \begin{bmatrix} I_p & 0 \\ -s_{21}(\lambda) & I_q \end{bmatrix} \end{aligned} \tag{4.40}$$

(7) $\|s_{12}(\lambda)\| < 1$ *and* $\|s_{21}(\lambda)\| < 1$ *at each point* $\lambda \in \mathbb{C}_+$.

(8) $\mathfrak{h}_W^+ = \mathfrak{h}_{s_{22}^{-1}}^+$.

(9) $\det W(\lambda) = \dfrac{\det s_{11}(\lambda)}{\det s_{22}(\lambda)}$ *at each point* $\lambda \in \mathfrak{h}_W^+$.

(10) $W \in \mathcal{P}^\circ(j_{pq}) \Longleftrightarrow \det s_{11}(\lambda) \not\equiv 0$.

(11) *If* $W \in \mathcal{P}^\circ(j_{pq})$, *then* $\mathfrak{h}_{W^{-1}}^+ = \mathfrak{h}_{s_{11}^{-1}}^+$ *and*

$$W(\lambda)^{-1} = \begin{bmatrix} I_p & 0 \\ s_{21}(\lambda) & I_q \end{bmatrix} \begin{bmatrix} s_{11}(\lambda)^{-1} & 0 \\ 0 & s_{22}(\lambda) \end{bmatrix} \begin{bmatrix} I_p & -s_{12}(\lambda) \\ 0 & I_q \end{bmatrix} \tag{4.41}$$

at every point $\lambda \in \mathfrak{h}_{W^{-1}}^+$.

Proof This is an easy consequence of Lemmas 2.3 and 2.7. □

Remark 4.21 *In view of formulas (4.2) and (4.3), formulas (4.39) and (4.40) may be reexpressed as*

$$S = \begin{bmatrix} I_p & -w_{12} \\ 0 & -w_{22} \end{bmatrix}^{-1} \begin{bmatrix} w_{11} & 0 \\ w_{21} & -I_q \end{bmatrix} \quad \text{and} \quad W = \begin{bmatrix} I_p & -s_{12} \\ 0 & -s_{22} \end{bmatrix}^{-1} \begin{bmatrix} s_{11} & 0 \\ s_{21} & -I_q \end{bmatrix}.$$

Corollary 4.22 *Let w_{ij} denote the blocks in the standard four block decomposition of a mvf $W \in \mathcal{P}(j_{pq})$. Then:*

(1) $\mathfrak{h}_W^+ = \mathfrak{h}_{w_{22}}^+$.

(2) $\|w_{22}(\lambda)\| \le \|W(\lambda)\| \le 3\|w_{22}(\lambda)\|$ *for every point* $\lambda \in \overline{\mathbb{C}_+} \cap \mathfrak{h}_W$.

If $W \in \mathcal{P}^\circ(j_{pq})$ and $S = PG(W)$, then

(3) $s_{11}(\lambda) = w_{11}(\lambda) - w_{12}(\lambda)w_{22}(\lambda)^{-1}w_{21}(\lambda)$ *for* $\lambda \in \mathfrak{h}_W^+$, $\overline{\mathbb{C}_+} \cap \mathfrak{h}_{W^{-1}} = \overline{\mathbb{C}_+} \cap \mathfrak{h}_{s_{11}^{-1}}$ *and*

$$\|s_{11}(\lambda)^{-1}\| \le \|W^{-1}(\lambda)\| \le 3\|s_{11}(\lambda)^{-1}\|$$

for every point $\lambda \in \overline{\mathbb{C}_+} \cap \mathfrak{h}_{W^{-1}}$.

If $W \in \mathcal{U}(j_{pq})$, then:

(4) $\mathfrak{h}_{W^\#}^+ = \mathfrak{h}_{w_{11}^\#}^+$.

(5) $\|w_{11}^\#(\lambda)\| \le \|W^\#(\lambda)\| \le 3\|w_{11}^\#(\lambda)\|$ *for every point* $\lambda \in \overline{\mathbb{C}_+} \cap \mathfrak{h}_{W^\#}$.

Proof Assertions (2), (3) and (5) follow from the corresponding inequalities in Lemma 2.7. Assertions (1) and (4) follow from (8) and (11) of Theorem 4.20, since $s_{22}^{-1} = w_{22}$ and $s_{11}^{-1} = w_{11}^\#$ when $U \in \mathcal{U}(j_{pq})$ (as is spelled out in Lemma 4.24). □

Corollary 4.23 *Let w_{ij} denote the blocks in the standard four block decomposition of a mvf $W \in \mathcal{P}(j_{pq})$. Then:*

(1) $W \in \widetilde{L}_2^{m\times m} \iff w_{22} \in \widetilde{L}_2^{q\times q}$.

(2) $W \in L_\infty^{m\times m} \iff w_{22} \in L_\infty^{q\times q}$.

If $W \in \mathcal{U}(j_{pq})$, then also

(3) $W \in \widetilde{L}_2^{m\times m} \iff w_{11} \in \widetilde{L}_2^{p\times p} \iff (I_p - s_{12}s_{12}^*)^{-1} \in \widetilde{L}_1^{p\times p} \iff (I_q - s_{21}s_{21}^*)^{-1} \in \widetilde{L}_1^{q\times q}$.

(4) $W \in L_\infty^{m\times m} \iff w_{11} \in L_\infty^{p\times p} \iff \|s_{12}\|_\infty < 1 \iff \|s_{21}\|_\infty < 1$.

Proof This is immediate from the inequalities in the preceding corollary and the fact that $W^{-1}(\lambda) = j_{pq}W^\#(\lambda)j_{pq}$ for every point $\lambda \in \mathfrak{h}_W \cap \mathfrak{h}_{W^\#}$ when $W \in \mathcal{U}(j_{pq})$. □

Lemma 4.24 *Let $W \in \mathcal{U}(j_{pq})$ and let $S = PG(W)$. Then:*

(1) $S \in \mathcal{S}_{in}^{m\times m}$.

(2) $s_{11}(\lambda) = w_{11}^{\#}(\lambda)^{-1}$ *for every point* $\lambda \in \mathfrak{h}_{w_{11}^{-\#}}$.

(3) $s_{12} = (w_{11}^{\#})^{-1} w_{21}^{\#}$ *and* $s_{21} = -w_{12}^{\#}(w_{11}^{\#})^{-1}$.

Conversely, if the lower $q \times q$ *diagonal block* s_{22} *of* $S \in \mathcal{S}_{in}^{m\times m}$ *satisfies the condition* $\det s_{22}(\lambda) \not\equiv 0$, *then* $S = PG(W)$ *for some mvf* $W \in \mathcal{U}(j_{pq})$.

Proof The first assertion (1) follows from Lemma 2.3, whereas the second follows from the identity

$$W(\lambda) j_{pq} W^{\#}(\lambda) = j_{pq}, \quad \text{for} \quad \lambda \in \mathfrak{h}_W \cap \mathfrak{h}_{W^{\#}}.$$

which implies that

$$W^{\#}(\lambda) = j_{pq} W(\lambda)^{-1} j_{pq}$$

and hence, in view of formula (4.41), that (2) and (3) hold. □

Lemma 4.25 *The block entries in the PG transform* S *of* $W \in \mathcal{U}(j_{pq})$ *are subject to the following two constraints:*

(1) $w_{11}^{\#}(\lambda) s_{11}(\lambda) = I_p$ *for every point* $\lambda \in \overline{\mathbb{C}_+} \cap \mathfrak{h}_{W^{\#}}$.
(2) $w_{22}(\lambda) s_{22}(\lambda) = I_q$ *for every point* $\lambda \in \overline{\mathbb{C}_+} \cap \mathfrak{h}_W$.

They are also subject to the following four inequalities, at every point $\lambda \in \mathbb{C}_+$, *with equality at at almost all points* $\lambda \in \mathbb{R}$:

(3) $I_q - s_{21}(\lambda) s_{21}(\lambda)^* \geq s_{22}(\lambda) s_{22}(\lambda)^*$.

(4) $I_q - s_{12}(\lambda)^* s_{12}(\lambda) \geq s_{22}(\lambda)^* s_{22}(\lambda)$.

(5) $I_p - s_{21}(\lambda)^* s_{21}(\lambda) \geq s_{11}(\lambda)^* s_{11}(\lambda)$.

(6) $I_p - s_{12}(\lambda) s_{12}(\lambda)^* \geq s_{11}(\lambda) s_{11}(\lambda)^*$.

Proof Assertions (1) and (2) are obtained in Lemma 4.24 and Theorem 4.20, respectively. The inequalities (3)–(6) are in force at every point $\lambda \in \mathbb{C}_+$ with equality a.e. on $\mathbb{R}$, since $S \in \mathcal{S}_{in}^{m\times m}$. □

4.6 Associated pairs of the first kind

In view of Theorem 4.20, the block diagonal entries in the PG transform of a mvf $W \in \mathcal{P}^{\circ}(j_{pq})$ satisfy the conditions

$$s_{11} \in \mathcal{S}^{p\times p}, \quad \det s_{11}(\lambda) \not\equiv 0, \quad s_{22} \in \mathcal{S}^{q\times q} \quad \text{and} \ \det s_{22}(\lambda) \not\equiv 0.$$

Therefore, s_{11} and s_{22} admit both left and right inner-outer factorizations. In particular we shall always write

$$s_{11}(\lambda) = b_1(\lambda)\varphi_1(\lambda) \qquad \text{and} \qquad s_{22}(\lambda) = \varphi_2(\lambda)b_2(\lambda), \tag{4.42}$$

where $b_1 \in \mathcal{S}_{in}^{p\times p}$, $\varphi_1 \in \mathcal{S}_{out}^{p\times p}$, $b_2 \in \mathcal{S}_{in}^{q\times q}$ and $\varphi_2 \in \mathcal{S}_{out}^{p\times p}$. We shall refer to the mvf's $b_1(\lambda)$ and $b_2(\lambda)$, which are determined up to a constant unitary factor (on the right for $b_1(\lambda)$ and on the left for $b_2(\lambda)$) as an **associated pair** of $W(\lambda)$ and shall write $\{b_1, b_2\} \in ap(W)$, for short. If $\{b_1, b_2\} \in ap(W)$, then $ap(W) = \{\{b_1u, vb_2\} : u$ and v are constant unitary matrices of sizes $p \times p$ and $q \times q\}$.

Lemma 4.26 *If $W \in \mathcal{P}^\circ(j_{pq})$ and if $\{b_1, b_2\} \in ap(W)$, then:*

(1) $\mathfrak{h}_W^+ = \mathfrak{h}_{b_2^{-1}}^+$ *and* $\mathfrak{h}_{W^{-1}}^+ = \mathfrak{h}_{b_1^{-1}}^+$.

(2) $W \in \mathcal{N}_+^{m\times m} \iff w_{22} \in \mathcal{N}_{out}^{q\times q} \iff b_2(\lambda) = constant.$

(3) $W^{-1} \in \mathcal{N}_+^{m\times m} \iff w_{11}^\# \in \mathcal{N}_{out}^{p\times p} \iff b_1(\lambda) = constant.$

Proof The first assertion follows from Corollary 4.22 and the definition of $ap(W)$. The remaining two assertions are a straightforward consequence of the definitions and formulas (4.39), (4.41) and Lemma 4.24. □

Lemma 4.27 *Let $W \in \mathcal{U}(j_{pq})$ and let b_j and φ_j, $j = 1, 2$, be defined for W as in (4.42). Then there exists a constant $\gamma \in \mathbb{C}$ with $|\gamma| = 1$ such that:*

$$\det \varphi_1(\lambda) = \gamma \ \det \varphi_2(\lambda) \tag{4.43}$$

for every point $\lambda \in \mathbb{C}_+$ and

$$\det W(\lambda) = \gamma \frac{\det b_1(\lambda)}{\det b_2(\lambda)} \tag{4.44}$$

for every point $\lambda \in \mathbb{C}_+$ at which $\det b_2(\lambda) \neq 0$.

Proof Lemma 4.25 implies that

$$\begin{aligned}
\det\{s_{22}(\mu)s_{22}(\mu)^*\} &= \det\{I_q - s_{21}(\mu)s_{21}(\mu)^*\} \\
&= \det\{I_p - s_{21}(\mu)^*s_{21}(\mu)\} \\
&= \det\{s_{11}(\mu)^*s_{11}(\mu)\}
\end{aligned}$$

for almost all points $\mu \in \mathbb{R}$ and hence that

$$|\det \varphi_2(\mu)| = |\det \varphi_1(\mu)|$$

for almost all points $\mu \in \mathbb{R}$.

This serves to complete the proof of (1), since a scalar outer function is determined uniquely by its modulus on $\mathbb{R}$, up to a multiplicative constant factor of modulus one. The second assertion is an immediate consequence of the first and item (9) in Theorem 4.20. □

Lemma 4.28 *Let W and W_1 belong to $\mathcal{P}^\circ(j_{pq})$, let $\{b_1, b_2\} \in ap(W)$ and $\{b_1^{(1)}, b_2^{(1)}\} \in ap(W_1)$ and suppose that $W_1^{-1}W \in \mathcal{P}^\circ(j_{pq})$. Then*

$$(b_1^{(1)})^{-1}b_1 \in \mathcal{S}_{in}^{p\times p} \quad \textit{and} \quad b_2(b_2^{(1)})^{-1} \in \mathcal{S}_{in}^{q\times q}.$$

Proof Let $W_2 = W_1^{-1}W$ and let $w_{ij}^{(k)}$ denote the ij block of W_k. Then, since

$$w_{11} = w_{11}^{(1)}w_{11}^{(2)} + w_{12}^{(1)}w_{21}^{(2)},$$

it follows readily from Lemma 4.24 that (in a self-evident notation for the entries in the corresponding PG transforms)

$$s_{11} = s_{11}^{(1)}\{I_p - s_{12}^{(2)}s_{21}^{(1)}\}^{-1}s_{11}^{(2)} \tag{4.45}$$

and hence, in terms of the adopted notation for the inner-outer factorization of s_{11} and $s_{11}^{(k)}$, that

$$b_1\varphi_1 = b_1^{(1)}\varphi_1^{(1)}\{I_p - s_{12}^{(2)}s_{21}^{(1)}\}^{-1}b_1^{(2)}\varphi_1^{(2)}.$$

Therefore,

$$(b_1^{(1)})^{-1}b_1\varphi_1 \in \mathcal{N}_+^{p\times p} \cap L_\infty^{p\times p} = H_\infty^{p\times p},$$

by the Smirnov maximum principle (Theorem 3.59), since $\{I_p - s_{12}^{(2)}s_{21}^{(1)}\}$ is outer in $H_\infty^{p\times p}$ by Lemma 3.54. This implies that $(b_1^{(1)})^{-1}b_1 \in \mathcal{S}_{in}^{p\times p}$.

A similar analysis of the formula

$$\begin{aligned} w_{22} &= w_{21}^{(1)}w_{12}^{(2)} + w_{22}^{(1)}w_{22}^{(2)} \\ &= w_{22}^{(1)}\{I_q - s_{21}^{(1)}s_{12}^{(2)}\}w_{22}^{(2)} \end{aligned}$$

yields the identity

$$s_{22} = s_{22}^{(2)}\{I_q - s_{21}^{(1)}s_{12}^{(2)}\}^{-1}s_{22}^{(1)}, \tag{4.46}$$

which leads easily to the second conclusion, much as before. □

Lemma 4.29 *Let the mvf's W and W_1 belong to $\mathcal{P}^\circ(j_{pq})$ and suppose that either $W_1^{-1}W \in \mathcal{P}^\circ(j_{pq})$ or $WW_1^{-1} \in \mathcal{P}^\circ(j_{pq})$. Then*

$$\mathfrak{h}_W^+ \subseteq \mathfrak{h}_{W_1}^+ \quad \text{and} \quad \mathfrak{h}_{W^{-1}}^+ \subseteq \mathfrak{h}_{W_1^{-1}}^+. \tag{4.47}$$

Proof Suppose first that $W_1^{-1}W \in \mathcal{P}^\circ(j_{pq})$ and let $\{b_1, b_2\} \in ap(W)$ and $\{b_1^{(1)}, b_2^{(1)}\} \in ap(W_1)$. Then Lemma 4.26 and Corollary 4.22 guarantee that

$$\mathfrak{h}_W^+ = \mathfrak{h}_{w_{22}}^+ = \mathfrak{h}_{s_{22}^{-1}}^+ = \mathfrak{h}_{b_2^{-1}}^+, \tag{4.48}$$

$$\mathfrak{h}_{W_1}^+ = \mathfrak{h}_{(b_2^{(1)})^{-1}}^+, \tag{4.49}$$

$$\mathfrak{h}_{W^{-1}}^+ = \mathfrak{h}_{s_{11}^{-1}}^+ = \mathfrak{h}_{b_1^{-1}}^+ \tag{4.50}$$

and

$$\mathfrak{h}_{W_1^{-1}}^+ = \mathfrak{h}_{(b_1^{(1)})^{-1}}^+. \tag{4.51}$$

Moreover, by Lemma 4.28,

$$(b_1^{(1)})^{-1} = ub_1^{-1} \quad \text{and} \quad (b_2^{(1)})^{-1} = b_2^{-1}v$$

with $u \in \mathcal{S}_{in}^{p\times p}$ and $v \in \mathcal{S}_{in}^{q\times q}$. Therefore,

$$\mathfrak{h}_{b_1^{-1}}^+ \subseteq \mathfrak{h}_{(b_1^{(1)})^{-1}}^+ \quad \text{and} \quad \mathfrak{h}_{b_2^{-1}}^+ \subseteq \mathfrak{h}_{(b_2^{(1)})^{-1}}^+. \tag{4.52}$$

The inclusions in (4.47) for the case $W_1^{-1}W \in \mathcal{P}^\circ(j_{pq})$ now follow from the inclusions in (4.52) and the equalities (4.48)–(4.51).

Finally, since $W \in \mathcal{P}^\circ(j_{pq}) \Longleftrightarrow W^\tau \in \mathcal{P}^\circ(j_{pq})$ the inclusions for the case $WW_1^{-1} \in \mathcal{P}^\circ(j_{pq})$ follow from the first case by considering transposes. □

Corollary 4.30 *Let $W = W_1W_2$, where the mvf's W_1 and W_2 belong to $\mathcal{P}^\circ(j_{pq})$. Then $W \in \mathcal{P}^\circ(j_{pq})$ and*

$$\mathfrak{h}_W^+ = \mathfrak{h}_{W_1}^+ \cap \mathfrak{h}_{W_2}^+ \quad \text{and} \quad \mathfrak{h}_{W^{-1}}^+ = \mathfrak{h}_{W_1^{-1}}^+ \cap \mathfrak{h}_{W_2^{-1}}^+. \tag{4.53}$$

Proof The inclusions in one direction follow from Lemma 4.29. The opposite inclusions are self-evident. □

Lemma 4.31 *Let $W = W_1W_2$, where W_1 and W_2 both belong to $\mathcal{P}^\circ(j_{pq})$ and let $\{b_1^{(j)}, b_2^{(j)}\} \in ap(W_j)$ for $j = 1, 2$ and $\{b_1, b_2\} \in ap(W)$. Then*

$$\det b_1 = \gamma_1 \det b_1^{(1)} \det b_1^{(2)} \quad \text{and} \quad \det b_2 = \gamma_2 \det b_2^{(1)} \det b_2^{(2)}, \tag{4.54}$$

where $\gamma_1, \gamma_2 \in \mathbb{C}$ and $|\gamma_1| = |\gamma_2| = 1$.

Proof Formula (4.45) implies that

$$\det s_{11} \det \{I_p - s_{12}^{(2)} s_{21}^{(1)}\} = \det s_{11}^{(1)} \det s_{11}^{(2)}. \tag{4.55}$$

Therefore, since the second factor on the left is an outer function in H_∞ by Lemma 3.54, the first equality in (4.54) follows by matching inner parts in formula (4.55). The second equality follows from (4.46) in much the same way. □

In the sequel we shall adopt the notion of associated pairs for mvf's $U \in \mathcal{P}^\circ(J)$ for arbirary signature matrices J that are unitarily equivalent to j_{pq}. Thus, if

$$J = V^* j_{pq} V$$

for some constant unitary matrix V, we shall say that $\{b_1(\lambda), b_2(\lambda)\}$ is an **associated pair of the first kind** for $U(\lambda)$, if $\{b_1, b_2\} \in ap(W)$, where $W(\lambda) = VU(\lambda)V^*$ and shall write

$$\{b_1, b_2\} \in ap_I(U).$$

Lemma 4.32 *The set $ap_I(U)$ depends upon the choice of the unitary matrix V only through extra unitary constant factors, i.e., $b_1(\lambda) \longrightarrow u_1 b_1(\lambda)$ and $b_2(\lambda) \longrightarrow b_2(\lambda) u_2$, where u_1 and u_2 are unitary matrices of sizes $p \times p$ and $q \times q$, respectively.*

Proof If $J = V_1^* j_{pq} V_1$, where V_1 is another unitary matrix, then, since $V_2 = V_1 V^*$ commutes with j_{pq}, it must be of the form $V_2 = \text{diag}\{u_1, u_2\}$, where u_1 and u_2 are unitary matrices of sizes $p \times p$ and $q \times q$, respectively. Consequently,

$$V_1 = \begin{bmatrix} u_1 & 0 \\ 0 & u_2 \end{bmatrix} V.$$

□

Lemma 4.33 *Let $U \in \mathcal{P}^\circ(J)$, $J \neq \pm I_m$, and let $\{b_1, b_2\} \in ap_I(U)$. Then*

$$U \in \mathcal{N}_+^{m\times m} \iff b_2(\lambda) \text{ is constant}$$

and

$$U^{-1} \in \mathcal{N}_+^{m\times m} \iff b_1(\lambda) \text{ is constant.}$$

Proof It suffices to consider the case $J = j_{pq}$. But then the conclusions are contained in Lemma 4.26. □

Lemma 4.34 *Let $U = U_1U_2$, where $U_k \in \mathcal{P}^\circ(J)$ for $k = 1, 2$ and $J \neq \pm I_m$. Then:*

(1) *$U \in \mathcal{N}_+^{m\times m}$ if and only if U_1 and U_2 both belong to $\mathcal{N}_+^{m\times m}$.*

(2) *$U^{-1} \in \mathcal{N}_+^{m\times m}$ if and only if U_1^{-1} and U_2^{-1} both belong to $\mathcal{N}_+^{m\times m}$.*

Proof It suffices to consider the case $J = j_{pq}$. But then the stated results follow for $W \in \mathcal{P}^\circ(j_{pq})$ from formulas (4.45), (4.46) and Lemmas 4.26 and 4.28. □

4.7 The classes $\mathcal{P}(J_p)$, $\mathcal{P}(j_p, J_p)$, $\mathcal{U}(J_p)$ and $\mathcal{U}(j_p, J_p)$

Let $m = 2p$ and let

$$A(\lambda) = \begin{bmatrix} a_{11}(\lambda) & a_{12}(\lambda) \\ a_{21}(\lambda) & a_{22}(\lambda) \end{bmatrix}$$

be a meromorphic $m \times m$ mvf in $\mathbb{C}_+$ with blocks $a_{ij}(\lambda)$ of size $p \times p$, and let

$$B(\lambda) = A(\lambda)\mathfrak{V} = \begin{bmatrix} b_{11}(\lambda) & b_{12}(\lambda) \\ b_{21}(\lambda) & b_{22}(\lambda) \end{bmatrix} \tag{4.56}$$

and

$$W(\lambda) = \mathfrak{V}A(\lambda)\mathfrak{V} = \begin{bmatrix} w_{11}(\lambda) & w_{12}(\lambda) \\ w_{21}(\lambda) & w_{22}(\lambda) \end{bmatrix}, \tag{4.57}$$

where $\mathfrak{V}$ is defined in formula (2.5). Then

$$A \in \mathcal{P}(J_p) \Longleftrightarrow B \in \mathcal{P}(j_p, J_p) \Longleftrightarrow W \in \mathcal{P}(j_p),$$

$$A \in \mathcal{P}^\circ(J_p) \Longleftrightarrow B \in \mathcal{P}^\circ(j_p, J_p) \Longleftrightarrow W \in \mathcal{P}^\circ(j_p),$$

and

$$A \in \mathcal{U}(J_p) \Longleftrightarrow B \in \mathcal{U}(j_p, J_p) \Longleftrightarrow W \in \mathcal{U}(j_p).$$

It turns out to be more convenient to work with the mvf's $B \in \mathcal{P}(j_p, J_p)$ than the mvf's $A \in \mathcal{P}(J_p)$. A mvf $B \in \mathcal{P}(j_p, J_p)$ is meromorphic in $\mathbb{C}_+$ and meets the constraint

$$B(\lambda)^* J_p B(\lambda) \leq j_p \quad \text{for every} \quad \lambda \in \mathfrak{h}_B^+. \tag{4.58}$$

If $B \in \mathcal{U}(j_p, J_p)$, then

$$B \in \mathcal{P}(j_p, J_p) \quad \text{and} \quad B(\mu)^* J_p B(\mu) = j_p \text{ for almost all points } \mu \in \mathbb{R}. \tag{4.59}$$

In view of (2.76), condition (4.58) is equivalent to the condition

$$B(\lambda) j_p B(\lambda)^* \leq J_p \quad \text{for every} \quad \lambda \in \mathfrak{h}_B^+, \tag{4.60}$$

whereas condition (4.59) is equivalent to the condition

$$B(\mu) j_p B(\mu)^* = J_p \quad \text{for almost all points } \mu \in \mathbb{R}. \tag{4.61}$$

In particular, (4.60) implies that:

$$b_{22}(\lambda) b_{22}(\lambda)^* - b_{21}(\lambda) b_{21}(\lambda)^* \geq 0 \quad \text{for } \lambda \in \mathfrak{h}_B^+, \tag{4.62}$$

whereas (4.58) implies that

$$b_{12}(\lambda)^* b_{22}(\lambda) + b_{22}(\lambda)^* b_{12}(\lambda) \geq I_p \quad \text{for } \lambda \in \mathfrak{h}_B^+, \tag{4.63}$$

and (4.59) implies that

$$b_{22}(\mu) b_{22}(\mu)^* - b_{21}(\mu) b_{21}(\mu)^* = 0 \quad \text{a.e. on } \mathbb{R}. \tag{4.64}$$

Lemma 4.35 *Let $B \in \mathcal{P}(j_p, J_p)$. Then:*

(1) $\det b_{22}(\lambda) \neq 0$ *for every point* $\lambda \in \mathfrak{h}_B^+$.

(2) *The mvf* $c_0 = b_{12} b_{22}^{-1}$ *belongs to* $\mathcal{C}^{p\times p}$.

(3) *The mvf* $\chi = b_{22}^{-1} b_{21}$ *belongs to* $\mathcal{S}^{p\times p}$.

(4) *The mvf* $\rho_i^{-1}(b_{22})^{-1}$ *belongs to* $H_2^{p\times p}$.

(5) $\rho_i^{-1}(b_{11} - b_{12} b_{22}^{-1} b_{21})$ *belongs to* $H_2^{p\times p}$.

If $B \in \mathcal{U}(j_p, J_p)$, then

(6) $\chi = b_{22}^{\#}(b_{21}^{\#})^{-1}$ *and* $\chi \in \mathcal{S}_{in}^{p\times p}$.

(7) $b_{11} - b_{12} b_{22}^{-1} b_{21} = -(b_{21}^{\#})^{-1}$.

(8) $\rho_i^{-1}(b_{21}^{\#})^{-1} \in H_2^{p\times p}$.

Proof Let $W = \mathfrak{V}B$. Then $W \in \mathcal{P}(j_p)$ and b_{22} can be expressed in terms of the blocks w_{ij} and s_{ij} of W and the PG transform S of W by the formulas

$$b_{22} = \frac{1}{\sqrt{2}}(w_{12} + w_{22}) = \frac{1}{\sqrt{2}}(I_p + s_{12}) w_{22}. \tag{4.65}$$

By Theorem 4.20, $s_{12} \in \mathcal{S}^{p\times p}$, $s_{12}(\lambda)^* s_{12}(\lambda) < I_p$ and $\det w_{22}(\lambda) \neq 0$ for $\lambda \in \mathfrak{h}_W^+$ and $s_{22} = w_{22}^{-1}$ belongs to $\mathcal{S}^{p\times p}$. Thus, (1) holds. Moreover, (2) holds because

$$c_0 = b_{12}b_{22}^{-1} = (-w_{12} + w_{22})(w_{12} + w_{22})^{-1} = (I_p - s_{12})(I_p + s_{12})^{-1} \quad (4.66)$$

and $s_{12} \in \mathcal{S}^{p\times p}$. Furthermore, the inequality (4.62) yields (3).

In view of (4.65), the mvf

$$b_{22}^{-1} = \sqrt{2}s_{22}(I_p + s_{12})^{-1} \quad \text{belongs to} \quad \mathcal{N}_+^{p\times p}, \qquad (4.67)$$

since $s_{22} \in \mathcal{S}^{p\times p}$ and, by Lemma 3.54, $(I_p + s_{12}) \in \mathcal{N}_{out}^{p\times p}$. Moreover, the inequalities

$$c_0 + c_0^* = (b_{22}^*)^{-1}(b_{12}^* b_{22} + b_{22}^* b_{12})b_{22}^{-1} \geq (b_{22}^*)^{-1} b_{22}^{-1}$$

imply that $\rho_i^{-1} b_{22}^{-1} \in L_2^{p\times p}$, since in view of condition (3.11), $\Re c_0 \in \widetilde{L}_1^{p\times p}$. Assertion (4) now follows by the Smirnov maximum principle, since $\rho_i^{-1} b_{22}^{-1} \in L_2^{p\times p} \cap \mathcal{N}_+^{p\times p}$.

Next, in view of formulas

$$\begin{bmatrix} b_{11}(\lambda) & b_{12}(\lambda) \\ b_{21}(\lambda) & b_{22}(\lambda) \end{bmatrix} = \frac{1}{\sqrt{2}} \begin{bmatrix} -w_{11}(\lambda) + w_{21}(\lambda) & -w_{12}(\lambda) + w_{22}(\lambda) \\ w_{11}(\lambda) + w_{21}(\lambda) & w_{12}(\lambda) + w_{22}(\lambda) \end{bmatrix} \qquad (4.68)$$

and (4.66), the mvf

$$\alpha = b_{11} - b_{12}b_{22}^{-1}b_{21}, \qquad (4.69)$$

may be reexpressed in terms of the blocks s_{ij} of S as

$$\begin{aligned} \alpha &= \frac{1}{\sqrt{2}}\{-w_{11} + w_{21} - (I_p - s_{12})(I_p + s_{12})^{-1}(w_{11} + w_{21})\} \\ &= \sqrt{2}\{-(I_p + s_{12})^{-1}w_{11} + s_{12}(I_p + s_{12})^{-1}w_{21}\} \\ &= \sqrt{2}(I_p + s_{12})^{-1}(-w_{11} + w_{12}w_{22}^{-1}w_{21}). \end{aligned}$$

Thus,

$$\alpha = -\sqrt{2}(I_p + s_{12})^{-1}s_{11} \qquad (4.70)$$

belongs to $\mathcal{N}_+^{p\times p}$, since $(I_p + s_{12}) \in \mathcal{N}_{out}^{p\times p}$ and $s_{11} \in \mathcal{S}^{p\times p}$. Moreover, as

$$\begin{aligned} \alpha\alpha^* &\leq 2(I_p + s_{12})^{-1}s_{11}s_{11}^*(I_p + s_{12})^{-*} \\ &\leq 2(I_p + s_{12})^{-1}(I_p - s_{12}s_{12}^*)(I_p + s_{12})^{-*} \\ &= 2\Re c_0 \end{aligned}$$

and $c_0 \in \mathcal{C}^{p\times p}$, it follows that $\alpha\alpha^* \in \widetilde{L}_1^{p\times p}$. Thus, $\rho_i^{-1}\alpha \in L_2^{p\times p} \cap \mathcal{N}_+^{p\times p}$ and the Smirnov maximum principle serves to complete the proof of (5).

If $B \in \mathcal{U}(j_p, J_p)$, then (6) follows from (3) and (4.64). Next, the identity

$$B(\lambda) j_p B^\#(\lambda) = J_p \quad \text{on} \quad \mathfrak{h}_B^+ \cap \mathfrak{h}_{B^\#}^+$$

implies that

$$b_{11}b_{21}^\# - b_{12}b_{22}^\# = -I_p \quad \text{on} \quad \mathfrak{h}_B^+ \cap \mathfrak{h}_{B^\#}^+$$

and hence, in view of (6), that

$$\begin{aligned} \alpha &= b_{11} - b_{12}b_{22}^{-1}b_{21} = b_{11} - b_{12}b_{22}^\#(b_{21}^\#)^{-1} \\ &= (b_{11}b_{21}^\# - b_{12}b_{22}^\#)(b_{21}^\#)^{-1} = -(b_{21}^\#)^{-1}, \end{aligned}$$

which justifies (7). Finally, (8) follows from (7) and (5). □

4.8 Associated pairs of the second kind

Let $B \in \mathcal{P}^\circ(j_p, J_p)$. Then the Schur complement factorization

$$B = \begin{bmatrix} I_p & b_{12}b_{22}^{-1} \\ 0 & I_p \end{bmatrix} \begin{bmatrix} b_{11} - b_{12}b_{22}^{-1}b_{21} & 0 \\ 0 & b_{22} \end{bmatrix} \begin{bmatrix} I_p & 0 \\ b_{22}^{-1}b_{21} & I_p \end{bmatrix},$$

can be reexpressed in terms of c_0, α and χ as

$$B = \begin{bmatrix} I_p & c_0 \\ 0 & I_p \end{bmatrix} \begin{bmatrix} \alpha & 0 \\ 0 & b_{22} \end{bmatrix} \begin{bmatrix} I_p & 0 \\ \chi & I_p \end{bmatrix}. \tag{4.71}$$

Consequently, $\det \alpha(\lambda) \not\equiv 0$, since

$$\det B = \det \alpha \cdot \det b_{22} \tag{4.72}$$

and $\det B(\lambda) \not\equiv 0$ by assumption. Thus,

$$\alpha \in \mathcal{N}_+^{p\times p} \quad \text{and} \quad \det \alpha(\lambda) \not\equiv 0 \text{ on } \mathfrak{h}_\alpha^+$$

and

$$b_{22}^{-1} \in \mathcal{N}_+^{p\times p} \quad \text{and} \quad \det \mathrm{b}_{22}(\lambda)^{-1} \not\equiv 0 \text{ on } \mathfrak{h}_{\mathrm{b}_{22}^{-1}}^+.$$

Therefore, $-\alpha$ and b_{22}^{-1} have inner-outer and outer-inner factorizations in $\mathcal{N}_+^{p\times p}$:

$$-\alpha = b_3\varphi_3 \quad \text{and} \quad \mathrm{b}_{22}^{-1} = \varphi_4\mathrm{b}_4, \tag{4.73}$$

where $b_j \in \mathcal{S}_{in}^{p\times p}$ and $\varphi_j = \mathcal{N}_{out}^{p\times p}$, $j = 1, 2$. If $B \in \mathcal{U}(j_p, J_p)$, then $\alpha = -(b_{21}^{\#})^{-1}$ and the formulas in (4.73) can be reexpressed as

$$(b_{21}^{\#})^{-1} = b_3\varphi_3 \quad \text{and} \quad b_{22}^{-1} = \varphi_4 b_4. \tag{4.74}$$

The pair $\{b_3, b_4\}$ is called an associated pair of the mvf B, and the set of all such pairs is denoted by $ap(B)$. Thus, $\{b_3, b_4\} \in ap(B)$. Moreover, if $\{b_3, b_4\}$ is a fixed associated pair of B, then $ap(B) = \{\{b_3 u,\ v b_4\} : u$ and v are unitary $p \times p$ matrices$\}$.

Now let $A \in \mathcal{P}^\circ(J_p)$ and let

$$B(\lambda) = A(\lambda)\mathfrak{V} \text{ and } \mathrm{W}(\lambda) = \mathfrak{V}\mathrm{A}(\lambda)\mathfrak{V}.$$

Then $B \in \mathcal{P}^\circ(j_p, J_p)$ and $W \in \mathcal{P}^\circ(j_p)$. Let $S = PG(W)$ and let b_{ij}, w_{ij} and s_{ij} denote the blocks of B, W and S, respectively.

The set $ap_I(A) = ap(W)$ of associated pairs of the first kind for A was defined earlier. Analogously, we define $ap_{II}(A) = ap(B)$ as the set of **associated pairs of the second kind** for A. These definitions can be extended to mvf's in the class $\mathcal{P}^\circ(J)$ based on a signature matrix J that is unitarily equivalent to J_p: If

$$J = V^* J_p V \text{ for some unitary matrix } V \in \mathbb{C}^{m\times m}, \tag{4.75}$$

then

$$\begin{aligned} U \in \mathcal{P}(J) \quad &\Longleftrightarrow \quad VUV^* \in \mathcal{P}(J_p) \\ U \in \mathcal{P}^\circ(J) \quad &\Longleftrightarrow \quad VUV^* \in \mathcal{P}^\circ(J_p) \\ U \in U(J) \quad &\Longleftrightarrow \quad VUV^* \in U(J_p). \end{aligned}$$

If $U \in \mathcal{P}^\circ(J)$, then the the pair $\{b_3, b_4\}$ is said to be an associated pair of the second kind for $U(\lambda)$ and we write $\{b_3, b_4\} \in ap_{II}(U)$ if $\{b_3, b_4\} \in ap_{II}(VUV^*)$. The set $ap_{II}(U)$ depends upon the choice of the unitary matrix V in (4.75).

Lemma 4.36 *Let $A \in \mathcal{P}^\circ(J_p)$ and let $W = \mathfrak{V}A\mathfrak{V}$. Let $\{b_1, b_2\} \in ap_I(A)$ and $\{b_3, b_4\} \in ap_{II}(A)$. Then*

$$\frac{1}{2}(I_p + s_{12})b_3 = b_1\varphi \quad \text{and} \quad b_4(I_p + s_{12})/2 = \psi b_2 \tag{4.76}$$

for some $\varphi \in \mathcal{S}_{out}^{p\times p}$ and $\psi \in \mathcal{S}_{out}^{p\times p}$, where $s_{12} = w_{12}w_{22}^{-1}$ and w_{ij} are the blocks of the mvf W.

Proof Formulas (4.70), (4.73) and (4.42) imply that

$$s_{11} = b_1\varphi_1 = \left((I_p + s_{12})/\sqrt{2}\right) b_3\varphi_3,$$

and hence that

$$\frac{1}{2}(I_p + s_{12})b_3 = \left(\frac{1}{\sqrt{2}}\right) b_1\varphi_1\varphi_3^{-1}.$$

Moreover, $\varphi = (1/\sqrt{2})\varphi_1\varphi_3^{-1} \in \mathcal{N}_{out}^{p\times p}$ because $\varphi_j \in \mathcal{N}_{out}^{p\times p}$. Therefore, since $\|\varphi\|_\infty = \|(I_p + s_{12})/2\|_\infty \leq 1$, the maximum principle guarantees that $\varphi \in \mathcal{S}_{out}^{p\times p}$.

The second assertion in (4.76) may be verified in much the same way via formulas (4.67), (4.73) and (4.42). □

Remark 4.37 *A stronger result on the connection between $ap_I(A)$ and $ap_{II}(A)$ will be obtained later in Lemma 7.68.*

Lemma 4.38 *Let $A \in \mathcal{P}^\circ(J_p)$ and let $\{b_1, b_2\} \in ap_I(A)$ and let $\{b_3, b_4\} \in ap_{II}(A)$. Then the following equivalences are valid:*

(1) $A \in \mathcal{N}_+^{m\times m} \iff b_2$ *is a constant matrix* $\iff b_4$ *is a constant matrix.*

(2) $A^{-1} \in \mathcal{N}_+^{m\times m} \iff b_1$ *is a constant matrix* $\iff b_3$ *is a constant matrix.*

Proof Since

$$\begin{bmatrix} I_p & c_0 \\ 0 & I_p \end{bmatrix} \in \mathcal{N}_{out}^{m\times m}, \quad \begin{bmatrix} I_p & 0 \\ \chi & I_p \end{bmatrix} \in \mathcal{N}_{out}^{m\times m},$$

$\alpha \in \mathcal{N}_+^{p\times p}$ and $b_{22}^{-1} \in \mathcal{N}_+^{p\times p}$, formula (4.71) implies that

$$B \in \mathcal{N}_+^{m\times m} \iff b_{22} \in \mathcal{N}_{out}^{m\times m} \iff b_4 \text{ is a constant matrix} \tag{4.77}$$

and

$$B^{-1} \in \mathcal{N}_+^{m\times m} \iff \alpha \in \mathcal{N}_{out}^{m\times m} \iff b_3 \text{ is a constant matrix.} \tag{4.78}$$

Moreover, the connections (4.76) imply that

$$b_3 \text{ is a constant matrix} \iff b_1 \text{ is a constant matrix}$$

and

$$b_4 \text{ is a constant matrix } \Longleftrightarrow b_2 \text{ is a constant matrix}$$

because $(\frac{1}{2})(I_p + s_{12}) \in \mathcal{S}_{out}^{p\times p}$. Thus, (1) and (2) are valid, since

$$A \in \mathcal{N}_+^{m\times m} \Longleftrightarrow B \in \mathcal{N}_+^{m\times m}$$

and

$$A^{-1} \in \mathcal{N}_+^{m\times m} \Longleftrightarrow B^{-1} \in \mathcal{N}_+^{m\times m}.$$

4.9 Singular and regular J-inner mvf's

A mvf $U \in \mathcal{U}(J)$ is said to be **singular** if it belongs to the class $\mathcal{N}_{out}^{m\times m}$. The class of singular J-inner mvf's will be denoted $\mathcal{U}_S(J)$.

Lemma 4.39 *Let $U \in \mathcal{U}(J)$, $J \neq \pm I_m$, and let $\{b_1, b_2\} \in ap_I(U)$. Then*

(1) *$U \in \mathcal{U}_S(J) \Longleftrightarrow b_1$ and b_2 are both constant unitary matrices.*

(2) *If $q = p$ and $\{b_3, b_4\} \in ap_{II}(U)$, then $U \in \mathcal{U}_S(J) \Longleftrightarrow b_3$ and b_4 are both constant unitary matrices.*

Proof Assertion (1) follows from the definition of the class $\mathcal{N}_{out}^{m\times m}$ and Lemma 4.26. Assertion (2) then follows from the connection (4.104) between pairs $\{b_1, b_2\} \in ap_I(A)$ and $\{b_3, b_4\} \in ap_{II}(A)$, where $A(\lambda) = VU(\lambda)V^*$ and V is a unitary matrix such that $J = V^* J_p V$. □

Lemma 4.40 *Let $U \in \mathcal{P}^\circ(J)$, $J \neq \pm I_m$, and let $S = PG(U)$. Then:*

(1) $U \in \mathcal{N}_+^{m\times m} \Longleftrightarrow (P + QS)/2 \in \mathcal{S}_{out}^{m\times m}$

(2) $U^{-1} \in \mathcal{N}_+^{m\times m} \Longleftrightarrow (Q + PS)/2 \in \mathcal{S}_{out}^{m\times m}$

(3) *If $U \in \mathcal{U}(J)$ and $J \neq \pm I_m$, then*

$$U \in \mathcal{U}_S(J) \Longleftrightarrow (P + QS)/2 \in \mathcal{S}_{out}^{m\times m} \quad \text{and} \quad (Q + PS)/2 \in \mathcal{S}_{out}^{m\times m}.$$

Proof Without loss of generality we may assume that $J = j_{pq}$. Then, in view of Theorem 3.50, the four block decompositions

$$P + QS = \begin{bmatrix} I_p & s_{12} \\ 0 & s_{22} \end{bmatrix} \quad \text{and} \quad Q + PS = \begin{bmatrix} s_{11} & s_{12} \\ 0 & I_q \end{bmatrix}$$

imply that

$$(P+QS)/2 \in \mathcal{S}_{out}^{m\times m} \Longleftrightarrow s_{22} \in \mathcal{S}_{out}^{q\times q} \quad \text{and}$$
$$(Q+PS)/2 \in \mathcal{S}_{out}^{m\times m} \Longleftrightarrow s_{11} \in \mathcal{S}_{out}^{p\times p}.$$

Therefore, (1) and (2) follow immediately from Lemma 4.33; and (3) follows from (1), (2) and Lemma 4.39. □

Lemma 4.41 *The equivalence*

$$U \in \mathcal{U}_S(\pm I_m) \Longleftrightarrow U^{\pm 1} \in \mathcal{S}_{in}^{m\times m} \cap \mathcal{S}_{out}^{m\times m}$$

exhibits the fact that if $J=\pm I_m$, then $\mathcal{U}_S(J)=\mathcal{U}_{const}(I_m)$ is equal to the class of constant $m\times m$ unitary matrices. Thus, if $\mathcal{U}_S(J)$ contains at least one nonconstant mvf, then $J\neq \pm I_m$.

Proof By definition, $\mathcal{U}_S(I_m) = \mathcal{S}_{in}^{m\times m} \cap \mathcal{N}_{out}^{m\times m} = \mathcal{S}_{in}^{m\times m} \cap \mathcal{S}_{out}^{m\times m}$. Therefore, by Theorem 3.50, $\det U \in \mathcal{S}_{in} \cap \mathcal{S}_{out}$ and hence as $U(\lambda)^*U(\lambda) \leq I_m$ in $\mathbb{C}_+$ and $\det U(\lambda)^*U(\lambda)=1$ in $\mathbb{C}_+$, it follows that $U(\lambda)$ is unitary at each point of $\mathbb{C}_+$ and consequently must be equal to a unitary constant matrix.

Since $U\in\mathcal{U}_S(-I_m) \Longleftrightarrow U^{-1}\in \mathcal{U}_S(I_m)$, the asserted conclusions are in force for $J=-I_m$ too. □

Lemma 4.42 *If $U\in\mathcal{U}_S(J)$, then* $\det U(\lambda)\equiv c$, *where c is a constant with $|c|=1$.*

Proof Let $\varphi(\lambda)=\det U(\lambda)$. Then, the result follows easily from the observation that $\varphi\in\mathcal{N}_{out}$, since $U^{\pm1}\in\mathcal{N}_+^{m\times m}$, and, at the same time, $|\varphi(\mu)|=1$ a.e. on $\mathbb{R}$. □

The simplest examples of a mvf $U\in\mathcal{U}_S(J)$ are the elementary BP factors of the third and fourth kind that are based on a constant $m\times m$ matrix E such that $E^2=0$ and $EJ\geq 0$:

$$U_\omega(\lambda) = I_m + \frac{E}{\pi i(\lambda-\omega)} = \exp\left\{\frac{E}{\pi i(\lambda-\omega)}\right\} \quad \text{for } \omega\in\mathbb{R}$$

and

$$U_\infty(\lambda) = I_m + i\lambda E = \exp\{i\lambda E\}.$$

The proof that these mvf's belong to $\mathcal{N}_{out}^{m\times m}$ follows from Remark 3.28.

Other important examples are the $2p \times 2p$ mvf's

$$U(\lambda) = \begin{bmatrix} I_p & c_s(\lambda) \\ 0 & I_p \end{bmatrix} \quad \text{and} \quad U(\lambda) = \begin{bmatrix} I_p & 0 \\ c_s(\lambda) & I_p \end{bmatrix},$$

which belong to $\mathcal{U}_S(J_p)$ when $c_s \in \mathcal{C}^{p\times p}$ and $\Re c_s(\mu) = 0$ for almost all points $\mu \in \mathbb{R}$.

Theorem 4.43 *Let $U = U_1U_2$, where $U_k \in \mathcal{U}(J)$ for $k = 1, 2$. Then $U \in \mathcal{U}_S(J)$ if and only if $U_1 \in \mathcal{U}_S(J)$ and $U_2 \in \mathcal{U}_S(J)$.*

Proof If $J = \pm I_m$, then there is nothing to check, since $\mathcal{U}_S(J) = \mathcal{U}_{const}(J)$. If $J \neq \pm I_m$, then the statement follows from Lemma 4.34. □

Theorem 4.44 *A finite or convergent infinite left or right product of singular J-inner mvf's is a singular J-inner mvf.*

Proof Let $U_j \in \mathcal{U}_S(J)$ for $j = 1, \dots, n$ and let $U(\lambda) = U_1(\lambda)\cdots U_n(\lambda)$. Then, by Theorem 4.43, $U \in \mathcal{U}_S(J)$. Convergent infinite products of singular J-inner mvf's are J-inner, by Corollary 4.62, which will be presented later. The proof that convergent infinite products are singular is not so obvious. It will follow from Theorem 4.60. □

An mvf $U \in \mathcal{U}(J)$ is said to belong to the class $\mathcal{U}_{\ell R}(J)$ of **left regular** J-inner mvf's if the factorization $U = U_1U_2$ with $U_2 \in \mathcal{U}(J)$ and $U_1 \in \mathcal{U}_S(J)$ implies that $U_1 \in \mathcal{U}_{const}(J)$. Similarly, an mvf $U \in \mathcal{U}(J)$ is said to belong to the class $\mathcal{U}_{rR}(J)$ of **right regular** J-inner mvf's if the factorization $U = U_1U_2$ with $U_1 \in \mathcal{U}(J)$ and $U_2 \in \mathcal{U}_S(J)$ implies that $U_2 \in \mathcal{U}_{const}(J)$.† Clearly,

$$U \in \mathcal{U}_{rR}(J) \iff U^\sim \in \mathcal{U}_{\ell R}(J).$$

The classes $\mathcal{U}_{rR}(j_{pq})$ and $\mathcal{U}_{rR}(J_p)$ are closely connected to the generalized interpolation problems in the classes $\mathcal{S}^{p\times q}$ and $\mathcal{C}^{p\times p}$, respectively, that will be considered in Chapter 7. Moreover, convergent right BP products of elementary factors of the first and second kind are connected with bitangential interpolation problems in the classes $\mathcal{S}^{p\times q}$ if $J = j_{pq}$ and $\mathcal{C}^{p\times p}$ if $J = J_p$. These connections will be used to obtain characterizations of the classes $\mathcal{U}_{rR}(j_{pq})$ and $\mathcal{U}_{\ell R}(j_{pq})$ in Chapter 7, and they were exploited in [Ar89] to establish the following result:

† The usage of left and right that is adopted here is not uniformly adhered to in the literature.

Theorem 4.45 *Let $U(\lambda)$ be a right BP product of elementary factors of the first and second kind, i.e., with poles that are not on $\mathbb{R}$. Then $U \in \mathcal{U}_{rR}(J)$. Moreover:*

(1) *If $J \neq \pm I_m$ and $\{b_1, b_2\} \in ap_I(U)$, then the inner mvf's b_1 and b_2 are both BP products.*

(2) *If $J \neq \pm I_m$, $q = p$ and $\{b_3, b_4\} \in ap_{II}(U)$, then the inner mvf's b_3 and b_4 are both BP products.*

Conversely, if $U \in \mathcal{U}_{rR}(J)$, $J \neq \pm I_m$ and either condition (1) or condition (2) is in force, then U is a right BP product of elementary factors of the first and second kind.

Remark 4.46 *A right BP product $U(\lambda)$ of elementary factors of the first kind belongs to the Smirnov class $\mathcal{N}_+^{m\times m}$ and, consequently, if $\{b_1, b_2\} \in ap_I(U)$ and $\{b_3, b_4\} \in ap_{II}(U)$, then $b_2(\lambda)$ and $b_4(\lambda)$ are both constant. A right BP product $U(\lambda)$ of elementary factors of the second kind belongs to the Smirnov class with respect to $\mathbb{C}_-$, i.e., $U^\# \in \mathcal{N}_+^{m\times m}$, and, consequently, if $\{b_1, b_2\} \in ap_I(U)$ and $\{b_3, b_4\} \in ap_{II}(U)$, then $b_1(\lambda)$ and $b_3(\lambda)$ are both constant.*

Remark 4.47 *In view of Theorem 4.45, necessary and sufficient conditions for a mvf $U \in \mathcal{U}(J)$, $J \neq \pm I_m$, to be a right BP product of elementary factors of the first and second kind may be obtained by combining any one of the necessary and sufficient conditions for $U \in \mathcal{U}_{rR}(J)$ (see Section 7.3) with Theorem 3.86, which gives necessary and sufficient conditions for an inner function $b(\lambda)$ to be a BP product. If $J = j_{pq}$, W is written in place of U and $\{b_1, b_2\} \in ap(W)$, then:*

(1) *b_2 is a BP product if and only if $\tau_{w_{22}}^+ = 0$ and*

$$\lim_{\nu\downarrow 0} \int_{\infty}^{\infty} \frac{\ln|\det w_{22}(\mu + i\nu)|}{1+\mu^2} d\mu = \int_{\infty}^{\infty} \frac{\ln|\det w_{22}(\mu)|}{1+\mu^2} d\mu. \tag{4.79}$$

(2) *b_1 is a BP product if and only if $\tau_{w_{11}}^- = 0$ and*

$$\lim_{\nu\downarrow 0} \int_{\infty}^{\infty} \frac{\ln|\det w_{11}(\mu - i\nu)|}{1+\mu^2} d\mu = \int_{\infty}^{\infty} \frac{\ln|\det w_{11}(\mu)|}{1+\mu^2} d\mu. \tag{4.80}$$

Remark 4.48 *In view of Lemma 4.41,*

$$J = \pm I_m \Longrightarrow \mathcal{U}_{rR}(J) = \mathcal{U}(J).$$

Theorem 4.49 *Every mvf $U \in \mathcal{U}(J)$ admits a pair of factorizations:*

$$U(\lambda) = U_1(\lambda)U_2(\lambda) \quad \text{where} \quad U_1 \in \mathcal{U}_{rR}(J) \quad \text{and} \quad U_2 \in \mathcal{U}_S(J)$$

and

$$U(\lambda) = U_1(\lambda)U_2(\lambda) \quad \textit{where} \quad U_1 \in \mathcal{U}_S(J) \quad \textit{and} \quad U_2 \in \mathcal{U}_{\ell R}(J).$$

These factorizations are unique up to constant J-unitary factors.

Proof The first assertion will be proved for $J = j_{pq}$ in Theorem 5.89 and then again by another method in Theorem 7.52. The proof is easily adapted to the case of general signature matrices $J \neq \pm I_m$ by considering $W = V^*UV$ with a constant $m \times m$ unitary matrix V such that $V^* j_{pq} V = J$. If $J = \pm I_m$, then, in view of Lemma 4.41 and Remark 4.48, the two indicated factorizations are both trivial.

The second factorization for $J \neq \pm I_m$ follows from the first factorization and the equivalences

$$\begin{aligned} U \in \mathcal{U}(J) &\iff U^\sim \in \mathcal{U}(J) \\ U \in \mathcal{U}_S(J) &\iff U^\sim \in \mathcal{U}_S(J) \\ U \in \mathcal{U}_{rR}(J) &\iff U^\sim \in \mathcal{U}_{\ell R}(J). \end{aligned}$$

□

Theorem 4.50 *Let $U(\lambda)$ be a left or right BP product of elementary factors $U_j \in \mathcal{U}(J)$. Then:*

(1) *$U \in \mathcal{U}(J)$.*

(2) *If the U_j are elementary factors of the first or second kind, then a left BP product belongs to the class $\mathcal{U}_{lR}(J)$ and a right BP product belongs to the class $\mathcal{U}_{rR}(J)$.*

(3) *If the U_j are elementary factors of the third or fourth kind, then $U \in \mathcal{U}_S(J)$.*

Proof If U is a finite BP product, then the theorem is self-evident. If U is an infinite BP product, then the asserted results follow from Corollary 4.62, which will be presented later, and Theorems 4.44 and 4.45. □

Necessary and sufficient conditions for a mvf $W \in \mathcal{U}(j_p)$ (recall that j_p is short for j_{pp}) to be right regular are presented in [Ar89] and [Ar95a]. A new characterization is presented in Section 6. A simple sufficient condition for a mvf $U \in \mathcal{U}(J)$ to be right and left regular will be presented in assertion (1) of Theorem 5.90.

The next theorem belongs in this section. But its proof depends upon Theorem 4.94, which will be established later.

Theorem 4.51 *Let $U \in \mathcal{U}(J)$, where $J \neq \pm I_m$ and let*

$$U = U_1 U_2 \quad \text{where } U_1 \in \mathcal{U}_{rR}(J),\ U_2 \in \mathcal{U}_S(J) \tag{4.81}$$

and

$$U = U_3 U_4 \quad \text{where } U_3 \in \mathcal{U}_S(J),\ U_4 \in \mathcal{U}_{\ell R}(J). \tag{4.82}$$

Then U_1 is a right BP product if and only if U_4 is a left BP product.

Proof Without loss of generality we may be assume that $J = j_{pq}$ and, in keeping with our usual conventions, write W in place of U and W_j in place of U_j. By Theorem 4.94, $ap(W) = ap(W_1)$ and $ap(W^\tau) = ap(W_4^\tau)$. Let W_1 be a right BP product of elementary factors of the first and second kind and let $\{b_1, b_2\} \in ap(W)$. Then, since $ap(W) = ap(W_1)$, Theorem 4.45 implies that b_1 and b_2 are both BP products and thus, by Remark 4.47, $\tau_{w_{22}}^+ = 0$, $\tau_{w_{11}}^- = 0$ and conditions (4.79) and (4.80) are both in force. Therefore, as the conditions in Remark 4.47 will also hold for the corresponding blocks of W^τ, it follows that if $\{\widetilde{b}_1, \widetilde{b}_2\} \in ap(W^\tau)$, then $\widetilde{b}_1$ and $\widetilde{b}_2$ will be BP products. Consequently, Theorem 4.45 implies that W_4^τ is a right BP product of elementary factors of the first and second kind. Therefore, W_4 is a left BP product of elementary factors of the first and second kind. The proof in the other direction is the same. □

Corollary 4.52 *Let $U \in \mathcal{U}(J)$, where $J \neq \pm I_m$. If U is a right (resp., left) BP product of elementary factors of the first and second kind, then the mvf U_4 in the factorization (4.82) (resp., U_1 in the factorization (4.81)) is a left (resp., right) BP product of elementary factors of the first and second kind.*

The next theorem is a much deeper result for the case $m = 2$.

Theorem 4.53 (V. E. Katsnelson) *Let $U(\lambda)$ be a 2×2 singular J-inner mvf. Then there exist convergent right and left infinite products of* *elementary BP factors of the first kind such that*

$$\left(\prod_{j=1}^{\overset{\curvearrowleft}{\infty}} U^r_{\omega_j}(\lambda)\right) U(\lambda) = \prod_{j=1}^{\overset{\curvearrowright}{\infty}} U^\ell_{\omega_j}(\lambda).$$

Proof See [Kat89] and [Kat90]. □

4.10 Domains of holomorphy of J inner mvf's

Theorem 4.54 *Let $U \in \mathcal{U}(J)$, $J \neq \pm I_m$, and let $\{b_1, b_2\} \in ap_I(U)$. Then*

$$\mathfrak{h}_U^+ = \mathfrak{h}^+_{b_2^\#}, \quad \mathfrak{h}_U^- = \mathfrak{h}^-_{b_1} \quad \text{and} \quad \mathfrak{h}_U \subseteq \mathfrak{h}_{b_2^\#} \cap \mathfrak{h}_{b_1}. \tag{4.83}$$

Moreover:

(1) *If U is entire, then b_1 and b_2 are entire.*

(2) *If $U \in \mathcal{U}_{rR}(J)$, then $\mathfrak{h}_U = \mathfrak{h}_{b_2^\#} \cap \mathfrak{h}_{b_1}$ and hence, U is entire if and only if b_1 and b_2 are entire.*

Proof It is enough to verify the theorem for $W \in \mathcal{U}(j_{pq})$ and $\{b_1, b_2\} \in ap(W)$. But then (4.83) follows from Lemma 4.26 and the fact that $W^{-1} = j_{pq} W^\# j_{pq}$. The verification of (2) is more complicated because it is necessary to check points on $\mathbb{R}$. Details may be found in [Ar90]. □

Corollary 4.55 *Let $U \in \mathcal{U}_{rR}(J)$, $J \neq \pm I_m$, and suppose that $\{b_1, b_2\} \in ap_I(U)$ and b_1 and b_2 are both entire mvf's. Then U is entire.*

Theorem 4.56 *Let $U \in \mathcal{U}(J)$, $J \neq \pm I_m$, let $q = p$ and let $\{b_3, b_4\} \in ap_{II}(U)$. Then*

$$\mathfrak{h}_U^+ = \mathfrak{h}^+_{b_4^\#}, \quad \mathfrak{h}_U^- = \mathfrak{h}^-_{b_3} \quad \text{and} \quad \mathfrak{h}_U \subseteq \mathfrak{h}_{b_4^\#} \cap \mathfrak{h}_{b_3}.$$

If U is entire, then b_3 and b_4 are entire. If $U \in \mathcal{U}_{rR}(J)$, then $\mathfrak{h}_U = \mathfrak{h}_{b_4^\#} \cap \mathfrak{h}_{b_3}$.

Proof The theorem follows from Lemma 4.36 and Theorem 4.54. □

Corollary 4.57 *Let $U \in \mathcal{U}_{rR}(J)$, $J \neq \pm I_m$, let $q = p$ and suppose that $\{b_3, b_4\} \in ap_{II}(U)$ are both entire mvf's. Then U is entire.*

4.11 Growth constraints on J-inner mvf's

Theorem 4.58 *Let $U \in \mathcal{U}(J)$ with J unitarily equivalent to j_{pq} and let $\{b_1, b_2\} \in ap_I(U)$. Then:*

(1) $\mathfrak{h}_U^+ = \mathbb{C}_+ \iff \mathfrak{h}_{b_2}^- = \mathbb{C}_-$ *and, if* $\mathfrak{h}_U^+ = \mathbb{C}_+$*, then* $\tau_U^+ = \tau_{b_2}^-$ *and* $\tau_U^+ \leq \delta_{b_2}^- \leq q\tau_U^+$.

(2) $\mathfrak{h}_U^- = \mathbb{C}_- \iff \mathfrak{h}_{b_1}^- = \mathbb{C}_-$ *and, if* $\mathfrak{h}_U^- = \mathbb{C}_-$*, then* $\tau_U^- = \tau_{b_1}^-$ *and* $\tau_U^- \leq \delta_{b_1}^- \leq p\tau_U^-$.

If $q = p$ and $\{b_3, b_4\} \in ap_{II}(U)$, then $\mathfrak{h}_{b_1} = \mathfrak{h}_{b_3}$, $\mathfrak{h}_{b_2} = \mathfrak{h}_{b_4}$ and:

(3) *If* $\mathfrak{h}_{b_3}^- = \mathbb{C}_-$*, then* $\tau_{b_1}^- = \tau_{b_3}^-$ *and* $\delta_{b_1}^- = \delta_{b_3}^-$.

(4) *If* $\mathfrak{h}_{b_4}^- = \mathbb{C}_-$*, then* $\tau_{b_2}^- = \tau_{b_4}^-$ *and* $\delta_{b_2}^- = \delta_{b_4}^-$.

Proof The identifications of the domains of holomorphy are established in Theorem 4.54. Assertions (1) and (2) may be verified for $W \in \mathcal{U}(j_{pq})$. Then, in view of the inequalities in (8) and (9) of Lemma 2.7,

$$\|s_{22}(\lambda)^{-1}\| \leq \|W(\lambda)\| \leq 3\|s_{22}(\lambda)^{-1}\| \quad \text{for every} \quad \lambda \in \mathfrak{h}_W^+$$

and

$$\|s_{11}(\lambda)^{-1}\| \leq \|W(\lambda)^{-1}\| \leq 3\|s_{11}(\lambda)^{-1}\| \quad \text{for every} \quad \lambda \in \mathfrak{h}_{W^{-1}}^+.$$

Consequently, by Lemma 3.112,

$$\mathfrak{h}_W^+ = \mathbb{C}_+ \Longrightarrow \tau_W^+ = \tau_{s_{22}^{-1}}^+ = \tau_{b_2^{-1}}^+ = \tau_{b_2}^- \quad \text{and} \quad \tau_{b_2}^- \leq \delta_{b_2}^- \leq q\tau_{b_2}^-,$$

whereas

$$\mathfrak{h}_W^- = \mathbb{C}_- \Longrightarrow \tau_W^- = \tau_{W^{-1}}^+ = \tau_{s_{11}^{-1}}^+ = \tau_{b_1^{-1}}^+ = \tau_{b_1}^- \quad \text{and} \quad \tau_{b_1}^- \leq \delta_{b_1}^- \leq p\tau_{b_1}^-.$$

This justifies (1) and (2).

If $q = p$, then assertions (3)–(6) may be verified for $A \in \mathcal{U}(J_p)$ by exploiting the connection (4.104) between the two sets of associated pairs $\{b_1, b_2\} \in ap_I(A)$ and $\{b_3, b_4\} \in ap_{II}(A)$ together with Lemma 3.54 and the fact that if $s \in \mathcal{S}^{p\times p}$, then $(I_p + s)/2 \in \mathcal{S}_{out}^{p\times p}$. □

4.12 Monotonic sequences in $\mathcal{P}^\circ(J)$

Lemma 4.59 *Let $U_1 \in \mathcal{P}^\circ(J)$, $U_2 \in \mathcal{P}^\circ(J)$ and let $U = U_1U_2$. Then $U \in \mathcal{P}^\circ(J)$ and*

$$\mathfrak{h}_U^+ = \mathfrak{h}_{U_1}^+ \cap \mathfrak{h}_{U_2}^+ \quad \text{and} \quad \mathfrak{h}_{U^{-1}}^+ = \mathfrak{h}_{U_1^{-1}}^+ \cap \mathfrak{h}_{U_2^{-1}}^+. \tag{4.84}$$

Proof There are three cases to consider: $J = I_m$, $J = -I_m$ and $J \neq \pm I_m$.

Case 1 If $J = I_m$, then the mvf's U_1, U_2 and U all belong to $\mathcal{S}^{m\times m}$ and hence $\mathfrak{h}_U^+ = \mathfrak{h}_{U_1}^+ = \mathfrak{h}_{U_2}^+ = \mathbb{C}_+$. Moreover, as

$$\det U(\lambda) = \det U_1(\lambda) \det U_2(\lambda)$$

it follows that $U \in \mathcal{P}^\circ(J)$ and

$$\det U(\lambda_0) \neq 0 \Longleftrightarrow \det U_1(\lambda_0) \neq 0 \quad \text{and} \quad \det U_2(\lambda_0) \neq 0.$$

Therefore, $\mathfrak{h}_{U^{-1}}^+ = \mathfrak{h}_{U_1^{-1}}^+ \cap \mathfrak{h}_{U_2^{-1}}^+$.

Case 2 If $J = -I_m$, then Case 1 is applicable to the mvf's $\widetilde{U} = U^{-1}$, $\widetilde{U}_1 = U_1^{-1}$ and $\widetilde{U}_2 = U_2^{-1}$.

Case 3 If $J \neq \pm I_m$, then without loss of generality it may be assumed that $J = j_{pq}$. Then the assertion of the lemma coincides with Corollary 4.30. □

A sequence $U_n \in \mathcal{P}^\circ(J), n \geq 1$, is called right (resp., left) **monotonic** if

$$U_n^{-1}U_{n+1} \in \mathcal{P}^\circ(J) \quad (\text{resp., } U_{n+1}U_n^{-1} \in \mathcal{P}^\circ(J)),\ n \geq 1.$$

If a sequence U_n is right (or left) monotonic, then, by Lemma 4.59,

$$\mathfrak{h}_{U_{n+1}}^+ \subseteq \mathfrak{h}_{U_n}^+ \quad \text{and} \quad \mathfrak{h}_{U_{n+1}^{-1}}^+ \subseteq \mathfrak{h}_{U_n^{-1}}^+.$$

Moreover, if the limit

$$U(\lambda) = \lim_{n\to\infty} U_n(\lambda) \text{ exists in } \mathbb{C}_+ \text{ except for at most a set of isolated points} \quad \text{and} \quad U \in \mathcal{P}^\circ(J), \tag{4.85}$$

then

$$\mathfrak{h}_U^+ \subseteq \bigcap_{n\geq 1} \mathfrak{h}_{U_n}^+ \quad \text{and} \quad \mathfrak{h}_{U^{-1}}^+ \subseteq \bigcap \mathfrak{h}_{U_n^{-1}}^+, \tag{4.86}$$

since

$$U_n^{-1}U \in \mathcal{P}^\circ(J) \quad \text{for } n \geq 1. \tag{4.87}$$

Moreover, the following assertion is valid.

Theorem 4.60 *Let $U_n \in \mathcal{P}^\circ(J)$, $n \geq 1$, be a right or left monotonic sequence that converges to a mvf $U \in \mathcal{P}^\circ(J)$ on $\mathfrak{h}_U^+$. Then*

$$\mathfrak{h}_U^+ = \bigcap_{n\geq 1} \mathfrak{h}_{U_n}^+ \quad \textit{and} \quad \mathfrak{h}_{U^{-1}}^+ = \bigcap_{n\geq 1} \mathfrak{h}_{U_n^{-1}}^+.$$

If J is unitarily equivalent to j_{pq} and if $\{b_1^{(n)}, b_2^{(n)}\} \in ap_I(U_n)$ are normalized by the conditions

$$b_1^{(n)}(\alpha) > 0 \quad \textit{and} \quad b_2^{(n)}(\alpha) > 0 \quad \textit{for } n = 1, 2, \ldots$$

at a fixed point $\alpha \in \mathfrak{h}_U^+ \cap \mathfrak{h}_{U^{-1}}^+$, then the limits

$$b_1(\lambda) = \lim_{n\uparrow\infty} b_1^{(n)}(\lambda) \quad \textit{and} \quad b_2(\lambda) = \lim_{n\uparrow\infty} b_2^{(n)}(\lambda)$$

exist at every point $\lambda \in \mathbb{C}_+$. Moreover, $\{b_1, b_2\} \in ap_I(U)$.

Proof It suffices to focus on right monotonicity.

If $J = I_m$, then $\mathfrak{h}_{U_n}^+ = \mathfrak{h}_U^+ = \mathbb{C}_+$ and the presumed monotonicity guarantees that

$$U_{n+1}(\lambda)U_{n+1}(\lambda)^* \leq U_n(\lambda)U_n(\lambda)^*, \quad \lambda \in \mathbb{C}_+.$$

Consequently,

$$U(\lambda)U(\lambda)^* \leq U_n(\lambda)U_n(\lambda)^*, \quad \lambda \in \mathbb{C}_+.$$

Let $s_n(\lambda) = \det U_n(\lambda)$, $s(\lambda) = \det U(\lambda)$ and let $\lambda_0 \in \bigcap_{n\geq 1} \mathfrak{h}_{U_n^{-1}}^+$. Then $s_n \in \mathcal{S}$ and $s_n(\lambda_0) \neq 0$ for $n \geq 1$; $s \in \mathcal{S}$ and, since $U \in \mathcal{P}^\circ(J)$, there exists a number r, $0 < r < |\lambda_0|$, such that $|s(\lambda)| > 0$ on the deleted disc $0 < |\lambda - \lambda_0| \leq r$. Therefore, $|s_n(\lambda)| > 0$ on the closed disc $0 \leq |\lambda - \lambda_0| \leq r$, since $|s_n(\lambda)| \geq |s(\lambda)|$ for $\lambda \in \mathbb{C}_+$ and $s_n(\lambda_0) \neq 0$ for $n = 1, 2, \ldots$. Thus, as $s_n(\lambda)$ tends uniformly to $s(\lambda)$ on the closed disc $0 \leq |\lambda - \lambda_0| \leq r$, a theorem of Hurwitz (which follows easily from Rouche's theorem; see e.g., p. 119 of [Ti60]) implies that $s(\lambda) \neq 0$ on this disc also, i.e., $\lambda_0 \in \mathfrak{h}_{U^{-1}}^+$. Thus,

$$\cap_{n=1}^{\infty} \mathfrak{h}_{U_n^{-1}}^+ \subseteq \mathfrak{h}_{U^{-1}}^+$$

and hence, as the opposite inclusion is available in (4.86), these two sets must coincide.

If $J = -I_m$, then $U_n^{-1} \in \mathcal{S}^{m\times m}$ and $U^{-1} \in \mathcal{S}^{m\times m}$, which implies that $\mathfrak{h}_{U^{-1}}^+ = \mathfrak{h}_{U_n^{-1}}^+ = \mathbb{C}_+$. That $\mathfrak{h}_U^+ = \mathfrak{h}_{U_n}^+$ follows by applying the preceding analysis to $S = U^{-1} \in \mathcal{P}^\circ(I_m)$, $S_n = U_n^{-1} \in \mathcal{P}^\circ(I_m)$.

If $J \neq \pm I_m$, then without loss of generality it may be assumed that $J = j_{pq}$ and then the asserted results follow from Lemmas 3.4 and 3.8 in [Ar97]. □

Theorem 4.61 *Let $\{U_n\}_{n=1}^{\infty}$ be a monotonic sequence of mvf's in $\mathcal{P}^\circ(J)$, let*

$$U(\lambda) = \lim_{n\uparrow\infty} U_n(\lambda) \quad for \quad \lambda \in \cap_{n=1}^{\infty} \left(\mathfrak{h}_{U_n}^+ \cap \mathfrak{h}_{U_n^{-1}}^+\right) \tag{4.88}$$

and assume that $U \in \mathcal{P}^\circ(J)$. Then there exists a subsequence $\{U_{n_k}\}_{k=1}^{\infty}$ such that

$$U(\mu) = \lim_{k\uparrow\infty} U_{n_k}(\mu) \quad a.e. \ on \ \mathbb{R}.$$

Proof See [Ar79b]. □

Corollary 4.62 *Let $\{U_n\}_{n=1}^{\infty}$ be a monotonic sequence of mvf's in $\mathcal{U}(J)$ and let $U \in \mathcal{P}^\circ(J)$ be the limit in (4.88). Then $U \in \mathcal{U}(J)$.*

Corollary 4.63 *A convergent infinite BP product $U(\lambda)$ belongs to the class $\mathcal{U}(J)$.*

4.13 Linear fractional transformations

In this section we consider left and right linear fractional transformations based on an $m \times m$ mvf

$$U(\lambda) = \begin{bmatrix} u_{11}(\lambda) & u_{12}(\lambda) \\ u_{21}(\lambda) & u_{22}(\lambda) \end{bmatrix} \tag{4.89}$$

that is meromorphic in $\mathbb{C}_+$ with blocks $u_{11}(\lambda)$ of size $p \times p$ and $u_{22}(\lambda)$ of size $q \times q$, respectively. The **right linear fractional transformation**

$$T_U^r[x] = \{u_{11}(\lambda)x(\lambda) + u_{12}(\lambda)\}\{u_{21}(\lambda)x(\lambda) + u_{22}(\lambda)\}^{-1} \tag{4.90}$$

acts in the set of $p \times q$ mvf's $\varepsilon(\lambda)$ that are meromorphic in $\mathbb{C}_+$ and belong to the set

$$\mathcal{D}(T_U^r) = \{x(\lambda) : \det[u_{21}(\lambda)x(\lambda) + u_{22}(\lambda)] \not\equiv 0 \ \text{ in } \ \mathbb{C}_+\},$$

i.e., to the domain of definition of T_U^r. The **left linear fractional transformation**

$$T_U^\ell[y] = \{y(\lambda)u_{12}(\lambda) + u_{22}(\lambda)\}^{-1}\{y(\lambda)u_{11}(\lambda) + u_{21}(\lambda)\} \tag{4.91}$$

acts in the set of $q \times p$ mvf's $y(\lambda)$ that are meromorphic in $\mathbb{C}_+$ and belong to the set

$$\mathcal{D}(T_U^\ell) = \{y(\lambda) : \det\{y(\lambda)u_{12}(\lambda) + u_{22}(\lambda)\} \not\equiv 0 \text{ in } \mathbb{C}_+\},$$

i.e., to the domain of definition of T_U^ℓ.

The notation

$$T_U^r[X] = \{T_U^r[x] : x \in X\} \quad \text{and} \quad T_U^\ell[Y] = \{T_U^\ell[y] : y \in Y\}$$

for subsets X of $\mathcal{D}(T_U^r)$ and subsets Y of $\mathcal{D}(T_U^\ell)$ will be useful.

Lemma 4.64 *Let $U(\lambda)$ be a meromorphic $m \times m$ mvf in $\mathbb{C}_+$ with block decomposition (4.89). Then:*

(1) $\mathcal{S}^{p\times q} \subseteq \mathcal{D}(T_U^r)$ *if and only if*

$$\det u_{22}(\lambda) \not\equiv 0 \text{ in } \mathfrak{h}_U^+, \quad \chi = u_{22}^{-1}u_{21} \in \mathcal{S}^{q\times p} \text{ and}$$
$$\chi(\lambda)^*\chi(\lambda) < I_p \quad \text{for each point } \lambda \in \mathbb{C}_+.$$

(2) $\mathcal{S}^{q\times p} \subseteq \mathcal{D}(T_U^\ell)$ *if and only if*

$$\det u_{22}(\lambda) \not\equiv 0 \text{ in } \mathfrak{h}_U^+, \quad s_{12} = u_{12}u_{22}^{-1} \in \mathcal{S}^{q\times p} \text{ and}$$
$$s_{12}(\lambda)^*s_{12}(\lambda) < I_q \quad \text{for each point } \lambda \in \mathbb{C}_+.$$

Proof The first assertion is immediate from Lemma 2.15. The second assertion follows from the first and the observation that

$$T_U^\ell[y] = (T_{U^\tau}^r[y^\tau])^\tau. \tag{4.92}$$

□

In the future, **we shall usually drop the superscript** r and write T_U instead of T_U^r.

Right and left transformations of $\mathcal{S}^{p\times q}$ into itself

Lemma 4.65 *Let $W \in \mathcal{P}(j_{pq})$. Then*

(1) $\mathcal{S}^{p\times q} \subseteq \mathcal{D}(T_W)$ *and* $T_W[\mathcal{S}^{p\times q}] \subseteq \mathcal{S}^{p\times q}$.

(2) $\mathcal{S}^{q\times p} \subseteq \mathcal{D}(T_W^\ell)$ *and* $T_W^\ell[\mathcal{S}^{q\times p}] \subseteq \mathcal{S}^{q\times p}$.

Proof Let s_{ij} denote the blocks of the PG transform S of W. Then Theorem 4.20 guarantees that $s_{12} \in \mathcal{S}^{p\times q}$, $s_{21} \in \mathcal{S}^{q\times p}$ and that

$$s_{12}(\lambda)^*s_{12}(\lambda) < I_q \quad \text{and} \quad s_{21}(\lambda)^*s_{21}(\lambda) < I_p \tag{4.93}$$

for each point $\lambda \in \mathbb{C}_+$. Therfore the first inclusions in (1) and (2) hold. On the other hand, if $\varepsilon \in \mathcal{S}^{p\times q}$, then $\varepsilon \in \mathcal{D}(T_W)$ and

$$(w_{21}\varepsilon + w_{22})^*(T_W[\varepsilon]^* T_W[\varepsilon] - I_q)(w_{21}\varepsilon + w_{22}) \\ = \begin{bmatrix} \varepsilon \\ I_q \end{bmatrix}^* (W^* j_{pq} W - j_{pq}) \begin{bmatrix} \varepsilon \\ I_q \end{bmatrix} \leq 0 \quad (4.94)$$

for each point $\lambda \in \mathfrak{h}_W^+$. Thus, $T_W[\varepsilon]$ is a holomorphic contractive mvf on $\mathfrak{h}_W^+$ and hence admits a unique extension to $\mathbb{C}_+$ that belongs to $\mathcal{S}^{p\times q}$. This completes the proof of the second inclusion in (1). The second inclusion in (2) may be verified in much the same way on the basis of the inequality

$$(\varepsilon w_{12} + w_{22})(T_W^\ell[\varepsilon] T_W^\ell[\varepsilon]^* - I_q)(\varepsilon w_{12} + w_{22})^* \\ = \begin{bmatrix} \varepsilon & I_q \end{bmatrix} (W j_{pq} W^* - j_{pq}) \begin{bmatrix} \varepsilon & I_q \end{bmatrix}^* \leq 0 \quad (4.95)$$

for $\varepsilon \in \mathcal{S}^{q\times p}$. □

Lemma 4.66 *Let $W \in \mathcal{U}(j_{pq})$. Then the assertions of Lemma 4.65 are in force. Moreover:*

(1) *If $p \geq q$, then $T_W[\mathcal{S}_{in}^{p\times q}] \subseteq \mathcal{S}_{in}^{p\times q}$ and $T_W^\ell[\mathcal{S}_{*in}^{q\times p}] \subseteq \mathcal{S}_{*in}^{q\times p}$.*

(2) *If $p \leq q$, then $T_W[\mathcal{S}_{*in}^{p\times q}] \subseteq \mathcal{S}_{*in}^{p\times q}$ and $T_W^\ell[\mathcal{S}_{in}^{q\times p}] \subseteq \mathcal{S}_{in}^{q\times p}$.*

Proof If $p \geq q$ and $\varepsilon \in \mathcal{S}_{in}^{p\times q}$, then inequality (4.94) will be an equality a.e. on $\mathbb{R}$, i.e., $T_W[\varepsilon]$ is an isometry a.e. on $\mathbb{R}$. Consequently, $T_W[\varepsilon] \in \mathcal{S}_{in}^{p\times q}$. The other three inclusions may be obtained similarly. □

Lemma 4.67 *Let $W \in \mathcal{P}(j_{pq})$, $W_1 \in \mathcal{P}^\circ(j_{pq})$ and suppose that $W_1^{-1}W \in \mathcal{P}(j_{pq})$. Then $T_W[\mathcal{S}^{p\times q}] \subseteq T_{W_1}[\mathcal{S}^{p\times q}]$.*

Proof Let $W_2 = W_1^{-1}W$. Then

$$T_W[\mathcal{S}^{p\times q}] \subseteq T_{W_1}[T_{W_2}[\mathcal{S}^{p\times q}]] \subseteq T_{W_1}[\mathcal{S}^{p\times q}],$$

since, $T_{W_2}[\mathcal{S}^{p\times q}] \subseteq \mathcal{S}^{p\times q}$ by Lemma 4.65. □

Lemma 4.68 *If $W_n \in \mathcal{P}^\circ(j_{pq})$, $n \geq 1$, is a monotonic sequence such that*

$$W(\lambda) = \lim_{n\uparrow\infty} W_n(\lambda) \quad \textit{for} \quad \lambda \in \bigcap_{n\geq 1} \mathfrak{h}_{W_n}^+ \quad \textit{and} \quad W \in \mathcal{P}^\circ(j_{pq}),$$

then

$$T_{W_{n+1}}[\mathcal{S}^{p\times q}] \subseteq T_{W_n}[\mathcal{S}^{p\times q} \quad and \quad T_W[\mathcal{S}^{p\times q}] = \bigcap_{n\geq 1} T_{W_n}[\mathcal{S}^{p\times q}].$$

Proof The first assertion follows from Lemma 4.67. Moreover, since $W_n^{-1}W_{n+k} \in \mathcal{P}^\circ(j_{pq})$ for $k \geq 1$, it follows that $W_n^{-1}W \in \mathcal{P}^\circ(j_{pq})$ and hence that $T_W[\mathcal{S}^{p\times q}] \subseteq \bigcap_{n\geq 1} T_{W_n}[\mathcal{S}^{p\times q}]$. Conversely, if $s \in \bigcap_{n\geq 1} T_{W_n}[\mathcal{S}^{p\times q}]$, then there exists a sequence of mvf's $\varepsilon_1, \varepsilon_2 \ldots \in \mathcal{S}^{p\times q}$ such that $s = T_{W_n}[\varepsilon_n]$ for every integer $n \geq 1$. Therefore, there exists a subsequence ε_{n_k} that converges to a limit $\varepsilon \in \mathcal{S}^{p\times q}$ at each point $\lambda \in \mathbb{C}_+$. Consequently,

$$s = T_{W_n}[\varepsilon_n] = \lim_{k\uparrow\infty} T_{W_{n_k}}[\varepsilon_{n_k}] = T_W[\varepsilon].$$

Thus, $s \in T_W[\mathcal{S}^{p\times q}]$. □

Remark 4.69 *If* $\varepsilon_n \in \mathcal{S}^{p\times q}$ *for* $n \geq 1$ *and*

$$s(\lambda) = \lim_{n\uparrow\infty} T_W[\varepsilon_n],$$

then there exists a subsequence $\{\varepsilon_{n_k}\}$, $k = 1, 2, \ldots$ *of* $\{\varepsilon_n\}$ *that converges to a mvf* $\varepsilon_0 \in \mathcal{S}^{p\times q}$ *at each point of* $\mathbb{C}_+$. *Therefore,*

$$s(\lambda) = \lim_{k\uparrow\infty} T_{W(\lambda)}[\varepsilon_{n_k}(\lambda)] = (T_W[\varepsilon_0])(\lambda)\,;$$

i.e., the set $T_W[\mathcal{S}^{p\times q}]$ *is a closed subspace of* $\mathcal{S}^{p\times q}$ *with respect to pointwise convergence.*

Transformations in $\mathcal{C}^{p\times p}$ and from $\mathcal{S}^{p\times p}$ into $\mathcal{C}^{p\times p}$

If $A \in \mathcal{P}(J_p)$, then it is convenient to consider linear fractional transformations based on the standard four block decompositions (with $p\times p$ blocks) of the mvf's

$$A(\lambda) = \mathfrak{V}W(\lambda)\mathfrak{V} \quad \text{and} \quad B(\lambda) = A(\lambda)\mathfrak{V}$$

together with that of $A(\lambda)$. It is readily checked that

$$\mathcal{S}^{p\times p} \cap \mathcal{D}(T_{\mathfrak{V}}) = \{s \in \mathcal{S}^{p\times p} : \det(I_p + s(\lambda)) \not\equiv 0\}$$

and

$$\mathcal{C}^{p\times p} = T_{\mathfrak{V}}[\mathcal{S}^{p\times p} \cap \mathcal{D}(T_{\mathfrak{V}})].$$

Consequently, in view of Lemma 4.65,

$$\mathcal{S}^{p\times p}\cap\mathcal{D}(T_B)=\{\varepsilon\in\mathcal{S}^{p\times p}:\,T_W[\varepsilon]\in\mathcal{D}(T_{\mathfrak{V}})\}$$

and

$$T_B[\mathcal{S}^{p\times p}\cap\mathcal{D}(T_B)]\;=\;T_{\mathfrak{V}}[T_W[\mathcal{S}^{p\times p}\cap\mathcal{D}(T_B)]]\subseteq\mathcal{C}^{p\times p}.$$

Let

$$\mathcal{C}(A)=T_B[\mathcal{S}^{p\times p}\cap\mathcal{D}(T_B)]. \tag{4.96}$$

Then it is easy to check that

$$T_A[\mathcal{C}^{p\times p}\cap\mathcal{D}(T_A)]\subseteq\mathcal{C}(A)\subseteq\mathcal{C}^{p\times p}\subset\mathcal{D}(T_{\mathfrak{V}}), \tag{4.97}$$

$$T_{\mathfrak{V}}[\mathcal{C}(A)]\subseteq T_W[\mathcal{S}^{p\times p}],\quad T_{\mathfrak{V}}[\mathcal{S}_{in}^{p\times p}\cap\mathcal{D}(T_{\mathfrak{V}})]\subseteq\mathcal{C}_{sing}^{p\times p}, \tag{4.98}$$

$$\mathring{\mathcal{C}}^{p\times p}\subset\mathcal{D}(T_A), \tag{4.99}$$

and, if $\tau\in\mathring{\mathcal{C}}^{p\times p}$ and $c=T_A[\tau]$, then $(\Re c)(\lambda)>0$ for every point $\lambda\in\mathbb{C}_+$. Moreover, it is readily seen that if $A\in\mathcal{U}(J_p)$, then also

$$T_A[\mathcal{C}_{sing}^{p\times p}\cap\mathcal{D}(T_A)]\subseteq T_B[\mathcal{S}_{in}^{p\times p}\cap\mathcal{D}(T_B)]\subseteq\mathcal{C}_{sing}^{p\times p}\subset\mathcal{D}(T_{\mathfrak{V}}) \tag{4.100}$$

and

$$T_{\mathfrak{V}B}[\mathcal{S}_{in}^{p\times p}\cap\mathcal{D}(T_B)]\subseteq T_W[\mathcal{S}_{in}^{p\times p}]. \tag{4.101}$$

Lemma 4.70 *Let $A\in\mathcal{P}(J_p)$, let $B=A\mathfrak{V}$ and let $\chi=b_{22}^{-1}b_{21}$. Then the following conditions are equivalent:*

(1) $I_p-\chi(\omega)\chi(\omega)^*>0$ *for at least one point* $\omega\in\mathbb{C}_+$.

(2) $I_p-\chi(\lambda)\chi(\lambda)^*>0$ *for every point* $\lambda\in\mathbb{C}_+$.

(3) $b_{22}(\omega)b_{22}(\omega)^*-b_{21}(\omega)b_{21}(\omega)^*>0$ *for at least one point* $\omega\in\mathfrak{h}_B^+$.

(4) $b_{22}(\lambda)b_{22}(\lambda)^*-b_{21}(\lambda)b_{21}(\lambda)^*>0$ *for every point* $\lambda\in\mathfrak{h}_B^+$.

(5) $\mathcal{S}^{p\times p}\subseteq\mathcal{D}(T_B)$.

Proof In view of Lemma 4.35, $\det b_{22}(\lambda)\neq0$ for $\lambda\in\mathfrak{h}_B^+$ and $\chi\in\mathcal{S}^{p\times p}$. Consequently, (1) is equivalent to (3). The equivalence (2) $\Longleftrightarrow$ (5) is valid by Lemma 4.64. The implications (2) $\Longrightarrow$ (1) and (4) $\Longrightarrow$ (3) are obvious. The converse implication (1) $\Longrightarrow$ (2) follows from Corollary 3.52. The implication (3) $\Longrightarrow$ (4) then follows from the formula $\chi=b_{22}^{-1}b_{21}$ and the implication (1) $\Longrightarrow$ (2). □

Lemma 4.71 *Let $A \in \mathcal{P}(J_p)$, $B = A\mathfrak{V}$ and $W = \mathfrak{V}A\mathfrak{V}$. Then*

$$T_{\mathfrak{V}}[T_A[\mathcal{C}^{p\times p} \cap \mathcal{D}(T_A)]] \subseteq T_W[\mathcal{S}^{p\times q}]. \tag{4.102}$$

If $s \in T_W[\mathcal{S}^{p\times q}]$, there is a sequence of mvf's $s_n \in T_{\mathfrak{V}A}[\mathcal{C}^{p\times p} \cap \mathcal{D}(T_A)]$ such that

$$s(\lambda) = \lim_{n\uparrow\infty} s_n(\lambda).$$

Moreover, if $A \in \mathcal{U}(J_p)$ and $s \in T_W[\mathcal{S}_{in}^{p\times q}]$, then the mvf's s_n may be chosen from $T_{\mathfrak{V}A}[\mathcal{C}_{sing}^{p\times p} \cap \mathcal{D}(T_A)]$.

Proof Let $\varepsilon \in \mathcal{S}^{p\times p}$, $\chi = b_{22}^{-1}b_{21}$, $\gamma \in \mathbb{C}$ and suppose that $|\gamma| = 1$. Then:

(a) $\gamma\varepsilon \in \mathcal{D}(T_{\mathfrak{V}}) \iff \det(\overline{\gamma}I_p + \varepsilon(\omega)) \neq 0$ for at least one (and hence every) point $\omega \in \mathbb{C}_+$.

(b) $\gamma\varepsilon \in \mathcal{D}(T_B) \iff \det(\overline{\gamma}I_p + \chi(\omega)\varepsilon(\omega)) \neq 0$ for at least one (and hence every) point $\omega \in \mathbb{C}_+$.

Now let $s = T_W[\varepsilon]$ for some mvf $\varepsilon \in \mathcal{S}^{p\times p}$. Then there exists a sequence of points $\gamma_n \in \mathbb{C}$ with $|\gamma_n| = 1$ such that

$$\lim \gamma_n = 1,\ \gamma_n\varepsilon \in \mathcal{D}(T_{\mathfrak{V}}) \cap \mathcal{D}(T_B),\ T_W[\gamma_n\varepsilon] \in T_{\mathfrak{V}A}[\mathcal{C}^{p\times p} \cap \mathcal{D}(T_A)]$$

and

$$s(\lambda) = \lim_{n\uparrow\infty} T_{W(\lambda)}[\gamma_n\varepsilon(\lambda)] = T_{W(\lambda)}[\varepsilon(\lambda)].$$

Moreover, if $A \in \mathcal{U}(J_p)$ and $\varepsilon \in \mathcal{S}_{in}^{p\times p}$, then

$$T_W[\gamma_n\varepsilon] \in \mathcal{S}_{in}^{p\times p} \quad \text{and} \quad T_{\mathfrak{V}}[T_W[\gamma_n\varepsilon]] \in \mathcal{C}_{sing}^{p\times p}.$$

□

Lemma 4.72 *Let $A \in \mathcal{P}(J_p)$, $A_1 \in \mathcal{P}^\circ(J_p)$ and $A_1^{-1}A \in \mathcal{P}(J_p)$. Then*

$$\mathcal{C}(A) \subseteq \mathcal{C}(A_1). \tag{4.103}$$

Proof Let $W = \mathfrak{V}A\mathfrak{V}$, $W_1 = \mathfrak{V}A_1\mathfrak{V}$, $B = A\mathfrak{V}$ and $B_1 = A_1\mathfrak{V}$. Then $W \in \mathcal{P}(j_p)$, $W_1 \in \mathcal{P}^\circ(j_p)$, and, by Lemma 4.67,

$$T_W[\mathcal{S}^{p\times p}] \subseteq T_{W_1}[\mathcal{S}^{p\times p}].$$

Consequently,

$$\mathcal{D}(T_B) = \mathcal{D}(T_{\mathfrak{V}}) \cap T_W[\mathcal{S}^{p\times p}] \subseteq \mathcal{D}(T_{\mathfrak{V}}) \cap T_{W_1}[\mathcal{S}^{p\times p}] = \mathcal{D}(T_{B_1}),$$

and, if $\varepsilon \in \mathcal{D}(T_B)$, then

$$T_B[\varepsilon] = T_{\mathfrak{V}}[T_W[\varepsilon]] \in T_{\mathfrak{V}}[\mathcal{D}(T_{\mathfrak{V}}) \cap T_{W_1}[\mathcal{S}^{p\times p}]] = \mathcal{C}(A_1).$$

□

4.14 Connections between the two kinds of associated pairs

The next lemma will be used to establish a connection between the two sets of associated pairs.

Lemma 4.73 *Let $A \in \mathcal{U}(J_p)$, let $B(\lambda) = A(\lambda)\mathfrak{V}$, $\chi = b_{22}^{-1}b_{21}$, $W(\lambda) = \mathfrak{V}A(\lambda)\mathfrak{V}$, $\{b_1, b_2\} \in ap_I(A)$ and $\{b_3, b_4\} \in ap_{II}(A)$. Then for every $s \in T_W[\mathcal{S}^{p\times p}] \cap \mathcal{D}(T_{\mathfrak{V}})$, there exist a pair of mvf's $\varphi_s \in \mathcal{S}_{out}^{p\times p}$ and $\psi_s \in \mathcal{S}_{out}^{p\times p}$ such that*

$$(1/2)(I_p + s)b_3 = b_1\varphi_s \quad \text{and} \quad b_4(I_p + s)/2 = \psi_s b_2. \tag{4.104}$$

If

$$I_p - \chi(\omega)\chi(\omega)^* > 0 \quad \text{for at least one point} \quad \omega \in \mathbb{C}_+, \tag{4.105}$$

then $T_W[\mathcal{S}^{p\times p}] \subset \mathcal{D}(T_{\mathfrak{V}})$ and (4.104) holds for every $s \in T_W[\mathcal{S}^{p\times p}]$.

Proof Let $W(\lambda) = \mathfrak{V}A(\lambda)\mathfrak{V}$, $B(\lambda) = \mathfrak{V}W(\lambda)$ and recall that if $\varepsilon \in \mathcal{S}^{p\times p}$, then

$$\begin{aligned} s = T_W[\varepsilon] &= (w_{11}\varepsilon + w_{12})(w_{21}\varepsilon + w_{22})^{-1} \\ &= (w_{11}^\# + \varepsilon w_{12}^\#)^{-1}(w_{21}^\# + \varepsilon w_{22}^\#). \end{aligned}$$

Consequently,

$$\begin{aligned} I_p + s = \sqrt{2}(b_{21}\varepsilon + b_{22})(w_{21}\varepsilon + w_{22})^{-1} \\ = \sqrt{2}(w_{11}^\# + \varepsilon w_{12}^\#)^{-1}(\varepsilon b_{22}^\# + b_{21}^\#). \end{aligned} \tag{4.106}$$

Then, by formulas (4.106), (3) and (6) of Lemma 4.35, (4.42), (4.74) and Lemmas 4.25 and 4.24, we obtain

$$b_4(I_p + s) = \sqrt{2}\varphi_4^{-1}(I_p + \chi\varepsilon)(I_p - s_{21}\varepsilon)^{-1}\varphi_2 b_2$$

and

$$(I_p + s)b_3 = \sqrt{2}b_1\varphi_1(I_p - \varepsilon s_{21})^{-1}(I_p + \varepsilon\chi)\varphi_3^{-1}.$$

By Lemma 3.54, the mvf's $I_p+\mathcal{X}\varepsilon$, $I_p+\varepsilon\mathcal{X}$, $I_p-s_{21}\varepsilon$ and $I_p-\varepsilon s_{21}$ all belong to the class $\mathcal{N}_{out}^{p\times p}$, since $\mathcal{X}\varepsilon$, $\varepsilon\mathcal{X}$, $s_{21}\varepsilon$ and εs_{21} belong to $\mathcal{S}^{p\times p}$ and

$$\det(I_p+\mathcal{X}\varepsilon)=\det(I_p+\varepsilon\mathcal{X})\not\equiv 0 \quad \text{if } s\in\mathcal{D}(T_{\mathfrak{V}})$$

and

$$\det(I_p-s_{21}\varepsilon)=\det(I_p-\varepsilon s_{21})\not\equiv 0,$$

by (7) of Theorem 4.20. Thus, as $\varphi_j\in\mathcal{N}_{out}^{p\times p}$ for $j=1,\ldots,4$, we see that (4.104) is valid with $\varphi_s,\psi_s\in\mathcal{N}_{out}^{p\times p}$. The supplementary conclusion, that in fact φ_s and ψ_s belong to $\mathcal{S}^{p\times p}$, then follows from the Smirnov maximum principle. The second statement then follows from Lemma 4.70. □

4.15 Redheffer transforms

Let $S\in\mathcal{S}^{m\times m}$ with blocks s_{ij}. The **right Redheffer transform** R_S^r is defined on $p\times q$ mvf's $x(\lambda)$ that are meromorphic in $\mathbb{C}_+$ and belong to the domain

$$\mathcal{D}(R_S^r)=\{x:\det(I_q-s_{21}(\lambda)x(\lambda))\not\equiv 0 \text{ in } \mathfrak{h}_x^+\} \tag{4.107}$$

by the formula

$$R_S^r[x]=s_{12}+s_{11}x(I_q-s_{21}x)^{-1}s_{22}, \tag{4.108}$$

and the **left Redheffer transform** R_S^ℓ is defined on $q\times p$ mvf's $y(\lambda)$ that are meromorphic in $\mathbb{C}_+$ and belong to the domain

$$\mathcal{D}(R_S^\ell)=\{y:\det(I_q+y(\lambda)s_{12}(\lambda))\not\equiv 0 \text{ in } \mathfrak{h}_y^+\} \tag{4.109}$$

by the formula

$$R_S^\ell[y]=-s_{21}+s_{22}(I_q+ys_{12})^{-1}ys_{11}. \tag{4.110}$$

The notation

$$R_S^r[X]=\{R_S^r[x]:x\in X\} \quad \text{and} \quad R_S^\ell[Y]=\{R_S^\ell[y]:y\in Y\}$$

for $X\subseteq\mathcal{D}(R_S^r)$ and $Y\subseteq\mathcal{D}(R_S^\ell)$ will be useful.

Theorem 4.74 *If $W\in\mathcal{P}(j_{pq})$ and $S=PG(W)$, then*

(1) $\mathcal{S}^{p\times q}\subset\mathcal{D}(R_S^r)$, $R_S^r[\mathcal{S}^{p\times q}]\subseteq\mathcal{S}^{p\times q}$ *and*

$$T_W^r[\varepsilon]=R_S^r[\varepsilon] \quad \textit{for every } \varepsilon\in\mathcal{S}^{p\times q}. \tag{4.111}$$

(2) $\mathcal{S}^{q\times p} \subset \mathcal{D}(R_S^\ell)$, $R_S^\ell[\mathcal{S}^{q\times p}] \subseteq \mathcal{S}^{q\times p}$ *and*

$$T_W^\ell[\varepsilon] = R_S^\ell[\varepsilon] \quad \text{for every } \varepsilon \in \mathcal{S}^{q\times p}. \tag{4.112}$$

(3) *If* $W \in \mathcal{U}(j_{pq})$, *then* $R_S^r[\mathcal{S}_{in}^{p\times q}] \subseteq \mathcal{S}_{in}^{p\times q}$ *if* $p \geq q$ *and* $R_S^r[\mathcal{S}_{*in}^{p\times q}] \subseteq \mathcal{S}_{*in}^{p\times q}$ *if* $p \leq q$. *Moreover,*

$$R_S^\ell[\mathcal{S}_{in}^{q\times p}] \subseteq \mathcal{S}_{in}^{q\times p} \quad \text{if } q \geq p \quad \text{and} \quad R_S^\ell[\mathcal{S}_{*in}^{q\times p}] \subseteq \mathcal{S}_{*in}^{q\times p} \quad \text{if } q \leq p.$$

Proof The inclusions $\mathcal{S}^{p\times q} \subset \mathcal{D}(R_S^r)$ and $\mathcal{S}^{q\times p} \subset \mathcal{D}(R_S^\ell)$ follow from the fact that if $W \in \mathcal{P}(j_{pq})$ and $S = PG(W)$, then $\det s_{22}(\lambda) \not\equiv 0$ in $\mathbb{C}_+$ and the inequalities (4.93) are in force for every point $\lambda \in \mathbb{C}_+$.

The remaining assertions are easy consequences of formula (2.58) and the observation that

$$R_S^\ell[y] = (R_{S_1}^r[y^\tau])^\tau, \quad \text{if} \quad S_1 = j_{pq} S^\tau j_{pq}.$$

□

In the future, **we shall drop the superscript** r and write R_S instead of R_S^r.

We remark that right (and left) Redheffer transforms may be defined just as in formulas (2.55) and (2.56) for mvf's $S \in \mathcal{S}^{(p+k)\times(r+q)}$ with blocks s_{11} and s_{22} that are not necessarily square. Then R_S is a mapping of $r \times k$ mvf's into the set of $p \times q$ mvf's and the analogue of Theorem 2.27 holds for mvf's.

4.16 Strongly regular J-inner mvf's

A mvf $W \in \mathcal{U}(j_{pq})$ is said to be a **right strongly regular** j_{pq}-inner mvf if

$$T_W[\mathcal{S}^{p\times q}] \cap \mathring{\mathcal{S}}^{p\times q} \neq \emptyset;$$

it is said to be a **left strongly regular** j_{pq}-inner mvf if

$$T_W^\ell[\mathcal{S}^{q\times p}] \cap \mathring{\mathcal{S}}^{q\times p} \neq \emptyset.$$

If $J \neq \pm I_m$ and $J = V^* j_{pq} V$ for some constant $m \times m$ unitary matrix V, then $U \in \mathcal{U}(J)$ is said to be a **right (left) strongly regular** J-inner mvf if $W(\lambda) = VU(\lambda)V^*$ is a right (left) strongly regular j_{pq}-inner mvf. This definition does not depend upon the choice of V: If $V_1^* j_{pq} V_1 = V_2^* j_{pq} V_2$, then the matrix $V_1 V_2^*$ is both unitary and j_{pq}-unitary and therefore, it must be of the form $V_1 V_2^* = \operatorname{diag}\{u_1, u_2\}$, where u_1 and u_2 are constant unitary matrices of sizes $p \times p$ and $q \times q$, respectively. Thus, the mvf's $W_1(\lambda) = V_1 U(\lambda) V_1^*$ and $W_2(\lambda) = V_2 U(\lambda) V_2^*$ are related by the formula

$$W_2(\lambda) = \begin{bmatrix} u_1^* & 0 \\ 0 & u_2^* \end{bmatrix} W_1(\lambda) \begin{bmatrix} u_1 & 0 \\ 0 & u_2 \end{bmatrix}$$

and, consequently,

$$T_{W_2}[\mathcal{S}^{p\times q}] = u_1^* T_{W_1}[\mathcal{S}^{p\times q}] u_2.$$

Therefore,

$$T_{W_2}[\mathcal{S}^{p\times q}] \cap \mathring{\mathcal{S}}^{p\times q} \neq \emptyset \Longleftrightarrow T_{W_1}[\mathcal{S}^{p\times q}] \cap \mathring{\mathcal{S}}^{p\times q} \neq \emptyset.$$

The classes of left and right strongly regular J-inner mvf's will be denoted $\mathcal{U}_{\ell sR}(J)$ and $\mathcal{U}_{rsR}(J)$, respectively. The convention $\mathcal{U}_{\ell sR}(\pm I_m) = \mathcal{U}_{rsR}(\pm I_m) = \mathcal{U}(\pm I_m)$ will be convenient.

Theorem 4.75 *The following inclusions hold:*

(1) $\mathcal{U}(J) \cap L_\infty^{m\times m}(\mathbb{R}) \subseteq \mathcal{U}_{rsR}(J) \cap \mathcal{U}_{\ell sR}(J)$.

(2) $\mathcal{U}_{rsR}(J) \cup \mathcal{U}_{\ell sR}(J) \subset \widetilde{L}_2^{m\times m}$.

(3) $\mathcal{U}(J) \cap \widetilde{L}_2^{m\times m} \subset \mathcal{U}_{\ell R}(J) \cap \mathcal{U}_{rR}(J)$.

Proof Corollary 4.23 justifies (1). To verify (2), it suffices to consider the case $J = j_{pq}$. Let $W \in \mathcal{U}_{rsR}(j_{pq})$ and let $s_{21} = -w_{22}^{-1} w_{21}$. Then there exists a mvf $\varepsilon \in \mathcal{S}^{p\times q}$ such that $s(\lambda) = T_W[\varepsilon]$ is strictly contractive: $\|s\|_\infty \leq \gamma < 1$. Let

$$c(\lambda) = \{I_q + s_{21}(\lambda)\varepsilon(\lambda)\}\{I_q - s_{21}(\lambda)\varepsilon(\lambda)\}^{-1}.$$

Then, since $c \in \mathcal{C}^{q\times q}$, it follows that $\Re c \in \widetilde{L}_1^{q\times q}$. The desired result now drops out easily from Corollary 4.23 and the inequality

$$\begin{aligned} \Re c(\mu) &= \{I_q - \varepsilon(\mu)^* s_{21}(\mu)^*\}^{-1}\{I_q - \varepsilon(\mu)^* s_{21}(\mu)^* s_{21}(\mu)\varepsilon(\mu)\} \\ &\quad \times \{I_q - s_{21}(\mu)\varepsilon(\mu)\}^{-1} \\ &\geq \{I_q - \varepsilon(\mu)^* s_{21}(\mu)^*\}^{-1}\{I_q - \varepsilon(\mu)^*\varepsilon(\mu)\}\{I_q - s_{21}(\mu)\varepsilon(\mu)\}^{-1} \\ &= w_{22}(\mu)^*\{I_q - s(\mu)^* s(\mu)\} w_{22}(\mu) \\ &\geq (1-\gamma^2) w_{22}(\mu)^* w_{22}(\mu). \end{aligned}$$

If $W \in \mathcal{U}_{\ell sR}(j_{pq})$, then $W^\tau \in \mathcal{U}_{rsR}(j_{pq})$ and hence $W^\tau \in \widetilde{L}_2^{m\times m}$, as proved above. The proof of (3) is postponed to Theorem 5.90. □

Theorem 4.76 *If $U \in \mathcal{U}_{rsR}(J) \cup \mathcal{U}_{\ell sR}(J)$, then $U \in \mathcal{U}_{\ell R}(J) \cap \mathcal{U}_{rR}(J)$, i.e.,*

$$\mathcal{U}_{rsR}(J) \cup \mathcal{U}_{\ell sR}(J) \subseteq \mathcal{U}_{\ell R}(J) \cap \mathcal{U}_{rR}(J).$$

Moreover, if $U \in \mathcal{U}_{rsR}(J)$ and admits a factorization of the form

$$U(\lambda) = U_1(\lambda)U_2(\lambda)U_3(\lambda)$$

with $U_i \in \mathcal{U}(J)$ for $i = 1, 2, 3$, then:

(1) $U_1 \in \mathcal{U}_{rsR}(J)$.

(2) $U_1U_2 \in \mathcal{U}_{rsR}(J)$.

(3) $U_i \in \mathcal{U}_{\ell R}(J) \cap \mathcal{U}_{rR}(J)$ *for* $i = 1, 2, 3$.

Proof The first assertion is immediate from the preceding two theorems.

To complete the proof, it suffices to consider $J = j_{pq}$ and then, in terms of our usual notation, we write W in place of U and W_i in place of U_i. Since

$$T_W[\mathcal{S}^{p\times q}] \subseteq T_{W_1W_2}[\mathcal{S}^{p\times q}] \subseteq T_{W_1}[\mathcal{S}^{p\times q}],$$

(1) and (2) are clear from the definition of right strong regularity. But these two conclusions imply that W_2 does not have a nonconstant singular j_{pq}-inner divisor on the left because otherwise it could be shifted to W_1, which would contradict (1). Similarly, W_3 does not have a left singular j_{pq}-inner factor because of (2). Moreover, in view of (2) and the already established fact that $W \in \mathcal{U}_{rR}(j_{pq})$, W_2 and W_3 do not have nonconstant singular j_{pq}-inner divisors on the right. □

Theorem 4.76 justifies the use of the terminology *strongly regular.*

4.17 Minus matrix valued functions

Our next objective is to develop analogues of the algebraic results that were presented in Sections 2.7 to 2.10 for meromorphic mvf's in $\mathbb{C}_+$. We shall show that the properties of the linear fractional transformations T_U based on $U \in \mathcal{P}(J)$ (resp., $U \in \mathcal{U}(J)$) for $J = j_{pq}$ and $J = J_p$ that are considered in this section serve to essentially characterize the classes $\mathcal{P}(J)$ and $\mathcal{U}(J)$ in the following sense: If a nondegenerate mvf $U(\lambda)$ that is meromorphic in $\mathbb{C}_+$ has these properties, then there exists a scalar function $\rho(\lambda)$ such that $\rho(\lambda)U(\lambda)$ belongs to $\mathcal{P}(J)$ (resp., $\rho U \in \mathcal{U}(J)$). These results are mostly adapted from the work of L. A. Simakova. It is first necessary to study minus mvf's.

An $m \times m$ mvf U is said to be a **minus mvf** with respect to a signature matrix J in $\mathbb{C}_+$ if U is meromorphic in $\mathbb{C}_+$ and $U(\lambda)$ is a minus matrix with respect to J for each point $\lambda \in \mathfrak{h}_U^+$. A minus mvf U is said to be nondegenerate if $\det U(\lambda) \not\equiv 0$ on $\mathfrak{h}_U^+$.

Lemma 4.77 *Let U be a nondegenerate minus mvf with respect to a signature matrix J in $\mathbb{C}_+$ and suppose that $J \neq \pm I_m$. Then there exists a scalar function $\rho(\lambda)$ that is meromorphic in $\mathbb{C}_+$ such that $\rho(\lambda) \not\equiv 0$ and $\rho U \in \mathcal{N}^{m\times m}$.*

Proof Without loss of generality we may assume that U is a nondegenerate minus mvf with respect to j_{pq} in $\mathbb{C}_+$, and, in keeping with our usual notation, replace U by W. Then, by Lemma 2.50 the mvf's

$$\alpha(\lambda) = w_{11}(\lambda) - w_{12}(\lambda)w_{22}(\lambda)^{-1}w_{21}(\lambda) \text{ and } \delta(\lambda) = w_{22}(\lambda)^{-1} \tag{4.113}$$

are well defined in terms of the standard four block decomposition of $W(\lambda)$, they are meromorphic in $\mathbb{C}_+$ and

$$s_{12}(\lambda) = w_{12}(\lambda)w_{22}(\lambda)^{-1} \quad \text{and} \quad s_{21}(\lambda) = -w_{22}(\lambda)^{-1}w_{21}(\lambda) \tag{4.114}$$

are holomorphic contractive mvf's on $\mathfrak{h}_W^+$. Therefore these two mvf's have unique holomorphic contractive extensions on $\mathbb{C}_+$, which we continue to denote $s_{12}(\lambda)$ and $s_{21}(\lambda)$, respectively. Thus,

$$s_{12} \in S^{p\times q} \quad \text{and} \quad s_{21} \in S^{q\times p}. \tag{4.115}$$

The auxiliary formula

$$W(\lambda) = \begin{bmatrix} I_p & s_{12}(\lambda) \\ 0 & I_q \end{bmatrix} \begin{bmatrix} \alpha(\lambda) & 0 \\ 0 & w_{22}(\lambda) \end{bmatrix} \begin{bmatrix} I_p & 0 \\ -s_{21}(\lambda) & I_q \end{bmatrix}$$

based on Schur complements, implies that

$$\det W(\lambda) = \det \alpha(\lambda) \cdot \det w_{22}(\lambda).$$

Since $\det W(\lambda) \not\equiv 0$ by assumption, there exists an entry $\alpha_{st}(\lambda)$ in $\alpha(\lambda)$ such that $\alpha_{st}(\lambda) \not\equiv 0$ in $\mathbb{C}_+$. Let

$$\rho(\lambda) = \alpha_{st}(\lambda)^{-1}. \tag{4.116}$$

We will show that $\rho W \in \mathcal{N}^{m\times m}$ with this choice of $\rho(\lambda)$. Let $\omega \in \mathfrak{h}_W^+ \cap \mathfrak{h}_{W^{-1}}$ be a fixed point. Then Theorem 2.51 implies that there exists a scalar $\rho_1(\omega) \neq 0$ such that $\widetilde{W}(\omega) = \rho_1(\omega)W(\omega)$ belongs to $\mathcal{P}_{const}(j_{pq})$. Let $\widetilde{S}(\omega)$ be the PG transform of $\widetilde{W}(\omega)$. Then $\widetilde{S}(\omega) \in \mathcal{S}_{const}^{m\times m}$ and in the standard four block decomposition of $\widetilde{W}(\omega)$,

$$\widetilde{s}_{11}(\omega) = \rho_1(\omega)\alpha(\omega) \in \mathcal{S}_{const}^{p\times p} \quad \text{and} \quad \widetilde{s}_{22}(\omega) = \rho_1(\omega)^{-1}\delta(\omega) \in \mathcal{S}_{const}^{q\times q}.$$

Consequently, the entries $\alpha_{kl}(\omega)$ of $\alpha(\omega)$ and $\delta_{ij}(\omega)$ of $\delta(\omega)$ are subject to the bounds

$$|\rho_1(\omega)\alpha_{kl}(\omega)| \leq 1 \text{ and } |\rho_1(\omega)^{-1}\delta_{ij}(\omega)| \leq 1,$$

and hence

$$\alpha_{kl}(\lambda)\delta_{ij}(\lambda) \in \mathcal{S}. \tag{4.117}$$

Moreover, the bounds

$$|\rho_1(\omega)\alpha_{st}(\omega)| \leq 1 \text{ and } \|\rho_1(\omega)^{-1}\delta(\omega)\| \leq 1$$

imply that $|\rho_1(\omega)| \leq |\rho(\omega)|$ and

$$\rho^{-1}(\omega)\delta(\omega) \in \mathcal{S}_{const}^{p\times p} \quad \text{for } \omega \in \mathfrak{h}_W^+ \cap \mathfrak{h}_{W^{-1}}^+.$$

Thus, $\rho^{-1}\delta$ has a unique holomorphic contractive extension to $\mathbb{C}_+$, i.e.,

$$\rho^{-1}\delta \in \mathcal{S}^{p\times p},$$

and hence

$$\rho\, w_{22} \in \mathcal{N}^{q\times q}.$$

Then, in view of (4.114) and (4.115),

$$\rho\, w_{12} \in \mathcal{N}^{p\times q} \quad \text{and} \quad \rho\, w_{21} \in \mathcal{N}^{q\times p}.$$

The inclusion (4.117) implies that

$$h_{kl} = \alpha_{kl}\delta \in H_\infty^{q\times q},$$

and consequently that

$$\rho\, \alpha_{kl} I_q = h_{kl}\rho\, w_{22} \in \mathcal{N}^{q\times q}$$

for every entry α_{kl} of α. Thus, $\rho\, \alpha \in \mathcal{N}^{p\times p}$ and therefore, $\rho\, w_{11} \in \mathcal{N}^{p\times p}$, since

$$w_{11} = \alpha + w_{12}w_{22}^{-1}w_{21} = \alpha + s_{12}w_{21}$$

and $\rho\, \alpha \in \mathcal{N}^{p\times p}$, $\rho\, w_{21} \in \mathcal{N}^{q\times p}$ and $s_{12} \in \mathcal{S}^{p\times q}$. □

Remark 4.78 *The function $\rho = \alpha_{st}^{-1}$ satisfies the needed condition $\rho W \in \mathcal{N}^{m\times m}$ and has an extra property:*

$$(\rho\, w_{22})^{-1} \in \mathcal{S}^{q\times q}. \tag{4.118}$$

Moreover, the PG transform S_ρ of the mvf ρW is a well defined holomorphic mvf on $\mathfrak{h}_W^+ \cap \mathfrak{h}_{W^{-1}}^+$ and its blocks $s_{ij}^{(\rho)}$ in the standard four block decomposition of S_ρ are such that:

$$\begin{aligned} s_{11}^{(\rho)} &= \rho\alpha \in \mathcal{N}^{p\times p}, \quad s_{12}^{(\rho)} = s_{12} \in \mathcal{S}^{p\times q}, \\ s_{21}^{(\rho)} &= s_{21} \in \mathcal{S}^{q\times p} \quad \textit{and} \quad s_{22}^{(\rho)} = (\rho\, w_{22})^{-1} \in \mathcal{S}^{q\times q}. \end{aligned}$$

Theorem 4.79 *Let U be a nondegenerate minus mvf with respect to a signature matrix J in $\mathbb{C}_+$ and suppose that $J \neq I_m$. Then there exists a scalar function ρ that is meromorphic in $\mathbb{C}_+$ such that $\rho U \in \mathcal{P}^\circ(J)$.*

Proof Without loss of generality, we may assume that U is a non-degenerate minus-mvf with respect to j_{pq} in $\mathbb{C}_+$ and then, in keeping with our usual notation, we write W in place of U. Moreover, in view of Lemma 4.77 and Remark 4.78, we can assume that $W \in \mathcal{N}^{m\times m}$ and $w_{22}^{-1} \in \mathcal{S}^{p\times p}$. Let $W(\mu)$ be the nontangential boundary values of W a.e. on $\mathbb{R}$ and let

$$G(\mu) = j_{pq} W(\mu)^* j_{pq} W(\mu) \quad \text{a.e. on } \mathbb{R}. \tag{4.119}$$

Then $\det W(\mu) \neq 0$ a.e. on $\mathbb{R}$, because $\det W \in \mathcal{N}$ and $\det W(\lambda) \not\equiv 0$ on $\mathfrak{h}_W^+$. Therefore $G(\mu)$ is invertible a.e. on $\mathbb{R}$, and hence, by Lemma 2.42, the eigenvalues of $G(\mu)$ are positive a.e. on $\mathbb{R}$ and taking into account their algebraic multiplicities, they may be indexed so that

$$0 < \lambda_1(\mu) \leq \lambda_2(\mu) \leq \cdots \leq \lambda_m(\mu) \quad \text{a.e. on } \mathbb{R}.$$

By Theorem 2.47, $c(\mu)W(\mu) \in \mathcal{P}_{const}(j_{pq})$ for some scalar $c(\mu)$ if and only if

$$\lambda_p(\mu) \leq |c(\mu)|^{-2} \leq \lambda_{p+1}(\mu). \tag{4.120}$$

All functions $\lambda_j(\mu), 1 \leq j \leq m$, are measurable on $\mathbb{R}$, because the coefficients of the characteristic polynomials $\det(\lambda I_m - W(\mu))$ of the matrices $W(\mu)$ are measurable functions of μ on $\mathbb{R}$ and the zeros $\lambda_j(\mu)$ of these polynomials are continuous functions of the coefficients. Moreover, in view of the bounds

$$\lambda_j(\mu) \leq \|W(\mu)\|^2 \quad \text{and} \quad \lambda_j(\mu)^{-1} \leq \|W(\mu)^{-1}\|^2 \quad \text{a.e. on } \mathbb{R}$$

and assertion (1) in Theorem 3.79,

$$\ln \lambda_j(\cdot) \in \widetilde{L}_1 \quad \text{for } 1 \leq j \leq m. \tag{4.121}$$

In particular,

$$\ln \lambda_p(\cdot) \in \widetilde{L}_1 \quad \text{and} \quad \ln \lambda_{p+1}(\cdot) \in \widetilde{L}_1. \tag{4.122}$$

Now let $f(\mu)$ be a measurable function on $\mathbb{R}$ such that

$$\lambda_p(\mu) \leq f(\mu) \leq \lambda_{p+1}(\mu) \quad \text{a.e. on } \mathbb{R}. \tag{4.123}$$

Then the inclusions (4.122) imply that

$$\ln f(\cdot) \in \widetilde{L}_1, \tag{4.124}$$

and hence, by Theorem 3.30, there exists a scalar function $\rho_e \in \mathcal{N}_{out}$ with nontangential boundary values $\rho_e(\mu)$ a.e. on $\mathbb{R}$ such that

$$|\rho_e(\mu)| = f(\mu)^{-2} \quad \text{a.e. on } \mathbb{R}. \tag{4.125}$$

In view of (2) of Theorem 2.47 and formulas (4.123) and (4.125), the nontangential boundary values $W(\mu)$ of the mvf

$$\widetilde{W}(\lambda) = \rho_e(\lambda) W(\lambda) \tag{4.126}$$

are j_{pq}-contractive a.e. on $\mathbb{R}$. Moreover, the blocks $\widetilde{S}_{ij}$ of the Potapov-Ginzburg transform $\widetilde{S}$ of $\widetilde{W}$ are such that

(1) $\widetilde{s}_{12} = \widetilde{w}_{12}\, \widetilde{w}_{22}^{-1} = w_{12} w_{22}^{-1}$, which belongs to $\mathcal{S}^{p\times q}$,

(2) $\widetilde{s}_{21} = -\widetilde{w}_{22}^{-1}\, \widetilde{w}_{21} = -w_{22} w_{21}^{-1}$, which belongs to $\mathcal{S}^{q\times p}$,

(3) $\widetilde{s}_{22} = \widetilde{w}_{22}^{-1} = \rho_e^{-1} w_{22}^{-1}$, which belongs to $\mathcal{S}^{p\times p}$.

The conclusions in (1) and (2) were discussed in the proof of Lemma 4.77; see (4.114). The conclusion in (3) follows from the Smirnov maximum principle, which is applicable because $\rho_e^{-1} \in \mathcal{N}_{out}$ and $w_{22}^{-1} \in \mathcal{S}^{q\times q}$ and, consequently, $\widetilde{s}_{22} \in \mathcal{N}_+^{p\times p}$ and $\|\widetilde{s}_{22}(\mu)\| \leq 1$ a.e. on $\mathbb{R}$.

The final step is to establish the existence of a scalar inner function $\beta_i(\lambda)$ such that

$$W_1 = \beta_i \widetilde{W} \in \mathcal{P}(j_{pq}). \tag{4.127}$$

To this end, let

$$\widetilde{\alpha} = \widetilde{w}_{11} - \widetilde{w}_{12}\widetilde{w}_{22}^{-1}\widetilde{w}_{21} (= \widetilde{s}_{11}) \quad \text{and} \quad \widetilde{\delta} = \widetilde{w}_{22}^{-1} (= \widetilde{s}_{22}).$$

Then it follows from the proof of Lemma 4.77 that

$$\widetilde{\alpha}_{st}\widetilde{\delta} \in \mathcal{S}^{q\times q}$$

for every entry $\widetilde{\alpha}_{st}$ of $\widetilde{\alpha}$. Consequently,

$$\widetilde{\alpha}_{st}\widetilde{\delta}_{kj} \in \mathcal{S}$$

for every entry $\widetilde{\alpha}_{st}$ of $\widetilde{\alpha}$ and every entry $\widetilde{\delta}_{kj}$ of $\widetilde{\delta}$. If $\widetilde{\alpha}_{st}(\lambda) \not\equiv 0$, then it admits a factorization of the form

$$\widetilde{\alpha}_{st} = \frac{b_{st}\ \varphi_{st}}{d_{st}} \tag{4.128}$$

in which $\varphi_{st} \in \mathcal{N}_{out}$ and b_{st} and d_{st} are coprime inner functions, i.e.,

$$\frac{b_{st}\ \varphi_{st}\ \widetilde{\delta}_{kj}}{d_{st}} \in \mathcal{S}.$$

Therefore,

$$\frac{\widetilde{\delta}_{kj}}{d_{st}} \in \mathcal{S}.$$

In view of Theorem 3.43, there exists a least common inner multiple β_i for all the d_{st} in the sense that $d_{st}^{-1}\beta_i \in \mathcal{S}_{in}$ for all $s, t = 1, \ldots, p$, and if $d_{st}^{-1}\beta \in \mathcal{S}_{in}$ for $s, t, \ldots, p$, then $\beta_i^{-1}\beta \in \mathcal{S}_{in}$. Thus,

$$\widetilde{\delta}_{kl}/\beta_i \in \mathcal{S}, \quad 1 \le k, l \le q.$$

Since $\widetilde{\delta} \in \mathcal{S}^{q\times q}$, $\beta_i^{-1}\widetilde{\delta} \in \mathcal{N}_+^{q\times q}$, $\beta_i\widetilde{\alpha} \in \mathcal{N}_+^{p\times p}$ and $\|\widetilde{\alpha}(\mu)\| \le 1$ a.e. on $\mathbb{R}$, the Smirnov maximum principle implies that

$$\beta_i^{-1}\widetilde{\delta} \in \mathcal{S}^{q\times q} \quad \text{and} \quad \beta_i\widetilde{\alpha} \in \mathcal{S}^{p\times p}.$$

Thus, the blocks of the PG transform S_1 of the mvf $W_1 = \beta_i\widetilde{W}$ all belong to the Schur class of appropriate size, and $\|S_1(\mu)\| \le 1$ a.e. on $\mathbb{R}$. Consequently, $S_1 \in \mathcal{S}^{m\times m}$, by the Smirnov maximum principle, and hence $W_1 \in \mathcal{P}(j_{pq})$. □

In view of Theorem 4.79, it is natural to look for a description of the set of all scalar functions $\rho(\lambda)$ such that $\rho U \in \mathcal{P}^\circ(J)$ when $U \in \mathcal{P}^\circ(J)$ and $J \neq \pm I_m$. Without loss of generality, we may assume that $J = j_{pq}$.

Theorem 4.80 (L. A. Simakova) *Let $W \in \mathcal{P}^\circ(j_{pq})$ and let $\lambda_j(\mu)$ denote the jth eigenvalue of $G(\mu) = j_{pq}W(\mu)^*j_{pq}W(\mu)$ in the ordering*

$\lambda_1(\mu) \leq \cdots \leq \lambda_m(\mu)$ a.e. on $\mathbb{R}$. Then the set of all scalar functions ρ such that $\rho W \in \mathcal{P}^\circ(j_{pq})$, is described by the formula

$$\rho = \frac{b}{d}\rho_e, \tag{4.129}$$

where

(1) *$\rho_e \in \mathcal{N}_{out}$ and is subject to the bounds*

$$\lambda_p(\mu) \leq |\rho_e(\mu)|^{-2} \leq \lambda_{p+1}(\mu) \quad \text{a.e. on } \mathbb{R}. \tag{4.130}$$

(2) *$b \in \mathcal{S}_{in}$ is a common inner divisor of all the entries in the mvf $\delta(\lambda)$ that is defined in (4.113).*

(3) *$d \in \mathcal{S}_{in}$ is a common inner divisor of all the entries in the mvf $\alpha(\lambda)$ that is defined in (4.113).*

Proof The identity

$$\rho I_m = (\rho W)W^{-1}$$

implies that $\rho \in \mathcal{N}$ when $W \in \mathcal{P}^\circ(j_{pq})$ and $\rho W \in \mathcal{P}^\circ(j_{pq})$, since $\mathcal{P}(j_{pq}) \subset \mathcal{N}^{m\times m}$. Thus, in view of (3.53), ρ admits a representation of the form (4.129), where $\rho_e \in \mathcal{N}_{out}$, $b \in \mathcal{S}_{in}$, $d \in \mathcal{S}_{in}$ and b and d are coprime inner functions. Moreover, ρ_e satisfies the condition (4.130), since $|\rho_e(\mu)| = |\rho(\mu)|$ a.e. on $\mathbb{R}$ and

$$\lambda_p(\mu) \leq |\rho(\mu)|^{-2} \leq \lambda_{p+1}(\mu) \quad \text{a.e. on } \mathbb{R}.$$

The next step is to verify (3). Since $W \in \mathcal{P}(j_{pq})$ and $\rho W \in \mathcal{P}(j_{pq})$, the mvf $\alpha \in \mathcal{S}^{p\times p}$ and $\rho\alpha \in \mathcal{S}^{p\times p}$. Therefore,

$$\alpha_{st} \in \mathcal{S} \quad \text{and} \quad \frac{b}{d}\rho_e\alpha_{st} \in \mathcal{S}$$

for every entry α_{st} of the mvf α. Thus,

$$\frac{b}{d}\alpha_{st} \in \mathcal{N}_+$$

and, since b and d are coprime,

$$\frac{\alpha_{st}}{d} \in \mathcal{N}_+.$$

The Smirnov maximum principle implies that

$$\frac{\alpha_{st}}{d} \in \mathcal{S},$$

i.e., d is an inner divisor of every entry α_{st} in α.

A similar argument based on the observation that

$$\delta \in \mathcal{S}^{q\times q} \text{ and } \rho^{-1}\delta \in \mathcal{S}^{q\times q}$$

leads to the conclusion that (2) is in force.

Conversely, if ρ is of the form (4.129) where ρ_e, b and d meet the conditions (1), (2) and (3), respectively, then, since $|\rho_e(\mu)| = |\rho(\mu)|$ a.e. on $\mathbb{R}$ and (1) is in force, Theorem 2.47 implies that $\rho(\mu)W(\mu)$ is j_{pq}-contractive a.e. on $\mathbb{R}$. The PG transform $\widetilde{S}(\mu)$ of $\rho(\mu)W(\mu)$ is a well defined contractive matrix a.e. on $\mathbb{R}$. Moreover, since $\rho W \in \mathcal{N}^{m\times m}$, $\widetilde{S}(\mu)$ is the nontangential boundary value of the PG transform $\widetilde{S}(\lambda)$ of $\widetilde{W}(\lambda) = \rho(\lambda)W(\lambda)$. The blocks $\widetilde{s}_{ij}$ of $\widetilde{S}$ are related to the blocks s_{ij} of the PG transform S of W by the formulas

$$\widetilde{s}_{11} = \rho s_{11}, \quad \widetilde{s}_{12} = s_{12}, \quad \widetilde{s}_{21} = s_{21} \quad \text{and} \quad \widetilde{s}_{22} = \rho^{-1} s_{22}.$$

Therefore, $\widetilde{s}_{12} \in \mathcal{S}^{p\times q}$ and $\widetilde{s}_{21} \in \mathcal{S}^{q\times p}$ since $W \in \mathcal{P}(j_{pq})$. Moreover, since $s_{11} = \alpha$ and $s_{22} = \delta$, the entries of $\widetilde{s}_{11}$ and $\widetilde{s}_{22}$ are

$$\frac{b}{d}\rho_e\alpha_{st} \quad \text{and} \quad \frac{d}{b}\rho_e^{-1}\delta_{jk},$$

respectively. Conditions (2) and (3) guarantee that

$$\frac{b}{d}\rho_e\alpha_{st} \in \mathcal{N}_+ \quad \text{and} \quad \frac{d}{b}\rho_e^{-1}\delta_{jk} \in \mathcal{N}_+.$$

Consequently,

$$\widetilde{s}_{11} \in \mathcal{N}_+^{p\times p} \quad \text{and} \quad \widetilde{s}_{22} \in \mathcal{N}_+^{q\times q}.$$

Thus, $\widetilde{S} \in \mathcal{N}_+^{m\times m}$ and the Smirnov maximum principle implies that $\widetilde{S} \in \mathcal{S}^{m\times m}$. Therefore, $\widetilde{W} \in \mathcal{P}(j_{pq})$. □

Remark 4.81 *In the setting of Theorem 4.80,*

$$\lambda_p(\mu) \le 1 \le \lambda_{p+1}(\mu) \quad a.e.\ on\ \mathbb{R}.$$

If $\lambda_p(\mu) = \lambda_{p+1}(\mu)$ a.e. on $\mathbb{R}$, then $\lambda_p(\mu) = 1$ a.e. on $\mathbb{R}$ and condition (4.127) is satisfied only for $\rho_e(\lambda) \equiv \exp(i\gamma)$, where $\gamma \in \mathbb{R}$. In this case, formula (4.129) reduces to

$$\rho(\lambda) = \frac{b(\lambda)}{d(\lambda)}.$$

Theorem 4.82 *If $U \in \mathcal{U}(J)$, $J \neq \pm I_m$ and $\{b_1, b_2\} \in ap_I(U)$, then $\rho U \in \mathcal{P}(J)$ for some scalar function ρ if and only if*

$$\rho = \frac{b}{d}, \tag{4.131}$$

where

$$b \in \mathcal{S}_{in}, \quad b^{-1}b_2 \in \mathcal{S}_{in}^{q\times q}, \quad d \in \mathcal{S}_{in} \quad and \quad d^{-1}b_1 \in \mathcal{S}_{in}^{p\times p}. \tag{4.132}$$

Moreover, for each such choice of ρ, $\rho U \in \mathcal{U}(J)$.

Proof We may assume that $J = j_{pq}$ and will write W in place of U. If $W \in \mathcal{U}(j_{pq})$ and $\{b_1, b_2\} \in ap(W)$, then

$$\alpha = s_{11} = b_1\varphi_1 \quad \text{and} \quad \delta = s_{22} = \varphi_2 b_2,$$

where $\varphi_1 \in \mathcal{S}_{out}^{p\times p}$ and $\varphi_2 \in \mathcal{S}_{out}^{q\times q}$. Therefore, if $b \in \mathcal{S}_{in}$, then

$$b^{-1}\alpha \in \mathcal{S}^{p\times p} \Longleftrightarrow b^{-1}b_1 \in \mathcal{S}^{p\times p},$$

and if $d \in \mathcal{S}_{in}$, then

$$d^{-1}\delta \in \mathcal{S}^{q\times q} \Longleftrightarrow d^{-1}b_2 \in \mathcal{S}^{q\times q}.$$

Moreover, since $G(\mu) = j_{pq}W(\mu)^* j_{pq}W(\mu) = I_m$ a.e. on $\mathbb{R}$, the characterization of ρ claimed in (4.131) and (4.132) follows from Remark 4.81).

Finally, since $\rho W \in \mathcal{P}(j_{pq})$ and $|\rho(\mu)| = 1$ a.e. on $\mathbb{R}$ for ρ of this form and $W \in \mathcal{U}(j_{pq})$, it follows that $\rho W \in \mathcal{U}(j_{pq})$ too. □

Corollary 4.83 *If $U \in \mathcal{E} \cap \mathcal{U}(J)$, $J \neq \pm I_m$, and $\{b_1, b_2\} \in ap_I(U)$, then $\rho U \in \mathcal{P}(J)$ for some scalar function ρ if and only if*

$$\rho(\lambda) = e^{i\gamma} e_\beta(\lambda), \quad where\ \overline{\gamma} = \gamma,\ \beta = \beta_2 - \beta_1, \tag{4.133}$$

and β_1 and β_2 are nonnegative numbers such that

$$e_{\beta_1}^{-1} b_1 \in \mathcal{S}_{in}^{p\times p} \quad and \quad e_{\beta_2}^{-1} b_2 \in \mathcal{S}_{in}^{q\times q}. \tag{4.134}$$

Moreover, for such a choice of ρ, $\rho U \in \mathcal{E} \cap \mathcal{U}(J)$.

Proof In view of Theorem 4.54, b_1 and b_2 are entire mvf's. Consequently, the corollary follows from Theorem 4.82 and the fact that a scalar inner divisor of an entire inner mvf is entire. □

Theorem 4.84 *If $A \in \mathcal{U}(J_p)$ and $\{b_3, b_4\} \in ap_{II}(A)$, then $\rho A \in \mathcal{P}(J_p)$ for some scalar function ρ if and only if*

$$\rho = \frac{b}{d}, \tag{4.135}$$

where

$$b \in \mathcal{S}_{in}, \quad b^{-1}b_4 \in \mathcal{S}_{in}^{p\times p}, \quad d \in \mathcal{S}_{in} \quad and \quad d^{-1}b_3 \in \mathcal{S}_{in}^{p\times p}. \tag{4.136}$$

Moreover, for each such choice of ρ, $\rho A \in \mathcal{U}(J_p)$.

Proof This is immediate from Theorem 4.82 and the connection (4.104) between the associated pairs of the first and second kind for $A \in \mathcal{U}(J_p)$. □

Remark 4.85 *In view of (4.74), condition (4.136) can be reformulated in terms of the blocks of the mvf $B(\lambda) = A(\lambda)\mathfrak{V}$:*

$$d \in \mathcal{S}_{in}, \quad d^{-1}(b_{21}^{\#})^{-1} \in \mathcal{N}_{+}^{p\times p}, \quad b \in \mathcal{S}_{in} \quad and \quad b^{-1}b_{22}^{-1} \in \mathcal{N}_{+}^{p\times p}. \tag{4.137}$$

Corollary 4.86 *If $A \in \mathcal{E} \cap \mathcal{U}(J_p)$ and $\{b_3, b_4\} \in ap_{II}(A)$, then $\rho A \in \mathcal{P}(J_p)$ for some scalar function ρ if and only if $\rho(\lambda)$ is of the form (4.133) where $\beta_1 \geq 0$, $\beta_2 \geq 0$ are such that*

$$e_{\beta_1}^{-1}b_3 \in \mathcal{S}_{in}^{p\times p} \quad and \quad e_{\beta_2}^{-1}b_4 \in \mathcal{S}_{in}^{p\times p}.$$

Moreover, for such a choice of ρ,

$$\rho A \in \mathcal{E} \cap \mathcal{U}(J_p).$$

Proof This is immediated from Corollary 4.83 and the connection (4.104) between the associated pairs of the first and second kind for $A \in \mathcal{U}(J_p)$. □

4.18 More on linear fractional transformations

Theorem 4.87 *Let W be a nondegenerate meromorphic $m \times m$ mvf in $\mathbb{C}_+$ and let*

$$\mathring{\mathcal{S}}_{const}^{p\times q} \subset \mathcal{D}(T_W) \quad and \quad T_W[\mathring{\mathcal{S}}_{const}^{p\times q}] \subseteq \mathcal{S}^{p\times q}. \tag{4.138}$$

Then there exists a scalar meromorphic function $\rho(\lambda)$ in $\mathbb{C}_+$ such that $\rho W \in \mathcal{P}^{\circ}(j_{pq})$. Moreover, if, in addition to the properties (4.138),

$$T_W \quad maps\ every\ isometric\ matrix\ in\ \mathbb{C}^{p\times q}\ into\ \mathcal{S}_{in}^{p\times q}\ if\ p \geq q$$

or

$$T_W \quad \textit{maps every coisometric matrix in } \mathbb{C}^{p\times q} \textit{ into } \mathcal{S}_{*in}^{p\times q} \textit{ if } p \le q,$$

then $\rho W \in \mathcal{U}(j_{pq})$ *with the same function* ρ.

Proof Lemma 2.49 guarantees that a nondegenerate meromorphic $m \times m$ mvf W that satisfies the conditions in (4.138) is a minus mvf with respect to j_{pq}. Then, since W is assumed to be nondegenerate, Theorem 4.79 implies that there exists a scalar meromorphic function $\rho(\lambda)$ in $\mathbb{C}_+$ such that $\rho W \in \mathcal{P}^\circ(j_{pq})$. The rest of the proof is the same as the proof of the corresponding statements in Theorem 2.52. □

Theorem 4.88 *Let* W *be a nondegenerate meromorphic* $m \times m$ *mvf in* $\mathbb{C}_+$ *and let*

$$\mathcal{S}^{p\times q} \subseteq \mathcal{D}(T_W) \quad \textit{and} \quad T_W[\mathcal{S}^{p\times q}] = \mathcal{S}^{p\times q}. \tag{4.139}$$

Then there exists a scalar meromorphic function $\rho(\lambda)$ *in* $\mathbb{C}_+$ *such that* $\rho(\lambda)W(\lambda) \in \mathcal{U}_{const}(j_{pq})$.

Proof By Theorem 4.87 there exists a scalar meromorphic function $\rho_1(\lambda)$ in $\mathbb{C}_+$ such that $\rho_1 W \in \mathcal{P}^\circ(j_{pq})$, and, since $T_{W^{-1}}[\mathcal{S}^{p\times q}] = \mathcal{S}^{p\times q}$, a scalar meromorphic function $\rho_2(\lambda)$ in $\mathbb{C}_+$ such that $\rho_2 W^{-1} \in \mathcal{P}^\circ(j_{pq})$. Thus,

$$|\rho_1(\lambda)|^2 W(\lambda)^* j_{pq} W(\lambda) \le j_{pq} \quad \text{for} \quad \lambda \in \mathfrak{h}_W^+ \cap \mathfrak{h}_{\rho_1}^+$$

and

$$|\rho_2(\lambda)|^2 j_{pq} \le W(\lambda)^* j_{pq} W(\lambda) \quad \text{for } \lambda \in \mathfrak{h}_W^+ \cap \mathfrak{h}_{\rho_2}^+.$$

Consequently,

$$(|\rho_2(\lambda)|^2 - |\rho_1(\lambda)|^{-2}) j_{pq} \le 0 \quad \text{for} \quad \lambda \in \mathfrak{h}_{\rho_1^{-1}}^+ \cap \mathfrak{h}_{\rho_2}^+.$$

Thus,

$$|\rho_2(\lambda)\rho_1(\lambda)| = 1 \text{ for} \quad \lambda \in \mathfrak{h}_{\rho_1}^+ \cap \mathfrak{h}_{\rho_2}^+;$$

i.e.,

$$\rho_2(\lambda) = e^{i\gamma}\rho_1(\lambda) \quad \text{for some } \gamma \in \mathbb{R}.$$

Thus, $W_0 = \rho_1 W$ and W_0^{-1} both belong to $\mathcal{P}(j_{pq})$ and $S_0 = PG(W_0)$ and $S_0^{-1} = PG(W_0^{-1})$ both belong to $\mathcal{S}^{m\times m}$. Therefore, $S_0 \in \mathcal{S}_{const}^{m\times m}$ and $W_0 \in \mathcal{U}_{const}(j_{pq})$. □

Theorem 4.89 *Let $W \in \mathcal{U}(j_{pq})$ and $W_1 \in \mathcal{P}^\circ(j_{pq})$, let $\{b_1, b_2\} \in ap(W)$ and suppose that*

$$T_W[\mathcal{S}^{p\times q}] \subseteq T_{W_1}[\mathcal{S}^{p\times q}]. \tag{4.140}$$

Then

(1) *$W_1 \in \mathcal{U}(j_{pq})$ and*

$$W = \frac{\beta_1}{\beta_2} W_1 W_2, \tag{4.141}$$

where $W_2 \in \mathcal{U}(j_{pq})$ and β_1 and β_2 are scalar inner functions such that

$$\beta_1^{-1} b_1 \in \mathcal{S}_{in}^{p\times p} \quad and \quad \beta_2^{-1} b_2 \in \mathcal{S}_{in}^{q\times q}. \tag{4.142}$$

(2) *If equality prevails in (4.140), then $W_2 \in \mathcal{U}_{const}(j_{pq})$.*

Proof Under the given assumptions the mvf $\widetilde{W}_2 = W_1^{-1} W$ is a nondegenerate meromorphic $m \times m$ mvf in $\mathbb{C}_+$ such that

$$\mathcal{S}^{p\times q} \subseteq \mathcal{D}(T_{\widetilde{W}_2}) \quad \text{and} \quad T_{\widetilde{W}_2}[\mathcal{S}^{p\times q}] \subseteq \mathcal{S}^{p\times q}.$$

Therefore, by Theorem 4.87, there exists a scalar meromorphic function $\rho(\lambda)$ in $\mathbb{C}_+$ such that

$$\rho \widetilde{W}_2 \in \mathcal{P}^\circ(j_{pq}).$$

Then $\rho W = W_1 W_2$ with $W_2 = \rho \widetilde{W}_2$, i.e., $W \in \mathcal{U}(j_{pq})$ and $\rho W \in \mathcal{P}^\circ(j_{pq})$; the rest follows from Theorems 4.82 and 4.88. □

Corollary 4.90 *If either W and $\widetilde{W}$ both belong to $\mathcal{U}(j_{pq}) \cap \mathcal{N}_+^{m\times m}$ or $W^\#$ and $\widetilde{W}^\#$ both belong to $\mathcal{U}(-j_{pq}) \cap \mathcal{N}_+^{m\times m}$, then*

$$T_W[\mathcal{S}^{p\times q}] = T_{\widetilde{W}}[\mathcal{S}^{p\times q}] \tag{4.143}$$

if and only if

$$\widetilde{W}(\lambda) = W(\lambda) V \quad \textit{for some } V \in \mathcal{U}_{const}(j_{pq}). \tag{4.144}$$

Proof Suppose first that (4.143) holds and that W and $\widetilde{W}$ both belong to $\mathcal{U}(j_{pq}) \cap \mathcal{N}_+^{m\times m}$. Then, β_2 is a constant in formula (4.141), since $\{b_1, I_q\} \in ap(W)$ by Lemma 4.33. However, this implies that $\widetilde{W} = \beta_1^{-1} W V^{-1}$ for some $V \in \mathcal{U}_{const}(J)$ and hence that $(\beta_1^{-1} b, \beta_1^{-1} I_q) \in ap(W)$. Therefore, since $\widetilde{W} \in \mathcal{N}_+^{m\times m}$, β_1 must be a constant. The proof of the second assertion is similar. □

Corollary 4.91 *Let W and W_1 both belong to $\mathcal{U}(j_{pq})$. Then*

$$T_W[\mathcal{S}^{p\times q}] = T_{W_1}[\mathcal{S}^{p\times q}] \quad \text{and} \quad ap(W) = ap(W_1)$$

if and only if

$$W_1(\lambda) = W(\lambda)U \quad \text{and} \quad U \in \mathcal{U}_{const}(j_{pq}).$$

Corollary 4.92 *Let $W \in \mathcal{E} \cap \mathcal{U}(j_{pq})$, $W_1 \in \mathcal{P}^\circ(j_{pq})$, $\{b_1, b_2\} \in ap(W)$ and suppose that*

$$T_W[\mathcal{S}^{p\times q}] \subseteq T_{W_1}[\mathcal{S}^{p\times q}]. \tag{4.145}$$

Then:

(1) *$W_1 \in \mathcal{U}(j_{pq})$ and*

$$W = \frac{e_{\beta_1}}{e_{\beta_2}} W_1 W_2, \tag{4.146}$$

where $W_2 \in \mathcal{U}(j_{pq})$, $\beta_1 \geq 0$, $\beta_2 \geq 0$,

$$e_{\beta_1}^{-1} b_1 \in \mathcal{E} \cap \mathcal{S}_{in}^{p\times p} \quad \text{and} \quad e_{\beta_2^{-1}} b_2 \in \mathcal{E} \cap \mathcal{S}_{in}^{q\times q}. \tag{4.147}$$

(2) *If equality prevails in (4.145), then $W_2 \in \mathcal{U}_{const}(j_{pq})$.*

Proof This is follows from Theorem 4.89 and and Corollary 4.83. □

Corollary 4.93 *If $W \in \mathcal{P}^\circ(j_{pq})$ is such that*

$$T_W[\mathcal{S}^{p\times q}] = \mathcal{S}^{p\times q},$$

then $W \in \mathcal{U}_{const}(j_{pq})$.

Proof This follows from Theorem 4.89, since $I_m \in \mathcal{U}(j_{pq})$, $\{I_p, I_q\} \in ap(I_m)$ and, under the given assumptions, $T_W[\mathcal{S}^{p\times q}] = T_{I_m}[\mathcal{S}^{p\times q}]$. □

Theorem 4.94 *Let W and W_1 both belong to $\mathcal{U}(j_{pq})$, let $\{b_1, b_2\} \in ap(W)$ and $\{b_1^{(1)}, b_2^{(1)}\} \in ap(W_1)$. Then the conditions*

$$(b_1^{(1)})^{-1} b_1 \in \mathcal{S}_{in}^{p\times p}, \quad b_2 (b_2^{(1)})^{-1} \in \mathcal{S}_{in}^{q\times q} \quad \text{and} \quad T_W[\mathcal{S}^{p\times q}] \subseteq T_{W_1}[\mathcal{S}^{p\times q}]$$

hold if and only if $W_1^{-1} W \in \mathcal{U}(j_{pq})$. Moreover, if $W_1^{-1} W \in \mathcal{U}(j_{pq})$, then

$$W_1^{-1} W \in \mathcal{U}_S(j_{pq}) \Longleftrightarrow ap(W_1) = ap(W). \tag{4.148}$$

Proof If the first set of conditions hold, then, by Theorem 4.89, there exist a mvf $W_2 \in \mathcal{U}(j_{pq})$ and a pair of scalar inner divisors β_1 and β_2 of b_1 and b_2, respectively (with no common divisors), such that

$$W = \frac{\beta_1}{\beta_2} W_1 W_2.$$

Therefore,

$$s_{11} = \frac{\beta_1}{\beta_2} s_{11}^{(1)} \{I_p - s_{12}^{(2)} s_{21}^{(1)}\}^{-1} s_{11}^{(2)},$$

just as in the verification of (4.45). But, in terms of the usual notation, this implies that

$$\frac{\beta_1}{\beta_2} b_1^{(2)} = \{I_p - s_{12}^{(2)} s_{21}^{(1)}\} (\varphi_1^{(1)})^{-1} (b_1^{(1)})^{-1} b_1 \varphi_1 (\varphi_1^{(2)})^{-1}, \tag{4.149}$$

which belongs to $L_\infty^{p\times p} \cap \mathcal{N}_+^{p\times p} = H_\infty^{p\times p}$. Therefore, each entry in the matrix $\beta_2^{-1} b_1^{(2)}$ belongs to H_∞ (since β_1 and β_2 have no common divisors) and hence $\beta_2^{-1} b_1^{(2)} \in \mathcal{S}_{in}^{p\times p}$.

In much the same way, formula

$$s_{22} = \frac{\beta_2}{\beta_1} s_{22}^{(2)} \{I_q - s_{21}^{(1)} s_{12}^{(2)}\}^{-1} s_{22}^{(1)} \tag{4.150}$$

implies that $\beta_1^{-1} b_2^{(2)} \in \mathcal{S}_{in}^{q\times q}$. Thus, by Theorem 4.82, $\beta_2^{-1}\beta_1 W_2 \in \mathcal{U}(j_{pq})$, as needed.

The converse is immediate from Lemmas 4.67 and 4.28.

Finally, since Formula (4.149) implies that

$$\frac{\beta_1}{\beta_2} b_1^{(2)} \text{ is constant} \quad \Longleftrightarrow \quad (b_1^{(1)})^{-1} b_1 \text{ is constant}$$

and formula (4.150) implies that

$$\frac{\beta_2}{\beta_1} b_2^{(2)} \text{ is constant} \quad \Longleftrightarrow \quad b_2 (b_2^{(1)})^{-1} \text{ is constant,}$$

the last assertion follows from Lemma 4.39. □

Next, a number of results on linear fractional transformations in $\mathcal{C}^{p\times p}$ and from $\mathcal{S}^{p\times p}$ into $\mathcal{C}^{p\times p}$ will be obtained by reduction to the linear fractional transformations in $\mathcal{S}^{p\times p}$ that were considered earlier.

Theorem 4.95 *Let $A(\lambda)$ be a nondegenerate meromorphic $m \times m$ mvf in $\mathbb{C}_+$ such that*

(1) $\mathring{\mathcal{C}}^{p\times p}_{const} \subset \mathcal{D}(T_A)$ *and* $T_A[\mathring{\mathcal{C}}^{p\times p}_{const}] \subset \mathcal{C}^{p\times p}$.

Then there exists a scalar meromorphic function $\rho(\lambda)$ *such that mvf* $\rho A \in \mathcal{P}^\circ(J_p)$. *Moreover, if in addition to (1), the condition*

(2) $T_A[(\mathcal{C}^{p\times p}_{sing})_{const}] \subseteq \mathcal{C}^{p\times p}_{sing}$

is in force, then $\rho A \in \mathcal{U}(J_p)$.

Proof Let $W(\lambda) = \mathfrak{V}A(\lambda)\mathfrak{V}$. Then, since $\mathring{\mathcal{C}}^{p\times p}_{const} = T_{\mathfrak{V}}[\mathring{\mathcal{S}}^{p\times p}_{const}]$, $\mathcal{C}^{p\times p}_{const} \subset \mathcal{D}(T_{\mathfrak{V}})$ and $T_{\mathfrak{V}}[\mathcal{C}^{p\times p}] \subseteq \mathcal{S}^{p\times p}$, the theorem follows from Theorem 4.87 and Lemma 4.71. □

Theorem 4.96 *Let* $A \in \mathcal{U}(J_p)$, $A_1 \in \mathcal{P}^\circ(J_p)$ *and* $\{b_3, b_4\} \in ap_{II}(A)$ *and suppose that*

$$\mathcal{C}(A) \subseteq \mathcal{C}(A_1). \tag{4.151}$$

Then

(1) $A_1 \in \mathcal{U}(J_p)$ *and there exist a mvf* $A_2 \in \mathcal{U}(J_p)$ *and a pair of scalar inner functions* β_1 *and* β_2 *such that*

$$A(\lambda) = \frac{\beta_2(\lambda)}{\beta_1(\lambda)} A_1(\lambda)A_2(\lambda)$$

and

$$\beta_1^{-1}b_3 \in \mathcal{S}^{p\times p}_{in} \quad \text{and} \quad \beta_2^{-1}b_4 \in \mathcal{S}^{p\times p}_{in}. \tag{4.152}$$

(2) *If equality prevails in (4.151), then* $A_2 \in \mathcal{U}_{const}(J_p)$.

Proof Let $W(\lambda) = \mathfrak{V}A(\lambda)\mathfrak{V}$, $W_1(\lambda) = \mathfrak{V}A_1(\lambda)\mathfrak{V}$, $B(\lambda) = A(\lambda)\mathfrak{V}$ and $B_1(\lambda) = A_1(\lambda)\mathfrak{V}$. Then $W \in \mathcal{U}(j_p)$ and $W_1 \in \mathcal{P}^\circ(j_p)$. Let $\{b_1, b_2\} \in ap(W)$, i.e., $\{b_1, b_2\} \in ap_I(A)$. Lemma 4.36 implies that property (4.152) is satisfied if and only if

$$\beta_1^{-1}b_1 \in \mathcal{S}^{p\times p}_{in} \quad \text{and} \quad \beta_2^{-1}b_2 \in \mathcal{S}^{p\times p}_{in}. \tag{4.153}$$

In view of Theorem 4.89, it is enough to verify that

$$T_W[\mathcal{S}^{p\times p}] \subseteq T_{W_1}[\mathcal{S}^{p\times p}],$$

but this follows from Lemma 4.71. □

Corollary 4.97 *Let* $A \in \mathcal{E} \cap \mathcal{U}(J_p)$ *and* $A_1 \in \mathcal{P}^\circ(J_p)$, *let* $\{b_3, b_4\} \in ap_{II}(A)$ *and suppose that*

$$\mathcal{C}(A) = \mathcal{C}(A_1).$$

Then $A_1 \in \mathcal{E} \cap \mathcal{U}(J_p)$ *and can be expressed in the form*

$$A_1(\lambda) = \frac{e_{\beta_2}(\lambda)}{e_{\beta_1}(\lambda)} A(\lambda)V,$$

where $e_{-\beta_1} b_3 \in \mathcal{S}_{in}^{p\times p}$, $e_{-\beta_2} b_4 \in \mathcal{S}_{in}$, $\beta_1 \geq 0$, $\beta_2 \geq 0$ *and* $V \in \mathcal{U}_{const}(J_p)$.

Theorem 4.98 *Let* A *and* A_1 *both belong to* $\mathcal{U}(J_p)$, *let* $\{b_3, b_4\} \in ap_{II}(A)$ *and* $\{b_3^{(1)}, b_4^{(1)}\} \in ap_{II}(A_1)$. *Then the conditions*

$$(b_3^{(1)})^{-1} b_3 \in \mathcal{S}_{in}^{p\times p}, \quad b_4 (b_4^{(1)})^{-1} \in \mathcal{S}_{in}^{p\times p} \quad \text{and} \quad \mathcal{C}(A) \subseteq \mathcal{C}(A_1)$$

hold if and only if $A_1^{-1} A \in \mathcal{U}(J_p)$.

Remark 4.99 *Theorem 4.98 remains valid if the condition* $\mathcal{C}(A) \subseteq \mathcal{C}(A_1)$ *is replaced by the condition*

$$T_A[\mathcal{C}^{p\times p} \cap \mathcal{D}(T_A)] \subseteq T_{A_1}[\mathcal{C}^{p\times p} \cap \mathcal{D}(T_{A_1})].$$

Corollary 4.100 *Let* A *and* A_1 *both belong to* $\mathcal{U}(J_p)$. *Then*

$$\mathcal{C}(A) = \mathcal{C}(A_1) \quad \text{and} \quad ap_{II}(A) = ap_{II}(A_1)$$

if and only if

$$A_1(\lambda) = A(\lambda)U \quad \text{and} \quad U \in \mathcal{U}_{const}(J_p).$$

Theorem 4.101 *Let* A *be a nondegenerate meromorphic* $m \times m$ *mvf in* $\mathbb{C}_+$, *let* $B = A\mathfrak{V}$ *and suppose that*

$$T_{\mathfrak{V}}[\mathcal{C}^{p\times p}] \subseteq \mathcal{D}(T_B) \quad \text{and} \quad \mathcal{C}(A) = \mathcal{C}^{p\times p}.$$

Then:

(1) *There exists a scalar function* $\rho(\lambda)$ *that is meromorphic in* $\mathbb{C}_+$ *such that* $\rho A \in \mathcal{U}_{const}(J)$.

(2) *If it is also assumed that* $A \in \mathcal{P}^\circ(J)$, *then* $A \in \mathcal{U}_{const}(J)$.

Proof Let $B(\lambda) = A(\lambda)\mathfrak{V}$ and let $W(\lambda) = \mathfrak{V}A(\lambda)\mathfrak{V}$. Then

$$\begin{aligned} \mathcal{C}^{p\times p} &= \mathcal{C}(A) = T_B[\mathcal{S}^{p\times p} \cap \mathcal{D}(T_B)] \\ &= T_{\mathfrak{V}}[T_W[\mathcal{S}^{p\times p} \cap \mathcal{D}(T_{\mathfrak{V}})]. \end{aligned}$$

Consequently,

$$T_W[\mathcal{S}^{p\times p} \cap \mathcal{D}(T_{\mathfrak{V}})] = \mathcal{S}^{p\times p} \cap \mathcal{D}(T_{\mathfrak{V}}),$$

and hence, in view of Lemma 4.71, the conditions in (4.139) are in force for W, since $\mathcal{S}^{p\times p}\cap\mathcal{D}(T_B)=\mathcal{S}^{p\times p}\cap\mathcal{D}(T_{\mathfrak{Y}})$ under the given assumptions. Thus, Theorem 4.88 and Corollary 4.93 are applicable. □

$T_W[\mathcal{S}^{p\times p}]=\mathcal{S}^{p\times p}$, since $\mathcal{S}^{p\times p}\cap\mathcal{D}(T_{\mathfrak{Y}})$ is dense in $\mathcal{S}^{p\times p}$ and $T_W[\mathcal{S}^{p\times p}]$ is closed with respect to pointwise convergence in $\mathcal{S}^{p\times p}$ and Theorem 4.88 and Corollary 4.93 are applicable. □

4.19 Bibliographical notes

Most of the information in this chapter that does not deal with multiplicative representations can be found in the papers [Ar89], [Ar95a] and the monograph [Dy89b]. The class $\mathcal{U}(J)$ of J-inner mvf's is a subclass of the class $\mathcal{P}^\circ(J)$ of J contractive mvf's, which was investigated by Potapov [Po60]. The discussion in this chapter of his results on multiplicative representations of mvf's in the class $\mathcal{P}^\circ(J)$ is adapted from the paper [Po60] and Chapter 3 of the monograph [Zol03].

The problem of establishing conditions for the convergence of infinite products of mvf's from $\mathcal{P}^\circ(J)$, or, equivalently, the convergence of monotone sequences of mvf's in $\mathcal{P}^\circ(J)$, is discussed in [Ar97].

The monograph [DF79] is an introduction to multiplicative integrals and applications to differential equations; for a more recent review of multi[licative integrals and multiplicative representations, see [GiSh94].

Remark 4.12 was established by Potopov [Po60].

The problem of recovering a mvf $H(t)$ that is subject to the constraints (4.33), given a mvf $U\in\mathcal{E}\cap\mathcal{U}(J)$ is known as the **inverse monodromy problem**. Potapov's theorem guarantees the existence of at least one solution. However, it is not unique unless additional constraints are imposed. If $J=I_m$ and $U\in\mathcal{U}_0(I_m)$, then Brodskii and Kisilevskii showed that the condition $\tau_U=\delta_U$ is necessary and sufficient for the existence of a unique solution in the class of solutions $H(t)\geq 0$ with trace $H(t)=1$ a.e. on $[0,d]$; see Section 30 in [Bro72]. If $m=2$, $U\in\mathcal{U}(J_1)$ and $\overline{U(\lambda)}=U(-\overline{\lambda})$, then L. de Branges showed that there is a unique solution in the class of solutions with trace $H(t)=1$ and $\overline{H(t)}=H(t)\geq 0$ a.e. on $[0,d]$; see [Br68a], and, for an expository account, [DMc76]. If the reality conditions $\overline{U(\lambda)}=U(-\overline{\lambda})$ and $\overline{H(t)}=H(t)$ are dropped, then the solution is not unique; see [ArD01a], and for additional information on the inverse monodromy problem, [ArD00a], [ArD00b] and the references cited therein.

Necessary and sufficient conditions for a 2×2 mvf $A \in \mathcal{U}(J_1)$ that is meromorphic in $\mathbb{C}$ to be the BP product of elementary factors of the first and second kind that are different from those in Remark 4.47 were obtained in [GuT03] and [GuT06], where necessary and sufficient conditions for A to belong to the class $A \in \mathcal{U}_{\ell sR}(J_1)$ are also obtained.

The identification of BP products of elementary factors of the third and fourth kind as resolvent matrices of bitangential interpolation problems with constraints at points on $\mathbb{R}$ or at ∞ was discussed in [Kov83], Chapter 8 of [Dy89b] and [BoD06]. Connections of the class $\mathcal{U}_S(J)$ with the Hamburger moment problem are discussed in [Kh96].

Theorem 4.17, which characterizes the class of mvf's $U \in \mathcal{P}^\circ(J)$ that admit a multiplicative integral representation of the form (4.37) is due to Potapov [Po60]. His proof was based on rational approximation of a given mvf $U \in \mathcal{P}^\circ(J)$ with properties (1)-(4) by a sequence of rational mvf's $U_n \in \mathcal{U}(J)$ with properties (1)-(4). He then represented each approximant $U_n(\lambda)$ as a finite product of normalized elementary BP factors and showed that this sequence of products tends to a multiplicative integral of the form (4.37); see e.g., [Zol03].

There is another proof of Theorem 4.17, in which, following M. S. Livsic, a mvf $B \in \mathcal{P}^\circ(J)$ that meets the conditions (1)–(4) is viewed as the characteristic mvf of a bounded linear operator A (in a Hilbert space) with real spectrum and rank $(A - A^*) = m$. The proof exploits the connection between the multiplicative representation of the characteristic mvf of such operators and the triangular representation of such operators in the Livsic model; see Brodskii [Bro72], Livsic [Liv73] and Zolotarev [Zol03]. Every convergent infinite product of elementary BP factors of the third and fourth kind can be written as a multiplicative integral with a discrete matrix valued measure. A much richer class of multiplicative integrals that are singular J-inner matrix valued functions was obtained by Katsnelson for the case $m = 2$. A number of other results on 2×2 singular J-inner mvf's may be found in [Kat89] and [Kat90].

Lemma 4.40 arose following a lecture by (and discussion with) S. Naboko in which the class of singular characteristic functions $U_A(\lambda)$ of nonselfadjoint operators A were defined by the properties of $S = PG(U_A)$ that are formulated in assertion (3) of the lemma.

A connection between 2×2 singular j_1-inner mvf's and a certain class of singular Nevanlinna type interpolation problems was established in [AFK95] on the basis of the work of Katsnelson [Kat89] and [Kat90].

The results in Section 4.17 are adapted from Simakova [Si74] and [Si75]. Extensions of these results to mvf's with extra properties (e.g., real, symmetric, symplectic) may be found in [Si03].

5
Reproducing kernel Hilbert spaces

5.1 Preliminaries

In this chapter we shall develop the basic properties of a number of RKHS's (reproducing kernel Hilbert spaces) with RK's (reproducing kernels) of a special form that were introduced and extensively exploited to resolve assorted problems in analysis by L. de Branges. In later chapters, RKHS methods will be used to describe the set of solutions to each of a number of completely indeterminate bitangential interpolation and extension problems.

Positive kernels

An $m \times m$ mvf $K_\omega(\lambda)$ that is defined on a set $\Omega \times \Omega$ is said to be a **positive kernel** on $\Omega \times \Omega$ if

$$\sum_{i,j=1}^{n} u_i^* K_{\omega_j}(\omega_i) u_j \geq 0 \tag{5.1}$$

for every positive integer n and every choice of points $\omega_1, \ldots, \omega_n$ in Ω and vectors $u_1, \ldots, u_n$ in $\mathbb{C}^m$.

Lemma 5.1 *Let $K_\omega(\lambda)$ be a positive kernel on $\Omega \times \Omega$. Then for every choice of $\alpha, \beta \in \Omega$:*

(1) $K_\alpha(\beta)^* = K_\beta(\alpha)$.

(2) $K_\alpha(\alpha) \geq 0$.

(3) $\|K_\alpha(\beta)\| \leq \|K_\alpha(\alpha)\|^{\frac{1}{2}} \|K_\beta(\beta)\|^{\frac{1}{2}}$.

(4) *Equality holds in (5.1) if and only if $\sum_{j=1}^{n} K_{\omega_j}(\lambda) u_j = 0$ for every point $\lambda \in \Omega$.*

(5) *If* $K_\alpha(\alpha)u = 0$ *for some point* $\alpha \in \Omega$ *and vector* $u \in \mathbb{C}^m$, *then* $K_\alpha(\lambda)u = 0$ *for every point* $\lambda \in \Omega$.

Proof Formula (5.1) implies that the matrix

$$\begin{bmatrix} u^*K_\alpha(\alpha)u & u^*K_\alpha(\beta)v \\ v^*K_\beta(\alpha)u & v^*K_\beta(\beta)v \end{bmatrix}$$

is positive semidefinite for arbitrary points α, β in Ω and arbitrary vectors u, v in $\mathbb{C}^m$. Thus, (1), (2) and the estimate

$$\begin{aligned} |v^*K_\alpha(\beta)u| &\leq \{u^*K_\alpha(\alpha)u\}^{\frac{1}{2}}\{v^*K_\beta(\beta)v\}^{\frac{1}{2}} \\ &\leq \|K_\alpha(\alpha)\|^{\frac{1}{2}}\|K_\beta(\beta)\|^{\frac{1}{2}}\|u\|\,\|v\|, \end{aligned}$$

which implies (3) and (5), are readily seen to be valid. Finally, another application of formula (5.1) implies that the matrix

$$\begin{bmatrix} \sum_{i,j=1}^n u_i^*K_{\omega_j}(\omega_i)u_j & \sum_{j=1}^n u_j^*K_{\omega_j}(\lambda)^*v \\ \sum_{j=1}^n v^*K_{\omega_j}(\lambda)u_j & v^*K_\lambda(\lambda)v \end{bmatrix}$$

is positive semidefinite for arbitrary points λ in Ω and arbitrary vectors v in $\mathbb{C}^m$ and hence that (4) is valid too. □

Basic definitions and properties

A Hilbert space $\mathcal{H}$ of $m \times 1$ vector valued functions defined on a set Ω is said to be a RKHS (**reproducing kernel Hilbert space**) if there exists an $m \times m$ mvf $K_\omega(\lambda)$ defined on $\Omega \times \Omega$ such that:

(1) The inclusion

$$K_\omega u \in \mathcal{H} \quad \text{(as a function of } \lambda) \tag{5.2}$$

holds for every choice of $\omega \in \Omega$ and $u \in \mathbb{C}^m$.

(2) The equality

$$\langle f, K_\omega u\rangle_{\mathcal{H}} = u^*f(\omega) \tag{5.3}$$

holds for every choice of $\omega \in \Omega$, $u \in \mathbb{C}^m$ and $f \in \mathcal{H}$.

Any mvf $K_\omega(\lambda)$ that meets these two conditions is said to be a RK (**reproducing kernel**) for $\mathcal{H}$. It is readily checked that:

(3) The kernel $K_\omega(\lambda)$ is a positive kernel.

(4) A RKHS has exactly one RK, i.e., the RK is unique.

(5) If $\|f_n - f\|_{\mathcal{H}} \longrightarrow 0$ as $n \uparrow \infty$, then $f_n(\omega) - f(\omega) \longrightarrow 0$ as $n \uparrow \infty$ at every point $\omega \in \Omega$.

Theorem 5.2 *If an $m \times m$ mvf $K_\omega(\lambda)$ is a positive kernel on $\Omega \times \Omega$, there is exactly one RKHS of $m \times 1$ vector valued functions on Ω with $K_\omega(\lambda)$ as its RK.*

Proof Let the $m \times m$ mvf $K_\omega(\lambda)$ be a positive kernel on $\Omega \times \Omega$ and let

$$\mathcal{L} = \left\{ \sum_{j=1}^{n} K_{\omega_j}(\lambda)u_j \; : \omega_j \in \Omega, \; u_j \in \mathbb{C}^m \, , \; n \geq 1 \right\} \tag{5.4}$$

be the linear manifold of finite sums of the indicated form endowed with the scalar product

$$\langle f, g \rangle = \sum_{i,j=1}^{n} v_i^* K_{\omega_j}(\omega_i) u_j, \tag{5.5}$$

when

$$f(\lambda) = \sum_{j=1}^{n} K_{\omega_j}(\lambda)u_j \quad \text{and} \quad g(\lambda) = \sum_{j=1}^{n} K_{\omega_j}(\lambda)v_j \tag{5.6}$$

for some $n \geq 1, \omega_j \in \Omega$ and $u_j, \; v_j \in \mathbb{C}^m, 1 \leq j \leq n$. The same sequence of points, $\omega_1 \ldots \omega_n$, may always be assumed in the representations (5.6) of the vvf's $f(\lambda)$ and $g(\lambda)$ by taking some of the $u_j = 0$ or $v_j = 0$, if needed.

The next step is to check that the inner product is well defined, or equivalently, that if either $f(\lambda) \equiv 0$ or $g(\lambda) \equiv 0$, then $\langle f, g \rangle = 0$. But if $f(\lambda) \equiv 0$ in Ω, then the following sequence of implications holds:

$$\begin{aligned} \sum_{j=1}^{n} K_{\omega_j}(\lambda)u_j \equiv 0 \quad &\Longrightarrow \quad \sum_{j=1}^{n} K_{\omega_j}(\omega_i)u_j = 0 \\ &\Longrightarrow \quad \sum_{j=1}^{n} v_i^* K_{\omega_j}(\omega_i)u_j = 0 \\ &\Longrightarrow \quad \langle f, g \rangle = \sum_{i,j=1}^{n} v_i^* K_{\omega_j}(\omega_i)u_j = 0. \end{aligned}$$

The linear manifold $\mathcal{L}$ with scalar product defined by formula (5.5) is a pre-Hilbert space. Moreover, if $\|f\|^2 = \langle f, f \rangle = 0$ for some vvf $f \in \mathcal{L}$, then, in view of item (4) in Lemma 5.1, $f(\lambda) \equiv 0$ on Ω.

Next, let $\{f_n\}$ be a fundamental sequence in $\mathcal{L}$. Then the formula

$$u^* f_n(\omega) = \langle f_n, K_\omega u \rangle$$

implies that $f_n(\omega)$ is a fundamental sequence of vectors in $\mathbb{C}^m$ for every fixed $\omega \in \Omega$. Let

$$f(\omega) = \lim_{n \to \infty} f_n(\omega), \quad \omega \in \Omega. \tag{5.7}$$

Thus, formula (5.7) defines exactly one $m \times 1$ vvf $f(\lambda)$ on Ω corresponding to each fundamental sequence $\{f_n\}$ in the pre-Hilbert space $\mathcal{L}$.

Let

$$f(\lambda) = \lim_{n \to \infty} f_n(\lambda) \quad \text{and} \quad g(\lambda) = \lim_{n \to \infty} g_n(\lambda), \quad \lambda \in \Omega,$$

where $\{f_n\}$ and $\{g_n\}$ are two fundamental sequences in $\mathcal{L}$. Define the scalar product for these two vvf's by formula

$$\langle f, g \rangle = \lim_{n \to \infty} \langle f_n, g_n \rangle.$$

This limit exists since $\{\langle f_n, g_n \rangle\}$, $n \geq 1$, is a fundamental sequence of points in $\mathbb{C}$. After such a completion a Hilbert space $\mathcal{H}$ of $m \times 1$ vvf's $g(\lambda)$ on Ω is obtained in which $\mathcal{L}$ is a dense manifold and, consequently, the first characteristic property (5.2) of a RKHS with RK $K_\omega(\lambda)$ holds. Moreover, since there exists a fundamental sequence $\{f_n\} \in \mathcal{L}$ such that $f(\lambda) = \lim_{n\uparrow\infty} f_n(\lambda)$ for every $f \in \mathcal{H}$, the evaluation

$$u^* f(\omega) = \lim_{n \to \infty} u^* f_n(\omega) = \lim_{n \to \infty} \langle f_n, K_\omega u \rangle = \langle f, K_\omega u \rangle.$$

for every choice of $\omega \in \Omega$ and $u \in \mathbb{C}^m$ implies that the second characteristic property (5.3) holds too.

The uniqueness of the RHKS $\mathcal{H}$ of $m \times 1$ vvf's with given RK $K_\omega(\lambda)$ follows from properties (5.2) and (5.3) of a RKHS. □

Lemma 5.3 *Let $\mathcal{H}$ be a RKHS of $p \times 1$ vvf's that are defined on a set Ω and let $K_\omega(\lambda)$ denote the RK of $\mathcal{H}$ and let*

$$\begin{aligned} \mathcal{N}_\omega &= \{u \in \mathbb{C}^p : K_\omega(\omega)u = 0\} \\ \mathcal{R}_\omega &= \{K_\omega(\omega)u : u \in \mathbb{C}^p\} \quad \textit{and} \\ \mathcal{F}_\omega &= \{f(\omega) : f \in \mathcal{H}\}. \end{aligned}$$

Then

$$\mathcal{R}_\omega = \mathcal{F}_\omega \quad \textit{and} \quad \mathcal{N}_\omega = \mathcal{F}_\omega^\perp$$

for every point $\omega \in \Omega$. In particular,

$$\textit{at each point } \omega \in \Omega, \quad \mathcal{F}_\omega = \mathbb{C}^p \iff K_\omega(\omega) > 0. \tag{5.8}$$

Proof It suffices to show that $\mathcal{N}_\omega = \mathcal{F}_\omega^\perp$, since $\mathcal{R}_\omega = \mathcal{N}_\omega^\perp$.

Suppose first that $u \in \mathcal{N}_\omega$. Then

$$|u^* f(\omega)| = |\langle f, K_\omega u\rangle_{\mathcal{H}}| \leq \|f\|_{\mathcal{H}} \|K_\omega u\|_{\mathcal{H}} = 0$$

for every $f \in \mathcal{H}$, since $\|K_\omega u\|_{\mathcal{H}}^2 = u^* K_\omega(\omega)u$. Therefore, $\mathcal{N}_\omega \subseteq \mathcal{F}_\omega^\perp$.

Conversely, if $u \in \mathcal{F}_\omega^\perp$, then $u^* K_\omega(\omega)v = 0$ for every $v \in \mathbb{C}^p$. But this implies that $u^* K_\omega(\omega) = 0$ and hence, since $K_\omega(\omega) = K_\omega(\omega)^*$ that $K_\omega(\omega)u = 0$, i.e., $u \in \mathcal{N}_\omega$. □

In the sequel we shall only be interested in RKHS's of vvf's that are holomorphic in some open subset Ω of the complex plane $\mathbb{C}$ and are invariant under the generalized backwards shift operator R_α which is defined for vvf's and mvf's in Chapter 3 by formula (3.69). In order to keep the typography simple, we shall not indicate the space in which R_α acts in the notation.

Lemma 5.4 *Let $\mathcal{H}$ be a RKHS of $m \times 1$ vvf's that are defined in an open nonempty subset Ω of $\mathbb{C}$ and suppose that $\mathcal{H}$ is R_α invariant for every point $\alpha \in \Omega$. Then the following two assertions are equivalent:*

(1) *$K_\omega(\omega) > 0$ for at least one point $\omega \in \Omega$.*

(2) *$K_\omega(\omega) > 0$ for every point $\omega \in \Omega$.*

Moreover, the sets $\mathcal{N}_\omega$, $\mathcal{R}_\omega$ and $\mathcal{F}_\omega$ defined in Lemma 5.3 are independent of the choice of the point $\omega \in \Omega$.

Proof Let $u \in \mathcal{N}_\beta$ for some point $\beta \in \Omega$. Then, since $\mathcal{N}_\beta = \mathcal{F}_\beta^\perp$ by Lemma 5.3,

$$u^* f(\beta) = 0 \quad \text{for every} \quad f \in \mathcal{H}.$$

Consequently,

$$\begin{aligned} 0 &= u^*(R_\alpha f)(\beta) \\ &= \frac{u^*f(\beta) - u^*f(\alpha)}{\alpha - \beta} \quad \text{for every } f \in \mathcal{H} \text{ and every } \alpha \in \Omega \setminus \{\beta\}. \end{aligned}$$

Therefore,

$$u^*f(\alpha) = 0 \quad \text{for every} \quad f \in \mathcal{H} \quad \text{and every} \quad \alpha \in \Omega.$$

Thus, $u \in \mathcal{F}_\alpha^\perp$, and as $\mathcal{F}_\alpha^\perp = \mathcal{N}_\alpha$, this proves that

$$\mathcal{N}_\beta \subseteq \mathcal{N}_\alpha \quad \text{for every} \quad \alpha \in \Omega.$$

Therefore, by symmetry,

$$\mathcal{N}_\beta = \mathcal{N}_\alpha \quad \text{for every point} \quad \alpha \in \Omega.$$

Moreover, as

$$K_\omega(\omega) > 0 \Longleftrightarrow \mathcal{N}_\omega = \{0\},$$

it follows that

$$\begin{aligned} K_\omega(\omega) > 0 \quad &\text{at one point} \quad \omega \in \Omega \\ &\Longleftrightarrow K_\omega(\omega) > 0 \quad \text{at every point } \omega \in \Omega. \end{aligned}$$

Since $\mathcal{R}_\omega = \mathcal{N}_\omega^\perp$ and $\mathcal{F}_\omega = \mathcal{R}_\omega$, the proof is complete. □

Theorem 5.5 *Let $\mathcal{H}$ be a RKHS of $m \times 1$ vector valued functions that are defined on a set Ω and let $K_\omega(\lambda)$ be its RK. Then an $m \times 1$ vvf $f(\lambda)$ that is defined on Ω belongs to $\mathcal{H}$ if and only if there exists a constant $\gamma \in \mathbb{R}$ such that*

$$\gamma^2 K_\omega(\lambda) - f(\lambda)f(\omega)^* \quad \textit{is a positive kernel.} \tag{5.9}$$

Moreover,

$$\|f\|_{\mathcal{H}}^2 = \inf\{\gamma^2 : \textit{(5.9) holds}\}.$$

Proof See e.g., pp. 16–17 of [Sai97]. □

Lemma 5.6 *Let $\mathcal{H}$ be a RKHS of $m \times 1$ vvf's on some nonempty open subset Ω of $\mathbb{C}$ with RK $K_\omega(\lambda)$ on $\Omega \times \Omega$. Then the two conditions*

(1) *$K_\omega(\lambda)$ is a holomorphic function of λ in Ω for every point $\omega \in \Omega$;*

(2) *the function* $\|K_\omega(\omega)\|$ *is bounded in every compact subset of* Ω*;*

are in force if and only if

(3) *every vvf* $f \in \mathcal{H}$ *is holomorphic in* Ω*.*

Proof If (1) and (2) are in force and $f \in \mathcal{H}$, then (3) may be verified by choosing a sequence $\{f_n\}$ of finite linear combinations of the form $f_n = \Sigma K_{\omega_j} u_j$ which approximate f in $\mathcal{H}$ and then invoking the inequality

$$\begin{aligned} |u^*\{f(\omega) - f_n(\omega)\}| &= |\langle f - f_n, K_\omega u\rangle_{\mathcal{H}}| \\ &\leq \|f - f_n\|_{\mathcal{H}} \|K_\omega(\omega)\|^{\frac{1}{2}} \|u\|. \end{aligned}$$

This does the trick since the f_n are holomorphic in Ω and tend to f uniformly on compact subsets of Ω.

Conversely, if (3) is in force, then (1) is selfevident and (2) follows from the uniform boundedness principle of Banach and Steinhaus; see, e.g., pp. 98–99 in [Ru74] for the latter. □

Corollary 5.7 *If, in the setting of the previous lemma, condition* (1) *is in force and the mvf* $K_\omega(\omega)$ *is continuous on* Ω*, then* (3) *is also in force.*

Theorem 5.8 *Let* $\mathcal{H}$ *be a RKHS of* $p \times 1$ *vvf's that are defined on a set* Ω *and suppose that* $\mathcal{H}$ *is invariant under the action of* R_α *for some point* $\alpha \in \Omega$*. Then* R_α *is a bounded linear operator in* $\mathcal{H}$*.*

Proof Under the assumptions of the theorem, R_α is a closed operator in $\mathcal{H}$: If $K_\omega(\lambda)$ is the RK of the RKHS $\mathcal{H}$ and if $h_n \to h$ and $R_\alpha h_n \to g$ in $\mathcal{H}$ when $n \to \infty$, then

$$\xi^* h_n(\lambda) = \langle h_n, K_\lambda \xi\rangle_{\mathcal{H}} \to \langle h, K_\lambda \xi\rangle_{\mathcal{H}} = \xi^* h(\lambda)$$

and

$$\xi^*(R_\alpha h_n)(\lambda) = \langle R_\alpha h_n, K_\lambda \xi\rangle_{\mathcal{H}} \to \langle g, K_\lambda \xi\rangle_{\mathcal{H}} = \xi^* g(\lambda)$$

for every $\lambda \in \Omega$ and $\xi \in \mathbb{C}^p$. Therefore, $g = R_\alpha h$, since $\xi^*(R_\alpha h_n)(\lambda) \to \xi^*(R_\alpha h)(\lambda)$ for every $\lambda \in \Omega$ and $\xi \in \mathbb{C}^p$. This completes the proof by a theorem of Banach (see e.g., Theorem 5.10 in [Ru74]), since R_α is a closed linear operator that is defined on the full space $\mathcal{H}$. □

5.2 Examples

Recall the notation

$$\rho_\omega(\lambda) = -2\pi i(\lambda - \bar{\omega}), \tag{5.10}$$

which will be used frequently.

Example 5.9 *The Hardy space H_2^p of $p \times 1$ holomorphic vvf's in $\Omega = \mathbb{C}_+$ equipped with the standard inner product in L_2^p:*

$$\langle g, h\rangle_{st} = \int_{-\infty}^{\infty} h(\mu)^* g(\mu) d\mu, \tag{5.11}$$

is perhaps the most familiar example of a RKHS. The RK for this space is

$$K_\omega(\lambda) = \frac{I_p}{\rho_\omega(\lambda)}, \tag{5.12}$$

defined on $\mathbb{C}_+ \times \mathbb{C}_+$.

Formula (5.3) is nothing more than Cauchy's formula for H_2^p. The reproducing kernel formulas lead easily to the evaluation described in the next lemma.

Lemma 5.10 *If $f \in H_\infty^{p\times q}$, $\xi \in \mathbb{C}^p$ and $\omega \in \mathbb{C}_+$, then*

$$\Pi_+ f^* \frac{\xi}{\rho_\omega} = f(\omega)^* \frac{\xi}{\rho_\omega}. \tag{5.13}$$

Proof Let $\eta \in \mathbb{C}^q$ and $\alpha \in \mathbb{C}_+$. Then

$$\left\langle \Pi_+ f^* \frac{\xi}{\rho_\omega}, \frac{\eta}{\rho_\alpha} \right\rangle_{st} = \left\langle \frac{\xi}{\rho_\omega}, f\frac{\eta}{\rho_\alpha} \right\rangle_{st} = \frac{\eta^* f(\omega)^* \xi}{\rho_\omega(\alpha)} = \left\langle f(\omega)^* \frac{\xi}{\rho_\omega}, \frac{\eta}{\rho_\alpha} \right\rangle_{st}.$$

□

The space H_2^p is R_α invariant for every point $\alpha \in \mathbb{C}_+$, since

$$R_\alpha f \in \mathcal{N}_+^p \cap L_2^p$$

for every $\alpha \in \mathbb{C}_+$ and every $f \in H_2^p$, and $\mathcal{N}_+^p \cap L_2^p = H_2^p$, by the Smirnov maximum principle. The same argument implies that $R_\alpha f \in H_2^p$ if $\alpha \in \mathbb{R} \cap \mathfrak{h}_f$, i.e., if the vvf f is holomorphic at the point α. Moreover, it is easy to check that

$$(R_\alpha K_\omega \xi)(\lambda) = 2\pi i K_\omega(\lambda) K_\omega(\alpha)\xi \tag{5.14}$$

for every point $\alpha \in \mathbb{C}_+$ and every vector $\xi \in \mathbb{C}^p$.

Example 5.11 *The Hardy space K_2^p of $p \times 1$ holomorphic vvf's in $\Omega = \mathbb{C}_-$ equipped with the standard inner product in L_2^p is an RKHS with RK*

$$K_\omega(\lambda) \;=\; -\frac{I_p}{\rho_\omega(\lambda)},$$

defined on $\mathbb{C}_- \times \mathbb{C}_-$.

The space K_2^p is R_α invariant for every point $\alpha \in \mathbb{C}_-$. This follows from the R_α invariance of H_2^p and the fact that

$$f \in K_2^p \Longleftrightarrow f^\# \in H_2^p.$$

Example 5.12 *If $s \in \mathcal{S}^{p\times q}$, then the kernel*

$$\Lambda_\omega(\lambda) = \frac{I_p - s(\lambda)s(\omega)^*}{\rho_\omega(\lambda)} \tag{5.15}$$

is positive on $\mathbb{C}_+ \times \mathbb{C}_+$ and hence, in view of Theorem 5.2, is the RK of exactly one RKHS which we shall refer to as $\mathcal{H}(s)$. Moreover, every vvf $f \in \mathcal{H}(s)$ is holomorphic in $\mathbb{C}_+$.

The asserted positivity of the kernel follows from the fact that the operator $\widetilde{M}_s$ of multiplication by the mvf s restricted to H_2^q is a contraction from H_2^q into H_2^p and formula (5.13) written as

$$(\widetilde{M}_s)^* \frac{\xi}{\rho_\omega} = s(\omega)^* \frac{\xi}{\rho_\omega}:$$

$$\begin{aligned}\sum_{i,j=1}^n \frac{\xi_i^* s(\omega_i)s(\omega_j)^*\xi_j}{\rho_{\omega_j}(\omega_i)} \;&=\; \left\|(\widetilde{M}_s)^* \sum_{j=1}^n \frac{\xi_j}{\rho_{\omega_j}}\right\|_{st}^2 \\ &\le \left\|\sum_{j=1}^n \frac{\xi_j}{\rho_{\omega_j}}\right\|_{st}^2 = \sum_{i,j=1}^n \frac{\xi_i^*\xi_j}{\rho_{\omega_j}(\omega_i)}\end{aligned}$$

for every choice of points $\omega_1, \ldots, \omega_n$ in $\mathbb{C}_+$ and vectors $\xi_1, \ldots, \xi_n$ in $\mathbb{C}^n$ and every integer $n \geq 1$. The vvf's in $\mathcal{H}(s)$ are holomorphic in $\mathbb{C}_+$, by Corollary 5.7.

Remark 5.13 *The space $\mathcal{H}(s)$ has two useful characterizations. The first, which originates with de Branges and Rovnyak (for the disc) [BrR66],* *characterizes* $\mathcal{H}(s)$ *as the set of* $g \in H_2^p$ *for which*

$$\kappa = \sup\{\|g + sh\|_{st}^2 - \|h\|_{st}^2 : \ h \in H_2^q\} \tag{5.16}$$

is finite. In this case

$$\kappa = \|g\|_{\mathcal{H}(s)}^2 \geq \|g\|_{st}^2 \quad \text{for every } g \in \mathcal{H}(s).$$

The second characterizes $\mathcal{H}(s)$ *as the range of the square root of the operator* $(I - \widetilde{M}_s\widetilde{M}_s^*)|_{H_2^p}$*. A good discussion of the connection between these two descriptions is given by Ando [An90]; see also Sarason [Sar94] and, for more on the second description, Fillmore and Williams [FW71]. These characterizations imply that* $\mathcal{H}(s)$ *is a subspace (not necessarily closed) of* H_2^p *and hence that every vvf* $f \in \mathcal{H}(s)$ *has nontangential boundary values* $f(\mu)$ *a.e. on* $\mathbb{R}$*. There is a natural analogue* $\mathcal{H}_*(s)$ *of* $\mathcal{H}(s)$ *in* K_2^q*; see, eg., [Dy03b].*

Theorem 5.14 *If* $s \in \mathcal{S}^{p\times q}$*, then:*

(1) $(R_\alpha s)\eta \in \mathcal{H}(s)$ *for every* $\alpha \in \mathbb{C}_+$ *and* $\eta \in \mathbb{C}^q$*.*

(2) *If* $\alpha_1, \ldots, \alpha_k \in \mathbb{C}_+$ *and* $\eta_1, \ldots, \eta_k \in \mathbb{C}^q$*, then*

$$\left\| \sum_{j=1}^k (R_{\alpha_j} s)\eta_j \right\|_{\mathcal{H}(s)}^2 \leq 4\pi^2 \sum_{i,j=1}^k \eta_i^* \frac{I_q - s(\alpha_i)^* s(\alpha_j)}{\rho_{\alpha_i}(\alpha_j)} \eta_j, \tag{5.17}$$

with equality if $s \in \mathcal{S}_{in}^{p\times q}$*.*

(3) *The space* $\mathcal{H}(s)$ *is* R_α *invariant for every* $\alpha \in \mathbb{C}_+$*.*

(4) *The inequality*

$$\|R_\alpha f\|_{\mathcal{H}(s)} \leq \frac{1}{\Im\alpha}\|f\|_{\mathcal{H}(s)}$$

is in force for every $\alpha \in \mathbb{C}_+$ *and every* $f \in \mathcal{H}(s)$*.*

(5) $\mathcal{H}(s) \subseteq H_2^p$ *as linear spaces and hence every* $f \in \mathcal{H}(s)$ *has nontangential boundary values* $f(\mu)$ *a.e. on* $\mathbb{R}$*.*

(6) *If* $s \in \mathcal{S}_{in}^{p\times q}$ *for some* $p \geq q$*, then*

$$\mathcal{H}(s) = H_2^p \ominus sH_2^q$$

and the inner product in $\mathcal{H}(s)$ *is just the standard inner product (5.11).*

Proof See Theorem 2.3 in [Dy89b]. □

Lemma 5.15 *If $s \in \mathcal{S}^{p\times q}$, then each of the vector spaces*

$$\text{range}\,\{I_p - s(\omega)s(\omega)^*\} \quad \textit{and} \quad \text{range}\,\{I_q - s(\omega)^*s(\omega)\}$$

is independent of the choice of the point $\omega \in \mathbb{C}_+$.

Proof The asserted conclusions follow by applying Lemma 5.4 to the RKHS's with RK's $\Lambda_\omega(\lambda)$ defined by s and $s^\sim$, respectively. □

Example 5.16 *Let $b \in \mathcal{S}_{*in}^{p\times q}$ for some $p \le q$ and let $\Omega = \mathbb{C}_-$. Then*

$$\mathcal{H}_*(b) = K_2^q \ominus b^\# K_2^p$$

is a RKHS of $q \times 1$ holomorphic vvf's in $\mathbb{C}_-$ with respect to the standard inner product in L_2^q with RK

$$K_\omega(\lambda) \;=\; \frac{b^\#(\lambda)b(\bar{\omega}) - I_q}{\rho_\omega(\lambda)}, \tag{5.18}$$

defined in $\mathbb{C}_- \times \mathbb{C}_-$. Moreover, $\mathcal{H}_*(b)$ is R_α invariant for every $\alpha \in \mathbb{C}_-$.

In view of the fact that

$$f \in \mathcal{H}_*(b) \Longleftrightarrow f \in K_2^q \quad \text{and} \quad bf \in H_2^p$$

it is readily seen that the space $\mathcal{H}_*(b)$ is R_α invariant for every $\alpha \in \mathbb{C}_-$, since $R_\alpha f \in K_2^q$ and

$$b(\lambda)(R_\alpha f)(\lambda) = \frac{1}{\lambda - \alpha} b(\lambda)f(\lambda) - \frac{b(\lambda)}{\lambda - \alpha} f(\alpha)$$

belongs to H_2^p for every choice of $\alpha \in \mathbb{C}_-$ and $f \in \mathcal{H}_*(b)$.

Example 5.17 *If $c \in \mathcal{C}^{p\times p}$, then the kernel*

$$k_\omega^c(\lambda) = 2\frac{c(\lambda) + c(\omega)^*}{\rho_\omega(\lambda)} \tag{5.19}$$

is positive on $\mathbb{C}_+ \times \mathbb{C}_+$ and hence, in view of Theorem 5.2, is the RK of exactly one RKHS which we shall refer to as $\mathcal{L}(c)$. Moreover, since the conditions of Corollary 5.7 are satisfied, every vvf $f \in \mathcal{L}(c)$ is holomorphic in $\mathbb{C}_+$.

The asserted positivity of the kernel is an easy consequence of the Riesz-Herglotz integral representation of a mvf $c \in \mathcal{C}^{p\times p}$: If

$$g = \sum_{j=1}^{n} k_{\omega_j}^c \xi_j, \quad g^\circ(\mu) = 2\sum_{j=1}^{n} \frac{\xi_j}{\rho_{\omega_j}(\mu)} \quad \text{and} \quad \xi = \sum_{j=1}^{n} \xi_j, \tag{5.20}$$

then

$$g(\lambda) = \frac{1}{\pi i}\int_{-\infty}^{\infty} \frac{1}{\mu - \lambda} d\sigma(\mu) g^\circ(\mu) + \frac{1}{\pi}\beta\xi$$

and

$$\begin{aligned} \sum_{i,j=1}^{n} \xi_i^* k_{\omega_j}^c(\omega_i)\xi_j &= \sum_{i,j=1}^{n} \frac{1}{\pi^2}\int_{-\infty}^{\infty} \frac{\xi_i^* d\sigma(\mu)\xi_j}{(\mu - \omega_i)(\mu - \overline{\omega_j})} + \frac{1}{\pi}\sum_{i,j=1}^{n} \xi_i^*\beta\xi_j \\ &= \int_{-\infty}^{\infty} g^\circ(\mu)^* d\sigma(\mu) g^\circ(\mu) + \frac{1}{\pi}\xi^*\beta\xi \\ &= \langle g, g\rangle_{\mathcal{L}(c)}. \end{aligned}$$

The main properties of $\mathcal{L}(c)$ are due to L. de Branges and may be found on pp. 9–13 of [Br68a]. In the sequel we shall only use the estimate in the next lemma, which is adapted from [Br63].

Lemma 5.18 *If $c \in \mathcal{C}^{p\times p}$ and the spectral function $\sigma(\mu)$ of c is such that $\sigma(\alpha + \varepsilon) = \sigma(\alpha - \varepsilon)$ for some $\alpha \in \mathbb{R}$ and $\varepsilon > 0$ and g is as in (5.20), then the kernel $k_\omega^c(\lambda)$ is positive on $\mathbb{C}_+ \cup \{\alpha\} \times \mathbb{C}_+ \cup \{\alpha\}$,*

$$R_\alpha g \in \mathcal{L}(c) \quad \textit{and} \quad \|R_\alpha g\|_{\mathcal{L}(c)} \le \varepsilon^{-1}\|g\|_{\mathcal{L}(c)}. \tag{5.21}$$

Proof If g is of the specified form, then

$$(R_\alpha g)(\lambda) = \frac{1}{\pi i}\int_{-\infty}^{\infty} \frac{1}{\mu - \lambda} d\sigma(u) \frac{1}{\mu - \alpha} g^\circ(\mu),$$

which belongs to $\mathcal{L}(c)$, since

$$\frac{1}{\mu - \alpha}\frac{1}{\mu - \overline{\omega_j}} = \frac{1}{\alpha - \overline{\omega_j}}\left\{\frac{1}{\mu - \alpha} - \frac{1}{\mu - \overline{\omega_j}}\right\}.$$

Moreover,

$$\begin{aligned} \|R_\alpha g\|_{\mathcal{L}(c)}^2 &= \int_{-\infty}^{\infty} \frac{1}{\mu - \alpha} g^\circ(\mu)^* d\sigma(\mu) \frac{1}{\mu - \alpha} g^\circ(\mu) \\ &\le \varepsilon^{-2}\int_{-\infty}^{\infty} g^\circ(\mu)^* d\sigma(\mu) g^\circ(\mu) \le \varepsilon^{-2}\|g\|_{\mathcal{L}(c)}^2. \end{aligned}$$

□

Example 5.19 *Let $U \in \mathcal{P}(J)$. Then*

$$K_\omega(\lambda) = \frac{J - U(\lambda)JU(\omega)^*}{\rho_\omega(\lambda)} \tag{5.22}$$

is a positive kernel on $\mathfrak{h}_U^+ \times \mathfrak{h}_U^+$ and hence, in view of Theorem 5.2, is the RK of exactly one RKHS of $m \times 1$ vvf's on $\mathfrak{h}_U^+$, which we shall refer to as $\mathcal{H}(U)$. In view of Corollary 5.7, these vvf's are holomorphic in $\mathfrak{h}_U^+$.

Theorem 5.20 *If $U \in \mathcal{P}(J)$, then:*

(1) *$(R_\alpha U)v \in \mathcal{H}(U)$ for every choice of $\alpha \in \mathfrak{h}_U^+$ and $v \in \mathbb{C}^m$.*

(2) *The inequality*

$$\left\| \sum_{j=1}^{\ell} (R_{\alpha_j} U) v_j \right\|_{\mathcal{H}(U)}^2 \leq 4\pi^2 \sum_{i,j=1}^{\ell} v_i^* \frac{J - U(\alpha_i)^* J U(\alpha_j)}{\rho_{\alpha_i}(\alpha_j)} v_j \tag{5.23}$$

holds for every choice of $\alpha_1 \cdots, \alpha_\ell \in \mathfrak{h}_U^+$ and $v_1, \ldots, v_\ell \in \mathbb{C}^m$.

(3) *The space $\mathcal{H}(U)$ is R_α invariant for every point $\alpha \in \mathfrak{h}_U^+$.*

(4) *The inequality*

$$\|R_\alpha f\|_{\mathcal{H}(U)} \leq \frac{1}{\Im\alpha}\{1 + 2^{-1}\|QU(\alpha)Q\|\}\|f\|_{\mathcal{H}(U)} \tag{5.24}$$

holds for every $f \in \mathcal{H}(U)$ and every $\alpha \in \mathfrak{h}_U^+$.

(5) *$\mathcal{H}(U) \subset \mathcal{N}^m$ and hence every $f \in \mathcal{H}(U)$ has nontangential boundary values.*

(6) *If $J \neq I_m$, $S = PG(U)$ and $L(\lambda) = (P - S(\lambda)Q)^{-1}$, then the formula*

$$f(\lambda) = L(\lambda)g(\lambda), \quad g \in \mathcal{H}(S), \tag{5.25}$$

defines a unitary operator from $\mathcal{H}(S)$ onto $\mathcal{H}(U)$.

Proof If $J = I_m$, then $\mathcal{P}(J) = \mathcal{S}^{m\times m}$, $\mathfrak{h}_U^+ = \mathbb{C}_+$ and (1)–(5) follow from Theorem 5.14.

If $J \neq I_m$, then Theorem 5.14 is applicable to $S = PG(U)$, since it belongs to $\mathcal{S}^{m\times m}$, and the kernels $K_\omega(\lambda)$ and $\Lambda_\omega(\lambda)$ that are defined by formulas

(5.22) and (5.15), respectively, are connected:

$$K_\omega(\lambda) = L(\lambda)\Lambda_\omega(\lambda)L(\omega)^* \quad \text{on } \mathfrak{h}_U^+ \times \mathfrak{h}_U^+.$$

Therefore, $K_\omega(\lambda)$ is a positive kernel on $\mathfrak{h}_U^+ \times \mathfrak{h}_U^+$. Moreover, (6) holds and the identities based on the two recipes for the PG transform (see (2.6)) imply that

$$\begin{aligned}(R_\alpha U)(\lambda) &= \frac{L(\lambda)(S(\lambda)P - Q) - (Q + PS(\alpha))\widetilde{L}(\alpha)}{\lambda - \alpha} \\ &= L(\lambda)\frac{(S(\lambda)P - Q)(P + QS(\alpha)) - (P - S(\lambda)Q)(Q + PS(\alpha))}{\lambda - \alpha}\widetilde{L}(\alpha) \\ &= L(\lambda)(R_\alpha S)(\lambda)\widetilde{L}(\alpha), \quad \text{where } \widetilde{L}(\lambda) = (P + QS(\lambda))^{-1},\end{aligned}$$

which serves to justify (1), thanks to (5.25).

The second assertion rests on the chain of inequalities

$$\begin{aligned}\left\| \sum_{j=1}^{\ell} (R_{\alpha_j} U)v_j \right\|_{\mathcal{H}(U)}^2 &= \left\| \sum_{j=1}^{\ell} (R_{\alpha_j} S)y_j \right\|_{\mathcal{H}(S)}^2 \\ &\leq 4\pi^2 \sum_{i,j=1}^{\ell} y_i^* \frac{I_m - S(\alpha_i)^* S(\alpha_j)}{\rho_{\alpha_i}(\alpha_j)} y_j,\end{aligned}$$

where $y_j = \widetilde{L}(\alpha_j))v_j$ for $j = 1, \ldots, \ell$, which coincides with the right hand side of (5.23).

Next, if $f \in \mathcal{H}(U)$ and $\alpha \in \mathfrak{h}_U^+$, the formula

$$(R_\alpha f)(\lambda) = L(\lambda)\{(R_\alpha g)(\lambda) + (R_\alpha S)(\lambda)QL(\alpha)g(\alpha)\} \tag{5.26}$$

displays the fact that $\mathcal{H}(U)$ is R_α invariant. Moreover, in view of (5.25) and (5.26)

$$\|R_\alpha f\|_{\mathcal{H}(U)} \leq \|R_\alpha g\|_{\mathcal{H}(S)} + \|(R_\alpha S)QL(\alpha)g(\alpha)\|_{\mathcal{H}(S)}.$$

Therefore, since

$$QL(\alpha) = Q(P - U(\alpha)Q) = -QU(\alpha)Q$$

and

$$g(\alpha)^* g(\alpha) = \left\langle g, \frac{g(\alpha)}{\rho_\alpha} \right\rangle_{st} \leq \|g\|_{st} \left\{ \frac{g(\alpha)^* g(\alpha)}{\rho_\alpha(\alpha)} \right\}^{1/2},$$

it follows that

$$\|(R_\alpha S)QL(\alpha)g(\alpha)\|_{\mathcal{H}(S)} \leq \frac{2\pi\|QL(\alpha)g(\alpha)\|}{\sqrt{\rho_\alpha(\alpha)}} \leq \frac{2\pi\|QU(\alpha)Q\|\|g\|_{st}}{\rho_\alpha(\alpha)}.$$

Thus, as

$$\|g\|_{st} \leq \|g\|_{\mathcal{H}(S)} \quad \text{and} \quad \|g\|_{\mathcal{H}(S)} = \|f\|_{\mathcal{H}(U)},$$

the preceding estimates combine to yield the bound in (4).

Assertion (5) follows from formula (5.25) and the fact that $L \in \mathcal{N}^{m\times m}$ and $g \in H_2^m$ if $g \in \mathcal{H}(S)$. □

5.3 A characterization of the spaces $\mathcal{H}(U)$

In this section we present a characterization of RKHS's of holomorphic vvf's with RK's of the form (5.34) in terms of R_α invariance and a constraint on the inner product that is formulated in terms of the structural identity (5.27), that is due to L. de Branges [Br63]. In de Branges original formulation it was assumed that the underlying domain Ω was subject to the constraint $\Omega \cap \mathbb{R} \neq \emptyset$. The relaxation of this condition to $\Omega \cap \overline{\Omega} \neq \emptyset$ is due to Rovnyak [Rov68].

Theorem 5.21 *Let $\mathcal{H}$ be a RKHS of holomorphic $m \times 1$ vector valued functions that are defined on an open subset Ω of $\mathbb{C}$ such that $\Omega \cap \overline{\Omega} \neq \emptyset$, let $J \in \mathbb{C}^{m\times m}$ be a signature matrix and suppose further that:*

(1) *$\mathcal{H}$ is invariant under R_α for every point $\alpha \in \Omega$.*

(2) *The de Branges identity*

$$\langle R_\alpha f, g\rangle_{\mathcal{H}} - \langle f, R_\beta g\rangle_{\mathcal{H}} - (\alpha - \overline{\beta})\langle R_\alpha f, R_\beta g\rangle_{\mathcal{H}} = 2\pi i g(\beta)^* J f(\alpha) \tag{5.27}$$

is in force for every pair of points $\alpha, \beta \in \Omega$ and every pair of functions $f, g \in \mathcal{H}$.

Then the RK $K_\omega(\lambda)$ of $\mathcal{H}$ meets the following conditions:

(1) *$J - \rho_\mu(\mu)K_\mu(\mu)$ is congruent to the matrix J for each point $\mu \in \Omega \cap \overline{\Omega}$.*

(2) *If $\mu \in \Omega \cap \overline{\Omega}$ is fixed and*

$$J - \rho_\mu(\mu)K_\mu(\mu) = T^* J T \tag{5.28}$$

for some invertible matrix $T \in \mathbb{C}^{m\times m}$, then

$$K_\omega(\lambda) = \frac{J - U(\lambda)JU(\omega)^*}{\rho_\omega(\lambda)} \tag{5.29}$$

on $\Omega \times \Omega$, where the mvf $U(\lambda)$ is uniquely specified by the formula

$$U(\lambda) = \{J - \rho_{\overline{\mu}}(\lambda)K_{\overline{\mu}}(\lambda)\}JT^* \tag{5.30}$$

up to a constant J-unitary factor on the right.

Proof The proof is broken into steps.

Step 1 *is to show that the mvf*

$$P_\omega(\lambda) = J - \rho_\omega(\lambda)K_\omega(\lambda) \tag{5.31}$$

satisfies the relation

$$P_\omega(\lambda) = P_{\overline{\mu}}(\lambda)JP_\mu(\mu)JP_{\overline{\mu}}(\omega)^* \tag{5.32}$$

for every pair of points $\lambda, \omega \in \Omega$.

The identity

$$\frac{K_\omega(\lambda) - K_\omega(\mu)}{\lambda - \mu} - \frac{K_\omega(\lambda) - K_{\overline{\mu}}(\lambda)}{\overline{\omega} - \mu} = 2\pi i K_{\overline{\mu}}(\lambda)JK_\omega(\mu)$$

is easily deduced from (5.27) by choosing $\alpha = \mu$, $\beta = \overline{\mu}$, $f = K_\omega u$, $g = K_\lambda v$ with $\lambda \in \Omega \setminus \{\mu\}$, $\omega \in \Omega \setminus \{\overline{\mu}\}$ and $u, v \in \mathbb{C}^m$. Then, upon cross multiplying by $(\lambda - \mu)(\overline{\omega} - \mu)$ and reexpressing in terms of $P_\omega(\lambda)$, it follows that

$$P_\omega(\lambda) = P_{\overline{\mu}}(\lambda)JP_\omega(\mu) \tag{5.33}$$

for every pair of points λ, $\omega \in \Omega$. Thus

$$P_\omega(\overline{\mu}) = P_{\overline{\mu}}(\overline{\mu})JP_\omega(\mu),$$

and, since $\rho_\mu(\overline{\mu}) = 0$,

$$J = P_\mu(\overline{\mu}) = P_{\overline{\mu}}(\overline{\mu})JP_\mu(\mu) = P_\mu(\mu)JP_{\overline{\mu}}(\overline{\mu}).$$

Consequently,

$$\begin{aligned} P_\omega(\mu) &= J^2 P_\omega(\mu) \\ &= P_\mu(\mu)JP_{\overline{\mu}}(\overline{\mu})JP_\omega(\mu) \\ &= P_\mu(\mu)JP_\omega(\overline{\mu}) \\ &= P_\mu(\mu)JP_{\overline{\mu}}(\omega)^*. \end{aligned}$$

Thus, upon substituting the last formula into (5.33), we obtain (5.32).

Step 2 *is to verify (5.28).*

It is convenient to let $M = P_\mu(\mu)$. Then $M = M^*$, M is invertible and $M^{-1} = JP_{\overline{\mu}}(\overline{\mu})J$. If $\mu \in \mathbb{R}$, then $M = J$ and we may choose $T = I_m$. If $\mu \notin \mathbb{R}$, then without loss of generality we may assume that $\mu \in \mathbb{C}_-$. Suppose further that $J \neq \pm I_m$, let $x \neq 0$ belong to the range of the orthogonal projection $P = (I_m + J)/2$ and let $y \neq 0$ belong to the range of the complementary projection $Q = (I_m - J)/2$. Then

$$\begin{aligned} \langle Mx, x\rangle &= \langle \{J - \rho_\mu(\mu)K_\mu(\mu)\}x, x\rangle \\ &= \langle x, x\rangle - \rho_\mu(\mu)\langle K_\mu(\mu)x, x\rangle \\ &> 0, \end{aligned}$$

since $\rho_\mu(\mu) < 0$ and $K_\mu(\mu) \geq 0$; and

$$\begin{aligned} \langle M^{-1}y, y\rangle &= \langle J\{J - \rho_{\overline{\mu}}(\overline{\mu})K_{\overline{\mu}}(\overline{\mu})\}Jy, y\rangle \\ &= -\langle y, y\rangle - \rho_{\overline{\mu}}(\overline{\mu})\langle K_{\overline{\mu}}(\overline{\mu})y, y\rangle \\ &< 0, \end{aligned}$$

since $\rho_{\overline{\mu}}(\overline{\mu}) > 0$ and $K_{\overline{\mu}}(\overline{\mu}) \geq 0$.

Now let

$$N = [X \quad M^{-1}Y]^* M [X \quad M^{-1}Y],$$

where $X \in \mathbb{C}^{m\times p}$, $Y \in \mathbb{C}^{m\times q}$, range X = range P, range Y = range Q, $p = \dim(\text{range}\, X)$ and $q = \dim(\text{range}\, Y)$. Then, since

$$N_{11} = X^*MX > 0, \quad N_{22} = Y^*M^{-1}MM^{-1}Y < 0,$$

$[X \quad M^{-1}Y]$ is invertible and $N = N^*$, it follows that M is congruent to the matrix

$$\begin{bmatrix} N_{11} & 0 \\ 0 & N_{22} - N_{12}^* N_{11}^{-1} N_{12} \end{bmatrix}$$

which has p positive eigenvalues, and q negative eigenvalues. Consequently, M is congruent to j_{pq}, which is congruent to J. Therefore, (5.28) holds when $J \neq \pm I_m$. The proof is easily modified to cover the cases $J = \pm I_m$. The details are left to the reader.

Step 3 *is to complete the proof.*

Formulas (5.28), (5.31) and (5.32) imply that

$$P_\omega(\lambda) = P_{\overline{\mu}}(\lambda)JT^*JTJP_{\overline{\mu}}(\omega)^* = U(\lambda)JU(\omega)^*$$

with

$$U(\lambda) = P_{\overline{\mu}}(\lambda)JT^*.$$

This serves to establish formulas (5.29) and (5.30).

It remains only to check uniqueness. But, if there exists a second $m \times m$ mvf $V(\lambda)$ that is holomorphic in Ω such that

$$K_\omega(\lambda) = \frac{J - V(\lambda)JV(\omega)^*}{\rho_\omega(\lambda)},$$

then

$$V(\mu)JV(\overline{\mu})^* = J = V(\overline{\mu})^*JV(\mu)$$

and

$$V(\lambda)JV(\omega)^* = U(\lambda)JU(\omega)^*$$

for every pair of points $\lambda, \omega \in \Omega$. Therefore,

$$V(\lambda) = V(\lambda)JV(\overline{\mu})^*JV(\mu) = U(\lambda)JU(\overline{\mu})^*JV(\mu) = U(\lambda)L,$$

where

$$L = JU(\overline{\mu})^*JV(\mu)$$

is independent of λ. Moreover, the development

$$J = V(\overline{\mu})^*JV(\mu) = L^*U(\overline{\mu})^*JU(\mu)L = L^*JL,$$

proves that L is J-unitary. □

Theorem 5.22 *Let $\mathcal{H}$ be a RKHS of $m \times 1$ vector valued functions that are holomorphic on a nonempty open subset Ω of $\mathbb{C}$ that is symmetric with respect to $\mathbb{R}$ and suppose that there exists an $m \times m$ signature matrix J and* *an $m \times m$ mvf $U(\lambda)$ that is holomorphic in Ω such that the RK $K_\omega(\lambda)$ can be expressed on $\Omega \times \Omega$ by the formula*

$$K_\omega(\lambda) = \begin{cases} \dfrac{J - U(\lambda)JU(\omega)^*}{\rho_\omega(\lambda)} & \text{if} \quad \lambda \neq \overline{\omega} \\ \dfrac{U'(\overline{\omega})JU(\omega)^*}{2\pi i} & \text{if} \quad \lambda = \overline{\omega} \end{cases} \tag{5.34}$$

Then

(1) *$\mathcal{H}$ is R_α invariant for every point $\alpha \in \Omega$.*

(2) *The de Branges identity (5.27) is valid for every choice of $f, g \in \mathcal{H}$ and $\alpha, \beta \in \Omega$.*

Proof The proof is broken into steps.

Step 1 *is to check that if $\lambda, \omega, \alpha, \in \Omega$, and $\alpha \neq \overline{\omega}$, then*

$$(R_\alpha K_\omega)(\lambda) = \frac{K_\omega(\lambda) - K_{\overline{\alpha}}(\lambda)JU(\alpha)JU(\omega)^*}{\overline{\omega} - \alpha}. \tag{5.35}$$

Let $N_\omega(\lambda) = J - U(\lambda)JU(\omega)^*$. Then, since

$$K_\omega(\lambda) - K_\omega(\alpha) = \left\{ \frac{1}{\rho_\omega(\lambda)} - \frac{1}{\rho_\omega(\alpha)} \right\} N_\omega(\lambda) + \frac{1}{\rho_\omega(\alpha)} \{N_\omega(\lambda) - N_\omega(\alpha)\},$$

and

$$J = U(\alpha)JU(\overline{\alpha})^* = U(\overline{\alpha})^* JU(\alpha),$$

the result follows from the evaluations

$$\left(R_\alpha \frac{1}{\rho_\omega} \right)(\lambda) = \left(\frac{1}{\overline{\omega} - \alpha} \right) \frac{1}{\rho_\omega(\lambda)}, \quad \text{for } \alpha \neq \overline{\omega}$$

and

$$\begin{aligned} \frac{N_\omega(\lambda) - N_\omega(\alpha)}{\lambda - \alpha} &= \frac{U(\alpha)JU(\omega)^* - U(\lambda)JU(\omega)^*}{\lambda - \alpha} \\ &= \left\{ \frac{U(\alpha)JU(\overline{\alpha})^* - U(\lambda)JU(\overline{\alpha})^*}{\lambda - \alpha} \right\} JU(\alpha)JU(\omega)^* \\ &= -2\pi i K_{\overline{\alpha}}(\lambda)JU(\alpha)JU(\omega)^*. \end{aligned}$$

Step 2 *is to verify the formula*

$$\langle (I + (\alpha - \overline{\beta})R_\alpha)f, (I + (\beta - \overline{\alpha})R_\beta)g \rangle_{\mathcal{H}} = \langle f, g \rangle_{\mathcal{H}} - \rho_\beta(\alpha) g(\beta)^* J f(\alpha) \tag{5.36}$$

for α, $\beta \in \Omega$ and sums of the form

$$f=\sum_{j=1}^{k} K_{\omega_j}u_j \quad \textit{and} \quad g=\sum_{i=1}^{\ell} K_{\gamma_i}v_i, \tag{5.37}$$

where the points ω_j, $\gamma_i \in \Omega$ and the vectors u_j, $v_i \in \mathbb{C}^m$.

The identity is verified first for $k=\ell=1$ by a straightforward, though lengthy, calculation that makes use of (5.34) and (5.35). The extension to finite sums is then immediate since inner products are sesquilinear.

Step 3 *If $f \in \mathcal{H}$ is as in Step 2 and $\alpha \in \Omega \setminus \mathbb{R}$, then*

$$\|R_\alpha f\|_{\mathcal{H}} \le \frac{1+\{1+2\pi|\alpha-\overline{\alpha}|\|K_\alpha(\alpha)\|^{1/2}\}}{|\alpha-\overline{\alpha}|}\|f\|_{\mathcal{H}}. \tag{5.38}$$

The inequalities

$$\begin{aligned} f(\alpha)^*f(\alpha) &= \langle f, K_\alpha f(\alpha)\rangle_{\mathcal{H}} \le \|f\|_{\mathcal{H}}\|K_\alpha f(\alpha)\|_{\mathcal{H}} \\ &= \|f\|_{\mathcal{H}}\{f(\alpha)^*K_\alpha(\alpha)f(\alpha)\}^{1/2} \end{aligned}$$

imply that

$$\|f(\alpha)\| \le \|f\|_{\mathcal{H}}\|K_\alpha(\alpha)\|^{1/2}.$$

Therefore, by formula (5.36) with $\alpha=\beta$ and $f=g$,

$$\|(I+(\alpha-\overline{\alpha})R_\alpha)f\|_{\mathcal{H}} \le \{1+2\pi|\alpha-\overline{\alpha}|\|K_\alpha(\alpha)\|^{1/2}\}\|f\|_{\mathcal{H}}.$$

The bound (5.38) then follows from the triangle inequality.

Step 4 *If $f \in \mathcal{H}$ is as in (5.37) and $[\alpha-\varepsilon,\alpha+\varepsilon] \subset \Omega \cap \mathbb{R}$ for some $\varepsilon > 0$ and $U(\mu)+U(\alpha)$ is invertible in this interval, then*

$$\|R_\alpha f\|_{\mathcal{H}(U)} \le \{\varepsilon^{-1}+\pi\|K_\alpha^U(\alpha)\|\}\|f\|_{\mathcal{H}(U)}. \tag{5.39}$$

Let $S=PG(U)$ and assume that $U(\alpha)=I_m$. (This involves no loss of generality because $U(\lambda)JU(\alpha)^*J$ meets this condition.) The mvf's

$$C=JT_{\mathfrak{V}}[S]J=(I_m-U)(I_m+U)^{-1}J \tag{5.40}$$

and

$$M(\lambda)=2(I_m+U(\lambda))^{-1}=I_m+C(\lambda)J \tag{5.41}$$

are holomorphic in an open set Ω_0 that contains $[\alpha-\varepsilon,\alpha+\varepsilon]$ and

$$\frac{C(\lambda)+C(\omega)^*}{-\pi i(\lambda-\overline{\omega})}=M(\lambda)K_\omega^U(\lambda)M(\omega)^*. \tag{5.42}$$

Thus, in view of Corollary 5.27 that will be presented in the next section, there exists a unique extension of $C(\lambda)$ into $\mathbb{C}_+$ that belongs to $\mathcal{C}^{m\times m}$ that will also be denoted by $C(\lambda)$. Formula (5.42) implies that

$$2\Re(C(\mu+i\nu)=M(\mu+i\nu)\{J-U(\mu+i\nu)JU(\mu+i\nu)^*\}M(\mu+i\nu)^*$$

and hence if the interval $[\alpha-\varepsilon,\alpha+\varepsilon]\subset\mathfrak{h}_U\cap\mathfrak{h}_M$, that the values of the spectral function $\sigma(\mu)$ of $C(\lambda)$ are equal at the endpoints of this interval. Thus,

$$\|R_\alpha g\|_{\mathcal{L}(C)}\leq\frac{1}{\varepsilon}\|g\|_{\mathcal{L}(C)}$$

for finite sums of the form $g=\sum K_{\omega_j}^C u_j$, by Lemma 5.18. Moreover, if f is as in (5.37), it is readily checked that

$$\|Mf\|^2_{\mathcal{L}(C)}=\|\sum_{j=1}^n K_{\omega_j}^C M(\omega_j)^{-*}u_j\|_{\mathcal{L}(C)}=\|f\|^2_{\mathcal{H}(U)}.$$

Therefore, since

$$(R_\alpha Mf)(\lambda)=M(\lambda)(R_\alpha f)(\lambda)+(R_\alpha M)(\lambda)f(\alpha)$$

and

$$(R_\alpha M)(\lambda)=-\pi i M(\lambda)K_\alpha^U(\lambda)J,$$

it follows that

$$\begin{aligned}\|R_\alpha f\|_{\mathcal{H}(U)} &= \|MR_\alpha f\|_{\mathcal{L}(C)}=\|R_\alpha Mf+\pi i MK_\alpha^U Jf(\alpha)\|_{\mathcal{L}(C)}\\ &\leq \frac{1}{\varepsilon}\|Mf\|_{\mathcal{L}(C)}+\pi\|K_\alpha^U Jf(\alpha)\|_{\mathcal{H}(U)},\end{aligned}$$

which yields (5.39).

Step 5 *is to complete the proof.*

Finite sums of the form considered in (5.37) are dense in $\mathcal{H}$ and the bounds in (5.36) and (5.39) insure that if a sequence f_n, $n=1,2,\ldots$ of finite sums of the form considered in (5.37) tend to a limit f in $\mathcal{H}$, then $R_\alpha f_n$ tends to a limit $\widetilde{f}$ in $\mathcal{H}$ when $\alpha\in\Omega$. Moreover, $\widetilde{f}$ can be identified as $R_\alpha f$, since norm convergence implies pointwise convergence in a RKHS. Thus, $\mathcal{H}$ is R_α invariant for every point $\alpha \in \Omega$ and the bounds on R_α are valid for arbitrary $f \in \mathcal{H}$ and formula (5.36), which is equivalent to the de Branges identity, holds for arbirary $f, g \in \mathcal{H}$ and $\alpha, \beta \in \Omega$. □

Lemma 5.23 *If $U \in \mathcal{P}(J)$, then*

$$\text{range}\,\{J - U(\omega)JU(\omega)^*\} \quad \textit{and} \quad \text{range}\,\{J - U(\omega)^*JU(\omega)\}$$

are independent of the choice of the point $\omega \in \mathfrak{h}_U^+$.

Proof The first assertion follows from Lemma 5.4; the second follows by applying the same lemma to the mvf $U_1(\lambda) = U(-\overline{\lambda})^*$. □

5.4 A finite dimensional example and extensions thereof

The conditions (1) R_α invariance and (2) the de Branges identity (5.27) in Theorem 5.21 are particularly transparent in the finite dimensional case. Thus, for example, if $F(\lambda)$ is an $m \times n$ meromorphic mvf in $\mathbb{C}$ with columns that are independent in the sense that $F(\lambda)u = 0$ for an infinite set of points that includes a limit point if and only if $u = 0$, then the space

$$\mathcal{H} = \{F(\lambda)u : u \in \mathbb{C}^n\} \tag{5.43}$$

endowed with the inner product

$$\langle Fu, Fv\rangle_{\mathcal{H}} = v^* P u \tag{5.44}$$

based on any positive definite $n \times n$ matrix P is a RKHS with RK

$$K_\omega(\lambda) = F(\lambda)P^{-1}F(\omega)^*. \tag{5.45}$$

If $\mathcal{H}$ is R_α invariant for some point $\alpha \in \mathfrak{h}_F$, then there exists an $n \times n$ matrix V such that

$$\frac{F(\lambda) - F(\alpha)}{\lambda - \alpha} = F(\lambda)V.$$

Thus,

$$F(\lambda) = F(\alpha)\{(I_n + \alpha V) - \lambda V\}^{-1},$$

i.e., R_α invariance forces $F(\lambda)$ to be a rational mvf. If it is also assumed that $F(\lambda) \to 0$ as $|\lambda| \to \infty$, then it is readily checked that V is invertible and hence that $F(\lambda)$ can be reexpressed as

$$F(\lambda) = \frac{C}{\sqrt{2\pi}}(\lambda I_n - A)^{-1} \tag{5.46}$$

where, $\cap_{j=0}^{n-1}\text{kernel}\, CA^j = \{0\}$ (i.e., (C, A) is observable). Thus, $\mathfrak{h}_F = \{\lambda \in \mathbb{C} : \det(\lambda I_n - A) \neq 0\}$ and, with the aid of the evaluation

$$(R_\alpha F)(\lambda) = -F(\lambda)(\alpha I_n - A)^{-1} \quad \text{for } \alpha \in \mathfrak{h}_F, \tag{5.47}$$

it is also easily verified that $\mathcal{H}$ is R_α invariant for every $\alpha \in \mathfrak{h}_F$ and that the de Branges identity (5.27) holds if and only if the positive matrix P that defines the inner product is a solution of the Lyapunov equation

$$A^*P - PA = iC^*JC. \tag{5.48}$$

Moreover, upon invoking formula (5.30) and letting $\mu \in \mathbb{R}$ tend to ∞, it is readily checked that

$$U(\lambda) = I_m - iC(\lambda I_n - A)^{-1}P^{-1}C^*J \tag{5.49}$$

up to a constant J-unitary multiplier on the right.

de Branges spaces are a convenient tool for solving tangential Nevanlinna-Pick problems. In particular:

Theorem 5.24 *Let $\xi_1, \ldots, \xi_n \in \mathbb{C}^p$, $\eta_1, \ldots, \eta_n \in \mathbb{C}^q$ and $\alpha_1, \ldots, \alpha_n \in \mathbb{C}_+$. Then there exists a mvf $s \in \mathcal{S}^{p\times q}$ such that $\xi_j^* s(\alpha_j) = \eta_j^*$ for $j = 1, \ldots, n$ if and only if the $n \times n$ matrix P with components*

$$p_{ij} = \frac{\xi_i^*\xi_j - \eta_i^*\eta_j}{\rho_{\alpha_j}(\alpha_i)}, \quad i, j = 1, \ldots, n,$$

is positive semidefinite.

Proof Suppose first that there exists a mvf $s \in \mathcal{S}^{p\times q}$ such that $\xi_j^* s(\alpha_j) = \eta_j^*$ for $j = 1, \ldots, n$. Then the necessity of the condition $P \geq 0$ is an immediate consequence of the inequalities

$$\begin{aligned} \left\| \sum_{j=1}^n c_j \rho_{\alpha_j}^{-1} \eta_j \right\|_{st}^2 &= \left\| \sum_{j=1}^n c_j \rho_{\alpha_j}^{-1} s(\alpha_j)^* \xi_j \right\|_{st}^2 = \left\| \Pi_+ M_{s^*} \sum_{j=1}^n c_j \rho_{\alpha_j}^{-1} \xi_j \right\|_{st}^2 \\ &\leq \left\| \sum_{j=1}^n c_j \rho_{\alpha_j}^{-1} \xi_j \right\|_{st}^2. \end{aligned}$$

Suppose next that $P > 0$, let $J = j_{pq}$, $v_j = \mathrm{col}\,[\xi_j, \eta_j]$ for $j = 1, \ldots, n$,

$$\begin{aligned} C &= -\frac{1}{i\sqrt{2\pi}} \begin{bmatrix} \xi_1 & \cdots & \xi_n \\ \eta_1 & \cdots & \eta_n \end{bmatrix} = -\frac{1}{i\sqrt{2\pi}}[v_1 \cdots v_n] \\ A &= \mathrm{diag}\,\{\overline{\alpha}_1, \ldots, \overline{\alpha}_n\}, \end{aligned}$$

$$F(\lambda) = \frac{C}{\sqrt{2\pi}}(\lambda I_n - A)^{-1} \quad \text{and} \quad W(\lambda) = I_m - iF(\lambda)P^{-1}C^* j_{pq}.$$

Then, since P is a positive definite solution of the Lyapunov equation (5.48) with $J = j_{pq}$, the space $\mathcal{H} = \mathrm{span}\,\{F(\lambda)u : u \in \mathbb{C}^n\}$ endowed with the inner product (5.44) is a de Branges space with RK

$$K_\omega^W(\lambda) = \frac{j_{pq} - W(\lambda) j_{pq} W(\omega)^*}{\rho_\omega(\lambda)}.$$

Moreover, if $f_j = \rho_{\alpha_j}^{-1} v_j$ denotes the jth column of F, then since

$$p_{k\ell} = \langle f_k, f_\ell \rangle_{\mathcal{H}} = \frac{v_\ell^* j_{pq} v_k}{\rho_{\alpha_k}(\alpha_\ell)} = \langle j_{pq} f_k, f_\ell \rangle_{st}$$

for $k, l = 1, \ldots, n$, it follows that

$$\begin{aligned} u^* f_k(\omega) &= \left\langle j_{pq} f_k, \frac{j_{pq} - W j_{pq} W(\omega)^*}{\rho_\omega} u \right\rangle_{st} \\ &= \left\langle j_{pq} f_k, \frac{j_{pq} u}{\rho_\omega} \right\rangle_{st} - \left\langle j_{pq} f_k, \frac{W j_{pq} W(\omega)^*}{\rho_\omega} u \right\rangle_{st} \\ &= u^* f_k(\omega) - \frac{u^* W(\omega) j_{pq} W(\alpha_k)^* j_{pq} v_k}{\rho_{\alpha_k}(\omega)} \end{aligned}$$

for every $\omega \in \mathbb{C}_+ \setminus \{\alpha_k\}$. Thus,

$$[\xi_k^* \quad -\eta_k^*] \begin{bmatrix} w_{11}(\alpha_k) & w_{12}(\alpha_k) \\ w_{21}(\alpha_k) & w_{22}(\alpha_k) \end{bmatrix} = v_k^* j_{pq} W(\alpha_k) = 0$$

and hence

$$\xi_k^* w_{12}(\alpha_k) w_{22}(\alpha_k)^{-1} = \eta_k^*,$$

which exhibits $s_{12} = w_{12} w_{22}^{-1}$ as a solution of the Nevanlinna-Pick interpolation problem if $P > 0$. A more detailed analysis exhibits $T_W[\mathcal{S}^{p\times q}]$ as the full set of solutions. The case $P \geq 0$ may be handled by a limiting argument. □

Corollary 5.25 *Let* $\omega_1, \ldots, \omega_k \in \mathbb{C}_+$ *and let* $s_{\omega_1} \ldots s_{\omega_k} \in \mathcal{S}^{p\times q}_{const}$. *Then there exists a mvf* $s \in \mathcal{S}^{p\times q}$ *such that* $s(\omega_j) = s_{\omega_j}$ *for* $j = 1, \ldots, k$ *if and only if*

$$\sum_{i,j=1}^{k} u_j^* \frac{I_p - s_{\omega_j} s_{\omega_k}^*}{\rho_{\omega_k}(\omega_j)} u_k \geq 0 \tag{5.50}$$

for every choice of vectors $u_1, \ldots, u_k \in \mathbb{C}^p$.

Proof This is a special case of Theorem 5.24 with $n = kp$,

$$C = -\frac{1}{i\sqrt{2\pi}} \begin{bmatrix} I_p & \cdots & I_p \\ s_{\omega_1}^* & \cdots & s_{\omega_k}^* \end{bmatrix} \quad \text{and} \quad A = \operatorname{diag}\{\overline{\omega_1} I_p, \ldots, \overline{\omega_k} I_p\}.$$

□

Let Ω be a subset of $\mathbb{C}_+$ and $s_\omega \in \mathbb{C}^{p\times q}$ for each point $\omega \in \Omega$ and let

$$\mathcal{S}(\Omega; s_\omega) = \{s \in \mathcal{S}^{p\times q} : s(\omega) = s_\omega \text{ for every point } \omega \in \Omega\}.$$

Lemma 5.26 $\mathcal{S}(\Omega; s_\omega) \neq \emptyset$ *if and only if*

$$\sum_{j,k=1}^{n} \xi_j^* \frac{I_p - s_{\omega_j} s_{\omega_k}^*}{\rho_{\omega_k}(\omega_j)} \xi_k \geq 0 \tag{5.51}$$

for every choice of points $\omega_1, \ldots, \omega_n \in \Omega$ *and vectors* $\xi_1, \ldots, \xi_n \in \mathbb{C}^p$.

Proof The necessity of condition (5.51) follows from the positivity of the kernel

$$\Lambda_\omega(\lambda) = \frac{I_p - s(\lambda)s(\omega)^*}{\rho_\omega(\lambda)}.$$

when $s \in \mathcal{S}^{p\times q}$. The sufficiency of this condition when Ω is a finite or countable set is a consequence of Corollary 5.25 and the sequential compactness of $\mathcal{S}^{p\times q}$.

If Ω is uncountable, then there exists a limit point $\omega_0 \in \mathbb{C}_+$ of the set Ω. Let $\omega_1, \omega_2, \ldots$ be a sequence of distict points in Ω that tend to this limit point ω_0. Then, in view of the preceding discussion, there exists a mvf $s \in \mathcal{S}^{p\times q}$ such that $s(\omega_j) = s_{\omega_j}$ for $j = 1, 2, \ldots$, and, since ω_0 is a limit point, there is only one such mvf $s \in \mathcal{S}^{p\times q}$. The same argument shows that if α is any other point of the set Ω, and if s is the unique mvf that was considered just above, then $s(\alpha) = s_\alpha$, since the sequence $\alpha, \omega_1, \omega_2, \ldots$ still tends to ω_0. □

Corollary 5.27 *If Ω is a subset of $\mathbb{C}_+$ and $c_\omega \in \mathcal{C}^{p\times p}_{const}$ for each point $\omega \in \Omega$, then the set $\mathcal{C}(\Omega; c_\omega)$ of mvf's $c \in \mathcal{C}^{p\times p}$ such that $c(\omega) = c_\omega$ for every point $\omega \in \Omega$ is nonempty if and only if*

$$\sum_{j,k=1}^{n} \xi_j^* \frac{I_p - c_{\omega_j} c_{\omega_k}^*}{\rho_{\omega_k}(\omega_j)} \xi_k \geq 0 \tag{5.52}$$

for every choice of points $\omega_1, \ldots, \omega_n \in \Omega$ and vectors $\xi_1, \ldots, \xi_n \in \mathbb{C}^p$.

Proof The necessity of condition (5.52) is clear from Example 5.17. Conversely, if this condition is satisfied and $s_\omega = PG(c_\omega)$, then the set $\{s_\omega : \omega \in \Omega\}$ satisfies the condition (5.51). Therefore, Lemma 5.26 guarantees that $\mathcal{S}(\omega; s_\omega) \neq \emptyset$. Thus, as $T_{\mathfrak{V}}[s_\omega] = c_\omega$, it is readily seen that $\mathcal{C}(\Omega; c_\omega) = T_{\mathfrak{V}}[\mathcal{S}(\Omega; s_\omega)]$ and hence that $\mathcal{C}(\Omega; c_\omega) \neq \emptyset$. □

Corollary 5.28 *If Ω is a subset of $\mathbb{C}_+$ and $U_\omega \in \mathcal{P}_{const}(J)$ for each point $\omega \in \Omega$, then the set $\mathcal{P}_J(\Omega; U_\omega)$ of mvf's $U \in \mathcal{P}(J)$ such that $U(\omega) = U_\omega$ for every $\omega \in \Omega$ is nonempty if and only if*

$$\sum_{j,k=1}^{n} v_j^* \frac{J - U_{\omega_j} J U_{\omega_k}^*}{\rho_{\omega_k}(\omega_j)} v_k \geq 0 \tag{5.53}$$

for every choice of points $\omega_1, \ldots, \omega_n \in \Omega$ and vectors $v_1, \ldots, v_n \in \mathbb{C}^m$.

Proof The necessity is easily checked directly from the identity

$$\left\| \sum_{k=1}^{n} K_{\omega_k}^U v_k \right\|_{\mathcal{H}(U)}^2 = \sum_{j,k=1}^{n} v_j^* \frac{J - U_{\omega_j} J U_{\omega_k}^*}{\rho_{\omega_k}(\omega_j)} v_k$$

which is valid for $U \in \mathcal{P}(J)$.

Conversely, if the condition (5.53) is met, then the set $\mathcal{S}(\Omega; S_\omega)$ with $S_\omega = PG(U_\omega)$ is nonempty by the preceding lemma, i.e., there exists a mvf $S \in \mathcal{S}^{m\times m}$ such that $S(\omega) = PG(U_\omega)$ for every point $\omega \in \Omega$. Thus,

$$\det(P + QS(\lambda)Q) \not\equiv 0,$$

and hence $U = PG(S)$ is well defined and $U(\omega) = PG(S_\omega) = U_\omega$, for every $\omega \in \Omega$, as claimed. □

Lemma 5.29 *Let J be an $m\times m$ signature matrix, let $U(\lambda)$ be an $m\times m$ mvf that is holomorphic in an open set $\Omega \subseteq \mathbb{C}_+$ such that the kernel $K_\omega^U(\lambda)$ that* *is defined by formula (5.34) is positive on $\Omega \times \Omega$. Then there exists a unique meromorphic extension $\widetilde{U} \in \mathcal{P}(J)$ of $U(\lambda)$ to a domain $\widetilde{\Omega} = \mathfrak{h}_{\widetilde{U}} \subseteq \mathbb{C}_+$. Morever, every vvf $h \in \mathcal{H}(U)$ has a unique extension $\widetilde{h} \in \mathcal{H}(\widetilde{U})$ to the domain $\widetilde{\Omega}$ and*

$$\langle h, g \rangle_{\mathcal{H}(U)} = \langle \widetilde{h}, \widetilde{g} \rangle_{\mathcal{H}(\widetilde{U})}.$$

Proof There are three cases to consider:

1. $J = I_m$**:** If $\omega_j \in \Omega$ for $j = 1, 2, \ldots$ is an infinite sequence of points in Ω with a limit point $\omega_0 \in \Omega$, then, since $\mathcal{P}(I_m) = \mathcal{S}^{m \times m}$, there exists a unique mvf $\widetilde{U} \in \mathcal{S}^{m \times m}$ such that $\widetilde{U}(\omega_j) = U(\omega_j)$ for $j = 1, 2, \ldots$. Moreover, the kernel $K_\omega^{\widetilde{U}}(\lambda)$ meets the conditions of the lemma.

2. $J = -I_m$**:** In this case, $U(\lambda)U(\lambda)^* \geq I_m$ for $\lambda \in \Omega$. Therefore, $U(\lambda)$ is invertible and the preceding argument is applicable to $S(\lambda) = U(\lambda)^{-1}$, since it belongs to $\mathcal{S}^{m \times m}$.

3. $J \neq \pm I_m$**:** The argument in the first paragraph is now applied to $S = PG(U)$. □

5.5 Extension of $\mathcal{H}(U)$ into $\mathbb{C}$ for $U \in \mathcal{P}^\circ(J)$

If $U \in \mathcal{P}^\circ(J)$ and

$$\Omega_+ = \{\lambda \in \mathfrak{h}_U^+ : \det U(\lambda) \neq 0\} \quad \text{and} \quad \Omega_- = \{\overline{\lambda} : \lambda \in \Omega_+\},$$

then U will be defined in Ω_- by the symmetry principle

$$U(\lambda) = JU^\#(\lambda)^{-1}J \quad \text{for } \lambda \in \Omega_-.$$

Let $\mathfrak{h}_U$ denote the largest open set in $\mathbb{C}$ to which U can be extended as a holomorphic mvf and let

$$\mathfrak{h}_U^- = \mathfrak{h}_U \cap \mathbb{C}_- \quad \text{and} \quad \mathfrak{h}_U^0 = \mathfrak{h}_U \cap \mathbb{R}. \tag{5.54}$$

Then $\mathfrak{h}_U^+ = \mathfrak{h}_U \cap \mathbb{C}_+ \supseteq \Omega_+$, $\mathfrak{h}_U^- \supseteq \Omega_-$ and

$$\omega \in \mathfrak{h}_U^0 \iff (\omega - \delta, \omega + \delta) \subset \mathfrak{h}_U \quad \text{and} \quad U(\mu)^* J U(\mu) = J$$
$$\text{for } \mu \in (\omega - \delta, \omega + \delta) \quad \text{for some } \delta > 0.$$

Moreover, U is holomorphic on the set

$$\Omega_U = \Omega_+ \cup \Omega_- \cup \mathfrak{h}_U^0 \tag{5.55}$$

and

$$U^\#(\lambda) J U(\lambda) = J \quad \text{for every point } \lambda \in \Omega_U. \tag{5.56}$$

Since U is J-contractive in $\mathfrak{h}_U^+$ and $-J$-contractive in $\mathfrak{h}_U^-$, the restrictions U_+ of U to $\mathbb{C}_+$ and U_- of U to $\mathbb{C}_-$ are meromorphic with bounded Nevanlinna characteristic in $\mathbb{C}_+$ and $\mathbb{C}_-$, respectively. Therefore, the limits

$$U_+(\mu) = \lim_{\nu \downarrow 0} U(\mu + i\nu) \quad \text{and} \quad U_-(\mu) = \lim_{\nu \downarrow 0} U(\mu - i\nu)$$

exist a.e. on $\mathbb{R}$ and, in view of (5.56), $U_-(\mu)^* J U_+(\mu) = J$ a.e. on $\mathbb{R}$. Thus a mvf $U \in \mathcal{P}^\circ(J)$ that is extended to $\mathbb{C}_-$ by the symmetry principle belongs to the class $\Pi^{m\times m}$ if and only it belongs to the class $\mathcal{U}(J)$.

Example 5.30 *Let $b(\lambda) = (\lambda - i)/(\lambda + i)$ for $\lambda \neq i$ and let $U(\lambda) = \mathrm{diag}\{b(\lambda)/2, 1\}$ in $\mathbb{C}_+$. Then $U \in \mathcal{P}^\circ(J)$, $\Omega_+ = \mathbb{C}_+ \setminus \{i\}$, $\Omega_- = \mathbb{C}_- \setminus \{-i\}$, $U_+ = U$, $U_-(\lambda) = \mathrm{diag}\{2b(\lambda), 1\}$. Thus, the mvf $U_-(\lambda)$, which is obtained by the symmetry principle is not the holomorphic extension of $U_+(\lambda)$.*

Theorem 5.31 *If $U \in \mathcal{P}^\circ(J)$, then:*

(1) *The formula (5.34) defines a positive kernel on $\mathfrak{h}_U \times \mathfrak{h}_U$. Every vvf in the RKHS $\mathcal{H}$ with RK $K_\omega(\lambda)$ is holomorphic on $\mathfrak{h}_U$.*

(2) *The RKHS $\mathcal{H}$ with RK $K_\omega(\lambda)$ is R_α invariant for every point $\alpha \in \Omega_U$.*

(3) *The de Branges identity (5.27) is in force for every choice of $\alpha, \beta \in \Omega_U$ and $f, g \in \mathcal{H}$.*

Proof To verify (1), it is necessary to show that

$$\sum_{s,t=1}^{n} x_t^* K_{\omega_s}(\omega_t) x_s \geq 0 \tag{5.57}$$

for every choice of points $\omega_1, \ldots, \omega_n \in \Omega_U$ and vectors $x_1, \ldots, x_n \in \mathbb{C}^m$. We shall treat the main case in which the sum (5.57) is considered with points in both Ω_+ and Ω_- and leave the other two cases to the reader. Without loss of generality, we may assume that $\omega_s \in \Omega_+$ for $s = 1, \ldots, k$ and $\omega_s \in \Omega_-$ for $s = k+1, \ldots, k+\ell$. Then $n = k + \ell$ and it is convenient to set $\alpha_s = \omega_s$

for $s = 1, \ldots, k$; and $\beta_t = \omega_{k+t}$ and $y_t = x_{k+t}$ for $t = 1, \ldots, \ell$. Then the sum (5.57) can be split into four pieces:

$$① = \sum_{i,j=1}^{k} x_j^* K_{\alpha_i}(\alpha_j) x_i, \quad ② = \sum_{t=1}^{\ell} \sum_{i=1}^{k} y_t^* K_{\alpha_i}(\beta_t) x_i,$$

$$③ = \sum_{j=1}^{\ell} \sum_{s=1}^{k} x_s^* K_{\beta_j}(\alpha_s) y_j \quad \text{and} \quad ④ = \sum_{t,j=1}^{\ell} y_t^* K_{\beta_j}(\beta_t) y_j.$$

Moreover, since

$$\begin{aligned} K_\beta(\alpha) &= \frac{J - U(\alpha) J U(\beta)^*}{\rho_\beta(\alpha)} \\ &= \frac{\{U(\overline{\beta}) - U(\alpha)\}}{-2\pi i(\alpha - \overline{\beta})} J U(\beta)^* \\ &= \frac{1}{2\pi i} \left(R_{\overline{\beta}} U \right)(\alpha) J U(\beta)^* \end{aligned}$$

for $\alpha \neq \overline{\beta}$,

$$③ = \sum_{j=1}^{\ell} \sum_{s=1}^{k} x_s^* \left(R_{\overline{\beta}_j} U \right)(\alpha_s) v_j \text{ and } ② = \sum_{t=1}^{\ell} \sum_{i=1}^{k} v_t^* \left(R_{\overline{\beta}_t} U \right)(\alpha_i)^* x_i,$$

with $2\pi i v_j = J U(\beta_j)^* y_j$ for $j = 1, \ldots, \ell$. On the other hand, since

$$\begin{aligned} ④ &= \sum_{t,j=1}^{\ell} y_t^* \left\{ \frac{J - U(\beta_t) J U(\beta_j)^*}{\rho_{\beta_j}(\beta_t)} \right\} y_j \\ &= \sum_{t,j=1}^{\ell} y_t^* U(\beta_t) J \left\{ \frac{U(\overline{\beta}_t)^* J U(\overline{\beta}_j) - J}{-2\pi i(\beta_t - \overline{\beta_j})} \right\} J U(\beta_j)^* y_j \\ &= 4\pi^2 \sum_{t,j=1}^{\ell} v_t^* \left\{ \frac{J - U(\overline{\beta}_t)^* J U(\overline{\beta}_j)}{\rho_{\overline{\beta}_t}(\overline{\beta}_j)} \right\} v_j \\ &\geq \left\| \sum_{j=1}^{\ell} (R_{\overline{\beta}_j} U) v_j \right\|^2, \end{aligned}$$

thanks to (5.23), it is readily seen that

$$①+②+③+④\geq \left\| \sum_{i=1}^{k} K_{\alpha_i} x_i + \sum_{j=1}^{\ell} \left(R_{\overline{\beta}_j} U \right) v_j \right\|^2 \geq 0.$$

Thus, the kernel $K_\omega(\lambda)$ is positive on $\Omega_U \times \Omega_U$ and hence also on $\mathfrak{h}_U \times \mathfrak{h}_U$. The rest of asserion (1) follows from Theorem 5.2 and Corollary 5.7, since $K_\omega(\omega)$ is continuous on $\mathfrak{h}_U$. Assertions (2) and (3) are immediate from Theorem 5.22, since Ω is symmetric with respect to $\mathbb{R}$. □

Remark 5.32 *If $U \in \mathcal{P}^\circ(J)$ has a holomorphic extension through some interval $(a, b) \subseteq \mathbb{R}$ and if $U(\mu)^* J U(\mu) = J$ for all points $\mu \in (a, b)$, then the assertions of Theorem 5.31 remain valid if Ω is replaced by the larger domain $\mathfrak{h}_U$.*

Theorem 5.33 *Let $\widetilde{\mathcal{H}}$ be a RKHS of $m \times 1$ vvf's on a subset $\widetilde{\Omega}$ of $\mathbb{C}$ with RK $K_\omega(\lambda)$ on $\widetilde{\Omega} \times \widetilde{\Omega}$ and let Ω be a nonempty subset of $\widetilde{\Omega}$. Then the set of restrictions $f = \widetilde{f}|_\Omega$ of $\widetilde{f} \in \widetilde{\mathcal{H}}$ to Ω is a RKHS $\mathcal{H}$ with RK that is the restriction of $K_\omega(\lambda)$ to $\Omega \times \Omega$ and the operator*

$$T_\Omega : \widetilde{f} \in \widetilde{\mathcal{H}} \longrightarrow \widetilde{f}|_\Omega,$$

is a partial isometry from $\widetilde{\mathcal{H}}$ onto $\mathcal{H}$ with $\ker T_\Omega = \{\widetilde{f} \in \widetilde{\mathcal{H}} : \widetilde{f}|_\Omega \equiv 0\}$. If Ω and $\widetilde{\Omega}$ are open and $\widetilde{\mathcal{H}}$ is a RKHS of holomorphic vvf's on $\widetilde{\Omega}$ and Ω has a nonempty intersection with each connected component of $\widetilde{\Omega}$, then T_Ω is unitary.

Proof In view of Theorem 5.2 and Corollary 5.7, the restriction of $K_\omega(\lambda)$ to $\Omega \times \Omega$ defines a unique RKHS $\mathcal{H}$ of holomorphic $m \times 1$ vvf's on Ω with the restricted $K_\omega(\lambda)$ as its RK. Moreover, with the help of Theorem 5.5, it is readily checked that

$$\|T_\Omega \widetilde{f}\|^2_{\mathcal{H}} \leq \|\widetilde{f}\|^2_{\widetilde{\mathcal{H}}} \quad \text{for every } \widetilde{f} \in \widetilde{\mathcal{H}},$$

i.e., T_Ω is a contraction. On the other hand, if $\omega_j \in \Omega$ and $u_j \in \mathbb{C}^m$ for $j = 1, \ldots, n$ and $\lambda \in \widetilde{\Omega}$, then

$$\widetilde{f}(\lambda) = \sum_{j=1}^{n} K_{\omega_j}(\lambda) u_j \quad \text{belongs to } \widetilde{\mathcal{H}}, \quad \widetilde{f}|_\Omega \in \mathcal{H}$$

and

$$\|\widetilde{f}\|_{\widetilde{\mathcal{H}}}^2 = \sum_{i,j=1}^{n} u_i^* K_{\omega_j}(\omega_i) u_i = \|\widetilde{f}|_{\mathcal{H}}\|_{\mathcal{H}}^2.$$

Then T_Ω maps a dense subspace of $\mathcal{H}$ isometrically onto the closure in $\widetilde{\mathcal{H}}$ of sums of the indicated form and is equal to zero on the orthogonal complement of this closure, as claimed. □

Theorem 5.34 *If $U \in \mathcal{P}^\circ(J)$ and $\mathcal{H} = \mathcal{H}(U)$ is the RKHS of holomorphic vvf's on $\mathfrak{h}_U$ that is considered in Theorem 5.31, then:*

(1) *The restrictions $f_\pm = f|_{\mathfrak{h}_U^\pm}$ belong to the Nevanlinna class of $m \times 1$ vvf's in $\mathbb{C}_\pm$, respectively, for every $f \in \mathcal{H}$.*

(2) *The limits $f_\pm(\mu) = \lim_{\nu \downarrow 0} f(\mu + i\nu)$ exist a.e. on $\mathbb{R}$ if $f \in \mathcal{H}$.*

(3) $U \in \mathcal{U}(J) \Longleftrightarrow \mathcal{H}(U) \subset \Pi^m$.

Proof (1) follows from Theorems 5.33 and 5.20 and (2) is immediate from (1). The condition $\mathcal{H}(U) \subset \Pi^m$ implies that that $f_+(\mu) = f_-(\mu)$ a.e. on $\mathbb{R}$ and hence, upon applying this to the vvf $f(\lambda) = K_\omega^U(\lambda)v$ for $\omega \in \mathfrak{h}_U \cap \mathfrak{h}_{U^{-1}}$ and $v \in \mathbb{C}^m$, that $U_+(\mu) = U_-(\mu)$ a.e. on $\mathbb{R}$. Thus, $U \in \mathcal{U}(J)$, since $J - U_-(\mu)JU_-(\mu)^* \leq 0$ and $J - U_+(\mu)JU_+(\mu)^* \geq 0$ a.e. on $\mathbb{R}$. The converse implication will be established later in Theorem 5.49. □

5.6 The space $\mathcal{H}(b)$ for $p \times p$ inner mvf's $b(\lambda)$

In this section the RKHS $\mathcal{H}(b)$ for $b \in \mathcal{S}_{in}^{p\times p}$ that was introduced earlier in Example 5.12 will be studied in more detail. Earlier, $\mathcal{H}(b)$ was defined as the RKHS of $p \times 1$ vvf's that are holomorphic in $\mathbb{C}_+$ with RK $\Lambda_\omega(\lambda)$ on $\mathbb{C}_+ \times \mathbb{C}_+$ defined by formula (5.15) with $s(\lambda) = b(\lambda)$. In view of assertion (5) of Theorem 5.14,

$$\mathcal{H}(b) = H_2^p \ominus bH_2^p$$

and the inner product in $\mathcal{H}(b)$ is just the standard inner product (5.11) in L_2^p.

Lemma 5.35 *If $b \in \mathcal{S}_{in}^{p\times p}$, then:*

(1) $b \in \Pi^{p\times p}$.

(2) $\mathcal{H}(b) \subseteq \Pi^p$.

Proof To verify the first assertion, recall that $b \in \mathcal{S}_{in}^{p\times p}$ admits a pseudocontinuation to the lower half plane $\mathbb{C}_-$ by means of the recipe

$$b(\lambda) = \{b^\#(\lambda)\}^{-1}$$

for every point $\lambda \in \mathbb{C}_-$ for which $b(\bar{\lambda})$ is invertible. Thus, as $b(\mu)$ is unitary for almost all points $\mu \in \mathbb{R}$,

$$\begin{aligned}\lim_{\varepsilon\downarrow 0} b(\mu - i\varepsilon) &= \lim_{\varepsilon\downarrow 0}\{b(\mu + i\varepsilon)^*\}^{-1} = \{b(\mu)^*\}^{-1} \\ &= b(\mu) = \lim_{\varepsilon\downarrow 0} b(\mu + i\varepsilon),\end{aligned}$$

for almost all points $\mu \in \mathbb{R}$.

Next, (2) follows from the fact that $\mathcal{H}(b) = H_2^p \cap bK_2^p$ and Corollary 3.107. □

Since every vvf $f \in \mathcal{H}(b)$ admits a pseudocontinuation to $\mathbb{C}_-$ that is holomorphic on the set $\mathfrak{h}_b$, the space $\mathcal{H}(b)$ may be viewed as a space of $p\times 1$ holomorphic vvf's on the set $\mathfrak{h}_b$ and not just on $\mathbb{C}_+$.

Lemma 5.36 *Let $b \in \mathcal{S}_{in}^{p\times p}$ and let the kernel $k_\omega^b(\lambda)$ be defined on $\mathfrak{h}_b \times \mathfrak{h}_b$ by the formula*

$$k_\omega^b(\lambda) = \begin{cases} \dfrac{I_p - b(\lambda)b(\omega)^*}{\rho_\omega(\lambda)} & \text{if} \quad \lambda \neq \overline{\omega} \\ \dfrac{b'(\overline{\omega})b(\omega)^*}{2\pi i} & \text{if} \quad \lambda = \overline{\omega} \end{cases}. \tag{5.58}$$

Then:

(1) *$k_\omega^b(\lambda)$ is a positive kernel on $\mathfrak{h}_b \times \mathfrak{h}_b$.*

(2) *$\|k_\omega^b(\omega)\|$ is continuous in $\mathfrak{h}_b$.*

Proof The first goal is to check that the kernel $k_\omega^b(\lambda)$ that is defined for $b \in \mathcal{S}_{in}^{p\times p}$ on $\mathfrak{h}_b \times \mathfrak{h}_b$ by formula (5.58) is positive. It is enough to check that this kernel is positive on the set $\Omega \times \Omega$, where $\Omega = \Omega_+ \cup \Omega_-$, $\Omega_+ = \{\lambda \in \mathbb{C}_+ : \det b(\lambda) \neq 0\}$ and $\Omega_-\{\overline{\lambda} \colon \lambda \in \Omega_+\}$, since this kernel is continuous on $\mathfrak{h}_b \times \mathfrak{h}_b$. But this follows from (1) of Theorem 5.31 with $J = I_p$. The second assertion is self-evident. □

Our next objective is to show that if $b \in \mathcal{S}_{in}^{p\times p}$, then

(1) $\cap_{f\in\mathcal{H}(b)}\mathfrak{h}_f = \mathfrak{h}_b$.

(2) The space of vvf's in $\mathcal{H}(b)$ extended by pseudocontinuation into all of $\mathfrak{h}_b$ coincides with the RKHS $\mathcal{H}$ of vvf's on $\mathfrak{h}_b$ with RK $k_\omega^b(\lambda)$ defined on $\mathfrak{h}_b \times \mathfrak{h}_b$.

Lemma 5.37 *Let $b \in \mathcal{S}_{in}^{p\times p}$. Then $R_\alpha b\xi \in \mathcal{H}(b)$ for every choice of $\alpha \in \mathfrak{h}_b$ and $\xi \in \mathbb{C}^p$.*

Proof The stated assertion is readily verified by direct calculation when $\alpha \in \mathbb{C}_+$. On the other hand, if $\alpha \in \mathbb{C}_- \cap \mathfrak{h}_b$, then $\overline{\alpha} \in \mathfrak{h}_b$, $b(\alpha)b(\overline{\alpha})^* = I_p$ and the desired conclusion follows from the identity

$$(R_\alpha b)(\lambda) = 2\pi i k_{\overline{\alpha}}^b(\lambda)b(\alpha), \tag{5.59}$$

which expresses the left hand side in terms of the RK of $\mathcal{H}(b)$.

If $\alpha \in \mathbb{R} \cap \mathfrak{h}_b$ and $\xi \in \mathbb{C}^p$, then another argument can be based on the observation that

$$R_\alpha b\xi = \frac{b(\lambda) - b(\alpha)}{\lambda - \alpha}\xi \in H_2^p$$

and

$$b^{-1}R_\alpha b\xi = \frac{I_p - b(\lambda)^{-1}b(\alpha)}{\lambda - \alpha}\xi = -\frac{b^\#(\lambda) - b^\#(\alpha)}{\lambda - \alpha}b(\alpha)\xi \in K_2^p.$$

Thus, $R_\alpha b\xi \in \mathcal{H}(b)$. □

Lemma 5.38 *Let $b \in \mathcal{S}_{in}^{p\times p}$ and let $d(\lambda) = \det b(\lambda)$. Then*

$$\mathfrak{h}_b = \mathfrak{h}_d \quad \text{and} \quad \mathcal{H}(b) \subseteq \mathcal{H}(dI_p). \tag{5.60}$$

Proof Clearly $\mathfrak{h}_b \subseteq \mathfrak{h}_d$. Moreover, since $b(\lambda)$ and $d(\lambda)$ are both holomorphic in $\mathbb{C}_+$, it remains only to check the opposite inclusion in $\mathbb{R} \cup \mathbb{C}_-$.

Suppose first that $\overline{\omega} \in \mathfrak{h}_d \cap \mathbb{C}_-$. Then, since $d(\omega)d(\overline{\omega})^* = 1$, $d(\omega) \neq 0$. Therefore, $b(\omega)$ is invertible and $\overline{\omega} \in \mathfrak{h}_b$ also.

Next, if $\omega \in \mathfrak{h}_d \cap \mathbb{R}$, then $d(\lambda)$ is holomorphic in a neighborhood of ω and hence (since $|d(\mu)| = 1$ for almost all points $\mu \in \mathbb{R}$), $|d(\lambda)| \geq \varepsilon > 0$ in a possibly smaller neighborhood of ω. Thus $b(\lambda)$ is holomorphic, invertible and bounded in the intersection of a small rectangle $\mathring{R}$ centered at ω with $\mathbb{C}_+ \cup \mathbb{C}_-$. The integral

$$\frac{1}{2\pi i}\int_\Gamma b(\zeta)(\zeta - \lambda)^{-1} d\zeta$$

around the boundary Γ of $\mathring{R}$ defines a holomorphic mvf in the interior of $\mathring{R}$ which coincides with $b(\lambda)$ in $\mathring{R} \cap (\mathbb{C}_+ \cup \mathbb{C}_-)$. Therefore, $b(\lambda)$ may be presumed to be holomorphic in $\mathring{R}$; see the proof of Lemma 5.39 below for more details and references.

Finally, the second assertion in (5.60) drops out easily from the observation that $db^{-1} \in \mathcal{S}_{in}^{p\times p}$ and hence that $dH_2^p \subseteq bH_2^p$. □

Lemma 5.39 *Let $b \in \mathcal{S}_{in}^{p\times p}$. Then $\omega \in \mathfrak{h}_b$ if and only if every $f \in \mathcal{H}(b)$ is holomorphic at ω, i.e.,*

$$\mathfrak{h}_b = \cap_{f\in\mathcal{H}(b)} \mathfrak{h}_f.$$

Proof Let $f \in \mathcal{H}(b)$. Then clearly

$$\mathfrak{h}_f^+ \;=\; \mathbb{C}_+ \;=\; \mathfrak{h}_b^+.$$

Moreover, since $f = bf_-$ for some mvf $f_- \in K_2^p$ and f_- is holomorphic in $\mathbb{C}_-$,

$$\mathfrak{h}_b^- \subseteq \mathfrak{h}_f^-$$

for every $f \in \mathcal{H}(b)$. On the other hand, the inclusion is reversed for $f = R_\alpha b\xi$, which belongs to $\mathcal{H}(b)$ for every $\alpha \in \mathfrak{h}_b$ and $\xi \in \mathbb{C}^p$ by Lemma 5.37.

Finally, if $\omega \in \mathbb{R} \cap \mathfrak{h}_b$, then $b(\lambda)$ is analytic in a small rectangle (with nonempty interior) centered at ω which intersects $\mathbb{R}$ in the interval $[a_1, a_2]$. Let $f \in \mathcal{H}(b)$. Then on the one hand, since $f \in H_2^p$, it follows that f is analytic in $\mathbb{C}_+$ and

$$\int_{a_1}^{a_2} \|f(\mu + i\nu) - f(\mu)\| d\mu \leq \left\{ (a_2 - a_1) \int_{-\infty}^{\infty} \|f(\mu + i\nu) - f(\mu)\|^2 d\mu \right\}^{\frac{1}{2}},$$

which tends to zero as $\nu \downarrow 0$. On the other hand the representation $f = bf_-$ with $f_- \in K_2^p$ insures that f is analytic in the intersection of the rectangle with $\mathbb{C}_-$. Moreover, it is not hard to show that

$$\int_{a_1}^{a_2} \|b(\mu - i\nu)f_-(\mu - i\nu) - b(\mu)f_-(\mu)\| d\mu \longrightarrow 0$$

as $\nu \downarrow 0$. Therefore, by a well known argument based on Cauchy's formula, it follows that every $f \in \mathcal{H}(b)$ can be continued analytically across the interval (a_1, a_2); see e.g., Carleman [Car67] and pp. 223–224 of Sz.-Nagy-Foias [SzNF70]. This proves that $\mathbb{R} \cap \mathfrak{h}_b \subseteq \mathbb{R} \cap \mathfrak{h}_f$ for every $f \in \mathcal{H}(b)$. The proof is now easily completed by resorting to functions of the form $R_\alpha b\xi$

with $\alpha \in \mathfrak{h}_b$ and $\xi \in \mathbb{C}^p$ in order to exhibit functions in $\mathcal{H}(b)$ for which the inclusion is reversed, just as before. □

Lemmas 5.35–5.39 are covered by the next theorem.

Theorem 5.40 *Let $b \in \mathcal{S}_{in}^{p\times p}$ and $\mathcal{H}(b) = H_2^p \ominus bH_2^p$. Then:*

(1) *Every vvf $f \in \mathcal{H}(b)$ admits a pseudocontinuation to $\mathbb{C}_-$, i.e.,*

$$\mathcal{H}(b) \subset \Pi^p.$$

(2) $\mathfrak{h}_b = \cap_{f\in\mathcal{H}(b)}\mathfrak{h}_f$.

Moreover,

(3) *The space $\mathcal{H}(b)$ is a RKHS of holomorphic $p \times 1$ vvf's on $\mathfrak{h}_b$ with respect to the standard inner product in $L_2^p(\mathbb{R})$ and the RK $k_\omega^b(\lambda)$ that is defined on $\mathfrak{h}_b \times \mathfrak{h}_b$ by formula (5.58).*

(4) *The space $\mathcal{H}(b)$ is R_α invariant for every point $\alpha \in \mathfrak{h}_b$.*

(5) *R_α is a bounded linear operator on $\mathcal{H}(b)$ for every $\alpha \in \mathfrak{h}_b$.*

(6) *$R_\alpha b\xi \in \mathcal{H}(b)$ for every point $\alpha \in \mathfrak{h}_b$ and every vector $\xi \in \mathbb{C}^p$.*

Proof Assertions (1), (2) and (5) are covered by the preceding lemmas. The verification of (4) for $\alpha \in \mathbb{C}_+$ is quick and easy. It rests on the observation that if $\delta_\alpha(\lambda) = \lambda - \alpha$, then

$$R_\alpha f \;=\; \Pi_+ \frac{f}{\delta_\alpha} \tag{5.61}$$

for $f \in H_2^p$ and $\alpha \in \mathbb{C}_+$. Thus, if $g \in \mathcal{H}(b)$ and $\alpha \in \mathbb{C}_+$, then $R_\alpha g \in \mathcal{H}(b)$, since $R_\alpha g \in H_2^p$ and

$$\langle R_\alpha g,\, bf\rangle_{st} = \langle \frac{g}{\delta_\alpha},\, bf\rangle_{st} = \langle g,\, b\frac{f}{\delta_{\bar{\alpha}}}\rangle_{st} = 0,$$

for every $f \in H_2^p$.

Next, if $\alpha \in \mathbb{C}_- \cap \mathfrak{h}_b$ and $g \in \mathcal{H}(b)$, then $b(\alpha)b(\overline{\alpha})^* = I_p$ and, by Lemma 5.39, $\alpha \in \mathfrak{h}_g$. Therefore, upon writing $g = bf$ with $f \in K_2^p$, it follows that $\alpha \in \mathfrak{h}_f$ and hence, from Lemma 5.37 and the formula

$$(R_\alpha g)(\lambda) = b(\lambda)(R_\alpha f)(\lambda) + (R_\alpha b)(\lambda)f(\alpha), \tag{5.62}$$

that $R_\alpha g \in bK_2^p$. The proof is now easily completed by checking that $R_\alpha g \in H_2^p$.

Finally, the argument for $\alpha \in \mathbb{C}_- \cap \mathfrak{h}_b$ works for $\alpha \in \mathbb{R} \cap \mathfrak{h}_b$ also in view of the fact that if $g \in H_2^p$ and $\alpha \in \mathbb{R} \cap \mathfrak{h}_g$, then $R_\alpha g \in H_2^p$ and, similarly, if $f \in K_2^p$ and $\alpha \in \mathbb{R} \cap \mathfrak{h}_f$, then $R_\alpha f \in K_2^p$.

It remains only to verify that the RKHS $\mathcal{H}$ with RK $k_\omega^b(\lambda)$ on $\mathfrak{h}_b \times \mathfrak{h}_b$ coincides with the space of vvf's $f \in \mathcal{H}(b)$ considered on $\mathfrak{h}_b$ instead of just $\mathbb{C}_+$. But this follows from Theorem 5.31. □

Lemma 5.41 *Let $b \in \mathcal{S}_{in}^{p\times p}$ and let $\mathcal{H}(b)$ denote the RKHS of $p \times 1$ vvf's on $\mathfrak{h}_b$ with RK $k_\omega^b(\lambda)$ that is defined on $\mathfrak{h}_b \times \mathfrak{h}_b$ by formula (5.58). Then the sets $\mathcal{N}_\omega$, $\mathcal{R}_\omega$ and $\mathcal{F}_\omega$ defined in Lemma 5.3 with $K_\omega(\lambda) = k_\omega^b(\lambda)$ are independent of the choice of the point $\omega \in \mathfrak{h}_b$. Moreover, the following conditions are equivalent:*

(1) *$k_\omega^b(\omega) > 0$ for at least one point $\omega \in \mathfrak{h}_b$.*

(2) *$k_\omega^b(\omega) > 0$ for every point $\omega \in \mathfrak{h}_b$.*

(3) *$\{f(\omega) : f \in \mathcal{H}(b)\} = \mathbb{C}^p$ for at least one point $\omega \in \mathfrak{h}_b$.*

(4) *$\{f(\omega) : f \in \mathcal{H}(b)\} = \mathbb{C}^p$ for every point $\omega \in \mathfrak{h}_b$.*

Proof The stated results follow from (4) of Theorem 5.40 and Lemma 5.4. □

5.7 The space $\mathcal{H}_*(b)$ for $p \times p$ inner mvf's $b(\lambda)$

If $b \in \mathcal{S}_{in}^{p\times p}$, then the space $\mathcal{H}_*(b)$ of $p\times 1$ vvf's on $\mathbb{C}_-$ considered in Example 5.16 may be extended to a space of vvf's on $\mathfrak{h}_{b^\#}$:

Theorem 5.42 *Let $b \in \mathcal{S}_{in}^{p\times p}$, let $\mathcal{H}_*(b) = K_2^p \ominus b^\# K_2^p$ and let the RK $\ell_\omega^b(\lambda)$ be defined on $\mathfrak{h}_{b^\#} \times \mathfrak{h}_{b^\#}$ by the formula*

$$\ell_\omega^b(\lambda) = \begin{cases} \dfrac{b^\#(\lambda)b^\#(\omega)^* - I_p}{\rho_\omega(\lambda)} & \text{if } \lambda \neq \overline{\omega} \\ \dfrac{1}{2\pi i} b'(\omega)^* b(\overline{\omega}) & \text{if } \lambda = \overline{\omega} \end{cases} . \tag{5.63}$$

Then:

(1) *Every vvf $g \in \mathcal{H}_*(b)$ admits a pseudocontinuation into $\mathbb{C}_+$ with bounded Nevanlinna characteristic, i.e., $\mathcal{H}_*(b) \subseteq \Pi^p$.*

(2) *$\mathfrak{h}_{b^\#} = \cap_{g\in\mathcal{H}_*(b)} \mathfrak{h}_g$.*

Moreover:

(3) *The space* $\mathcal{H}_*(b)$ *of* $p \times 1$ *holomorphic vvf's on* $\mathfrak{h}_{b^\#}$ *is a RKHS with respect to the standard inner product in* L_2^p, *and the RK* $\ell_\omega^b(\lambda)$ *is defined by formula (5.63).*

(4) *The space* $\mathcal{H}_*(b)$ *is* R_α *invariant for every point* $\alpha \in \mathfrak{h}_{b^\#}$.

(5) R_α *is a bounded linear operator on* $\mathcal{H}_*(b)$ *for every* $\alpha \in \mathfrak{h}_{b^\#}$.

(6) $R_\alpha b^\# \xi \in \mathcal{H}_*(b)$ *for every point* $\alpha \in \mathfrak{h}_{b^\#}$ *and vector* $\xi \in \mathbb{C}^p$.

Proof Let $f_\sim(\lambda) = \overline{f(\overline{\lambda})}$. Then $\mathfrak{h}_{b^\tau} = \mathfrak{h}_b$ and:

(a) $f \in \mathcal{H}_*(b) \Longleftrightarrow f_\sim \in \mathcal{H}(b^\tau)$.

(b) $\alpha \in \mathfrak{h}_{b^\#} \Longleftrightarrow \overline{\alpha} \in \mathfrak{h}_{b^\tau}$.

(c) $(R_\alpha f)_\sim(\lambda) = (R_{\overline{\alpha}} f_\sim)(\lambda)$.

(d) $(R_\alpha b^\# \xi)_\sim(\lambda) = (R_{\overline{\alpha}} b^\tau)(\lambda)\xi$.

(e) If $\omega \in \mathfrak{h}_{b^\#}$, then $\overline{\omega} \in \mathfrak{h}_b$ and

$$(\ell_\omega^b)_\sim(\lambda) = \frac{I_p - b^\tau(\lambda)b^\tau(\overline{\omega})^*}{-2\pi i(\lambda - \omega)} = k_{\overline{\omega}}^{b^\tau}(\lambda) \quad \text{if } \lambda \neq \omega.$$

Thus, all the stated assertions follow from the corresponding assertions in Theorem 5.40 applied to the mvf b^τ in place of b. □

Lemma 5.43 *Let* $b \in \mathcal{S}_{in}^{p\times p}$. *Then the operator* M_b *of multiplication by the mvf* $b(\mu)$ *is unitary in* L_2^p *and maps the subspace* $\mathcal{H}_*(b)$ *onto the subspace* $\mathcal{H}(b)$.

Proof The proof is easy and is left to the reader. □

5.8 The space $\mathcal{H}(U)$ for $m \times m$ J-inner mvf's $U(\lambda)$

Lemma 5.44 *Let* $U \in \mathcal{U}(J)$, *where* J *is an* $m \times m$ *signature matrix. Then the kernel* $K_w^U(\lambda)$ *that is defined on* $\mathfrak{h}_U \times \mathfrak{h}_U$ *by formula (5.34) is positive on* $\mathfrak{h}_U \times \mathfrak{h}_U$.

Proof This follows from Theorem 5.31. □

In view of Theorem 5.2 and Corollary 5.7, there is a unique RKHS $\mathcal{H}$ of holomorphic vvf's on $\mathfrak{h}_U$ with RK $K_\omega^U(\lambda)$ on $\mathfrak{h}_U \times \mathfrak{h}_U$. Consequently,

$$\langle f, K_\omega^U \xi\rangle_{\mathcal{H}} = \xi^* f(w) \quad \text{for every } f \in \mathcal{H},\ \xi \in \mathbb{C}^m \text{ and } \omega \in \mathfrak{h}_U. \tag{5.64}$$

At the same time, since $\mathcal{U}(J) \subset \mathcal{P}(J)$, to each mvf $U \in \mathcal{U}(J)$ there corresponds a RKHS $\mathcal{H}(U)$ of vvf's that are holomorphic in $\mathfrak{h}_U^+$ with RK $K_\omega^U(\lambda)$ on $\mathfrak{h}_U^+ \times \mathfrak{h}_U^+$ that is defined by formula (5.34).

Let

$$L(\lambda) = (P - S(\lambda)Q)^{-1} \quad \text{for } \lambda \in \Omega, \tag{5.65}$$

where $J \neq \pm I_m$, $S = PG(U)$ and

$$\Omega = \begin{cases} \mathfrak{h}_U \cap \mathfrak{h}_S & \text{if } U \in \mathcal{U}(J) \\ \mathfrak{h}_U^+ & \text{if } U \in \mathcal{P}(J) \setminus \mathcal{U}(J) \end{cases}.$$

Then

$$K_\omega^U(\lambda) = L(\lambda)K_\omega^S(\lambda)L(\omega)^* \quad \text{on } \Omega \times \Omega. \tag{5.66}$$

Theorem 5.45 *Let J be an $m \times m$ signature matrix, $U \in \mathcal{U}(J)$, $S = PG(U)$,*

$$G_\ell(\mu) = P + U(\mu)QU(\mu)^* \quad a.e. \text{ on } \mathbb{R}$$

and let $L(\lambda)$ be defined by (5.65). Then:

(1) *Every vvf $f \in \mathcal{H}(U)$ belongs to Π^m and, consequently, has nontangential limit values a.e. on $\mathbb{R}$.*

(2) *The formula*

$$f(\lambda) = L(\lambda)g(\lambda), \quad g \in \mathcal{H}(S) \tag{5.67}$$

defines a unitary operator, acting from $\mathcal{H}(S)$ onto $\mathcal{H}(U)$, i.e., $f \in \mathcal{H}(U)$ if and only if $(P - SQ)f \in H_2^m$ and $(Q - S^\# P)f \in K_2^m$ and, if $f \in \mathcal{H}(U)$, then

$$\| f \|_{\mathcal{H}(U)}^2 = \| (P - SQ)f \|_{st}^2 = \|G_\ell^{-1/2} f\|_{st}^2.$$

Proof If $J = I_m$, then $S = U$ and there is nothing to prove. If $J \neq I_m$, then the identities

$$\sum_{j=1}^n K_{\omega_j}^U(\lambda)\xi_j = L(\lambda) \sum_{j=1}^n K_{\omega_j}^S(\lambda)L(\omega_j)^*\xi_j \tag{5.68}$$

and

$$\sum_{i,j=1}^n \xi_i^* K_{w_j}^U(w_i)\xi_i = \sum_{i,j=l}^n \xi_i^* L(\omega_i)K_{\omega_j}^S(\omega_i)L(\omega_j)^*\xi_j, \tag{5.69}$$

which are in force for every choice of points $\omega_i, \ldots, \omega_n$ in $\mathfrak{h}_U \cap \mathfrak{h}_S$ and vectors $\xi_1, \ldots, \xi_n$ in $\mathbb{C}^p$, imply that formula (5.67) defines an isometric operator from the linear manifold $\mathcal{L}_1$ of vvf's $g \in \mathcal{H}(S)$ of the form

$$g(\lambda) = \sum_{j=1}^{n} K_{\omega_j}^S(\lambda) L(\omega_j)^* \xi_j \tag{5.70}$$

that is dense in $\mathcal{H}(S)$ into the linear manifold $\mathcal{L}_2$ of vvf's $f \in \mathcal{H}(U)$ of the form

$$f(\lambda) = \sum_{i,j=1}^{n} K_{\omega_j}^U(\lambda)\xi_j, \tag{5.71}$$

which is dense in $\mathcal{H}(U)$, since

$$\langle f, f \rangle_{\mathcal{H}(U)} = \langle g, g \rangle_{\mathcal{H}(S)} = \langle g, g \rangle_{st}.$$

Next to check that the unitary extension of the isometric operator is given by (5.67), let $g \in \mathcal{H}(S)$ and $g_n \in \mathcal{L}_1$ be such that $\lim_{n\to\infty} g_n = g$ in the space $\mathcal{H}(S)$, i.e., in the L_2^m metric, and let $f_n(\lambda) = L(\lambda) g_n(\lambda)$. Then $f_n \in \mathcal{L}_2$ and $\lim_{n\to\infty} f_n = f$ in the space $\mathcal{H}(U)$ for some $f \in \mathcal{H}(U)$. Thus,

$$\begin{aligned} \xi^* f(w) &= \langle f, K_\omega^U \xi \rangle_{\mathcal{H}(U)} = \lim_{n\to\infty} \langle f_n, K_\omega^U \xi \rangle_{\mathcal{H}(U)} \\ &= \lim_{n\to\infty} \langle L g_n, K_\omega^U \xi \rangle_{\mathcal{H}(U)} = \lim_{n\to\infty} \langle g_n, K_\omega^S L(w)^* \xi \rangle_{\mathcal{H}(S)} \\ &= \langle g, K_\omega^S L(w)^* \xi \rangle_{\mathcal{H}(S)} = \xi^* L(w) g(w) \end{aligned}$$

for every $w \in \mathfrak{H}_U \bigcap \mathfrak{h}_S$ and $\xi \in \mathbb{C}^m$. Consequently, $f(\lambda) = L(\lambda) g(\lambda)$. for $\lambda \in \mathfrak{h}_S \bigcap \mathfrak{h}_U$. Thus, formula (5.67) defines a unitary operator acting from $\mathcal{H}(S)$ onto $\mathcal{H}(U)$. It is easily checked that

$$(P - S(\mu)Q)^*(P - S(\mu)Q) = G_\ell(\mu)^{-1} \quad \text{a.e. on } \mathbb{R}.$$

Assertion (1) follows from (5.67), since $L \in \Pi^{m\times m}$ and (as shown in Theorem 5.40) $\mathcal{H}(S) \subset \Pi^m$. □

Corollary 5.46 *If $U \in \mathcal{U}(J)$ is rational and $S = PG(U)$, then*

$$\dim \mathcal{H}(U) = \dim \mathcal{H}(S) < \infty.$$

Proof This is immediate from Theorem 5.45 and Lemma 3.47. □

Remark 5.47 *If $U \in \mathcal{P}(J)$ and $S = PG(U)$, then formula (5.66) is valid in $\mathfrak{h}_U^+ \times \mathfrak{h}_U^+$ and multiplication by L is a unitary operator from $\mathcal{H}(S)$ onto $\mathcal{H}(U)$:*

$$\|(P - SQ)^{-1}g\|_{\mathcal{H}(U)} = \|g\|_{\mathcal{H}(S)} \quad \text{for every } g \in \mathcal{H}(s).$$

Since $\mathcal{H}(S) \subseteq H_2^m$, this implies that $\mathcal{H}(U) \subset \mathcal{N}^m$ and hence that every $f \in \mathcal{H}(U)$ has nontangential boundary values $f(\mu)$ a.e. on $\mathbb{R}$. If $(P - SQ) \in \mathcal{N}_{out}^{m\times m}$, then $\mathcal{H}(U) \subset \mathcal{N}_+^m$.

Lemma 5.48 *Let $U \in \mathcal{U}(J)$, $J \neq \pm I_m$, let $S = PG(U)$ and let*

$$L_*(\lambda) = (S^\#(\lambda)P - Q)^{-1}. \tag{5.72}$$

Then the formula

$$f(\lambda) = L_*(\lambda)h(\lambda) \quad \text{with} \quad h \in \mathcal{H}_*(S) \tag{5.73}$$

defines a unitary map L_* from $\mathcal{H}_*(S)$ onto $\mathcal{H}(U)$.

Proof Since $S \in \mathcal{S}_{in}^{m\times m}$

$$L_*(\lambda) = L(\lambda)S(\lambda), \tag{5.74}$$

and, by Lemma 5.43, the operator M_S of multiplication by S maps $\mathcal{H}_*(S)$ isometrically onto $\mathcal{H}(S)$, whereas M_L defines a unitary operator from $\mathcal{H}(S)$ onto $\mathcal{H}(U)$, by Theorem 5.45. □

Theorem 5.49 *Let $U \in \mathcal{U}(J)$. Then:*

(1) *Every vvf $f \in \mathcal{H}(U)$ admits a pseudocontinuation to $\mathbb{C}_-$, i.e.,*
$$\mathcal{H}(U) \subset \Pi^m.$$

(2) $\mathfrak{h}_U = \cap_{f\in\mathcal{H}(U)}\mathfrak{h}_f$.

Moreover, in view of (1) and (2):

(3) *The space $\mathcal{H}(U)$ is a RKHS of holomorphic $m \times 1$ vvf's on $\mathfrak{h}_U$ with RK $k_\omega^U(\lambda)$ that is defined on $\mathfrak{h}_U \times \mathfrak{h}_U$ by formula (5.34).*

(4) *The space $\mathcal{H}(U)$ is R_α invariant for every point $\alpha \in \mathfrak{h}_U$.*

(5) *The operator R_α is bounded in $\mathcal{H}(U)$ for every $\alpha \in \mathfrak{h}_U$.*

(6) *$(R_\alpha U)\xi \in \mathcal{H}(U)$ for every point $\alpha \in \mathfrak{h}_U$ and vector $\xi \in \mathbb{C}^m$.*

Proof The cases $J = \pm I_m$ are covered by Theorems 5.40 and 5.42. Suppose, therefore, that $J \neq \pm I_m$, let $S = PG(U)$ and let $L(\lambda)$ be defined by formula (5.65). Then (1) is established in Theorem 5.45.

Next, fix a point $\omega \in \mathfrak{h}_U$ at which U is invertible. Then, since $K_\omega u \in \mathcal{H}(U)$ for every vector $u \in \mathbb{C}^m$, the formula

$$\rho_\omega(\lambda)K_\omega(\lambda) = J - U(\lambda)JU(\omega)^*$$

clearly implies that

$$\bigcap_{f\in\mathcal{H}(U)} \mathfrak{h}_f \subseteq \mathfrak{h}_U.$$

The proof of the opposite inclusion is more complicated. Formulas (4.2), (4.3) and (5.65) imply that

$$\mathfrak{h}_U \cap \overline{\mathbb{C}_+} = \mathfrak{h}_S \cap \mathfrak{h}_L \cap \overline{\mathbb{C}_+} \tag{5.75}$$

and hence, in view of formula (5.67) and Theorem 5.40, that

$$\mathfrak{h}_U \cap \overline{\mathbb{C}_+} \subseteq \mathfrak{h}_f \cap \overline{\mathbb{C}_+} \quad \text{for every } f \in \mathcal{H}(U).$$

In much the same way, Lemma 5.48 implies that

$$\mathfrak{h}_U \cap \mathbb{C}_- = \mathfrak{h}_{S^\#} \cap \mathfrak{h}_{L_*} \cap \mathbb{C}_- = \mathfrak{h}_{L_*} \cap \mathbb{C}_- \tag{5.76}$$

and hence that

$$\mathfrak{h}_U \cap \mathbb{C}_- \subseteq \mathfrak{h}_f \cap \mathbb{C}_- \quad \text{for every } f \in \mathcal{H}(U),$$

which serves to complete the proof of (2).

Assertion (4) now follows from formulas (5.75), (5.67), (5.26) and Theorem 5.40 if $\alpha \in \mathfrak{h}_U \cap \overline{\mathbb{C}_+}$, and from formulas (5.76), (5.73), Theorem 5.42, Lemma 5.48 and the identity

$$(R_\alpha f)(\lambda) = L_*(\lambda)\{(R_\alpha h)(\lambda) - (R_\alpha S^\#)(\lambda)Ph(\alpha)\},$$

if $\alpha \in \mathfrak{h}_U \cap \mathbb{C}_-$. Assertion (5) follows from (4) and Theorem 5.8.

To verify (6), suppose first that $\alpha \in \mathfrak{h}_U \cap \overline{\mathbb{C}_+}$. Then the formula

$$(R_\alpha U)(\lambda) = L(\lambda)(R_\alpha S)(\lambda)(P + QS(\alpha))^{-1},$$

which is established in the proof of Theorem 5.20 for α and λ in $\mathfrak{h}_U^+$, is still valid for $\alpha \in \mathfrak{h}_S \cap \mathfrak{h}_L \cap \overline{\mathbb{C}_+}$ (the same proof works even if $\alpha \in \mathfrak{h}_S \cap \mathfrak{h}_L \cap \mathbb{R}$).

Therefore, by Theorems 5.45 and 5.40, $(R_\alpha U)u \in \mathcal{H}(U)$ for every vector $u \in \mathbb{C}^m$. Now suppose that $\alpha \in \mathfrak{h}_U \cap \mathbb{C}_-$. Then $\alpha \in \mathfrak{h}_{L_*} \cap \mathbb{C}_-$, and, since

$$U(\lambda) = (S^\#(\lambda)P - Q)^{-1}(P - S^\#(\lambda)Q) = (QS^\#(\lambda) + P)(PS^\#(\lambda) + Q)^{-1},$$

the formula

$$(R_\alpha U)(\lambda) = -L_*(\lambda)(R_\alpha S^\#)(\lambda)(PS^\#(\alpha) + Q)^{-1}$$

and Lemmas 5.48 and 5.43 imply that $(R_\alpha U)u \in \mathcal{H}(U)$ for every vector $u \in \mathbb{C}^m$. □

Theorem 5.50 *Let $U \in \mathcal{U}(J)$ and let $\mathcal{L}$ be a closed subspace of $\mathcal{H}(U)$ that is R_α invariant for every point $\alpha \in \mathfrak{h}_U$. Then there exists a mvf $U_1 \in \mathcal{U}(J)$ such that $\mathfrak{h}_{U_1} \supseteq \mathfrak{h}_U$, $\mathcal{L} = \mathcal{H}(U_1)$ and $U_1^{-1}U \in \mathcal{U}(J)$. Moreover, the space $\mathcal{H}(U_1)$ is isometrically included in $\mathcal{H}(U)$, and*

$$\mathcal{H}(U) = \mathcal{H}(U_1) \oplus U_1\mathcal{H}(U_2), \quad \text{where} \quad U_2 = U_1^{-1}U. \tag{5.77}$$

Proof Let u_j denote the jth column of I_m for $j = 1, \ldots, m$ and let $\omega \in \mathfrak{h}_U$. Then, since the formula $u_j^* f(\omega)$ defines a bounded linear functional on $\mathcal{L}$, the Riesz representation theorem guarantees the existence of a set of vectors $\{g_{j,\omega}\} \in \mathcal{L}$, $j = 1, \ldots, m$, such that

$$u_j^* f(\omega) = \langle f, g_{j,\omega} \rangle_{\mathcal{H}(U)} \quad \text{for} \quad j = 1, \ldots, m \quad \text{and for every} \quad f \in \mathcal{L}.$$

Thus, the $m \times m$ mvf

$$G_\omega(\lambda) = \begin{bmatrix} g_{1,\omega}(\lambda) & \cdots & g_{m,\omega}(\lambda) \end{bmatrix}$$

defines a RK on $\mathcal{L}$, i.e.,

$$G_\omega u \in \mathcal{L} \quad \text{and} \quad u^* f(\omega) = \langle f, G_\omega u \rangle_{\mathcal{H}(U)} \quad \text{for every } u \in \mathbb{C}^m.$$

Consequently, Theorem 5.21 is applicable to the RKHS $\mathcal{L}$ with RK $G_\omega(\lambda)$ on $\mathfrak{h}_U \times \mathfrak{h}_U$, and hence $\mathcal{L} = \mathcal{H}(U_1)$ for some $m \times m$ mvf $U_1 \in \mathcal{P}(J)$ that is holomorphic in $\mathfrak{h}_U$; and $\mathcal{H}(U_1)$ is isometrically included in $\mathcal{H}(U)$. Moreover, the evaluations

$$|\langle G_\alpha u, K_\alpha u \rangle|^2 \leq \langle G_\alpha u, G_\alpha u \rangle \, \langle K_\alpha u, K_\alpha u \rangle$$

imply that

$$\{u^* G_\alpha(\alpha) u\}^2 \leq u^* G_\alpha(\alpha) u \, u^* K_\alpha(\alpha) u$$

for every $\alpha \in \mathfrak{h}_U$ and every $u \in \mathbb{C}^m$, and hence that

$$u^*\{J - U_1(\alpha)JU_1(\alpha)^*\}u \le u^*\{J - U(\alpha)JU(\alpha)^*\}u$$

for every $\alpha \in \mathfrak{h}_U^+$ and every $u \in \mathbb{C}^m$. Therefore, $U \in \mathcal{U}(J)$ and

$$u^*U(\alpha)JU(\alpha)^*u \le u^*U_1(\alpha)JU_1(\alpha)^*u$$

for every $\alpha \in \mathfrak{h}_U^+$ and every $u \in \mathbb{C}^m$. Thus, $U_1^{-1}U \in \mathcal{U}(J)$. Moreover, since $\mathcal{L} = \mathcal{H}(U_1)$ is isometrically included in $\mathcal{H}(U)$, (5.77) follows from the observation that

$$\begin{aligned} 0 &= \langle f, K_\omega^U v - K_\omega^{U_1} v\rangle_{\mathcal{H}(U)} \\ &= \left\langle f, U_1 \left\{ \frac{J - U_2 J U_2(\omega)^*}{\rho_\omega} \right\} U_1(\omega)^* v \right\rangle_{\mathcal{H}(U)} \end{aligned}$$

for every choice of $f \in \mathcal{H}(U_1)$, $\omega \in \mathfrak{h}_{U_1} \cap \mathfrak{h}_{U_2}$ and $v \in \mathbb{C}^m$. □

Proof of Theorem 4.7 If $U \in \mathcal{U}(J)$ is rational, then, in view of Corollary 5.46 and Theorem 5.49, $\dim \mathcal{H}(U) = r < \infty$ and $\mathcal{H}(U)$ is R_α invariant for $\alpha \in \mathfrak{h}_U$. Therefore, there exists a nonzero vvf $f \in \mathcal{H}(U)$ such that $R_\alpha f = \mu f$ for some $\mu \in \mathbb{C}$. It is readily checked that

$$f(\lambda) = \begin{cases} \frac{u}{\rho_\omega(\lambda)} & \text{for some } \omega \in \mathbb{C} \text{ and } u \in \mathbb{C}^m \text{ if } \mu \neq 0 \\ u & \text{for some } u \in \mathbb{C}^m \text{ if } \mu = 0. \end{cases}$$

Therefore, since $X_1 = \{cf : c \in \mathbb{C}\}$ is a one dimensional R_α invariant subspace of $\mathcal{H}(U)$, Theorem 5.50 guarantees that $X_1 = \mathcal{H}(U_1)$, where U_1 is a primary BP factor that is uniquely determined by X_1 up to a constant J unitary factor on the right. Moreover,

$$\mathcal{H}(U) = \mathcal{H}(U_1) \oplus U_1\mathcal{H}(\widetilde{U}_2), \quad \text{where } \widetilde{U}_2 = U_1^{-1}U$$

belongs to $\mathcal{U}(J)$ and $\dim\mathcal{H}(\widetilde{U}_2) = r - 1$. If $r > 1$, then this algorithm can be repeated $r - 1$ more times to yield the factorization $U(\lambda) = U_1(\lambda)\cdots U_r(\lambda)\widetilde{U}_{r+1}$, where the first r factors are primary BP factors and $\widetilde{U}_{r+1} \in \mathcal{U}_{const}(J)$. Thus, as

$$U(\lambda) = \left\{I + i(\lambda - \alpha)C_\alpha(I - (\lambda - \alpha)R_\alpha)^{-1}C_\alpha^* J\right\} U(\alpha) \quad \text{for } \alpha \in \mathfrak{h}_U \cap \mathbb{R},$$

where $C_\alpha f = \sqrt{2\pi} f(\alpha)$, $C_\alpha^* v = \sqrt{2\pi} K_\alpha v$, $f \in \mathcal{H}(U)$ and $v \in \mathbb{C}^m$, and this realization of $U(\lambda)$ is mininimal, a theorem of Kalman [Kal65] guarantees that the McMillan degree of $U(\lambda)$ is equal to r. (Variants of the proof of this theorem may be found, e.g., in [AlD86] and on pp. 418–424 in [Dy07].) □

Lemma 5.51 *If $U \in \mathcal{P}^\circ(J)$ and $\omega \in \mathfrak{h}_U \cap \mathfrak{h}_{U^{-1}}$, then the formula*

$$U(\lambda) = \{J - \rho_\omega(\lambda) K_\omega^U(\lambda)\} U(\omega)^{-*} J \tag{5.78}$$

defines $U(\lambda)$ in terms of $K_\omega^U(\lambda)$ up to a J-unitary constant factor on the right.

Proof Formula (5.78) follows directly from the definition of the RK. If $U_1 \in \mathcal{P}^\circ(J)$ is such that $\omega \in \mathfrak{h}_{U_1} \cap \mathfrak{h}_{U_1^{-1}}$ and $K_\omega^{U_1}(\lambda) = K_\omega^U(\lambda)$, then $U_1(\lambda) J U_1(\omega)^* = U(\lambda) J U(\omega)^*$. Thus, $U(\omega)^{-1} U_1(\omega) J U_1(\omega)^* U(\omega)^{-*} = J$, and the proof is complete. □

If $U \in \mathcal{U}_0(J)$, then (5.78) evaluated at $\omega = 0$ simplifies to

$$U(\lambda) = J + 2\pi i \lambda K_0^U(\lambda) J. \tag{5.79}$$

Theorem 5.52 (L. de Branges) *If U, U_1, $U_2 \in \mathcal{U}(J)$ and $U = U_1 U_2$, then $\mathcal{H}(U_1)$ sits contractively in $\mathcal{H}(U)$, i.e., $\mathcal{H}(U_1) \subseteq \mathcal{H}(U)$ (as linear spaces) and*

$$\|f\|_{\mathcal{H}(U)} \le \|f\|_{\mathcal{H}(U_1)} \quad \textit{for every } f \in \mathcal{H}(U_1).$$

The inclusion is isometric if and only if

$$\mathcal{H}(U_1) \cap U_1 \mathcal{H}(U_2) = \{0\}. \tag{5.80}$$

If the condition (5.80) is in force, then

$$\mathcal{H}(U) = \mathcal{H}(U_1) \oplus U_1 \mathcal{H}(U_2). \tag{5.81}$$

Proof Let

$$\begin{aligned} \mathcal{L} &= \{f + U_1 g : f \in \mathcal{H}(U_1) \text{ and } g \in \mathcal{H}(U_2)\}, \\ \mathcal{L}_0 &= \{h \in \mathcal{H}(U_2) : U_1 h \in \mathcal{H}(U_1)\} \end{aligned}$$

and let

$$\|f + U_1 g\|_{\mathcal{L}}^2 = \inf\{\|f - U_1 h\|_{\mathcal{H}(U_1)}^2 + \|g + h\|_{\mathcal{H}(U_2)}^2 : \quad h \in \mathcal{L}_0\}.$$

The strategy of the proof is to identify $\mathcal{L}$ with $\mathcal{H}(U)$ with the help of the identity

$$K_\omega^U(\lambda) = K_\omega^{U_1}(\lambda) + U_1(\lambda) K_\omega^{U_2}(\lambda) U_1(\omega)^*. \tag{5.82}$$

The stated claim then drops out easily:

$$\mathcal{H}(U_1) \subseteq \mathcal{L} = \mathcal{H}(U)$$

and

$$\begin{aligned} \|f\|^2_{\mathcal{H}(U)} &= \|f\|^2_{\mathcal{L}} = \inf\{\|f - U_1 h\|^2_{\mathcal{H}(U_1)} + \|h\|^2_{\mathcal{H}(U_2)} : h \in \mathcal{L}_0\} \\ &\leq \|f\|^2_{\mathcal{H}(U_1)} \end{aligned} \tag{5.83}$$

for every $f \in \mathcal{H}(U_1)$. Thus, the inclusion $\mathcal{H}(U_1) \subseteq \mathcal{H}(U)$ is isometric if $\mathcal{L}_0 = \{0\}$. Conversely, if this inclusion is isometric, then by Theorem 5.50 applied to $\mathcal{L} = \mathcal{H}(U_1)$, there exists a mvf $\widetilde{U_1} \in \mathcal{U}(J)$ such that $\mathcal{H}(\widetilde{U_1}) = \mathcal{H}(U_1)$, $\widetilde{U_1}^{-1} U \in \mathcal{U}(J)$ and $\mathcal{H}(U) = \mathcal{H}(\widetilde{U_1}) \oplus \widetilde{U_1}\mathcal{H}(\widetilde{U_1}^{-1} U)$. However, Theorem 5.21 implies that $\widetilde{U_1} = U_1 V$ for some $V \in \mathcal{U}_{const}(J)$. Thus,

$$\mathcal{H}(U) = \mathcal{H}(U_1 V) \oplus U_1 V \mathcal{H}(V^{-1} U_1^{-1} U) = \mathcal{H}(U_1) \oplus U_1 V \mathcal{H}(U_2).$$

The identification of $\mathcal{L}$ with $\mathcal{H}(U)$ is broken into a number of steps.

Step 1. *$\mathcal{L}$ is an inner product space.*

Clearly $\mathcal{L}$ is a vector space. Suppose next that

$$\|f + U_1 g\|_{\mathcal{L}} = 0$$

for some choice of $f \in \mathcal{H}(U_1)$ and $g \in \mathcal{H}(U_2)$. Then there exists a sequence of vvf's $h_j \in \mathcal{L}_0$ such that

$$\|f - U_1 h_j\|^2_{\mathcal{H}(U_1)} + \|g + h_j\|^2_{\mathcal{H}(U_2)} \leq 1/j.$$

Thus, $U_1 h_j \to f$ in $\mathcal{H}(U_1)$ and $h_j \to -g$ in $\mathcal{H}(U_2)$, and so, as the convergence is pointwise in $\mathfrak{h}_{U_1} \cap \mathfrak{h}_{U_2}$,

$$f + U_1 g = 0$$

in $\mathfrak{h}_{U_1} \cap \mathfrak{h}_{U_2}$, i.e., $\|f + U_1 g\|_{\mathcal{L}} > 0$ if $f + U_1 g \not\equiv 0$. Moreover, it is readily checked from the definition that

$$\|\alpha(f + U_1 g)\|_{\mathcal{L}} = |\alpha|\, \|f + U_1 g\|_{\mathcal{L}} \tag{5.84}$$

for every $\alpha \in \mathbb{C}$.

The next objective is to check that the parallelogram law holds. To this end let $p_1 = f_1 + U_1 g_1$ and $p_2 = f_2 + U_1 g_2$ be any two elements in $\mathcal{L}$. Then there exists a pair of sequences $\{h_j\}$ and $\{k_j\}$ in $\mathcal{L}_0$ such that

$$\|p_1\|_{\mathcal{L}}^2 = \lim_{j\to\infty} \{\|f_1 - U_1 h_j\|_{\mathcal{H}(U_1)}^2 + \|g_1 + h_j\|_{\mathcal{H}(U_2)}^2\}$$

and

$$\|p_2\|_{\mathcal{L}}^2 = \lim_{j\to\infty} \{\|f_2 - U_1 k_j\|_{\mathcal{H}(U_1)}^2 + \|g_2 + k_j\|_{\mathcal{H}(U_2)}^2\}.$$

Moreover, by the parallelogram law in $\mathcal{H}(U_1)$ and $\mathcal{H}(U_2)$, the sum of the terms in the curly brackets on the right hand side of the last two formulas is equal to

$$\frac{1}{2}\{\|f_1 + f_2 - U_1(h_j + k_j)\|_{\mathcal{H}(U_1)}^2 + \|g_1 + g_2 + h_j + k_j\|_{\mathcal{H}(U_2)}^2 + \|f_1 - f_2 - U_1(h_j - k_j)\|_{\mathcal{H}(U_1)}^2 + \|g_1 - g_2 + h_j - k_j\|_{\mathcal{H}(U_2)}^2\}$$

which in turn is clearly bounded from below by

$$\frac{1}{2}\{\|p_1 + p_2\|_{\mathcal{L}}^2 + \|p_1 - p_2\|_{\mathcal{L}}^2\}.$$

Therefore,

$$2\|p_1\|_{\mathcal{L}}^2 + 2\|p_2\|_{\mathcal{L}}^2 \geq \|p_1 + p_2\|_{\mathcal{L}}^2 + \|p_1 - p_2\|_{\mathcal{L}}^2. \tag{5.85}$$

However, upon reexpressing (5.85) in terms of $p_1' = p_1 + p_2$ and $p_2' = p_1 - p_2$, it is readily seen that the opposite inequality to (5.85) is also valid, and hence that in fact equality prevails in (5.85). Thus the parallelogram law is established, and therefore, as is well known, $\mathcal{L}$ admits an inner product and $\| \ \|_{\mathcal{L}}$ satisfies the triangle inequality.

Step 2. *For every finite set of points $\omega_1, \ldots, \omega_k \in \mathfrak{h}_{U_1} \cap \mathfrak{h}_{U_2}$ and vectors $\xi_1, \ldots, \xi_k \in \mathbb{C}^m$, $\Sigma K_{\omega_j}^U \xi_j$ belongs to $\mathcal{L}$ and*

$$\|\Sigma_j K_{\omega_j}^U \xi_j\|_{\mathcal{L}}^2 = \Sigma_{i,j} \xi_i^* K_{\omega_j}^U(\omega_i)\xi_j = \|\Sigma_j K_{\omega_k}^U \xi_j\|_{\mathcal{H}(U)}^2.$$

Let $\eta_j = U_1(\omega_j)^* \xi_j$. Then, by (5.82),

$$\Sigma_j K_{\omega_j}^U(\lambda)\xi_j = \Sigma_j K_{\omega_j}^{U_1}(\lambda)\xi_j + U_1(\lambda)\Sigma_j K_{\omega_j}^{U_2}(\lambda)\eta_j$$

belongs to $\mathcal{L}$. The rest is plain from the evaluation

$$\begin{aligned}
&\|\Sigma_j K_{\omega_j}^{U_1}\xi_j - U_1 h\|_{\mathcal{H}(U_1)}^2 + \|\Sigma_j K_{\omega_j}^{U_2}\eta_j + h\|_{\mathcal{H}(U_2)}^2 \\
&= \|\Sigma_j K_{\omega_j}^{U_1}\xi_j\|_{\mathcal{H}(U_1)}^2 + \|U_1 h\|_{\mathcal{H}(U_1)}^2 - 2\Re\Sigma_j\eta_j^* h(\omega_j) \\
&\quad + \|\Sigma_j K_{\omega_j}^{U_2}\eta_j\|_{\mathcal{H}(U_2)}^2 + \|h\|_{\mathcal{H}(U_2)}^2 + 2\Re\Sigma_j\eta_j^* h(\omega_j) \\
&= \|\Sigma_j K_{\omega_j}^{U}\xi_j\|_{\mathcal{H}(U)}^2 + \|U_1 h\|_{\mathcal{H}(U_1)}^2 + \|h\|_{\mathcal{H}(U_2)}^2,
\end{aligned}$$

which is valid for every $h \in \mathcal{L}_0$.

Step 3. *If $u \in \mathcal{L}$, then*

$$\langle u, K_\omega^U \xi\rangle_{\mathcal{L}} = \xi^* u(\omega) \tag{5.86}$$

for every point $\omega \in \mathfrak{h}_{U_1} \cap \mathfrak{h}_{U_2}$ and every $\xi \in \mathbb{C}^m$.

Suppose $u = f + U_1 g$ with $f \in \mathcal{H}(U_1)$ and $g \in \mathcal{H}(U_2)$, and let $\eta = U_1(\omega)^*\xi$. Then

$$\|u + K_\omega^U \xi\|_{\mathcal{L}}^2 = \inf\{\|f + K_\omega^{U_1}\xi - U_1 h\|_{\mathcal{H}(U_1)}^2 + \|g + K_\omega^{U_2}\eta + h\|_{\mathcal{H}(U_2)}^2 : h \in \mathcal{L}_0\}.$$

But the term in curly brackets is readily seen to be equal to

$$\begin{aligned}
&\|f - U_1 h\|_{\mathcal{H}(U_1)}^2 + \|K_\omega^{U_1}\xi\|_{\mathcal{H}(U_1)}^2 + 2\Re\xi^*\{f(\omega) - U_1(\omega)h(\omega)\} \\
&\qquad + \|g + h\|_{\mathcal{H}(U_2)}^2 + \|K_\omega^{U_2}\eta\|_{\mathcal{H}(U_2)}^2 + 2\Re\eta^*\{g(\omega) + h(\omega)\}.
\end{aligned}$$

Thus the terms involving $h(\omega)$ cancel out and the indicated infimum is thus seen to be equal to

$$\|u\|_{\mathcal{L}}^2 + \xi^* K_\omega^{U_1}(\omega)\xi + \eta^* K_\omega^{U_2}(\omega)\eta + 2\Re\xi^* u(\omega)$$

which, in view of (5.82) and Step 2, serves to establish the identity

$$\|u + K_\omega^U \xi\|_{\mathcal{L}}^2 = \|u\|_{\mathcal{L}}^2 + \|K_\omega^U \xi\|_{\mathcal{L}}^2 + 2\Re\xi^* u(\omega).$$

The rest is plain.

Step 4. $\mathcal{H}(U) = \mathcal{L}$.

Let $u \in \mathcal{H}(U)$. Then there exist finite linear combinations of reproducing kernels

$$\sum_{j=1}^n K_{\omega_j}^U \xi_j = \sum_{j=1}^n K_{\omega_j}^{U_1}\xi_j + U_1 \sum_{j=1}^n K_{\omega_j}^{U_2}\eta_j = f_n + U_1 g_n$$

such that

$$\|u - f_n - U_1 g_n\|_{\mathcal{H}(U)} \to 0 \text{ as } n \to \infty.$$

But now, by the calculations in Step 2 with $h = 0$, it follows that

$$\|f_n + U_1 g_n\|^2_{\mathcal{H}(U)} = \|f_n\|^2_{\mathcal{H}(U_1)} + \|g_n\|^2_{\mathcal{H}(U_2)},$$

and hence that f_n tends to a limit f in $\mathcal{H}(U_1)$ and g_n tends to a limit g in $\mathcal{H}(U_2)$. Therefore,

$$u = f + U_1 g$$

belongs to $\mathcal{L}$ and

$$\begin{aligned}
&\|u\|^2_{\mathcal{H}(U)} = \|f\|^2_{\mathcal{H}(U_1)} + \|g\|^2_{\mathcal{H}(U_2)} \\
&= \inf_{h\in\mathcal{L}_0} \lim_{n\to\infty}\{\|f_n\|^2_{\mathcal{H}(U_1)} + \|U_1 h\|^2_{\mathcal{H}(U_1)} + \|g_n\|^2_{\mathcal{H}(U_2)} + \|h\|^2_{\mathcal{H}(U_2)}\} \\
&= \inf_{h\in\mathcal{L}_0} \lim_{n\to\infty}\{\|f_n - U_1 h\|^2_{\mathcal{H}(U_1)} + \|g_n + h\|^2_{\mathcal{H}(U_2)}\} \\
&= \inf_{h\in\mathcal{L}_0}\{\|f - U_1 h\|^2_{\mathcal{H}(U_1)} + \|g + h\|^2_{\mathcal{H}(U_2)}\} \\
&= \|u\|^2_{\mathcal{L}}.
\end{aligned}$$

Thus $\mathcal{H}(U)$ sits isometrically inside $\mathcal{L}$ and is therefore a closed subspace of $\mathcal{L}$. On the other hand, it follows readily from (5.86) that $u = 0$ is the only element of $\mathcal{L}$ which is orthogonal to $\mathcal{H}(U)$. Thus $\mathcal{L} = \mathcal{H}(U)$ as asserted, and so the proof of both the step and the theorem is complete. □

Formula (5.83) implies that the inclusion $\mathcal{H}(U_1) \subseteq \mathcal{H}(U)$ is isometric if $\mathcal{L}_0 = \{0\}$. Conversely, if this inclusion is isometric, then by Theorem 5.50 applied to $\mathcal{L} = \mathcal{H}(U_1)$, there exists a mvf $\widetilde{U_1} \in \mathcal{U}(J)$ such that $\mathcal{H}(\widetilde{U_1}) = \mathcal{H}(U_1)$, $\widetilde{U_1}^{-1} U \in \mathcal{U}(J)$ and $\mathcal{H}(U) = \mathcal{H}(\widetilde{U_1}) \oplus \widetilde{U_1}\mathcal{H}(\widetilde{U_1}^{-1} U)$. However, Theorem 5.21 implies that $\widetilde{U_1} = U_1 V$ for some $V \in \mathcal{U}_{const}(J)$. Thus,

$$\mathcal{H}(U) = \mathcal{H}(U_1 V) \oplus U_1 V \mathcal{H}(V^{-1} U_1^{-1} U) = \mathcal{H}(U_1) \oplus U_1 V \mathcal{H}(U_2).$$

We remark that the space $\mathcal{L}_0$ endowed with the norm

$$\|h\|^2_{\mathcal{L}_0} = \|h\|^2_{\mathcal{H}(U_2)} + \|U_1 h\|^2_{\mathcal{H}(U_1)}$$

is referred to by de Branges as the **overlapping space** [dB4], [dB8].

A simple example which illustrates the last theorem is obtained by setting

$$U_\alpha(\lambda) = I_m - \alpha c(\lambda) v v^* J$$

where $\alpha \geq 0$, $c(\lambda)$ is a scalar Carathéodory function and $v^*Jv = 0$. Then it is readily checked that

$$U_{\alpha+\beta} = U_\alpha U_\beta$$

and that

$$K_\omega^{U_\alpha}(\lambda) = \frac{J - U_\alpha(\lambda)JU_\alpha(\omega)^*}{\rho_\omega(\lambda)} = 2\alpha k_\omega^c(\lambda)vv^*$$

in which $k_\omega^c(\lambda)$ is given by (5.19). If $c \in \mathcal{C}_{sing}^{p\times p}$, then $U_\alpha \in \mathcal{U}(J)$ and

$$K_\omega^{U_\alpha}(\lambda) = \frac{\alpha}{\gamma} K_\omega^{U_\gamma}(\lambda) \quad \text{for} \quad 0 < \alpha \leq \gamma.$$

Therefore,

$$\begin{aligned} &\| \sum_j K_{\omega_j}^{U_\alpha}(\lambda)\xi_j \|_{\mathcal{H}(U_\gamma)}^2 \\ &= \left(\frac{\alpha}{\gamma}\right)^2 \| \sum_j K_{\omega_j}^{U_\gamma}(\lambda)\xi_j \|_{\mathcal{H}(U_\gamma)}^2 = \left(\frac{\alpha}{\gamma}\right)^2 \sum_{i,j} \xi_i^* K_{\omega_j}^{U_\gamma}(\omega_i)\xi_j \\ &= \left(\frac{\alpha}{\gamma}\right) \sum_{i,j} \xi_i^* K_{\omega_j}^{U_\alpha}(\omega_i)\xi_j = \left(\frac{\alpha}{\gamma}\right) \| \sum_j K_{\omega_j}^{U_\alpha}(\lambda)\xi_j \|_{\mathcal{H}(U_\alpha)}^2. \end{aligned}$$

But this in turn implies that $\mathcal{H}(U_\alpha)$ sits contractively in $\mathcal{H}(U_\gamma)$ for $0 \leq \alpha \leq \gamma$, with isometry if and only if $\alpha = \gamma$.

Theorem 5.53 *If $U \in \mathcal{U}(J)$, then $\{\xi \in \mathbb{C}^m : U\xi \in \mathcal{H}(U)\}$ is a J-neutral subspace of $\mathbb{C}^m$.*

Proof If $U\xi$ and $U\eta$ belong to $\mathcal{H}(U)$ and if α and $\overline{\alpha}$ belong to $\mathfrak{h}_U$, then the identity

$$(R_\alpha U)(\lambda) = 2\pi i K_{\overline{\alpha}}(\lambda)JU(\alpha)$$

implies that

$$\langle R_\alpha U\xi, U\eta\rangle_{\mathcal{H}(U)} = 2\pi i\eta^* J\xi \quad \text{and} \quad \langle U\xi, R_{\overline{\alpha}} U\eta\rangle_{\mathcal{H}(U)} = -2\pi i\eta^* J\xi.$$

Therefore, formula (5.27) with $f = U\xi$, $g = U\eta$ and $\beta = \overline{\alpha}$ implies that

$$\eta^* J\xi = 0,$$

as desired. □

Lemma 5.54 *Let $U \in \mathcal{U}(J)$ and let $\mathcal{H}(U)$ denote the RKHS of $m \times 1$ vvf's on $\mathfrak{h}_U$ with RK $K_\omega^U(\lambda)$ that is defined on $\mathfrak{h}_U \times \mathfrak{h}_U$ by formula (5.34). Then the sets $\mathcal{N}_\omega$, $\mathcal{R}_\omega$ and $\mathcal{F}_\omega$ defined in Lemma 5.3 with $K_\omega(\lambda) = K_\omega^U(\lambda)$ are independent of the choice of the point $\omega \in \mathfrak{h}_b$. Moreover, the following conditions are equivalent:*

(1) $K_\omega^U(\omega) > 0$ *for at least one point* $\omega \in \mathfrak{h}_U$.

(2) $K_\omega^U(\omega) > 0$ *for every point* $\omega \in \mathfrak{h}_U$.

(3) $\{f(\omega) : f \in \mathcal{H}(U)\} = \mathbb{C}^m$ *for at least one point* $\omega \in \mathfrak{h}_U$.

(4) $\{f(\omega) : f \in \mathcal{H}(U)\} = \mathbb{C}^m$ *for every point* $\omega \in \mathfrak{h}_U$.

Proof This follows from Theorem 5.49 and Lemma 5.4. □

5.9 The space $\mathcal{H}(W)$ for $W \in \mathcal{U}(j_{pq})$

Description of $\mathcal{H}(W)$

If $W \in \mathcal{U}(j_{pq})$ and $S = PG(W)$, then formula (5.65) may be written in terms of the standard four block decompositions of S and W as

$$L(\lambda) = \begin{bmatrix} I_p & -s_{12} \\ 0 & -s_{22} \end{bmatrix}^{-1} = \begin{bmatrix} I_p & -w_{12} \\ 0 & -w_{22} \end{bmatrix} \tag{5.87}$$

and the description of the space $\mathcal{H}(U)$ given in Theorem 5.45 yields the following description of $\mathcal{H}(W)$:

Theorem 5.55 *Let $W \in \mathcal{U}(j_{pq})$ and let s_{ij}, $i, j = 1, 2$ be the blocks of $S = PG(W)$ in the four block decomposition that is conformal with j_{pq}. Then*

$$\mathcal{H}(W) = \left\{ \begin{bmatrix} f_1 \\ f_2 \end{bmatrix} : \begin{array}{c} f_1 - s_{12} f_2 \in H_2^p \\ s_{22} f_2 \in H_2^q \end{array} \quad \text{and} \quad \begin{array}{c} s_{11}^* f_1 \in K_2^p \\ f_2 - s_{12}^* f_1 \in K_2^q \end{array} \right\} \tag{5.88}$$

and

$$f = \begin{bmatrix} f_1 \\ f_2 \end{bmatrix} \Longrightarrow \langle f, f \rangle_{\mathcal{H}(W)} = \left\langle \begin{bmatrix} I_p & -s_{12} \\ -s_{12}^* & I_q \end{bmatrix} f, f \right\rangle_{st}. \tag{5.89}$$

Proof This is immediate from (2) of Theorem 5.45 and formula (5.87). □

Corollary 5.56 *If $W \in \mathcal{U}(j_{pq})$ and $f \in \mathcal{H}(W)$ with components f_1 and f_2 as in (5.88), then*

$$f \in L_2^m \iff f_1 \in L_2^p \iff f_2 \in L_2^q.$$

Proof This is immediate from the description of $\mathcal{H}(W)$ given in (5.88). □

Recovering W from the RK $K_\omega^W(\lambda)$

Let

$$K_\omega^W(\lambda) = K_\omega(\lambda) = \begin{bmatrix} K_\omega^{11}(\lambda) & K_\omega^{12}(\lambda) \\ K_\omega^{21}(\lambda) & K_\omega^{22}(\lambda) \end{bmatrix} \tag{5.90}$$

be the four block decomposition of the RK $K_\omega^W(\lambda)$ with blocks $K_\omega^{11}(\lambda)$ of size $p \times p$ and $K_\omega^{22}(\lambda)$ of size $q \times q$ for the mvf $W \in \mathcal{U}(j_{pq})$. The mvf $W(\lambda)$ is defined by the RK $K_\omega(\lambda)$ up to a j_{pq}- unitary constant factor on the right by Lemma 5.51. In view of Lemma 2.17, this constant factor can be chosen so that

$$w_{11}(\omega) > 0, \quad w_{22}(\omega) > 0 \quad \text{and} \quad w_{21}(\omega) = 0 \tag{5.91}$$

at a fixed point $\omega \in \mathfrak{h}_W \cap \mathfrak{h}_{W^{-1}} \cap \overline{\mathbb{C}_+}$.

Theorem 5.57 *Let $W_1 \in \mathcal{U}(j_{pq})$ and let $\omega \in \mathfrak{h}_W \cap \mathfrak{h}_{W^{-1}} \cap \overline{\mathbb{C}_+}$. Then there exists exactly one mvf $W \in \mathcal{U}(j_{pq})$ such that $\mathcal{H}(W) = \mathcal{H}(W_1)$ and W meets the normalization condition (5.91). Moreover:*

(1) *If $\omega \in \mathfrak{h}_W \cap \mathfrak{h}_{W^{-1}} \cap \mathbb{R}$, then*

$$W(\lambda) = I_m - \rho_\omega(\lambda) K_\omega^{W_1}(\lambda) j_{pq} \quad \textit{for } \lambda \in \mathfrak{h}_W. \tag{5.92}$$

(2) *If $\omega \in \mathfrak{h}_W \cap \mathfrak{h}_{W^{-1}} \cap \mathbb{C}_+$, then*

$$W(\lambda) = \{j_{pq} - \rho_\omega(\lambda) K_\omega^{W_1}(\lambda) j_{pq}\} W(\omega)^{-*} j_{pq} \quad \textit{for} \quad \lambda \in \mathfrak{h}_W, \tag{5.93}$$

where

$$W(\omega) = \begin{bmatrix} w_{11}(\omega) & w_{12}(\omega) \\ 0 & w_{22}(\omega) \end{bmatrix}, \tag{5.94}$$

$$w_{11}(\omega) = \{I_p + \rho_\omega(\omega) K_{\overline{\omega}}^{11}(\overline{\omega})\}^{-1/2}, \tag{5.95}$$

$$w_{22}(\omega) = \{I_q + \rho_\omega(\omega) K_\omega^{22}(\omega)\}^{-1/2} \tag{5.96}$$

and

$$w_{12}(\omega) = \rho_\omega(\omega) K_\omega^{12}(\omega) w_{22}(\omega)^{-1}. \tag{5.97}$$

Proof Formulas (5.96) and (5.97) follow easily upon substituting the normalization (5.91) into the formula

$$W(\lambda) j_{pq} W(\omega)^* = j_{pq} - \rho_\omega(\lambda) K_\omega(\lambda), \tag{5.98}$$

setting $\lambda = \omega$ and matching blocks. This also leads to formula (5.92) (upon taking taking note of the fact that $W(\omega)$ is j_{pq}-unitary if $\omega \in \mathbb{R}$, which forces $W(\omega) = I_m$) as well as a lengthy formula for $w_{11}(\omega)$. Formula (5.95) is obtained from the 11 block of the formula

$$W(\overline{\omega}) j_{pq} W(\overline{\omega})^* = j_{pq} + \rho_\omega(\omega) K_{\overline{\omega}}(\overline{\omega})$$

and the observation that $w_{12}(\overline{\omega}) = 0$ and $w_{11}(\overline{\omega}) = w_{11}(\omega)^{-1}$. The last two evaluations come from the formulas

$$W(\overline{\omega}) j_{pq} W(\omega)^* = j_{pq} \quad \text{and} \quad w_{21}(\omega) = 0.$$

□

Remark 5.58 *The blocks $w_{21}(\omega)$ and $w_{22}(\omega)$ are uniquely defined by formulas (5.96) and (5.97) for points $\omega \in \mathfrak{h}_W^+$ even if $W(\omega)$ is not invertible. The invertibility of $W(\omega)$ is only needed to justify formula (5.95).*

Matrix balls $\mathcal{B}_W(\omega)$

In this subsection we shall show that if $W \in \mathcal{U}(j_{pq})$ and $\omega \in \mathfrak{h}_W^+ \cap \mathfrak{h}_{W^{-1}}^+$, then the set

$$\mathcal{B}_W(\omega) = \{s(\omega) : s \in T_W[\mathcal{S}^{p\times q}]\} \tag{5.99}$$

is a matrix ball with positive semiradii and we shall present formulas for its center and its left and right semiradii in terms of the reproducing kernel $K_\omega^W(\omega)$.

Theorem 5.59 *If $W \in \mathcal{U}(j_{pq})$ and $\omega \in \mathfrak{h}_W^+$, then the set $\mathcal{B}_W(\omega)$ defined by formula (5.99) is a matrix ball with left and right semiradii $R_\ell(\omega) \geq 0$ and $R_r(\omega) > 0$ and center $s_c(\omega) \in \mathcal{S}_{const}^{p\times q}$:*

$$\mathcal{B}_W(\omega) = \{s_c(\omega) + R_\ell(\omega) \varepsilon R_r(\omega) : \varepsilon \in \mathcal{S}_{const}^{p\times q}\}, \tag{5.100}$$

where

$$s_c(\omega) = \rho_\omega(\omega) K_\omega^{12}(\omega)\{I_q + \rho_\omega(\omega) K_\omega^{22}(\omega)\}^{-1} \tag{5.101}$$

and

$$R_r(\omega) = \{I_q + \rho_\omega(\omega) K_\omega^{22}(\omega)\}^{1/2}. \tag{5.102}$$

If $\{b_1, b_2\} \in ap(W)$, *then*

$$\frac{\det R_\ell(\omega)}{\det R_r(\omega)} = \frac{|\det b_1(\omega)|}{|\det b_2(\omega)|} = |\det W(\omega)| \tag{5.103}$$

and

$$R_\ell(\omega) > 0 \Longleftrightarrow \omega \in \mathfrak{h}_W^+ \cap \mathfrak{h}_{W^{-1}}^+. \tag{5.104}$$

Moreover, if $\omega \in \mathfrak{h}_W^+ \cap \mathfrak{h}_{W^{-1}}^+$, *then*

$$R_\ell(\omega) = \{I_p + \rho_\omega(\omega) K_{\overline{\omega}}^{11}(\overline{\omega})\}^{-1/2}. \tag{5.105}$$

Proof Since $T_W[\mathcal{S}^{p\times q}] = T_{WW_0}[\mathcal{S}^{p\times q}]$ for $W \in \mathcal{U}(j_{pq})$ and $W_0 \in \mathcal{U}_{const}(j_{pq})$, we can, without loss of generality, consider $T_W[\mathcal{S}^{p\times q}]$ under the assumption that $W(\omega)$ meets the normalization conditions (5.91). Then $\mathcal{B}_W(\omega)$ is clearly a matrix ball with center

$$s_c(\omega) = w_{12}(\omega) w_{22}(\omega)^{-1} \tag{5.106}$$

and semiradii

$$R_\ell(\omega) = w_{11}(\omega) \quad \text{and} \quad R_r(\omega) = w_{22}(\omega)^{-1}. \tag{5.107}$$

The stated formulas for the center and the semiradii are now easily obtained from Theorem 5.57. Finally, to obtain formula (5.103), it suffices to consider the case that $W(\lambda)$ is subject to the normalization conditions (5.91) at the point ω. Then the semiradii can be reexpressed in terms of the blocks in the PG transform $S = PG(W)$ by the formulas

$$R_\ell(\omega) = w_{11}(\omega) = w_{11}^{\#}(\omega)^{-1} = s_{11}(\omega) = b_1(\omega)\varphi_1(\omega)$$

and

$$R_r(\omega) = w_{22}(\omega)^{-1} = s_{22}(\omega) = \varphi_2(\omega) b_2(\omega).$$

The next step is to invoke the identities

$$s_{11}(\mu)s_{11}(\mu)^* = I_p - s_{12}(\mu)s_{12}(\mu)^* \quad \text{a.e. on } \mathbb{R}$$

and

$$s_{22}^*(\mu)s_{22}(\mu) = I_q - s_{12}(\mu)^* s_{12}(\mu) \quad \text{a.e. on } \mathbb{R}$$

to verify that

$$\begin{aligned} |\det \varphi_1(\mu)|^2 &= \det (I_p - s_{12}(\mu)s_{12}(\mu)^*) \\ &= \det (I_q - s_{12}(\mu)^* s_{12}(\mu)) = |\det \varphi_2(\mu)|^2 \quad \text{a.e. on } \mathbb{R} \end{aligned}$$

and hence that $|\det \varphi_1(\omega)| = |\det \varphi_2(\omega)|$. Formula (5.103) now drops out upon combining formulas. □

Remark 5.60 *The center of the matrix ball $\mathcal{B}_W(\omega)$ is uniquely defined. However, the semiradii are only unique up to a multiplicative positive scalar constant: i.e., $\mathcal{B}_W(\omega)$ does not change if $R_\ell(\omega)$ is replaced by $\delta R_\ell(\omega)$ and $R_r(\omega)$ is replaced by $\delta^{-1}R_r(\omega)$ for any choice of $\delta > 0$. If $R_\ell(\omega) > 0$, then there exist only one pair of semiradii $R_\ell(\omega)$ and $R_r(\omega)$ of the ball $\mathcal{B}_W(\omega)$ that satisfy the relation (5.103). Consequently, in view of formulas (5.103)–(5.107), the matrix ball $\mathcal{B}_W(\omega)$ and the number* det $W(\omega)$ *serve to uniquely determine the matrix $W(\omega)$ normalized by the condition (5.91) for a mvf $W \in \mathcal{U}(j_{pq})$.*

5.10 The de Branges space $\mathcal{B}(\mathfrak{E})$

A $p \times 2p$ mvf

$$\mathfrak{E}(\lambda) = [E_-(\lambda) \quad E_+(\lambda)], \tag{5.108}$$

with $p \times p$ blocks $E_\pm$ that are meromorphic in $\mathbb{C}_+$ and meet the conditions

$$\det E_+(\lambda) \not\equiv 0 \quad \text{and} \quad \chi \overset{def}{=} E_+^{-1}E_- \in \mathcal{S}_{in}^{p\times p} \tag{5.109}$$

will be called a **de Branges matrix**. If $\mathfrak{E}$ is a de Branges matrix, then the kernel

$$K_\omega^{\mathfrak{E}}(\lambda) = -\frac{\mathfrak{E}(\lambda)j_p\mathfrak{E}(\omega)^*}{\rho_\omega(\lambda)} = \frac{E_+(\lambda)E_+(\omega)^* - E_-(\lambda)E_-(\omega)^*}{\rho_\omega(\lambda)} \tag{5.110}$$

is positive on $\mathfrak{h}_{\mathfrak{E}}^+ \times \mathfrak{h}_{\mathfrak{E}}^+$, since $\chi \in \mathcal{S}_{in}^{p\times p}$ and

$$K_\omega^{\mathfrak{E}}(\lambda) = E_+(\lambda)k_\omega^\chi(\lambda)E_+(\omega)^* \quad \text{on } \mathfrak{h}_{\mathfrak{E}}^+ \times \mathfrak{h}_{\mathfrak{E}}^+. \tag{5.111}$$

Therefore, by Theorem 5.2, there is exactly one RKHS $\mathcal{B}(\mathfrak{E})$ with RK $K_\omega^{\mathfrak{E}}(\lambda)$ associated with each de Branges matrix; it will be called a **de Branges**

space. Moreover, since the kernel $K_\omega^{\mathfrak{E}}(\lambda)$ is a holomorphic function of λ on $\mathfrak{h}_{\mathfrak{E}}^+$ for every fixed $\omega \in \mathfrak{h}_{\mathfrak{E}}^+$ and $K_\omega^{\mathfrak{E}}(\omega)$ is continuous on $\mathfrak{h}_{\mathfrak{E}}^+$, Corollary 5.7 guarantees that every vvf $f \in \mathcal{B}(\mathfrak{E})$ is holomorphic in $\mathfrak{h}_{\mathfrak{E}}^+$. A vvf $f \in \mathcal{B}(\mathfrak{E})$ may have a holomorphic extension onto a larger set in $\mathbb{C}_+$ than $\mathfrak{h}_{\mathfrak{E}}^+$. Thus, for example, if ω_1 and ω_2 are two distinct points in $\mathbb{C}_+$ and if

$$E_-(\lambda) = I_p, \quad \text{and} \quad E_+(\lambda) = \frac{(\lambda - \overline{\omega_1})}{(\lambda - \omega_1)} \frac{(\lambda - \overline{\omega_2})}{(\lambda - \omega_2)} \quad \text{and} \quad f = \frac{1}{\lambda - \omega_1},$$

then

$$\mathcal{B}(\mathfrak{E}) = \left\{ \frac{\alpha}{\lambda - \omega_1} + \frac{\beta}{\lambda - \omega_2} : \alpha, \beta \in \mathbb{C} \right\}$$

with respect to the standard inner product and

$$\mathfrak{h}_f^+ = \mathbb{C}_+ \setminus \{\omega_1\} \quad \text{whereas} \quad \mathfrak{h}_{\mathfrak{E}}^+ = \mathbb{C}_+ \setminus \{\omega_1, \omega_2\}.$$

Lemma 5.61 *Let $\mathfrak{E}$ be a de Branges matrix, let $\chi = E_+^{-1} E_-$ and suppose that*

$$K_\omega^\chi(\omega) > 0 \quad \textit{for at least one (and hence every) point} \quad \omega \in \mathbb{C}_+. \tag{5.112}$$

Then

$$\bigcap_{f \in \mathcal{B}(\mathfrak{E})} \mathfrak{h}_f^+ = \mathfrak{h}_{\mathfrak{E}}^+. \tag{5.113}$$

Proof In view of Lemma 5.41, $\det\,[I_p - \chi(\lambda)\chi(\omega)^*] \neq 0$ for every choice of $\lambda, \omega \in \mathbb{C}_+$ and consequently, $\mathfrak{h}_f^+ = \mathfrak{h}_{\mathfrak{E}}^+$ for vvf's of the form

$$f(\lambda) = K_\omega^{\mathfrak{E}}(\lambda) u$$

with $\omega \in \mathfrak{h}_{E_+}^+ \cap \mathfrak{h}_{(E_+)^{-1}}^+$, $u \in \mathbb{C}^p$ and $u \neq 0$. Thus, the condition (5.112) implies (5.113). □

Lemma 5.62 *Let $\mathfrak{E}$ be a de Branges matrix and let $\chi = E_+^{-1} E_-$. Then the formula*

$$(Tf)(\lambda) = E_+(\lambda)^{-1} f(\lambda) \tag{5.114}$$

defines a unitary operator T from $\mathcal{B}(\mathfrak{E})$ onto $\mathcal{H}(\chi)$. Moreover,

$$\begin{aligned} f \in \mathcal{B}(\mathfrak{E}) &\iff E_+^{-1} f \in \mathcal{H}(\chi) \\ &\iff E_-^{-1} f \in \mathcal{H}_*(\chi) \\ &\iff E_+^{-1} f \in H_2^p \text{ and } E_-^{-1} f \in K_2^p. \end{aligned} \tag{5.115}$$

Proof The proof follows easily from the definitions. □

Lemma 5.63 *Let $\mathfrak{E} = [E_+ \quad E_-]$ be a de Branges matrix. Then the following three conditions are equivalent:*

(1) *$E_+(\lambda)$ has a limit $E_+(\mu)$ as λ tends nontangentially to $\mu \in \mathbb{R}$ at almost all points $\mu \in \mathbb{R}$ and* $\det E_+(\mu) \neq 0$ *a.e. on $\mathbb{R}$.*

(2) *$E_-(\lambda)$ has a limit $E_-(\mu)$ as λ tends nontangentially to $\mu \in \mathbb{R}$ at almost all points $\mu \in \mathbb{R}$ and* $\det E_-(\mu) \neq 0$ *a.e. on $\mathbb{R}$.*

(3) *$\mathfrak{E}(\lambda)$ has a limit $\mathfrak{E}(\mu)$ as λ tends nontangentially to $\mu \in \mathbb{R}$ at almost all points $\mu \in \mathbb{R}$ and* $\operatorname{rank} \mathfrak{E}(\mu) = p$ *a.e. on $\mathbb{R}$.*

Proof The implications (3) $\Longrightarrow$ (1), (3) $\Longrightarrow$ (2) and the equivalence (1) $\Longleftrightarrow$ (2) follow from the formula $E_-(\lambda) = E_+(\lambda)\chi(\lambda)$, since $\chi \in \mathcal{S}_{in}^{p\times p}$. Next, if (1) and (2) are in force, then the nontangential convergence asserted in (3) clearly takes place. The rank condition is then immediate from the inequality

$$E_+(\mu)E_+(\mu)^* \leq \mathfrak{E}(\mu)\mathfrak{E}(\mu)^*,$$

which is valid a.e. on $\mathbb{R}$. □

The conditions (1), (2) and (3) in Lemma 5.63 are met if any one (and hence every one) of the three equivalent conditions

$$(i)\ E_+ \in \Pi^{p\times p}, \quad (ii)\ E_- \in \Pi^{p\times p}, \quad (iii)\ \mathfrak{E} \in \Pi^{p\times 2p} \tag{5.116}$$

is in force. The conditions (1)–(3) are also met if the de Branges matrix $\mathfrak{E}(\lambda)$ is holomorphic on $\mathbb{R}$ except for a set of Lebesgue measure zero and $\operatorname{rank} \mathfrak{E} = p$ (in the sense defined in Chapter 3) and hence in particular if $\mathfrak{E}(\lambda)$ is meromorphic on the full complex plane $\mathbb{C}$ and $\operatorname{rank} \mathfrak{E} = p$.

Theorem 5.64 *If any one (and hence every one) of the equivalent conditions (1)–(3) in Lemma 5.63 is in force, then:*

(1) *The identity*

$$E_+(\mu)E_+(\mu)^* = E_-(\mu)E_-(\mu)^* \tag{5.117}$$

is in force for almost all points $\mu \in \mathbb{R}$.

(2) *The mvf*

$$\Delta_{\mathfrak{E}}(\mu) = E_+(\mu)^{-*}E_+(\mu)^{-1} = E_-(\mu)^{-*}E_-(\mu)^{-1} \tag{5.118}$$

is well defined a.e. on $\mathbb{R}$.

(3) *Every vvf* $f(\lambda)$ *in* $\mathcal{B}(\mathfrak{E})$ *has a nontangential limit* $f(\mu)$ *as* λ *tends nontangentially to* $\mu \in \mathbb{R}$ *at almost all points* $\mu \in \mathbb{R}$.

(4) *If* $f \in \mathcal{B}(\mathfrak{E})$, *then*

$$\|f\|^2_{\mathcal{B}(\mathfrak{E})} = \|E_+^{-1}f\|^2_{st} = \int_{-\infty}^{\infty} f(\mu)^*\Delta_{\mathfrak{E}}f(\mu)d\mu. \tag{5.119}$$

Proof The theorem follows from the equivalences (5.115) and the fact that the operator T defined in (5.114) is a unitary map of $\mathcal{B}(\mathfrak{E})$ onto $\mathcal{H}(\chi)$. □

Theorem 5.65 *If a de Branges matrix* $\mathfrak{E} = [E_- \quad E_+]$ *belongs to* $\Pi^{p\times 2p}$, *then:*

(1) $f \in \Pi^p$ *for every* $f \in \mathcal{B}(\mathfrak{E})$, *i.e.,*

$$\mathcal{B}(\mathfrak{E}) \subset \Pi^p. \tag{5.120}$$

(2) *The inclusion*

$$\mathfrak{h}_f \supseteq \mathfrak{h}_{\mathfrak{E}} \tag{5.121}$$

is in force for every $f \in \mathcal{B}(\mathfrak{E})$.

(3) *The space* $\mathcal{B}(\mathfrak{E})$ *of vvf's is a RKHS of holomorphic* $p \times 1$ *vvf's on* $\mathfrak{h}_{\mathfrak{E}}$ *with RK* $K_w^{\mathfrak{E}}(\lambda)$, *defined on* $\mathfrak{h}_{\mathfrak{E}} \times \mathfrak{h}_{\mathfrak{E}}$ *by the formula*

$$K_\omega^{\mathfrak{E}}(\lambda) = \begin{cases} \dfrac{E_+(\lambda)E_+(\omega)^* - E_-(\lambda)E_-(\omega)^*}{\rho_\omega(\lambda)} & \text{if } \lambda \neq \overline{\omega} \\ -\dfrac{1}{2\pi i}\{E_+'(\overline{\omega})E_+(\omega)^* - E_-'(\overline{\omega})E_-(\omega)^*\} & \text{if } \lambda = \overline{\omega}. \end{cases} \tag{5.122}$$

Proof Assertion (1) follows from the relation (5.114) between the vvf's of the spaces $\mathcal{B}(\mathfrak{E})$ and $\mathcal{H}(\chi)$. Assertion (2) follows from (3). In view of formula (5.111), the kernel $K_\omega^{\mathfrak{E}}(\lambda)$ defined by formula (5.122)is clearly positive on $\Omega \times \Omega$, where $\Omega = \mathfrak{h}_{\mathfrak{E}} \bigcap \mathfrak{h}_\chi$. However, since Ω is dense in $\mathfrak{h}_{\mathfrak{E}}$ and $K_\omega^{\mathfrak{E}}(\lambda)$ is continuous on $\mathfrak{h}_{\mathfrak{E}} \times \mathfrak{h}_{\mathfrak{E}}$, the kernel $K_\omega^{\mathfrak{E}}(\lambda)$ is in fact positive on the larger set $\mathfrak{h}_{\mathfrak{E}} \times \mathfrak{h}_{\mathfrak{E}}$. Thus, by Theorem 5.2, there exists a unique RKHS $\mathcal{H}$ of $p \times 1$ vvf's on $\mathfrak{h}_{\mathfrak{E}}$ with RK $K_\omega^{\mathfrak{E}}(\lambda)$ that is defined on $\mathfrak{h}_{\mathfrak{E}} \times \mathfrak{h}_{\mathfrak{E}}$ by formula (5.122).

Moreover, since the mvf $K_\omega^{\mathfrak{E}}(\lambda)$ is holomorphic on $h_{\mathfrak{E}}$ for every $\omega \in \mathfrak{h}_{\mathfrak{E}}$ and the mvf $K_\omega^{\mathfrak{E}}(\omega)$ is continuous on $\mathfrak{h}_{\mathfrak{E}}$, the vvf's in $\mathcal{H}$ are holomorphic on $\mathfrak{h}_{\mathfrak{E}}$ by Corollary 5.7.

Next, to verify (3), let $\mathcal{L}$ denote the linear manifold of vvf's f of the form

$$\left\{ f = \sum_{j=1}^{n} K_{\omega_j}^{\mathfrak{E}}(\lambda) u_j, \text{ where } \omega_j \in \mathfrak{h}_{\mathfrak{E}} \bigcap \mathfrak{h}_\chi \text{ and } u_j \in \mathbb{C}^p,\ 1 \le j \le n \right\},$$

and note that for every such vvf f,

$$g = E_+^{-1} f = \sum_{j=1}^{n} K_{\omega_j}^{\chi}(\lambda) E_+(\omega_j)^* u_j$$

belongs to $\mathcal{H}(\chi)$ by Theorem 5.40 and

$$\|f\|_{\mathcal{H}}^2 = \|g\|_{\mathcal{H}(\chi)}^2 = \|g\|_{st}^2.$$

Consequently, $f \in \mathcal{B}(\mathfrak{E})$ and $\|f\|_{\mathcal{B}(\mathfrak{E})}^2 = \|f\|_{\mathcal{H}}^2$. Therefore, $\mathcal{H} = \mathcal{B}(\mathfrak{E})$, since $\mathcal{L}$ is dense in $\mathcal{H}$. □

Lemma 5.66 *Let $\mathfrak{E} = [E_- \quad E_+]$ be a de Branges matrix and let $\chi = E_+^{-1} E_-$. Then the following conditions are equivalent:*

(1) *The inequality $K_\omega^{\mathfrak{E}}(\omega) > 0$ holds for at least one point $\omega \in \mathfrak{h}_{\mathfrak{E}}^+$.*

(2) *The inequality $K_\omega^{\mathfrak{E}}(\omega) > 0$ holds for every point $\omega \in \mathfrak{h}_{\mathfrak{E}}^+$.*

(3) *The equality $\{f(\omega) : f \in \mathcal{B}(\mathfrak{E})\} = \mathbb{C}^p$ holds for at least one point $\omega \in \mathfrak{h}_{\mathfrak{E}}^+$.*

(4) *The equality $\{f(\omega) : f \in \mathcal{B}(\mathfrak{E})\} = \mathbb{C}^p$ holds for every point $\omega \in \mathfrak{h}_{\mathfrak{E}}^+$.*

(5) *The inequality $k_\omega^\chi(\omega) > 0$ holds for at least one point $\omega \in \mathfrak{h}_\chi$.*

(6) *The inequality $k_\omega^\chi(\omega) > 0$ holds for every point $\omega \in \mathfrak{h}_\chi$.*

(7) *The equality $\{f(\omega) : f \in \mathcal{H}(\chi)\} = \mathbb{C}^p$ holds for at least one point $\omega \in \mathfrak{h}_\chi$.*

(8) *The equality $\{f(\omega) : f \in \mathcal{H}(\chi)\} = \mathbb{C}^p$ holds for every point $\omega \in \mathfrak{h}_\chi$.*

Moreover, if $\mathfrak{E} \in \Pi^{p\times 2p}$, then the preceding eight equivalences hold with $\mathfrak{h}_{\mathfrak{E}}$ in place of $\mathfrak{h}_{\mathfrak{E}}^+$ in (1)–(4).

Proof The last four equivalences follow from Lemma 5.41, since $\chi \in \mathcal{S}_{in}^{p\times p}$. The implication (1) $\Longrightarrow$ (2) follows from (5) $\Longrightarrow$ (6) and formula (5.111), which expresses the kernel $K_\omega^{\mathfrak{E}}(\lambda)$ in terms of the kernel $k_\omega^{\chi}(\lambda)$. The converse implication (2) $\Longrightarrow$ (1) is self-evident. The rest follows from Lemma 5.41. □

5.11 Regular de Branges matrices $\mathfrak{E}$ and spaces $\mathcal{B}(\mathfrak{E})$

Let $\mathfrak{E}$ be a de Branges matrix. Then the space $\mathcal{B}(\mathfrak{E})$ will be called a **regular de Branges space** if it is R_α invariant for every point $\alpha \in \mathfrak{h}_{\mathfrak{E}}^+$; $\mathfrak{E}$ will be called a **regular de Branges matrix** if

$$\mathfrak{E} \in \Pi^{p\times 2p}, \quad \rho_\alpha^{-1} E_+^{-1} \in H_2^{p\times p} \quad \text{and} \quad \rho_{\overline{\alpha}}^{-1} E_-^{-1} \in K_2^{p\times p}, \tag{5.123}$$

for at least one (and hence every) point $\alpha \in \mathbb{C}_+$.

Lemma 5.67 *If $\mathfrak{E} = [E_- \quad E_+]$ is a de Branges matrix, then the following implications are in force:*

(a) *$\mathfrak{E}$ is a regular de Branges matrix* $\Longrightarrow$

(b) *$\mathcal{B}(\mathfrak{E})$ is a regular de Branges space* $\Longrightarrow$

(c) *$\mathcal{B}(\mathfrak{E})$ is R_α invariant for at least one point $\alpha \in \mathfrak{h}_{\mathfrak{E}}^+$;*

i.e., (a) $\Longrightarrow$ (b) $\Longrightarrow$ (c). If

$$K_\omega^{\mathfrak{E}}(\omega) > 0 \quad \text{for at least one (and hence every) point } \omega \in \mathfrak{h}_{\mathfrak{E}}^+, \tag{5.124}$$

then (c) $\Longrightarrow$ (a) and hence (a) $\Longleftrightarrow$ (b) $\Longleftrightarrow$ (c).

Proof If $\mathfrak{E} = [E_- \quad E_+]$ is a de Branges matrix and $f = E_+ g$, then $f \in \mathcal{B}(\mathfrak{E}) \Longleftrightarrow g \in \mathcal{H}(\chi)$, and

$$(R_\alpha f)(\lambda) = E_+(\lambda)(R_\alpha g)(\lambda) + (R_\alpha E_+)(\lambda) g(\alpha) \tag{5.125}$$

for every $\alpha \in \mathfrak{h}_{\mathfrak{E}}^+$. Therefore, since $R_\alpha g \in \mathcal{H}(\chi)$ for every $\alpha \in \mathbb{C}_+$,

$$R_\alpha f \in \mathcal{B}(\mathfrak{E}) \Longleftrightarrow (R_\alpha E_+)(\lambda) g(\alpha) \in \mathcal{B}(\mathfrak{E}), \tag{5.126}$$

i.e., in view of Lemma 5.62, $R_\alpha f \in \mathcal{B}(\mathfrak{E})$ if and only if

$$E_+^{-1}(R_\alpha E_+) u \in H_2^p \quad \text{and} \quad E_-^{-1}(R_\alpha E_-) u \in K_2^p \quad \text{for } u = g(\alpha). \tag{5.127}$$

However, if (a) is in force, then Lemma 3.60 and the second condition in (5.123) imply that

$$(R_\alpha E_+^{-1})v \in H_2^p \quad \text{for every } v \in \mathbb{C}^p.$$

The choice $v = E_+(\alpha)g(\alpha)$ justifies the first condition in (5.127); the second is immediate from the third condition in (5.123). Thus, (a) $\Longrightarrow$ (b) $\Longrightarrow$ (c).

Suppose next that the extra constraint (5.124) is in force. Then the other seven conditions in Lemma 5.66 are also in force. In particular,

$$\{g(\alpha) : g \in \mathcal{H}(\chi)\} = \mathbb{C}^p \quad \text{for every } \alpha \in \mathbb{C}_+. \tag{5.128}$$

Moreover, since (5.124) implies that $E_+(\omega)^* E_+(\omega) > 0$ at every point $\omega \in \mathfrak{h}_{\mathfrak{E}}^+$, it follows that the mvf $E_+(\lambda)^{-1}$ is holomorphic in $\mathfrak{h}_{\mathfrak{E}}^+$. Thus, if (c) is in force for some point $\alpha \in \mathfrak{h}_{\mathfrak{E}}^+$, then the two conditions in (5.127) hold for every $u \in \mathbb{C}^p$ and hence (5.123) holds for the given point $\alpha \in \mathfrak{h}_{\mathfrak{E}}^+$. However, this suffices to insure that (5.123) holds for every point $\alpha \in \mathbb{C}_+$ and hence that *(c)* $\Longrightarrow$ *(a)*. □

Remark 5.68 *If $\mathfrak{E} = [E_- \quad E_+]$ is a regular de Branges matrix. then $\mathcal{B}(\mathfrak{E})$ is a RKHS of $p \times 1$ holomorphic vvf's on $\mathfrak{h}_{\mathfrak{E}}$ and its RK, $K_\omega^{\mathfrak{E}}(\lambda)$, is defined on $\mathfrak{h}_{\mathfrak{E}} \times \mathfrak{h}_{\mathfrak{E}}$ by formula (5.122). Moreover, the proof of Lemma 5.67 is easily adapted to show that $\mathcal{B}(\mathfrak{E})$ is R_α invariant for every point $\alpha \in \mathfrak{h}_{\mathfrak{E}} \cap \mathfrak{h}_\chi$.*

A simple example

A simple example of a regular de Branges matrix is obtained by setting

$$\mathfrak{E}(\lambda) = \begin{bmatrix} E_-(\lambda) & E_+(\lambda) \end{bmatrix} = \begin{bmatrix} b_3(\lambda) & b_4^\#(\lambda) \end{bmatrix},$$

for any choice of $b_3 \in \mathcal{S}_{in}^{p\times p}$ and $b_4 \in \mathcal{S}_{in}^{p\times p}$. Then,

$$\chi = E_+^{-1} E_- = b_4 b_3$$

and the RK

$$K_\omega^{\mathfrak{E}}(\lambda) = \frac{b_4^\#(\lambda) b_4^\#(\omega)^* - b_3(\lambda) b_3(\omega)^*}{\rho_\omega(\lambda)} = \ell_\omega^{b_4}(\lambda) + k_\omega^{b_3}(\lambda).$$

Thus,

$$\mathcal{B}(\mathfrak{E}) = \mathcal{H}_*(b_4) \oplus \mathcal{H}(b_3),$$

which reduces to $\mathcal{H}(b)$ with $b = b_3$ if $b_4(\lambda)$ is constant and to $\mathcal{H}_*(b)$ with $b = b_4$ if $b_3(\lambda)$ is constant. If $b_3 = e_\alpha I_p$ and $b_4 = e_\beta I_p$ with $\alpha \geq 0$ and $\beta \geq 0$, then

$$\mathcal{B}(\mathfrak{E}) = \left\{ \int_{-\beta}^{\alpha} e^{i\lambda t} h(t) dt : h \in L_2^p([-\beta, \alpha]) \right\}$$

is the Paley-Wiener space of entire functions $\widehat{h}(\lambda)$ of exponential type that belong to $L_2^p(\mathbb{R})$ and are subject to the growth constraints $\tau^+(\widehat{h}) \leq \beta$ and $\tau^-(\widehat{h}) \leq \alpha$.

Theorem 5.69 *Let $\mathfrak{E} = [E_- \quad E_+]$ be a de Branges matrix, let $\chi = E_+^{-1} E_-$ and suppose that*

$$0 \in \mathfrak{h}_{\mathfrak{E}} \quad \text{and} \quad E_+(0) = E_-(0) = I_p. \tag{5.129}$$

Then:

(1) *$\mathfrak{E}$ is a regular de Branges matrix if and only if $\mathfrak{E} \in \Pi^{p \times 2p}$ and $R_0 E_+ \xi \in \mathcal{B}(\mathfrak{E})$ and $R_0 E_- \xi \in \mathcal{B}(\mathfrak{E})$ for every $\xi \in \mathbb{C}^p$.*

(2) *If $\mathfrak{E}$ is a regular de Branges matrix, then $\mathcal{B}(\mathfrak{E})$ is R_0 invariant.*

(3) *If $\mathcal{B}(\mathfrak{E})$ is R_0 invariant and $-i\chi'(0) > 0$, then $\mathfrak{E}$ is a regular de Branges matrix,*

Proof In view of Theorem 5.40, $0 \in \mathfrak{h}_g$ for every $g \in \mathcal{H}(\chi)$, since $0 \in \mathfrak{h}_\chi$. Consequently, $0 \in \mathfrak{h}_f$ for every $f \in \mathcal{B}(\mathfrak{E})$. Assertion (1) now follows from Lemmas 3.60 and 5.62 and Theorem 5.40. The proof of (2) is similar to the proof of Lemma 5.67.

Finally, suppose that the assumptions in (3) hold. Then, in view of Lemma 5.41, $\{g(0) : g \in \mathcal{H}(\chi)\} = \mathbb{C}^p$, and, by Theorem 5.40, $R_0 \chi \xi \in \mathcal{H}(\chi)$ for every $\xi \in \mathbb{C}^p$. Thus, formula (5.125) with $\alpha = 0$ implies that $E_+^{-1} R_0 E_+ \xi \in \mathcal{H}(\chi)$ for every $\xi \in \mathbb{C}^p$. Therefore, E_+ and $E_- = E_+ \chi$ both belong to $\Pi^{p \times p}$ and $R_0 E_\pm \xi \in \mathcal{B}(\mathfrak{E})$ for every $\xi \in \mathbb{C}^p$. Thus, (3) follows from (1). □

5.12 Connections between A and $\mathfrak{E}$

This section focuses on the connections between the mvf $A \in \mathcal{U}(J_p)$ and the de Branges matrix $\mathfrak{E}$ that is defined by the formula

$$\mathfrak{E}(\lambda) = [E_-(\lambda) \quad E_+(\lambda)] = \sqrt{2}[0 \quad I_p]\, A(\lambda) \mathfrak{V} \tag{5.130}$$

or, equivalently, in terms of the blocks a_{ij} of A and b_{ij} of $B = A\mathfrak{V}$,

$$E_- = a_{22} - a_{21} = \sqrt{2}b_{21} \quad \text{and} \quad E_+ = a_{22} + a_{21} = \sqrt{2}b_{22}. \tag{5.131}$$

Lemma 5.70 *The components $E_\pm(\lambda)$ of the mvf $\mathfrak{E}(\lambda)$ defined by formula (5.130) enjoy the following properties:*

(1) $E_\pm \in \Pi^{p\times p}$.

(2) $E_+(\lambda)E_+(\lambda)^* - E_-(\lambda)E_-(\lambda)^* = -\mathfrak{E}(\lambda)j_p\mathfrak{E}(\lambda)^* \geq 0$ *in* $\mathfrak{h}^+_{\mathfrak{E}}$.

(3) $E_+(\mu)E_+(\mu)^* - E_-(\mu)E_-(\mu)^* = -\mathfrak{E}(\mu)j_p\mathfrak{E}(\mu)^* = 0$ *for almost all points* $\mu \in \mathbb{R}$.

(4) $\det E_+(\omega) \neq 0$ *for at least one point* $\omega \in \mathfrak{h}^+_{\mathfrak{E}}$, *i.e.*, $\operatorname{rank} E_+ = p$.

(5) *The mvf* $\chi(\lambda) = E_+(\lambda)^{-1}E_-(\lambda)$ *belongs to the class* $\mathcal{S}_{in}^{p\times p}$.

(6) $\rho_i^{-1}E_+^{-1} \in H_2^{p\times p}$ *and* $\rho_{-i}^{-1}E_-^{-1} \in K_2^{p\times p}$.

(7) $\Delta(\mu) = E_+(\mu)^{-*}E_+(\mu)^{-1} = E_-(\mu)^{-*}E_-(\mu)^{-1}$ *a.e. on* $\mathbb{R}$.

(8) $\Delta \in \widetilde{L}_1^{p\times p}$.

Proof (1) is immediate from the fact that $A \in \Pi^{m\times m}$, while (2) and (3) are easy consequences of the fact that $A \in \mathcal{U}(J_p)$ and $\mathfrak{V}J_p\mathfrak{V} = j_p$.

Next, if assertion (4) is false, then $\det E_+(\lambda) \equiv 0$ on $\mathfrak{h}^+_{\mathfrak{E}}$, which in turn leads to the following sequence of implications:

$$\begin{aligned} \det \mathfrak{E}(\lambda)\mathfrak{E}(\lambda)^* \equiv 0 \text{ on } \mathfrak{h}^+_{\mathfrak{E}} &\implies \det A(\lambda) \equiv 0 \text{ on } \mathfrak{h}_A \\ &\implies \det A(\mu) = 0 \text{ a.e on } \mathbb{R}. \end{aligned}$$

But on the other hand, $|A(\mu)| = 1$ a.e on $\mathbb{R}$, since $A \in \mathcal{U}(J_p)$. Consequently, (4) holds and the mvf's $E_+(\lambda)^{-1}$ and $\chi(\lambda)$ are meromorphic in $\mathbb{C}_+$. Thus (5) follows from (2) and (3).

Assertion (6) is equivalent to the pair of assertions (4) and (8) in Lemma 4.35, since $E_+ = \sqrt{2}b_{22}$ and $E_- = \sqrt{2}b_{21}$.

Assertions (7) and (8) follow from (5) and (6). □

Theorem 5.71 *If $A \in \mathcal{U}(J_p)$, then the mvf $\mathfrak{E} = [E_- \quad E_+]$ that is defined by formula (5.130) is a regular de Branges matrix. Conversely, if $\mathfrak{E} = [E_- \quad E_+]$ is a regular de Branges matrix, then a family of mvf's $A \in \mathcal{U}(J_p)$ exists such that formula (5.130) holds.*

Proof The first part of the theorem follows from Lemma 5.70. The converse statement will be established in Theorem 9.18. □

In Section 9.4 it will be shown that every regular de Branges matrix $\mathfrak{E}$ may be obtained from the bottom block row of some mvf $A \in \mathcal{U}(J_p)$ by formula (5.130). Moreover, the set of mvf's $A \in \mathcal{U}(J_p)$ that correspond to a given regular de Branges matrix in this way will be described.

Corollary 5.72 *Let $A \in \mathcal{U}(J_p)$ and let the mvf $\mathfrak{E} = [E_- \quad E_+]$ be defined by the bottom $p \times 2p$ block row $[a_{21} \quad a_{22}]$ of the mvf A by formula (5.130). Then $\mathfrak{E}$ is a regular de Branges matrix and $\mathcal{B}(\mathfrak{E})$ is a regular de Branges space.*

Proof This follows from Theorem 5.71 and (1) of Lemma 5.67. □

5.13 A factorization and parametrization of mvf's $A \in \mathcal{U}(J_p)$.

In this section we shall present some results that are adapted from [ArD01b] on the factorization and parametrization of mvf's $A \in \mathcal{U}(J_p)$ with prescribed blocks $a_{21}(\lambda)$ and $a_{22}(\lambda)$. These blocks will be specified via formula (5.130) in terms of a de Branges matrix $\mathfrak{E} = [E_+ \quad E_-]$ that satisfies the auxiliary conditions (5.123), and the mvf $c = T_A[I_p]$.

Theorem 5.73 *If $A \in \mathcal{U}(J_p)$, $\mathfrak{E}$ is given by (5.130) and $c = T_A[I_p]$, then:*

(1) *$\mathfrak{E}$ is a regular de Branges matrix and $c \in \Pi \cap \mathcal{C}^{p\times p}$.*

(2) *The mvf A can be recovered from $\mathfrak{E}$ and c by the formula*

$$A = \frac{1}{\sqrt{2}} \begin{bmatrix} -c^{\#}E_- & cE_+ \\ E_- & E_+ \end{bmatrix} \mathfrak{V}. \tag{5.132}$$

(3) *The mvf c admits an essentially unique decomposition of the form*

$$c = c_s + c_a \tag{5.133}$$

with components $c_s \in \mathcal{C}_{sing}^{p\times p}$ and $c_a \in \mathcal{C}_a^{p\times p}$. Moreover, both of the components c_s and c_a belong to the class $\Pi^{p\times p}$, $c_s + c_s^{\#} = 0$ and

$$c_a(\lambda) = i\alpha + \frac{1}{\pi i}\int_{-\infty}^{\infty}\left\{\frac{1}{\mu-\lambda} - \frac{\mu}{1+\mu^2}\right\} E_+(\mu)^{-*}E_+(\mu)^{-1}d\mu \quad \text{for } \lambda \in \mathbb{C}_+ \tag{5.134}$$

and some Hermitian matrix $\alpha \in \mathbb{C}^{p\times p}$.

(4) *The given mvf* $A \in \mathcal{U}(J_p)$ *admits a factorization of the form*

$$A(\lambda) = A_s(\lambda) A_a(\lambda), \tag{5.135}$$

where

$$A_s(\lambda) = \begin{bmatrix} I_p & c_s(\lambda) \\ 0 & I_p \end{bmatrix} \tag{5.136}$$

and

$$A_a(\lambda) = \frac{1}{\sqrt{2}} \begin{bmatrix} -c_a^{\#}(\lambda)E_-(\lambda) & c_a(\lambda)E_+(\lambda) \\ E_-(\lambda) & E_+(\lambda) \end{bmatrix} \mathfrak{V}. \tag{5.137}$$

Moreover, $A_s \in \mathcal{U}_S(J_p)$, $A_a \in \mathcal{U}(J_p)$ *and the mvf's* c_a *and* A_a *are uniquely determined by* $\mathfrak{E}$ *up to an additive constant* $i\alpha$ *in (5.134) and a corresponding left constant multiplicative factor in (5.137) that is of the form*

$$\begin{bmatrix} I_p & i\alpha \\ 0 & I_p \end{bmatrix} \quad \textit{with} \quad \alpha = \alpha^* \in \mathbb{C}^{p\times p}. \tag{5.138}$$

Conversely, if $\mathfrak{E}$ *is a regular de Branges matrix and if* $c = c_s + c_a$, *where* c_s *is any mvf from the class* $\mathcal{C}_{sing}^{p\times p}$ *and* c_a *is defined by formula (5.134), then* $c_a \in \Pi^{p\times p}$; *the mvf* A *that is defined by formulas (5.132)–(5.134) belongs to the class* $\mathcal{U}(J_p)$; *and (5.130) holds.*

Proof The first assertion in (1) follows from Lemma 5.70; the second from the formula $c = T_A[I_p]$, since $A \in \mathcal{U}(J_p)$ and $\mathcal{U}(J_p) \subset \Pi^{m\times m}$. Formula (5.132) and the converse implications of the theorem follow from Theorem 9.18, which will be established in Chapter 9. Assertion (3) follows from the equality

$$\Re c(\mu) = \Re c_a(\mu) = E_+(\mu)^{-*}E_+(\mu)^{-1} \quad \text{a.e. on } \mathbb{R}.$$

Finally assertion (4) follows from formulas (5.132) and (5.133). □

An mvf $A \in \mathcal{U}(J_p)$ is called **perfect** if $T_A[I_p] \in \mathcal{C}_a^{p\times p}$. Recall that $c \in \mathcal{C}_a^{p\times p}$ if and only if (a) $\lim_{\nu\uparrow\infty} \nu^{-1}c(i\nu) = 0$ and (b) the spectral function $\sigma(\mu)$ is locally absolutely continuous. Moreover (a) and (b) hold if and only if (3.13) holds.

If $A \in \mathcal{E} \cap \mathcal{U}(J)$, then (b) is automatically in force and hence A is perfect if and only if $c = T_A[I_p]$ meets condition (a).

Theorem 5.74 *If $A \in \mathcal{U}(J_p)$ is perfect and $\mathfrak{E}$ is given by (5.130), then $\mathfrak{h}_A = \mathfrak{h}_E$. Moreover, $\mathfrak{E}$ is rational (resp., meromorphic, entire) if and only if A is rational (resp., meromorphic, entire).*

Proof The inclusion $\mathfrak{h}_A \subseteq \mathfrak{h}_{\mathfrak{E}}$ is obvious. The proof of the opposite inclusion rests on the fact that if $A \in \mathcal{U}(J_p)$ is perfect, then $c = T_A[I_p]$ belongs to $\Pi \cap \mathcal{C}_a^{p\times p}$ and A can be recovered from $\mathfrak{E}$ and c by formula (5.132) up to a J_p-unitary constant factor on the left that does not effect $\mathfrak{h}_A$. Moreover, as

$$c = -c^{\#} + 2(E_+^{\#})^{-1}E_+^{-1} = -c^{\#} + 2(E_-^{\#})^{-1}E_-^{-1}, \tag{5.139}$$

the entries in the top block row of A are given by the formulas

$$cE_+ = -c^{\#}E_+ + 2(E_+^{\#})^{-1} \quad \text{and} \quad -c^{\#}E_- = cE_- - 2(E_-^{\#})^{-1}. \tag{5.140}$$

Suppose now that $\lambda \in \mathfrak{h}_{\mathfrak{E}}^-$. Then $c^{\#}$, E_- and E_+ are all holomorphic in a neighborhood of λ as is cE_+ thanks to the first formula in (5.140) and (6) of Lemma 5.70.

Similarly if $\lambda \in \mathfrak{h}_{\mathfrak{E}}^+$, then $\lambda \in \mathfrak{h}_A^+$ thanks to to the second formula in (5.140). Finally, if $\lambda \in \mathfrak{h}_{\mathfrak{E}} \cap \mathbb{R}$, then c and $c^{\#}$ are holomorphic in a neighborhood of λ by Remark 3.7, whereas (6) of Lemma 5.70 guarantees that E_+^{-1} and E_-^{-1} have no poles on $\mathbb{R}$. Therefore, $\mathfrak{h}_A = \mathfrak{h}_{\mathfrak{E}}$ and hence A is entire if and only if $\mathfrak{E}$ is entire.

If A is rational (resp., meromorphic) then formula (5.132) clearly implies that $\mathfrak{E}$ is rational (resp., meromorphic). Conversely, if $\mathfrak{E}$ is rational, then formula (5.134) implies that c is rational in $\mathbb{C}_+$. Therefore, its boundary values and its extension to $\mathbb{C}_-$ via (5.139) are also rational, as is $c^{\#}$. Thus A is rational. Similarly, if $\mathfrak{E}$ is meromorphic in $\mathbb{C}$, then (6) of Lemma 5.70 and formula (5.140) imply that A is meromorphic on $\mathbb{C} \setminus \mathbb{R}$ and $\mathbb{R} \subset \mathfrak{h}_{\mathfrak{E}}$. Thus, in view of Remark 3.7, c and $c^{\#}$ are also holomorphic on $\mathbb{R}$. Therefore. A is meromorphic on $\mathbb{C}$. □

Theorem 5.75 *Let $\mathfrak{E} = [E_- \quad E_+]$ be a regular de Branges matrix and let $\Delta_{\mathfrak{E}}(\mu) = E_+(\mu)^{-*}E_+(\mu)^{-1}$ a.e. on $\mathbb{R}$. Then:*

(1) *There exists a perfect mvf $A \in \mathcal{U}(J_p)$ such that (5.130) holds. Moreover, A is uniquely defined by $\mathfrak{E}$ up to a left constant J_p-unitary factor of the form (5.138) by formulas (5.134) and (5.137). There is only one such perfect mvf A_a for which $c_a(i) > 0$: if $\alpha = 0$ in (5.134).*

(2) *If $\mathfrak{E}$ satisfies the conditions (5.129), then there is exactly one perfect mvf $A \in \mathcal{U}_0(J_p)$ for which (5.130) holds. It is given by formula (5.132), where*

$$c(\lambda) = I_p + \frac{\lambda}{\pi i} \int_{-\infty}^{\infty} \frac{1}{\mu(\mu - \lambda)} \{\Delta_{\mathfrak{E}}(\mu) - I_p\} d\mu. \tag{5.141}$$

Proof The first assertion follows from Theorem 5.73 with $c = c_a$ and $A = A_a$ given by formulas (5.134) and (5.137), respectively.

Assertion (2) follows from Theorem 5.74 and the identity

$$c(\lambda) - I_p = i\alpha + \frac{1}{\pi i} \int_{-\infty}^{\infty} \left\{ \frac{\lambda}{\mu(\mu - \lambda)} + \frac{1}{\mu(1 + \mu^2)} \right\} \{\Delta_{\mathfrak{E}}(\mu) - I_p\} d\mu$$

by choosing

$$\alpha = \frac{1}{\pi} \int_{-\infty}^{\infty} \frac{1}{\mu(1 + \mu^2)} \{\Delta_{\mathfrak{E}}(\mu) - I_p\} d\mu.$$

□

In view of Theorem 5.73, formula (5.130) defines a one to one correspondence between perfect mvf's $A \in \mathcal{U}_0(J_p)$ and regular de Branges matrices $\mathfrak{E} = [E_- \quad E_+]$ with the properties (5.129).

The next theorem describes the connection between the corresponding RKHS's $\mathcal{H}(A)$ and $\mathcal{B}(\mathfrak{E})$.

Theorem 5.76 *Let $A \in \mathcal{U}(J_p)$ and let $\mathfrak{E}$, c_a, c_s, A_a and A_s be defined by A as in Theorem 5.73. Let U_2 denote the operator that is defined on $\mathcal{H}(A)$ by the formula*

$$(U_2 f)(\lambda) = \sqrt{2}[0 \quad I_p] f(\lambda) \quad \text{for} \quad f \in \mathcal{H}(A). \tag{5.142}$$

Then:

(1) $\mathcal{H}(A_s) = \{f \in \mathcal{H}(A) : (U_2 f)(\lambda) \equiv 0\}$, *i.e.,*

$$\mathcal{H}(A_s) = \ker U_2. \tag{5.143}$$

(2) *The orthogonal complement of $\mathcal{H}(A_s)$ in $\mathcal{H}(A)$ is equal to $A_s\mathcal{H}(A_a)$, i.e.,*

$$\mathcal{H}(A) = \mathcal{H}(A_s) \oplus A_s\mathcal{H}(A_a). \tag{5.144}$$

(3) *The operator U_2 is a partial isometry from $\mathcal{H}(A)$ onto $\mathcal{B}(\mathfrak{E})$ with*

kernel $\mathcal{H}(A_s)$, i.e., U_2 maps $\mathcal{H}(A) \ominus \mathcal{H}(A_s)$ isometrically onto $\mathcal{B}(\mathfrak{E})$. Moreover,

$$\mathcal{H}(A_s) = \begin{matrix} \mathcal{L}(c_s) \\ \oplus \\ \{0\} \end{matrix} \tag{5.145}$$

and $\mathcal{L}(c_s)$ is the RKHS with RK

$$k_\omega^{c_s}(\lambda) = \frac{c_s(\lambda) + c_s(\omega)^*}{\rho_\omega(\lambda)}. \tag{5.146}$$

(4) *The operator U_2 is unitary from $\mathcal{H}(A)$ onto $\mathcal{B}(\mathfrak{E})$ if and only if the mvf A is perfect.*

Proof Let $N_2^* = [0 \quad I_p]$, $\mathcal{H}_2(A) = \ker U_2$ and observe that

$$\begin{aligned} 2N_2^* K_\omega^A(\lambda) N_2 &= \rho_\omega(\lambda)^{-1} 2\{N_2^* J_p N_2 - N_2^* B(\lambda)\mathfrak{V} J_p \mathfrak{V}^* B(\omega)^* N_2\} \\ &= \rho_\omega(\lambda)^{-1}\{0 - \mathfrak{E}(\lambda) j_p \mathfrak{E}(\omega)^*\} \qquad (5.147) \\ &= K_\omega^{\mathfrak{E}}(\lambda). \end{aligned}$$

Moreover, since

$$U_2 K_\omega^A(\lambda) N_2 \eta = \frac{1}{\sqrt{2}} K_\omega^{\mathfrak{E}}(\lambda)\eta, \tag{5.148}$$

the operator U_2 maps the linear manifold

$$\mathcal{L}_A = \left\{ \sum_{i=1}^n K_{\omega_i}^A(\lambda) N_2 \eta_i : \ \omega_i \in \mathfrak{h}_A, \ \eta_i \in \mathbb{C}^p \ \text{ and } \ n \geq 1 \right\} \tag{5.149}$$

onto the linear manifold

$$\mathcal{L}_{\mathfrak{E}} = \left\{ \frac{1}{\sqrt{2}} \sum_{i=1}^n K_{\omega_i}^{\mathfrak{E}}(\lambda)\eta_i : \ \omega_i \in \mathfrak{h}_A, \ \ \eta_i \in \mathbb{C}^p \ \text{ and } \ n \geq 1 \right\}.$$

Furthermore, the formula

$$\begin{aligned} 2\left\| \sum_{i=1}^n K_{\omega_i}^A N_2 \eta_i \right\|_{\mathcal{H}(A)}^2 &= 2\sum_{i,j=1}^n \eta_j^* N_2^* K_{\omega_i}^A(\omega_j) N_2 \eta_i \\ &= \sum_{i,j=1}^n \eta_j^* K_{\omega_i}^{\mathfrak{E}}(\omega_j)\eta_i = \left\| \sum_{i=1}^n K_{\omega_i}^{\mathfrak{E}} \eta_i \right\|_{\mathcal{B}(\mathfrak{E})}^2 \qquad (5.150) \end{aligned}$$

exhibits U_2 as an isometry from $\mathcal{L}_A$ into $\mathcal{B}(\mathfrak{E})$:

$$\|f\|^2_{\mathcal{H}(A)} = \|U_2 f\|^2_{\mathcal{B}(\mathfrak{E})} \quad \text{for} \quad f \in \mathcal{L}_A. \tag{5.151}$$

Therefore, since $\mathcal{L}_A^\perp = \mathcal{H}_2(A)$ and the set $U_2\mathcal{L}_A$ is dense in $\mathcal{B}(\mathfrak{E})$, the operator U_2 is a partial isometry from $\mathcal{H}(A) = \overline{\mathcal{L}_A} \oplus \mathcal{L}_A^\perp$ onto $\mathcal{B}(\mathfrak{E})$ with kernel $\mathcal{H}_2(A)$.

Next, it follows readily from formula (5.136) that

$$K_\omega^{A_s}(\lambda) = \frac{1}{\rho_\omega(\lambda)} \begin{bmatrix} c_s(\lambda) + c_s(\omega)^* & 0 \\ 0 & 0 \end{bmatrix}$$

and hence that

$$N_2^* K_\omega^{A_s}(\lambda) = [0 \quad 0].$$

Thus, $\mathcal{H}(A_s) \subseteq \mathcal{H}_2(A)$ (as linear spaces).

Finally, to obtain the opposite inclusion, observe that since $\mathcal{H}_2(A)$ is a closed subset of $\mathcal{H}(A)$ that is R_α invariant for every point $\alpha \in \mathfrak{h}_A$, Theorem 5.50 guarantees that there exists a pair of mvf's $A_j \in \mathcal{U}(J_p)$, $j = 1, 2$, such that

$$\mathcal{H}_2(A) = \mathcal{H}(A_1), \quad A(\lambda) = A_1(\lambda)A_2(\lambda)$$

and

$$\mathcal{H}(A) = \mathcal{H}(A_1) \oplus A_1\mathcal{H}(A_2).$$

Moreover, upon writing

$$A_1(\lambda) = A_s^{(1)}(\lambda)A_a^{(1)}(\lambda) \quad \text{and} \quad A_2(\lambda) = A_s^{(2)}(\lambda)A_a^{(2)}(\lambda)$$

in terms of the formulas given in Theorem 5.73 and observing that

$$\begin{aligned} N_2^* K_\omega^{A_1}(\lambda) = 0 \quad &\Longleftrightarrow \quad N_2^*\{A_1(\overline{\omega}) - A_1(\lambda)\}J_p A_1(\omega)^* = 0 \\ &\qquad \text{for } \lambda, \omega, \overline{\omega} \in \mathfrak{h}_{A_1} \\ &\Longleftrightarrow \quad N_2^*\{A_1(\overline{\omega}) - A_1(\lambda)\} = 0 \text{ for } \lambda, \overline{\omega} \in \mathfrak{h}_{A_1} \\ &\Longleftrightarrow \quad N_2^*\{A_a^{(1)}(\lambda) - A_a^{(1)}(\omega)\} = 0 \text{ for } \lambda, \omega \in \mathfrak{h}_{A_1}, \end{aligned}$$

we see that the bottom block row of $A_a^{(1)}(\lambda)$ is constant. Therefore, by formulas (5.137) and (5.134), $A_a^{(1)}(\lambda)$ is a J_p–unitary constant matrix, which may be taken equal to I_m. Thus

$$A(\lambda) = A_s^{(1)}(\lambda)A_s^{(2)}(\lambda)A_a^{(2)}(\lambda)$$

from which it follows that

$$A_s(\lambda) = A_s^{(1)}(\lambda) A_s^{(2)}(\lambda)$$

and hence that

$$\mathcal{H}(A_1) \subseteq \mathcal{H}(A_s),$$

as needed to complete the proof of (3). Finally, (4) is immediate from (1)–(3). □

Remark 5.77 *It is readily seen that the mvf $A_s(\lambda)$ in the factorization (5.135) belongs to the class $\mathcal{U}_S(J_p)$, since $c_s \in \mathcal{N}_+^{p\times p}$. Thus,*

$$A \in \mathcal{U}_{\ell R}(J_p) \Longrightarrow A_s(\lambda) \text{ is constant.} \tag{5.152}$$

The converse is not true. In order to have $A \in \mathcal{U}_{\ell R}(J_p)$, a second condition is also needed; see Section 7.4. The inclusion

$$\mathcal{U}_{\ell sR} \cup \mathcal{U}_{rsR} \subseteq \mathcal{U}_{\ell R} \cap \mathcal{U}_{rR}$$

guarantees that if $A \in \mathcal{U}_{\ell sR}$ or $A \in \mathcal{U}_{rsR}$, then the operator U_2 is a unitary operator from $\mathcal{H}(A)$ onto $\mathcal{B}(\mathfrak{E})$.

5.14 A description of $\mathcal{H}(W) \cap L_2^m$

The classes $\mathcal{U}_S(J)$, $\mathcal{U}_{rR}(J)$ and $\mathcal{U}_{srR}(J)$ of singular, right regular and strongly right regular J-inner mvf's will be characterized in terms of the linear manifold $\mathcal{L}_U = \mathcal{H}(U) \cap L_2^m(\mathbb{R})$. Since $\mathcal{H}(U) \subset L_2^m(\mathbb{R})$ if $J = \pm I_m$, only the case $J \neq \pm I_m$ is of interest and thus, it suffices to focus on $J = j_{pq}$ and $\mathcal{H}(W) \cap L_2^m(\mathbb{R})$ for $W \in \mathcal{U}(j_{pq})$.

To every $W \in \mathcal{U}(j_{pq})$ we associate a pair of operators Γ_{11}, Γ_{22} and a pair of linear manifolds $\mathcal{L}_W^+$, $\mathcal{L}_W^-$ which are defined in terms of $\{b_1, b_2\} \in ap(W)$ and $s_{12} = T_W[0]$:

$$\Gamma_{11} = \Pi_{\mathcal{H}(b_1)} M_{s_{12}}|_{H_2^q}, \quad \Gamma_{22} = \Pi_- M_{s_{12}}|_{\mathcal{H}_*(b_2)} \tag{5.153}$$

$$\mathcal{L}_W^+ = \left\{ \begin{bmatrix} g \\ \Gamma_{11}^* g \end{bmatrix} : g \in \mathcal{H}(b_1) \right\} \text{ and } \mathcal{L}_W^- = \left\{ \begin{bmatrix} \Gamma_{22} h \\ h \end{bmatrix} : h \in \mathcal{H}_*(b_2) \right\}. \tag{5.154}$$

Lemma 5.78 *If $W \in \mathcal{U}(j_{pq})$, then:*

(1) $\mathcal{L}_W^+ = \mathcal{H}(W) \cap H_2^m$.

(2) *The formula*

$$\left\| \begin{bmatrix} g \\ \Gamma_{11}^* g \end{bmatrix} \right\|_{\mathcal{H}(W)}^2 = \left\langle \begin{bmatrix} I & -\Gamma_{11} \\ -\Gamma_{11}^* & I \end{bmatrix} \begin{bmatrix} g \\ \Gamma_{11}^* g \end{bmatrix}, \begin{bmatrix} g \\ \Gamma_{11}^* g \end{bmatrix} \right\rangle_{st}$$

$$= \langle (I - \Gamma_{11}\Gamma_{11}^*)g, g\rangle_{st}$$

holds for every $g \in \mathcal{H}(b_1)$. In this formula, $\Gamma_{11}^* = \Pi_+ M_{s_{12}^*}|_{\mathcal{H}(b_1)}$.

(3) *The formulas $\Gamma_{11} = \Pi_+ M_s|_{H_2^q}$ and*

$$\left\langle \begin{bmatrix} I_p & -s \\ -s^* & I_q \end{bmatrix} \begin{bmatrix} g \\ \Gamma_{11}^* g \end{bmatrix}, \begin{bmatrix} g \\ \Gamma_{11}^* g \end{bmatrix} \right\rangle_{st} = \left\| \begin{bmatrix} g \\ \Gamma_{11}^* g \end{bmatrix} \right\|_{\mathcal{H}(W)}^2$$

hold for every mvf $s \in T_W[\mathcal{S}^{p\times q}]$ and every $g \in \mathcal{H}(b_1)$.

Proof. If $f = \mathrm{col}(f, g)$ belongs to $\mathcal{H}(W) \cap H_2^m$, then (5.88) implies that $h = \Pi_+ s_{12}^* g$ and $s_{11}^* g \in K_2^p$. Thus, as $s_{11} = b_1\varphi_1$ and $\varphi_1 \in \mathcal{S}_{out}^{p\times p}$, it is readily checked that g is orthogonal to $b_1 H_2^p$ with respect to the standard inner product and hence that $g \in \mathcal{H}(b_1)$. Therefore $\mathcal{H}(W)\cap H_2^m \subseteq \mathcal{L}_W^+$. The opposite inclusion is even easier.

Since the verification of the formulas in (2) follows easily from (5.89), we turn next to (3). Let $\varepsilon \in \mathcal{S}^{p\times q}$. Then

$$\begin{aligned} T_W[\varepsilon] - T_W[0] &= (w_{11}\varepsilon + w_{12})(w_{21}\varepsilon + w_{22})^{-1} - w_{12}w_{22}^{-1} \\ &= \{w_{11}\varepsilon + w_{12} - w_{12}w_{22}^{-1}(w_{21}\varepsilon + w_{22})\}(w_{21}\varepsilon + w_{22})^{-1} \\ &= (w_{11} - w_{12}w_{22}^{-1}w_{21})\varepsilon(I_q - s_{21}\varepsilon)^{-1}s_{22} \\ &= w_{11}^{\#}\varepsilon(I_q - s_{21}\varepsilon)^{-1}s_{22} \\ &= s_{11}\varepsilon(I_q - s_{21}\varepsilon)^{-1}s_{22} \\ &= b_1\varphi_1\varepsilon(I_q - s_{21}\varepsilon)^{-1}\varphi_2 b_2 \\ &= b_1\psi b_2, \end{aligned}$$

where $\psi = \varphi_1\varepsilon(I_q - s_{21}\varepsilon)^{-1}\varphi_2$ belongs to $\mathcal{N}_+^{p\times q}$, since $(I_q - s_{21}\varepsilon)^{-1}$ belongs to $\mathcal{N}_+^{p\times q}$, by Lemma 3.54. Thus,

$$s(\lambda) - s_{12}(\lambda) = b_1(\lambda)\psi(\lambda)b_2(\lambda) \tag{5.155}$$

for some choice of $\psi \in H_\infty^{p\times q}$ and hence $\Gamma_{11} = \Pi_{\mathcal{H}(b_1)} M_s|_{H_2^q}$ and

$$\Pi_+(s^* - s_{12}^*)g = \Pi_+ b_2^* \psi^* b_1^* g = \Pi_+ b_2^* \psi^* \Pi_+ b_1^* g = 0 \tag{5.156}$$

for $g \in \mathcal{H}(b_1)$ and $s \in T_W[\mathcal{S}^{p\times q}]$. The asserted formula is now easily checked by direct computation. □

Lemma 5.79 *If $W \in \mathcal{U}(j_{pq})$, then:*

(1) $\mathcal{L}_W^- = \mathcal{H}(W) \cap K_2^m$.

(2) *The formula*

$$\left\| \begin{bmatrix} \Gamma_{22}h \\ h \end{bmatrix} \right\|_{\mathcal{H}(W)}^2 = \left\langle \begin{bmatrix} I & -\Gamma_{22} \\ -\Gamma_{22}^* & I \end{bmatrix} \begin{bmatrix} \Gamma_{22}h \\ h \end{bmatrix}, \begin{bmatrix} \Gamma_{22}h \\ h \end{bmatrix} \right\rangle_{st}$$
$$= \langle (I - \Gamma_{22}^*\Gamma_{22})h, h\rangle_{st}$$

is valid for every $h \in \mathcal{H}_*(b_2)$.

(3) *The formulas $\Gamma_{22} = \Pi_- M_s|_{\mathcal{H}_*(b_2)}$ and*

$$\left\langle \begin{bmatrix} I_p & -s \\ -s^* & I_q \end{bmatrix} \begin{bmatrix} \Gamma_{22}h \\ h \end{bmatrix}, \begin{bmatrix} \Gamma_{22}h \\ h \end{bmatrix} \right\rangle_{st} = \langle (I - \Gamma_{22}^*\Gamma_{22})h, h\rangle_{st}$$

are valid for for every mvf $s \in T_W[\mathcal{S}^{p\times q}]$ and every $h \in \mathcal{H}_*(b_2)$.

Proof The proof is much the same as the proof of Lemma 5.78. The verification of the third assertion rests on formula (5.155) and the evaluation

$$\Pi_-\{(s - s_{12})h\} = \Pi_- b_1 \psi b_2 h = \Pi_- b_1 \psi \Pi_- b_2 h = 0 \tag{5.157}$$

for $h \in \mathcal{H}_*(b_2)$. Thus

$$\Gamma_{22}h - sh = -\Pi_+ sh$$

and

$$-s^*\Gamma_{22}h + h = -s^*\Pi_- sh + h.$$

The rest is straightforward. □

We turn next to the sum

$$\mathcal{L}_W = \mathcal{L}_W^- \dotplus \mathcal{L}_W^+ \tag{5.158}$$

and its closure $\overline{\mathcal{L}_W}$ in $\mathcal{H}(W)$. The sum is direct because $\mathcal{L}_W^- \subseteq (K_2^m)$ and $\mathcal{L}_W^+ \subseteq H_2^m$. However, it is not an orthogonal sum in $\mathcal{H}(W)$. The inner product between elements in the two spaces can best be expressed in terms of the operator

$$\Gamma_{12} = \Pi_{\mathcal{H}(b_1)} M_{s_{12}}|_{\mathcal{H}_*(b_2)}. \tag{5.159}$$

Since

$$\Pi_{\mathcal{H}(b_1)} b_1 \psi b_2|_{\mathcal{H}_*(b_2)} = 0 \tag{5.160}$$

for $\psi \in H_\infty^{p \times q}$, it follows readily from (5.155), (5.156) and (5.157), respectively, that the operators Γ_{12}, Γ_{11} and Γ_{22} do not change if s_{12} is replaced by any $s \in T_W[\mathcal{S}^{p \times q}]$.

Lemma 5.80 *If $W \in \mathcal{U}(j_{pq})$, $g \in \mathcal{H}(b_1)$ and $h \in \mathcal{H}_*(b_2)$, then:*

(1) $\mathcal{L}_W = \mathcal{H}(W) \cap L_2^m$.

(2) $\left\langle \begin{bmatrix} g \\ \Gamma_{11}^* g \end{bmatrix}, \begin{bmatrix} \Gamma_{22} h \\ h \end{bmatrix} \right\rangle_{\mathcal{H}(W)} = -\langle \Gamma_{12}^* g, h \rangle_{st}$.

(3) $\left\langle \begin{bmatrix} I_p & -s \\ -s^* & I_q \end{bmatrix} \begin{bmatrix} g \\ \Gamma_{11}^* g \end{bmatrix}, \begin{bmatrix} \Gamma_{22} h \\ h \end{bmatrix} \right\rangle_{st} = -\langle \Gamma_{12}^* g, h \rangle_{st}$

and $\Gamma_{12} = \Pi_{\mathcal{H}(b_1)} M_s|_{\mathcal{H}_*(b_2)}$ for every choice of $s \in T_W[\mathcal{S}^{p \times q}]$.

Proof It is readily checked with the aid of Theorem 5.55 that if $f \in \mathcal{H}(W) \cap L_2^m$ is decomposed as $f = f_1 + f_2$ with $f_1 \in H_2^m$ and $f_2 \in K_2^m$, then $f_1 \in \mathcal{H}(W) \cap H_2^m$ and $f_2 \in \mathcal{H}(W) \cap K_2^m$. Therefore, in view of Lemmas 5.78 and 5.79, we see that $\mathcal{H}(W) \cap L_2^m \subseteq \mathcal{L}_W$. This serves to establish (1), since the opposite inclusion is self-evident. The rest is straightforward. □

The next theorem summarizes the main conclusions from the preceding three lemmas.

Theorem 5.81 *Let $W \in \mathcal{U}(j_{pq})$. Then*

$$\Gamma_{11} = \Pi_{\mathcal{H}(b_1)} M_s|_{H_2^q}, \quad \Gamma_{22} = \Pi_- M_s|_{\mathcal{H}_*(b_2)} \quad \textit{and}$$

$$\Gamma_{12} = \Pi_{\mathcal{H}(b_1)} M_s|_{\mathcal{H}_*(b_2)} \quad \textit{for every choice of } s \in T_W[\mathcal{S}^{p \times q}], \tag{5.161}$$

and:

(1) $\mathcal{L}_W = \mathcal{H}(W) \cap L_2^m$.

(2) *Every $f \in \mathcal{L}_W$ has a unique decomposition of the form*

$$f = \begin{bmatrix} g \\ \Gamma_{11}^* g \end{bmatrix} + \begin{bmatrix} \Gamma_{22} h \\ h \end{bmatrix} \quad \textit{with } g \in \mathcal{H}(b_1) \quad \textit{and} \quad h \in \mathcal{H}_*(b_2). \tag{5.162}$$

(3) *If f is as in (2), then*

$$\|f\|^2_{\mathcal{H}(W)} = \left\langle \begin{bmatrix} I - \Gamma_{11}\Gamma_{11}^* & -\Gamma_{12} \\ -\Gamma_{12}^* & I - \Gamma_{22}^*\Gamma_{22} \end{bmatrix} \begin{bmatrix} g \\ h \end{bmatrix}, \begin{bmatrix} g \\ h \end{bmatrix} \right\rangle_{st} . \tag{5.163}$$

(4) *If f is as in (2) and $s \in T_W[\mathcal{S}^{p\times q}]$, then*

$$\|f\|^2_{\mathcal{H}(W)} = \left\langle \begin{bmatrix} I_p & -s \\ -s^* & I_q \end{bmatrix} f, f \right\rangle_{st} . \tag{5.164}$$

Corollary 5.82 *If $W \in \mathcal{U}(j_{pq})$, then $\mathcal{L}_W$ is R_α invariant for every point $\alpha \in \mathfrak{h}_W$.*

Proof This is immediate from the identification of $\mathcal{L}_W$ in (1) of the theorem, since $\mathcal{H}(W)$ is R_α invariant for every point $\alpha \in \mathfrak{h}_W$ and $f \in \mathcal{H}(W)$ is holomorphic at all such points α, by Theorem 5.49. Moreover, if $f \in \mathcal{H}(W) \cap L_2^m$ and $\alpha \in \mathfrak{h}_W$, then $R_\alpha f \in L_2^m$. Thus, $R_\alpha f \in \mathcal{L}_W$. □

Lemma 5.83 *Let $U \in \mathcal{U}(J)$. Then:*

(1) *$U \in \mathcal{N}_+^{m\times m}$ if and only if $\mathcal{H}(U) \subseteq \mathcal{N}_+^m$.*

(2) *$U \in \mathcal{N}_-^{m\times m}$ if and only if $\mathcal{H}(U) \subseteq \mathcal{N}_-^m$, where $f \in \mathcal{N}_-^{m\times k} \iff f^\# \in \mathcal{N}_+^{k\times m}$.*

Proof Clearly it suffices to prove both statements for the special case that $J = j_{pq}$. Suppose first that $W \in \mathcal{N}_+^{m\times m}$ and let $f = \mathrm{col}(f, g)$ belong to $\mathcal{H}(W)$. Then $f \in \mathcal{N}_+^m$, since $s_{22}h \in H_2^q$ and $g - s_{12}h \in H_2^p$, by Theorem 5.55, and $s_{22} \in \mathcal{S}_{out}^{q\times q}$ when $W \in \mathcal{N}_+^{m\times m}$. The other direction follows from the formula for the reproducing kernel. This completes the proof of (1). The proof of (2) is similar. □

Corollary 5.84 *If $W \in \mathcal{U}(j_{pq}) \cap \mathcal{N}_+^{m\times m}$, then*

$$\mathcal{L}_W^+ = \mathcal{H}(W) \cap L_2^m = \mathcal{H}(W) \cap H_2^m .$$

If $W \in \mathcal{U}(j_{pq}) \cap \mathcal{N}_-$, then

$$\mathcal{L}_W^- = \mathcal{H}(W) \cap L_2^m = \mathcal{H}(W) \cap K_2^m .$$

Proof By the Smirnov maximum principle,

$$\mathcal{N}_+ \cap L_2^m = H_2^m \quad \text{and (by symmetry)} \quad \mathcal{N}_- \cap L_2^m = K_2^m .$$

□

5.15 Characterizations of the classes $\mathcal{U}_S(J)$, $\mathcal{U}_{rR}(J)$ and $\mathcal{U}_{rsR}(J)$

Let

$$\mathcal{H}(b_1, b_2) = \begin{array}{c} \mathcal{H}(b_1) \\ \oplus \\ \mathcal{H}_*(b_2) \end{array}.$$

Theorem 5.81 guarantees that the operator

$$L_W = \begin{bmatrix} I & \Gamma_{22} \\ \Gamma_{11}^* & I \end{bmatrix} : \mathcal{H}(b_1, b_2) \to \mathcal{H}(W) \tag{5.165}$$

is well defined, injective and bounded. The operator

$$\Delta_W = \begin{bmatrix} I - \Gamma_{11}\Gamma_{11}^* & -\Gamma_{12} \\ -\Gamma_{12}^* & I - \Gamma_{22}^*\Gamma_{22} \end{bmatrix} : \mathcal{H}(b_1, b_2) \to \mathcal{H}(b_1, b_2) \tag{5.166}$$

that appears in formula (5.163) is also bounded and (since $\Delta_W = L_W^* L_W$) nonnegative. We turn next to characterizations of the indicated subclasses of $\mathcal{U}(J)$ in terms of the properties of the set

$$\mathcal{L}_U = \mathcal{H}(U) \cap L_2^m.$$

Extensive use will be made of the next lemma and Theorem 5.50.

Lemma 5.85 *Let $W \in \mathcal{U}(j_{pq})$ and let $\overline{\mathcal{L}_W}$ denote the closure of $\mathcal{H}(W) \cap L_2^m$ in $\mathcal{H}(W)$. Then there exists an essentially unique mvf $W_1 \in \mathcal{U}(j_{pq})$ such that*

$$\overline{\mathcal{L}_W} = \mathcal{H}(W_1). \tag{5.167}$$

Moreover:

(1) $W_1^{-1}W \in \mathcal{U}_S(j_{pq})$.

(2) *$\mathcal{H}(W_1)$ is included isometrically in $\mathcal{H}(W)$, i.e.,*

$$\|f\|_{\mathcal{H}(W_1)} = \|f\|_{\mathcal{H}(W)} \quad \text{for every } f \in \mathcal{H}(W_1).$$

(3) *$\mathcal{L}_W = \mathcal{L}_{W_1}$ and $ap(W) = ap(W_1)$.*

Proof By Corollary 5.82, $\mathcal{L}_W$ is R_α invariant for every point $\alpha \in \mathfrak{h}_W$. Therefore, since R_α is a bounded operator in $\mathcal{H}(W)$ for each such point α, $\overline{\mathcal{L}_W}$ is also invariant under R_α. Thus, by Theorem 5.50, there exists an essentially unique mvf $W_1 \in \mathcal{U}(j_{pq})$ such that (5.167) holds, (2) holds and $W_1^{-1}W \in \mathcal{U}(J_{pq})$. Consequently,

$$\mathcal{L}_{W_1} = \mathcal{H}(W_1) \cap L_2^m \subseteq \mathcal{H}(W) \cap L_2^m = \mathcal{L}_W \subseteq \mathcal{L}_{W_1}.$$

Therefore,

$$\mathcal{L}_{W_1} = \mathcal{L}_W.$$

Let $\{b_1^{(1)}, b_2^{(1)}\} \in ap(W_1)$ and let $s \in T_W[\mathcal{S}^{p\times q}]$. Then, since $W_1^{-1}W \in \mathcal{U}(j_{pq})$, $s \in T_{W_1}[\mathcal{S}^{p\times q}]$. Therefore, $\mathcal{L}_{W_1}$ admits a description as in (2) of Theorem 5.81 in terms of the operators $\Gamma_{ij}^{(1)}$ that are defined just as in formulas (5.153) and (5.159), with the same s, but with $\{b_1^{(1)}, b_2^{(1)}\}$ in place of $\{b_1, b_2\}$. Thus, as $\mathcal{L}_{W_1} = \mathcal{L}_W$, it follows that $\mathcal{H}(b_1^{(1)}) = \mathcal{H}(b_1)$ and $\mathcal{H}_*(b_2^{(1)}) = \mathcal{H}_*(b_2)$ and hence that $b_1^{(1)} = b_1 v_1$ and $b_2^{(1)} = v_2 b_2$ for some pair of unitary matrices $v_1 \in \mathbb{C}^{p\times p}$ and $v_2 \in \mathbb{C}^{q\times q}$. This completes the proof of (3) and justifies the formulas

$$\Gamma_{ij}^{(1)} = \Gamma_{ij} \quad \text{for} \quad i, j = 1, 2. \tag{5.168}$$

In particular, $\{b_1, b_2\} \in ap(W_1)$, and hence, in view of Theorem 4.94, (1) holds. □

Theorem 5.86 *Let $U \in \mathcal{U}(J)$. Then:*

(1) $U \in \mathcal{U}_S(J) \Longleftrightarrow \mathcal{H}(U) \cap L_2^m = \{0\}$.

(2) $U \in \mathcal{U}_{rR}(J) \Longleftrightarrow \mathcal{H}(U) \cap L_2^m$ *is dense in* $\mathcal{H}(U)$.

(3) $U \in \mathcal{U}_{rsR}(J) \Longleftrightarrow \mathcal{H}(U) \subset L_2^m$.

Proof Suppose first that $J \neq \pm I_m$. Then, there is no loss of generality in assuming that $J = j_{pq}$. Consequently, Lemma 4.39 and the description of $\mathcal{L}_W$ that is furnished in Theorem 5.81 supply the key ingredients in the following sequence of equivalences if $W \in \mathcal{U}(J_{pq})$ and $\{b_1, b_2\} \in ap(W)$:

$$\begin{aligned} \mathcal{L}_W = \{0\} \quad &\Longleftrightarrow \quad \mathcal{H}(b_1) = \{0\} \quad \text{and} \quad \mathcal{H}_*(b_2) = \{0\} \\ &\Longleftrightarrow \quad b_1(\lambda) = \text{constant} \quad \text{and} \quad b_2(\lambda) = \text{constant} \\ &\Longleftrightarrow \quad W \in \mathcal{U}_S(j_{pq}). \end{aligned}$$

Moreover, since $\overline{\mathcal{L}_W} = \mathcal{H}(W_1)$ for some mvf $W_1 \in \mathcal{U}(j_{pq})$ and $W_1^{-1}W \in \mathcal{U}_S(j_{pq})$ by Lemma 5.85, the formula $W = W_1(W_1^{-1}W)$ implies that

$$\overline{\mathcal{L}_W} \neq \mathcal{H}(W) \Longleftrightarrow W_1^{-1}W \notin \mathcal{U}_{const}(j_{pq}) \Longleftrightarrow W \notin \mathcal{U}_{rR}(j_{pq}).$$

This completes the proof of (1) and (2) if $J \neq \pm I_m$. However, if $J = \pm I_m$, then (1) and (2) are self-evident, because $\mathcal{U}(J) = \mathcal{U}_{rR}(J) \subset L_2^m$.

Assertion (3) is listed here to ease the comparison with (1) and (2). It will be justified in Theorem 5.92. □

Corollary 5.87 *Let $W \in \mathcal{U}(j_{pq})$ and let W_1 be the mvf that is considered in Lemma 5.85. Then $W_1 \in \mathcal{U}_{rR}(j_{pq})$.*

Proof This follows from the equality $\overline{\mathcal{L}_{W_1}} = \mathcal{H}(W_1)$ and Theorem 5.86. □

Theorem 5.88 *Let $W \in \mathcal{U}(j_{pq})$, let $\{b_1, b_2\} \in ap(W)$ and let $s_{12} = T_W[0]$. Then $W(\lambda)$ is not right regular if and only if there exists a nonzero element $f \in \mathcal{H}(W)$ such that*

$$\begin{bmatrix} I_p & -s_{12} \\ -s_{12}^* & I_q \end{bmatrix} f \in \begin{array}{c} b_1 H_2^p \\ \oplus \\ b_2^* K_2^q \end{array}. \tag{5.169}$$

Proof If $W(\lambda)$ is not right regular, then by Theorems 5.81 and 5.86, there exists a nonzero $f \in \mathcal{H}(W)$ such that

$$\left\langle \begin{bmatrix} I_p & -s_{12} \\ -s_{12}^* & I_q \end{bmatrix} f, \begin{bmatrix} g \\ \Gamma_{11}^* g \end{bmatrix} + \begin{bmatrix} \Gamma_{22} h \\ h \end{bmatrix} \right\rangle_{st} = 0 \tag{5.170}$$

for every choice of $g \in \mathcal{H}(b_1)$ and $h \in \mathcal{H}_*(b_2)$. Therefore, since

$$\begin{bmatrix} I_p & -s_{12} \\ -s_{12}^* & I_q \end{bmatrix} f \in \begin{array}{c} H_2^p \\ \oplus \\ K_2^q \end{array}, \quad \Gamma_{22} h \in K_2^p \quad \text{and} \quad \Gamma_{11}^* g \in H_2^q,$$

the condition (5.170) is readily seen to be equivalent to the condition

$$\left\langle \begin{bmatrix} I_p & -s_{12} \\ -s_{12}^* & I_q \end{bmatrix} f, \begin{bmatrix} g \\ h \end{bmatrix} \right\rangle_{st} = 0$$

for all such g and h. But this is equivalent to the asserted condition (5.169). □

Theorem 5.89 *Let $U \in \mathcal{U}(J)$. Then:*

(1) *U admits a right regular-singular factorization*

$$U = U_1 U_2 \quad \text{with} \quad U_1 \in \mathcal{U}_{rR}(J) \quad \text{and} \quad U_2 \in \mathcal{U}_S(J) \tag{5.171}$$

that is unique up to multiplication by a constant J-unitary factor V on the right of U_1 and V^{-1} on the left of U_2.

(2) $\overline{\mathcal{L}_U} = \overline{\mathcal{L}_{U_1}} = \mathcal{H}(U_1)$.

(3) $\mathcal{H}(U_1)$ *is isometrically included in* $\mathcal{H}(U)$.

Proof If $U \in \mathcal{U}(J)$ and $J = \pm I_m$, then the theorem is self-evident, because $\mathcal{U}(J) = \mathcal{U}_{rR}(J)$, $\mathcal{U}_S(J) = \mathcal{U}_{const}(J)$ and $\mathcal{L}_U(J) = \mathcal{H}(U)$.

If $J \neq \pm I_m$, then it suffices to verify the theorem for $J = j_{pq}$. In this setting, Lemma 5.85 and Corollary 5.87 guarantee the existence of a right regular-singular factorization $W = W_1 W_2$ as in (1) and that (2) and (3) hold. Therefore, it remains only to check the essential uniqueness of the factorization considered in (1). But if $W = W_3 W_4$ with $W_3 \in \mathcal{U}_{rR}(j_{pq})$ and $W_4 \in \mathcal{U}_S(j_{pq})$, then, by the same arguments that were used to prove (2) of Theorem 5.86, $\overline{\mathcal{L}_W} = \overline{\mathcal{L}_{W_3}} = \mathcal{H}(W_3)$. Thus, as $\overline{\mathcal{L}_W} = \mathcal{H}(W_1)$, by Lemma 5.85, it follows that $\mathcal{H}(W_3) = \mathcal{H}(W_1)$ and hence that $W_3 = W_1 V$ for some $V \in \mathcal{U}_{const}(j_{pq})$. □

Theorem 5.90 *Let* $U \in \mathcal{U}(J)$. *Then:*

(1) $U \in \widetilde{L_2^m} \Longrightarrow U \in \mathcal{U}_{rR} \cap \mathcal{U}_{\ell R}$.

(2) $U \in \mathcal{U}_{rsR} \cup \mathcal{U}_{\ell sR} \Longrightarrow U \in \widetilde{L_2^m}$.

(3) $U \in \mathcal{U}_{rsR} \cup \mathcal{U}_{\ell sR} \Longrightarrow U \in \mathcal{U}_{rR} \cap \mathcal{U}_{\ell R}$.

Proof If $U \in \widetilde{L_2^m}$, then $K_\omega^U v \in L_2^m$ for every choice of $v \in \mathbb{C}^m$ and $\omega \in \mathfrak{h}_U$. Therefore, $\mathcal{L}_U$ is dense in $\mathcal{H}(U)$ and hence $U \in \mathcal{U}_{rR}(J)$, by Theorem 5.86. The same argument applied to $U^\sim$ implies that $U^\sim \in \mathcal{U}_{rR}(J)$ and hence that $U \in \mathcal{U}_{\ell R}(J)$. Thus, (1) holds, and the proof is complete, since (2) coincides with Theorem 4.75 and (3) is immediate from (1) and (2). □

Let

$$\Gamma_W = \begin{bmatrix} \Gamma_{11} & \Gamma_{12} \\ 0 & \Gamma_{22} \end{bmatrix} : \begin{array}{c} H_2^p \\ \oplus \\ \mathcal{H}_*(b_2) \end{array} \longrightarrow \begin{array}{c} \mathcal{H}(b_1) \\ \oplus \\ K_2^q \end{array}, \tag{5.172}$$

where $\{b_1, b_2\} \in ap(W)$ for $W \in \mathcal{U}(j_{pq})$ and the operators Γ_{ij} are defined by formulas (5.153) and (5.159).

Lemma 5.91 *If* $W \in \mathcal{U}(j_{pq})$, *then the following are equivalent:*

(1) $W \in \mathcal{U}_{rsR}(j_{pq})$.

(2) $W \in \mathcal{U}_{rR}(j_{pq})$ *and* $\|\Gamma_W\| < 1$.

(3) $\Delta_W \geq \varepsilon I$ *on* $\mathcal{H}(b_1, b_2)$ *for some* $\epsilon > 0$.

Proof The implication (1) $\Longrightarrow$ (2) will be proved in Theorem 5.92. The converse implication will follow from Theorem 7.54. Moreover, the equivalence of (2) and (3) is well known. □

Lemma 5.91 exhibits a useful characterization of the class $\mathcal{U}_{rsR}(j_{pq})$. The implication (2) $\Longrightarrow$ (1) will be used in the proof of Theorem 5.92.

Theorem 5.92 *Let $U \in \mathcal{U}(J)$. Then the following are equivalent:*

(1) *$U \in \mathcal{U}_{rsR}(J)$.*

(2) *There exist a pair of constants $\gamma_2 \geq \gamma_1 > 0$ such that*

$$\gamma_1 \|f\|_{st} \leq \|f\|_{\mathcal{H}(U)} \leq \gamma_2 \|f\|_{st} \quad \textit{for every} \quad f \in \mathcal{H}(U). \tag{5.173}$$

(3) *$\mathcal{H}(U)$ is a closed subspace of L_2^m.*

(4) *$\mathcal{H}(U) \subset L_2^m$.*

Proof If $J = \pm I_m$, then $\mathcal{H}(U)$ is included isometrically in L_2^m and the theorem is self evident. If $J \neq \pm I_m$, then it is enough to verify the theorem for $J = j_{pq}$. But if $W \in \mathcal{U}_{rsR}(j_{pq})$, then, since $\mathcal{U}_{rsR}(j_{pq}) \subseteq \mathcal{U}_{rR}(j_{pq})$ by Theorem 5.90, Lemma 5.85 guarantees that $\overline{\mathcal{L}_W} = \mathcal{H}(W)$. Moreover, by Theorem 5.81, $\|f\|_{\mathcal{H}(W)}$ can be evaluated by formula (5.164) when $f \in \mathcal{L}_W$, which, upon choosing $s \in T_W[\mathcal{S}^{p\times q}]$ with $\|s\|_\infty < \delta$ and $\delta < 1$ leads easily to the bounds

$$(1-\delta)\|f\|_{st}^2 \leq \|f\|_{\mathcal{H}(W)}^2 \leq (1+\delta)\|f\|_{st}^2 \quad \text{for} \quad f \in \mathcal{L}_W.$$

Moreover, since $\mathcal{L}_W$ is dense in $\mathcal{H}(W)$, the inequalities extend to $\mathcal{H}(W)$. Thus, (1) $\Longrightarrow$ (2), and clearly (2) $\Longrightarrow$ (3) and (3) $\Longrightarrow$ (4).

It remains only to check that (4) $\Longrightarrow$ (1). Towards this end, let $\mathcal{H}(W) \subset L_2^m$. Then $\mathcal{L}_W = \mathcal{H}(W)$, $W \in \mathcal{U}_{rR}(j_{pq})$ by Theorem 5.86, and the operator L_W defined in (5.165) is a bounded invertible operator from $\mathcal{H}(b_1, b_2)$ onto $\mathcal{H}(W)$. Therefore, by a theorem of Banach, L_W has a bounded inverse. Thus, in view of Theorem 5.81,

$$\left\| L_W \begin{bmatrix} g \\ h \end{bmatrix} \right\|_{\mathcal{H}(W)}^2 = \left\langle \Delta_W \begin{bmatrix} g \\ h \end{bmatrix}, \begin{bmatrix} g \\ h \end{bmatrix} \right\rangle_{st}$$

Δ_W is positive and has a bounded inverse. But this is equivalent to the fact that the operator Γ_W defined by (5.172) is strictly contractive, i.e., $\|\Gamma_W\| < 1$. Thus,

$$\mathcal{H}(W) \subset L_2^m \Longrightarrow W \in \mathcal{U}_{rR}(j_{pq}) \quad \text{and} \quad \|\Gamma_W\| < 1,$$

and hence $W \in \mathcal{U}_{rsR}(j_{pq})$ by Lemma 5.91. □

5.16 Regular J-inner mvf's

A J-inner mvf U is said to belong to the class $\mathcal{U}_R(J)$ of **regular** J-inner mvf's if it is both right and left regular, i.e.,

$$\mathcal{U}_R(J) = \mathcal{U}_{rR}(J) \cap \mathcal{U}_{\ell R}(J).$$

Theorem 5.93 *The following inclusions hold:*

(1) $\mathcal{U}(J) \cap \widetilde{L}_2^{m\times m} \subseteq \mathcal{U}_R(J)$.

(2) $\mathcal{U}_{rsR}(J) \cup \mathcal{U}_{\ell sR}(J) \subseteq \mathcal{U}_R(J) \cap \widetilde{L}_2^{m\times m}$.

(3) $\mathcal{U}_{rsR}(J) \cup \mathcal{U}_{\ell sR}(J) \subseteq \mathcal{U}_R(J)$.

Proof This is just a reformulation of Theorem 5.90. □

Theorem 5.94 *Let $U = U_1U_2$, where $U_j \in \mathcal{U}(J)$ for $j = 1, 2$. Then:*

(1) *If $U \in \mathcal{U}_{rsR}(J)$, then $U_1 \in \mathcal{U}_{rsR}(J)$ and $U_2 \in \mathcal{U}_R(J)$.*

(2) *If $U \in \mathcal{U}_{\ell sR}(J)$, then $U_1 \in \mathcal{U}_R(J)$ and $U_2 \in \mathcal{U}_{\ell sR}(J)$.*

Proof Suppose first that $U \in \mathcal{U}_{rsR}(J)$. Then, by Theorems 5.52 and 5.86,

$$\mathcal{H}(U_1) \subseteq \mathcal{H}(U) \subseteq L_2^m$$

and $U_1 \in \mathcal{U}_{rsR}$. Moreover,

$$U \in \mathcal{U}_{rsR}(J) \Longrightarrow U \in \mathcal{U}_{rR}(J) \Longrightarrow U_2 \in \mathcal{U}_{rR}(J),$$

where the first implication follows from Theorem 5.93 and the second implication follows from the definition of the class $\mathcal{U}_{rR}(J)$. Next, to verify that $U_2 \in \mathcal{U}_{\ell R}(J)$, let $U_2 = U_3U_4$, where $U_3 \in \mathcal{U}_S(J)$ and $U_4 \in \mathcal{U}(J)$. Then $U = U_1U_3U_4$ and, by the preceding argument, $U_1U_3 \in \mathcal{U}_{rsR}(J)$. Therefore, $U_1U_3 \in \mathcal{U}_{rR}(J)$ and hence, U_3 is a constant matrix. Thus, $U_2 \in \mathcal{U}_R(J)$.

Since (2) follows from (1) by considering $U^\sim = U_2^\sim U_1^\sim$, the proof is complete. □

5.17 Formulas for $W(\lambda)$ when $W \in \mathcal{U}_{rsR}(j_{pq})$

Let $W \in \mathcal{U}_{rsR}(j_{pq})$. Let $\{b_1, b_2\} \in ap(W)$, $s_{12} = T_W[0]$ and let the operators Γ_{11}, Γ_{12} and Γ_{22} be defined by formulas (5.153) and (5.159). Then, by Theorem 5.92, $\mathcal{H}(W) \subset L_2^m$ and, consequently, Theorem 5.81 provides a description of $\mathcal{H}(W)$. Moreover, if $d_j(\lambda) = \det b_j(\lambda)$, $j = 1, 2$, then

$$\mathcal{H}(b_1) \subseteq \mathcal{H}(d_1 I_p), \quad \mathcal{H}_*(b_2) \subseteq \mathcal{H}_*(d_2 I_q),$$

$$\overline{\Gamma_{11}^* \mathcal{H}(b_1)} \subseteq \mathcal{H}(d_1 I_p), \quad \overline{\Gamma_{22} \mathcal{H}_*(b_2)} \subseteq \mathcal{H}_*(d_2 I_q)$$

and it is easy to check that both subspaces $\overline{\Gamma_{11}^* \mathcal{H}(b_1)}$ and $\overline{\Gamma_{22} \mathcal{H}_*(b_2)}$ are R_α invariant for every $\alpha \in \mathfrak{h}_W$. Thus, in view of Theorem 3.38 and Corollary 3.42,

$$\overline{\Gamma_{11}^* \mathcal{H}(b_1)} = \mathcal{H}(\widehat{b}_1) \quad \text{and} \quad \overline{\Gamma_{22} \mathcal{H}_*(b_2)} = \mathcal{H}_*(\widehat{b}_2) \tag{5.174}$$

for some $\widehat{b}_1 \in \mathcal{S}_{in}^{p \times p}$ and $\widehat{b}_2 \in \mathcal{S}_{in}^{q \times q}$. □

Theorem 5.95 *Let $W \in U_{rsR}(j_{pq})$, let $\{b_1, b_2\} \in ap(W)$, and $s_{12} = T_W[0]$. Let the operators Γ_{11}, Γ_{22} and Γ_{12} be defined by the formulas (5.153) and (5.159) and let the operators L_W and Δ_W be defined by formulas (5.165) and (5.166). Then Δ_W and L_W are bounded and boundedly invertible operators, the adjoint L_W^* of L_W acts from $\mathcal{H}(W)$ onto $\mathcal{H}(b_1, b_2)$,*

$$\Delta_W = L_W^* L_W > 0 \tag{5.175}$$

and the RK of the RKHS $\mathcal{H}(W)$ is given by the formula

$$K_\omega^W \begin{bmatrix} \xi \\ \eta \end{bmatrix} = L_W \Delta_W^{-1} \begin{bmatrix} k_\omega^{b_1} \xi + \Gamma_{11} k_\omega^{\widehat{b}_1} \eta \\ \Gamma_{22}^* \ell_\omega^{\widehat{b}_2} \xi + \ell_\omega^{b_2} \eta \end{bmatrix}, \quad \text{for } \xi \in \mathbb{C}^p \text{ and } \eta \in \mathbb{C}^q, \tag{5.176}$$

where $k_\omega^{b_1}$ and $\ell_\omega^{b_2}$ are the RK's of the RKHS's $\mathcal{H}(b_1)$ and $\mathcal{H}_*(b_2)$, and $k_\omega^{\widehat{b}_1}$ and $\ell_\omega^{\widehat{b}_2}$ are the RK's of the RKHS's $\mathcal{H}(\widehat{b}_1)$ and $\mathcal{H}_*(\widehat{b}_2)$, respectively.

Proof In the proof of Theorem 5.92 it is shown that the operator Δ_W is positive, bounded and boundedly invertible in the space $\mathcal{H}(b_1, b_2)$ and that L_W is a bounded, invertible and boundedly invertible operator from

$\mathcal{H}(b_1, b_2)$ onto $\mathcal{H}(W)$. It is readily checked that for every point $\omega \in \mathfrak{h}_W$

$$\begin{aligned}\left\langle L_W \begin{bmatrix} g \\ h \end{bmatrix}, K_\omega^W \begin{bmatrix} \xi \\ \eta \end{bmatrix} \right\rangle_{\mathcal{H}(W)} &= [\xi^* \quad \eta^*] \left(L_W \begin{bmatrix} g \\ h \end{bmatrix} \right)(\omega) \\ &= \langle g, k_\omega^{b_1} \xi \rangle_{st} + \langle \Gamma_{22} h, \ell_\omega^{\widehat{b}_2} \xi \rangle_{st} + \langle \Gamma_{11}^* g, k_\omega^{\widehat{b}_1} \rangle_{st} \\ &\quad + \langle h, \ell_\omega^{b_2} \eta \rangle_{st} \\ &= \left\langle \begin{bmatrix} g \\ h \end{bmatrix}, \begin{bmatrix} k_\omega^{b_1} \xi + \Gamma_{11} k_\omega^{\widehat{b}_1} \eta \\ \Gamma_{22}^* \ell_\omega^{\widehat{b}_2} \xi + \ell_\omega^{b_2} \eta \end{bmatrix} \right\rangle_{st} .\end{aligned}$$

At the same time, by assertion (3) of Theorem 5.81,

$$\left\langle L_W \begin{bmatrix} g \\ h \end{bmatrix}, K_\omega^W \begin{bmatrix} \xi \\ \eta \end{bmatrix} \right\rangle_{\mathcal{H}(W)} = \left\langle \begin{bmatrix} g \\ h \end{bmatrix}, \Delta_W L_W^{-1} K_\omega^W \begin{bmatrix} \xi \\ \eta \end{bmatrix} \right\rangle_{st},$$

which also serves to justify (5.175). Thus,

$$\left\langle \begin{bmatrix} g \\ h \end{bmatrix}, \begin{bmatrix} k_\omega^{b_1} \xi + \Gamma_{11} k_\omega^{\widehat{b}_1} \eta \\ \Gamma_{22}^* \ell_\omega^{\widehat{b}_2} \xi + \ell_\omega^{b_2} \eta \end{bmatrix} \right\rangle_{st} = \left\langle \begin{bmatrix} g \\ h \end{bmatrix}, \Delta_W L_W^{-1} K_\omega^W \begin{bmatrix} \xi \\ \eta \end{bmatrix} \right\rangle_{st}$$

for every $g \in \mathcal{H}(b_1)$ and $h \in \mathcal{H}(b_2)$, which yields (5.176). □

Formula (5.176) may be rewritten in the following form. Consider the set $\mathcal{H}^m(b_1, b_2)$ of $m \times m$ mvf's $F = \begin{bmatrix} f_1 & f_2 & \cdots & f_m \end{bmatrix}$ with columns $f_j \in \mathcal{H}(b_1, b_2)$ for $1 \leq j \leq m$, as the orthogonal sum of m copies of the space $\mathcal{H}(b_1, b_2)$ and the set $\mathcal{H}^m(W)$ of $m \times m$ mvf's $K = \begin{bmatrix} k_1 & k_2 & \cdots & k_m \end{bmatrix}$ with columns $k_j \in \mathcal{H}(W)$, for $1 \leq j \leq m$, as the orthogonal sum of m copies of the space $\mathcal{H}(W)$. Let the operators

$$\mathbf{\Delta}_W : \mathcal{H}^m(b_1, b_2) \to \mathcal{H}^m(b_1, b_2) \text{ and } L_W : \mathcal{H}^m(b_1, b_1) \to \mathcal{H}^m(W)$$

act on these spaces of $m \times m$ mvf's columns by column:

$$\mathbf{\Delta}_W \begin{bmatrix} f_1 & f_2 & \cdots & f_m \end{bmatrix} = \begin{bmatrix} \Delta_W f_1 & \Delta_W f_2 & \cdots & \Delta_W f_m \end{bmatrix}$$

and

$$L_W \begin{bmatrix} f_1 & f_2 & \cdots & f_m \end{bmatrix} = \begin{bmatrix} L_W f_1 & L_W f_2 & \cdots & L_W f_m \end{bmatrix}.$$

Analogously, let the operators $\mathbf{\Gamma}_{11}$ and $\mathbf{\Gamma}_{22}^*$ act on $p \times p$ and $q \times q$ mvf's respectively, column by column. Then the mvf

$$F_\omega^W(\lambda) = \begin{bmatrix} k_\omega^{b_1}(\lambda) & (\mathbf{\Gamma}_{11} k_\omega^{\widehat{b}_1})(\lambda) \\ (\mathbf{\Gamma}_{22}^* \ell_\omega^{\widehat{b}_2})(\lambda) & \ell_\omega^{b_2}(\lambda) \end{bmatrix} \in \mathcal{H}^m(b_1, b_2) \tag{5.177}$$

and formula (5.176) may be rewritten in the form

$$K_\omega^W(\lambda) = (L_W \Delta_W^{-1} F_\omega^W)(\lambda) \tag{5.178}$$

for every $\omega \in \mathfrak{h}_W$.

The mvf $W(\lambda)$ is defined by the RK $K_\omega(\lambda)$ of the RKHS $\mathcal{H}(W)$ up to a constant j_{pq}-unitary right factor and

$$W(\lambda) j_{pq} W(\omega)^* = j_{pq} - \rho_\omega(\lambda) K_\omega(\lambda). \tag{5.179}$$

If $\omega \in \mathfrak{h}_W^+ \cap \mathfrak{h}_{W^{-1}}^+$, then there is only one mvf $W(\lambda)$ that meets the normalization condition (5.91) at the point ω, and $W(\omega)$ for this W may be expressed in terms of $K_\omega(\omega)$ and $K_{\overline{\omega}}(\overline{\omega})$ by formulas (5.94)–(5.97).

If $0 \in \mathfrak{h}_W$, then $W(\lambda)$ may be normalized at the point $\omega = 0$ by the condition $W(0) = I_m$. This normalized $W(\lambda)$ is defined uniquely by the formula

$$W(\lambda) = I_m + 2\pi i \lambda \, K_0^W(\lambda) j_{pq}, \tag{5.180}$$

where

$$K_0^W(\lambda) = (L_W \mathbf{\Delta}_W^{-1} F_0^W)(\lambda) \tag{5.181}$$

and

$$F_0^W = \begin{bmatrix} k_0^{b_1}(\lambda) & (\mathbf{\Gamma}_{11} k_0^{\widehat{b}_1})(\lambda) \\ (\mathbf{\Gamma}_{22} \ell_0^{\widehat{b}_2})(\lambda) & \ell_0^{b_2}(\lambda) \end{bmatrix}. \tag{5.182}$$

Since $\mathfrak{h}_b \subseteq \mathfrak{h}_W$ and $\mathfrak{h}_{\widehat{b}_j} \subseteq \mathfrak{h}_W$, the mvf's b_j and $\widehat{b}_j$ may also be normalized by the conditions $b_1(0) = \widehat{b}_1(0) = I_p$ and $b_2(0) = \widehat{b}_2(0) = I_q$. Then

$$\begin{aligned} k_0^{b_1}(\lambda) &= \frac{I_p - b_1(\lambda)}{-2\pi i \lambda}, \qquad k_0^{\widehat{b}_1}(\lambda) = \frac{I_q - \widehat{b}_1(\lambda)}{-2\pi i \lambda}, \\ \ell_0^{b_2}(\lambda) &= \frac{b_2^\#(\lambda) - I_q}{-2\pi i \lambda} \quad \text{and} \quad \ell_0^{\widehat{b}_2}(\lambda) = \frac{\widehat{b}_2^\#(\lambda) - I_p}{-2\pi i \lambda}. \end{aligned} \tag{5.183}$$

If $b_2(\lambda)$=constant, i.e., if $W \in \mathcal{N}_+^{m \times m}$, then $\mathcal{H}_*(b_2) = \{0\}$ and the preceding formulas simplify:

$$L_W = \begin{bmatrix} I \\ \Gamma_{11}^* \end{bmatrix} : \mathcal{H}(b_1) \to \mathcal{H}(W), \tag{5.184}$$

and

$$\Delta_W = I - \Gamma_{11} \Gamma_{11}^* : \mathcal{H}(b_1) \to \mathcal{H}(b_1), \tag{5.185}$$

where

$$\Gamma_{11} = P_{\mathcal{H}(b_1)} M_s|_{H_2^q}, \; s \in T_W[\mathcal{S}^{p\times q}], \tag{5.186}$$

$$F_\omega^W(\lambda) = [k_\omega^{b_1}(\lambda) \quad (\mathbf{\Gamma}_{11} k_\omega^{\widehat{b}_1})(\lambda)]. \tag{5.187}$$

Analogously, if $b_1(\lambda)$=constant, i.e., if $W \in \mathcal{N}_-^{m\times m}$, then $\mathcal{H}(b_1) = 0$,

$$L_W = \begin{bmatrix} \Gamma_{22} \\ I \end{bmatrix} : \mathcal{H}_*(b_2) \to \mathcal{H}(W), \tag{5.188}$$

and

$$\Delta_W = I - \Gamma_{22}^*\Gamma_{22} : \mathcal{H}_*(b_2) \to \mathcal{H}_*(b_2), \tag{5.189}$$

where

$$\Gamma_{22} = \Pi_- M_s|_{\mathcal{H}_*(b_2)}, \; s \in T_W[\mathcal{S}^{p\times q}], \tag{5.190}$$

$$F_\omega^W(\lambda) = [(\mathbf{\Gamma}_{22}^* \ell_\omega^{\widehat{b}_2})(\lambda) \quad \ell_\omega^{b_2}(\lambda)]. \tag{5.191}$$

Remark 5.96 *In formula (5.177) the mvf's $\widehat{b}_1$ and $\widehat{b}_2$ may be replaced by any other pair of inner mvf's $b_5 \in \mathcal{S}_{in}^{p\times p}$ and $b_6 \in \mathcal{S}_{in}^{q\times q}$, such that $\mathcal{H}(\widehat{b}_1) \subseteq \mathcal{H}(b_5)$, $\mathcal{H}_*(\widehat{b}_2) \subseteq \mathcal{H}_*(b_6)$ and $\mathfrak{h}_{b_5} \cap \mathfrak{h}_{b_6^{-1}} \supseteq h_W$. In particular, it is possible to choose $b_5 = (\det b_1)I_p$ and $b_6 = (\det b_2)I_q$.*

Remark 5.97 *Theorem 5.57 guarantees that there is exactly one mvf $W \in \mathcal{U}(j_{pq})$ for which (5.179) and the normalization conditions (5.91) hold at a point $\omega \in \mathfrak{h}_W^+ \cap \mathfrak{h}_{W^{-1}}^+ \cap \overline{\mathbb{C}_+}$ hold. The formulas that are given in Theorem 5.57 for this normalized mvf $W(\lambda)$ in terms of the RK may be combined with the preceding analysis to obtain formulas for the normalized mvf $W(\lambda)$ in terms of the operators Γ_{ij}.*

5.18 A description of $\mathcal{H}(A)$ when $A \in \mathcal{U}_{rsR}(J_p)$

In this section we shall explain how to obtain a description of the space $\mathcal{H}(A)$ for $A \in \mathcal{U}_{rsR}(J_p)$ from the description of the space $\mathcal{H}(W)$ that was given for $W \in \mathcal{U}_{rsR}(j_{pq})$ in Theorem 5.81. The proof will be obtained by applying this theorem to $W(\lambda) = \mathfrak{V}A(\lambda)\mathfrak{V}$. (In this case $p = q$.) The description of $\mathcal{H}(A)$ will depend upon the connection between $\{b_1, b_2\} \in ap_I(A)$ and $\{b_3, b_4\} \in ap_{II}(A)$ (see (5.196) below) and the interplay between the operators Γ_{11}, Γ_{22} and Γ_{12} defined in terms of $\{b_1, b_2\} \in ap(W)$ and $s \in T_W[\mathcal{S}^{p\times q}]$ by formulas

(5.153) and (5.159) that were used to describe $\mathcal{H}(W)$ and the operators Φ_{ij} that are defined in analogous fashion by the formulas

$$\Phi_{11} = \Pi_{\mathcal{H}(b_3)} M_c|_{H_2^p}, \quad \Phi_{22} = \Pi_- M_c|_{\mathcal{H}_*(b_4)}, \quad \Phi_{12} = \Pi_{\mathcal{H}(b_3)} M_c|_{\mathcal{H}_*(b_4)} \tag{5.192}$$

for $c \in \mathcal{C}(A) \cap \mathring{\mathcal{C}}^{p\times p}$. The operators Γ_{ij} and Φ_{ij} are independent of the considered choices of the mvf's $s(\lambda)$ and $c(\lambda)$.

Theorem 5.98 *Let $A \in \mathcal{U}_{rsR}(J_p)$, let $\mathfrak{E}(\lambda) = \sqrt{2}N_2^* A(\lambda)\mathfrak{V}$ and $\{b_3, b_4\} \in ap_{II}(A)$. Then*

$$\mathcal{B}(\mathfrak{E}) = \mathcal{H}_*(b_4) \oplus \mathcal{H}(b_3) \tag{5.193}$$

as linear spaces of vvf's, but not as Hilbert spaces (unless $E_+E_+^\# = I_p$) and there exist a pair of positive constants γ_1 and γ_2 such that

$$\gamma_1 \|f\|_{st} \le \|f\|_{\mathcal{B}(\mathfrak{E})} \le \gamma_2 \|f\|_{st} \tag{5.194}$$

for every $f \in \mathcal{B}(\mathfrak{E})$.

Proof By Theorem 5.92, there exist a pair of positive constants α_1 and α_2 such that

$$\alpha_1 \|f\|_{st} \le \|f\|_{\mathcal{H}(A)} \le \alpha_2 \|f\|_{st}$$

for every $f \in \mathcal{H}(A)$. Moreover, since $A \in \mathcal{U}_{rsR}(J_p)$,

$$\|f\|_{\mathcal{H}(A)} = \sqrt{2}\, \|N_2^* f\|_{\mathcal{B}(\mathfrak{E})},$$

by Theorem 5.76. Therefore,

$$\alpha_1 \|N_2^* f\|_{st} \le \sqrt{2}\, \|N_2^* f\|_{\mathcal{B}(\mathfrak{E})}$$

for every $f \in \mathcal{H}(A)$. This supplies the first inequality in (5.194) (with $\gamma_1 = \alpha_1/\sqrt{2}$) and exhibits the inclusion $\mathcal{B}(\mathfrak{E}) \subset L_2^p(\mathbb{R})$.

The verification of (5.193) rests on the fact that

$$f \in \mathcal{B}(\mathfrak{E}) \iff E_+^{-1} f \in H_2^p \quad \text{and } E_-^{-1} f \in K_2^p$$

and the factorizations

$$E_+^{-1} = \varphi_4 b_4 \quad \text{and} \quad (E_-^\#)^{-1} = b_3 \varphi_3 \tag{5.195}$$

with $b_j \in \mathcal{S}_{in}^{p\times p}$ and $(\lambda+i)^{-1}\varphi_j \in \mathcal{N}_{out}^{p\times p} \cap H_2^{p\times p}$ for $j = 3, 4$. Thus, as $(\mu+i)^{-1}\varphi_4(\mu)$ is outer in $H_2^{p\times p}$, the following equivalences hold for $f \in L_2^p$:

$$\begin{aligned} E_+^{-1} f \in H_2^p &\iff \left\langle \varphi_4 b_4 f, \left(\frac{f_+}{\mu+i}\right)^* \right\rangle_{st} = 0 \text{ for all } f_+ \in H_\infty^{1\times p} \\ &\iff \left\langle b_4 f, \left(\frac{\varphi_4}{\mu+i}\right)^* f_+^* \right\rangle_{st} = 0 \text{ for all } f_+ \in H_\infty^{1\times p} \\ &\iff f \text{ is orthogonal to } b_4^* K_2^p. \end{aligned}$$

Similar considerations lead to the conclusion that

$$E_-^{-1} f \in K_2^p \iff f \text{ is orthogonal to } b_3 H_2^p.$$

This serves to complete the proof of (5.193).

Next, let T denote the (embedding) operator that maps $f \in \mathcal{B}(\mathfrak{E})$ onto $f \in \mathcal{H}_*(b_3) \oplus \mathcal{H}(b_4)$. By the first inequality in (5.194), T is also bounded and one to one. Therefore, by a theorem of Banach, T has a bounded inverse. Consequently, the second inequality in (5.194) must also hold for some $\gamma_2 > 0$. □

Lemma 5.99 *Let $A \in \mathcal{U}_{rsR}(J_p)$, let $\{b_1, b_2\} \in ap_I(A)$, $\{b_3, b_4\} \in ap_{II}(A)$, $W = \mathfrak{V}A\mathfrak{V}$ and let Γ_{11} and Γ_{22} be defined as in (5.153). Then*

$$(I + \Gamma_{11}^*)\mathcal{H}(b_1) = \mathcal{H}(b_3), \quad (I + \Gamma_{22})\mathcal{H}_*(b_2) = \mathcal{H}_*(b_4) \tag{5.196}$$

and

$$\mathcal{H}(A) = \left\{ \begin{bmatrix} -(I - \Gamma_{11}^*)g + (I - \Gamma_{22})h \\ (I + \Gamma_{11}^*)g + (I + \Gamma_{22})h \end{bmatrix} : g \in \mathcal{H}(b_1) \text{ and } h \in \mathcal{H}_*(b_2) \right\}. \tag{5.197}$$

Proof Formula (5.197) follows from the relations

$$K_\omega^A(\lambda) = \mathfrak{V}K_\omega^W(\lambda)\mathfrak{V} \quad \text{and} \quad \mathcal{H}(A) = \mathfrak{V}\mathcal{H}(W)$$

between the RK's of the RKHS's $\mathcal{H}(A)$ and $\mathcal{H}(W)$ and the description of $\mathcal{H}(W)$ that is given in Theorem 5.81, because $\mathcal{L}_W = \mathcal{H}(W)$, since $W \in \mathcal{U}_{rsR}(J_p)$. Then (5.196) follows from (5.197) and Theorem 5.98. □

Lemma 5.100 *Let $A \in \mathcal{U}_{rsR}(J_p)$. Then*

$$\Phi_{11}^* = (I - \Gamma_{11}^*)(I + \Gamma_{11}^*)^{-1}|_{\mathcal{H}(b_3)} \quad \text{and}$$

$$\Phi_{22} = (I - \Gamma_{22})(I + \Gamma_{22})^{-1}|_{\mathcal{H}_*(b_4)}. \tag{5.198}$$

Proof If $c \in \mathcal{C}(A) \cap \mathring{\mathcal{C}}^{p\times p}$, then $c = -I_p + 2(I_p + s)^{-1}$, where $s \in T_W[\mathcal{S}^{p\times q}] \cap \mathring{\mathcal{S}}^{p\times p}$. Consequently,

$$\begin{aligned}\Phi_{11}^* &= \Pi_+ M_{c^*}|_{\mathcal{H}(b_3)} = \Pi_+\{-I + 2M_{(I_p+s^*)^{-1}}\}|_{\mathcal{H}(b_3)} \\ &= \{-I + 2\Pi_+ M_{(I_p+s^*)^{-1}}\}|_{\mathcal{H}(b_3)}.\end{aligned}$$

Thus, in order to complete the proof of the first equality in (5.198), it is enough to show that

$$(I + \Gamma_{11}^*)^{-1} g = \Pi_+(I_p + s^*)^{-1} g \tag{5.199}$$

for every $g \in \mathcal{H}(b_3)$. Since $g = b_3 g_-$ for some $g_- \in K_2^p$, it follows from Lemma 4.73 that

$$\Pi_+(I_p + s^*)^{-1} g = \Pi_+(I_p + s^*)^{-1} b_3 g_- = \frac{1}{2}\Pi_+ b_1 \varphi_s^{-*} g_-,$$

where $\varphi_s^{-1} \in H_\infty^{p\times p}$. This last conclusion stems from the Smirnov maximum principle, which is applicable since $\varphi_s^{-1} \in \mathcal{N}_+^{p\times p}$ and is equal to the bounded mvf $2b_3^*(I_p + s)^{-1} b_1$ a.e. on $\mathbb{R}$. This proves that $\Pi_+(I_p + s^*)^{-1} g \in \mathcal{H}(b_1)$. Moreover,

$$\begin{aligned}(I + \Gamma_{11}^*)\Pi_+(I_p + s^*)^{-1} g &= \Pi_+(I_p + s^*)\Pi_+(I_p + s^*)^{-1} g \\ &= \Pi_+(I_p + s^*)(I_p + s^*)^{-1} g \\ &= g.\end{aligned}$$

This justifies (5.199). Then,

$$\Phi_{22} = \Pi_-\{-I + 2M_{(I_p+s)^{-1}}\}|_{\mathcal{H}_*(b_4)}$$

and so, to verify the second equality in (5.198), we need to show that

$$(I + \Gamma_{22})^{-1} h = \Pi_-(I_p + s)^{-1} h \tag{5.200}$$

for every $h \in \mathcal{H}_*(b_4)$. But, writing $h = b_4^* h_+$ for some $h_+ \in H_2^p$ and invoking Lemma 4.73, we see that

$$\Pi_-(I_p + s)^{-1} h = \Pi_-(I_p + s)^{-1} b_4^* h_+ = \frac{1}{2}\Pi_- b_2^* \psi_s^{-1} h_+.$$

Moreover, since

$$\psi_s^{-1} = 2b_2(I_p + s)^{-1} b_4^*$$

belongs to $L_\infty^{p\times p} \cap \mathcal{N}_+$, it follows from the Smirnov maximum principle that $\psi_s^{-1} \in H_\infty^{p\times p}$. Consequently, $\Pi_- b_2^* \psi_s^{-1} h_+$ belongs to $\mathcal{H}_*(b_2)$ and

$$\begin{aligned}(I+\Gamma_{22})\Pi_-(I_p+s)^{-1}h &= \Pi_-(I_p+s)\Pi_-(I_p+s)^{-1}b_4^*h_+ \\ &= \Pi_-(I_p+s)(I_p+s)^{-1}b_4^*h_+ \\ &= \Pi_- b_4^* h_+ = h,\end{aligned}$$

as required. □

Lemma 5.101 *In the setting of this section,*

$$\langle \Phi_{12}h, g\rangle_{st} = -2\langle \Gamma_{12}(I+\Gamma_{22})^{-1}h, (I+\Gamma_{11}^*)^{-1}g\rangle_{st} \quad \textit{for every}$$
$$g \in \mathcal{H}(b_3) \textit{ and } h \in \mathcal{H}_*(b_4). \tag{5.201}$$

Proof In view of the identifications (5.196), $(I+\Gamma_{11}^*)^{-1}g \in \mathcal{H}(b_1)$ and $(I+\Gamma_{22})^{-1}h \in \mathcal{H}_*(b_2)$. Therefore,

$$\langle \Gamma_{12}(I+\Gamma_{22})^{-1}h, (I+\Gamma_{11}^*)^{-1}g\rangle_{st} = \langle M_s(I+\Gamma_{22})^{-1}h, (I+\Gamma_{11}^*)^{-1}g\rangle_{st}.$$

Next, invoking (5.199) and (5.200), we can write the term on the right as

$$\begin{aligned}&\langle s\Pi_-(I_p+s)^{-1}h, \Pi_+(I_p+s^*)^{-1}g\rangle_{st} \\ &= \langle (I_p+s)\Pi_-(I_p+s)^{-1}h, \Pi_+(I_p+s^*)^{-1}g\rangle_{st} \\ &= -\langle (I_p+s)\Pi_+(I_p+s)^{-1}h, (I_p+s^*)^{-1}g\rangle_{st} \\ &= -\langle \Pi_+(I_p+s)^{-1}h, g\rangle_{st} = -\langle (I_p+s)^{-1}h, g\rangle_{st} \\ &= -(\frac{1}{2})\langle (c+I_p)h, g\rangle_{st} = -(\frac{1}{2})\langle ch, g\rangle_{st} = -(\frac{1}{2})\langle \Phi_{12}h, g\rangle_{st},\end{aligned}$$

as claimed. □

Next, in order to obtain formulas for the RK $K_\omega^A(\lambda)$ of the RKHS $\mathcal{H}(A)$ in terms of the operators Φ_{ij}, it is convenient to introduce the operators

$$\Delta_A = 2\Re \begin{bmatrix} \Phi_{11}|_{\mathcal{H}(b_3)} & \Phi_{12} \\ 0 & \Pi_{\mathcal{H}_*(b_4)}\Phi_{22} \end{bmatrix} : \begin{array}{c}\mathcal{H}(b_3)\\ \oplus \\ \mathcal{H}_*(b_4)\end{array} \longrightarrow \begin{array}{c}\mathcal{H}(b_3)\\ \oplus \\ \mathcal{H}_*(b_4)\end{array} \tag{5.202}$$

and

$$L_A = \begin{bmatrix} -\Phi_{11}^* & \Phi_{22} \\ I & I \end{bmatrix} : \begin{array}{c}\mathcal{H}(b_3)\\ \oplus \\ \mathcal{H}_*(b_4)\end{array} \longrightarrow \mathcal{H}(A). \tag{5.203}$$

Theorem 5.102 *Let $A \in \mathcal{U}_{rsR}(J_p)$, $B(\lambda) = A(\lambda)\mathfrak{V}$ and let the operators Φ_{11}, Φ_{22} and Φ_{12} be defined by formula (5.192), where $c \in \mathcal{C}(A) \cap H_\infty^{p\times p}$ and $\{b_3, b_4\} \in ap_{II}(A)$. Then*

$$\mathcal{H}(A) = \left\{ L_A \begin{bmatrix} g \\ h \end{bmatrix} : \ g \in \mathcal{H}(b_3) \ \text{ and } \ h \in \mathcal{H}_*(b_4) \right\}. \tag{5.204}$$

Moreover, if

$$f = L_A \begin{bmatrix} g \\ h \end{bmatrix} \quad \text{for some } g \in \mathcal{H}(b_3) \text{ and } h \in \mathcal{H}_*(b_4), \tag{5.205}$$

then

$$\|f\|^2_{\mathcal{H}(A)} = \langle \Delta_A f, f\rangle_{st} = \langle (c+c^*)(g+h), (g+h)\rangle_{st}. \tag{5.206}$$

Proof The description of $\mathcal{H}(A)$ in (5.204) follows from Lemmas 5.99 and 5.100. Moreover, in view of (5.196),

$$g \in \mathcal{H}(b_3) \Longleftrightarrow \widetilde{g} = (I + \Gamma_{11}^*)^{-1} g \quad \text{belongs to } \mathcal{H}(b_1)$$

and

$$h \in \mathcal{H}_*(b_4) \Longleftrightarrow \widetilde{h} = (I + \Gamma_{22})^{-1} h \quad \text{belongs to } \mathcal{H}_*(b_2).$$

Thus, as

$$\begin{bmatrix} g \\ h \end{bmatrix} = \sqrt{2}\mathfrak{V} L_W \begin{bmatrix} \widetilde{g} \\ \widetilde{h} \end{bmatrix} \quad \text{and} \quad \|f\|_{\mathcal{H}(A)} = \|\mathfrak{V} f\|_{\mathcal{H}(W)}$$

when $W = \mathfrak{V}A\mathfrak{V}$, it is readily seen that

$$\begin{aligned} \left\| L_A \begin{bmatrix} g \\ h \end{bmatrix} \right\|^2_{\mathcal{H}(A)} &= \left\| \mathfrak{V} L_A \begin{bmatrix} g \\ h \end{bmatrix} \right\|^2_{\mathcal{H}(W)} = 2 \left\| L_W \begin{bmatrix} \widetilde{g} \\ \widetilde{h} \end{bmatrix} \right\|^2_{\mathcal{H}(W)} \\ &= 2 \left\langle \Delta_W \begin{bmatrix} \widetilde{g} \\ \widetilde{h} \end{bmatrix}, \begin{bmatrix} \widetilde{g} \\ \widetilde{h} \end{bmatrix} \right\rangle_{st} \\ &= \left\langle \begin{bmatrix} I - \Gamma_{11}\Gamma_{11}^* & -\Gamma_{12} \\ -\Gamma_{12}^* & I - \Gamma_{22}^*\Gamma_{22} \end{bmatrix} \begin{bmatrix} \widetilde{g} \\ \widetilde{h} \end{bmatrix}, \begin{bmatrix} \widetilde{g} \\ \widetilde{h} \end{bmatrix} \right\rangle_{st}. \end{aligned}$$

This leads easily to formula (5.206) with the help of the identities

$$2\langle (I - \Gamma_{11}\Gamma_{11}^*)\widetilde{g}, \widetilde{g}\rangle_{st} = \langle (\Phi_{11} + \Phi_{11}^*) g, g\rangle_{st},$$

$$2\langle (I - \Gamma_{22}^*\Gamma_{22})\widetilde{h}, \widetilde{h}\rangle_{st} = \langle (\Phi_{22} + \Phi_{22}^*) h, h\rangle_{st}$$

and, by (5.201),

$$-2\langle\Gamma_{12}\widetilde{h},\widetilde{g}\rangle_{st} = \langle\Phi_{12}h, g\rangle_{st}. \qquad \Box$$

Lemma 5.103 *Let $A \in \mathcal{U}_{rsR}(J_p)$, $\dot{A}(\lambda) = J_pA(\lambda)J_p$ and let $\{b_3, b_4\} \in ap_{II}(A)$ and $\{\dot{b}_3, \dot{b}_4\} \in ap_{II}(\dot{A})$. Then*

$$\Phi_{11}^*\mathcal{H}(b_3) = \mathcal{H}(\dot{b}_3) \quad \textit{and} \quad \Phi_{22}\mathcal{H}_*(b_4) = \mathcal{H}_*(\dot{b}_4). \tag{5.207}$$

Proof Under the given assumptions, $\dot{A} \in \mathcal{U}_{rsR}(J_p)$ because

$$c \in \mathcal{C}(A) \cap \mathring{\mathcal{C}}^{p\times p} \Longleftrightarrow c^{-1} \in \mathcal{C}(\dot{A}) \cap \mathring{\mathcal{C}}^{p\times p},$$

and, in view of formula (5.204),

$$[0 \quad I_p]\mathcal{H}(\dot{A}) = \{\dot{g} + \dot{h} : \dot{g} \in \mathcal{H}(\dot{b}_3) \quad \text{and} \quad h \in \mathcal{H}_*(\dot{b}_4)\}.$$

On the other hand, since $K_\omega^{\dot{A}}(\lambda) = J_pK_\omega^A(\lambda)J_p$, formula (5.204) implies that

$$[0 \quad I_p]\mathcal{H}(\dot{A}) = \{-\Phi_{11}^*g + \Phi_{22}h : g \in \mathcal{H}(b_3) \quad \text{and} \quad h \in \mathcal{H}_*(b_4)\}.$$

The identities in (5.207) now follow easily, since $\Phi_{11}^*g \in H_2^p$ and $\Phi_{22}h \in K_2^p$. $\Box$

Theorem 5.104 *Let $A \in \mathcal{U}_{rsR}(J_p)$, $\dot{A}(\lambda) = J_pA(\lambda)J_p$, $\{b_3, b_4\} \in ap_{II}(A)$, $\{\dot{b}_3, \dot{b}_4\} \in ap_{II}(\dot{A})$, let the operators Δ_A and L_A be defined by formulas (5.202) and (5.203) and let $b_5(\lambda)$ and $b_6(\lambda)$ be a pair of mvf's in $\mathcal{S}_{in}^{p\times p}$ such that*

$$\dot{b}_3^{-1}b_5 \in \mathcal{S}_{in}^{p\times p} \quad \textit{and} \quad b_6\dot{b}_4^{-1} \in \mathcal{S}_{in}^{p\times p}$$

and $\mathfrak{h}_{b_3} \cap \mathfrak{h}_{b_4^{-1}} \subseteq \mathfrak{h}_{b_5} \cap \mathfrak{h}_{b_6^{-1}}$. Then Δ_A and L_A are bounded invertible operators with bounded inverses and, for every point $\omega \in \mathfrak{h}_A$, the RK $K_\omega^A(\lambda)$ for the space $\mathcal{H}(A)$ is given by the formula

$$K_\omega^A(\lambda)\begin{bmatrix}\xi \\ \eta\end{bmatrix} = L_A(\Delta_A)^{-1}\begin{bmatrix}-\Phi_{11}k_\omega^{b_5} & k_\omega^{b_3} \\ \Phi_{22}^*\ell_\omega^{b_6} & \ell_\omega^{b_4}\end{bmatrix}\begin{bmatrix}\xi \\ \eta\end{bmatrix}. \tag{5.208}$$

Proof Since $A \in \mathcal{U}_{rsR}(J_p)$, we may assume that $c \in \mathcal{C}(A) \cap \mathring{\mathcal{C}}^{p\times p}$ in formula (5.198) and hence that there exists a pair of numbers $\delta_2 > \delta_1 > 0$ such that

$$\delta_1 I_p \le c(\lambda) + c(\lambda)^* \le \delta_2 I_p \quad \text{for every point } \lambda \in \mathbb{C}_+. \tag{5.209}$$

Thus, if $g \in \mathcal{H}(b_3)$ and $h \in \mathcal{H}_*(b_4)$, then,

$$\|g + h\|_{st}^2 = \|g\|_{st}^2 + \|h\|_{st}^2$$

and, in view of formulas (5.205) and (5.206),

$$\delta_1 \|g+h\|_{st}^2 \leq \left\| L_A \begin{bmatrix} g \\ h \end{bmatrix} \right\|_{\mathcal{H}(A)}^2 \leq \delta_2 \|g+h\|_{st}^2.$$

Therefore, L_A is a bounded one to one map of $\mathcal{H}(b_3) \oplus \mathcal{H}_*(b_4)$ onto $\mathcal{H}(A)$ with a bounded inverse $(L_A)^{-1}$:

$$\|L_A^{-1}\| \leq \delta_1^{-1/2} \quad \text{and} \quad \|\Delta_A^{-1}\| \leq \delta_1^{-1}.$$

Now let $f \in \mathcal{H}(A)$ and let

$$(L_A)^{-1} f = \begin{bmatrix} g \\ h \end{bmatrix} \in \begin{array}{c} \mathcal{H}(b_3) \\ \oplus \\ \mathcal{H}_*(b_4) \end{array}.$$

Then, for every choice of $\xi, \eta \in \mathbb{C}^p$ and every point $\omega \in \mathfrak{h}_A$,

$$\begin{aligned} [\xi^* \quad \eta^*] f(\omega) &= \left\langle f, K_\omega^A \begin{bmatrix} \xi \\ \eta \end{bmatrix} \right\rangle_{\mathcal{H}(A)} \\ &= \left\langle \Delta_A (L_A)^{-1} f, (L_A)^{-1} K_\omega^A \begin{bmatrix} \xi \\ \eta \end{bmatrix} \right\rangle_{st} \\ &= \left\langle \begin{bmatrix} g \\ h \end{bmatrix}, \Delta_A (L_A)^{-1} K_\omega^A \begin{bmatrix} \xi \\ \eta \end{bmatrix} \right\rangle_{st}, \end{aligned}$$

by formulas (5.206) and (5.202). On the other hand, in view of (5.203),

$$\begin{aligned} &[\xi^* \quad \eta^*] f(\omega) = [\xi^* \quad \eta^*] \left(L_A \begin{bmatrix} g \\ 0 \end{bmatrix} \right)(\omega) + [\xi^* \quad \eta^*] \left(L_A \begin{bmatrix} 0 \\ h \end{bmatrix} \right)(\omega) \\ &= \left\langle \begin{bmatrix} -\Phi_{11}^* g \\ g \end{bmatrix}, \begin{bmatrix} k_\omega^{b_5} \xi \\ k_\omega^{b_3} \eta \end{bmatrix} \right\rangle_{st} + \left\langle \begin{bmatrix} \Phi_{22} h \\ h \end{bmatrix}, \begin{bmatrix} \ell_\omega^{b_6} \xi \\ \ell_\omega^{b_4} \eta \end{bmatrix} \right\rangle_{st} \\ &= \langle g, -\Phi_{11} k_\omega^{b_5} \xi + k_\omega^{b_3} \eta \rangle_{st} + \langle h, \Phi_{22}^* \ell_\omega^{b_6} \xi + \ell_\omega^{b_4} \eta \rangle_{st} \\ &= \left\langle \begin{bmatrix} g \\ h \end{bmatrix}, \begin{bmatrix} -\Phi_{11} k_\omega^{b_5} \xi + k_\omega^{b_3} \eta \\ \Phi_{22}^* \ell_\omega^{b_6} \xi + \ell_\omega^{b_4} \eta \end{bmatrix} \right\rangle_{st} \end{aligned}$$

for every choice of $g \in \mathcal{H}(b_3)$, $h \in \mathcal{H}_*(b_4)$ and $\xi, \eta \in \mathbb{C}^p$. Thus, upon comparing the two formulas for $[\xi^* \quad \eta^*] f(\omega)$, we obtain formula (5.208). □

We remark that the exhibited lower bound for Δ_A depends heavily on the "full operator"; the symbol

$$\begin{bmatrix} c+c^* & c \\ c^* & c+c^* \end{bmatrix}$$

may be a singular matrix (e.g., $c = (1+i\sqrt{3})a$ with $a \in \mathbb{R}$).

Formula (5.208) may be rewritten in a form that is analogous to formula (5.178):

$$K_\omega^A(\lambda) = (L_A \Delta_A^{-1} F_\omega^A)(\lambda), \tag{5.210}$$

where the operators

$$L_A : \mathcal{H}^m(b_3, b_4) \to \mathcal{H}^m(A) \quad \text{and} \quad \Delta_A : \mathcal{H}^m(b_3, b_4) \to \mathcal{H}^m(b_3, b_4)$$

act on the columns of $m \times m$ mvf's and

$$F_\omega^A(\lambda) = \begin{bmatrix} -(\Phi_{11} k_\omega^{b_5})(\lambda) & k_\omega^{b_3}(\lambda) \\ (\Phi_{22}^* \ell_\omega^{b_6})(\lambda) & \ell_\omega^{b_4}(\lambda) \end{bmatrix} \in \mathcal{H}^m(b_3, b_4).$$

Then the mvf $A(\lambda)$ is defined up to a constant right J_p-unitary factor by the formula

$$A(\lambda) J_p A(\omega)^* = J_p - \rho_\omega(\lambda) K_\omega^A(\lambda).$$

If $A \in \mathcal{U}_0(J_p)$, then

$$A(\lambda) = I_m + 2\pi i \lambda K_0(\lambda) J_p.$$

Theorem 5.105 *Let $A \in \mathcal{U}_{rsR}(J_p)$, $\{b_3, b_4\} \in ap_{II}(A)$, $\mathfrak{E}(\lambda) = \sqrt{2} N_2^* A(\lambda) \mathfrak{V}$ and let the operator Δ_A be defined by formula (5.202). Then*

$$\eta_2^* K_\omega^{\mathfrak{E}}(\lambda) \eta_1 = 2 \left\langle \Delta_A^{-1} \begin{bmatrix} k_\omega^{b_3} \eta_1 \\ \ell_\omega^{b_4} \eta_1 \end{bmatrix}, \begin{bmatrix} k_\lambda^{b_3} \eta_2 \\ \ell_\lambda^{b_4} \eta_2 \end{bmatrix} \right\rangle_{st} \tag{5.211}$$

for every pair of points $\lambda, \omega \in \mathfrak{h}_A$ and every pair of vectors $\eta_1, \eta_2 \in \mathbb{C}^p$. Moreover, if $c \in \mathcal{C}(A) \cap \mathring{\mathcal{C}}^{p \times p}$, then there exist numbers $\gamma_1 > 0$, $\gamma_2 > \gamma_1$ such that

$$\gamma_1 I_p \le (\Re c)(\mu) \le \gamma_2 I_p \tag{5.212}$$

for almost all points $\mu \in \mathbb{R}$, and for every such choice of γ_1 and γ_2, and every point $\omega \in \mathfrak{h}_A$,

$$\gamma_2^{-1} \{ k_\omega^{b_3}(\omega) + \ell_\omega^{b_4}(\omega) \} \le K_\omega^{\mathfrak{E}}(\omega) \le \gamma_1^{-1} \{ k_\omega^{b_3}(\omega) + \ell_\omega^{b_4}(\omega) \}. \tag{5.213}$$

Proof Formula (5.211) follows from formulas (5.208) and (5.203) and the identity

$$\eta_2^* K_\omega^{\mathfrak{E}}(\lambda)\eta_1 = 2\left\langle K_\omega^A \begin{bmatrix} 0 \\ \eta_1 \end{bmatrix}, K_\lambda^A \begin{bmatrix} 0 \\ \eta_2 \end{bmatrix} \right\rangle_{\mathcal{H}(A)}.$$

Next, the equality (5.206) and the bounds exhibited in (5.212)imply that $2\gamma_1 I \leq \Delta_A \leq 2\gamma_2 I$. Therefore,

$$(2\gamma_2)^{-1} I \leq (\Delta_A)^{-1} \leq (2\gamma_1)^{-1} I$$

and hence, by formula (5.211),

$$\gamma_2^{-1} \left\| \begin{matrix} k_\omega^{b_3}\eta \\ \ell_\omega^{b_4}\eta \end{matrix} \right\|_{st}^2 \leq \eta^* K_\omega^{\mathfrak{E}}(\omega)\eta \leq \gamma_1^{-1} \left\| \begin{matrix} k_\omega^{b_3}\eta \\ \ell_\omega^{b_4}\eta \end{matrix} \right\|_{st}^2.$$

The asserted inequalities (5.213) now drop out easily from the evaluation

$$\left\| \begin{matrix} k_\omega^{b_3}\eta \\ \ell_\omega^{b_4}\eta \end{matrix} \right\|_{st}^2 = \eta^* \{k_\omega^{b_3}(\omega) + \ell_\omega^{b_4}(\omega)\}\eta. \qquad \square$$

We are primarily interested in formula (5.210) for the case $\omega = 0$. With a slight abuse of notation, it can be reexpressed in the following more convenient form:

Theorem 5.106 *If, in the setting of Theorem 5.105, it is also assumed that $0 \in \mathfrak{h}_A$, then*

$$K_0^A = L_A \begin{bmatrix} \widehat{u}_{11} & \widehat{u}_{12} \\ \widehat{u}_{21} & \widehat{u}_{22} \end{bmatrix}, \tag{5.214}$$

where the $\widehat{u}_{ij} = \widehat{u}_{ij}(\lambda)$ are $p \times p$ mvf's that are obtained as the solutions of the system of equations

$$\Delta_A \begin{bmatrix} \widehat{u}_{11} & \widehat{u}_{12} \\ \widehat{u}_{21} & \widehat{u}_{22} \end{bmatrix} = \begin{bmatrix} -\Phi_{11} k_0^{b_5} & k_0^{b_3} \\ \Phi_{22}^* \ell_0^{b_6} & \ell_0^{b_4} \end{bmatrix} \tag{5.215}$$

and the operators in formulas (5.214) and (5.215) act on the indicated matrix arrays column by column. In particular, the columns of $\widehat{u}_{11}(\lambda)$ and $\widehat{u}_{12}(\lambda)$ belong to $\mathcal{H}(b_3)$ and the columns of $\widehat{u}_{21}(\lambda)$ and $\widehat{u}_{22}(\lambda)$ belong to $\mathcal{H}_*(b_4)$.

5.19 Bibliographical notes

Theorem 5.2 seems to be essentially due to Aronszajn and Moore; see [Arn50] for a scalar version and [Sch64] for more information on RKHS'.

The RKHS's $\mathcal{H}(U)$ and $\mathcal{B}(\mathfrak{E})$ were introduced by L. de Branges, who developed the theory of these spaces and applied it to a number of problems in analysis, including inverse problems for canonical differential systems for $m = 2$. de Branges' early papers focused on entire $\mathcal{J}_1$-inner mvf's $U(\lambda)$ and spaces of scalar entire inner functions $\mathcal{B}(\mathfrak{E})$; see [Br68a] and the references cited therein. His later papers treated entire and meromorphic matrix and operator valued functions; see [Br63], [Br65], [Br68b] and, for spaces $\mathcal{H}(S)$ of Schur class operator valued functions in the open unit disc, [BrR66].

An introduction to the theory of the de Branges spaces $\mathcal{B}(\mathfrak{E})$ of scalar entire functions and its application to prediction theory for stationary Gaussian processes and inverse spectral problems for the Feller-Krein string equation is presented in the monograph [DMc76]. Additional information on direct problems for Feller-Krein type string equations is presented in [Dy70].The monographs [An90] and [ADRS97] also contain useful information on de Branges spaces.

Connections of the de Branges RKHS's $\mathcal{H}(U)$ with the Krein theory of symmetric operators in Hilbert space with finite defect indices (m, m) are presented in [AlD84] and [AlD85].

Theorems 5.21 and 5.22 are essentially due to L. de Branges [Br63]. In de Branges' original formulation it was assumed that the domain Ω of the RKHS included at least one point of $\mathbb{R}$. The relaxation of this requirement to the weaker constraint $\Omega \cap \overline{\Omega} \neq \emptyset$ is due to Rovnyak [Rov68]. A formulation for a wider class of domains that includes analogous results for the disc that are due to Ball [Ba75] is given in [AlD93]. The identification of $\mathcal{H}(U)$ with maximal domain with a mvf $U \in \mathcal{P}^\circ(J)$ in Theorem 5.31 seems to be new.

The verification of the unitary similarity (5.67) between $\mathcal{H}(U)$ and $\mathcal{H}(S)$ for $U \in \mathcal{H}(U)$ ans $S = PG(U)$, and the use of this equivalence to obtain the characterization Theorem 5.55 of $\mathcal{H}(W)$ for $W \in \mathcal{U}(j_{pq})$ is based on Theorems 2.4 and 2.7 of [Dy89b]. This characterization of $\mathcal{H}(W)$ was obtained by other methods by Z. Arova [Ara97].

There are many ways to establish Theorem 5.24. The presented method via finite dimensional reproducing kernel Hilbert spaces is adapted from the treatments in [Dy89b] and [Dy98]; more details on the limiting argument may be found in [Dy89b]. Limiting arguments can be avoided by passing to Riccati equations as in [Dy03a]. The identification of the de Branges identity in finite dimensional R_α invariant spaces with solutions of the Lyapunov equation seems to have been first noted in [Dy89a]; a better proof is furnished in [Dy94a] and [Dy98].

6

Operator nodes and passive systems

In this chapter the RKHS's $\mathcal{H}(U)$ and $\mathcal{B}(\mathfrak{E})$ will used to construct functional models for operator nodes and passive systems. The following notation will be useful:

$$
\begin{aligned}
\mathcal{L}(X,Y) &= \text{the set of bounded linear operators from a Hilbert space } X \\
&\qquad\qquad \text{into a Hilbert space } Y; \\
\mathcal{L}(X) &= \mathcal{L}(X,X); \\
\mathcal{D}(A) &= \text{the domain of a linear operator } A; \\
A|_{\mathcal{M}} &= \text{the restriction of a linear operator to a subspace } \mathcal{M}; \\
\Lambda_A &= \{\lambda \in \mathbb{C} : (I - \lambda A)^{-1} \in \mathcal{L}(X)\}; \\
\rho(A) &= \{\lambda \in \mathbb{C} : (\lambda I - A)^{-1} \in \mathcal{L}(X)\}.
\end{aligned}
$$

6.1 Characteristic mvf's of Livsic-Brodskii nodes

Let $\mathcal{LB}(J)$ denote the class of mvf's U that are meromorphic on $\mathbb{C} \setminus \mathbb{R}$ such that:

(1) The restriction of U to $\mathfrak{h}_U^+$ belongs to the class $\mathcal{P}^\circ(J)$.

(2) $0 \in \mathfrak{h}_U$ and $U(0) = I_m$.

(3) There exists a $\delta > 0$ such that $(-\delta, \delta) \subset \mathfrak{h}_U$ and $U(\mu)^* J U(\mu) = J$ for all points $\mu \in (-\delta, \delta)$.

(4) The mvf U does not have a holomorphic extension to an open set Ω that contains $\mathfrak{h}_U$ properly.

Clearly,

$$\mathcal{LB}(J) \supset \mathcal{U}_0(J). \tag{6.1}$$

The class $\mathcal{LB}(J)$ coincides with the class of characteristic mvf's of simple LB (Livsic-Brodskii) J-nodes, which plays a fundamental role in the spectral theory of bounded linear nonselfadjoint operators with finite dimensional imaginary part.

A set $\Sigma = (A, C; X, Y)$ of Hilbert spaces X and Y and operators $A \in \mathcal{L}(X)$ and $C \in \mathcal{L}(X, Y)$ is called an LB J-node if

$$A - A^* = iC^*JC, \tag{6.2}$$

where $J \in \mathcal{L}(Y)$ is both unitary and selfajoint.

An LB J-node is called **simple** if

$$\bigcap_{n\geq 0} \ker CA^n = \{0\}. \tag{6.3}$$

It is readily checked that Σ is simple if and only if

$$\bigvee_{n\geq 0} (A^*)^n C^* Y = X \quad \text{or, equivalently,} \quad \bigvee_{\lambda\in\Omega} (I - \lambda A^*)^{-1} C^* Y = X$$

for some neighborhood Ω of zero. In view of (6.2), these conditions are also valid if A^* is replaced by A. The operator valued function

$$U_\Sigma(\lambda) = I_Y + i\lambda C(I_X - \lambda A)^{-1} C^* J \quad \text{for } \lambda \in \Lambda_A \tag{6.4}$$

is called the **characteristic function** of the LB J-node Σ. If Y is an m-dimensional linear space, then $\mathcal{L}(Y)$ can be identified with the set of $m \times m$ matrices that define these operators in a fixed orthornormal basis in Y. In this case we can assume without loss of generality that $Y = \mathbb{C}^m$ and $\mathcal{L}(Y) = \mathbb{C}^{m\times m}$, respectively. Then J is an $m \times m$ signature matrix and $U_\Sigma(\lambda)$ is an $m \times m$ mvf that will be called the **characteristic mvf** of the LB J-node Σ.

Theorem 6.1 *Let $U \in \mathcal{LB}(J)$, $\mathring{X} = \mathcal{H}(U)$, $\mathring{A} = R_0$ and $\mathring{C}f = \sqrt{2\pi} f(0)$ for $f \in \mathcal{H}(U)$. Then $\mathring{\Sigma} = (\mathring{A}, \mathring{C}; \mathring{X}, \mathbb{C}^m)$ is a simple LB J-node. Moreover, $\Lambda_{\mathring{A}} = \mathfrak{h}_U$ and $U_{\mathring{\Sigma}}(\lambda) \equiv U(\lambda)$ on $\mathfrak{h}_U$.*

Proof Since

$$\xi^* \mathring{C} f = \sqrt{2\pi}\xi^* f(0) = \sqrt{2\pi}\langle f, K_0^U \xi\rangle_{\mathcal{H}(U)} \quad \text{for } f \in \mathcal{H}(U) \quad \text{and} \quad \xi \in \mathbb{C}^m,$$

the adjoint operator $\mathring{C}^* \in \mathcal{L}(\mathbb{C}^m, \mathcal{H}(U))$ is given by the formula

$$(\mathring{C}^*\xi)(\lambda) = \sqrt{2\pi} K_0^U(\lambda)\xi \quad \text{for } \lambda \in \mathfrak{h}_U \quad \text{and} \quad \xi \in \mathbb{C}^m. \tag{6.5}$$

In view of Remark 5.32, the RKHS $\mathcal{H}(U)$ is R_0 invariant and the de Branges identity (5.27) for $\alpha = \beta = 0$:

$$\langle (R_0 - R_0^*)f, g\rangle_{\mathcal{H}(U)} = 2\pi i g(0)^* J f(0) \quad \text{for } f, g \in \mathcal{H}(U), \tag{6.6}$$

is equivalent to the operator identity

$$\mathring{A} - \mathring{A}^* = i\mathring{C}^* J \mathring{C}.$$

Moreover, since

$$\mathring{C}\mathring{A}^n f = \sqrt{2\pi}\frac{f^{(n)}(0)}{n!} \quad \text{for every } f \in \mathcal{H}(U),$$

it is readily seen that $\cap_{n\geq 0}\ker \mathring{C}\mathring{A}^n = \{0\}$ and hence that the node $\mathring{\Sigma}$ is simple.

It remains only to check that $\Lambda_{\mathring{A}} = \mathfrak{h}_U$ and that $U_{\mathring{\Sigma}}(\lambda) = U(\lambda)$ for every $\lambda \in \mathfrak{h}_U$. The formula

$$((I - \lambda R_0)^{-1} f)(\omega) = \frac{\omega f(\omega) - \lambda f(\lambda)}{\omega - \lambda} = f(\omega) + \lambda (R_\lambda f)(\omega)$$
$$\text{for } \lambda \in \mathfrak{h}_U \cap \mathfrak{h}_{U^{-1}},\ \omega \in \mathfrak{h}_U,\ \lambda \neq \omega, \tag{6.7}$$

implies that $(I - \lambda R_0)^{-1} \in \mathcal{L}(\mathcal{H}(U))$ for each point $\lambda \in \mathfrak{h}_U \cap \mathfrak{h}_{U^{-1}}$, i.e., $\mathfrak{h}_U \cap \mathfrak{h}_{U^{-1}} \subseteq \Lambda_{\mathring{A}}$. It also implies that

$$\mathring{C}(I - \lambda\mathring{A})^{-1} f = \sqrt{2\pi} f(\lambda) \quad \text{for } \lambda \in \mathfrak{h}_U \cap \mathfrak{h}_{U^{-1}} \tag{6.8}$$

and hence that

$$\mathring{C}(I - \lambda\mathring{A})^{-1}\mathring{C}^* J\xi = 2\pi K_0^U(\lambda) J\xi \quad \text{for } \lambda \in \mathfrak{h}_U \cap \mathfrak{h}_{U^{-1}} \quad \text{and} \quad \xi \in \mathbb{C}^m.$$

Thus,

$$U_{\mathring{\Sigma}}(\lambda) = I_m + 2\pi i\lambda K_0^U(\lambda) J = U(\lambda) \quad \text{for } \lambda \in \mathfrak{h}_U \cap \mathfrak{h}_{U^{-1}}$$

and, consequently $U_{\mathring{\Sigma}}$ satisfies properties (1)–(3) in the list furnished above. Therefore, as $\Lambda_{\mathring{A}} = \mathfrak{h}_{U_{\mathring{\Sigma}}}$ by definition, and $\mathfrak{h}_U$ satisfies property (4) in the list, it follows that $\Lambda_{\mathring{A}} \subseteq \mathfrak{h}_U$ and hence that $U_{\mathring{\Sigma}}(\lambda) = U(\lambda)$ in $\Lambda_{\mathring{A}}$. Moreover, $\Lambda_{\mathring{A}} \cap \mathbb{R} = \mathfrak{h}_U \cap \mathbb{R}$, since $\mathfrak{h}_{U^{-1}} \cap \mathfrak{h}_U \cap \mathbb{R} = \mathfrak{h}_U \cap \mathbb{R}$.

Suppose next that $\omega \in \mathfrak{h}_U$ and $\omega \notin \mathbb{R}$. Then there exists a $\delta > 0$ such that the mvf $F(\lambda) = U(1/\lambda)$ is holomorphic in the disc $\mathcal{B}_\delta(1/\omega) = \{\lambda \in \mathbb{C} : |\lambda - 1/\omega| < \delta\}$ and $\mathcal{B}_\delta(1/\omega) \setminus \{1/\omega\} \subset \mathfrak{h}_U \cap \mathfrak{h}_{U^{-1}}$. Therefore,

$$F(\lambda) = I + i\mathring{C}(\lambda I - \mathring{A})^{-1}\mathring{C}^* J \quad \text{in } \mathcal{B}_\delta(1/\omega) \setminus \{1/\omega\}.$$

Thus, if Γ_r is circular contour of radius $0 < r < \delta$ centered at $1/\omega$ and directed counterclockwise, then

$$\int_{\Gamma_r} \zeta^k \mathring{C}(\zeta I - \mathring{A})^{-1}\mathring{C}^* d\zeta = -i \int_{\Gamma_r} \zeta^k (F(\zeta) - I) J d\zeta = 0.$$

Therefore,

$$\int_{\Gamma_r} \mathring{C}\mathring{A}^{k+1}(\zeta I - \mathring{A})^{-1}\mathring{C}^* d\zeta = \int_{\Gamma_r} \mathring{C}\mathring{A}^k\{\zeta(\zeta I - \mathring{A})^{-1} - I\}\mathring{C}^* d\zeta = 0$$

for $k = 0, 1, \ldots$. Consequently,

$$\int_{\Gamma_r} \langle \mathring{C}\mathring{A}^{k+j}(\zeta I - \mathring{A})^{-1}\mathring{C}^* u, v\rangle_{st} d\zeta =$$

$$\int_{\Gamma_r} \langle (\zeta I - \zeta\mathring{A})^{-1}\mathring{A}^j\mathring{C}^* u, (\mathring{A}^*)^k \mathring{C}^* v\rangle_{\mathcal{H}(U)} d\zeta = 0$$

for $u, v \in \mathbb{C}^m$ and $j, k = 0, 1, \ldots$. Thus, as the node is simple,

$$\int_{\Gamma_r} \langle (\zeta I - \mathring{A})^{-1} g, h\rangle_{\mathcal{H}(U)} d\zeta = 0 \quad \text{for every } 0 < r < \delta,$$

for dense sets of vectors $g \in \mathcal{H}(U)$ and $h \in \mathcal{H}(U)$ and hence, by passing to limits, for every pair of vectors $g, h \in \mathcal{H}(U)$. But this in turn implies that the Riesz projector (see e.g., Chapter 11 of [RSzN55])

$$\int_{\Gamma_r} (\zeta I - \mathring{A})^{-1} d\zeta = 0,$$

and hence that the operator $(\omega^{-1}I - \mathring{A})$ has a bounded inverse, i.e., $\omega \in \Lambda_{\mathring{A}}$. □

Theorem 6.2 *Let $U \in \mathcal{LB}(J)$ and let $\mathcal{H}(U)$ be the RKHS of holomorphic vvf's defined on $\mathfrak{h}_U$ with RK $K_\omega(\lambda)$ defined on $\mathfrak{h}_U \times \mathfrak{h}_U$ by formula (5.34). Then:*

(1) *$R_\alpha \in \mathcal{L}(\mathcal{H}(U))$ for every point $\alpha \in \mathfrak{h}_U$.*

(2) *The de Branges identity holds for every pair of vectors* $f, g \in \mathcal{H}(U)$ *and every pair of points* $\alpha, \beta \in \mathfrak{h}_U$.

(3) $\mathfrak{h}_U = \Lambda_{\mathring{A}}$ *for* $\mathring{A} = R_0$ *and*

$$R_\alpha = R_0(I - \alpha R_0)^{-1} \quad \text{for every } \alpha \in \mathfrak{h}_U. \tag{6.9}$$

Proof Let $\mathring{\Sigma} = (R_0, \mathring{C}; \mathcal{H}(U), \mathbb{C}^m)$ be the simple LB J-node considered in Theorem 6.1. Then formula (6.8) implies that

$$\begin{aligned}(R_\alpha f)(\lambda) &= (\lambda - \alpha)^{-1}\sqrt{2\pi}\left\{\mathring{C}(I - \lambda\mathring{A})^{-1} - \mathring{C}(I - \alpha\mathring{A})^{-1}\right\} f \\ &= \sqrt{2\pi}\mathring{C}(I - \lambda\mathring{A})^{-1}\mathring{A}(I - \alpha\mathring{A})^{-1} f \end{aligned} \tag{6.10}$$

for every $f \in \mathcal{H}(U)$ and $\alpha \in \Lambda_{\mathring{A}}$. Thus,

$$R_\alpha f = \mathring{A}(I - \alpha\mathring{A})^{-1} f \quad \text{for } f \in \mathcal{H}(U) \text{ and } \alpha \in \Lambda_{\mathring{A}}.$$

Consequently, $\mathcal{H}(U)$ is R_α invariant for every $\alpha \in \mathfrak{h}_U$; formula (6.9) holds; and the de Branges identity hold for $f, g \in \mathcal{H}(U)$ and $\alpha, \beta \in \mathfrak{h}_U$. □

Since the LB J-node $\mathring{\Sigma}$ is uniquely defined by the mvf $U \in \mathcal{LB}(J)$, we shall also denote it as

$$\Sigma_U = (R_0, C_U; \mathcal{H}(U), \mathbb{C}^m),$$

where it is understood that R_0 acts in $\mathcal{H}(U)$ and that C_U maps $f \in \mathcal{H}(U)$ into $\sqrt{2\pi} f(0)$.

There exists another model of a simple LB J-node with given characteristic mvf $U \in \mathcal{LB}(J)$ that is due to Livsic that rests on the following lemma:

Lemma 6.3 *If* $U \in \mathcal{P}(J)$, *and* $\det(I_m + U(\lambda)) \not\equiv 0$ *in* $\mathbb{C}_+$, *then the mvf* $c = JT_{\mathfrak{V}}[PG(U)]J$ *belongs to* $\mathcal{C}^{m\times m}$. *Moreover, if* $U = U_\Sigma$ *is the characteristic mvf of an LB J-node* $\Sigma = (A, C; X, \mathbb{C}^m)$, *then*

$$2c(\lambda) = -i\lambda C(I - \lambda A_R)^{-1}C^*, \quad \text{where} \quad A_R = (A + A^*)/2. \tag{6.11}$$

Proof This is verified by straightforward calculation with the aid of formulas (5.40) and (5.42). □

Theorem 6.4 *If* $U \in \mathcal{LB}(J)$, $\widetilde{U}(\lambda) = U(-1/\lambda)$, *and* $c = JT_{\mathfrak{V}}[PG(\widetilde{U})]J$, *then* $\lim_{\mu\uparrow\infty} \widetilde{U}(\mu) = I_m$ *and there is an interval* $[a, b] \subset \mathbb{R}$ *such that:*

(1) $\mathbb{R} \setminus [a, b] \subset \mathfrak{h}_{\widetilde{U}}$ *and* $\widetilde{U}^* J \widetilde{U} = J$ *for* $\mu \in \mathbb{R} \setminus [a, b]$.

(2) *$c \in \mathcal{C}^{p\times p}$ and admits a representation of the form*

$$c(\lambda) = \frac{1}{\pi i}\int_a^b \frac{d\sigma(\mu)}{\mu - \lambda} \quad \text{for } \lambda \in \mathbb{C} \setminus [a, b].$$

(3) *If $X = L_2^m(d\sigma, [a, b])$,*

$$(A_1 f)(\mu) = \mu f(\mu), \quad (Cf)(\mu) = \frac{1}{\sqrt{\pi}} J \int_a^b d\sigma(\mu) f(\mu) \quad \text{for } f \in X,$$

and $A = A_1 + (1/2)iC^*JC$, then $\Sigma = (A, C; X, \mathbb{C}^m)$ is a simple LB J-node with characteristic mvf $U_\Sigma(\lambda) = U(\lambda)$.

Proof Assertion (1) follows directly from the fact that $U \in \mathcal{LB}(J)$. Thus, in view of formula (5.42) (applied to $\widetilde{U}$), the Stieltjes inversion formula implies that the spectral function $\sigma(\mu)$ of c is constant for $\mu < a$ and $\mu > b$. Assertion (2) then follows from the general representation formula (3.4), since $c(\infty) = 0$.

Since $A_1 = A_1^*$, it is readily checked that Σ is an LB J-node. Moreover, Σ is simple because

$$\cap_{n\geq 0} \ker CA^n = \cap_{n\geq 0} \ker CA_1^n \quad \text{and} \quad \cap_{n\geq 0} \ker CA_1^n = \{0\},$$

since vector valued polynomials are dense in $L_2^m(d\sigma, [a, b])$.

Finally, to identify $U_\Sigma(-1/\lambda)$, let

$$c_\Sigma(\lambda) = (I_m - U_\Sigma(-1/\lambda))(I_m + U_\Sigma(-1/\lambda))^{-1} J$$

Then, it is readily checked that

$$2c_\Sigma(\lambda)v = iC(\lambda I - A_1)^{-1}C^* v = \frac{2}{\pi i}\int_a^b \frac{d\sigma(\mu)}{\mu - \lambda} v \quad \text{for every } v \in \mathbb{C}^m,$$

and hence that $c_\Sigma(\lambda)$ coincides with the mvf $c(\lambda)$ that was defined in terms of $U(-1/\lambda)$. Therefore, $U(\lambda) = U_\Sigma(\lambda)$, as claimed. □

Two LB J-nodes $\Sigma_j = (A_j, C_j; X_j, Y)$, $j = 1, 2$, are called **similar** if there exists an invertible operator $T \in \mathcal{L}(X_1, X_2)$ with $T^{-1} \in \mathcal{L}(X_2, X_1)$ such that

$$A_2 = TA_1T^{-1} \quad \text{and} \quad C_2 = C_1T^{-1}.$$

If there exists such a T which is unitary, then Σ_1 and Σ_2 are said to be **unitarily similar** and T is called a **unitary similarity operator** from Σ_1 to Σ_2. If one of two unitary similar nodes is simple, then the other is simple and there is only one such T.

Theorem 6.5 *If the characteristic functions U_{Σ_1} and U_{Σ_2} of two simple LB J-nodes coincide in a neighborhood of zero, then $\Sigma_1 = (A_1, C_1; X_1, Y_1)$ is unitarily similar to $\Sigma_2 = (A_2, C_2; X_2, Y_2)$ and hence $\Lambda_{A_1} = \Lambda_{A_2}$ and $U_{\Sigma_1}(\lambda) \equiv U_{\Sigma_2}(\lambda)$ on Λ_{A_1}.*

Proof Since U_{Σ_1} and U_{Σ_2} coincide in a neighborhood Ω of zero,

$$C_1(I_{X_1} - \lambda A_1)^{-1}(I_{X_1} - \overline{\omega} A_1^*)^{-1} C_1^* = C_2(I_{X_2} - \lambda A_2)^{-1}(I_{X_2} - \overline{\omega} A_2^*)^{-1} C_2^*$$

for $\lambda, \overline{\omega} \in \Omega$. Therefore,

$$C_1 A_1^j (A_1^*)^k C_1^* = C_2 A_2^j (A_2^*)^k C_2^* \quad \text{for } j, k = 0, 1, \dots. \tag{6.12}$$

The identities in (6.12) imply that the formula

$$T \sum_{k=0}^{n} (A_1^*)^k C_1^* y_k = \sum_{k=0}^{n} (A_2^*)^k C_2^* y_k, \quad y_k \in Y \quad \text{for } k = 1, \dots, n,$$

defines an isometric operator that extends by continuity to an isometric operator from

$$\bigvee_{k \geq 0} (A_1^*)^k C_1^* Y \quad \text{onto} \quad \bigvee_{k \geq 0} (A_2^*)^k C_2^* Y \quad \text{such that} \quad T C_1^* = C_2^*.$$

If the LB J-nodes Σ_1 and Σ_2 are simple, then this extension is unitary from X_1 onto X_2, Moreover, in this case the formulas

$$(T A_1^*)(A_1^*)^k C_1^* y = (A_2^*)^{k+1} C_2^* y = (A_2^* T)(A_1^*)^k C_1^* y$$

imply that $T A_1^* = A_2^* T$. □

Theorem 6.6 *If $\Sigma = (A, C; X, \mathbb{C}^m)$ is an LB J-node, then its characteristic mvf $U_\Sigma(\lambda)$ satisfies conditions (1)–(3) in the definition of the class $\mathcal{LB}(J)$ and the operator*

$$(T_\Sigma x)(\lambda) = (2\pi)^{-1/2} C(I - \lambda A)^{-1} x \tag{6.13}$$

is a partial isometry from X onto the space $\mathcal{H}(U)$ based on $U = U_\Sigma$ (with $\mathfrak{h}_U = \Lambda_A$),

$$\ker T_\Sigma = \bigcap_{n \geq 0} \ker(C A^n).$$

Moreover, T_Σ is a unitary similarity operator from the node Σ to the node $\Sigma_U = (R_0, C_U; \mathcal{H}(U), \mathbb{C}^m)$ based on the mvf $U = U_\Sigma$ if and only if Σ is a simple node. If Σ is a simple LB J-node, then $U_\Sigma \in \mathcal{LB}(J)$,

Proof A straightforward calculation based on (6.2) and (6.4) yields the identity

$$J - U_\Sigma(\lambda)JU_\Sigma(\omega)^* = -i(\lambda - \overline{\omega})C(I - \lambda A)^{-1}(I - \overline{\omega}A^*)^{-1}C^* \qquad (6.14)$$

for every pair of points $\lambda, \omega \in \Lambda_A$. Thus, U_Σ is holomorphic on Λ_A and satisfies conditions (1)–(3) in the definition of the class $\mathcal{LB}(J)$. Formula (6.14) implies that

$$\sum_{j=1}^{n} K_{\omega_j}(\lambda)v_j = \frac{1}{\sqrt{2\pi}}C(I - \lambda A)^{-1}x = T_\Sigma x,$$

where

$$x = \frac{1}{\sqrt{2\pi}}\sum_{j=1}^{n}(I - \overline{\omega_j}A)^{-1}C^*v_j,$$

$\omega_1, \ldots, \omega_n \in \Lambda_A$ and $v_1, \ldots, v_n \in \mathbb{C}^m$. Therefore,

$$\begin{aligned}\|T_\Sigma x\|^2_{\mathcal{H}(U)} &= \sum_{i,j=1}^{n} v_i^* K_{\omega_j}(\omega_i)v_i \\ &= \frac{1}{2\pi}\sum_{i,j=1}^{n} v_i^* C(I - \omega_i A)^{-1}(I - \overline{\omega_j}A)^{-1}C^*v_j = \|x\|^2_X.\end{aligned}$$

Thus, T_Σ maps

$$X_1 = \bigvee_{\omega \in \Lambda_A} (I - \overline{\omega}A^*)^{-1}C^*\mathbb{C}^m$$

isometrically onto $\mathcal{H}(U)$. Therefore, since

$$X \ominus X_1 = \cap_{n\geq 0}\ker(CA^n) = \ker T_\Sigma,$$

T_Σ is a partial isometry from X onto $\mathcal{H}(U)$ and T_Σ is unitary if and only if $X = X_1$, i.e., if and only if the LB J-node Σ is simple.

The fact that U_Σ satisfies (4) too if Σ is simple follows from Theorems 6.1 and 6.5.

The formulas $T_\Sigma Ax = R_0 T_\Sigma x$ and $Cx = C_U T_\Sigma x$ for $x \in X$ are equivalent to the formula

$$C(I - \lambda A)^{-1}x = C_U(I - \lambda R_0)^{-1}T_\Sigma x = \sqrt{2\pi}f(\lambda)$$

for $f(\lambda) = T_\Sigma x$, which is equivalent to (6.13). □

The characteristic function $U_{\Sigma_*}(\lambda)$ of the dual LB J-node $\Sigma_* = (-A^*, C; X, Y)$ is equal to $U_\Sigma^\sim(\lambda)$ and the operator

$$(T_{\Sigma_*}x)(\lambda) = \frac{1}{\sqrt{2\pi}} C(I + \lambda A^*)^{-1} x \quad \text{from } X \text{ into } \mathcal{H}(U_{\Sigma_*})$$

is related to the operator T_Σ by the formula

$$(T_{\Sigma_*}x)(\lambda) = (U_\Sigma)^\sim(\lambda)(T_\Sigma x)(-\lambda).$$

Lemma 6.7 *If $\Sigma = (A, C; X, Y)$ is an LB J-node, X_1 is a closed subspace of X and*

$$A_1 = P_{X_1} A|_{X_1} \quad \text{and} \quad C_1 = C|_{X_1}, \tag{6.15}$$

then $\Sigma_1 = (A_1, C_1; X_1, Y)$ is an LB J-node.

Proof Formulas (6.15) and (6.2) imply that

$$A_1 - A_1^* = P_{X_1}(A - A_1^*)|_{X_1} = iP_{X_1} C^* JC|_{X_1} = iC_1^* JC_1.$$

□

The LB J-node Σ_1 that is defined in Lemma 6.7 is called the **projection of the LB J-node Σ onto the subspace X_1**.

Theorem 6.8 *Let $\Sigma = (A, C; X, Y)$ be an LB J-node, let X_1 be a closed subspace of X that is invariant under A, let $X_2 = X \ominus X_1$ and let $\Sigma_j = (A_j, C_j; X_j, Y)$ be the projections of Σ onto X_j for $j = 1, 2$. Then $X = X_1 \oplus X_2$ and:*

(1) *The operators A and C can be expressed as*

$$A = \begin{bmatrix} A_1 & iC_1^* JC_2 \\ 0 & A_2 \end{bmatrix} \quad \text{and} \quad C = [C_1 \quad C_2]. \tag{6.16}$$

(2) $\Lambda_A = \Lambda_{A_1} \cap \Lambda_{A_2}$.

(3) *The characteristic functions of Σ_j are divisors of the characteristic function of Σ:*

$$U_\Sigma(\lambda) = U_{\Sigma_1}(\lambda) U_{\Sigma_2}(\lambda) \quad \text{for } \lambda \in \Lambda_A; \tag{6.17}$$

and, if $x_1 \in X_1$ and $x_2 \in X_2$, then

$$\left(T_\Sigma \begin{bmatrix} x_1 \\ x_2 \end{bmatrix}\right)(\lambda) = (T_{\Sigma_1}x_1)(\lambda) + U_{\Sigma_1}(\lambda)(T_{\Sigma_2}x_2)(\lambda). \tag{6.18}$$

(4) *If Σ is simple, then Σ_1 and Σ_2 are simple LB J-nodes, the two components in the sum in (6.18) are orthogonal in $\mathcal{H}(U_\Sigma)$ and*

$$\mathcal{H}(U_\Sigma) = \mathcal{H}(U_{\Sigma_1}) \oplus U_{\Sigma_1}\mathcal{H}(U_{\Sigma_2}).$$

Conversely, if $\Sigma_j = (A_j, C_j; X_j, Y)$, $j = 1, 2$ are two LB J-nodes, and if $X = X_1 \oplus X_2$ and the operators A and C are defined by the formulas in (6.16), then $\Sigma = (A, C; X, Y)$ is an LB J-node and X_1 is invariant under A.

Proof Under the first set of given assumptions,

$$A = \begin{bmatrix} A_1 & A_{12} \\ 0 & A_2 \end{bmatrix} \quad \text{and} \quad C = [C_1 \quad C_2]$$

for some operator $A_{12} \in \mathcal{L}(X_2, X_1)$. Thus,

$$A - A^* = \begin{bmatrix} A_1 - A_1^* & A_{12} \\ -A_{12}^* & A_2 - A_2^* \end{bmatrix} = i \begin{bmatrix} C_1^* \\ C_2^* \end{bmatrix} J[C_1 \quad C_2],$$

which implies that $A_{12} = iC_1^* JC_2$ and hence that (1) holds.

Assertions (2) and (4) follow from the block triangular structure of A; and (3) is a straightforward computation.

Conversely, if A and C are defined on $X = X_1 \oplus X_2$ in terms of the operators from the two LB J-nodes $\Sigma_1 = (A_1, C_1; X_1, Y)$ and $\Sigma_2 = (A_2, C_2; X_2, Y)$ by the formulas in (6.16), then

$$A - A^* = \begin{bmatrix} A_1 - A_1^* & iC_1^* JC_2 \\ -iC_2^* JC_1 & A_2 - A_2^* \end{bmatrix} = i \begin{bmatrix} C_1^* \\ C_2^* \end{bmatrix} J[C_1 \quad C_2] = iC^* JC,$$

i.e., $\Sigma = (A, C; X, Y)$ is an LB J-node, and $AX_1 \subseteq X_1$. □

The LB J-node $\Sigma = (A, C; X, Y)$ that is connected with the LB J-nodes $\Sigma_j = (A_j, C_j; X_j, Y)$ by the formulas in (6.16) is called the **product of Σ_1 with Σ_2** and is written $\Sigma = \Sigma_1 \times \Sigma_2$.

Remark 6.9 *The product $\Sigma_1 \times \Sigma_2$ of two simple LB J-nodes may not be simple, even if $J = I$; see, e.g., Theorem 2.3 in [Bro72]. Necessary and sufficient conditions for $\Sigma_1 \times \Sigma_2$ to be simple in terms of the characteristic functions U_{Σ_1} and U_{Σ_2} for $J = I$ are given in [SzNF70] and [Shv70]. Analogous conditions for $J \neq I$ are necessary but not sufficient for the product of two simple nodes to be simple; see [Ve91a].*

Lemma 6.10 *If $U \in \mathcal{LB}(J)$ admits a factorization $U(\lambda) = U_1(\lambda)U_2(\lambda)$ in $\mathfrak{h}_U^+$ with factors $U_1, U_2 \in \mathcal{P}(J)$, then U_1 and U_2 have unique holomorphic extensions to $\mathfrak{h}_{U_1} \supseteq \mathfrak{h}_U$ and $\mathfrak{h}_{U_2} \supseteq \mathfrak{h}_U$, respectively, and for these extensions $U_1(\lambda)U_1(0)^{-1} \in \mathcal{LB}(J)$ and $U_2(\lambda)U_1(0)^{-1} \in \mathcal{LB}(J)$.*

Proof In view of the equality (5.82) for $\lambda, \omega \in \mathfrak{h}_U^+$ and Theorem 5.5, the RKHS $\mathcal{H}(U_1)$ of vvf's on $\mathfrak{h}_{U_1}^+$ sits contractively in the RKHS $\mathcal{H}(U)$ of vvf's on $\mathfrak{h}_U^+$ and hence every vvf $f \in \mathcal{H}(U_1)$ has a unique holomorphic extension onto $\mathfrak{h}_U$. Then U_1 has a unique holomorphic extensions to $\mathfrak{h}_U$ that meets the stated conditions. The same conclusion may be obtained for U_2 by considering $U^\sim$. □

In view of Lemma 6.10 we can (and will) assume that both of the factors in every such factorization $U(\lambda) = U_1(\lambda)U_2(\lambda)$ of $U \in \mathcal{LB}(J)$ also belong to $\mathcal{LB}(J)$. The divisor U_1 of U in this factorization is called **left regular** in the Livsic-Brodskii sense if the product $\Sigma = \Sigma_1 \times \Sigma_2$ of two simple LB J-nodes with characteristic mvf's $U_{\Sigma_1} = U_1$ and $U_{\Sigma_2} = U_2$ is a simple LB J-node. Thus, if Σ is a simple LB J-node and $\Sigma = \Sigma_1 \times \Sigma_2$, then U_{Σ_1} is a left regular divisor of U_Σ. But not every divisor $U_1 \in \mathcal{LB}(J)$ in the factorization $U_\Sigma = U_1U_2$ is left regular; see [Bro72].

Theorem 6.11 *If $U \in \mathcal{U}_0(J)$, then the factors U_1 and U_2 in a right regular-singular factorization (5.171) may be uniquely specified by choosing them in $\mathcal{U}_0(J)$ and U_1 is a left regular divisor of U in the Livsic-Brodskii sense. Moreover, if $\Sigma = (A, C; X, \mathbb{C}^m)$ is a simple LB J-node with characteristic mvf $U_\Sigma = U$,*

$$X_1 = \overline{\{x \in X : T_\Sigma x \in L_2^m\}} \quad \text{and} \quad X_2 = X \ominus X_1,$$

then $\Sigma = \Sigma_1 \times \Sigma_2$, where Σ_1 and Σ_2 are the projections of the node Σ onto X_1 and X_2, respectively, and $U_j = U_{\Sigma_j}$ for $j = 1, 2$.

Proof This follows from Theorems 5.89 and 6.8 and the fact that X_1 is an invariant subspace of A. □

Theorem 6.12 *If $\Sigma = (A, C; X, Y)$ is an LB J-node and $X_1 = \cap_{n \geq 0} \ker(CA^n)$, then:*

(1) *$AX_1 \subseteq X_1$, $CX_1 = \{0\}$ and $A_1 = A|_{X_1}$ is a selfadjoint operator in X_1, i.e., the projection $\Sigma_1 = (A_1, 0; X_1, Y)$ of Σ onto X_1 is an LB J-node in which A_1 is selfadjoint.*

(2) *The space* $X_2 = X \ominus X_1$ *is invariant under* A *and the projection* $\Sigma_2 = (A_2, C_2; X_2, Y)$ *of* Σ *onto* X_2 *is a simple LB J-node.*

(3) $\Sigma = \Sigma_1 \times \Sigma_2 = \Sigma_2 \times \Sigma_1$.

(4) $U_{\Sigma_1}(\lambda) = I_Y$ *for* $\lambda \in \Lambda_{A_1}$, $U_{\Sigma_2}(\lambda) = U_\Sigma(\lambda)$ *for* $\lambda \in \Lambda_A$ *and* $\Lambda_A \cap (\mathbb{C} \setminus \mathbb{R}) = \Lambda_{A_2} \cap (\mathbb{C} \setminus \mathbb{R})$.

Proof Assertion (1) and the invariance of X_2 under A are easy (with the help of (6.2)) and also serve to justify the formula

$$CA^n(x_1 + x_2) = C_1 A_1^n x_1 + C_2 A_2^n x_2 \quad \text{when } x_1 \in X_1 \text{ and } x_2 \in X_2.$$

But this in turn implies that $CA^n x_2 = 0 \Longleftrightarrow C_2 A_2^n x_2 = 0$ and hence that Σ_2 is simple. Moreover, as the last formula is equivalent to

$$CA^n \begin{bmatrix} x_1 \\ x_2 \end{bmatrix} = [C_1 \quad C_2] \begin{bmatrix} A_1^n & 0 \\ 0 & A_2^n \end{bmatrix} \begin{bmatrix} x_1 \\ x_2 \end{bmatrix},$$

(3) and (4) follow easily from the definitions. □

Corollary 6.13 *If* $\Sigma = (A, C; X, Y)$ *is an LB J-node and* A *has no selfadjoint part, then* Σ *is a simple node.*

Proof This is immediate from Theorem 6.12. □

Remark 6.14 *If* $A \in \mathcal{L}(X)$ *is a selfadjoint injective operator,* $Y = X \oplus X$, $J = \mathrm{diag}\{I_X, -I_X\}$ *and* $C = \mathrm{col}(I_X, I_X)$, *then* $A - A^* = 0 = iC^*JC$, *i.e.,* $\Sigma = (A, C; X, Y)$ *is an LB J-node. Moreover, since* A *is injective, and* $CAx = 0 \Longrightarrow Ax = 0$, Σ *is simple.*

If $\Sigma = (A, C; X, \mathbb{C}^m)$ is a simple LB J-node,with characteristic mvf U_Σ, then, in view of Theorems 6.6 and 5.49, $U_\Sigma \in \mathcal{U}_0(J)$ if and only if

$$T_\Sigma x = \frac{1}{\sqrt{2\pi}} C(I - \lambda A)^{-1} x \quad \text{belongs to } \Pi^m \text{ for every } x \in X.$$

Moreover, if $U_\Sigma \in \mathcal{U}_0(J)$, then

$$\{x \in X : (T_\Sigma x)(\mu) \in L_2^m\} \quad \text{is dense in } X \Longleftrightarrow U_\Sigma \in \mathcal{U}_{rR}(J) \tag{6.19}$$

and

$$\{x \in X : (T_\Sigma x)(\mu) \in L_2^m\} = \{0\} \Longleftrightarrow U_\Sigma \in \mathcal{U}_S(J). \tag{6.20}$$

In order to keep the notation simple, the same symbol R_0 is used for the backward shift in both $\mathcal{H}(A)$ and $\mathcal{B}(\mathfrak{E})$.

Theorem 6.15 *Let $\mathfrak{E} = [E_- \quad E_+]$ be a regular de Branges matrix such that (5.129) holds and let $A \in \mathcal{U}_0(J_p)$ be the unique perfect J_p-inner mvf such that $\mathfrak{E} = \sqrt{2}[I_p \quad 0]A\mathfrak{V}$. Let $\mathring{\Sigma} = (R_0, C_A; \mathcal{H}(A), \mathbb{C}^m)$ be the simple LB J-node with characteristic mvf $U_{\mathring{\Sigma}}(\lambda) = A(\lambda)$. Then the unitary operator U_2 from $\mathcal{H}(A)$ onto $\mathcal{B}(\mathfrak{E})$ that is defined by formula (5.142) is a unitary similarity operator from the node $\mathring{\Sigma}$ onto the node $\Sigma_{\mathfrak{E}} = (R_0, C_{\mathfrak{E}}; \mathcal{B}(\mathfrak{E}), \mathbb{C}^m)$, where $C_{\mathfrak{E}}g = \sqrt{2\pi}(U_2^* g)(0)$ for $g \in \mathcal{B}(\mathfrak{E})$. Moreover, if*

$$G_{\pm}(\lambda) = (R_0 E_{\pm})(\lambda), \quad G(\lambda) = [G_+(\lambda) + G_-(\lambda) \quad G_+(\lambda) - G_-(\lambda)]$$

and $\Delta_{\mathfrak{E}}(\mu)$ is given by formula (5.118), then:

(1) *$G_+\xi$ and $G_-\xi$ belong to $\mathcal{B}(\mathfrak{E})$ for every $\xi \in \mathbb{C}^p$.*

(2) *The operator $C_{\mathfrak{E}}$ from $\mathcal{B}(\mathfrak{E})$ into $\mathbb{C}^m$ may be defined by the formula*

$$C_{\mathfrak{E}}g = \frac{\sqrt{\pi}}{2\pi i}\int_{-\infty}^{\infty} G(\mu)^* \Delta_{\mathfrak{E}}(\mu) g(\mu) d\mu. \tag{6.21}$$

(3) *The adjoint U_2^* of the unitary operator U_2 is given by the formula*

$$(U_2^* g)(\lambda) = \frac{1}{\sqrt{2}2\pi i}\int_{-\infty}^{\infty} G(\mu)^* \Delta_{\mathfrak{E}}(\mu) \frac{\lambda g(\lambda) - \mu g(\mu)}{\lambda - \mu} d\mu. \tag{6.22}$$

Proof If $A \in \mathcal{U}_0(J_p)$ is perfect, then, in view of Theorem 5.76, the operator U_2 from $\mathcal{H}(A)$ onto $\mathcal{B}(\mathfrak{E})$ that is defined by formula (5.142) is unitary. Moreover,

$$U_2 R_0|_{\mathcal{H}(A)} = R_0|_{\mathcal{B}(\mathfrak{E})} U_2$$

and

$$\begin{aligned} U_2 K_0^A(\lambda)u &= U_2 \frac{J_p - A(\lambda)J_p}{-2\pi i\lambda} u = U_2 \frac{\mathfrak{V} - A(\lambda)\mathfrak{V}}{-2\pi i\lambda} j_p \mathfrak{V} u \\ &= \frac{1}{2\pi i}[G_-(\lambda) \quad G_+(\lambda)] j_p \mathfrak{V} u \quad \text{for } u \in \mathbb{C}^m. \end{aligned}$$

Therefore, if $f \in \mathcal{H}(A)$, then

$$\begin{aligned} u^* f(0) &= \langle f, K_0^A u\rangle_{\mathcal{H}(A)} = \langle U_2 f, U_2 K_0^A u\rangle_{\mathcal{B}(\mathfrak{E})} \\ &= \frac{u^*}{\sqrt{2}2\pi i}\int_{-\infty}^{\infty} G(\mu)^* \Delta_{\mathfrak{E}}(\mu)(U_2 f)(\mu) d\mu \end{aligned}$$

and hence, as

$$u^*C_{\mathfrak{E}}U_2f = u^*C_Af = \sqrt{2\pi}u^*f(0),$$

that (6.21) holds for every $g \in \mathcal{B}(\mathfrak{E})$.

In view of Theorem 6.6, the inverse operator U_2^* from $\mathcal{B}(\mathfrak{E})$ onto $\mathcal{H}(A)$ is given by the formula

$$\begin{aligned}(U_2^*g)(\lambda) &= (2\pi)^{-1/2}C_{\mathfrak{E}}(I-\lambda R_0)^{-1}g \\ &= \frac{1}{\sqrt{2}2\pi i}\int_{-\infty}^{\infty} G(\mu)^*\Delta_{\mathfrak{E}}(\mu)((I-\lambda R_0)^{-1}g)(\mu)d\mu,\end{aligned}$$

which coincides with (6.22). □

Remark 6.16 *If $A \in \mathcal{U}(J_p)$ is perfect and $c = T_A[I_p]$, then there is another formula for U_2^* that is valid even if $0 \notin \mathfrak{h}_A$: If $g \in \mathcal{B}(\mathfrak{E})$, then*

$$(U_2^*g)(\lambda) = \frac{1}{\sqrt{2}}\begin{bmatrix} c(\lambda)g(\lambda) - \frac{1}{\pi i}\int_{-\infty}^{\infty}\frac{\Re c(\mu)}{\mu-\lambda}g(\mu)d\mu \\ g(\lambda)\end{bmatrix} \quad \text{for } \lambda \in \mathfrak{h}_A. \tag{6.23}$$

This follows from Theorem 3.1 in [AlD84]; see also Theorem 2.12 in [ArD05a].

Theorem 6.17 *Let $\mathfrak{E} = [E_- \quad E_+]$ be a regular de Branges matrix such that (5.129) holds and let $\chi = E_+^{-1}E_-$. Then χ is the characteristic mvf of the following two simple LB I-nodes:*

(1) *$\Sigma_\chi = (R_0, C_\chi; \mathcal{H}(\chi), \mathbb{C}^p)$, where $C_\chi = \sqrt{2\pi}h(0)$ for $h \in \mathcal{H}(\chi)$ and R_0 acts in $\mathcal{H}(\chi)$.*

(2) *$\Sigma_1 = (A_1, C_1; \mathcal{B}(\mathfrak{E}), \mathbb{C}^p)$, where*

$$(A_1g)(\lambda) = \frac{g(\lambda) - E_+(\lambda)g(0)}{\lambda} \quad \text{and} \quad C_1g = \sqrt{2\pi}g(0)$$

for $g \in \mathcal{B}(\mathfrak{E})$. Moreover, $T : h \longrightarrow E_+h$ is the unitary similarity operator from Σ_χ to Σ_1.

Proof Assertion (1) holds because $\chi \in \mathcal{U}_0(I_p)$. The rest follows from Lemma 5.62, since the operator

$$T : h \in \mathcal{H}(\chi) \longrightarrow g = E_+h \in \mathcal{B}(\mathfrak{E})$$

is unitary,

$$A_1 g = TR_0 h = E_+(\lambda)\frac{h(\lambda)-h(0)}{\lambda} = \frac{g(\lambda)-E_+(\lambda)g(0)}{\lambda}$$
$$= (R_0 g)(\lambda) - G_+(\lambda)g(0) = (R_0 g)(\lambda) - \frac{1}{\sqrt{2\pi}}G_+(\lambda)C_1 g \qquad (6.24)$$

and

$$C_1 g = C_\chi T^{-1} g = C_\chi E_+^{-1} g = \sqrt{2\pi} g(0).$$

□

We remark that formula (6.24) exhibits the operator A_1 as a finite dimensional perturbation of the operator R_0 in the space $\mathcal{B}(\mathfrak{E})$.

6.2 Connections with systems theory

LB J-nodes also arise naturally in the study of linear continuous time invariant conservative systems that are described by a system of equations of the form

$$\begin{aligned} -i\frac{dx}{dt} &= Ax(t) + Bu(t) \\ y(t) &= Cx(t) + Du(t) \quad \text{for } t \geq 0, \end{aligned} \qquad (6.25)$$

where $x(t)$, $u(t)$ and $y(t)$ belong to the Hilbert spaces X, U and Y, respectively; $A \in \mathcal{L}(X)$, $B \in \mathcal{L}(Y, X)$, $C \in \mathcal{L}(X, U)$ and $D \in \mathcal{L}(U, Y)$.

The system (6.25) will be denoted $\mathfrak{S} = (A, B, C, D; X, U, Y)$. Let

$$X^c_{\mathfrak{S}} = \bigvee_{n\geq 0} A^n B \quad \text{and} \quad X^o_{\mathfrak{S}} = \bigvee_{n\geq 0} (A^*)^n C^* Y.$$

The system $\mathfrak{S}$ is said to be

controllable if $X^c_{\mathfrak{S}} = X$; **observable** if $X^o_{\mathfrak{S}} = X$;

and **minimal** if it is both controllable and observable; it is called **simple** if

$$X^c_{\mathfrak{S}} \bigvee X^o_{\mathfrak{S}} = X.$$

The condition for observability is equivalent to the condition

$$\bigcap_{n\geq 0} \ker C A^n = \{0\}.$$

If $x(0) = 0$, then the Fourier transforms of the functions $u(t)$ and $y(t)$ are connected:

$$\widehat{y}(\lambda) = T_{\mathfrak{S}}(\lambda)\widehat{u}(\lambda), \quad \text{where} \quad T_{\mathfrak{S}}(\lambda) = D + C(\lambda I - A)^{-1}B \tag{6.26}$$

for $\lambda \in \rho(A)$. The function $T_{\mathfrak{S}}(\lambda)$ is called the **transfer function** of the system $\mathfrak{S}$.

Two systems $\mathfrak{S}_j = (A_j, B_j, C_j, D_j; X_j, U_j, Y_j)$, $j = 1, 2$, with $Y_1 = U_2$ may be connected in **cascade** to obtain the system $\mathfrak{S} = (A, B, C, D; X, U, Y)$, in which $U = U_1$, $Y = Y_2$, $X = X_1 \oplus X_2$ and the input $u(t) = u_1(t)$, the state $x(t) = \text{col}\,(x_1(t), x_2(t))$, the output $y(t) = y_2(t)$ and $u_2(t) = y_1(t)$. The transfer functions of the systems $\mathfrak{S}_j$ are related to the transfer function of $\mathfrak{S}$ by the formula

$$T_{\mathfrak{S}}(\lambda) = T_{\mathfrak{S}_2}(\lambda)T_{\mathfrak{S}_1}(\lambda) \quad \text{for } \lambda \in \rho(A_1) \cap \rho(A_2).$$

A system $\mathfrak{S} = (A, B, C, D; X, U, Y)$ is said to be (J_U, J_Y)-**conservative** with respect to a pair of signature operators $J_U \in \mathcal{L}(U)$ and $J_Y \in \mathcal{L}(Y)$ if

$$\frac{d}{dt}\|x(t)\|^2 = \langle J_U u(t), u(t)\rangle_U - \langle J_Y y(t), y(t)\rangle_Y \tag{6.27}$$

for every $t \geq 0$, when $x(0) \in X$ and $u(t) \in U$ and an analogous condition is satisfied for mvf's $u(t)$, $x(t)$ and $y(t)$ associated with the adjoint system

$$\mathfrak{S}_* = (-A^*, -iC^*, -iB^*, D^*; X, Y, U).$$

These two conditions are equivalent to the constraints

$$\begin{bmatrix} -i(A - A^*) - C^* J_Y C & -iB - C^* J_Y D \\ iB^* - D^* J_Y C & J_U - D^* J_Y D \end{bmatrix} = 0 \tag{6.28}$$

and

$$\begin{bmatrix} -i(A - A^*) - BJ_U B^* & -C^* - iBJ_U D^* \\ -C + iDJ_U B^* & J_Y - DJ_U D^* \end{bmatrix} = 0, \tag{6.29}$$

which in turn is equivalent to the three conditions

(1) $D^* J_Y D = J_U$ and $DJ_U D^* = J_Y$.

(2) $B = iC^* J_Y D$.

(3) $A - A^* = iC^* J_Y C$.

Thus, there is a one to one correspondence between

LB J-nodes $\Sigma = (A, C; X, Y)$ and

$$J\text{-conservative systems } \mathfrak{S} = (A, iC^*J, C, I_Y; X, Y, Y)$$

and

$$T_{\mathfrak{S}}(\lambda) = U_\Sigma(1/\lambda) \quad \text{for } \lambda \in \rho(A). \tag{6.30}$$

Moreover, if Σ_j corresponds to $\mathfrak{S}_j$ for $j = 1, 2$, then $\Sigma = \Sigma_1 \times \Sigma_2$ corresponds to the cascade connection of $\mathfrak{S}_2$ and $\mathfrak{S}_1$.

A system $\mathfrak{S} = (A, B, C, D; X, U, Y)$ is said said to be (J_U, J_Y)-**passive** if the equality in (6.27) and its analogue for $\mathfrak{S}_*$ is replaced by $\leq$, or, equivalently, if the equalities in (6.28) and (6.29) are both replaced by $\geq$.

A (J_U, J_Y) **conservative** (resp., **passive**) system with $J_U = I_U$ and $J_Y = I_Y$ is called a **conservative** (resp., **passive**) **scattering system** and its transfer function is called a **scattering matrix**. Thus, the system $\mathfrak{S} = (A, B, C, D; X, U, Y)$ is a conservative scattering system if $A - A^* = iC^*C$, $B = iC^*D$ and D is unitary. The scattering matrix of a passive scattering system is defined by the formula

$$T_{\mathfrak{S}}(\lambda) = \{I + iC(\lambda I - A)^{-1}C^*\}D \quad \text{for } \lambda \in \rho(A). \tag{6.31}$$

If the equalities (6.28) and (6.29)are in force, then the properties of observability, controllability, simplicity and minimality are equivalent to those of the system (6.25).

In view of (6.2), the main operator in an LB J-node is **dissipative** if $J = I$ and **accumulative** if $J = -I$. Correspondingly, following Brodskii, LB I-nodes and LB $(-I)$ nodes are referred to as dissipative and accumulative nodes, respectively. This terminology is not used in this monograph because these nodes correspond to conservative but not dissipative or accumulative systems. Thus, each LB I-node corresponds to a linear continuous time invariant conservative scattering system (6.25) with $U = Y$, $D = I_Y$ and $B = iC^*$, and $U_\Sigma(1/\lambda)$ is equal to the scattering matrix of this system. Moreover, every such conservative scattering system can be imbedded into the **Lax-Phillips** framework in which the evolution of the state is described by a group of unitary dilations of the semigroup e^{itA}:

$$e^{itA} = P_X U(t)|_X \quad \text{for } t \geq 0;$$

for details see [AdAr66], [St05] and [Zol03].

The Lax-Phillips framework was used in [ArNu96] to generalize the results on conservative and passive scattering systems discussed above to a setting in which one or more of the operators may be unbounded. In this case it is convenient to rewrite the second equation in (6.25) as

$$y(t) = N \begin{bmatrix} x(t) \\ u(t) \end{bmatrix} \quad \text{for } t \geq 0,$$

where N is a linear operator from $X \oplus U$ into Y and to impose the following restrictions on the operators A, B and N:

(1) iA is the generator of a continuous semigroup $T(t) = e^{itA}$ for $t \geq 0$;

(2) $B \in \mathcal{L}(X, X_-)$;

(3) $N \in \mathcal{L}(\mathcal{D}(N), Y)$;

where X_- is the space of continuous linear functionals on the space $X_+ = \mathcal{D}(A^*)$, the domain of A^* with the graph norm; $\mathcal{D}(N) = \{\mathrm{col}\,(x, u) : \widehat{A}x + Bu \in X\}$ with the graph norm of $[\widehat{A} \quad B]|_{\mathcal{D}(N)}$ and $\widehat{A} = (A^*)^* \in \mathcal{L}(X, X_-)$, the adjoint of $A^* \in \mathcal{L}(X_+, X)$.

We remark that A^* and $\widehat{A}$ are contractive operators between the indicated spaces and that $\widehat{A}$ is a natural extension of A.

The controllability and observability subspaces for the system $\mathfrak{S} = (A, B, N; X, U, Y)$ are defined as

$$X^c_{\mathfrak{S}} = \bigvee_{\lambda \in \Lambda_{\widehat{A}} \cap \mathbb{C}_+} (I - \lambda \widehat{A})^{-1} BU \quad \text{and} \quad X^o_{\mathfrak{S}} = \bigvee_{\lambda \in \Lambda_{\widehat{A}^*} \cap \mathbb{C}_-} (I - \lambda \widehat{A})^*)^{-1} C^* Y,$$

where $C = N|_{\mathcal{D}(A)}$. The definitions of controllable, observable, simple and minimal systems are the same as before, but with respect to these subspaces. The transfer function of $\mathfrak{S}$ is defined by the formula

$$T_{\mathfrak{S}}(\lambda) = N \begin{bmatrix} (\lambda I - \widehat{A})^{-1} B \\ I_U \end{bmatrix} \quad \text{for } \lambda \in \rho(\widehat{A}).$$

A system $\mathfrak{S} = (A, B, N; X, U, Y)$ is said to be a conservative (resp., passive) scattering system if the equality (resp., inequality $\leq$ in place of $=$) (6.27) holds with $J_U = I_U$ and $J_Y = I_Y$ for every admissible $x(0) \in X$ and $u(t) \in U$ and analogous constraints are in force for the adjoint system $\mathfrak{S}_* = (-A^*, -iC^*, N_*; X, Y, U)$, where N_* is defined so that $T_{\mathfrak{S}}(\lambda) = T^{\sim}_{\mathfrak{S}_*}(\lambda)$.

This condition is equivalent to the equality (resp., inequality $\geq$) for the quadratic forms

$$\|u\|_U^2 - \|N \begin{bmatrix} x \\ u \end{bmatrix}\|_Y^2 = i\langle \widehat{A}x + Bu, x\rangle_X - i\langle x, \widehat{A}x + Bu\rangle_X \tag{6.32}$$

for col $(x, u) \in \mathcal{D}(N)$ and a second constraint analogous to the first for the adjoint system $\mathfrak{S}_*$. The transfer function for a passive scattering system is called a scattering matrix. In this setting, a simple conservative scattering system need not be minimal.

The colligation $\Sigma = (A, B, N; X, U, Y)$ that corresponds to a conservative scattering system $\mathfrak{S} = (A, B, N; X, U, Y)$ will be called a generalized LB I-node. The characteristic function $U_\Sigma(\lambda)$ of such a generalized LB I-node is defined by the formula

$$U_\Sigma(\lambda) = N \begin{bmatrix} \lambda(I - \lambda\widehat{A})^{-1}B \\ I_U \end{bmatrix} \quad \text{for } \lambda \in \Lambda_{\widehat{A}}, \tag{6.33}$$

and hence $T_{\mathfrak{S}}(\lambda) = U_\Sigma(1/\lambda)$. This node Σ is called simple if the corresponding system $\mathfrak{S}$ is simple.

The following facts are known for scattering systems and generalized LB I-nodes:

(1) The main operator A of a passive scattering system is a maximal dissipative operator.
(2) Every maximal dissipative operator A in a Hilbert space X may be imbedded as the main operator in a generalized LB I-node $\Sigma = (A, B, N; X, U, Y)$, which is equivalent to a conservative scattering system $\mathfrak{S}$. Morever, the system $\mathfrak{S}$ is simple if and only if A is a simple dissipative operator (i.e., A has no selfadjoint part).
(3) If $T_{\mathfrak{S}}(\lambda)$ the scattering matrix of a passive system, then the restriction $U(\lambda)$ of $T_{\mathfrak{S}}(1/\lambda)$ to $\mathbb{C}_+$ belongs to he Schur class $\mathcal{S}(U, Y)$ of holomorphic contractive $\mathcal{L}(U, Y)$ valued functions in $\mathbb{C}_+$.
(4) Every function $U \in \mathcal{S}(U, Y)$ may be represented as the characteristic function U_Σ of a generalized simple LB I-node, which is defined by $U(\lambda)$ up to unitary similarity.
(5) If $\mathfrak{S} = (A, B, N; X, U, Y)$ is a conservative scattering system and if

$$X_\circ = \overline{P_{X^o_{\mathfrak{S}}} X^c_{\mathfrak{S}}} \quad \text{and} \quad X_\bullet = \overline{P_{X^c_{\mathfrak{S}}} X^o_{\mathfrak{S}}}$$

then appropriately defined restrictions

$$\mathfrak{S}_\circ = (A_\circ, B_\circ, N_\circ; X_\circ, U, Y) \quad \text{and} \quad \mathfrak{S}_\bullet = (A_\bullet, B_\bullet, N_\bullet; X_\bullet, U, Y)$$

of $\mathfrak{S}$ onto the subspaces $X_\circ$ and $X_\bullet$ are minimal passive scattering systems with scattering matrices that coincide with $T_{\mathfrak{S}}(\lambda)$ in $\mathbb{C}_-$.

(6) The systems $\mathfrak{S}_\circ$ and $\mathfrak{S}_\bullet$ may be characterized as extremal systems in the set of all minimal passive scattering realizations of a given scattering matrix $T(\lambda)$ with $T(1/\lambda) \in \mathcal{S}(U, Y)$ in the sense that

$$\|x_\circ(t)\|_{X_\circ} \leq \|x_{min}(t)\|_{X_{min}} \leq \|x_\bullet(t)\|_{X_\bullet}$$

if $x_\circ(0) = 0$, $x_{min}(0) = 0$, $x_\bullet(0) = 0$ and the admissible input data $u(t)$ is the same for all three systems. The systems $\mathfrak{S}_\circ$ and $\mathfrak{S}_\bullet$ are defined by $T(\lambda)$ up to unitary similarity and are called **minimal optimal** and **minimal $*$-optimal**, respectively.

Thus for every $U \in \mathcal{S}(U, Y)$, $U(1/\lambda)$ may be realized as the restriction to $\mathbb{C}_-$ of the scattering matrix of either a simple conservative scattering system $\mathfrak{S}$ that is defined by $U(\lambda)$ up to unitary similarity, or a minimal passive scattering system that is defined by $U(\lambda)$ up to weak similarity.

Every rational mvf $U \in \mathcal{S}^{p\times q}$ coincides with the characteristic mvf $U_\Sigma(\lambda)$ of a simple generalized LB I-node $\Sigma = (A, B, N; X, \mathbb{C}^q, \mathbb{C}^p)$ in $\mathbb{C}_+$. However, $\dim X < \infty$ if and only if $U \in \mathcal{S}_{in}^{p\times p}$. In this case, $\dim X = \deg U$, $N = [C \quad D]$, the quadratic equality (6.32) and its analogue for $\mathfrak{S}_*$ is equivalent to the operator identities (6.28) and (6.29) with $J_U = I_q$ and $J_Y = I_p$, and formula (6.33) can be rewritten as

$$U_\Sigma(\lambda) = D + \lambda C(I - \lambda A)^{-1} B, \tag{6.34}$$

where $B = iC^* D$. If $U \notin \mathcal{S}_{in}^{p\times p}$, then $\dim X = \infty$, A is unbounded and $\Lambda_A \cap \mathbb{R} = \emptyset$. Nevertheless, there exists a restriction

$$\mathfrak{S}_{min} = (A_{min}, B_{min}, [C_{min} \quad D_{min}]; X_{min}, \mathbb{C}^q, \mathbb{C}^p)$$

of a conservative system $\mathfrak{S} = (A, B, N; X, \mathbb{C}^q, \mathbb{C}^p)$ to a subspace X_{min} of X such that $\dim X_{min} = \deg U$, the operators in $\mathfrak{S}_{min}$ satisfy (6.28) with $\geq$ in place of $=$, $J_U = I_q$, $J_Y = I_p$; and $T_{\mathfrak{S}_{min}}$ is defined by the operators in $\mathfrak{S}_{min}$ by formula (6.31). Moreover, $T_{\mathfrak{S}_{min}}(1/\lambda) = U(\lambda)$ in $\mathbb{C}$ (and not just in $\mathbb{C}_+$). In particular, the choices $\mathfrak{S}_\circ$ and $\mathfrak{S}_\bullet$ yield minimal optimal and minimal $*$-optimal passive scattering system realizations for $U(\lambda)$,

6.3 Bibliographical notes

In the mid forties M. S. Livsic introduced the notion of the characteristic function of an operator $A \in \mathcal{L}(X)$ by the formula

$$U_A(\lambda) = \{I + i|A_I|^{1/2}(\lambda I - A)^{-1}|A_I|^{1/2} J_A\}|_{Y_A}, \tag{6.35}$$

where $A_I = 2\Im A = (A - A^*)/i$, $Y_A = \overline{A_I X}$ and $J_A = \mathrm{sign} A_I|_{Y_A}$; see, e.g., [Liv54] and the references cited therein. Livsic discovered that:

(1) $U_A(1/\lambda)^* J U_A(1/\lambda) \le J$ for $\lambda \in \Lambda_A \cap \mathbb{C}_+$ with equality for $\lambda \in \Lambda_A \cap \mathbb{R}$.

(2) If A is unitarily similar to B, then $U_A(\lambda) = U_B(\lambda)$.

(3) If $U_A(\lambda) = U_B(\lambda)$ in a neighborhood of zero and A and B are simple in the sense that $\cap \ker\{A_I A^n\} = \cap \ker\{B_I B^n\} = \{0\}$, then A is unitarily similar to B.

(4) If A is simple and $\dim Y_A = m < \infty$, then the mvf $U_A(1/\lambda)$ belongs to the class $\mathcal{LB}(J)$.

(5) The characteristic function can be identified as the transfer function of a system, as in (6.25). Moreover, in a number of concrete problems in scattering theory, circuit theory and quantum mechanics the operator A in the evolution semigroup $T(t) = e^{itA}$ is dissipative and the characteristic function of A coincides with the Heisenberg scattering matrix.

(6) The reduction of A to triangular form is connected with the existence of a monotonically increasing system of invariant subspaces, which, in turn correspond to the resolution of the characteristic matrix function $U_A(\lambda)$ into factors and to the decomposition of an open system with main operator A into a cascade of simpler systems.

The connections referred to in (6) stimulated the development of multiplicative representations of mvf's in the class $\mathcal{P}^\circ(J)$ by Potapov [Po60]. Brodskii introduced the operator node that we call an LB J-node and its characteristic function to simplify these connections. It is easily seen that $U_A(1/\lambda)$ is the characteristic function of the LB J_A-node $\Sigma_A = (A, |A_I|^{1/2}; X, Y_A)$. A detailed discussion of LB J-nodes may be found in [Bro72]. The argument based on Riesz projections in last part of the proof of Theorem 6.1 is adapted from the proof of Theorem 9.3 in [Bro72].

Theorem 6.4 was generalized by E. R. Tsekanovskii and Yu. L. Shmulyan in [TsS77] to the setting of operator valued functions $U(\lambda)$ and $c(\lambda)$ and possibly unbounded operators A and was used to develop the theory of generalized LB J-nodes and the Krein theory of resolvent matrices. A number of other authors, including A. V. Shtraus [Sht60], A. N. Kochubei [Ko80] have contributed to the theory of generalized LB J-nodes; see, e.g., the references cited in [TsS77] and in [Ku96].

Passive minimal realizations of rational mvf's in $\mathcal{S}^{p\times q}$, $\mathcal{C}^{p\times p}$, $\mathcal{P}(J)$ and other classes by purely algebraic methods that do not depend upon operator theory were considered independently in control theory and passive network theory; see, e.g., [Bel68], [DBN71] and [Kal63a]. A minimal realization for a rational transfer mvf $T(\lambda)$ that is holomorphic at infinity is defined by four matrices A, B, C and $D = T(\infty)$, up to similarity, i.e., up to replacing the first three matrices by $R^{-1}AR$, $R^{-1}B$ and CR, respectively. If a minimal realization of $T(\lambda)$ is given, then a passive minimal realization may be obtained by choosing the similarity matrix R as a positive definite square root of a solution of the appropriate KYP (Kalman-Yakubovich-Popov) inequality. If $U \in \mathcal{S}^{p\times q}$ is rational, then the KYP inequality is

$$\begin{bmatrix} -i(PA - A^*P) - C^*C & -iPB - C^*D \\ iB^*P - D^*C & I - D^*D \end{bmatrix} \geq 0.$$

Moreover, if $P > 0$ is a solution of this KYP inequality, then the minimal system $\mathfrak{S} = (A, B, C, D; \mathbb{C}^r, \mathbb{C}^q, \mathbb{C}^p)$ is a minimal passive scattering system with respect to the inner product $\langle x, x\rangle_P = \langle Px, x\rangle_{\mathbb{C}^r}$ in the state space $X = \mathbb{C}^r$, where $r = \text{degree}\, T(\lambda) = \text{degree}\, U(\lambda)$. There exist a unique pair of extremal solutions $P_\circ$ and $P_\bullet$ of the KYP inequality such that $0 < P_\circ \leq P \leq P_\bullet$. If $P = P_\circ$ or $P = P_\bullet$, then the corresponding systems $\mathfrak{S}_\circ$ and $\mathfrak{S}_\bullet$ will be minimal optimal and minimal $*$-optimal, respectively.

KYP inequalities were used to study the absolute stability problem by the Lyapunov method in control theory, first for finite dimensional spaces (see [Pop61], [Yak62], [Kal63b] and [Pop73]) and subsequently in Hilbert spaces, where one or more of the operators may be unbounded (see [Yak74], [Yak75] and [LiYa6]). In control theory, the quadratic functionals $\langle Px, x\rangle_X$ are called storage or Lyapunov functions; the extremal functions $\langle P_\circ x, x\rangle_X$ and $\langle P_\bullet x, x\rangle_X$ are called available storage and required supply functions; see [Wi72a], [Wi72b], [Pop73]; see also [Kai74] for applications to stochastic processes. Solutions P of the KYP inequality that may be unbounded and/or have unbounded inverses were considered in [AKP05] and [ArSt05]. In particular, criteria for $P_\circ = P_\bullet$ and $P_\bullet \le cP_\circ$ in terms of constraints on the scattering matrix are presented there.

An observable optimal discrete time scattering system $\mathfrak{S} = (A, B, C, D; X, U, Y)$ with scattering matrix

$$S(\lambda) = D + \lambda C(I - \lambda A)^{-1}B \quad \text{where} \quad \begin{bmatrix} A & B \\ C & D \end{bmatrix} \text{ is coisometric}$$

for a given function $S \in \mathcal{S}(U, Y)$ (with respect to $\mathbb{D}$) was obtained by L. de Branges with $A = R_0$ and state space $X = \mathcal{H}(S)$; see [BrR66] and Theorem 5.3 and Corollary 5.6 in [An90]. The restriction of $\mathfrak{S}$ to the controllability space $X^c_{\mathfrak{S}}$ yields a minimal optimal realization $\mathfrak{S}_\circ$ of $S(\lambda)$.

Connections of the theory of characteristic functions of operators and operator nodes with the theory of conservative and passive systems are discussed in [Hel74], [Ar74b], [Ar79c], [BaC91] and [Ar95b]. The presented results on conservative and passive scattering systems are adapted from [ArNu96] (which includes proofs); for additional information, see also [ArNu00], [Ar00b], [ArNu02] and [St05].

Passive minimal realizations of rational mvf's in the classes $\mathcal{S}^{p\times q}$ and $\mathcal{C}^{p\times p}$ may also be obtained via the minimal Darlington representations that are described in Chapter 9.

If $P > 0$ in (5.48), then $\Sigma = (P^{1/2}AP^{-1/2}, CP^{-1/2}; \mathbb{C}^n, \mathbb{C}^m)$ is an LB $(-J)$-node with characteristic mvf $U_\Sigma(\lambda) = U(1/\lambda)$, where $U(\lambda)$ is defined by formula (5.49). Operator nodes based on operator identities in Hilbert space analogous to (5.48) were used extensively by L. A. Sakhnovich in his study of interpolation, moment problems and canonical systems; see, e.g., [Sak93], [Sak97] and [Sak99].

The functional model of a simple LB J-node with state space $\mathcal{H}(U)$ in Theorem 6.1 was considered in [AlD84] for $U \in \mathcal{U}_0(J)$. Analogous models for simple J-unitary nodes where the state space X is the RKHS $\mathcal{H}(U)$ of holomorphic vvf's in a domain $\Omega \subseteq \mathbb{D}$ that contains the point zero are well known; see, e.g., [ADRS97] and the references cited therein. A generalization that drops this restriction may be found in [ArSt07]. The functional model with state space $\mathcal{B}(\mathfrak{E})$ presented in Theorem 6.15 seems to be new, though equivalent functional models for LB J-nodes with characteristic mvf's that are meromorphic in $\mathbb{C}$ were considered earlier by L. de Branges [Br68a],

[Br68b] and subsequently by L. Golinskii and I Mikhailova [GoM97], L. A. Sakhnovich [Sak97] and G. M. Gubreev [Gu00a], [Gu00b].

The operator A is said to have **absolutely continuous spectrum** (resp., **singular spectrum**) if the condition on the left of (6.19) (resp., (6.20)) is met; see, e.g., [Na76] and [Na78] for a study of these classes. Operator valued generalizations of Theorems 5.89 and 6.11 were obtained in [Ve91a], [Ve91b] and [Tik02].

7
Generalized interpolation problems

In this chapter and the next, we will discuss a number of interpolation problems:

(1) **GSIP**, the generalized Schur interpolation problem.
(2) **GCIP**, the generalized Carathéodory interpolation problem.
(3) **GKEP**, the generalized Krein extension problem.
(4) **NP**, the Nehari problem.

If these problems are completely indeterminate in a sense that will be defined below, then the set of solutions of each of the first three problems may be described in terms of a linear fractional transformation that is based on a right regular J-inner mvf, with $J = j_{pq}$ for the GSIP and $J = J_p$ for the GCIP and the GKEP. Moreover, the J-inner mvf's that correspond to GKEP's are entire. Conversely, every mvf $U \in \mathcal{U}_{rR}(J)$ corresponds in this way to a completely indeterminate GSIP or GCIP, according as $J = j_{pq}$ or $J = J_p$, and every entire mvf $U \in \mathcal{U}_{rR}(J_p)$ corresponds to a completely indeterminate GKEP.

The GKEP may be considered as a special case of the GCIP, which in turn, may be reduced to a GSIP and may be viewed as a special case of the NP. The set of solutions of a completely indeterminate NP can be expressed as the image of a linear fractional transformation of the class $\mathcal{S}^{p\times q}$ that is based on a mvf that belongs to the class $\mathfrak{M}_{rR}(j_{pq})$ of right regular γ-generating matrices, which will be defined below.

Two sided connections also exist between the class $\mathcal{U}_{rsR}(J)$ of right strongly regular J-inner mvf's ($\mathfrak{M}_{rsR}$ of right strongly regular γ-generating matrices) and strictly completely indeterminate interpolation problems. These connections will be exploited in Chapter 10 to establish alternative criteria for the strong regularity of J-inner mvf's (γ-generating matrices) in terms of the Treil-Volberg matrix version of the Muckenhoupt $(A)_2$ condition.

The final sections of this chapter will be devoted to a discussion of the generalized Sarason problem.

7.1 The Nehari problem

In this section we review a number of results on the Nehari problem, largely without proof. Let $f \in L_\infty^{p\times q}$ and let $\Gamma(f)$ denote the linear operator from H_2^q into K_2^p that is defined by the rule

$$\Gamma(f) = \Pi_- M_f|_{H_2^q}, \tag{7.1}$$

where M_f denotes the operator of multiplication by f from L_2^q into L_2^p and Π_- denotes the orthogonal projection of L_2^p onto K_2^p.

Clearly,

$$\Gamma(f) \in \mathcal{L}(H_2^q, K_2^p) \quad \text{and} \quad \|\Gamma(f)\| \le \|f\|_\infty. \tag{7.2}$$

Lemma 7.1 *Let f_1 and f_2 be two mvf's from $L_\infty^{p\times q}$. Then*

$$\Gamma(f_1) = \Gamma(f_2) \iff f_1 - f_2 \in H_\infty^{p\times q}.$$

Proof The implication $\Longleftarrow$ is self-evident. Conversely, if $\Gamma(f_1) = \Gamma(f_2)$ and $h = f_1 - f_2$, then $h \in L_\infty^{p\times q}$ and $\Gamma(h) = 0$. Consequently, $M_h H_2^q \subseteq H_2^p$, which is equivalent to the assertion that $h \in H_\infty^{p\times q}$, by Lemma 3.47. □

Let $V_+(t)$, $t \ge 0$ denote the semigroup of operators of multiplication by $e_t = \exp(i\mu t)$ in H_2^q and let $V_-(t)$, $t \ge 0$ denote the semigroup of operators of multiplication by e_{-t} in K_2^p.

Theorem 7.2 *Let $\Gamma \in \mathcal{L}(H_2^q, K_2^p)$. Then there exists a mvf $f \in L_\infty^{p\times q}$ such that $\Gamma = \Gamma(f)$ if and only if*

$$\Gamma V_+(t) = V_-(t)^*\Gamma. \tag{7.3}$$

Moreover, if this condition is satisfied, then there exists at least one mvf $f \in L_\infty^{p\times q}$ such that

$$\Gamma = \Gamma(f) \quad \text{and} \quad \|f\|_\infty = \|\Gamma\|, \tag{7.4}$$

i.e.,

$$\|\Gamma\| = \min\{\|f\|_\infty : f \in L_\infty^{p\times q} \quad \text{and} \quad \Gamma(f) = \Gamma\}. \tag{7.5}$$

Proof It is readily checked that if $f \in L_\infty^{p\times q}$, $\Gamma = \Gamma(f)$ and $g \in H_2^q$, then

$$\begin{aligned} \Pi_- f e_t g &= \Pi_- e_t f g = \Pi_- e_t(\Pi_- + \Pi_+) f g \\ &= \Pi_- e_t \Pi_- f g = \Pi_- e_t \Gamma g, \end{aligned}$$

i.e., (7.3) holds; and $\|\Gamma\| \le \|f\|_\infty$. The converse lies deeper. If $\|\Gamma\| < 1$, then Theorem 7.45 implies that there exists a mvf $f \in L_\infty^{p\times q}$ with $\|f\|_\infty = 1$ such that $\Gamma = \Gamma(f)$. Thus, if $\|\Gamma\| = 1$ and $0 < \rho < 1$, then there exists a function $f_\rho \in L_\infty^{p\times q}$ with $\|f_\rho\|_\infty = 1$ such that $\rho\Gamma = \Gamma(f_\rho)$. Consequently, there exists a sequence of points $\rho_n \uparrow 1$ as $n \uparrow \infty$ such that $f_{\rho_n} \to f$ weakly in $L_\infty^{p\times q}$ as $\rho_n \uparrow 1$ and hence, $\Gamma = \Gamma(f)$ and $\|f\|_\infty \le 1$. Therefore, $\|f\|_\infty \le \|\Gamma\|$. □

An operator $\Gamma \in \mathcal{L}(H_2^q, K_2^p)$ that satisfies the condition (7.3) is called a **Hankel operator**.

The Nehari problem NP(Γ): *Given a Hankel operator* Γ*, define the set*

$$\mathcal{N}(\Gamma) = \{f \in L_\infty^{p\times q} : \Gamma(f) = \Gamma \quad \textit{and} \quad \|f\|_\infty \le 1\}. \tag{7.6}$$

The mvf's $f \in \mathcal{N}(\Gamma)$ *are called solutions of the NP*(Γ).

In view of Theorem 7.2,

$$\mathcal{N}(\Gamma) \ne \emptyset \iff \|\Gamma\| \le 1. \tag{7.7}$$

Moreover, if $\|\Gamma\| \le 1$, then

$$\|\Gamma\| = \min\{\|f\|_\infty : f \in \mathcal{N}(\Gamma)\}.$$

Remark 7.3 *Theorem 7.2 and other results on the NP may be obtained from the corresponding analogues for mvf's on the circle* $\mathbb{T}$ *by replacing the independent variable*

$$\psi(\zeta) = i\frac{1-\zeta}{1+\zeta}, \quad \zeta \in \mathbb{T} \quad \textit{that maps } \mathbb{T} \textit{ onto } \mathbb{R} \cup \infty,$$

which transforms mvf's $f(\mu)$ *on* $\mathbb{R}$ *to* $f(\psi(\zeta))$ *on* $\mathbb{T}$.

If Γ is a Hankel operator, then NP(Γ) is called

(1) **determinate** if it has exactly one solution.

(2) **indeterminate** if it has more than one solution.

(3) **completely indeterminate** if for every nonzero vector $\eta \in \mathbb{C}^q$ there exist at least two solutions $f_1 \in \mathcal{N}(\Gamma)$, $f_2 \in \mathcal{N}(\Gamma)$ such that $\|(f_1 - f_2)\eta\|_\infty > 0$.

(4) **strictly completely indeterminate** if it has at least one solution $f \in \mathcal{N}(\Gamma)$ such that $\|f\|_\infty < 1$.

It is readily checked that if NP(Γ) is strictly completely indeterminate, then it is automatically completely indeterminate: If $f_1 \in \mathcal{N}(\Gamma)$ and $\|f_1\|_\infty < 1$, then $f_2 = f_1 + \xi\eta^* \in \mathcal{N}(\Gamma)$ for every choice of nonzero vectors $\xi \in \mathbb{C}^p$ and $\eta \in \mathbb{C}^q$ such that $\|\xi\eta^*\| \leq 1 - \|f_1\|_\infty$. Moreover,

$$\|(f_2 - f_1)\eta\|_\infty = \|\xi\eta^*\eta\| = \eta^*\eta\|\xi\| > 0.$$

Lemma 7.4 *If Γ is a Hankel operator, then*

$$NP(\Gamma) \quad \text{is strictly completely indeterminate} \iff \|\Gamma\| < 1. \tag{7.8}$$

Proof The assertion is immediate from Theorem 7.2. □

Let $\omega \in \mathbb{C}_+$ and let

$$\mathfrak{A}_+(\omega) = \left\{ \frac{\eta}{\rho_\omega} : \eta \in \mathbb{C}^q \right\} \cap (I - \Gamma^*\Gamma)^{1/2} H_2^q \tag{7.9}$$

and

$$\mathfrak{A}_-(\omega) = \left\{ \frac{\xi}{\rho_{\overline{\omega}}} : \xi \in \mathbb{C}^p \right\} \cap (I - \Gamma\Gamma^*)^{1/2} K_2^p. \tag{7.10}$$

Theorem 7.5 *Let Γ be a Hankel operator with $\|\Gamma\| \leq 1$. Then:*

(1) *The numbers* $\dim \mathfrak{A}_+(\omega)$ *and* $\dim \mathfrak{A}_-(\omega)$ *are independent of the choice of the point $\omega \in \mathbb{C}_+$.*

(2) *The NP(Γ) is determinate if and only if*

$$\mathfrak{A}_+(\omega) = \{0\} \quad or \quad \mathfrak{A}_-(\omega) = \{0\} \tag{7.11}$$

for at least one (and hence every) point $\omega \in \mathbb{C}_+$.

(3) *The NP(Γ) is completely indeterminate if and only if*

$$\dim \mathfrak{A}_+(\omega) = q \quad and \quad \dim \mathfrak{A}_-(\omega) = p \tag{7.12}$$

for at least one (and hence every) point $\omega \in \mathbb{C}_+$. Moreover, the two conditions in (7.12) are equivalent.

Proof This follows with the help of Remark 7.3 from the analogous results for mvf's on $\mathbb{T}$ that are established in [AAK71a] and Theorem 4 of [Ar89]. □

Remark 7.6 *It is useful to note that if $\omega \in \mathbb{C}_+$, then*

$$\dim \mathfrak{A}_+(\omega) = q \Longleftrightarrow \left\{ \frac{\eta}{\rho_\omega} : \eta \in \mathbb{C}^q \right\} \subseteq (I - \Gamma^*\Gamma)^{1/2} H_2^q \tag{7.13}$$

and

$$\dim \mathfrak{A}_-(\omega) = p \Longleftrightarrow \left\{ \frac{\xi}{\rho_{\overline{\omega}}} : \xi \in \mathbb{C}^p \right\} \subseteq (I - \Gamma\Gamma^*)^{1/2} K_2^p. \tag{7.14}$$

Lemma 7.7 *Let A be a linear operator in a Hilbert space $\mathcal{H}$ such that $0 \leq A \leq I$, let $x \in \mathcal{H}$ and let*

$$\lim_{t \uparrow 1} \langle (I - tA)^{-1} x, x \rangle = \kappa. \tag{7.15}$$

Then $\kappa < \infty$ if and only if $x \in$ range $(I - A)^{1/2}$.

Proof Suppose first that $\kappa < \infty$. Then, since the bound

$$\begin{aligned} |\langle y, x \rangle_{\mathcal{H}}| &= |\langle (I - tA)^{1/2} y, (I - tA)^{-1/2} x \rangle_{\mathcal{H}}| \\ &\leq \|(I - tA)^{1/2} y\|_{\mathcal{H}} \, \|(I - tA)^{-1/2} x\|_{\mathcal{H}} \\ &\leq \{\langle (I - A)y, y \rangle_{\mathcal{H}} + (1 - t)\langle Ay, y \rangle_{\mathcal{H}}\}^{1/2} \, \kappa^{1/2} \end{aligned}$$

is valid for every t in the interval $0 \leq t < 1$, it follows that

$$|\langle y, x \rangle_{\mathcal{H}}| \leq \kappa^{1/2} \|(I - A)^{1/2} y\|_{\mathcal{H}} \quad \text{for every} \quad y \in \mathcal{H}.$$

Thus, the linear functional

$$\varphi((I - A)^{1/2} y) = \langle y, x \rangle_{\mathcal{H}}$$

is well defined and bounded on the range of $(I - A)^{1/2}$. Therefore, by the Riesz representation theorem, there exists a vector $u \in \mathcal{H}$ such that

$$\varphi((I - A)^{1/2} y) = \langle (I - A)^{1/2} y, u \rangle_{\mathcal{H}}.$$

Thus,

$$\langle y, x \rangle_{\mathcal{H}} = \langle y, (I - A)^{1/2} u \rangle_{\mathcal{H}}$$

for every $y \in \mathcal{H}$, and hence $x = (I - A)^{1/2} u$. This completes the proof that if $\kappa < \infty$, then $x \in$ range $(I - A)^{1/2}$. The converse is easy and is left to the reader. □

Remark 7.8 *Let $\omega \in \mathbb{C}_+, \eta \in \mathbb{C}_q$ and $\xi \in \mathbb{C}^p$. Then, by Lemma 7.7,*

$$\frac{\eta}{\rho_\omega} \in \mathfrak{A}_+(\omega) \Longleftrightarrow \lim_{\delta \uparrow 1} \left\langle (I - \delta\Gamma^*\Gamma)^{-1} \frac{\eta}{\rho_\omega}, \frac{\eta}{\rho_\omega} \right\rangle_{st} < \infty$$

and

$$\frac{\xi}{\rho_{\overline{\omega}}} \in \mathfrak{A}_-(\omega) \Longleftrightarrow \lim_{\delta \uparrow 1} \left\langle (I - \delta\Gamma\Gamma^*)^{-1} \frac{\xi}{\rho_{\overline{\omega}}}, \frac{\xi}{\rho_{\overline{\omega}}} \right\rangle_{st} < \infty.$$

Consequently the condition (7.12) is equivalent to the following:

$$\lim_{\delta \uparrow 1} \left\langle (I - \delta\Gamma^*\Gamma)^{-1} \frac{\eta}{\rho_\omega}, \frac{\eta}{\rho_\omega} \right\rangle_{st} < \infty \quad \text{for every} \quad \eta \in \mathbb{C}^q \tag{7.16}$$

and

$$\lim_{\delta \uparrow 1} \left\langle (I - \delta\Gamma\Gamma^*)^{-1} \frac{\xi}{\rho_{\overline{\omega}}}, \frac{\xi}{\rho_{\overline{\omega}}} \right\rangle_{st} < \infty \quad \text{for every} \quad \xi \in \mathbb{C}^p \tag{7.17}$$

for at least one (and hence every) point $\omega \in \mathbb{C}_+$. The conditions (7.11) are equivalent to

$$\lim_{\delta \uparrow 1} \left\langle (I - \delta\Gamma^*\Gamma)^{-1} \frac{\eta}{\rho_\omega}, \frac{\eta}{\rho_\omega} \right\rangle_{H_2^q} = \infty \quad \text{for every nonzero vector } \eta \in \mathbb{C}^q \tag{7.18}$$

or

$$\lim_{\delta \uparrow 1} \left\langle (I - \delta\Gamma\Gamma^*)^{-1} \frac{\xi}{\rho_{\overline{\omega}}}, \frac{\xi}{\rho_{\overline{\omega}}} \right\rangle_{K_2^p} = \infty \quad \text{for every nonzero vector } \xi \in \mathbb{C}^p, \tag{7.19}$$

respectively.

Remark 7.9 *If 1 is a singular value of a contractive Hankel operator Γ, i.e., if $\|\Gamma x\| = \|x\|$ for some nonzero vector x, then $NP(\Gamma)$ is not completely indeterminate. Consequently, if the $NP(\Gamma)$ is completely indeterminate, then*

$$I - \Gamma^*\Gamma > 0 \quad \text{and} \quad I - \Gamma\Gamma^* > 0 \tag{7.20}$$

and, in view of Theorem 7.5 and Remark 7.6, the inclusions

$$\frac{I_q}{\rho_\omega} \in (I - \Gamma^*\Gamma)^{1/2} H_2^{q\times q} \quad \text{and} \quad \frac{I_p}{\rho_{\overline{\omega}}} \in (I - \Gamma\Gamma^*)^{1/2} K_2^{p\times p} \tag{7.21}$$

are necessary and sufficient for the $NP(\Gamma)$ to be completely indeterminate.

7.2 γ-generating matrices

Let $\mathfrak{M}_r(j_{pq})$ denote the class of $m \times m$ mvf's $\mathfrak{A}(\mu)$ on $\mathbb{R}$ of the form

$$\mathfrak{A}(\mu) = \begin{bmatrix} \mathfrak{a}_{11}(\mu) & \mathfrak{a}_{12}(\mu) \\ \mathfrak{a}_{21}(\mu) & \mathfrak{a}_{22}(\mu) \end{bmatrix} \tag{7.22}$$

with blocks $\mathfrak{a}_{11}$ of size $p \times p$ and $\mathfrak{a}_{22}$ of size $q \times q$ such that

(1) $\mathfrak{A}(\mu)$ is a measurable mvf on $\mathbb{R}$ that is j_{pq}-unitary a.e. on $\mathbb{R}$.

(2) $\mathfrak{a}_{22}(\mu)$ and $\mathfrak{a}_{11}(\mu)^*$ are the boundary values of mvf's $\mathfrak{a}_{22}(\lambda)$ and $\mathfrak{a}_{11}^{\#}(\lambda)$ that are holomorphic in $\mathbb{C}_+$ and, in addition,

$$(\mathfrak{a}_{22})^{-1} \in \mathcal{S}_{out}^{q\times q} \quad \text{and} \quad (\mathfrak{a}_{11}^{\#})^{-1} \in \mathcal{S}_{out}^{p\times p}. \tag{7.23}$$

(3_r) The mvf

$$s_{21}(\mu) = -\mathfrak{a}_{22}(\mu)^{-1}\mathfrak{a}_{21}(\mu) = -\mathfrak{a}_{12}(\mu)^*(\mathfrak{a}_{11}(\mu)^*)^{-1} \tag{7.24}$$

is the boundary value of a mvf $s_{21}(\lambda)$ that belongs to the class $\mathcal{S}^{q\times p}$.

The mvf's in the class $\mathfrak{M}_r(j_{pq})$ are called **right γ-generating matrices**. They play a fundamental role in the study of the matrix Nehari problem: if the NP(Γ) is completely indeterminate, then

$$\mathcal{N}(\Gamma) = T_{\mathfrak{A}}[\mathcal{S}^{p\times q}] \quad \text{for some } \mathfrak{A} \in \mathfrak{M}_r(j_{pq}).$$

Let $\mathfrak{M}_\ell(j_{pq})$ denote the class of $m \times m$ mvf's $\mathfrak{A}(\mu)$ on $\mathbb{R}$ of the form (7.22) that meet the conditions (1) and (2) that are stated above for $\mathfrak{M}_r(j_{pq})$ and (in place of (3_r))

(3_ℓ) The mvf

$$s_{12}(\mu) = \mathfrak{a}_{12}(\mu)\mathfrak{a}_{22}(\mu)^{-1} = (\mathfrak{a}_{11}(\mu)^*)^{-1}\mathfrak{a}_{21}(\mu)^* \tag{7.25}$$

is the boundary value of a mvf $s_{12}(\lambda)$ that belongs to the class $\mathcal{S}^{p\times q}$. The mvf's in the class $\mathfrak{M}_\ell(j_{pq})$ will be called **left γ-generating matrices**. This class was introduced and briefly discussed in Section 7.3 of [ArD01b].

The standard four block decompositions

$$\mathfrak{A} = \begin{bmatrix} \mathfrak{a}_- & \mathfrak{b}_- \\ \mathfrak{b}_+ & \mathfrak{a}_+ \end{bmatrix}, \quad \text{if } \mathfrak{A} \in \mathfrak{M}_r(j_{pq}) \tag{7.26}$$

and

$$\mathfrak{A} = \begin{bmatrix} \mathfrak{d}_- & \mathfrak{c}_+ \\ \mathfrak{c}_- & \mathfrak{d}_+ \end{bmatrix}, \quad \text{if } \mathfrak{A} \in \mathfrak{M}_\ell(j_{pq}), \tag{7.27}$$

in which the indices $\pm$ indicate that the corresponding blocks are nontangential limits of mvf's in the Nevanlinna class in $\mathbb{C}_\pm$, will be used frequently.

We remark that, in view of property (1),

$$\varepsilon \in \mathcal{S}^{p\times q} \Longrightarrow f(\mu) = T_{\mathfrak{A}}[\varepsilon] \quad \text{belongs to} \quad L_\infty^{p\times q} \text{ and } \|f\|_\infty \le 1.$$

Let

$$T_{\mathfrak{A}}[\mathcal{S}^{p\times q}] = \{T_{\mathfrak{A}}[\varepsilon] : \varepsilon \in \mathcal{S}^{p\times q}\}.$$

A mvf $\mathfrak{A} \in \mathfrak{M}_r(j_{pq})$ is said to be:

(1) **right singular** if $T_{\mathfrak{A}}[\mathcal{S}^{p\times q}] \subseteq \mathcal{S}^{p\times q}$.

(2) **right regular** if the factorization $\mathfrak{A} = \mathfrak{A}_1\mathfrak{A}_2$ with a factor $\mathfrak{A}_1 \in \mathfrak{M}_r(j_{pq})$ and a right singular factor $\mathfrak{A}_2 \in \mathfrak{M}_r(j_{pq})$ implies that $\mathfrak{A}_2$ is constant.

(3) **right strongly regular** if there is an $f \in T_{\mathfrak{A}}[\mathcal{S}^{p\times q}]$ with $\|f\|_\infty < 1$.

These three subclasses of $\mathfrak{M}_r(j_{pq})$ will be designated $\mathfrak{M}_{rS}(j_{pq})$, $\mathfrak{M}_{rR}(j_{pq})$ and $\mathfrak{M}_{rsR}(j_{pq})$, respectively. The corresponding subclasses $\mathfrak{M}_{\ell S}(j_{pq})$, $\mathfrak{M}_{\ell R}(j_{pq})$ and $\mathfrak{M}_{\ell sR}(j_{pq})$ of $\mathfrak{M}_\ell(j_{pq})$ are defined analogously:

A mvf $\mathfrak{A} \in \mathfrak{M}_\ell$ is said to be:

(1) **left singular** if $T_{\mathfrak{A}}^\ell[\mathcal{S}^{q\times p}] \subseteq \mathcal{S}^{q\times p}$.

(2) **left regular** if $\mathfrak{A} = \mathfrak{A}_2\mathfrak{A}_1$ with $\mathfrak{A}_j \in \mathfrak{M}_\ell(j_{pq})$ for $j = 1, 2$ and $\mathfrak{A}_2$ is left singular, then $\mathfrak{A}_2$ is constant.

(3) **left strongly regular** if there exists an $f \in T_{\mathfrak{A}}^\ell[\mathcal{S}^{q\times p}]$ with $\|f\|_\infty < 1$.

In view of the equivalences

$$\begin{aligned}
f = T_{\mathfrak{A}}^\ell[\varepsilon] \quad &\Longleftrightarrow \quad f^\sim = T_{\mathfrak{A}^\sim}[\varepsilon^\sim] \Longleftrightarrow f^\tau = T_{\mathfrak{A}^\tau}[\varepsilon^\tau] \\
\mathfrak{A} \in \mathfrak{M}_\ell(j_{pq}) \quad &\Longleftrightarrow \quad \mathfrak{A}^\sim \in \mathfrak{M}_r(j_{pq}) \Longleftrightarrow \mathfrak{A}^\tau \in \mathfrak{M}_r(j_{pq}) \\
\mathfrak{A} \in \mathfrak{M}_{\ell S}(j_{pq}) \quad &\Longleftrightarrow \quad \mathfrak{A}^\sim \in \mathfrak{M}_{rS}(j_{pq}) \Longleftrightarrow \mathfrak{A}^\tau \in \mathfrak{M}_{rS}(j_{pq}) \\
\mathfrak{A} \in \mathfrak{M}_{\ell R}(j_{pq}) \quad &\Longleftrightarrow \quad \mathfrak{A}^\sim \in \mathfrak{M}_{rR}(j_{pq}) \Longleftrightarrow \mathfrak{A}^\tau \in \mathfrak{M}_{rR}(j_{pq}) \\
\mathfrak{A} \in \mathfrak{M}_{\ell sR}(j_{pq}) \quad &\Longleftrightarrow \quad \mathfrak{A}^\sim \in \mathfrak{M}_{rsR}(j_{pq}) \Longleftrightarrow \mathfrak{A}^\tau \in \mathfrak{M}_{rsR}(j_{pq}),
\end{aligned}$$

there is a correspondence between results on mvf's in the class $\mathfrak{M}_r(j_{pq})$ and mvf's in the class $\mathfrak{M}_\ell(j_{pq})$. Thus, only results on right γ-generating matrices and right regular J-inner mvf's will be presented in detail. In those cases when a dual result for left γ-generating matrices or left regular J-inner mvf's is presented, the proof will be omitted.

A mvf $W \in \mathcal{U}(j_{pq})$ will be identified with its boundary values $W(\mu)$, which are defined a.e. on $\mathbb{R}$.

Lemma 7.10 *The following conditions are equivalent:*

(1) $\mathfrak{A} \in \mathfrak{M}_r(j_{pq})$ *and* $T_{\mathfrak{A}}[0] \in \mathcal{S}^{p\times q}$.

(2) $\mathfrak{A} \in \mathfrak{M}_r(j_{pq})$ *and* $T_{\mathfrak{A}}[\mathcal{S}^{p\times q}] \subseteq \mathcal{S}^{p\times q}$, *i.e.,* $\mathfrak{A} \in \mathfrak{M}_{rS}(j_{pq})$.

(3) $\mathfrak{A} \in \mathcal{U}_S(j_{pq})$.

(4) $\mathfrak{A} \in \mathfrak{M}_r(j_{pq})$ *and* $T_{\mathfrak{A}}[\varepsilon] \in \mathcal{S}^{p\times q}$ *for some* $\varepsilon \in \mathcal{S}^{p\times q}$.

Moreover, the classes $\mathfrak{M}_{rS}(j_{pq})$ *and* $\mathfrak{M}_{\ell S}(j_{pq})$ *can (and will) be identified with* $\mathcal{U}_S(j_{pq})$.

Proof If (1) holds, then $S = PG(\mathfrak{A})$ is a measurable mvf on $\mathbb{R}$ that is unitary a.e. on $\mathbb{R}$. Moreover, since $s_{11} = (\mathfrak{a}_-^{\#})^{-1} \in \mathcal{S}_{out}^{p\times p}$, $s_{21} = -\mathfrak{a}_+^{-1}\mathfrak{b}_+ \in \mathcal{S}^{q\times p}$ and $s_{22} = \mathfrak{a}_+^{-1} \in \mathcal{S}_{out}^{q\times q}$ by definition of the class $\mathfrak{M}_r(j_{pq})$, and $s_{12} = \mathfrak{b}_-\mathfrak{a}_+^{-1} \in \mathcal{S}^{p\times q}$ by assumption, it follows that $S \in H_\infty^{m\times m}$. Consequently, $S \in \mathcal{S}_{in}^{m\times m}$, by the maximum principle. Therefore, $\mathfrak{A} = PG(S)$ is j_{pq}-inner, and, since $\mathfrak{a}_{22} \in \mathcal{N}_{out}^{q\times q}$ and $\mathfrak{a}_{11}^{\#} \in \mathcal{N}_{out}^{p\times p}$, $\mathfrak{A} \in \mathcal{U}_S(j_{pq})$, i.e., (1) $\Longrightarrow$ (3).

The implications (3) $\Longrightarrow$ (2), (2) $\Longrightarrow$ (1) and (3) $\Longrightarrow$ (4) are obvious. The implication (4) $\Longrightarrow$ (1) will be established later.

The final statement follows from the equivalences

$$\mathfrak{A} \in \mathfrak{M}_{rS} \Longleftrightarrow \mathfrak{A}^\tau \in \mathfrak{M}_{rS} \quad \text{and} \quad U \in \mathcal{U}_S(j_{pq}) \Longleftrightarrow U^\tau \in \mathcal{U}_S(j_{pq})$$

and the equivalence of (2) with (3). □

If either $\mathfrak{A} \in \mathfrak{M}_r(j_{pq})$ or $\mathfrak{A} \in \mathfrak{M}_l(j_{pq})$, then the $m \times m$ mvf $S = PG(\mathfrak{A})$ is unitary a.e. on $\mathbb{R}$ and its blocks can be expressed in terms of the blocks of $\mathfrak{A}$ by the formula

$$\begin{bmatrix} s_{11}(\mu) & s_{12}(\mu) \\ s_{21}(\mu) & s_{22}(\mu) \end{bmatrix} = \begin{bmatrix} \mathfrak{a}_{11}(\mu)^{-*} & \mathfrak{a}_{12}(\mu)\mathfrak{a}_{22}(\mu)^{-1} \\ -\mathfrak{a}_{22}(\mu)^{-1}\mathfrak{a}_{21}(\mu) & \mathfrak{a}_{22}(\mu)^{-1} \end{bmatrix} \text{ a.e. on } \mathbb{R}.$$

If either $\mathfrak{A} \in \mathfrak{M}_r(j_{pq})$ or $\mathfrak{A} \in \mathfrak{M}_l(j_{pq})$ and $\mathfrak{a}_{ij}(\mu)$ are the blocks, considered in the decomposition (7.22) of $\mathfrak{A}$, the notations

$$s_{12}(\mu) = \mathfrak{a}_{12}(\mu)\mathfrak{a}_{22}^{-1}(\mu), \quad s_{21}(\mu) = -\mathfrak{a}_{22}^{-1}(\mu)\mathfrak{a}_{21}(\mu), \tag{7.28}$$

$$\Delta_r(\mu) = \begin{bmatrix} I_p & -s_{21}(\mu)^* \\ -s_{21}(\mu) & I_q \end{bmatrix}, \quad \Delta_\ell(\mu) = \begin{bmatrix} I_p & s_{12}(\mu) \\ s_{12}(\mu)^* & I_q \end{bmatrix} \tag{7.29}$$

will be useful.

Lemma 7.11 *Let $\mathfrak{A} \in \mathfrak{M}_r(j_{pq}) \cup \mathfrak{M}_\ell(j_{pq})$, let $m = p+q$ and let the mvf's s_{21}, s_{12}, Δ_r and Δ_ℓ be defined by formulas (7.28) and (7.29). Then the following conditions are equivalent:*

(1) $\mathfrak{A} \in L_\infty^{m\times m}$.

(2) $\|s_{12}\|_\infty < 1$.

(3) $\|s_{21}\|_\infty < 1$.

(4) $\Delta_r^{-1} \in L_\infty^{m\times m}$.

(5) $\Delta_\ell^{-1} \in L_\infty^{m\times m}$.

(6) $\|T_{\mathfrak{A}}[\varepsilon]\|_\infty < 1$ *for at least one matrix* $\varepsilon \in \mathcal{S}_{const}^{p\times q}$.

(7) $\|T_{\mathfrak{A}}[\varepsilon]\|_\infty < 1$ *for every mvf* $\varepsilon \in \mathring{\mathcal{S}}^{p\times q}$.

Proof Since $\mathfrak{A}(\mu)$ is j_{pq}-unitary a.e. on $\mathbb{R}$, the blocks $\mathfrak{a}_{11}(\mu)$ and $\mathfrak{a}_{22}(\mu)$ are invertible a.e. on $\mathbb{R}$ and the equalities

$$\begin{aligned} \mathfrak{a}_{11}(\mu)\mathfrak{a}_{11}(\mu)^* &= (I_p - s_{12}(\mu)s_{12}(\mu)^*)^{-1} \\ \mathfrak{a}_{22}(\mu)^*\mathfrak{a}_{22}(\mu) &= (I_q - s_{21}(\mu)s_{21}(\mu)^*)^{-1} \\ \mathfrak{a}_{22}(\mu)\mathfrak{a}_{22}(\mu)^* &= (I_q - s_{12}(\mu)^*s_{12}(\mu))^{-1} \end{aligned}$$

are in force a.e. on $\mathbb{R}$. Next, upon invoking the formula

$$\|(I_q - A^*A)^{-1}\| = (1 - \|A\|^2)^{-1},$$

which is valid for every $A \in \mathbb{C}^{p\times q}$ with $\|A\| < 1$, it is readily seen that

$$\mathfrak{a}_{11} \in L_\infty^{p\times p} \iff \|s_{12}\|_\infty < 1 \iff \mathfrak{a}_{22} \in L_\infty^{q\times q} < 1 \iff \|s_{21}\|_\infty < 1.$$

Moreover,

$$\mathfrak{a}_{22} \in L_\infty^{q\times q} \Longrightarrow \mathfrak{a}_{12} \in L_\infty^{q\times p} \quad \text{and} \quad \mathfrak{a}_{11} \in L_\infty^{p\times p} \Longrightarrow \mathfrak{a}_{21} \in L_\infty^{p\times q},$$

since

$$\mathfrak{a}_{12} = s_{12}\mathfrak{a}_{22}, \ \mathfrak{a}_{21} = s_{12}^*\mathfrak{a}_{11}, \ \|s_{12}\|_\infty < 1 \text{ and } \|s_{21}\|_\infty < 1.$$

Thus, the equivalences (1) $\Longleftrightarrow$ (2) $\Longleftrightarrow$ (3) are established. Moreover, (1) $\Longleftrightarrow$ (4) $\Longleftrightarrow$ (5), then follows from Schur complements.

Next, let $\varepsilon \in \mathcal{S}^{p\times q}$ and $f_\varepsilon = T_{\mathfrak{A}}[\varepsilon]$ and suppose that (3) holds. Then the identity

$$I_q - f_\varepsilon^* f_\varepsilon = \mathfrak{a}_{22}^{-*}(I_q - s_{21}\varepsilon)^{-*}(I_q - \varepsilon^*\varepsilon)(I_q - s_{21}\varepsilon)^{-1}\mathfrak{a}_{22}^{-1} \tag{7.30}$$

is in force and implies that

$$\|f_\varepsilon\|_\infty < 1 \Longleftrightarrow \|\varepsilon\|_\infty < 1.$$

Thus, (3) $\Longrightarrow$ (7). Since the implication (7) $\Longrightarrow$ (6) is obvious, it remains only to check that (6) $\Longrightarrow$ (1). Suppose therefore that (6) holds, i.e., that $\|f_\varepsilon\|_\infty < 1$ for some constant $p \times q$ contractive matrix ε. Then, by formula (7.30), $\|\varepsilon\| < 1$. Thus, the matrix

$$V_\varepsilon = \begin{bmatrix} (I_p - \varepsilon\varepsilon^*)^{-1/2} & \varepsilon(I_q - \varepsilon^*\varepsilon)^{-1/2} \\ \varepsilon^*(I_p - \varepsilon\varepsilon^*)^{-1/2} & (I_q - \varepsilon^*\varepsilon)^{-1/2} \end{bmatrix}$$

is a constant j_{pq}-inner matrix with the property $T_{V_\varepsilon}[0] = \varepsilon$. Let

$$\mathfrak{A}_\varepsilon(\mu) = \mathfrak{A}(\mu)V_\varepsilon.$$

Then the formulas

$$T_{\mathfrak{A}_\varepsilon}[0_{p\times q}] = T_{\mathfrak{A}}[\varepsilon] = f_\varepsilon$$

imply that $\|T_{\mathfrak{A}_\varepsilon}[0_{p\times q}]\| < 1$ and hence, upon applying the implication (2) $\Longrightarrow$ (1) to the mvf $\mathfrak{A}_\varepsilon$, that $\mathfrak{A}_\varepsilon \in L_\infty^{m\times m}$. Thus, as $\mathfrak{A}_\varepsilon \in \mathfrak{M}_r$ and

$$\mathfrak{A}_\varepsilon \in L_\infty^{m\times m} \Longleftrightarrow \mathfrak{A} \in L_\infty^{m\times m},$$

the proof is complete. □

Corollary 7.12 *If* $\mathfrak{A} \in \mathfrak{M}_r(j_{pq}) \cap L_\infty^{m\times m}$, *then* $\mathfrak{A} \in \mathfrak{M}_{rsR}(j_{pq})$. *If* $\mathfrak{A} \in \mathfrak{M}_\ell(j_{pq}) \cap L_\infty^{m\times m}$, *then* $\mathfrak{A} \in \mathfrak{M}_{\ell sR}(j_{pq})$.

Proof The first assertion is immediate from the preceding lemma. The second assertion follows from the first and the equivalences

$$\mathfrak{A} \in \mathfrak{M}_\ell(j_{pq}) \iff \mathfrak{A}^\sim \in \mathfrak{M}_r(j_{pq})$$
$$\mathfrak{A} \in \mathfrak{M}_{\ell sR}(j_{pq}) \iff \mathfrak{A}^\sim \in \mathfrak{M}_{rsR}(j_{pq})$$
$$\mathfrak{A} \in L_\infty^{m\times m} \iff \mathfrak{A}^\sim \in L_\infty^{m\times m}.$$

□

Lemma 7.13 *Let $\mathfrak{A} \in \mathfrak{M}_r(j_{pq}) \cup \mathfrak{M}_\ell(j_{pq})$, let $m = p+q$ and let the mvf's s_{21}, s_{12}, Δ_r and Δ_ℓ be defined by formulas (7.28) and (7.29). Then the following conditions are equivalent:*

(1) $\mathfrak{A} \in \widetilde{L}_2^{m\times m}$.

(2) $(I_p - s_{21}^* s_{21})^{-1} \in \widetilde{L}_1^{p\times p}$.

(3) $(I_p - s_{12} s_{12}^*)^{-1} \in \widetilde{L}_1^{p\times p}$.

(4) $\Delta_r^{-1} \in \widetilde{L}_1^{m\times m}$.

(5) $\Delta_\ell^{-1} \in \widetilde{L}_1^{m\times m}$.

Moreover, if any one (and hence everyone) of these conditions is in force, then $\mathfrak{A} \in \mathfrak{M}_r \Longrightarrow \mathfrak{A} \in \mathfrak{M}_{rR}$ and $\mathfrak{A} \in \mathfrak{M}_\ell \Longrightarrow \mathfrak{A} \in \mathfrak{M}_{\ell R}$

Proof The proof of the equivalence of (1)–(5) is much the same as for Lemma 7.11. The last assertion follows from Lemma 7.36. □

Our next objective is to **parametrize** the mvf's $\mathfrak{A} \in \mathfrak{M}_r(j_{pq})$.

Lemma 7.14 *Let $\mathfrak{A} \in \mathfrak{M}_r(j_{pq})$ and let the mvf $s(\mu) = s_{21}(\mu)$ be defined by formula (7.24). Then*

$$s \in \mathcal{S}^{q\times p} \quad \text{and} \quad \ln \det\{I_q - ss^*\} \in \widetilde{L}_1. \tag{7.31}$$

Conversely, if a mvf $s(\lambda)$ satisfies the conditions (7.31), then there exists a mvf $\mathfrak{A} \in \mathfrak{M}_r(j_{pq})$ such that $s(\mu) = -\mathfrak{a}_{22}(\mu)^{-1}\mathfrak{a}_{12}(\mu)$ a.e. on $\mathbb{R}$. Moreover, this mvf $\mathfrak{A}(\mu)$ is uniquely defined by $s(\lambda)$ up to a left block diagonal j_{pq}-unitary multiplier by the formula

$$\mathfrak{A}(\mu) = \begin{bmatrix} \mathfrak{a}_-(\mu) & \mathfrak{b}_-(\mu) \\ \mathfrak{b}_+(\mu) & \mathfrak{a}_+(\mu) \end{bmatrix} = \begin{bmatrix} \mathfrak{a}_-(\mu) & 0 \\ 0 & \mathfrak{a}_+(\mu) \end{bmatrix} \Delta_r(\mu), \tag{7.32}$$

where $\mathfrak{a}_+(\mu)$ *and* $\mathfrak{a}_-(\mu)$ *are the essentially unique solutions of the factorization problems*

$$\mathfrak{a}_+^{-1}\mathfrak{a}_+^{-*} = I_q - ss^* \quad a.e. \text{ on } \mathbb{R} \text{ with } \mathfrak{a}_+^{-1} \in \mathcal{S}_{out}^{q\times q}, \tag{7.33}$$

$$\mathfrak{a}_-^{-1}\mathfrak{a}_-^{-*} = I_p - s^*s \quad a.e. \text{ on } \mathbb{R} \text{ with } \mathfrak{a}_-^{-\#} \in \mathcal{S}_{out}^{p\times p}, \tag{7.34}$$

$$\mathfrak{b}_+ = -\mathfrak{a}_+ s, \quad \mathfrak{b}_- = -\mathfrak{a}_- s^* \tag{7.35}$$

and Δ_r *is defined by formula (7.29) with* $s_{21} = s$.

Proof Let $\mathfrak{A} \in \mathfrak{M}_r(j_{pq})$ and $s = s_{21}$, where s_{21} is defined by formula (7.24). Then $s \in \mathcal{S}^{p\times p}$ and the blocks of the mvf $\mathfrak{A}(\mu)$ satisfy the equalities in (7.33)–(7.35). Moreover $\varphi_2(\lambda) = \mathfrak{a}_+(\lambda)^{-1}$ and $\varphi_1(\lambda) = a_-^{\#}(\lambda)^{-1}$ are solutions of the factorization problems

$$\varphi_1(\mu)^*\varphi_1(\mu) = I_q - s(\mu)^*s(\mu) \quad \text{a.e. on } \mathbb{R}, \text{ with } \varphi_1 \in \mathcal{S}_{out}^{p\times p}, \tag{7.36}$$

and

$$\varphi_2(\mu)\varphi_2(\mu)^* = I_p - s(\mu)s(\mu)^* \quad \text{a.e. on } \mathbb{R}, \text{ with } \varphi_2 \in \mathcal{S}_{out}^{q\times q}. \tag{7.37}$$

Consequently, the second condition in (7.31) holds by Theorem 3.78.

Conversely, if the conditions in (7.31) are in force, then the factorization problems (7.36) and (7.37) are solvable, by the Zasukhin-Krein theorem. Let

$$\mathfrak{a}_-(\mu) = \varphi_1(\mu)^{-*}, \quad \mathfrak{a}_+(\mu) = \varphi_2(\mu)^{-1}, \quad \mathfrak{b}_+(\mu) = -\varphi_2(\mu)^{-1}s(\mu)$$
$$\text{and} \quad \mathfrak{b}_-(\mu) = -\varphi_1(\mu)^{-*}s(\mu)^* \quad \text{a.e. on } \mathbb{R}. \tag{7.38}$$

Then the mvf $\mathfrak{A}(\mu)$ defined by formula (7.32) is a right γ-generating matrix. Moreover, since the solutions of the factorization problems (7.36) and (7.37) are uniquely defined up to constant unitary multipliers u and v:

$$\varphi_1 \longrightarrow u\varphi_1 \quad \text{and} \quad \varphi_2 \longrightarrow \varphi_2 v,$$

the preceding analysis shows that the blocks of the mvf $\mathfrak{A}(\mu)$ are defined by $s(\lambda)$ up to constant multipliers

$$\mathfrak{a}_-(\mu) \longrightarrow u\mathfrak{a}_-(\mu), \quad \mathfrak{a}_+(\mu) \longrightarrow v\mathfrak{a}_+(\mu),$$

$$\mathfrak{b}_-(\mu) \longrightarrow u\mathfrak{b}_-(\mu), \quad \mathfrak{b}_+(\mu) \longrightarrow v\mathfrak{b}_+(\mu),$$

i.e., the mvf $\mathfrak{A}(\mu)$ is defined by $s(\lambda)$ up to a constant block diagonal j_{pq}-unitary (and unitary) multiplier

$$\mathfrak{A}(\mu) \longrightarrow \begin{bmatrix} u & 0 \\ 0 & v \end{bmatrix} \mathfrak{A}(\mu). \tag{7.39}$$

□

Lemma 7.15 *Let* $\mathfrak{A} \in \mathfrak{M}_{rsR}(j_{pq}) \cup \mathfrak{M}_{\ell sR}(j_{pq})$. *Then* $\mathfrak{A} \in \widetilde{L}_2^{m\times m}$.

Proof If $\mathfrak{A} \in \mathfrak{M}_{rsR}(j_{pq})$, then there exists an $\varepsilon \in \mathcal{S}^{p\times q}$ such that the mvf

$$f = T_{\mathfrak{A}}[\varepsilon]$$

is strictly contractive: $\|f\|_\infty = \delta < 1$. Then, since $s_{21}\varepsilon \in \mathcal{S}^{q\times q}$ and $\|(s_{21}\varepsilon)(\lambda)\| < 1$ for every point $\lambda \in \mathbb{C}_+$, the mvf

$$c = (I_q + s_{21}\varepsilon)(I_q - s_{21}\varepsilon)^{-1}$$

belongs to the Carathéodory class $\mathcal{C}^{q\times q}$. Consequently, $(\Re c) \in \widetilde{L}_1^{q\times q}$. Thus, as

$$\begin{aligned}
&(\Re c)(\mu) \\
&= \{I_q - (s_{21}\varepsilon)(\mu)^*\}^{-1}\{I_q - (s_{21}\varepsilon)(\mu)^*(s_{21}\varepsilon)(\mu)\}\{I_q - (s_{21}\varepsilon)(\mu)\}^{-1} \\
&\geq \{I_q - (s_{21}\varepsilon)(\mu)^*\}^{-1}\{I_q - \varepsilon(\mu)^*\varepsilon(\mu)\}\{I_q - (s_{21}\varepsilon)(\mu)\}^{-1} \\
&= \mathfrak{a}_+(\mu)^*\{I_q - f(\mu)^*f(\mu)\}\mathfrak{a}_+(\mu) \geq (1-\delta^2)\mathfrak{a}_+(\mu)^*\mathfrak{a}_+(\mu) \\
&= (1-\delta^2)\{I_q - s_{21}(\mu)s_{21}(\mu)^*\}^{-1}
\end{aligned}$$

for almost all points $\mu \in \mathbb{R}$, the asserted result follows from Lemma 7.13 if $\mathfrak{A} \in \mathfrak{M}_{rsR}(j_{pq})$.

If $\mathfrak{A} \in \mathfrak{M}_{\ell sR}(j_{pq})$, then $\mathfrak{A}^\tau \in \mathfrak{M}_{rsR}(j_{pq})$ and consequently, $\mathfrak{A}^\tau \in \widetilde{L}_2^{m\times m}$ and hence $\mathfrak{A} \in \widetilde{L}_2^{m\times m}$. □

The next theorem clarifies the connection between the class $\mathfrak{M}_{rR}(j_{pq})$ (resp., $\mathfrak{M}_{rsR}(j_{pq})$) and the completely indeterminate (resp., strictly completely indeterminate) Nehari problems. We begin with three lemmas, which are of interest in their own right.

Lemma 7.16 $\mathfrak{M}_{rsR}(j_{pq}) \subset \mathfrak{M}_{rR}(j_{pq})$ *and* $\mathfrak{M}_{\ell sR}(j_{pq}) \subset \mathfrak{M}_{\ell R}(j_{pq})$.

Proof This follows from Lemmas 7.15 and 7.13. □

Lemma 7.17 *Let $\mathfrak{A}_1 \in \mathfrak{M}_r(j_{pq})$ and let*

$$\mathfrak{A}_2 = \mathfrak{A}_1 W, \quad \textit{where} \quad W \in \mathcal{U}_S(j_{pq}). \tag{7.40}$$

Then $\mathfrak{A}_2 \in \mathfrak{M}_r(j_{pq})$ and

$$T_{\mathfrak{A}_2}[\mathcal{S}^{p\times q}] \subseteq T_{\mathfrak{A}_1}[\mathcal{S}^{p\times q}]. \tag{7.41}$$

Conversely, if $\mathfrak{A}_j \in \mathfrak{M}_r(j_{pq})$ for $j = 1, 2$, and (7.41) holds, then (7.40) is in force.

Proof Let the mvf's $\mathfrak{A}_1 \in \mathfrak{M}_r(j_{pq})$, $W \in \mathcal{U}_S(j_{pq})$, $\mathfrak{A}_2 = \mathfrak{A}_1 W$ and their PG transforms have block decompositions

$$\mathfrak{A}_j = \begin{bmatrix} \mathfrak{a}_{11}^{(j)} & \mathfrak{a}_{12}^{(j)} \\ \mathfrak{a}_{21}^{(j)} & \mathfrak{a}_{22}^{(j)} \end{bmatrix} \quad S_j = \begin{bmatrix} s_{11}^{(j)} & s_{12}^{(j)} \\ s_{21}^{(j)} & s_{22}^{(j)} \end{bmatrix} = PG(\mathfrak{A}_j),\ j = 1, 2,$$

$$W = \begin{bmatrix} w_{11} & w_{12} \\ w_{21} & w_{22} \end{bmatrix} \quad \text{and} \quad S = \begin{bmatrix} s_{11} & s_{12} \\ s_{21} & s_{22} \end{bmatrix} = PG(W),$$

respectively. Then the equality

$$\mathfrak{a}_{22}^{(2)} = \mathfrak{a}_{21}^{(1)} w_{12} + \mathfrak{a}_{22}^{(1)} w_{22} = \mathfrak{a}_{22}^{(1)}\{I_q - s_{21}^{(1)} s_{12}\} w_{22} \tag{7.42}$$

implies that $\mathfrak{a}_{22}^{(2)} \in \mathcal{N}_{out}^{q\times q}$, since

$$\mathfrak{A}^{(1)} \in \mathfrak{M}_r \Longrightarrow \mathfrak{a}_{22}^{(1)} \in \mathcal{N}_{out}^{q\times q}, \quad W \in \mathcal{U}_S(j_{pq}) \Longrightarrow w_{22} \in \mathcal{N}_{out}^{q\times q}$$

and, in view of Lemma 3.54 and the fact that $s_{21}^{(1)} \in \mathcal{S}^{q\times p}$, $s_{12} \in \mathcal{S}^{p\times q}$ and $\|s_{12}(\lambda)\| < 1$ for $\lambda \in \mathbb{C}_+$,

$$I_q - s_{21}^{(1)} s_{21} \in \mathcal{N}_{out}^{q\times q}.$$

Moreover,

$$\mathfrak{a}_{12}^{(2)}(\mu)^* \mathfrak{a}_{12}^{(2)}(\mu) - \mathfrak{a}_{22}^{(2)}(\mu)^* \mathfrak{a}_{22}^{(2)}(\mu) = -I_q \quad \text{a.e. on } \mathbb{R}, \tag{7.43}$$

since $\mathfrak{A}_2(\mu)$ is j_{pq}-unitary a.e. on $\mathbb{R}$. Consequently, $\mathfrak{a}_{22}^{(2)}(\mu)^{-*}\mathfrak{a}_{22}^{(2)}(\mu)^{-1} \leq I_q$ a.e. on $\mathbb{R}$, which implies that $(\mathfrak{a}_{22}^{(2)})^{-1} \in \mathcal{S}_{out}^{q\times q}$, by the Smirnov maximum principle. In much the same way, the equality

$$\mathfrak{a}_{11}^{(2)} = \mathfrak{a}_{11}^{(1)} w_{11} + \mathfrak{a}_{12}^{(1)} w_{21} = \mathfrak{a}_{11}^{(1)}\{I_p - s_{21}^{(1)*} s_{21}^*\} w_{11}, \tag{7.44}$$

which is valid a.e. on $\mathbb{R}$, implies that $(\mathfrak{a}_{11}^{(2)})^{-\#} \in \mathcal{S}_{out}^{p\times p}$. Furthermore,

$$\mathfrak{a}_{21}^{(2)} = \mathfrak{a}_{21}^{(1)} w_{11} + \mathfrak{a}_{22}^{(1)} w_{21} \in \mathcal{N}_+^{q\times p},$$

since $\mathfrak{a}_{21}^{(1)} \in \mathcal{N}_+^{q\times p}$, $\mathfrak{a}_{22}^{(1)} \in \mathcal{N}_+^{q\times q}$ and $W \in \mathcal{N}_+^{m\times n}$. Moreover, $s_{21}^{(2)} \in \mathcal{N}_+^{q\times p}$, since $s_{21}^{(2)} = -(\mathfrak{a}_{22}^{(2)})^{-1}\mathfrak{a}_{21}^{(2)}$, $\mathfrak{a}_{21}^{(2)} \in \mathcal{N}_+^{p\times p}$ and $(\mathfrak{a}_{22}^{(2)})^{-1} \in \mathcal{S}^{q\times q}$. The Smirnov maximum principle and the supplementary inequality

$$s_{21}^{(2)}(\mu)s_{21}^{(2)}(\mu)^* \leq I_q \quad \text{a.e. on } \mathbb{R}$$

imply that $s_{21}^{(2)} \in \mathcal{S}^{q\times p}$. Thus, $\mathfrak{A}_2 \in \mathfrak{M}_r(j_{pq})$. Moreover, (7.41) holds because

$$T_{\mathfrak{A}_2}[\mathcal{S}^{p\times p}] = T_{\mathfrak{A}_1}[T_W[\mathcal{S}^{p\times q}]] \quad \text{and} \quad T_W[\mathcal{S}^{p\times q}] \subseteq \mathcal{S}^{p\times q}.$$

Conversely, let $\mathfrak{A}_j \in \mathfrak{M}(j_{pq})_r$, $j = 1, 2$, let $W = \mathfrak{A}_1^{-1}\mathfrak{A}_2$ and assume that (7.41) holds. Then $W(\mu)$ is j_{pq}-unitary a.e. on $\mathbb{R}$ and

$$T_W[\mathcal{S}^{p\times q}] \subseteq \mathcal{S}^{p\times q}. \tag{7.45}$$

To prove that $W \in \mathcal{U}_S(j_{pq})$, let $S_j = [s_{ik}^{(j)}]$ and $S = [s_{ik}]$ denote the PG transforms of $\mathfrak{A}_j$ and W, respectively. Then, since $W = PG(S)$ and $S(\mu)$ is unitary a.e. on $\mathbb{R}$, it is enough to prove that $S \in \mathcal{N}_+^{m\times m}$, $s_{11} \in \mathcal{S}_{out}^{p\times p}$ and $s_{22} \in \mathcal{S}_{out}^{q\times q}$.

In view of (7.45), the mvf $s_{12} = T_W[0]$ belongs to $\mathcal{S}^{p\times q}$. Then the equality (7.42) implies that $w_{22} \in \mathcal{N}_{out}^{q\times q}$ and hence that $s_{22} \in \mathcal{N}_{out}^{q\times q}$. Moreover, $s_{22} \in \mathcal{S}_{out}^{q\times q}$, by the Smirnov maximum principle. In the same way, the equality (7.44) implies that $w_{11}^{\#} \in \mathcal{N}_{out}^{p\times p}$ and then that $s_{11} \in \mathcal{S}_{out}^{p\times p}$.

To complete the proof of the lemma, it remains to show that $s_{21} \in \mathcal{N}_+^{q\times p}$. To this end, let $0 < \gamma < 1$ and let $u \in \mathbb{C}^{p\times q}$ be isometric if $p \geq q$ and coisometric if $p < q$. Then the evaluation

$$\begin{aligned}
T_W[\gamma u] - T_W[0] &= (w_{11}\gamma u + w_{12})(w_{21}\gamma u + w_{22})^{-1} - w_{12}w_{22}^{-1} \\
&= (w_{11} - w_{12}w_{22}^{-1}w_{21})\gamma u(w_{21}\gamma u + w_{22}^{-1} \\
&= (w_{11}^{\#})^{-1}\gamma u(w_{22}^{-1}w_{21}\gamma u + I_q)^{-1}w_{22}^{-1} \\
&= s_{11}\gamma u(I_q - s_{21}\gamma u)^{-1}s_{22}
\end{aligned}$$

implies that $u(I_q - \gamma s_{21}u)^{-1} \in \mathcal{N}_+^{p\times q}$. Thus, if $p \geq q$, then $u^*u = I_q$, $(I_q - \gamma s_{21}u)^{-1} \in \mathcal{N}_+^{q\times q}$, and, as

$$\|(I_q - \gamma s_{21}(\mu)u)^{-1}\| \leq (1-\gamma)^{-1} \quad \text{a.e. on} \quad \mathbb{R},$$

the Smirnov maximum principle implies that $(I_q - \gamma s_{21}u)^{-1} \in H_\infty^{q\times q}$. Moreover, the Poisson integral representation of mvf's in the class $H_\infty^{q\times q}$ implies that $(I_q - \gamma s_{21}u) \in \mathcal{C}^{q\times q}$, since $\Re(I_q - \gamma s_{21}(\mu)u)^{-1} \geq 0$ a.e. on $\mathbb{R}$. Consequently, $s_{21}u \in \mathcal{N}_+^{q\times q}$. Therefore, $s_{21} \in \mathcal{N}_+^{q\times p}$, since u is an arbitrary isometric $p \times q$ matrix.

The case $p < q$ then follows from the identity

$$u(I_q - \gamma s_{21}u)^{-1} = (I_q - \gamma u s_{21})^{-1}u. \qquad \square$$

Lemma 7.18 *Let $\mathfrak{A}_j \in \mathfrak{M}_r(j_{pq})$ for $j = 1, 2$. Then*

$$T_{\mathfrak{A}_1}[\mathcal{S}^{p\times q}] = T_{\mathfrak{A}_2}[\mathcal{S}^{p\times q}] \tag{7.46}$$

if and only if

$$\mathfrak{A}_1(\mu) = \mathfrak{A}_2(\mu)W \quad a.e.\ on\ \mathbb{R} \tag{7.47}$$

for some constant j_{pq}-unitary matrix W.

Proof If $\mathfrak{A}_j \in \mathfrak{M}_r(j_{pq})$ for $j = 1, 2$ and (7.46) holds, then, by Lemma 7.17, the identity (7.47) is in force and $W^{\pm 1} \in \mathcal{U}_S(j_{pq})$. Therefore, $W(\lambda)$ is j_{pq}-unitary at all points $\lambda \in \mathbb{C}_+$ and hence (as the PG transform of $W(\lambda)$ is unitary in $\mathbb{C}_+$) must be constant. Conversely, if (7.47) holds, it is readily seen that (7.46) holds, since

$$T_{\mathfrak{A}_1}[\mathcal{S}^{p\times q}] = T_{\mathfrak{A}_2}[T_W[\mathcal{S}^{p\times q}]] \quad \text{and} \quad T_W[\mathcal{S}^{p\times q}] = \mathcal{S}^{p\times q}$$

when $W \in \mathcal{U}_{const}(j_{pq})$. $\square$

Lemma 7.19 *If $\mathfrak{A} \in \mathfrak{M}_r(j_{pq})$ and $T_{\mathfrak{A}}[\mathcal{S}^{p\times q}] = \mathcal{S}^{p\times q}$, then $\mathfrak{A}(\mu)$ is a constant j_{pq}-unitary matrix.*

Proof The conclusion is immediate from Lemma 7.18 with $\mathfrak{A}_1 = \mathfrak{A}$ and $\mathfrak{A}_2 = I_m$. $\square$

Theorem 7.20 *Let Γ be a Hankel operator such that the $NP(\Gamma)$ is completely indeterminate. Then there exists a mvf $\mathfrak{A} \in \mathfrak{M}_r(j_{pq})$ such that*

$$\mathcal{N}(\Gamma) = T_{\mathfrak{A}}[\mathcal{S}^{p\times q}]. \tag{7.48}$$

This mvf is defined essentially uniquely by Γ up to a constant j_{pq}-unitary multiplier on the right and is automatically right regular, i.e., $\mathfrak{A} \in \mathfrak{M}_{rR}(j_{pq})$. Moreover,

$$\mathfrak{A} \in \mathfrak{M}_{rsR}(j_{pq}) \Longleftrightarrow \|\Gamma\| < 1. \tag{7.49}$$

Proof A description of the form (7.48) for completely indeterminate NP's considered on the circle $\mathbb{T}$ was obtained in [AAK68] for the case $p = q = 1$ and then for arbitrary p and q in [Ad73a]. The asserted uniqueness of the mvf $\mathfrak{A} \in \mathfrak{M}_r(j_{pq})$ in formula (7.48) follows from Lemma 7.18. The fact that this mvf belongs to the class $\mathfrak{M}_{rR}(j_{pq})$ will be established in Theorem 7.22. The equivalence (7.49) follows from Theorem 7.2. □

Lemma 7.21 *Let $\mathfrak{A} \in \mathfrak{M}_r(j_{pq})$, $\varepsilon \in \mathcal{S}^{p\times q}$ and let*

$$\varphi_+^\varepsilon = \mathfrak{a}_+ + \mathfrak{b}_+\varepsilon \quad and \quad \varphi_-^\varepsilon = \mathfrak{a}_- + \mathfrak{b}_-\varepsilon^\#. \tag{7.50}$$

Then:

(1) $\varphi_+^\varepsilon \in \mathcal{N}_{out}^{q\times q}$ *and* $(\varphi_-^\varepsilon)^\# \in \mathcal{N}_{out}^{p\times p}$.

(2) $\rho_\omega^{-1}(\varphi_+^\varepsilon)^{-1} \in H_2^{q\times q}$ *and* $\rho_\omega^{-1}(\varphi_-^{\varepsilon\#})^{-1} \in H_2^{p\times p}$ *for every point* $\omega \in \mathbb{C}_+$.

If $\mathfrak{A} \in \mathfrak{M}_{rsR}(j_{pq})$, then also

(3) $\rho_\omega^{-1}\varphi_+^\varepsilon \in H_2^{q\times q}$ *and* $\rho_\omega^{-1}(\varphi_-^\varepsilon)^\# \in H_2^{p\times p}$ *for every point* $\omega \in \mathbb{C}_+$.

Proof Consider $s_{21} = -\mathfrak{a}_+^{-1}\mathfrak{b}_+$ for the mvf $\mathfrak{A} \in \mathfrak{M}_r(j_{pq})$. Then

$$\varphi_+^\varepsilon = \mathfrak{a}_+(I_q - s_{21}\varepsilon) \quad \text{and} \quad (\varphi_-^\varepsilon)^\# = (I_p - \varepsilon s_{21})\mathfrak{a}_-^\#$$

and the statement (1) holds by Lemma 3.54, which is applicable to $I_q - s_{21}\varepsilon$ and $I_p - \varepsilon s_{21}$, since $s_{21}\varepsilon \in \mathcal{S}^{q\times q}$, $\varepsilon s_{21} \in \mathcal{S}^{p\times p}$ and $\|s_{21}(\mu)\| < 1$ a.e. on $\mathbb{R}$. Furthermore, $c_\varepsilon(\lambda) = (I_q + s_{21}(\lambda)\varepsilon(\lambda))^{-1}$ belongs to $\mathcal{C}^{q\times q}$ and

$$\begin{aligned}
\varphi_+^\varepsilon(\mu)^{-1}\varphi_+^\varepsilon(\mu)^{-*} &= c_\varepsilon(\mu)\mathfrak{a}_+(\mu)\mathfrak{a}_+(\mu)^{-*}c_\varepsilon(\mu)^* \\
&= c_\varepsilon(\mu)(I_q - s_{21}(\mu)s_{21}(\mu)^*)c_\varepsilon(\mu)^* \\
&\le c_\varepsilon(\mu)(I_q - s_{21}(\mu)\varepsilon(\mu)\varepsilon(\mu)^*s_{21}(\mu)^*)c_\varepsilon(\mu)^* \\
&\le \Re c_\varepsilon(\mu) \quad \text{a.e. on } \mathbb{R}.
\end{aligned}$$

Thus, as $\Re c_\varepsilon \in \widetilde{L}_1^{q\times q}$, it follows that $(\varphi_+^\varepsilon)^{-1} \in \widetilde{L}_1^{q\times q} \cap \mathcal{N}_{out}^{q\times q}$ and hence, by the Smirnov maximum principle, that $\rho_\omega^{-1}(\varphi_+^\varepsilon)^{-1} \in H_2^{q\times q}$ for every point $\omega \in \mathbb{C}_+$. The second inclusion in (2) may be checked in much the same way.

Let $\mathfrak{A} \in \mathfrak{M}_{rsR}(j_{pq})$. Then $\mathfrak{A}(\mu) \in \widetilde{L}_2^{m\times m}$ by Lemma 7.15. Consequently,

$$\rho_\omega^{-1}\varphi_+^\varepsilon \in L_2^{m\times m} \quad \text{and} \quad \rho_\omega^{-1}(\varphi_-^\varepsilon)^\# \in L_2^{m\times m}.$$

Then (3) follows from (1) and the Smirnov maximum principle. □

Theorem 7.22 *Let $\mathfrak{A} \in \mathfrak{M}_r(j_{pq})$ and let $\Gamma = \Gamma(f)$ for some $f \in T_{\mathfrak{A}}[\mathcal{S}^{p\times q}]$. Then:*

(1) *The Hankel operator Γ does not depend upon the choice of f, i.e.,*

$$T_{\mathfrak{A}}[\mathcal{S}^{p\times q}] \subseteq \mathcal{N}(\Gamma). \tag{7.51}$$

(2) *$NP(\Gamma)$ is completely indeterminate.*

(3) *$T_{\mathfrak{A}}[\mathcal{S}^{p\times q}] = \mathcal{N}(\Gamma)$ if and only if $\mathfrak{A} \in \mathfrak{M}_{rR}(j_{pq})$.*

(4) *$\mathfrak{A} \in \mathfrak{M}_{rR}(j_{pq})$ and $\|\Gamma\| < 1$ if and only if $\mathfrak{A} \in \mathfrak{M}_{rsR}(j_{pq})$.*

Proof . We have already observed that if $f \in T_{\mathfrak{A}}[S^{p\times q}]$ and $\mathfrak{A} \in \mathfrak{M}_r(j_{pq})$, then $f \in L_\infty^{p\times q}$ and $\|f\|_\infty \leq 1$. Let $\mathfrak{A} \in \mathfrak{M}_r(j_{pq})$ and let $f_j = T_{\mathfrak{A}}[\varepsilon_j]$ for $\varepsilon_j \in \mathcal{S}^{p\times q}$, $j = 1, 2$. Then, invoking the left linear fractional representation of the mvf $f_2(\mu)$:

$$f_2(\mu) = \{\varepsilon_2(\mu)\mathfrak{b}_-(\mu)^* + \mathfrak{a}_-(\mu)^*\}^{-1}\{\varepsilon_2(\mu)\mathfrak{a}_+(\mu)^* + \mathfrak{b}_+(\mu)^*\}, \tag{7.52}$$

and the right linear fractional representation of the mvf $f_1(\mu)$:

$$f_1(\mu) = \{\mathfrak{a}_-(\mu)\varepsilon_1(\mu) + \mathfrak{b}_-(\mu)\}\{\mathfrak{b}_+(\mu)\varepsilon_1(\mu) + \mathfrak{a}_+(\mu)\}^{-1}, \tag{7.53}$$

it is readily checked that

$$\begin{aligned} f_2(\mu) - f_1(\mu) = \{\varepsilon_2(\mu)\mathfrak{b}_-(\mu)^* + \mathfrak{a}_-(\mu)^*\}^{-1} \\ \times \{\varepsilon_2(\mu) - \varepsilon_1(\mu)\}\{\mathfrak{b}_+(\mu)\varepsilon_1(\mu) + \mathfrak{a}_+(\mu)\}^{-1}. \end{aligned} \tag{7.54}$$

Then Lemma 7.21 and the Smirnov maximum principle imply that $f_2 - f_1 \in H_\infty^{p\times q}$ and consequently, $\Gamma(f_2) = \Gamma(f_1)$. Thus, (1) is proved.

To verify (2), let $\varepsilon_1 = 0_{p\times q}$ in (7.53) and let $\varepsilon_2 = \varepsilon$ be a constant isometric (resp,. coisometric) matrix if $p \geq q$ (resp., $p < q$). Then for $\eta \in \mathbb{C}^q$, $\eta \neq 0$:

$$(f_2 - f_1)\eta = \mathfrak{a}_-^{-*}(I_p - \varepsilon s_{21}^*)^{-1}\varepsilon\mathfrak{a}_+^{-1}\eta.$$

Consequently,

$$(f_2(\mu) - f_1(\mu))\eta = 0 \quad \text{a.e. on } \mathbb{R} \Longleftrightarrow \varepsilon\mathfrak{a}_+(\mu)^{-1}\eta = 0 \quad \text{a.e. on } \mathbb{R}.$$

Therefore, it suffices to exhibit a matrix $\varepsilon \in \mathcal{S}_{const}^{p\times q}$ such that $\|\varepsilon\mathfrak{a}_+^{-1}\eta\|_\infty > 0$. If $p \geq q$, then every isometric matrix $\varepsilon \in \mathbb{C}^{p\times q}$ satisfies this condition. If $p < q$, then there exists at least one coisometric matrix $\varepsilon \in \mathbb{C}^{p\times q}$ that satisfies this condition, because otherwise $P_{\mathcal{L}}\mathfrak{a}_+^{-1}\eta = 0$ a.e. on $\mathbb{R}$ for every orthoprojection $\varepsilon^*\varepsilon = P_{\mathcal{L}}$ onto $\mathcal{L} \subseteq \mathbb{C}^q$ with $\dim \mathcal{L} = p$. But then $\mathfrak{a}_+^{-1}(\mu)\eta = 0$ a.e. on $\mathbb{R}$. Thus, (2) holds.

Suppose next that equality holds in (7.51) and let $\mathfrak{A} = \mathfrak{A}_1 W$, where $\mathfrak{A}_1 \in \mathfrak{M}_r(j_{pq})$ and $W \in \mathcal{U}_S(j_{pq})$. Then

$$T_{\mathfrak{A}}[\mathcal{S}^{p\times p}] = T_{\mathfrak{A}_1}[T_W[\mathcal{S}^{p\times q}]] \subseteq T_{\mathfrak{A}_1}[\mathcal{S}^{p\times q}].$$

Therefore, by the preceding analysis, $T_{\mathfrak{A}_1}[\mathcal{S}^{p\times q}] \subseteq \mathcal{N}(\Gamma) = T_{\mathfrak{A}}[\mathcal{S}^{p\times q}]$. Consequently, $T_{\mathfrak{A}_1}[\mathcal{S}^{p\times q}] = T_{\mathfrak{A}}[\mathcal{S}^{p\times q}]$ and hence, by Lemma 7.18, $W \in \mathcal{U}_{const}(j_{pq})$, i.e., $\mathfrak{A} \in \mathfrak{M}_{rR}(j_{pq})$.

To verify (3), assume that $\mathfrak{A} \in \mathfrak{M}_{rR}(j_{pq})$, $f \in T_{\mathfrak{A}}[\mathcal{S}^{p\times q}]$ and $\Gamma = \Gamma(f)$. Then the NP(Γ) is completely indeterminate and consequently, by Theorem 7.20, there exists a mvf $\mathfrak{A}_1 \in \mathfrak{M}_r(j_{pq})$ such that $T_{\mathfrak{A}_1}[\mathcal{S}^{p\times q}] = \mathcal{N}(\Gamma)$. Thus,

$$T_{\mathfrak{A}}[\mathcal{S}^{p\times q}] \subseteq T_{\mathfrak{A}_1}[\mathcal{S}^{p\times q}] = \mathcal{N}(\Gamma)$$

and hence, by Lemma 7.17, $\mathfrak{A} = \mathfrak{A}_1 W$ for some $W \in \mathcal{U}_S(j_{pq})$. However, as $\mathfrak{A} \in \mathfrak{M}_{rR}(j_{pq})$, it follows that $W \in \mathcal{U}_{const}(j_{pq})$, and thus

$$T_{\mathfrak{A}}[\mathcal{S}^{p\times q}] = \mathcal{N}(\Gamma).$$

This completes the proof of (3).

Let $\mathfrak{A} \in \mathfrak{M}_{rsR}(j_{pq})$. Then $\mathfrak{A} \in \mathfrak{M}_{rR}(j_{pq})$ by Lemma 7.16. Moreover, there exists a mvf $f \in T_{\mathfrak{A}}[\mathcal{S}^{p\times q}]$ with $\|f\|_\infty < 1$ and, consequently, $\|\Gamma(f)\| < 1$, since $\|\Gamma(f)\| \le \|f\|_\infty$. Conversely, if $\mathfrak{A} \in \mathfrak{M}_{rR}(j_{pq})$ and $\|\Gamma(f)\| < 1$, for some $f \in T_{\mathfrak{A}}[\mathcal{S}^{p\times q}]$, then $T_{\mathfrak{A}}[\mathcal{S}^{p\times q}] = \mathcal{N}(\Gamma(f))$ and hence $\mathfrak{A} \in \mathfrak{M}_{rsR}(j_{pq})$ by Theorem 7.20. Thus, (4) is proved. □

Theorem 7.23 *If $\mathfrak{A} \in \mathfrak{M}_r(j_{pq})$ and the mvf s_{12} is defined by the first formula in (7.28), then:*

$$\mathfrak{A} \in \mathfrak{M}_{rsR}(j_{pq}) \Longleftrightarrow \|\Gamma(s_{12})\| < 1 \quad \text{and at least one (and hence each)}$$

of the five conditions considered in Lemma 7.13 holds.
If $\mathfrak{A} \in \mathfrak{M}_\ell(j_{pq})$ and the mvf s_{21} is defined by the second formula in (7.28), then:

$$\mathfrak{A} \in \mathfrak{M}_{rsR}(j_{pq}) \Longleftrightarrow \|\Gamma(s_{21})\| < 1 \text{ and at least one (and hence each)}$$

of the five conditions considered in Lemma 7.13 holds.

Proof If $\mathfrak{A} \in \mathfrak{M}_{rR}(j_{pq})$ then the implication $\Longrightarrow$ follows from Theorem 7.22 and Lemma 7.15, whereas the implication $\Longleftarrow$ follows from the same theorem and Lemma 7.13.

The proof of the case $\mathfrak{A} \in \mathfrak{M}_l(j_{pq})$ then follows from the equivalence

$$\mathfrak{A} \in \mathfrak{M}_{lsR}(j_{pq}) \Longleftrightarrow \mathfrak{A}^\tau \in \mathfrak{M}_{rsR}(j_{pq}). \qquad \square$$

Theorem 7.24 *Every mvf* $\mathfrak{A} \in \mathfrak{M}_r(j_{pq})$ *admits a factorization*

$$\mathfrak{A} = \mathfrak{A}_1\mathfrak{A}_2, \quad \text{with} \quad \mathfrak{A}_1 \in \mathfrak{M}_{rR}(j_{pq}) \quad \text{and} \quad \mathfrak{A}_2 \in \mathcal{U}_S(j_{pq}) \tag{7.55}$$

that is unique up to constant j_{pq}*-unitary multipliers:*

$$\mathfrak{A}_1 \longrightarrow \mathfrak{A}_1 V \quad \text{and} \quad \mathfrak{A}_2 \longrightarrow V^{-1}\mathfrak{A}_2 \quad \text{with} \quad V \in \mathcal{U}_{const}(j_{pq}). \tag{7.56}$$

Proof Let $\mathfrak{A} \in \mathfrak{M}_r(j_{pq})$ and consider $f \in T_{\mathfrak{A}}[\mathcal{S}^{p\times q}]$ and $\Gamma = \Gamma(f)$. Then, by Theorems 7.22 and 7.20, the NP(Γ) is completely indeterminate and there exists a mvf $\mathfrak{A}_1 \in \mathfrak{M}_{rR}(j_{pq})$ such that

$$T_{\mathfrak{A}}[\mathcal{S}^{p\times q}] \subseteq \mathcal{N}(\Gamma) = T_{\mathfrak{A}_1}[\mathcal{S}^{p\times q}].$$

Moreover, by Lemma 7.17, $\mathfrak{A}$ admits a factorization of the form (7.55), which is unique up to the transformations (7.56), by Theorem 7.20. $\square$

The next theorem establishes a connection between the classes $\mathfrak{M}_r(j_{pq})$, $\mathfrak{M}_{rR}(j_{pq})$, $\mathfrak{M}_{rsR}(j_{pq})$ and the classes $\mathcal{U}(j_{pq})$, $\mathcal{U}_{rR}(j_{pq})$, $\mathcal{U}_{rsR}(j_{pq})$, respectively.

Lemma 7.25 *If* $\mathfrak{A} \in \mathfrak{M}_r(j_{pq})$, *then*

$$T_{\mathfrak{A}}[0] \in \Pi^{p\times q} \Longleftrightarrow T_{\mathfrak{A}}^{\ell}[0] \in \Pi^{q\times p} \Longleftrightarrow \mathfrak{A} \in \Pi^{m\times m}.$$

Proof Let $s = s_{21} \in \Pi^{q\times p}$. Then, by Theorem 3.110, the equalities in (7.33)–(7.35) imply that the outer mvf's $\mathfrak{a}_-^\#$ and $\mathfrak{a}_+$ belong to the classes $\Pi^{p\times p}$ and $\Pi^{q\times q}$, respectively. Thus, $\mathfrak{a}_- \in \Pi^{p\times p}$, $s_{21}^\# \in \Pi^{p\times q}$ and, as $\mathfrak{b}_+ = -\mathfrak{a}_+ s_{21}$, it follows that $\mathfrak{b}_+ \in \Pi^{q\times p}$ and $\mathfrak{b}_- \in \Pi^{p\times q}$. Therefore, $\mathfrak{A} \in \Pi^{m\times m}$. In much the same way, the identities

$$\begin{aligned} I_p - f_{12}(\mu)f_{12}(\mu)^* &= \mathfrak{a}_-(\mu)^{-*}\mathfrak{a}_-(\mu)^{-1}, \\ I_q - f_{12}(\mu)^* f_{12}(\mu) &= \mathfrak{a}_+(\mu)^{-*}\mathfrak{a}_+(\mu)^{-1}, \end{aligned}$$

$$\mathfrak{b}_- = f_{12}\mathfrak{a}_+ \qquad \text{and} \qquad \mathfrak{b}_+ = f_{12}^*\mathfrak{a}_-$$

for $f_{12} = T_{\mathfrak{A}}[0]$, imply that $\mathfrak{A} \in \Pi^{m\times m}$ when $f_{12} \in \Pi^{p\times q}$. The remaining implications are self-evident. $\square$

We recall that a denominator of a mvf $f \in \mathcal{N}^{p\times q}$ is a pair $\{b_1, b_2\}$ of mvf's such that $b_1 \in \mathcal{S}_{in}^{p\times p}$, $b_2 \in \mathcal{S}_{in}^{q\times q}$ and $b_1 f b_2 \in \mathcal{N}_+^{p\times q}$. The set of denominators of f is denoted den f.

Theorem 7.26 *Let $\mathfrak{A} \in \Pi \cap \mathfrak{M}_r(j_{pq})$, $\{b_1, b_2\} \in \operatorname{den} T_{\mathfrak{A}}[0]$ and let*

$$W(\mu) = \begin{bmatrix} b_1(\mu) & 0 \\ 0 & b_2(\mu)^{-1} \end{bmatrix} \mathfrak{A}(\mu) \quad a.e.\ on\ \mathbb{R}. \tag{7.57}$$

Then

$$W \in \mathcal{U}(j_{pq}) \quad and \quad \{b_1, b_2\} \in ap(W). \tag{7.58}$$

Conversely, if (7.58) holds and the mvf $\mathfrak{A}$ is defined by formula (7.57), then $\mathfrak{A} \in \Pi \cap \mathfrak{M}_r(j_{pq})$ and $\{b_1, b_2\} \in \operatorname{den} T_{\mathfrak{A}}[0]$.

Proof Let $\mathfrak{A} \in \Pi \cap \mathfrak{M}_r(j_{pq})$, let $\{b_1, b_2\} \in \operatorname{den} T_{\mathfrak{A}}[0]$ and let W be defined by formula (7.57). Then W is j_{pq}-unitary a.e on $\mathbb{R}$ and hence $S = PG(W)$ is unitary a.e. on $\mathbb{R}$. Next, to show that $S \in \mathcal{S}_{in}^{m\times m}$, let

$$S = PG(W) = \begin{bmatrix} s_{11} & s_{12} \\ s_{21} & s_{22} \end{bmatrix} \quad \text{and} \quad \mathfrak{A} = \begin{bmatrix} \mathfrak{a}_{11} & \mathfrak{a}_{12} \\ \mathfrak{a}_{21} & \mathfrak{a}_{22} \end{bmatrix};$$

and recall that, in view of the Smirnov maximum principle, it is enough to check that $S \in \mathcal{N}_+^{m\times m}$. To this point, we know that $s_{12} \in \mathcal{N}_+^{p\times q}$, since $s_{12} = b_1 T_{\mathfrak{A}}[0] b_2$ and $\{b_1, b_2\} \in \operatorname{den} T_{\mathfrak{A}}[0]$. Furthermore, $s_{22} \in \mathcal{S}^{q\times q}$, since $s_{22} = \mathfrak{a}_{22}^{-1} b_2$ and $\mathfrak{a}_{22}^{-1} \in \mathcal{S}_{out}^{q\times q}$; $s_{11} \in \mathcal{S}^{p\times p}$, since $s_{11} = b_1(\mathfrak{a}_{11}^{\#})^{-1}$ and $(\mathfrak{a}_{11}^{\#})^{-1} \in \mathcal{S}_{out}^{p\times p}$; and $s_{21} = -\mathfrak{a}_{22}^{-1}\,\mathfrak{a}_{21} \in S^{q\times p}$. Thus, $W \in \mathcal{U}(j_{pq})$ and $\{b_1, b_2\} \in ap(W)$.

The converse assertion follows from the definitions of the set $ap(W)$, the class $\mathfrak{M}_r(j_{pq})$ and the equality $T_W[0] = b_1 T_{\mathfrak{A}}[0] b_2$. □

Theorem 7.26 yields a one to one correspondence between the class of mvf's $W \in \mathcal{U}(j_{pq})$ with a given associated pair $\{b_1, b_2\} \in ap(W)$ and the class of mvf's $\mathfrak{A} \in \Pi \cap \mathfrak{M}_r(j_{pq})$ with $\{b_1, b_2\} \in \operatorname{den} T_{\mathfrak{A}}[0]$.

Theorem 7.27 *If the mvf's $\mathfrak{A} \in \Pi \cap \mathfrak{M}_r(j_{pq})$ and $W \in \mathcal{U}(j_{pq})$ are connected by formula (7.57), then*

$$\mathfrak{A} \in \mathfrak{M}_{rR}(j_{pq}) \iff W \in \mathcal{U}_{rR}(j_{pq}) \tag{7.59}$$

and

$$\mathfrak{A} \in \mathfrak{M}_{rsR}(j_{pq}) \iff W \in \mathcal{U}_{rsR}(j_{pq}). \tag{7.60}$$

Proof Let $\mathfrak{A} \in \mathfrak{M}_{rR}(j_{pg})$ and let $W \in \mathcal{U}(j_{pq})$ be connected with $\mathfrak{A}$ by relation (7.57). Let $W = W_1 W_2$, where $W_1 \in \mathcal{U}(j_{pq})$, $W_2 \in \mathcal{U}_S(j_{pq})$. Then $\{b_1, b_2\} \in ap(W_1)$ by Theorem 4.94, and hence, if

$$\mathfrak{A}_1 = \begin{bmatrix} b_1^{-1} & 0 \\ 0 & b_2 \end{bmatrix} W_1, \text{ then } \mathfrak{A}_1 \in \mathfrak{M}_r(j_{pq}).$$

Therefore,

$$\mathfrak{A} = \mathfrak{A}_1 W_2, \quad \text{where } \mathfrak{A}_1 \in \mathfrak{M}_r(j_{pq}) \text{ and } W_2 \in \mathcal{U}_S(j_{pq}).$$

Consequently, W_2 is a constant matrix, since $\mathfrak{A} \in \mathfrak{M}_{rR}(j_{pq})$. Thus, $W \in \mathcal{U}_{rR}(j_{pq})$.

Conversely, if $W \in \mathcal{U}_{rR}(j_{pq})$, $\{b_1, b_2\} \in ap(W)$ and (7.57) holds for some mvf $\mathfrak{A} \in \mathfrak{M}_r(j_{pq})$, then $\mathfrak{A} = \mathfrak{A}_1 W_2$, with $\mathfrak{A}_1 \in \mathfrak{M}_{rR}(j_{pq})$ and $W_2 \in \mathcal{U}_S(j_{pq})$, and

$$W = \begin{bmatrix} b_1 & 0 \\ 0 & b_2^{-1} \end{bmatrix} \mathfrak{A}_1 W_2.$$

Thus, as

$$T_{\mathfrak{A}}[S^{p\times q}] \subseteq T_{\mathfrak{A}_1}[S^{p\times q}],$$

it follows that

$$f - f_{12} \in H_\infty^{p\times q} \text{ for every } f \in T_{\mathfrak{A}_1}[S^{p\times q}].$$

Consequently, $\{b_1, b_2\} \in \text{den} f$ for every $f \in \mathbf{T}_{\mathfrak{A}_1}[S^{p\times q}]$. In particular, $\{b_1, b_2\} \in \text{den} f_0$, where $f_0 = T_{\mathfrak{A}_1}[0_{p\times q}]$ and

$$W_1 = \begin{bmatrix} b_1 & 0 \\ 0 & b_2^{-1} \end{bmatrix} \mathfrak{A}_1 \in \mathcal{U}(j_{pq}) \text{ and } W = W_1 W_2.$$

Thus, $W_2(\lambda)$ is constant, since $W \in \mathcal{U}_{rR}(j_{pq})$. Moreover, since $f \in T_{\mathfrak{A}_1}[S^{p\times q}] \Longleftrightarrow b_1 f b_2 \in T_W[S^{p\times q}]$, (7.60) follows from Lemma 7.16, (7.59), Theorem 7.22, and the fact that $\|f\|_\infty < 1$ if and only if $\|b_1 f b_2\|_\infty < 1$. □

Theorem 7.28 *If $W \in \mathcal{U}(j_{pq})$, then the following conditions are equivalent:*

(1) $W \in L_\infty^{m\times m}$.

(2) $T_W[0] \in \mathring{\mathcal{S}}^{p\times q}$.

(3) $T_W^\ell[0] \in \mathring{\mathcal{S}}^{q\times p}$.

(4) $T_W[\mathcal{S}_{const}^{p\times q}] \cap \mathring{\mathcal{S}}^{p\times q} \neq \emptyset$.

(5) $T_W[\mathring{\mathcal{S}}^{p\times q}] \subseteq \mathring{\mathcal{S}}^{p\times q}$.

Proof This follows from Theorem 7.26 and Lemma 7.11. □

Corollary 7.29 $\mathcal{U}(J) \cap L_\infty^{m\times m} \subseteq \mathcal{U}_{rsR}(J) \cap \mathcal{U}_{\ell sR}(J)$.

Proof If $J \neq \pm I_m$, then U is unitarily equivalent to a mvf $W \in \mathcal{U}(j_{pq})$ and then, since $U \in \mathcal{U}_{rsR}(J) \Longleftrightarrow W \in \mathcal{U}_{rsR}(j_{pq})$ and $U \in \mathcal{U}_{\ell sR}(J) \Longleftrightarrow W \in \mathcal{U}_{\ell sR}(j_{pq})$, the assertion follows from Theorem 7.28. If $J = \pm I_m$, then the assertion is true by definition. □

Theorem 7.30 *Let $A \in \mathcal{U}(J_p)$. Then the following conditions are equivalent:*

(1) $A \in L_\infty^{m\times m}$.

(2) $T_A[I_p] \in \mathring{\mathcal{C}}^{p\times p}$.

(3) $T_{A^\sim}[I_p] \in \mathring{\mathcal{C}}^{p\times p}$.

(4) $T_A[\mathcal{C}_{const}^{p\times p}] \cap \mathring{\mathcal{C}}^{p\times p} \neq \emptyset$.

(5) $T_A[\mathring{\mathcal{C}}^{p\times p}] \subseteq \mathring{\mathcal{C}}^{p\times p}$.

Proof The assertions follow from the fact that $\mathring{\mathcal{C}}^{p\times p} = T_{\mathfrak{V}}[\mathring{\mathcal{S}}^{p\times p}]$ and that $A \in \mathcal{U}(J_p) \Longleftrightarrow \mathfrak{V}A\mathfrak{V} \in \mathcal{U}(j_p)$. □

Examples to show that the inclusions $\mathfrak{M}_r(j_p) \cap L_\infty^{m\times m} \subseteq \mathfrak{M}_{rsR}(j_p)$ and $\mathcal{U}_r(j_p) \cap L_\infty^{m\times m} \subseteq \mathcal{U}_{rsR}(j_p)$ are proper when $p = 1$ will be presented in Section 10.4.

7.3 Criteria for right regularity

Lemma 7.14 yields a parametrization of the mvf's $\mathfrak{A} \in \mathfrak{M}_r(j_{pq})$. If $q = p$, then a second parametrization of the mvf's $\mathfrak{A} \in \mathfrak{M}_r(j_p)$ may be obtained by setting

$$c = T_{\mathfrak{V}}[s], \quad \text{where } s = s_{21} = -\mathfrak{a}_{22}^{-1}\mathfrak{a}_{21}. \tag{7.61}$$

Then $c \in \mathcal{C}^{p\times p}$ and the second condition in (7.31) is equivalent to the requirement that

$$\ln \det(\Re c) \in \widetilde{L}_1. \tag{7.62}$$

Let

$$\Delta(\mu) = 2\{c(\mu) + c(\mu)^*\}^{-1} \quad \text{a.e. on } \mathbb{R}. \tag{7.63}$$

Since $\Re c \in \widetilde{L}_1^{p\times p}$ when $c \in \mathcal{C}^{p\times p}$, Theorem 3.78 guarantees that there exist essentially unique solutions of the factorization problems

$$\Delta(\mu) = \varphi_-(\mu)^*\varphi_-(\mu) \quad \text{and} \quad \Delta(\mu) = \varphi_+(\mu)^*\varphi_+(\mu) \tag{7.64}$$

such that $(\rho_i\varphi_-^{\#})^{-1}$ and $(\rho_i\varphi_+)^{-1}$ are outer mvf's in $H_2^{p\times p}$. Then

$$\mathfrak{A}(\mu) = \frac{1}{2}\begin{bmatrix} \varphi_-(\mu)\{I_p + c(\mu)^*\} & -\varphi_-(\mu)\{I_p - c(\mu)^*\} \\ -\varphi_+(\mu)\{I_p - c(\mu)\} & \varphi_+(\mu)\{I_p + c(\mu)\} \end{bmatrix} \quad \text{a.e. on } \mathbb{R} \quad (7.65)$$

and every mvf $\mathfrak{A} \in \mathfrak{M}_r(j_p)$ may be parametrized by this formula, up to a constant block diagonal unitary factor on the left; (7.65) may be obtained from (7.32) and the formulas (7.33)–(7.35), (7.61), (7.63) and (7.64).

Our next objective is to obtain a new factorization of mvf's $\mathfrak{A} \in \mathfrak{M}_r(j_p)$ that is based on the parametrization formula (7.65) and the Riesz-Herglotz integral representation formula for the mvf's $c \in \mathcal{C}^{p\times p}$, which yields the decomposition (3.14)–(3.16).

Theorem 7.31 *Let $\mathfrak{A} \in \mathfrak{M}_r(j_p)$ be parametrized by formula (7.65) and let the mvf $c(\lambda) \in \mathcal{C}^{p\times p}$ considered in this formula be expressed in the form (3.14). Then the formula for $\mathfrak{A}(\mu)$ can be reexpressed in the following equivalent ways:*

$$\mathfrak{A}(\mu) = \mathfrak{A}_a(\mu) + \frac{1}{2}\begin{bmatrix} \varphi_-(\mu)c_s(\mu)^* & \varphi_-(\mu)c_s(\mu)^* \\ \varphi_+(\mu)c_s(\mu) & \varphi_+(\mu)c_s(\mu) \end{bmatrix}, \quad (7.66)$$

$$\mathfrak{A}(\mu) = \mathfrak{A}_a(\mu) + \frac{1}{2}\begin{bmatrix} -\varphi_-(\mu)c_s(\mu) & -\varphi_-(\mu)c_s(\mu) \\ \varphi_+(\mu)c_s(\mu) & \varphi_+(\mu)c_s(\mu) \end{bmatrix} \quad (7.67)$$

and

$$\mathfrak{A}(\mu) = \mathfrak{A}_a(\mu)\mathfrak{A}_s(\mu), \quad (7.68)$$

where

$$\mathfrak{A}_a(\mu) = \frac{1}{2}\begin{bmatrix} \varphi_-(\mu)\{I_p + c_a(\mu)^*\} & -\varphi_-(\mu)\{I_p - c_a(\mu)^*\} \\ -\varphi_+(\mu)\{I_p - c_a(\mu)\} & \varphi_+(\mu)\{I_p + c_a(\mu)\} \end{bmatrix} \quad (7.69)$$

and

$$\mathfrak{A}_s(\mu) = \begin{bmatrix} I_p & 0 \\ 0 & I_p \end{bmatrix} + \frac{1}{2}\begin{bmatrix} -c_s(\mu) & -c_s(\mu) \\ c_s(\mu) & c_s(\mu) \end{bmatrix} \quad (7.70)$$

a.e. on $\mathbb{R}$. Moreover, $\mathfrak{A}_a \in \mathfrak{M}_r(j_p)$ and $\mathfrak{A}_s \in \mathcal{U}_S(j_p)$.

Proof This is an immediate consequence of the stated formulas and the fact that $c_s(\mu) = -c_s(\mu)^*$ a.e. on $\mathbb{R}$. □

Corollary 7.32 *The class $\mathcal{U}_{const}(j_p)$ is parametrized by the formulas*

$$\mathfrak{A} = \mathfrak{A}_a \mathfrak{A}_s, \tag{7.71}$$

$$\mathfrak{A}_a = \begin{bmatrix} u & 0 \\ 0 & v \end{bmatrix} \begin{bmatrix} (\delta^{\frac{1}{2}} + \delta^{-\frac{1}{2}})/2 & -(\delta^{\frac{1}{2}} - \delta^{-\frac{1}{2}})/2 \\ -(\delta^{\frac{1}{2}} - \delta^{-\frac{1}{2}})/2 & (\delta^{\frac{1}{2}} + \delta^{\frac{1}{2}})/2 \end{bmatrix} \tag{7.72}$$

and

$$\mathfrak{A}_s = \begin{bmatrix} I_p - i\gamma & -i\gamma \\ i\gamma & I_p + i\gamma \end{bmatrix}, \tag{7.73}$$

where u, v, δ and γ are arbitrary $p \times p$ matrices such that

$$u^*u = v^*v = I_p, \quad \delta > 0 \quad \textit{and} \quad \gamma^* = \gamma. \tag{7.74}$$

Proof The class of mvf's $\mathfrak{A} \in \mathfrak{M}_r(j_{pq})$ which are constant on $\mathbb{R}$ coincides with the class of constant j_{pq}-unitary matrices. Therefore the sought for parametrization can be obtained by specializing formulas (7.69) and (7.70) to the case where $\mathfrak{A}(\mu)$ is constant. In particular, this assumption implies that $\varphi_-(\mu)$, $\varphi_+(\mu)$, $\Delta(\mu)$, $c_a(\mu)$ and $c_s(\mu)$ are all constant. Upon setting

$$\Delta(\mu) = \delta > 0,$$

it follows that

$$\varphi_-(\mu) = u\delta^{\frac{1}{2}}, \quad \varphi_+(\mu) = v\delta^{\frac{1}{2}}, \quad c_a(\mu) = \delta^{-1} \quad \text{and} \quad c_s(\mu) = i2\gamma,$$

where u, v, δ and γ are subject to (7.74). The rest is immediate from formulas (7.68)–(7.70). □

Let $g(\mu)$ be a $p \times p$ measurable mvf that is unitary a.e. on $\mathbb{R}$ and admits a factorization of the form

$$g(\mu) = \psi_-(\mu)\psi_+(\mu)^{-1}, \tag{7.75}$$

where

$$\rho_\omega^{-1}(\psi_-^{\#})^{-1} \in H_2^{p\times p} \quad \text{and} \quad \rho_\omega^{-1}\psi_+^{-1} \in H_2^{p\times p} \tag{7.76}$$

for at least one (and hence every) point $\omega \in \Omega_+$. Then we shall say that **index**$\{g\} = 0$ if for any other pair of mvf's $\widetilde{\psi}_-$, $\widetilde{\psi}_+$ with the properties (7.75) and (7.76) the equalities

$$\widetilde{\psi}_-(\mu) = \psi_-(\mu)\kappa \quad \text{and} \quad \widetilde{\psi}_+(\mu) = \psi_+(\mu)\kappa \tag{7.77}$$

hold a.e. on $\mathbb{R}$ for some invertible constant $p \times p$ matrix κ.

We remark that this automatically forces $\psi_-^\#(\lambda)$ and $\psi_+(\lambda)$ to be outer mvf's.

Theorem 7.33 *Let $\mathfrak{A} \in \mathfrak{M}_r(j_p)$ be expressed in the form*

$$\mathfrak{A}(\mu) = \mathfrak{A}_a(\mu)\mathfrak{A}_s(\mu)$$

considered in Theorem 7.31 and let

$$c(\lambda) = c_a(\lambda) + c_s(\lambda) \tag{7.78}$$

be defined by formula (7.61). Then the following statements are equivalent:

(1) $\mathfrak{A} \in \mathfrak{M}_{rR}(j_p)$.

(2) $\mathfrak{A}_a \in \mathfrak{M}_{rR}(j_p)$ *and* $\mathfrak{A}_s(\mu)$ *is constant.*

(3) $\text{index}\{T_{\mathfrak{A}}[I_p]\} = 0$ *and* $\mathfrak{A}_s(\mu)$ *is constant.*

(4) $\text{index}\{T_{\mathfrak{A}}[I_p]\} = 0$ *and* $c_s(\lambda)$ *is constant.*

(5) $\text{index}\{T_{\mathfrak{A}}[\varepsilon]\} = 0$ *for every constant unitary* $p \times p$ *matrix* ε *and* $\mathfrak{A}_s(\mu)$ *is constant.*

Proof The implications (1) $\Longleftrightarrow$ (2), (3) $\Longleftrightarrow$ (4) and (5) $\Longrightarrow$ (3) are self-evident; (1) $\Longrightarrow$ (5) is established in [AAK68] for $p = 1$ and in [Ad73a] for $p > 1$. Finally, (3) $\Longrightarrow$ (2), by Theorem 5.4 of [ArD01b]. □

Theorem 7.33 is not directly applicable to mvf's $\mathfrak{A} \in \mathfrak{M}_{rR}(j_{pq})$ when $p \neq q$. The next two lemmas present two different ways of embedding a mvf $\mathfrak{A} \in \mathfrak{M}_{rR}(j_{pq})$ into a mvf $\mathfrak{A}_1 \in \mathfrak{M}_{rR}(j_{p_1})$ with the property that $\mathfrak{A} \in \mathfrak{M}_{rR}(j_{pq}) \Longleftrightarrow \mathfrak{A}_1 \in \mathfrak{M}_{rR}(j_{p_1})$.

Lemma 7.34 *Let*

$$\mathfrak{A} = \begin{bmatrix} \mathfrak{a}_- & \mathfrak{b}_- \\ \mathfrak{b}_+ & \mathfrak{a}_+ \end{bmatrix} \in \mathfrak{M}_r(j_{pq}) \tag{7.79}$$

and, supposing that $p \neq q$, *let*

$$k = |p - q|, \quad p^\circ = \max\{p, q\}$$

and

$$\mathfrak{A}^\circ = \begin{bmatrix} \mathfrak{a}_-^\circ & \mathfrak{b}_-^\circ \\ \mathfrak{b}_+^\circ & \mathfrak{a}_+^\circ \end{bmatrix}, \tag{7.80}$$

where

$$\left.\begin{array}{rclrcl} \mathfrak{a}_-^\circ & = & \mathfrak{a}_-, & \mathfrak{b}_-^\circ & = & [0_{p\times k} \quad \mathfrak{b}_-], \\ \mathfrak{b}_+^\circ & = & \begin{bmatrix} 0_{k\times p} \\ \mathfrak{b}_+ \end{bmatrix}, & \mathfrak{a}_+^\circ & = & \begin{bmatrix} I_k & 0 \\ 0 & \mathfrak{a}_+ \end{bmatrix} \end{array}\right\} \textit{if } p > q \tag{7.81}$$

and

$$\left.\begin{array}{rclrcl} \mathfrak{a}_-^\circ & = & \begin{bmatrix} \mathfrak{a}_- & 0 \\ 0 & I_k \end{bmatrix}, & \mathfrak{b}_-^\circ & = & \begin{bmatrix} 0_{k\times q} \\ \mathfrak{b}_- \end{bmatrix} \\ \mathfrak{b}_+^\circ & = & [0_{q\times k} \quad \mathfrak{b}_+], & \mathfrak{a}_+^0 & = & \mathfrak{a}_+ \end{array}\right\} \textit{if } p < q. \tag{7.82}$$

Then

(1) $\mathfrak{A}^\circ \in \mathfrak{M}_r(j_{p^\circ})$.

(2) $\mathfrak{A} \in \mathfrak{M}_{rR}(j_{pq})$ *if and only if* $\mathfrak{A}^\circ \in \mathfrak{M}_{rR}(j_{p^\circ})$.

(3) $\mathfrak{A} \in \mathfrak{M}_{rsR}(j_{pq})$ *if and only if* $\mathfrak{A}^\circ \in \mathfrak{M}_{rsR}(j_{p^\circ})$.

Proof Suppose first that $p > q$. Then it is readily checked that the identities

$$\begin{aligned} (\mathfrak{a}_-^\circ)^*\mathfrak{a}_-^\circ - (\mathfrak{b}_+^\circ)^*\mathfrak{b}_+^\circ &= \mathfrak{a}_-^*\mathfrak{a}_- - \mathfrak{b}_+^*\mathfrak{b}_+ = I_p, \\ (\mathfrak{a}_-^\circ)^*\mathfrak{b}_-^\circ - (\mathfrak{b}_+^\circ)^*\mathfrak{a}_+^\circ &= 0 \quad \text{and} \\ (\mathfrak{b}_-^\circ)^*\mathfrak{b}_-^\circ - (\mathfrak{a}_+^\circ)^*\mathfrak{a}_+^\circ &= \begin{bmatrix} -I_k & 0_{k\times q} \\ 0_{q\times k} & \mathfrak{b}_-^*\mathfrak{b}_- - \mathfrak{a}_+^*\mathfrak{a}_+ \end{bmatrix} = -I_p \end{aligned}$$

hold a.e. on $\mathbb{R}$ and hence that $\mathfrak{A}^\circ$ is j_p-unitary a.e. on $\mathbb{R}$. Moreover, as the blocks $\mathfrak{a}_\pm^\circ$ and $\mathfrak{b}_\pm^\circ$ of $\mathfrak{A}^\circ$ inherit the properties (7.23) and (7.24) from the blocks $\mathfrak{a}_\pm$ and $\mathfrak{b}_\pm$ of $\mathfrak{A}$, it follows that $\mathfrak{A}^\circ \in \mathfrak{M}_r(j_p)$.

Next, to prove (2), let

$$s = T_{\mathfrak{A}}[0_{p\times q}], \quad s^\circ = T_{\mathfrak{A}^\circ}[0_{p\times p}], \quad \Gamma = \Gamma(s) \quad \text{and} \quad \Gamma^\circ = \Gamma(s^\circ). \tag{7.83}$$

Then, in view of Theorem 7.22, it suffices to show that

(a) $\mathfrak{A}^\circ \in \mathfrak{M}_{rR}(j_p) \Longrightarrow \mathcal{N}(\Gamma) \subseteq T_{\mathfrak{A}}[\mathcal{S}^{p\times q}]$

and

(b) $\mathfrak{A} \in \mathfrak{M}_{rR}(j_{pq}) \Longrightarrow \mathcal{N}(\Gamma^\circ) \subseteq T_{\mathfrak{A}^\circ}[\mathcal{S}^{p\times p}]$.

We first focus on the case $p > q$. To establish (a), let $\mathfrak{A}^\circ \in \mathfrak{M}_{rR}(j_p)$ and $f \in \mathcal{N}(\Gamma)$. Then, since $f^\circ = [0_{p\times k} \quad f]$ belongs to $\mathcal{N}(\Gamma^\circ)$, $f^\circ = T_{\mathfrak{A}^\circ}[\varepsilon^\circ]$ for some $\varepsilon^\circ \in \mathcal{S}^{p\times p}$, by Theorem 7.22. A direct calculation based on the block decomposition $\varepsilon^\circ = [\varepsilon_1 \quad \varepsilon]$ with $\varepsilon_1 \in \mathcal{S}^{p\times k}$ and $\varepsilon \in \mathcal{S}^{p\times q}$ yields the formula

$$T_{\mathfrak{A}^\circ}[\varepsilon^\circ] = [(\mathfrak{a}_- - T_{\mathfrak{A}}[\varepsilon]\,\mathfrak{b}_+)\varepsilon_1 \quad T_{\mathfrak{A}}[\varepsilon]]. \tag{7.84}$$

Thus, $f = T_{\mathfrak{A}}[\varepsilon]$ belongs to $T_{\mathfrak{A}}[\mathcal{S}^{p\times q}]$, as claimed.

Next, to establish (b), let $\mathfrak{A} \in \mathfrak{M}_{rR}(j_{pq})$ and $f^\circ \in \mathcal{N}(\Gamma^\circ)$. Then $f^\circ = [f_1 \quad f]$ where $f \in \mathcal{N}(\Gamma)$, $f_1 \in \mathcal{S}^{p\times k}$ and $\|f^\circ\|_\infty \le 1$. Therefore since $\mathcal{N}(\Gamma) = T_{\mathfrak{A}}[\mathcal{S}^{p\times q}]$ by Theorem 7.22, $f = T_{\mathfrak{A}}[\varepsilon]$ for some $\varepsilon \in \mathcal{S}^{p\times q}$. Let

$$\varepsilon^\circ = T_{(\mathfrak{A}^\circ)^{-1}}[f^\circ].$$

Then $\|\varepsilon^\circ\|_\infty \le 1$ and, in terms of the block decomposition

$$\varepsilon^\circ = [\varepsilon_1 \quad \varepsilon] \quad \text{and} \quad f^\circ = [f_1 \quad f] = T_{\mathfrak{A}^\circ}[\varepsilon^\circ] = [(\mathfrak{a}_- - f\mathfrak{b}_+)\varepsilon_1 \quad f].$$

Thus, $(\mathfrak{a}_- - f\mathfrak{b}_+)\varepsilon_1 = f_1 \in \mathcal{S}^{p\times k}$. But, with the aid of the formula

$$f = T_{\mathfrak{A}}[\varepsilon] = (\varepsilon\mathfrak{b}_-^* + \mathfrak{a}_-^*)^{-1}\,(\varepsilon\mathfrak{a}_+^* + \mathfrak{b}_+^*), \tag{7.85}$$

it is readily checked that $\mathfrak{a}_- - f\mathfrak{b}_+ = (\varepsilon\mathfrak{b}_-^* + \mathfrak{a}_-^*)^{-1}$ and hence that

$$\varepsilon_1(\mu) = \{\varepsilon(\mu)\mathfrak{b}_-(\mu)^* + \mathfrak{a}_-(\mu)^*\}f_1(\mu) \quad \text{a.e. on } \mathbb{R}.$$

Consequently, $\varepsilon^\circ \in \mathcal{N}_+^{p\times p}$ and, as it also meets the bound

$$\|\varepsilon^\circ(\mu)\| \le 1 \quad \text{a.e on } \mathbb{R},$$

the Smirnov maximum principle guarantees that $\varepsilon^\circ \in \mathcal{S}^{p\times p}$. This completes the proof of (b) and assertion (2) for the case $p > q$. The verification of (2) for the case $p < q$ is similar and is left to the reader.

Finally, assertion (3) follows from (2) and Theorem 7.22 and the fact that $\|\Gamma^\circ\| = \|\Gamma\|$. □

Remark 7.35 *In view of Lemma 7.34, Theorem 7.33 applied to the mvf* $\mathfrak{A}^\circ$ *yields a number of equivalent conditions for* $\mathfrak{A} \in \mathfrak{M}_{rR}(j_{pq})$ *even if* $p \ne q$.

Another useful embedding of the mvf $\mathfrak{A} \in \mathfrak{M}_r(j_{pq})$ is described below.

Lemma 7.36 *Let* $\mathfrak{A} \in \mathfrak{M}_r(j_{pq})$, $m = p + q$ *and define the* $2m \times 2m$ *mvf*

$$\widetilde{\mathfrak{A}}(\mu) = \begin{bmatrix} \widetilde{\mathfrak{a}}_-(\mu) & \widetilde{\mathfrak{b}}_-(\mu) \\ \widetilde{\mathfrak{b}}_+(\mu) & \widetilde{\mathfrak{a}}_+(\mu) \end{bmatrix}, \tag{7.86}$$

where

$$\widetilde{\mathfrak{a}}_- = \begin{bmatrix} \mathfrak{a}_- & 0 \\ 0 & I_q \end{bmatrix}, \qquad \widetilde{\mathfrak{b}}_- = \begin{bmatrix} 0_{p\times p} & \mathfrak{b}_- \\ 0_{q\times p} & 0_{q\times q} \end{bmatrix},$$
$$\widetilde{\mathfrak{b}}_+ = \begin{bmatrix} 0_{p\times p} & 0_{p\times q} \\ \mathfrak{b}_+ & 0_{q\times q} \end{bmatrix}, \quad \widetilde{\mathfrak{a}}_+ = \begin{bmatrix} I_p & 0 \\ 0 & \mathfrak{a}_+ \end{bmatrix}, \tag{7.87}$$

and $\mathfrak{a}_\pm$ and $\mathfrak{b}_\pm$ are the blocks of $\mathfrak{A}$, considered in (7.79). Then:

(1) $\widetilde{\mathfrak{A}} \in \mathfrak{M}_r(j_m)$.

(2) $\mathfrak{A} \in \mathfrak{M}_{rR}(j_{pq})$ *if and only if* $\widetilde{\mathfrak{A}} \in \mathfrak{M}_{rR}(j_m)$.

(3) $\mathfrak{A} \in \mathfrak{M}_{rsR}(j_{pq})$ *if and only if* $\widetilde{\mathfrak{A}} \in \mathfrak{M}_{rsR}(j_m)$.

Proof A straightforward calculation yields the identities

$$\begin{aligned}\widetilde{\mathfrak{A}}^* j_m \widetilde{\mathfrak{A}} &= \begin{bmatrix} \mathfrak{a}_-^*\mathfrak{a}_- - \mathfrak{b}_+^*\mathfrak{b}_+ & 0 & 0 & \mathfrak{a}_-^*\mathfrak{b}_- - \mathfrak{b}_+^*\mathfrak{a}_+ \\ 0 & I_q & 0 & 0 \\ 0 & 0 & -I_p & 0 \\ \mathfrak{b}_-^*\mathfrak{a}_- - \mathfrak{a}_+^*\mathfrak{b}_+ & 0 & 0 & \mathfrak{b}_-^*\mathfrak{b}_- - \mathfrak{a}_+^*\mathfrak{a}_+ \end{bmatrix} \\ &= j_m \quad \text{a.e on } \mathbb{R},\end{aligned}$$

where the variable μ has been supressed to save space. Therefore, since $(\widetilde{\mathfrak{a}}_-^{\#})^{-1} \in \mathcal{S}_{out}^{m\times m}$, $(\widetilde{\mathfrak{a}}_+)^{-1} \in \mathcal{S}_{out}^{m\times m}$ and $\widetilde{\mathfrak{a}}_+^{-1}\widetilde{\mathfrak{b}} \in \mathcal{S}^{m\times m}$, (1) is established.

Next, let

$$\Gamma = \Gamma(f_0) \quad \text{and} \quad \widetilde{\Gamma} = \Gamma(\widetilde{f}_0), \tag{7.88}$$

where

$$f_0 = T_{\mathfrak{A}}[0_{p\times q}] \quad \text{and} \quad \widetilde{f}_0 = T_{\widetilde{\mathfrak{A}}}[0_{m\times m}]. \tag{7.89}$$

Then

$$\widetilde{f}_0 = \begin{bmatrix} 0_{p\times p} & f_0 \\ 0_{q\times p} & 0_{q\times q} \end{bmatrix} \quad \text{and} \quad \widetilde{\Gamma} = \begin{bmatrix} 0 & \Gamma \\ 0 & 0 \end{bmatrix} \tag{7.90}$$

and hence

$$\|\widetilde{\Gamma}\| = \|\Gamma\|.$$

Thus, in view of Theorem 7.22, (3) follows from (2).

The verification of (2) rests on Theorem 7.22 and the observation that if

$$\widetilde{\varepsilon} = \begin{bmatrix} \varepsilon_{11} & \varepsilon_{12} \\ \varepsilon_{21} & \varepsilon_{22} \end{bmatrix} \tag{7.91}$$

is the four block decomposition of a mvf $\widetilde{\varepsilon} \in L_\infty^{m\times m}$ with $\|\widetilde{\varepsilon}\|_\infty \le 1$ and diagonal blocks of sizes $p \times p$ and $q \times q$, respectively, then, with the help of formulas (7.87) and (2.27), the mvf

$$\widetilde{f} = T_{\widetilde{\mathfrak{A}}}[\widetilde{\varepsilon}] = \begin{bmatrix} \mathfrak{a}_-\varepsilon_{11} & \mathfrak{a}_-\varepsilon_{12} + \mathfrak{b}_- \\ \varepsilon_{21} & \varepsilon_{22} \end{bmatrix} \begin{bmatrix} I_p & 0_{p\times q} \\ \mathfrak{b}_+\varepsilon_{11} & \mathfrak{b}_+\varepsilon_{12} + \mathfrak{a}_+ \end{bmatrix}^{-1} \tag{7.92}$$

can be reduced to

$$\widetilde{f} = \begin{bmatrix} (\varepsilon_{12}\mathfrak{b}_-^* + \mathfrak{a}_-^*)^{-1}\varepsilon_{11} & T_{\mathfrak{A}}[\varepsilon_{12}] \\ \varepsilon_{21} - \varepsilon_{22}(\mathfrak{b}_+\varepsilon_{12} + \mathfrak{a}_+)^{-1}\mathfrak{b}_+\varepsilon_{11} & \varepsilon_{22}(\mathfrak{b}_+\varepsilon_{12} + \mathfrak{a}_+)^{-1} \end{bmatrix}.$$

To establish (2), suppose first that $\mathfrak{A} \in \mathfrak{M}_{rR}(j_{pq})$ and let $\widetilde{f} \in \mathcal{F}(\widetilde{\Gamma})$. Then, in view of (7.25),

$$\widetilde{f} = \begin{bmatrix} f_{11} & f \\ f_{21} & f_{22} \end{bmatrix},$$

where $f_{11} \in \mathcal{S}^{p\times p}$, $f_{21} \in \mathcal{S}^{q\times p}$, $f_{22} \in \mathcal{S}^{q\times q}$ and $f \in \mathcal{N}(\Gamma)$. Therefore, by Theorem 7.22, there exists a mvf $\varepsilon \in \mathcal{S}^{p\times q}$ such that $f = T_{\mathfrak{A}}[\varepsilon]$. Moreover,

$$\widetilde{\varepsilon} = T_{(\widetilde{\mathfrak{A}})^{-1}}[\widetilde{f}] \quad \text{belongs to } L_\infty^{m\times m} \text{ and } \|\widetilde{\varepsilon}\|_\infty \le 1.$$

Therefore, since $f = T_{\mathfrak{A}}[\varepsilon]$ for some $\varepsilon \in \mathcal{S}^{p\times q}$, it follows from the two block decompositions of $\widetilde{f}$ that $T_{\mathfrak{A}}[\varepsilon] = T_{\mathfrak{A}}[\varepsilon_{12}]$ and hence that $\varepsilon = \varepsilon_{12}$. Thus, $\varepsilon_{12} \in \mathcal{S}^{p\times q}$ and $\varepsilon_{11} = (\varepsilon\mathfrak{a}_-^\# + \mathfrak{b}_-^\#)f_{11} \in \mathcal{N}_+^{p\times p}$, $\varepsilon_{22} = f_{22}(\mathfrak{b}_+\varepsilon + \mathfrak{a}_+) \in \mathcal{N}_+^{q\times q}$ and $\varepsilon_{21} = f_{21} + \varepsilon_{22}(\mathfrak{b}_+\varepsilon_{12} + \mathfrak{a}_+)^{-1}b_+\varepsilon_{11} \in \mathcal{N}_+^{q\times p}$. Therefore, by the Smirnov maximum principle, $\widetilde{\varepsilon} \in \mathcal{S}^{m\times m}$. Thus, $\widetilde{f} \in T_{\widetilde{\mathfrak{A}}}[\mathcal{S}^{m\times m}]$ and consequently, by Theorem 7.22, $\widetilde{\mathfrak{A}} \in \mathfrak{M}_{rR}(j_m)$.

To establish the converse, let $\widetilde{\mathfrak{A}} \in \mathfrak{M}_{rR}(j_m)$, $f \in \mathcal{N}(\Gamma)$ and set

$$\widetilde{f} = \begin{bmatrix} 0_{p\times p} & f \\ 0_{q\times p} & 0_{q\times q} \end{bmatrix}. \tag{7.93}$$

Then, since $f \in \mathcal{N}(\Gamma)$, Theorem 7.22 guarantees that $\widetilde{f} = T_{\widetilde{\mathfrak{A}}}[\widetilde{\varepsilon}]$ for some mvf $\widetilde{\varepsilon} \in \mathcal{S}^{m\times m}$. Thus, upon invoking the four block decomposition (7.91) of $\widetilde{\varepsilon}$ and comparing formulas (7.92) and (7.93), it follows that $f = T_{\mathfrak{A}}[\varepsilon_{12}]$ with $\varepsilon_{12} \in \mathcal{S}^{p\times q}$ (and $\varepsilon_{11} = 0_{p\times p}$, $\varepsilon_{21} = 0_{q\times p}$ and $\varepsilon_{22} = 0_{q\times q}$). Thus, Theorem 7.22 guarantees that $\mathfrak{A} \in \mathfrak{M}_{rR}(j_{pq})$. □

Lemma 7.37 *Let $\mathfrak{A} \in \mathfrak{M}_r(j_{pq})$ be given by formula (7.79), let $\widetilde{\mathfrak{A}}$ be defined by formulas(7.86) and (7.87) and set*

$$\widetilde{c} = (I_m - \widetilde{s})(I_m + \widetilde{s})^{-1}, \quad \textit{where} \quad \widetilde{s} = -\widetilde{\mathfrak{a}}_+^{-1}\widetilde{\mathfrak{b}}_+ \tag{7.94}$$

then $\widetilde{c} \in \mathcal{C}^{m\times m} \cap H_\infty^{m\times m}$ *(and, consequently, its spectral function is absolutely continuous, i.e., the component* $\widetilde{c}_s$ *of* $\widetilde{c}$ *in the decomposition (7.78) with* $\widetilde{c}$ *in place of* c *is a constant* $m \times m$ *matrix).*

Proof In view of formula (7.87),

$$\widetilde{s} = -\widetilde{\mathfrak{a}}_+^{-1}\widetilde{\mathfrak{b}}_+ = \begin{bmatrix} 0_{p\times p} & 0_{p\times q} \\ s & 0_{q\times q} \end{bmatrix}, \quad \text{where } s = -\mathfrak{a}_+^{-1}\mathfrak{b}_+ \in \mathcal{S}^{q\times p}.$$

Consequently,

$$\widetilde{c} = \begin{bmatrix} I_p & 0 \\ -s & I_q \end{bmatrix}\begin{bmatrix} I_p & 0 \\ s & I_q \end{bmatrix}^{-1} = \begin{bmatrix} I_p & 0 \\ -2s & I_q \end{bmatrix}$$

belongs to $H_\infty^{m\times m}$. □

Theorem 7.38 *Let* $\mathfrak{A} \in \mathfrak{M}_r(j_{pq})$ *be given by formula (7.79) and let* $\widetilde{\mathfrak{A}} \in \mathfrak{M}_r(j_m)$ *be defined by formulas (7.86) and (7.87). Then the following assertions are equivalent:*

(1) $\mathfrak{A} \in \mathfrak{M}_{rR}(j_{pq})$.

(2) $\operatorname{index}\{T_{\widetilde{\mathfrak{A}}}[I_m]\} = 0$.

(3) $\operatorname{index}\{\widetilde{T}_{\widetilde{\mathfrak{A}}}[\widetilde{\varepsilon}]\} = 0$ *for every constant unitary* $m \times m$ *matrix* $\widetilde{\varepsilon}$.

Proof The theorem follows from Lemmas 7.36 and 7.37 and Theorem 7.33 applied to the mvf $\widetilde{\mathfrak{A}}(\mu)$. □

Lemma 7.39 *Let* $\mathfrak{A} \in \mathfrak{M}_r(j_{pq})$ *be given by formula (7.32), let* $\widetilde{\mathfrak{A}}$ *be defined by formulas (7.86) and (7.87), let* $\Delta_r(\mu)$ *be defined by formula (7.29) and let* $\widetilde{g} = T_{\widetilde{\mathfrak{A}}}[I_m]$. *Then:*

(1)

$$\Delta_r(\mu)^{-1} = \widetilde{\psi}_+(\mu)^*\widetilde{\psi}_+(\mu) = \widetilde{\psi}_-(\mu)^*\widetilde{\psi}_-(\mu), \tag{7.95}$$

where

$$\widetilde{\psi}_+ = \begin{bmatrix} I_p & 0 \\ \mathfrak{b}_+(\mu) & \mathfrak{a}_+(\mu) \end{bmatrix} j_{pq} \quad \textit{and} \tag{7.96}$$

$$\widetilde{\psi}_-(\mu) = \begin{bmatrix} \mathfrak{a}_-(\mu) & \mathfrak{b}_-(\mu) \\ 0 & I_q \end{bmatrix} j_{pq} \tag{7.97}$$

(2) $\widetilde{g} = \widetilde{\psi}_- \widetilde{\psi}_+^{-1}$.

(3) $\widetilde{\psi}_+^{-1} \in H_\infty^{m\times m}$ *and* $(\widetilde{\psi}_-^{\#})^{-1} \in H_\infty^{m\times m}$.

(4) *If* $\Delta_r^{-1} \in \widetilde{L}_1^{m\times m}$, *then* $\frac{\widetilde{\psi}_+}{\rho_i} \in H_2^{m\times m}$, $\frac{\widetilde{\psi}_-^{\#}}{\rho_i} \in H_2^{m\times m}$, index $\widetilde{g} = 0$, *and* $\mathfrak{A} \in \mathfrak{M}_{rR}(j_{pq})$.

Proof (1)–(3) and the first two assertions of (4) are easy; the last two follow from Lemma 7.13 and Theorem 7.38. □

7.4 Criteria for left regularity

The following dual versions of Lemma 7.14 and Theorems 7.26 and 7.27 are useful.

Lemma 7.40 *Let*

$$\mathfrak{A} \in \mathfrak{M}_\ell(j_{pq}) \quad \text{and} \quad s = T_{\mathfrak{A}}[0]. \tag{7.98}$$

Then

$$s \in \mathcal{S}^{p\times q} \quad \text{and} \quad \ln\det\{I_p - ss^*\} \in \widetilde{L}_1. \tag{7.99}$$

Conversely, if a mvf s meets the conditions in (7.99), then there exists a mvf $\mathfrak{A}$ such that (7.98) holds. Moreover, this mvf is uniquely defined up to a constant block diagonal j_{pq}-unitary matrix multiplier on the right by the formula

$$\mathfrak{A}(\mu) = \Delta_\ell(\mu) \begin{bmatrix} \mathfrak{d}_-(\mu) & 0 \\ 0 & \mathfrak{d}_+(\mu) \end{bmatrix} \quad \text{a.e. on } \mathbb{R}, \tag{7.100}$$

where Δ_ℓ is defined by formula (7.29) with $s_{12} = s$ and the mvf's $\mathfrak{d}_-$ and $\mathfrak{d}_+$ are solutions of the factorization problems

$$\mathfrak{d}_-(\mu)^{-*}\mathfrak{d}_-(\mu)^{-1} = I_p - s(\mu)s(\mu)^* \quad \text{a.e. on } \mathbb{R}, \quad \mathfrak{d}_-^{-\#} \in \mathcal{S}_{out}^{p\times p}, \tag{7.101}$$

and

$$\mathfrak{d}_+(\mu)^{-*}\mathfrak{d}_+(\mu)^{-1} = I_q - s(\mu)^*s(\mu) \quad \text{a.e. on } \mathbb{R}, \quad \mathfrak{d}_+^{-1} \in \mathcal{S}_{out}^{q\times q}, \tag{7.102}$$

respectively.

Theorem 7.41 *Let $\mathfrak{A} \in \Pi \cap \mathfrak{M}_\ell(j_{pq})$, $\{b_5, b_6\} \in \operatorname{den} T^\ell_{\mathfrak{A}}[0_{q\times p}]$ and let*

$$W(\lambda) = \mathfrak{A}(\lambda) \begin{bmatrix} b_5(\lambda) & 0 \\ 0 & b_6(\lambda)^{-1} \end{bmatrix}. \tag{7.103}$$

Then

$$W \in \mathcal{U}(j_{pq}) \quad \text{and} \quad \{b_5^\sim, b_6^\sim\} \in ap(W^\sim). \tag{7.104}$$

Conversely, if (7.104) is in force and the mvf $\mathfrak{A}(\lambda)$ is defined by formula (7.103), then $\mathfrak{A} \in \Pi \cap \mathfrak{M}_\ell(j_{pq})$ and $\{b_5, b_6\} \in \operatorname{den} T^\ell_{\mathfrak{A}}[0_{q\times p}]$.

Theorem 7.42 *Let the mvf's $W \in \mathcal{U}(j_{pq})$ and $\mathfrak{A} \in \Pi\cap\mathfrak{M}_\ell(j_{pq})$ be connected by formula (7.103). Then*

$$W \in \mathcal{U}_{\ell R}(j_{pq}) \iff \mathfrak{A} \in \mathfrak{M}_{\ell R}(j_{pq})$$

$$W \in \mathcal{U}_{\ell sR}(j_{pq}) \iff \mathfrak{A} \in \mathfrak{M}_{\ell sR}(j_{pq}).$$

Dual versions of the parametrization formula (7.65) and the factorization formula (7.68) for mvf's $\mathfrak{A} \in \mathfrak{M}_\ell(j_p)$ exist. They are based on the decomposition

$$z(\lambda) = z_s(\lambda) + z_a(\lambda)$$

of the mvf $z = T_{\mathfrak{V}\mathfrak{A}}[0_{p\times p}]$ into components $z_s \in \mathcal{C}^{p\times p}_{sing}$ and $z_a \in \mathcal{C}^{p\times p}_a$ with $z_a(i) > 0$. Additional information on dual formulas is furnished on pp. 284–289 of [ArD01b].

7.5 Formulas for $\mathfrak{A}$ when $\|\Gamma\| < 1$

Lemma 7.43 *If $\mathfrak{A} \in \mathfrak{M}_r(j_{pq})$ and $\omega \in \mathbb{C}_+$, then $\mathfrak{A}$ admits a unique factorization of the form*

$$\mathfrak{A}(\mu) = \mathfrak{A}^\circ_\omega(\mu) V_\omega, \tag{7.105}$$

where $V_\omega \in \mathcal{U}_{const}(j_{pq})$ and $\mathfrak{A}^\circ_\omega \in \mathfrak{M}_r(j_{pq})$ is normalized by the conditions

$$\mathfrak{a}^\circ_-(\overline{\omega}) > 0,\ \mathfrak{b}^\circ_-(\overline{\omega}) = 0,\ \mathfrak{b}^\circ_+(\omega) = 0 \text{ and } \mathfrak{a}^\circ_+(\omega) > 0. \tag{7.106}$$

Proof If the matrix V_ω^{-1} is expressed in terms of the parameters $k \in \mathcal{S}_{const}^{p\times q}$, $u \in \mathcal{U}_{const}(I_p)$ and $v \in \mathcal{U}_{const}(I_q)$ in (2.19), then (7.105) is equivalent to the system of equations

$$\mathfrak{a}_-^\circ = (\mathfrak{a}_- + \mathfrak{b}_- k^*)(I_p - kk^*)^{-\frac{1}{2}}u, \quad \mathfrak{b}_-^\circ = (\mathfrak{a}_- k + \mathfrak{b}_-)(I_q - k^*k)^{-\frac{1}{2}}v$$

$$\mathfrak{b}_+^\circ = (\mathfrak{b}_+ + \mathfrak{a}_+ k^*)(I_p - kk^*)^{-\frac{1}{2}}u, \quad \mathfrak{a}_+^\circ = (\mathfrak{b}_+ k + \mathfrak{a}_+)(I_q - k^*k)^{-\frac{1}{2}}v.$$

Moreover, since $\mathfrak{A} \in \mathfrak{M}_r(j_{pq})$,

$$s_{21} = -\mathfrak{a}_+^{-1}\mathfrak{b}_+ = -\mathfrak{b}_-^\#(\mathfrak{a}_-^\#)^{-1} \text{ belongs to } \mathcal{S}^{q\times p} \tag{7.107}$$

and $\|s_{21}(\omega)\| < 1$ for every point $\omega \in \mathbb{C}_+$. Thus,

$$\mathfrak{b}_+^\circ(\omega) = 0 \Longleftrightarrow \mathfrak{b}_-^\circ(\overline{\omega}) = 0 \Longleftrightarrow k = s_{21}(\omega)^*,$$

and hence, if $k = -s_{21}(\omega)^*$, then

$$\mathfrak{a}_-^\circ(\overline{\omega}) = \mathfrak{a}_-(\overline{\omega})(I_p - kk^*)^{\frac{1}{2}}u \quad \text{and} \quad \mathfrak{a}_+^\circ(\omega) = \mathfrak{a}_+(\omega)(I_q - k^*k)^{\frac{1}{2}}v.$$

Thus, as u and v are unitary matrices, $\mathfrak{a}_-^\circ(\overline{\omega}) > 0$ and $\mathfrak{a}_+^\circ(\overline{\omega}) > 0$ if and only if

$$\begin{aligned}
\mathfrak{a}_-^\circ(\overline{\omega}) &= \{\mathfrak{a}_-(\overline{\omega})(I_p - kk^*)\mathfrak{a}_-(\overline{\omega})^*\}^{\frac{1}{2}},\\
\mathfrak{a}_+^\circ(\omega) &= \{\mathfrak{a}_+(\omega)(I_q - k^*k)\mathfrak{a}_+(\omega)^*\}^{\frac{1}{2}},\\
u &= (I_p - kk^*)^{-\frac{1}{2}}\mathfrak{a}_-(\overline{\omega})^{-1}\mathfrak{a}_-^\circ(\overline{\omega}) \quad \text{and}\\
v &= (I_q - k^*k)^{-\frac{1}{2}}\mathfrak{a}_+(\omega)^{-1}\mathfrak{a}_+^\circ(\omega).
\end{aligned}$$

□

Remark 7.44 *If there is a point $\omega \in \mathbb{R} \cap \mathfrak{h}_U$, then $\mathfrak{A}(\omega) \in \mathcal{U}_{const}(j_{pq})$ and $\mathfrak{A}$ admits a factorization of the form (7.105) with $V_\omega = \mathfrak{A}(\omega)$ and $\mathfrak{A}_\omega^\circ = I_m$.*

If Γ is a Hankel operator from H_2^q into K_2^p, then Γ and Γ^* will be defined on mvf's $h = [h_2 \cdots h_\ell]$ in $H_2^{q\times\ell}$ and $[g_1 \cdots g_k]$ in $K_2^{p\times k}$ column by column, i.e., by the rules

$$\Gamma[h_1 \quad \cdots \quad h_\ell] = [\Gamma h_1 \quad \cdots \quad \Gamma h_\ell]$$

and

$$\Gamma^*[g_1 \ \cdots \ g_k] = [\Gamma^* g_1 \ \cdots \ \Gamma^* g_k].$$

Correspondingly, $I - \Gamma^*\Gamma$ and $I - \Gamma\Gamma^*$ act on mvf's in $H_2^{q\times q}$ and $K_2^{p\times p}$ by the rules

$$(I - \Gamma^*\Gamma)[h_1 \ \cdots \ h_q] = [(I - \Gamma^*\Gamma)h_1 \ \cdots \ (I - \Gamma^*\Gamma)h_q]$$

and

$$(I - \Gamma\Gamma^*)[g_1 \ \cdots \ g_p] = [(I - \Gamma^*\Gamma)g_1 \ \cdots \ (I - \Gamma\Gamma^*)g_p].$$

Theorem 7.45 *If the $NP(\Gamma)$ is strictly completely indeterminate, i.e, if $\|\Gamma\| < 1$, then*

$$\mathcal{N}(\Gamma) = T_{\mathfrak{A}_\omega^\circ}[\mathcal{S}^{p\times q}], \tag{7.108}$$

where $\mathfrak{A}_\omega^\circ \in \mathfrak{M}_{rsR}(j_{pq})$ is normalized at the point $\omega \in \mathbb{C}_+$ by the conditions (7.106) and its blocks are uniquely specified by the formulas

$$\mathfrak{a}_+^\circ(\mu) = \rho_\omega(\mu)\{(I - \Gamma^*\Gamma)^{-1}\rho_\omega^{-1}I_q\}(\mu)\alpha_\omega, \tag{7.109}$$

$$\mathfrak{b}_+^\circ(\mu) = \rho_{\overline{\omega}}(\mu)\{\Gamma^*(I - \Gamma\Gamma^*)^{-1}\rho_{\overline{\omega}}^{-1}I_p\}(\mu)\beta_\omega, \tag{7.110}$$

$$\mathfrak{b}_-^\circ(\mu) = \rho_\omega(\mu)\{\Gamma(I - \Gamma^*\Gamma)^{-1}\rho_\omega^{-1}I_q\}(\mu)\alpha_\omega, \tag{7.111}$$

$$\mathfrak{a}_-^\circ(\mu) = \rho_{\overline{\omega}}(\mu)\{(I - \Gamma\Gamma^*)^{-1}\rho_{\overline{\omega}}^{-1}I_p\}(\mu)\beta_\omega, \tag{7.112}$$

where

$$\alpha_\omega = \{\rho_\omega(\omega)((I - \Gamma^*\Gamma)^{-1}\rho_\omega^{-1}I_q)(\omega)\}^{-1/2} \tag{7.113}$$

and

$$\beta_\omega = \{\rho_{\overline{\omega}}(\overline{\omega})((I - \Gamma\Gamma^*)^{-1}\rho_{\overline{\omega}}^{-1}I_p)(\overline{\omega})\}^{-1/2}. \tag{7.114}$$

Proof Theorem 7.20 and Lemma 7.43 guarantee the existence of a normalized mvf $\mathfrak{A}_\omega^\circ \in \mathfrak{M}_{rsR}(j_{pq})$ such that (7.105) holds. Moreover, since $\mathfrak{A}_\omega^\circ \in \widetilde{L}^{m\times m}$ by Lemma 7.15, the Smirnov maximum principle implies that

$$\rho_\omega^{-1}\mathfrak{a}_+^\circ \in H_2^{q\times q} \quad \text{and} \quad \rho_\omega^{-1}\mathfrak{b}_+^\circ \in H_2^{q\times p},$$

and, by consideration of $(\rho_{\overline{\omega}}^{-1}\mathfrak{a}_-^\circ)^\#$ and $(\rho_{\overline{\omega}}^{-1}\mathfrak{b}_-^\circ)^\#$, that

$$\rho_{\overline{\omega}}^{-1}\mathfrak{a}_-^\circ \in K_2^{p\times p} \quad \text{and} \quad \rho_{\overline{\omega}}^{-1}\mathfrak{b}_-^\circ \in K_2^{p\times q}.$$

Let

$$f_\varepsilon = T_{\mathfrak{A}_\omega^\circ}[\varepsilon] \quad \text{for } \varepsilon \in \mathcal{S}^{p\times q}.$$

Then

$$\Gamma\rho_\omega^{-1}(\mathfrak{b}_+^\circ\varepsilon+\mathfrak{a}_+^\circ)=\Pi_-f_\varepsilon\rho_\omega^{-1}(\mathfrak{b}_+^\circ\varepsilon+\mathfrak{a}_+^\circ)=\Pi_-\rho_\omega^{-1}(\mathfrak{a}_-^\circ\varepsilon+\mathfrak{b}_-^\circ).$$

Therefore,

$$\Gamma\rho_\omega^{-1}\mathfrak{a}_+^\circ=\Pi_-\rho_\omega^{-1}\mathfrak{b}_-^\circ=\rho_\omega^{-1}\mathfrak{b}_-^\circ, \tag{7.115}$$

since $\mathfrak{b}_-^\circ(\overline{\omega})=0$, and

$$\Gamma\rho_\omega^{-1}\mathfrak{b}_+^\circ=\rho_\omega^{-1}\{\mathfrak{a}_-^\circ-\mathfrak{a}_-^\circ(\overline{\omega})\}. \tag{7.116}$$

Moreover, since $\mathfrak{A}_\omega^\circ$ is j_{pq}-unitary a.e. on $\mathbb{R}$,

$$f_\varepsilon=(\varepsilon(\mathfrak{b}_-^\circ)^*+(\mathfrak{a}_-^\circ)^*)^{-1}(\varepsilon(\mathfrak{a}_+^\circ)^*+(\mathfrak{b}_+^\circ)^*).$$

Therefore,

$$f_\varepsilon^*=(\mathfrak{a}_+^\circ\varepsilon^*+\mathfrak{b}_+^\circ)(\mathfrak{b}_-^\circ\varepsilon^*+\mathfrak{a}_-^\circ)^{-1}$$

and

$$\Gamma^*\rho_{\overline{\omega}}^{-1}(\mathfrak{b}_-^\circ\varepsilon^*+\mathfrak{a}_-^\circ)=\Pi_+\rho_{\overline{\omega}}^{-1}f_\varepsilon^*(\mathfrak{b}_-^\circ\varepsilon^*+\mathfrak{a}_-^\circ)^{-1}=\Pi_+\rho_{\overline{\omega}}^{-1}(\mathfrak{a}_+^\circ\varepsilon^*+\mathfrak{b}_+^\circ)$$

and hence

$$\Gamma^*\rho_{\overline{\omega}}^{-1}\mathfrak{a}_-^\circ=\rho_{\overline{\omega}}^{-1}\mathfrak{b}_+^\circ, \tag{7.117}$$

since $\mathfrak{b}_+^\circ(\omega)=0$, and

$$\Gamma^*\rho_{\overline{\omega}}^{-1}\mathfrak{b}_-^\circ=\rho_{\overline{\omega}}^{-1}(\mathfrak{a}_+^\circ-\mathfrak{a}_+^\circ(\omega)). \tag{7.118}$$

Thus, as

$$\rho_{\overline{\omega}}(\lambda)/\rho_\omega(\lambda)=b_\omega(\lambda)=\frac{\lambda-\omega}{\lambda-\overline{\omega}}=1+\frac{\overline{\omega}-\omega}{\lambda-\overline{\omega}}=1-\frac{2\pi i(\overline{\omega}-\omega)}{\rho_\omega(\lambda)},$$

and

$$\Pi_+b_\omega\Pi_-f_\varepsilon^*\rho_{\overline{\omega}}^{-1}\mathfrak{b}_-^\circ=\rho_\omega^{-1}c$$

with $c\in\mathbb{C}_2^{q\times q}$, it follows that

$$\begin{aligned}\Gamma^*\Gamma\rho_\omega^{-1}\mathfrak{a}_+^\circ &= \Gamma^*\rho_\omega^{-1}\mathfrak{b}_-^\circ=\Pi_+b_\omega f_\varepsilon^*\rho_{\overline{\omega}}^{-1}\mathfrak{b}_-^\circ\\ &= \Pi_+b_\omega\Gamma^*\rho_{\overline{\omega}}^{-1}\mathfrak{b}_-^\circ+\Pi_+b_\omega\Pi_-f_\varepsilon^*\rho_{\overline{\omega}}^{-1}\mathfrak{b}_-^\circ\\ &= b_\omega\Gamma^*\rho_{\overline{\omega}}^{-1}\mathfrak{b}_-^\circ+\rho_\omega^{-1}c\\ &= b_\omega\frac{\mathfrak{a}_+^\circ-\mathfrak{a}_+(\omega)}{\rho_{\overline{\omega}}}+\frac{c}{\rho_\omega}.\end{aligned}$$

Therefore,

$$(I - \Gamma^*\Gamma)\rho_\omega^{-1}\mathfrak{a}_+^\circ = \rho_\omega^{-1}c_1 \tag{7.119}$$

for some matrix $c_1 \in \mathbb{C}^{q\times q}$. In much the same way, it follows that

$$\begin{aligned} \Gamma\Gamma^*\rho_\omega^{-1}\mathfrak{a}_-^\circ &= \Gamma\rho_{\overline{\omega}}^{-1}\mathfrak{b}_+^\circ = \Pi_- b_\omega^{-1} f_\varepsilon \rho_\omega^{-1}\mathfrak{b}_+^\circ \\ &= \Pi_- b_\omega^{-1}\Gamma\rho_\omega^{-1}\mathfrak{b}_+^\circ + \rho_{\overline{\omega}}^{-1}d \\ &= b_\omega^{-1}\Gamma^*\rho_\omega^{-1}\mathfrak{b}_+^\circ + \rho_{\overline{\omega}}^{-1}d \end{aligned}$$

for some $d \in \mathbb{C}p \times p_2$, and hence that

$$\Gamma\Gamma^*\rho_{\overline{\omega}}^{-1}\mathfrak{a}_-^\circ = \rho_{\overline{\omega}}^{-1}\{\mathfrak{a}_-^\circ(\lambda) - \mathfrak{a}_-^\circ(\overline{\omega})\} + \rho_{\overline{\omega}}^{-1}h_1(\omega)$$

and

$$(I - \Gamma\Gamma^*)\rho_{\overline{\omega}}^{-1}\mathfrak{a}_-^\circ = \rho_{\overline{\omega}}^{-1}c_2 \tag{7.120}$$

for some matrix $c_2 \in \mathbb{C}^{p\times p}$.

Next, since $\mathfrak{A}_\omega^\circ$ is j_{pq}-unitary a.e. on $\mathbb{R}$, the auxiliary evaluation

$$\begin{aligned} \rho_\omega(\omega)^{-1}I_q &= \int_{-\infty}^{\infty} \frac{\mathfrak{a}_+^\circ(\mu)^*\mathfrak{a}_+^\circ(\mu) - \mathfrak{b}_-^\circ(\mu)^*\mathfrak{b}_-^\circ(\mu)}{|\rho_\omega(\mu)|^2} d\mu \\ &= \langle (I - \Gamma^*\Gamma)^{-1}\rho_\omega^{-1}c_1,\, (I - \Gamma^*\Gamma)^{-1}\rho_\omega^{-1}c_1\rangle_{st} \\ &\quad -\langle \Gamma(I - \Gamma^*\Gamma)^{-1}\rho_\omega^{-1}c_1,\, \Gamma(I - \Gamma^*\Gamma)^{-1}\rho_\omega^{-1}c_1\rangle_{st} \\ &= \langle (I - \Gamma^*\Gamma)^{-1}\rho_\omega^{-1}c_1,\, (I - \Gamma^*\Gamma)(I - \Gamma^*\Gamma)^{-1}\rho_\omega^{-1}c_1\rangle_{st} \\ &= \langle \rho_\omega^{-1}a_+^\circ,\, \rho_\omega^{-1}c_1\rangle_{st} = \rho_\omega(\omega)^{-1}c_1^* a_+^\circ(\omega) \end{aligned}$$

implies that $c_1 = (\mathfrak{a}_+^\circ(\omega)^{-1}$ is positive definite and that

$$c_1 = \{\rho_\omega(\omega)\langle (I - \Gamma^*\Gamma)^{-1}\rho_\omega^{-1}I_q,\, \rho_\omega^{-1}I_q\rangle_{st}\}^{\frac{1}{2}}. \tag{7.121}$$

Similarly,

$$c_2 = \{-\rho_{\overline{\omega}}(\overline{\omega})\langle (I - \Gamma\Gamma^*)^{-1}\rho_{\overline{\omega}}^{-1}I_p,\, \rho_{\overline{\omega}}^{-1}I_p\rangle_{st}\}^{-\frac{1}{2}}. \tag{7.122}$$

The asserted formulas now follow easily from formulas (7.115)–(7.122). □

Remark 7.46 *If the $NP(\Gamma)$ is completely indeterminate, then, in view of Remark 7.9, formulas (7.109)–(7.112) may be written as (even if $\|\Gamma\| = 1$)*

$$\mathfrak{a}_+^\circ(\lambda) = \rho_\omega(\lambda)\{\langle (I - \Gamma^*\Gamma)^{-1/2}\rho_\omega^{-1}I_q,\, (I - \Gamma^*\Gamma)^{-1/2}\rho_\lambda^{-1}I_q\rangle_{st}\alpha_\omega \tag{7.123}$$

and

$$\mathfrak{b}_+^\circ(\lambda) = \rho_{\overline{\omega}}(\lambda)\langle \Gamma^*(I - \Gamma\Gamma^*)^{-1/2}\rho_{\overline{\omega}}^{-1}I_p, (I - \Gamma^*\Gamma)^{-1/2}\rho_\lambda^{-1}I_q\rangle_{st}\beta_\omega \tag{7.124}$$

for $\lambda \in \mathbb{C}_+$; *and*

$$\mathfrak{b}_-^\circ(\lambda) = \rho_\omega(\lambda)\langle \Gamma(I - \Gamma^*\Gamma)^{-1/2}\rho_\omega^{-1}I_q, (I - \Gamma\Gamma^*)^{-1/2}\rho_\lambda^{-1}I_q\rangle_{st}\alpha_\omega \tag{7.125}$$

and

$$\mathfrak{a}_-^\circ(\lambda) = \rho_{\overline{\omega}}(\lambda)\langle (I - \Gamma\Gamma^*)^{-1/2}\rho_{\overline{\omega}}^{-1}I_p, (I - \Gamma\Gamma^*)^{-1/2}\rho_\lambda^{-1}I_p\rangle\beta_\omega, \tag{7.126}$$

for $\lambda \in \mathbb{C}_-$.

7.6 The generalized Schur interpolation problem

In this section we shall study the following interpolation problem for mvf's in the Schur class $\mathcal{S}^{p\times q}$, which we shall refer to as the GSIP, an acronym for the **generalized Schur interpolation problem**:

GSIP$(b_1, b_2; s^\circ)$: *Given mvf's* $b_1 \in \mathcal{S}_{in}^{p\times p}$, $b_2 \in \mathcal{S}_{in}^{q\times q}$ *and* $s^\circ \in \mathcal{S}^{p\times q}$, *describe the set*

$$\mathcal{S}(b_1, b_2; s^\circ) = \{s \in \mathcal{S}^{p\times q} : (b_1)^{-1}(s - s^\circ)(b_2)^{-1} \in H_\infty^{p\times q}\}. \tag{7.127}$$

The mvf's $s(\lambda)$ in this set are called solutions of this problem. The classical Schur and Nevanlinna-Pick interpolation problems for mvf's in the Schur class $\mathcal{S}^{p\times q}$ and their tangential and bitangential generalizations are all special cases of the GSIP, if there exists at least one solution s° to the problem. Moreover, every GSIP may be easily reduced to an NP.

The GSIP$(b_1, b_2; s^\circ)$ is said to be

(1) **determinate** if it has exactly one solution.

(2) **indeterminate** if it has more than one solution.

(3) **completely indeterminate** if for every nonzero vector $\eta \in \mathbb{C}^q$ there exist at least two solutions $s_1 \in \mathcal{S}(b_1, b_2; s^\circ)$, $s_2 \in \mathcal{S}(b_1, b_2; s^\circ)$ such that $\|(s_1 - s_2)\eta\|_\infty > 0$.

(4) **strictly completely indeterminate** if there exists at least one solution $s \in \mathcal{S}(b_1, b_2; s^\circ)$ such that $\|s\|_\infty < 1$.

It is readily seen that there is a simple connection between the set $\mathcal{S}(b_1, b_2; s^\circ)$ and the set $\mathcal{N}(\Gamma)$ based on the Hankel operator

$$\Gamma = \Gamma(f^\circ), \quad \text{where} \quad f^\circ(\mu) = b_1(\mu)^* s^\circ(\mu) b_2(\mu)^* \quad \text{a.e. on } \mathbb{R}, \tag{7.128}$$

that is given by the equivalence

$$s \in \mathcal{S}(b_1, b_2; s^\circ) \iff b_1^* s b_2^* \in \mathcal{N}(\Gamma). \tag{7.129}$$

Moreover, a GSIP$(b_1, b_2; s^\circ)$ is determinate, indeterminate, completely indeterminate, or strictly completely indeterminate if and only if the corresponding NP(Γ) belongs to the same class.

It is also easily checked that

$$s_1 \in \mathcal{S}(b_1, b_2; s^\circ) \iff \mathcal{S}(b_1, b_2; s^\circ) = \mathcal{S}(b_1, b_2; s_1),$$

and hence that the given data for the GSIP is really b_1, b_2 and the Hankel operator defined by formula (7.128).

Theorem 7.47 *Let $b_1 \in \mathcal{S}_{in}^{p\times p}$, $b_2 \in \mathcal{S}_{in}^{q\times q}$ and let Γ be a Hankel operator. Then there exists at least one mvf f° such that $b_1 f^\circ b_2 \in \mathcal{S}^{p\times q}$ and $\Gamma = \Gamma(f^\circ)$ if and only if*

$$\|\Gamma\| \le 1 \quad \text{and} \quad \Gamma M_{b_2} H_2^q \subseteq \mathcal{H}_*(b_1). \tag{7.130}$$

Moreover, if (7.130) is in force and $s^\circ(\mu) = b_1(\mu) f^\circ(\mu) b_2(\mu)$ a.e. on $\mathbb{R}$, then

$$\mathcal{S}(b_1, b_2; s^\circ) = \{s : b_1^* s b_2^* \in \mathcal{N}(\Gamma)\} = \{b_1 f b_2 : f \in \mathcal{N}(\Gamma)\}$$

and

$$\|\Gamma\| = \min\{\|s\|_\infty : s \in \mathcal{S}(b_1, b_2; s^\circ)\}. \tag{7.131}$$

Proof Let Γ be the Hankel operator corresponding to the GSIP $(b_1, b_2; s^\circ)$ that is defined by formula (7.128). Then $\|\Gamma\| \le 1$ and

$$\Gamma M_{b_2} H_2^q = \Pi_- M_{b_1}^* M_{s^\circ} H_2^q \subseteq \mathcal{H}_*(b_1),$$

i.e., (7.130) holds. Conversely, let $b_1 \in \mathcal{S}_{in}^{p\times p}$, $b_2 \in \mathcal{S}_{in}^{q\times q}$ and let Γ be a Hankel operator that meets the conditions in (7.130). Then $\mathcal{N}(\Gamma) \neq \emptyset$. Let $f^\circ \in \mathcal{N}(\Gamma)$. Then

$$\Gamma M_{b_2} H_2^q = \Pi_- M_{f^\circ} M_{b_2} H_2^q \subseteq \mathcal{H}_*(b_1).$$

Consequently,

$$M_{b_1} M_{f^\circ} M_{b_2} H_2^q \subseteq H_2^p.$$

Let $s^\circ(\mu) = b_1(\mu)f^\circ(\mu)b_2(\mu)$ a.e. on $\mathbb{R}$. Then $s^\circ \in L_\infty^{p\times q}$, $\|s^\circ\|_\infty \le 1$ and $M_{s^\circ}H_2^q \subseteq H_2^p$ and hence, $s^\circ \in \mathcal{S}^{p\times q}$, by Lemma 3.47. The equivalence (7.129) between the GSIP$(b_1, b_2; s^\circ)$ and $\mathcal{N}(\Gamma)$ for this choice of s° now drops out easily: if

$$f(\mu) = b_1(\mu)^* s(\mu) b_2(\mu)^* \quad \text{a.e. on } \mathbb{R},$$

then

$$\begin{aligned} f \in \mathcal{N}(\Gamma) &\Longleftrightarrow f - f^\circ \in H_\infty^{p\times q} \quad \text{and } \|f\|_\infty \le 1 \\ &\Longleftrightarrow b_1^{-1}(s - s^\circ)b_2^{-1} \in H_\infty^{p\times q} \quad \text{and } s \in \mathcal{S}^{p\times q} \\ &\Longleftrightarrow s \in \mathcal{S}(b_1,\, b_2;\, s^\circ). \end{aligned}$$

Formula 7.131 follows from Theorem 7.2. □

Connections between the classes $\mathcal{U}(j_{pq})$, $\mathcal{U}_{rR}(j_{pq})$ and $\mathcal{U}_{rsR}(j_{pq})$ and the GSIP that are analogous to the connections between the classes $\mathfrak{M}_r(j_{pq})$, $\mathfrak{M}_{rR}(j_{pq})$, $\mathfrak{M}_{rsR}(j_{pq})$ and the NP also exist.

Theorem 7.48 *Let $b_1 \in \mathcal{S}_{in}^{p\times p}$, $b_2 \in \mathcal{S}_{in}^{q\times q}$ and $s^\circ \in \mathcal{S}^{p\times q}$ be a given set of mvf's such that the GSIP$(b_1, b_2; s^\circ)$ is completely indeterminate. Then there exists an essentially unique (i.e., up to a right j_{pq}-unitary constant factor) mvf $\mathring{W} \in \mathcal{U}(j_{pq})$ for which both of the following two conditions hold:*

(1) $T_{\mathring{W}}[\mathcal{S}^{p\times q}] = \mathcal{S}(b_1, b_2; s^\circ)$.

(2) $\{b_1, b_2\} \in ap(\mathring{W})$.

If $W \in \mathcal{P}^\circ(j_{pq})$, then $T_W[\mathcal{S}^{p\times q}] = \mathcal{S}(b_1, b_2; s^\circ)$, if and only if $W \in \mathcal{U}(j_{pq})$ and is of the form

$$W(\lambda) = \frac{\beta_2(\lambda)}{\beta_1(\lambda)}\mathring{W}(\lambda)V, \tag{7.132}$$

where the $\beta_k(\lambda)$ are scalar inner divisors of the $b_k(\lambda)$ for $k = 1, 2$, and V is a constant j_{pq}-unitary matrix. These mvf's $W(\lambda)$ automatically belong to the class $\mathcal{U}_{rR}(j_{pq})$. Moreover,

$$\begin{aligned} &W \in \mathcal{U}_{rsR}(j_{pq}) \\ &\qquad \Longleftrightarrow \text{the GSIP}(b_1, b_2; s^\circ) \textit{ is strictly completely indeterminate.} \end{aligned}$$

Proof The NP(Γ) that corresponds to a completely indeterminate GSIP$(b_1, b_2; s^\circ)$ via (7.128) and (7.129) is also completely indeterminate and

$$\mathcal{N}(\Gamma) = T_{\mathfrak{A}}[\mathcal{S}^{p\times q}]$$

for some mvf $\mathfrak{A} \in \mathfrak{M}_{rR}(j_{pq})$. Thus, if

$$\mathring{W} = \begin{bmatrix} b_1 & 0 \\ 0 & b_2^{-1} \end{bmatrix} \mathfrak{A},$$

then

$$\mathcal{S}(b_1, b_2; s^\circ) = T_{\mathring{W}}\left[\mathcal{S}^{p\times q}\right].$$

Consequently, if $f_{12} = T_{\mathfrak{A}}[0]$, then, by Theorem 7.47, the mvf

$$b_1 f_{12} b_2 = T_{\mathring{W}}[0]$$

belongs to $\mathcal{S}^{p\times q}$. Therefore, $f_{12} \in \Pi^{p\times q}$, $\{b_1, b_2\} \in \operatorname{den} f_{12}$, $\mathfrak{A} \in \Pi^{m\times m}$ by Lemma 7.25, and, by Theorem 7.26,

$$\mathring{W} \in \mathcal{U}(j_{pq}) \quad \text{and} \quad \{b_1, b_2\} \in ap(\mathring{W}).$$

Moreover, $\mathring{W} \in \mathcal{U}_{rR}(j_{pq})$, thanks to Theorem 7.27, since $\mathfrak{A} \in \mathfrak{M}_{rR}(j_{pq})$; and the assertions related to (7.132) follow from Theorem 4.89. □

A mvf $W \in \mathcal{P}^\circ(j_{pq})$ such that $\mathcal{S}(b_1, b_2; s^\circ) = T_W[\mathcal{S}^{p\times q}]$ is called a **resolvent matrix** of the GSIP$(b_1, b_2; s^\circ)$.

Remark 7.49 *In view of Theorems 7.48 and 4.89, every completely indeterminate GSIP$(b_1, b_2; s^\circ\}$ has exactly one resolvent matrix $\mathring{W}$ that is normalized at a point $\omega \in \mathfrak{h}^+_{b_2^{-1}}$ by the conditions*

$$\mathring{w}_{11}(\omega) > 0, \quad \mathring{w}_{21}(\omega) = 0 \quad \textit{and} \quad \mathring{w}_{22}(\omega) > 0 \tag{7.133}$$

$$\left(\textit{or by the condition} \quad \mathring{W}(\omega) = I_m \ \textit{if}\ \omega \in \mathfrak{h}_{b_1} \cap \mathfrak{h}_{b_2^{-1}} \cap \mathbb{R}\right)$$

such that $\{b_1, b_2\} \in ap(\mathring{W})$. Moreover, every resolvent matrix W of this problem belongs automatically to the class $\mathcal{U}_{rR}(j_{pq})$, and the set of all resolvent matrices W of this problem is described by formula (7.132). A formula for the resolvent matrices W of this problem with $\{b_1, b_2\} \in ap(W)$ will be presented in Section 7.8.

Theorem 7.50 *Let $W \in \mathcal{U}(j_{pq})$, let $\{b_1, b_2\} \in ap(W)$ and let $s^\circ \in T_W[\mathcal{S}^{p\times q}]$. Then the GSIP$(b_1, b_2; s^\circ)$ is completely indeterminate and*

$$T_W[\mathcal{S}^{p\times q}] \subseteq \mathcal{S}(b_1, b_2; s^\circ). \tag{7.134}$$

Moreover, equality prevails in (7.134) if and only if $W \in \mathcal{U}_{rR}(j_{pq})$.

Proof Let $W \in \mathcal{U}(j_{pq})$, $\{b_1, b_2\} \in ap(W)$ and $s^\circ \in T_W[\mathcal{S}^{p\times q}]$, let the mvf $\mathfrak{A}$ be defined by formula (7.57) and let $f^\circ = b_1^{-1}s^\circ b_2^{-1}$ and $\Gamma = \Gamma(f^\circ)$. Then $\{b_1, b_2\} \in \text{den}\, f^\circ$ and $\mathfrak{A} \in \Pi \cap \mathfrak{M}_r(j_{pq})$ by Theorem 7.26. Moreover, by Theorem 7.22, $\mathcal{N}(\Gamma) \supseteq T_{\mathfrak{A}}[\mathcal{S}^{p\times q}]$, $NP(\Gamma)$ is completely indeterminate and

$$\mathcal{N}(\Gamma) = T_{\mathfrak{A}}[\mathcal{S}^{p\times q}] \iff \mathfrak{A} \in \mathfrak{M}_{rR}(j_{pq}),$$

the GSIP$(b_1, b_2; s^\circ)$ is completely indeterminate and

$$T_W[\mathcal{S}^{p\times q}] = b_1 T_{\mathfrak{A}}[\mathcal{S}^{p\times q}]b_2 \subseteq b_1\mathcal{N}(\Gamma)b_2 = \mathcal{S}(b_1, b_2; s^\circ).$$

Moreover, formula (7.134) holds with equality if and only if $W \in \mathcal{U}_{rR}(j_{pq})$. □

Theorem 7.51 *Let $W \in \mathcal{U}(j_{pq})$, let $\{b_1, b_2\} \in ap(W)$ and let $b_1^{(1)} \in \mathcal{S}_{in}^{p\times p}$ and $b_2^{(1)} \in \mathcal{S}_{in}^{q\times q}$ be any pair of matrix valued inner functions such that $(b_1^{(1)})^{-1}b_1 \in \mathcal{S}_{in}^{p\times p}$ and $b_2(b_2^{(1)})^{-1} \in \mathcal{S}_{in}^{q\times q}$. Then there exists an essentially unique mvf $W_1 \in \mathcal{U}_{rR}(j_{pq})$ such that*

(1) *$\{b_1^{(1)}, b_2^{(1)}\} \in ap(W_1)$ and*

(2) *$W_1^{-1}W \in \mathcal{U}(j_{pq})$.*

Moreover, $W_1^{-1}W \in \mathcal{U}_S(j_{pq})$ if and only if $\{b_1^{(1)}, b_2^{(1)}\} \in ap(W)$.

Proof Under the stated conditions, let $s^\circ \in T_W[\mathcal{S}^{p\times q}]$. Then, by Theorem 7.50, the GSIP$(b_1, b_2; s^\circ)$ is completely indeterminate and

$$T_W[\mathcal{S}^{p\times q}] \subseteq \mathcal{S}(b_1, b_2; s^\circ). \tag{7.135}$$

Moreover, since

$$\mathcal{S}(b_1, b_2; s^\circ) \subseteq \mathcal{S}(b_1^{(1)}, b_2^{(1)}; s^\circ), \tag{7.136}$$

the GSIP$(b_1^{(1)}, b_2^{(1)}; s^\circ)$ is completely indeterminate too. By Theorem 7.48, there exists an essentially unique mvf $W_1 \in U_{rR}(j_{pq})$ such that

$$\mathcal{S}(b_1^{(1)}, b_2^{(1)}; s^\circ) = T_{W_1}[\mathcal{S}^{p\times q}] \text{ and } \{b_1^{(1)}, b_2^{(1)}\} \in ap(W_1). \tag{7.137}$$

Conditions (7.135)–(7.137) imply that

$$T_W[\mathcal{S}^{p\times q}] \subseteq T_{W_1}[\mathcal{S}^{p\times q}]$$

and hence, Theorem 4.94 implies that $W_1^{-1}W \in \mathcal{U}(j_{pq})$ and $W_1^{-1}W \in \mathcal{U}_S(j_{pq})$ if and only if $\{b_1^{(1)}, b_2^{(1)}\} \in ap(W)$. □

Theorem 7.52 *Every mvf* $W \in \mathcal{U}(j_{pq})$ *admits a factorization*

$$W = W_1 W_2, \quad \text{with} \quad W_1 \in \mathcal{U}_{rR}(j_{pq}) \quad \text{and} \quad W_2 \in \mathcal{U}_S(j_{pq}) \tag{7.138}$$

that is unique up to constant j_{pq}*-unitary multipliers*

$$W_1 \longrightarrow W_1 V \quad \text{and} \quad W_2 \longrightarrow V^{-1} W_2 \quad \text{with} \quad V \in \mathcal{U}_{const}(j_{pq}). \tag{7.139}$$

Proof Let $W \in \mathcal{U}(j_{pq})$, $s^\circ \in T_W[\mathcal{S}^{p\times q}]$ and $\{b_1, b_2\} \in ap(W)$. Then, by Theorems 7.48 and 7.50, the GSIP$(b_1, b_2; s^\circ)$ is completely indeterminate, $\mathcal{S}(b_1, b_2; s^\circ) = T_{W_1}[\mathcal{S}^{p\times q}]$ for some essentially unique mvf $W_1 \in \mathcal{U}_1(j_{pq})$ and $T_W[\mathcal{S}^{p\times q}] \subseteq T_{W_1}[\mathcal{S}^{p\times q}]$. Moreover, since $T_{W_1}[\mathcal{S}^{p\times q}] = \mathcal{S}(b_1, b_2; s^\circ)$ and $ap(W) = ap(W_1)$, Theorem 7.51 guarantees that the mvf $W_2 = W_1^{-1} W$ belongs to he class $\mathcal{U}_S(j_{pq})$, i.e., that a factorization of the form (7.138) exists and that it is unique, up to the transformations indicated in (7.139). □

7.7 Generalized Sarason problems in $\mathcal{S}^{p\times q}$ and $H_\infty^{p\times q}$

The GSIP$(b_1, b_2; s^\circ)$ is formulated in terms of the three mvf's b_1, b_2 and s°. However, the set $\mathcal{S}(b_1, b_2; s^\circ)$ of solutions to this problem depends only upon the mvf's $b_1 \in \mathcal{S}_{in}^{p\times p}$, $b_2 \in \mathcal{S}_{in}^{q\times q}$ and the three operators

$$\Gamma_{11} = P_{\mathcal{H}(b_1)} M_s|_{H_2^q}, \; \Gamma_{22} = \Pi_- M_s|_{\mathcal{H}_*(b_2)} \text{ and}$$
$$\Gamma_{12} = P_{\mathcal{H}(b_1)} M_s|_{\mathcal{H}_*(b_2)}, \quad \text{with } s = s^\circ. \tag{7.140}$$

In fact these operators are independent of the choice of $s \in \mathcal{S}(b_1, b_2; s^\circ)$. Moreover, $\mathcal{S}(b_1, b_2; s^\circ)$ coincides with the set of $s \in \mathcal{S}^{p\times q}$ for which (7.140) holds. This observation leads to the following **generalized Sarason problem** in the class $\mathcal{S}^{p\times q}$ in which the given data are $b_1 \in \mathcal{S}_{in}^{p\times p}$, $b_2 \in \mathcal{S}_{in}^{q\times q}$ and a block upper triangular operator

$$X = \begin{bmatrix} \Gamma_{11} & \Gamma_{12} \\ 0 & \Gamma_{22} \end{bmatrix} : \begin{array}{c} H_2^q \\ \oplus \\ \mathcal{H}_*(b_2) \end{array} \longrightarrow \begin{array}{c} \mathcal{H}(b_1) \\ \oplus \\ K_2^p \end{array}. \tag{7.141}$$

GSP$(b_1, b_2; X; \mathcal{S}^{p\times q})$: *Given* $b_1 \in \mathcal{S}_{in}^{p\times p}$, $b_2 \in \mathcal{S}_{in}^{q\times q}$ *and a block triangular operator* X *of the form (7.141), find a mvf* $s \in \mathcal{S}^{p\times q}$ *such that (7.140) holds.*

Let $\mathcal{S}(b_1, b_2; X)$ denote the set of solutions to the GSP$(b_1, b_2; X; \mathcal{S}^{p\times q})$.

If $\mathcal{S}(b_1, b_2; X) \neq \emptyset$ and $s^\circ \in \mathcal{S}(b_1, b_2; X)$, then

$$\mathcal{S}(b_1, b_2; X) = \mathcal{S}(b_1, b_2; s^\circ). \tag{7.142}$$

Moreover, there is also a one to one correspondence between $\mathcal{S}(b_1, b_2; X)$ and the set of solutions $\mathcal{N}(\Gamma)$ to the Nehari problem based on the Hankel operator

$$\Gamma = \Pi_- M_{f^\circ}|_{H_2^q} \quad \text{with} \quad f^\circ = b_1^* s^\circ b_2^*, \tag{7.143}$$

i.e.,

$$s \in \mathcal{S}(b_1, b_2; X) \Longleftrightarrow b_1^* s b_2^* \in \mathcal{N}(\Gamma). \tag{7.144}$$

There is a two sided connection between the operators X and the Hankel operator Γ considered in the equivalence (7.144) that may be written in in terms of the blocks Γ_{ij}, $1 \leq i \leq j \leq 2$, of X in (7.141) and the operator Γ:

$$\Gamma = M_{b_1^*}\{\Gamma_{11}\Pi_+ + (\Gamma_{22} + \Gamma_{12})P_{\mathcal{H}_*(b_2)}\}M_{b_2^*}|_{H_2^q}, \tag{7.145}$$

and, conversely,

$$\begin{aligned} \Gamma_{11} &= M_{b_1}\Gamma M_{b_2}|_{H_2^q}, \quad \Gamma_{22} = \Pi_- M_{b_1}\Gamma M_{b_2}|_{\mathcal{H}_*(b_2)}, \\ \Gamma_{12} &= P_{\mathcal{H}(b_1)} M_{b_1}\Gamma M_{b_2}|_{\mathcal{H}_*(b_2)}. \end{aligned} \tag{7.146}$$

By Theorem 7.2, $\mathcal{N}(\Gamma) \neq \emptyset$ if and only if Γ is a contractive Hankel operator from H_2^q into K_2^p, i.e., if and only if (7.3) holds and $\|\Gamma\| \leq 1$. This leads to the following criteria for the solvability of a GSP in the class $\mathcal{S}^{p\times q}$.

Theorem 7.53 *Let $b_1 \in \mathcal{S}_{in}^{p\times p}$, $b_2 \in \mathcal{S}_{in}^{q\times q}$ and let X be a bounded linear block triangular operator with blocks Γ_{ij}, $1 \leq i \leq j$, as in (7.141). Then $\mathcal{S}(b_1, b_2; X) \neq \emptyset$ if and only if the operator X is contractive and the operator Γ from H_2^q into K_2^p defined by formula (7.145) is a Hankel operator, i.e., it has property (7.3).*

Proof Since $\|\Gamma\| = \|X\|$ the theorem follows from Theorem 7.2 and the fact that

$$\mathcal{S}(b_1, b_2; X) \neq \emptyset \Longleftrightarrow \mathcal{N}(\Gamma) \neq \emptyset. \qquad \square$$

The generalized Sarason problem can also be formulated in $H_\infty^{p\times q}$, with a pair of mvf's $b_1 \in \mathcal{S}_{in}^{p\times p}$, $b_2 \in \mathcal{S}_{in}^{p\times p}$ and a bounded linear upper block triangular operator X of the form (7.141) as given data.

GSP$(b_1, b_2; X; H_\infty^{p\times q})$: *Given* $b_1 \in \mathcal{S}_{in}^{p\times p}$, $b_2 \in \mathcal{S}_{in}^{q\times q}$ *and a block triangular operator* X *of the form (7.141), find a mvf* $s \in H_\infty^{p\times q}$ *such that the relations (7.140) hold.*

Let $H_\infty(b_1, b_2; X)$ denote the set of solutions to the GSP$(b_1, b_2; X; H_\infty^{p\times q})$.

Theorem 7.54 *A GSP*$(b_1, b_2; X; H_\infty^{p\times q})$ *is solvable if and only if the operator* Γ, *defined by the formula (7.145) is a Hankel operator. Moreover, if the problem is solvable, then*

$$\|X\| = \min\{\|s\|_\infty : s \in H_\infty(b_1, b_2; X)\}. \tag{7.147}$$

Proof If $X = 0$ then $\Gamma = 0$ and $H_\infty(b_1, b_2; X) = H_\infty^{p\times q}$. Suppose therefore that $X \neq 0$ and Γ is defined by formula (7.145). If $H_\infty(b_1, b_2; X) \neq \emptyset$ and $s \in H_\infty(b_1, b_2; X)$, i.e., if $s \in H_\infty^{p\times q}$ and the relations (7.140) hold, then formulas (7.140) and (7.145) imply that Γ is a Hankel operator and $\|s\|_\infty \geq \|X\|$.

To establish the converse, let Γ be a Hankel operator and let $\Gamma_1 = \|X\|^{-1}\Gamma$ and $X_1 = \|X\|^{-1}X$. then Γ_1 is a Hankel operator with $\|X_1\| = 1$. Therefore, by Theorem 7.53, $\mathcal{S}(b_1, b_2; X_1) \neq \emptyset$. Let $s_1 \in \mathcal{S}(b_1, b_2; X_1)$ and let $s = \|X\|s_1$. Then $s \in H_\infty(b_1, b_2; X)$ and

$$\|s\|_\infty = \|X\| \, \|s_1\|_\infty \leq \|X\|.$$

Therefore, $\|s\|_\infty = \|X\|$, since $\|s\|_\infty \geq \|X\|$ when $s \in H_\infty(b_1, b_2; X)$. □

Corollary 7.55 *The GSP*$(b_1, b_2; X; \mathcal{S}^{p\times q})$ *is strictly completely indeterminate if and only if the operator* Γ *defined by formula (7.145) is a Hankel operator and* $\|X\| < 1$.

Remark 7.56 *Theorems 7.53 and 7.54 are equivalent, i.e., Theorem 7.53 follows from Theorem 7.54 and conversely. Moreover,*

$$\begin{aligned} \mathcal{S}(b_1, b_2; X) &= \{s \in H_\infty(b_1, b_2; X) : \|s\|_\infty \leq 1\} \\ &\subseteq H_\infty(b_1, b_2; X) \end{aligned}$$

and, if $X \neq 0$, *then*

$$\mathcal{S}(b_1, b_2; \|X\|^{-1}X) \neq \emptyset \Longleftrightarrow H_\infty(b_1, b_2; X) \neq \emptyset.$$

A solvable GSP in the class $\mathcal{S}^{p\times q}$ is determinate, indeterminate, completely indeterminate or strictly completely indeterminate, if and only if the corresponding GSIP is such, and results that are obtained for one problem are immediately transferrable to the other. In particular, Theorem 7.48 may be reformulated as follows:

Theorem 7.57 *If the* $\text{GSP}(b_1, b_2; X; \mathcal{S}^{p\times q})$ *is completely indeterminate, then there exists a mvf* W *such that*

$$\mathcal{S}(b_1, b_2; X) = T_W[\mathcal{S}^{p\times q}], \tag{7.148}$$

where $W \in \mathcal{U}(j_{pq})$ *and* $\{b_1, b_2\} \in ap(W)$. *Moreover,* $W \in \mathcal{U}_{rR}(j_{pq})$ *and is uniquely defined by the given data* b_1, b_2, X *up to a constant* j_{pq}*-unitary factor on the right, which may be uniquely specified by imposing the normalization conditions*

$$w_{11}(\omega) > 0, \quad , w_{22}(\omega) > 0 \quad \text{and} \quad w_{21}(\omega) = 0 \tag{7.149}$$

at a point $\omega \in \mathfrak{h}_{b_1} \cap \mathfrak{h}_{b_2^\#} \cap \overline{\mathbb{C}_+}$.
If $\omega \in \mathfrak{h}_{b_1} \cap \mathfrak{h}_{b_2^\#} \cap \mathbb{R}$, *then (7.149) holds if and only if* $W(\omega) = I_m$.

Proof In view of Theorem 4.54, $\mathfrak{h}_{b_1} \cap \mathfrak{h}_{b_2^\#} = \mathfrak{h}_W$, since $W \in \mathcal{U}_{rR}(j_{pq})$; and (7.149) holds if and only if $\det W(\omega) \neq 0$, i.e., if and only if $\omega \in \mathfrak{h}_{W^{-1}}$. If $\omega \in \mathfrak{h}_{b_1} \cap \mathfrak{h}_{b_2^\#} \cap \mathfrak{h}_{W^{-1}}$, then a mvf $W \in \mathcal{U}_{const}(j_{pq})$ that meets the normalization conditions (7.149) may be obtained by Lemma 2.17. □

7.8 Resolvent matrices for the GSP in the Schur class

An $m \times m$ mvf $W(\lambda)$ that is meromorphic in $\mathbb{C}_+$ with $\det W(\lambda) \not\equiv 0$ is said to be a **resolvent matrix** for the $\text{GSP}(b_1, b_2; X; \mathcal{S}^{p\times q})$ if $\mathcal{S}(b_1, b_2; X) = T_W[\mathcal{S}^{p\times q}]$. In view of the identification $\mathcal{S}(b_1, b_2; s^\circ) = \mathcal{S}(b_1, b_2; X)$ when $s^\circ \in \mathcal{S}(b_1, b_2; X)$, Remark 7.49 is applicable to the resolvent matrices of a completely indeterminate $\text{GSP}(b_1, b_2; X)$. If $\|X\| < 1$, and only in this case, the $\text{GSP}(b_1, b_2; X)$ is strictly completely indeterminate, its resolvent matrices $W \in \mathcal{U}_{rsR}(j_{pq})$, and a resolvent matrix W such that $\{b_1, b_2\} \in ap(W)$ may be obtained in terms of the given data by formulas (5.165), (5.166) and (5.177)–(5.179). If $0 \in \mathfrak{h}_{b_1} \cap \mathfrak{h}_{b_2}$ then there exists exactly one normalized resolvent matrix W with $0 \in \mathfrak{h}_W$ and $W(0) = I_m$ such that $\{b_1, b_2\} \in ap(W)$; it is given by formulas (5.180)–(5.183). In these formulas, the mvf's $\widehat{b}_1 \in \mathcal{S}_{in}^{p\times p}$ and $\widehat{b}_2 \in \mathcal{S}_{in}^{q\times q}$ may be replaced by any mvf's $b_5 \in \mathcal{S}^{p\times p}$ and $b_6 \in \mathcal{S}^{q\times q}$ such that $(\widehat{b}_1)^{-1}b_5 \in \mathcal{S}_{in}^{p\times p}$ and $b_6(\widehat{b}_2)^{-1} \in \mathcal{S}_{in}^{q\times q}$ and $\mathfrak{h}_{b_5} \cap \mathfrak{h}_{b_6^\#} \supseteq \mathfrak{h}_{b_1} \cap \mathfrak{h}_{b_2^\#}$. In particular, we may choose $b_5 = \det b_1 I_p$ and $b_6 = \det b_2 I_q$.

The next step is to derive formulas for this resolvent matrix when the problem is strictly completely indeterminate, i.e., in view of Theorem 7.54, when $\|X\| < 1$ and the operator Γ defined by formula (7.145) is a Hankel operator. In this case $W \in \mathcal{U}_{rsR}(j_{pq})$ and consequently formulas (5.177)–(5.179) for W are applicable with

$$L_W = L_X = \begin{bmatrix} I & \Gamma_{22} \\ \Gamma_{11}^* & I \end{bmatrix} : \mathcal{H}(b_1, b_2) \to \mathcal{H}(W) \tag{7.150}$$

and

$$\Delta_W = \Delta_X = \begin{bmatrix} I - \Gamma_{11}\Gamma_{11}^* & -\Gamma_{12} \\ -\Gamma_{12}^* & I - \Gamma_{22}^*\Gamma_{22} \end{bmatrix} : \mathcal{H}(b_1, b_2) \to \mathcal{H}(b_1, b_2), \tag{7.151}$$

where Γ_{ij} is now taken from X. The value $W(\omega)$ of the mvf $W(\lambda)$ normalized by the condition (7.149) at a point $\omega \in \mathfrak{h}_{b_1}^+ \cap \mathfrak{h}_{b_2^\#}^+$ is obtained from formulas (5.94)–(5.97).

Lemma 7.58 *If X is of the form (7.141) and Δ_X is defined by formula (7.151), then $\|X\| \le 1$ if and only if $\Delta_X \ge 0$ and*

$$\|X\| < 1 \iff \Delta_X \ge \delta I \text{ for some } \delta > 0. \tag{7.152}$$

Let $b \in \mathcal{S}_{in}^{p\times p}$,

$$V(t) = M_{e_t}|_{H_2^q} \quad \text{and} \quad V_1(t) = P_{\mathcal{H}(b)} M_{e_t}|_{\mathcal{H}(b)} \tag{7.153}$$

be two semigroups of contractive operators in the spaces H_2^q and $\mathcal{H}(b)$, respectively. Then Theorem 7.2 yields the following results:

Theorem 7.59 *$H_\infty(b, I_q; X) \neq \emptyset$ if and only if*

$$V_1(t)X = XV(t) \quad \text{holds for every} \quad t \ge 0. \tag{7.154}$$

Moreover, if this condition is in force, then

$$\min\{\|h\|_\infty : h \in H_\infty(b, I_q; X)\} = \|X\|, \tag{7.155}$$

i.e., $\|h\|_\infty \ge \|X\|$ and there exists at least one $h \in H_\infty(b, I_q; X)$ with $\|h\|_\infty = \|X\|$.

Proof This is immediate from Theorems 7.2 and 7.53, since $\mathcal{H}_*(b_2) = \{0\}$ when $b_2 = I_q$. $\square$

Corollary 7.60 *$\mathcal{S}(b, I_q; X) \neq \emptyset$ if and only if (7.154) is in force for every $t \geq 0$ and $\|X\| \leq 1$.*

Remark 7.61 *If $\mathcal{S}(b, I_q; X) \neq \emptyset$ and $s^\circ \in \mathcal{S}(b, I_q; X)$, then*

$$\mathcal{S}(b, I_q; X) = \mathcal{S}(b, I_q; s^\circ) = \{s \in \mathcal{S}^{p\times q} : b^{-1}(s - s^\circ) \in H_\infty^{p\times q}\},$$

i.e., the GSP($b, I_q; X; \mathcal{S}^{p\times q}$) is a one sided GSIP, with conditions imposed on the left. Consequently, all the results that were obtained earlier for the GSIP with $b_1 = b$ and $b_2 = I_q$ are applicable to a solvable GSP($b, I_q; X; \mathcal{S}^{p\times q}$). A number of them will be formulated below without proof.

The connection

$$X = M_b\Gamma, \quad \text{where } \Gamma = \Gamma(f) = \Pi_- M_f|_{H_2^q}, \ f = b^*h \quad \text{and} \quad h \in H_\infty^{p\times q}, \quad (7.156)$$

between the operator X and the Hankel operator Γ implies that

$$(I - \Gamma^*\Gamma)^{1/2} = (I - X^*X)^{1/2} \quad (7.157)$$

and

$$(I - \Gamma\Gamma^*)^{1/2} = \{M_{b^*}(I - XX^*)^{1/2}M_b P_{\mathcal{H}_*(b)} + P_{b^*K_2^p}\}|_{K_2^p}. \quad (7.158)$$

Consequently, the subspaces $\mathfrak{A}_+(\omega)$ and $\mathfrak{A}_-(\omega)$ that were defined for $\omega \in \mathbb{C}_+$ in terms of the contractive Hankel operator Γ by formulas (7.9) and (7.10) may be reexpressed in terms of the operator X by the formulas

$$\mathfrak{A}_+(\omega) = \left\{\frac{\eta}{\rho_\omega} : \eta \in \mathbb{C}^q\right\} \cap (I - X^*X)^{1/2}H_2^q \quad (7.159)$$

and

$$\mathfrak{A}_-(\omega) = \left\{\frac{\xi}{\rho_{\overline{\omega}}} : \xi \in \mathbb{C}^p\right\} \cap (b^*K_2^p \oplus M_{b^*}(I - XX^*)^{1/2}\mathcal{H}(b)). \quad (7.160)$$

Moreover, setting

$$\mathfrak{A}'_-(\omega) = \{R_\omega b\eta : \eta \in \mathbb{C}^p\} \cap (I - XX^*)^{1/2}\mathcal{H}(b), \quad (7.161)$$

it is easy to check that $\dim \mathfrak{A}_-(\omega) = \dim \mathfrak{A}'_-(\omega)$.

In keeping with the definitions introduced above for the GSIP, the GSP$(b, I_q; X; \mathcal{S}^{p\times q})$ is said to be

(1) **determinate**, if there is only one solution to this problem;

(2) **completely indeterminate**, if for every $\eta \in \mathbb{C}^q$ there are at least two solutions $s_1, s_2 \in \mathcal{S}(b, I_q; X)$ such that $s_1(\lambda)\eta \not\equiv s_2(\lambda)\eta$;

(3) **strictly completely indeterminate**, if $\mathcal{S}(b, I_q; X) \cap \mathring{\mathcal{S}}^{p\times q} \neq \emptyset$.

Theorem 7.62 *Let $\mathcal{S}(b, I_q; X) \neq \emptyset$. Then:*

(1) *The GSP$(b, I_q; X; \mathcal{S}^{p\times q})$ is determinate if and only if $\mathfrak{A}_+(\omega) = \{0\}$ or $\mathfrak{A}'_-(\omega) = \{0\}$ for at least one (and in fact for every) point $\omega \in \mathbb{C}_+$.*

(2) *The GSP$(b, I_q; X; \mathcal{S}^{p\times q})$ is completely indeterminate if and only if $\dim \mathfrak{A}_+(\omega) = q$ and $\dim \mathfrak{A}'_-(\omega) = p$ for at least one (and in fact for every) point $\omega \in \mathbb{C}_+$.*

(3) *The GSP$(b, I_q; X; \mathcal{S}^{p\times q})$ is strictly completely indeterminate if and only if $\|X\| < 1$.*

Remark 7.63 *It is useful to note that*

$$\dim \mathfrak{A}_+(\omega) = q \Longleftrightarrow \left\{ \frac{\eta}{\rho_\omega} : \eta \in \mathbb{C}^q \right\} \subseteq \operatorname{range}(I - X^*X)^{1/2} \tag{7.162}$$

and

$$\dim \mathfrak{A}'_-(\omega) = p \Longleftrightarrow \{R_\omega b\xi : \xi \in \mathbb{C}^p\} \subseteq \operatorname{range}(I - XX^*)^{1/2}. \tag{7.163}$$

If (7.162) is in force, then $I - XX^* > 0$, whereas if (7.163) is in force, then $I - X^*X > 0$. Both of these conditions fail if 1 is a singular value of X.

Theorem 7.64 *If the GSP$(b, I_q; X; \mathcal{S}^{p\times q})$ is completely indeterminate, then there exists a unique mvf $W \in \mathcal{U}(j_{pq})$ with $\{b, I_q\} \in ap(W)$, up to an arbitrary constant j_{pq}-unitary multiplier on the right, such that*

$$\mathcal{S}(b, I_q; X) = T_W[\mathcal{S}^{p\times q}]; \tag{7.164}$$

and this mvf W automatically belongs to the class $\mathcal{U}_{rR}(j_{pq})$. Moreover, $W \in \mathcal{U}_{rsR}(j_{pq})$ if and only if the GSP$(b, I_q; X; \mathcal{S}^{p\times q})$ is strictly completely indeterminate.

Theorem 7.65 *Let $W \in \mathcal{U}(j_{pq})$, let $\{b, I_q\} \in ap(W)$, $s^\circ \in T_W[\mathcal{S}^{p\times q}]$ and $X = P_{\mathcal{H}(b)} M_{s^\circ}|_{H_2^q}$. Then:*

(1) *X does not depend upon the choice of s° in $T_W[\mathcal{S}^{p\times q}]$.*

(2) *The GSP$(b, I_q; X; \mathcal{S}^{p\times q})$ is completely indeterminate.*

(3) *$\mathcal{S}(b, I_q; X) \subseteq T_W[\mathcal{S}^{p\times q}]$, with equality if and only if $W \in \mathcal{U}_{rR}(j_{pq})$.*

Theorem 7.66 *Let $W \in \mathcal{U}(j_{pq})$, let $\{b, I_q\} \in ap(W)$, $s^\circ \in T_W[\mathcal{S}^{p\times q}]$ and $X = P_{\mathcal{H}(b)} M_{s^\circ}|_{H_2^q}$. Then:*

(1) $\mathcal{H}(W) \cap L_2^m = \left\{ \begin{bmatrix} g \\ X^* g \end{bmatrix} : g \in \mathcal{H}(b) \right\}$.

(2) $W \in \mathcal{U}_{rR}(j_{pq}) \Longleftrightarrow \left\{ \begin{bmatrix} g \\ X^* g \end{bmatrix} : g \in \mathcal{H}(b) \right\}$ *is dense in* $\mathcal{H}(W)$.

(3) $W \in \mathcal{U}_{rsR}(j_{pq}) \Longleftrightarrow \left\{ \begin{bmatrix} g \\ X^* g \end{bmatrix} : g \in \mathcal{H}(b) \right\} = \mathcal{H}(W)$. *Moreover, in this case*

$$\left\| \begin{bmatrix} g \\ X^* g \end{bmatrix} \right\|_{\mathcal{H}(W)} = \left\langle j_{pq} \begin{bmatrix} g \\ X^* g \end{bmatrix}, \begin{bmatrix} g \\ X^* g \end{bmatrix} \right\rangle_{st} = \langle (I - XX^*) g, g \rangle_{st}.$$

(4) *If $W \in \mathcal{U}_{rsR}(j_{pq})$, then it is uniquely specified by the normalization condition (5.91) and the formulas (5.93)–(5.97) with*

$$K_\omega(\lambda) = \left(\begin{bmatrix} I \\ X^* \end{bmatrix} (I - XX^*)^{-1} \begin{bmatrix} k_\omega^b & X k_\omega^{\widehat{b}} \end{bmatrix} \right)(\lambda),$$

where $\widehat{b} \in \mathcal{S}_{in}^{p\times p}$ is such that $\overline{X\mathcal{H}(b)} = \mathcal{H}(\widehat{b})$.

7.9 The generalized Carathéodory interpolation problem

In this section we shall study the following interpolation problem for mvf's in the Carathéodory class $\mathcal{C}^{p\times p}$, which we shall refer to as the GCIP, an acronym for **generalized Carathéodory interpolation problem**:

GCIP$(b_3, b_4; c^\circ)$: *Given mvf's $b_3 \in \mathcal{S}_{in}^{p\times p}$, $b_4 \in \mathcal{S}_{in}^{p\times p}$ and $c^\circ \in \mathcal{C}^{p\times p}$, describe the set*

$$\mathcal{C}(b_3, b_4; c^\circ) = \{c \in \mathcal{C}^{p\times p} : (b_3)^{-1}(c - c^\circ)(b_4)^{-1} \in \mathcal{N}_+^{p\times p}\}. \tag{7.165}$$

The mvf's $c(\lambda)$ in this set are called solutions of this problem. The classical Carathéodory and Nevanlinna-Pick interpolation problems for mvf's in the Carathéodory class $\mathcal{C}^{p\times p}$, and their tangential and bitangential generalizations, may be reformulated as GCIP's, if there exists at least one solution c° to the problem.

The difference between the formulation of the GSIP and the GCIP is due to the fact that although $\mathcal{S}^{p\times q} \subseteq H_\infty^{p\times q}$, there are mvf's $c \in \mathcal{C}^{p\times p}$ that do not belong to $H_\infty^{p\times p}$, for instance $c(\lambda) = -i\lambda I_p$. It turns out that the space $\mathcal{N}_+^{p\times p}$, which contains $\mathcal{C}^{p\times p}$, is a reasonable substitute for $H_\infty^{p\times p}$ in the formulation of the GCIP. Thus, for example, if $\omega_1, \ldots, \omega_n$ is a distinct set of points in $\mathbb{C}_+$, and if $s^\circ \in \mathcal{S}^{p\times q}$, $c^\circ \in \mathcal{C}^{p\times p}$ and $b(\lambda) = \prod_{j=1}^n (\lambda - \omega_j)/(\lambda - \overline{\omega_j})$, then:

$$\begin{aligned} \{s \in \mathcal{S}^{p\times q} : s(\omega_j) = s^\circ(\omega_j) \quad \text{for} \quad j = 1, \ldots, n\} \\ = \{s \in \mathcal{S}^{p\times q} : b^{-1}(s - s^\circ) \in H_\infty^{p\times q}\} \end{aligned}$$

and

$$\begin{aligned} \{c \in \mathcal{C}^{p\times p} : c(\omega_j) = c^\circ(\omega_j) \quad \text{for} \quad j = 1, \ldots, n\} \\ = \{c \in \mathcal{C}^{p\times p} : b^{-1}(c - c^\circ) \in \mathcal{N}_+^{p\times p}\}. \end{aligned}$$

Consequently, in this example, $s(\lambda)$ (resp., $c(\lambda)$) is a solution of the Nevanlinna-Pick problem in the class $\mathcal{S}^{p\times q}$ (resp., $\mathcal{C}^{p\times p}$) with prescribed values $s^\circ(\omega_1), \ldots, s^\circ(\omega_n)$ (respectively $c^\circ(\omega_1), \ldots, c^\circ(\omega_n)$) at the points $\omega_1, \ldots, \omega_n$.

Bitangential Nevanlinna-Pick problems in the class $\mathcal{C}^{p\times p}$ with a solution $c^\circ(\lambda)$ correspond to a GCIP$(b_3, b_4; c^\circ)$ in which the $p \times p$ inner mvf's $b_3(\lambda)$ and $b_4(\lambda)$ are BP products.

The GCIP$(b_3, b_4; c^\circ)$ is said to be

(1) **determinate** if it has exactly one solution;

(2) **indeterminate** if it has more than one solution;

(3) **completely indeterminate** if for every nonzero vector $\xi \in \mathbb{C}^p$ there exist at least two solutions $c_1 \in \mathcal{C}(b_3, b_4; c^\circ)$, $c_2 \in \mathcal{C}(b_3, b_4; c^\circ)$ such that $c_1(\lambda)\xi \not\equiv c_2(\lambda)\xi$ in $\mathbb{C}_+$;

(4) **strictly completely indeterminate** if there exists at least one solution $c \in \mathcal{C}(b_3, b_4; c^\circ)$ such that $c \in H_\infty^{p\times p}$ and $\Re c(\lambda) \geq \delta I_p$ for some $\delta > 0$,

or, in other words, if

$$\mathcal{C}(b_3, b_4; c^\circ) \cap \mathring{\mathcal{C}}^{p\times p} \neq \emptyset. \tag{7.166}$$

Remark 7.67 *The GCIP($b_3, b_4; c^\circ$) is completely indeterminate if at some point $\omega \in \mathbb{C}_+$ at which $b_3(\omega)$ and $b_4(\omega)$ are invertible,*

$$\{c(\omega)\xi - c^\circ(\omega)\xi : c \in \mathcal{C}(b_3, b_4; c^\circ)\} \neq \{0\} \tag{7.167}$$

for every nonzero vector $\xi \in \mathbb{C}^p$.

7.10 Connections between the GSIP and the GCIP

The next lemma serves to connect the GCIP with the GSIP.

Lemma 7.68 *Let $c^\circ \in \mathcal{C}^{p\times p}$ and let $s^\circ = T_{\mathfrak{V}}[c^\circ]$. Then the conditions*

$$\frac{1}{2}b_1^{-1}(I_p + s^\circ)b_3 \in \mathcal{S}_{out}^{p\times p} \quad \text{and} \quad \frac{1}{2}b_4(I_p + s^\circ)b_2^{-1} \in \mathcal{S}_{out}^{p\times p} \tag{7.168}$$

serve to define one of the pairs $\{b_1, b_2\}$ and $\{b_3, b_4\}$ of $p \times p$ inner mvf's in terms of the other, up to constant unitary multipliers. Moreover, for any two such pairs, the conditions

$$\frac{1}{2}b_1^{-1}(I_p + s)b_3 \in \mathcal{S}_{out}^{p\times p} \quad \text{and} \quad \frac{1}{2}b_4(I_p + s)b_2^{-1} \in \mathcal{S}_{out}^{p\times p} \tag{7.169}$$

are satisfied for every mvf $s \in \mathcal{S}(b_1, b_2; s^\circ) \cap \mathcal{D}(T_{\mathfrak{V}})$ and

$$\mathcal{C}(b_3, b_4; c^\circ) = T_{\mathfrak{V}}[\mathcal{S}(b_1, b_2; s^\circ) \cap \mathcal{D}(T_{\mathfrak{V}})]. \tag{7.170}$$

Proof The proof rests on the essential uniqueness of inner-outer factorizations and outer-inner factorizations of mvf's $f \in \mathcal{S}^{p\times p}$ and $f \in \mathcal{N}_+^{p\times p}$ with $\det f(\lambda) \not\equiv 0$ in $\mathbb{C}_+$. These results are applicable because $\frac{1}{2}(I_p + s^\circ) \in \mathcal{S}^{p\times p}$ and $\det(I_p + s^\circ) \not\equiv 0$ in $\mathbb{C}_+$ implies that $\frac{1}{2}(I_p + s^\circ) \in \mathcal{S}_{out}^{p\times p}$ and $2(I_p + s^\circ)^{-1} \in \mathcal{N}_{out}^{p\times p}$. Thus, b_3 and b_4 are essentially uniquely defined by s°, b_1 and b_2:

b_3 is a minimal right denominator of $b_1^{-1}(I_p + s^\circ)$ and
b_4 is a minimal left denominator of $(I_p + s^\circ)b_2^{-1}$.

Since $b_3^{-1}(I_p + s^\circ)^{-1}b_1 \in \mathcal{N}_{out}^{p\times p}$ and $b_2(I_p + s^\circ)^{-1}b_4^{-1} \in \mathcal{N}_{out}^{p\times p}$, b_1 and b_2 are essentially uniquely defined by b_3, b_4 and s°:

b_1 is a minimal right denominator of $b_3^{-1}(I_p + s^\circ)$ and
b_2 is a minimal left denominator of $(I_p + s^\circ)b_4^{-1}$.

Let $s \in S(b_1, b_2; s^\circ) \cap \mathcal{D}(T_{\mathfrak{V}})$. Then

$$b_1^{-1}(I_p + s)b_3 = b_1^{-1}(I_p + s^\circ)b_3 + b_1^{-1}(s - s^\circ)b_3.$$

Since $b_1^{-1}(I_p+s)b_3 \in H_\infty^{p\times p}$, b_3 is a minimal right denominator of $b_1^{-1}(I_p+s)$ if and only if b_3 is a minimal right denominator of $b_1^{-1}(I_p+s^\circ)$. Analogously, since

$$b_4(I_p + s)b_2^{-1} = b_4(I_p + s^\circ)b_2^{-1} + b_4(s - s^\circ)b_2^{-1},$$

b_4 is a minimal left denominator of $(I_p+s)b_2^{-1}$ if and only if b_4 is a minimal left denominator of $(I_p + s^\circ)b_2^{-1}$. Moreover, if $s = T_{\mathfrak{V}}[c]$ and $c \in \mathcal{C}^{p\times p}$, then

$$I_p + c = 2(I_p + s)^{-1}, \quad I_p + c^\circ = 2(I_p + s^\circ)^{-1}$$

and, consequently, if $s \in S(b_1, b_2; s^\circ) \cap \mathcal{D}(T_{\mathfrak{V}})$, then

$$\begin{aligned} 2b_3^{-1}(c - c^\circ)b_4^{-1} &= 4b_3^{-1}\{(I_p + s)^{-1} - (I_p + s^\circ)^{-1}\}b_4^{-1} \\ &= 4b_3^{-1}\{(I_p + s)^{-1}(s^\circ - s)(I_p + s^\circ)^{-1}\}b_4^{-1} \\ &= -\varphi_s^{-1}b_1^{-1}(s - s^\circ)b_2^{-1}\psi_{s^\circ}^{-1}, \end{aligned}$$

where $\varphi_s = b_1^{-1}(I_p+s)b_3/2$ and $\psi_s = b_4^{-1}(I_p+s)b_2/2$ are the mvf's in $S_{out}^{p\times q}$ that are considered in (7.169). Thus,

$$T_{\mathfrak{V}}[\mathcal{S}(b_1, b_2; s^\circ) \cap \mathcal{D}(T_{\mathfrak{V}})] \subseteq \mathcal{C}(b_3, b_4; c^\circ). \tag{7.171}$$

Conversely, (7.169) is equivalent to the fact that

$$\varphi_s = b_1^{-1}(I_p + c)^{-1}b_3 \in \mathcal{S}_{out}^{p\times p} \quad \text{and} \quad \psi_s = b_4(I_p + c)^{-1}b_2^{-1} \in \mathcal{S}_{out}^{p\times p}$$

if $c \in \mathcal{C}(b_3, b_4; c^\circ)$. Moreover, since

$$b_1^{-1}(s - s^\circ)b_2^{-1} = -2\varphi_s b_3^{-1}(c - c^\circ)b_4^{-1}\psi_{s^\circ},$$

$$T_{\mathfrak{V}}[\mathcal{C}(b_3, b_4; c^\circ) \subseteq \mathcal{S}(b_1, b_2; s^\circ) \cap \mathcal{D}(T_{\mathfrak{V}}). \tag{7.172}$$

The inclusions (7.171) and (7.172) yield the equality (7.170). □

Theorem 7.69 *Let $A \in \mathcal{U}(J_p)$, $\{b_3, b_4\} \in ap_{II}(A)$ and $c^\circ \in \mathcal{C}(A)$. Then:*

(1) *The GCIP$(b_3, b_4; c^\circ)$ is completely indeterminate.*

(2) $\mathcal{C}(A) \subseteq \mathcal{C}(b_3, b_4; c^\circ)$.

(3) *Equality prevails in (2) if and only if $A \in \mathcal{U}_{rR}(J_p)$.*

(4) $A \in \mathcal{U}_{rsR}$ *if and only if the GCIP*$(b_3, b_4; c^\circ)$ *is strictly completely indeterminate and equality prevails in (2).*

Proof This is a consequence of the preceding lemma, the relations between the classes $\mathcal{U}(J_p)$ and $\mathcal{U}(j_p)$, $\mathcal{U}_{rR}(J_p)$ and $\mathcal{U}_{rR}(j_p)$, and $\mathcal{U}_{rsR}(J_p)$ and $\mathcal{U}_{rsR}(j_p)$, respectively. □

Theorem 7.70 *Let the GCIP*$(b_3, b_4; c^\circ)$ *be completely indeterminate. Then there exists a mvf* $\mathring{A} \in \mathcal{U}(J_p)$ *such that:*

(1) $\mathcal{C}(b_3, b_4; c^\circ) = \mathcal{C}(\mathring{A})$.

(2) $\{b_3, b_4\} \in ap_{II}(\mathring{A})$.

Such a mvf $\mathring{A}(\lambda)$ *is defined up to a constant right* J_p*-unitary multiplier and is automatically right regular, i.e.,* $\mathring{A} \in \mathcal{U}_{rR}(J_p)$.

If $A \in \mathcal{P}(J_p)$, *then* $\mathcal{C}(b_3, b_4; c^\circ) = \mathcal{C}(A)$ *if and only if* $A \in \mathcal{U}_{rR}(J_p)$ *and is of the form*

$$A(\lambda) = \frac{\beta_4(\lambda)}{\beta_3(\lambda)}\mathring{A}(\lambda)V,$$

where $\beta_j(\lambda)$ *is a scalar inner divisor of* $b_j(\lambda)$, $j = 3, 4$ *and* $V \in \mathcal{U}_{const}(J_p)$.

The GCIP$(b_3, b_4; c^\circ)$ *is strictly completely indeterminate if and only if* $\mathring{A} \in \mathcal{U}_{rsR}(J_p)$.

Proof This is immediate from Lemma 7.68 and the analogous result for the GSIP. □

Analogues of Theorems 7.51 and 7.52 hold for mvf's $A \in \mathcal{U}(J_p)$.

Lemma 7.71 *Let* $b_3, b_4 \in \mathcal{S}_{in}^{p\times p}$ *and let* $\chi_1(\lambda) = b_4(\lambda)b_3(\lambda)$ *satisfy the condition*

$$\chi_1(\omega)\chi_1(\omega)^* < I_p \quad \text{for at least one (and hence every) point } \omega \in \mathbb{C}_+. \tag{7.173}$$

Then the GCIP $(b_3, b_4; 0_{p\times p})$ *is determinate, i.e.,*

$$\mathcal{C}(b_3, b_4; 0_{p\times p}) = \{0_{p\times p}\}.$$

Proof If $c^\circ(\lambda) = 0_{p\times p}$, then $s^\circ = T_{\mathfrak{V}}[0] = I_p$ and, in view of Lemma 7.68, we can (and will) choose $b_1 = b_3$ and $b_2 = b_4$. Thus, in order to complete the proof, it suffices to show that under condition (7.173) $\mathcal{S}(b_3, b_4; I_p) = \{I_p\}$. Let $f^\circ(\mu) = \chi_1(\mu)^*$ a.e. on $\mathbb{R}$ and let $\Gamma = \Gamma(f^\circ)$. Then, by Theorem 7.47,

$$\mathcal{S}(b_3, b_4; I_p) = \{b_3 f b_4 : f \in \mathcal{N}(\Gamma)\}.$$

Consequently,

$$\mathcal{S}(b_3, b_4; I_p) = I_p \iff \mathcal{N}(\Gamma) = \{f^\circ\}.$$

In the present setting,

$$(I - \Gamma^*\Gamma)^{1/2} H_2^p = \chi_1 H_2^p$$

and hence, for $\omega \in \mathbb{C}_+$,

$$\frac{\xi}{\rho_\omega} \in (I - \Gamma^*\Gamma)^{1/2} H_2^p \iff \frac{\xi}{\rho_\omega} \in \chi_1 H_2^p \iff \frac{\xi - \chi_1\chi_1(\omega)^*\xi}{\rho_\omega} \in \chi_1 H_2^p.$$

However, since

$$\frac{\xi - \chi_1\chi_1(\omega)^*\xi}{\rho_\omega} \in H_2^p \ominus \chi_1 H_2^p,$$

the assumption

$$\frac{\xi}{\rho_\omega} \in \chi_1 H_2^p \quad \text{implies that} \quad \frac{\xi - \chi_1(\lambda)\chi_1(\omega)^*\xi}{\rho_\omega} = 0$$

for every point $\lambda \in \mathbb{C}_+$. In particular, $\xi^*(I_p - \chi_1(\omega)\chi_1(\omega)^*)\xi = 0$, which implies that $\xi = 0$, since $\chi_1(\omega)\chi_1(\omega)^* < I_p$. □

Remark 7.72 *If $A \in \mathcal{U}_{rsR}(J_p)$, $\{b_3, b_4\} \in ap_{II}(A)$, (7.173) holds and $B = A\mathfrak{V}$, then $\mathcal{S}^{p\times p} \subset \mathcal{D}(T_B)$. This follows from Lemmas 4.70, 5.3 and Theorem 5.98.*

7.11 Generalized Sarason problems in $\mathring{C}^{p\times p}$

The given data for a generalized Sarason problem in the class $\mathring{C}^{p\times p}$ is a pair of mvf's $b_3, b_4 \in \mathcal{S}_{in}^{p\times p}$ and a bounded linear block triangular operator

$$X = \begin{bmatrix} \Phi_{11} & \Phi_{12} \\ 0 & \Phi_{22} \end{bmatrix} : \begin{matrix} H_2^p \\ \oplus \\ \mathcal{H}_*(b_4) \end{matrix} \longrightarrow \begin{matrix} \mathcal{H}(b_3) \\ \oplus \\ K_2^p \end{matrix} . \tag{7.174}$$

GSP$(b_3, b_4; X; \mathring{C}^{p\times p})$: *Given $b_1 \in \mathcal{S}_{in}^{p\times p}$, $b_2 \in \mathcal{S}_{in}^{q\times q}$ and a block triangular operator X of the form (7.174), find a mvf $c \in \mathring{C}^{p\times p}$ such that*

$$\Phi_{11} = P_{\mathcal{H}(b_3)} M_c|_{H_2^p}, \quad \Phi_{22} = \Pi_- M_c|_{\mathcal{H}_*(b_4)} \quad \text{and}$$
$$\Phi_{12} = P_{\mathcal{H}(b_3)} M_c|_{\mathcal{H}_*(b_4)}. \tag{7.175}$$

Let $\mathring{\mathcal{C}}(b_3, b_4; X)$ denote the set of solutions of this problem. If this problem is solvable and $c^\circ \in \mathring{\mathcal{C}}(b_3, b_4; X)$, then it is easy to verify that

$$\mathring{\mathcal{C}}(b_3, b_4; X) = \{c \in \mathring{\mathcal{C}}^{p\times p} : b_3^{-1}(c - c^\circ)b_4^{-1} \in H_\infty^{p\times p}\}. \tag{7.176}$$

Moreover, since both of the mvf's c and c° in (7.176) belong to $H_\infty^{p\times p}$,

$$b_3^{-1}(c - c^\circ)b_4^{-1} \in H_\infty^{p\times p} \Longleftrightarrow b_3^{-1}(c - c^\circ)b_4^{-1} \in \mathcal{N}_+^{p\times p}$$

by the Smirnov maximum principle. Thus,

$$\mathring{\mathcal{C}}(b_3, b_4; X) = H_\infty(b_3, b_4; X) \cap \mathring{\mathcal{C}}^{p\times p} \tag{7.177}$$

and, if $c^\circ \in \mathring{\mathcal{C}}(b_3, b_4; X)$, then

$$\mathring{\mathcal{C}}(b_3, b_4; X) = \mathcal{C}(b_3, b_4; c^\circ) \cap \mathring{\mathcal{C}}^{p\times p} \tag{7.178}$$

and the GCIP$(b_3, b_4; c^\circ)$ is strictly completely indeterminate. Conversely, if the GCIP$(b_3, b_4; c^\circ)$ is strictly completely indeterminate, then the right hand side of (7.178) is nonempty, since c° may be chosen to belong to $\mathring{\mathcal{C}}^{p\times p}$. Moreover, the operator X is well defined by formulas (7.174) and (7.175) with $c \in$ GCIP$(b_3, b_4; c^\circ) \cap H_\infty^{p\times p}$ and X is independent of the choice of such a c. Consequently, $\mathring{\mathcal{C}}(b_3, b_4; X) \neq \emptyset$ and (7.178) holds. Thus, formula (7.178) exhibits a two sided connection between strictly completely indeterminate GCIP's and solvable GSP's in $\mathring{\mathcal{C}}^{p\times p}$.

The operators

$$\Gamma_X = M_{b_3^*}\{\Phi_{11}\Pi_+ + (\Phi_{22} + \Phi_{12})P_{\mathcal{H}_*(b_4)}\}M_{b_4^*}|_{H_2^p} \tag{7.179}$$

from H_2^p into K_2^p and

$$\Delta_X = 2\Re \begin{bmatrix} \Phi_{11}|_{\mathcal{H}(b_3)} & \Phi_{12} \\ 0 & P_{\mathcal{H}_*(b_4)}\Phi_{22} \end{bmatrix} : \mathcal{H}(b_3, b_4) \to \mathcal{H}(b_3, b_4), \tag{7.180}$$

where

$$\mathcal{H}(b_3, b_4) = \begin{array}{c} \mathcal{H}(b_3) \\ \oplus \\ \mathcal{H}_*(b_4), \end{array}$$

will be useful.

Theorem 7.73 *A GSP$(b_3, b_4; X; \mathring{\mathcal{C}}^{p\times p})$ is solvable if and only if:*

(1) *The operator Γ_X, defined by formula (7.179) is a Hankel operator.*

(2) *The operator* Δ_X*, defined by formula (7.180) is positive and has a bounded inverse, i.e.,*

$$\Delta_X \geq \delta I \text{ for some } \delta > 0. \tag{7.181}$$

Proof Suppose first that $\mathring{\mathcal{C}}(b_3, b_4; X) \neq \emptyset$. Then, in view of (7.177), $H_\infty(b_3, b_4; X) \neq \emptyset$. Therefore, by Theorem 7.54, Γ_X is a Hankel operator, i.e., (1) holds. Moreover, if $c \in \mathring{\mathcal{C}}(b_3, b_4; X)$, then $2(\Re c)(\lambda) \geq \delta I_p$ for some $\delta > 0$ and consequently,

$$\left\langle \Delta_X \begin{bmatrix} g \\ h \end{bmatrix}, \begin{bmatrix} g \\ h \end{bmatrix} \right\rangle_{st} = \langle (c + c^*)(g + h), (g + h) \rangle_{st} \geq \delta \left\langle \begin{bmatrix} g \\ h \end{bmatrix}, \begin{bmatrix} g \\ h \end{bmatrix} \right\rangle_{st}$$

for every choice of $g \in \mathcal{H}(b_3)$ and $h \in \mathcal{H}_*(b_4)$, i.e., (2) holds.

It remains to show that (1) and (2) imply that $\mathring{\mathcal{C}}(b_3, b_4; X) \neq \emptyset$. The verification of this implication is much more complicated.

One approach is based on the connection between the GCIP$(b_3, b_4; c^\circ)$ and the GSIP $(b_1, b_2; s^\circ)$ considered in the previous section, which yields a connection between GSP$(b_1, b_2; X_1; \mathcal{S}^{p\times p})$ and GSP$(b_3, b_4; X; \mathcal{C}^{p\times p})$. The connection between the data of each of these two GSP's may by found in formulas (5.196), (5.198) and (5.201); for more details see the first appendix in [ArD05a].

Condition (1) is equivalent to restricting the operator Γ defined by the blocks Γ_{ij} of X_1 by formula (7.145) to be a Hankel operator, whereas condition (2) is equivalent to the condition $\|X_1\| < 1$. Thus, if (1) and (2) are in force, then by Theorem 7.54, the corresponding GSP$(b_1, b_2; X_1; \mathcal{S}^{p\times p})$ is strictly completely indeterminate, i.e.,

$$\mathcal{S}(b_1, b_2; X_1) \cap \mathring{\mathcal{S}}^{p\times p} \neq \emptyset.$$

Consequently, $\mathring{\mathcal{C}}(b_3, b_4; X) \neq \emptyset$ and, in fact,

$$\mathring{\mathcal{C}}(b_3, b_4; X) = T_{\mathfrak{V}}[\mathcal{S}(b_1, b_2; X_1) \cap \mathring{\mathcal{S}}^{p\times p}]. \tag{7.182}$$

□

Formula (7.182) leads to a description of the set $\mathring{\mathcal{C}}(b_3, b_4; X)$ of solutions of a solvable GSP in $\mathring{\mathcal{C}}^{p\times p}$ and, at the same time, of the set $\mathcal{C}(b_3, b_4; c^\circ)$ of solutions of a strictly completely indeterminate GCIP.

Theorem 7.74 *Let* $b_3, b_4 \in \mathcal{S}_{in}^{p\times p}$ *and let* X *be the block upper triangular operator defined by formula (7.174) and assume that conditions (1) and (2)* *of Theorem 7.73 are satisfied for the GSP*$(b_3, b_4; X; \mathring{\mathcal{C}}^{p\times p})$. *Then there exists a mvf* $A(\lambda)$ *which is uniquely specified by the conditions*

(1) $A \in \mathcal{U}_{rsR}(J_p)$,

(2) $\{b_3, b_4\} \in ap_{II}(A)$,

(3) $\mathring{\mathcal{C}}(b_3, b_4; X) = \mathcal{C}(A) \cap \mathring{\mathcal{C}}^{p\times p}$,

up to a constant J_p*-unitary multiplier on the right, which is obtained from the formulas*

$$A(\lambda)J_pA(\omega)^*J_p = I_m - \rho_\omega(\lambda)K_\omega^A(\lambda)J_p, \tag{7.183}$$

where $\omega \in \mathfrak{h}_{b_3} \cap \mathfrak{h}_{b_4^\#}$ *is fixed and* $K_\omega^A(\lambda)$ *may be obtained from the given data by the formulas that are presented in (5.210) and Theorem 5.104. Moreover, if* $c^\circ \in \mathcal{C}(A) \cap \mathring{\mathcal{C}}^{p\times p}$, *then the GCIP*$(b_3, b_4; c^\circ)$ *is strictly completely indeterminate and*

$$\mathcal{C}(b_3, b_4; c^\circ) = \mathcal{C}(A).$$

Proof Conditions (1) and (2) of Theorem 7.73 are satisfied if and only if $\mathring{\mathcal{C}}(b_3, b_4; X) \neq \emptyset$. In this case $\mathring{\mathcal{C}}(b_3, b_4; X) = \mathcal{C}(b_3, b_4; c^\circ)$, where $c^\circ \in \mathring{\mathcal{C}}(b_3, b_4; X)$. The rest of the theorem follows from Theorems 7.69 and 7.70. □

An mvf $A \in \mathcal{P}^\circ(J_p)$ is called a **resolvent matrix** of the GCIP$(b_3, b_4; c^\circ)$ if

$$\mathcal{C}(b_3, b_4; c^\circ) = \mathcal{C}(A),$$

and $B(\lambda) = A(\lambda)\mathfrak{V}$ is called the B**-resolvent matrix** of this problem. If

$$\mathring{\mathcal{C}}(b_3, b_4; X) = \mathcal{C}(A) \cap \mathring{\mathcal{C}}^{p\times p},$$

then A is called a resolvent matrix of the GSP$(b_3, b_4; X)$, and formulas for it are furnished in Theorem 7.74, whereas Theorem 7.70 describes the set of all resolvent matrices of a completely indeterminate GCIP$(b_3, b_4; c^\circ)$.

If $b_3 = I_p$ or $b_4 = I_p$, then the Sarason problem reduces to a simpler one sided problem. Thus, for example, if $b_3 = b \in \mathcal{S}_{in}^{p\times p}$, $b_4 = I_p$ and $X \in \mathcal{L}(H_2^p, \mathcal{H}(b))$, then the GSP$(b, I_p; X; \mathring{\mathcal{C}}^{p\times p})$ is to find a mvf c such that

$$c \in \mathring{\mathcal{C}}^{p\times p} \quad \text{and} \quad X = P_{\mathcal{H}(b)}M_c|_{H_2^p}.$$

In this setting the operator Δ_X is defined by the formula

$$\Delta_X = X|_{\mathcal{H}(b)} + (X|_{\mathcal{H}(b)})^*,$$

and the following theorems are special cases of the preceding results.

Theorem 7.75 *$\mathring{\mathcal{C}}(b, I_p; X) \neq \emptyset$ if and only if conditions (7.154) and (7.181) are in force.*

Theorem 7.76 *If $\mathring{\mathcal{C}}(b, I_p; X) \neq \emptyset$, then there exists a unique mvf $A \in \mathcal{U}_{rsR}(J_p)$ with $\{b, I_p\} \in ap_{II}(A)$, up to an arbitrary constant J_p-unitary multiplier on the right such that*

$$\mathring{\mathcal{C}}(b, I_p; X) = \mathcal{C}(A) \cap \mathring{\mathcal{C}}^{p\times p}. \tag{7.184}$$

Theorem 7.77 *Let $A \in \mathcal{U}_{rsR}(J_p)$, $\{b, I_p\} \in ap_{II}(A)$, $c^\circ \in \mathcal{C}(A) \cap \mathring{\mathcal{C}}^{p\times p}$ and $X = P_{\mathcal{H}(b)} M_{c^\circ}|_{H_2^p}$. Then X does not depend upon the choice of c° and (7.181) holds. Moreover,*

$$\mathcal{H}(A) = \left\{ L_X g = \begin{bmatrix} -X^* g \\ g \end{bmatrix} : g \in \mathcal{H}(b) \right\}, \tag{7.185}$$

$L_X \in \mathcal{L}(\mathcal{H}(b), \mathcal{H}(A))$ and, for every vvf $g \in \mathcal{H}(b)$,

$$\left\| \begin{bmatrix} -X^* g \\ g \end{bmatrix} \right\|_{\mathcal{H}(A)} = \langle \Delta_X g, g \rangle_{st} = -\left\langle J_p \begin{bmatrix} -X^* g \\ g \end{bmatrix}, \begin{bmatrix} -X^* g \\ g \end{bmatrix} \right\rangle_{st}, \tag{7.186}$$

and the mvf $A(\lambda)$ may be recovered from the formulas (7.183), where $K_\omega^A(\lambda) = (L_X \Delta_X^{-1} F_\omega^X)(\lambda)$, $b^{-1} b_5 \in \mathcal{S}^{p\times p}$ and

$$F_\omega^X(\lambda) = \begin{bmatrix} -(X k_\omega^{b_5})(\lambda) & k_\omega^b(\lambda) \end{bmatrix}.$$

7.12 Bibliographical notes

Theorem 7.2 is an analogue on the line $\mathbb{R}$ of a theorem for infinite block Hankel matrices $[\gamma_{i+j-1}]$, $i, j = 1, 2, \ldots$ with symbols that are mvf's on the unit circle $\mathbb{T} = \{\zeta \in \mathbb{C} : |\zeta| = 1\}$ that was first established by Nehari [Ne57] for scalar valued functions and then extended to the setting of mvf's and operator valued functions by Page [Pa70], see also [AAK71a] and [Ad73a]. One way to establish Theorems 7.2 and 7.20 is based on the notion of a scattering suboperator of a unitary coupling of the semigroups of simple isometric operators $V_-(t)$ and $V_+(t)$, as was done in the circle case in [AAK68] for scalar functions and [Ad73a] for operator valued functions. In this way a description of $\mathcal{N}(\Gamma)$ in the indeterminate case as the Redheffer transform of Schur class operator valued functions was obtained in [Ad73b] for mvf's and subsequently in [Kh00] for operator valued functions. A complete exposition of these results and a number of others (including the Nehari-Takagi problem) are presented by Peller in Chapter 5 of his monograph [Pe03]; the scalar case and assorted applications are treated in [Ni02].

Theorems 7.2 and 7.20 and a number of other results on the NP may also be established by purely function theoretical methods, as in Stray [St81] for the scalar case and [AFK98] for the matrix case. In the latter, formulas (7.123)–(7.126) were obtained by considering the operators $r\Gamma$ with $0 < r < 1$ in place of Γ in formulas (7.109)–(7.114) and then letting $r \uparrow 1$.

A more general version of the GSIP in which the condition in (7.127) is replaced by the constraint $\{s \in \mathcal{S}^{p\times q} : b_1^*(s - s^\circ)b_2^* \in H_\infty^{k\times\ell}\}$, where $b_1 \in \mathcal{S}_{in}^{p\times k}$ and $b_2 \in \mathcal{S}_{*in}^{\ell\times q}$, as well as its operator valued version were studied by A. Kheifets in [Kh90]. Kheifets obtained his results as an application of the abstract interpolation problem of [KKY87]. In the latter, the set of solutions was expressed as a Redheffer transform based on the four blocks of a Schur class operator valued function. This description was obtained by reducing the problem to the problem of describing the set of scattering matrices of the unitary extensions of an isometric operators and then using the parametrization formula for such a set that was presented in [ArG83] and later with complete proofs in [ArG94]. This approach arose as a development of M. G. Krein's theory of resolvent matrices of isometric and symmetric operators [Kr49]; see also [GoGo97]. The same strategy was used by M. Morán [Mo93] to describe the set of solutions of the commutant lifting problem, which yields yet another approach to the problems considered in this chapter; see [SzNF70], [FoFr90] and [FFGK98]. A good review of this circle of ideas may be found in [Arc94].

V. P. Potapov developed another method for resolving interpolation and extension problems for mvf's in $\mathcal{S}^{p\times q}$ and $\mathcal{C}^{p\times p}$ that was based on reducing the problem to the solution of certain inequality that he called the **fundamental matrix inequality**, which arose as an application of his theory of J-contractive mvf's; see [KoPo74], [Kov83], and for a good expository survey of this approach [Kat95]. This method was developed and applied to a variety of interpolation and extension problems by a number of authors, see, e.g., [Nu77], [Nu81], [KKY87], [Sak97] and [BoD98].

The formulas for resolvent matrices for the GSIP and the GCIP in the strictly completely indeterminate case are taken from Theorem 7.1 in [ArD02a] and Theorem 3.12 in [ArD05a], respectively. The RKHS methods that were used to obtain these formulas follow and partially extend the methods that were introduced earlier to study interpolation and extension problems in [Dy89a], [Dy89b], [Dy98], [Dy01a], [Dy01b], [Dy03a] and [Dy03b].

Analogues of the classes $\mathfrak{M}_r(j_{pq})$, $\mathfrak{M}_{rR}(j_{pq})$ and $\mathfrak{M}_{rS}(j_{pq})$ for mvf's on the unit circle $\mathbb{T}$ were introduced and investigated in [Ar89] in connection with the completely indeteterminate Nehari problem on $\mathbb{T}$. The corresponding **left** classes $\mathfrak{M}_\ell(j_{pq})$, $\mathfrak{M}_{\ell R}(j_{pq})$ and $\mathfrak{M}_{\ell S}(j_{pq})$ and the classes $\mathfrak{M}_{rsR}(j_{pq})$ and $\mathfrak{M}_{\ell sR}(j_{pq})$ were introduced and studied later in [ArD01b].

The pair $s_{22} = \mathfrak{a}_+^{-1}$ and $s_{21} = -\mathfrak{a}_+^{-1}\mathfrak{b}_+$ that corresponds to a mvf $\mathfrak{A} \in \mathfrak{M}_r(j_{pq})$ for $p = q = 1$ was studied by by A. Kheifets in response to a question raised by D. Sarason in [Sar89].

The material in Sections 7.10 and 7.9 is adapted from [Ar93]. Theorem 7.31 is connected with Theorem 5.2 in [ArD01b].

Lemma 7.15 was established in [Ar89] for mvf's on $\mathbb{T}$.

Theorem 7.31 is connected with Theorem 5.2 in [ArD01b]. The embedding defined in Lemma 7.36 was used in the study of the matrix Nehari problem in [AFK98] and then subsequently in [ArD01b].

Analogues of the classes $\mathcal{U}_{rR}(J)$ and $\mathcal{U}_S(J)$ for mvf's in $\mathbb{D}$ were introduced in [Ar84a], [Ar88] and [Ar90] for $J = j_{pq}$ and in [Ar93] for $J = J_p$ to study the GSIP and the GCIP in $\mathbb{D}$, respectively; see also the review [BulM98].

The GSIP is a special case of the model matching problem that is studied in control theory; see, e.g., the monographs [Fr87], [Fe98] and [FFGK98]; the latter discusses the connection of this problem and a number of others with the commutant lifting theorem. Computational issues for a number of interpolation problems are discussed in [DvdV98]. The monograph [BGR91] is a useful reference for rational interpolation problems.

8
Generalized Krein extension problems

In this chapter we shall focus on the GCIP$(b_3, b_4; c^\circ)$ in the special setting when $b_3(\lambda)$ and $b_4(\lambda)$ are entire inner mvf's. Each such interpolation problem is equivalent to a bitangential Krein extension problem in the class $\mathcal{G}_\infty^{p\times p}$ of helical $p\times p$ mvf's on the interval $(-\infty,\infty)$. This extension problem and two analogous extension problems in the class of positive definite mvf's and in the class of accelerants that were considered by Krein will also be discussed.

8.1 Helical extension problems

A mvf $g(t)$ is said to belong to the class $\mathcal{G}_\infty^{p\times p}$ of **helical** $p\times p$ mvf's on the interval $(-\infty,\infty)$ if it meets the following three conditions:

(1) $g(t)$ is a continuous $p\times p$ mvf on the interval $(-\infty,\infty)$.

(2) $g(t)^* = g(-t)$ for every t in the interval $(-\infty,\infty)$.

(3) The kernel

$$k(t,s) = g(t-s) - g(t) - g(-s) + g(0) \tag{8.1}$$

is positive on $[0,\infty)\times[0,\infty)$.

The class $\mathcal{G}_a^{p\times p}$ of helical $p\times p$ mvf's on the closed interval $[-a,a]$ is defined for $a>0$ by the same set of three conditions except that the finite closed intervals $[-a,a]$ and $[0,a]$ are considered in place of the intervals $(-\infty,\infty)$ and $[0,\infty)$, respectively.

The classical **Krein helical extension problem** is:

HEP$(g^\circ; a)$: *Given a mvf* $g^\circ \in \mathcal{G}_a^{p\times p}$, $0 < a < \infty$, *describe the set*

$$\mathcal{G}(g^\circ; a) = \{g \in \mathcal{G}_\infty^{p\times p} : g(t) = g^\circ(t) \quad \text{for every} \quad t \in [-a, a]\}. \tag{8.2}$$

Theorem 8.1 (M. G. Krein) *A* $p \times p$ *mvf* $g(t)$ *belongs to the class* $\mathcal{G}_a^{p\times p}$ *if and only if it admits a representation of the form*

$$g(t) = -\beta + it\alpha + \frac{1}{\pi}\int_{-\infty}^{\infty}\left\{e^{-i\mu t} - 1 + \frac{i\mu t}{1+\mu^2}\right\}\frac{d\sigma(\mu)}{\mu^2} \tag{8.3}$$

on the interval $(-a, a)$, *where* $\alpha = \alpha^*$ *and* $\beta = \beta^*$ *are constant* $p\times p$ *matrices and* $\sigma(\mu)$ *is a nondecreasing* $p \times p$ *mvf on* $\mathbb{R}$ *such that*

$$\int_{-\infty}^{\infty}(1+\mu^2)^{-1}d(\text{trace }\sigma(\mu)) < \infty. \tag{8.4}$$

Proof A proof for the case $p = 1$, based on Krein's unpublished Lecture Notes is given in Chapter 3 of [GoGo97]. This proof can be extended to cover the matrix case. □

Theorem 8.1 implies that the HEP$(g^\circ; a)$ is always solvable i.e., the set $\mathcal{G}(g^\circ; a)$ is not empty and can be obtained by extending the set of all representations (8.3) for $g(t)$ from $(-a, a)$ to the full axis $\mathbb{R}$. Moreover, Theorem 8.1 leads easily to a growth estimate for mvf's $g \in \mathcal{G}_\infty^{p\times p}$.

Corollary 8.2 *If* $g \in \mathcal{G}_\infty^{p\times p}$, *then*

$$\|g(t)\| = O(t^2) \quad \text{as} \quad t \to \infty. \tag{8.5}$$

Proof In view of Theorem 8.1, it suffices to check that the integral in formula (8.3) meets the asserted growth condition. But clearly the integrand

$$\left\{e^{-i\mu t} - 1 + \frac{i\mu t}{1+\mu^2}\right\}\frac{1}{\mu^2} = \left\{\frac{①+②}{1+\mu^2}\right\},$$

where

$$① = \{e^{-i\mu t} - 1\}$$

and

$$② = \left\{\frac{e^{-i\mu t} - 1 + i\mu t}{\mu^2}\right\} = \int_0^t (s-t)e^{-i\mu s}ds.$$

The rest is self-evident from (8.4), since

$$|①| \leq 2 \quad \text{and} \quad |②| \leq t^2/2$$

for $\mu \in \mathbb{R}$. □

The next theorem exhibits an important connection between the classes $\mathcal{G}_\infty^{p\times p}$ and $\mathcal{C}^{p\times p}$.

Theorem 8.3 *There is a one to one correspondence between mvf's $c(\lambda)$ in the class $\mathcal{C}^{p\times p}$ and mvf's $g(t)$ in the subclass*

$$\mathcal{G}_\infty^{p\times p}(0) = \{g \in \mathcal{G}_\infty^{p\times p} : g(0) \leq 0\} \tag{8.6}$$

that is defined by the formula

$$c_g(\lambda) = \lambda^2 \int_0^\infty e^{i\lambda t} g(t)dt \quad \text{for} \quad \lambda \in \mathbb{C}_+. \tag{8.7}$$

Moreover, if $\widetilde{g}^\circ \in \mathcal{G}(g^\circ; a)$ for a given mvf $g^\circ \in \mathcal{G}_a^{p\times p}$ with $g^\circ(0) \leq 0$ and $c^\circ = c_{\widetilde{g}^\circ}$, then

$$\mathcal{C}(e_a I_p, I_p; c^\circ) = \{c_g : g \in \mathcal{G}(g^\circ; a)\}. \tag{8.8}$$

Proof The preceding corollary guarantees that the integral in (8.7) below exists and is holomorphic in $\mathbb{C}_+$ for every choice of $g \in \mathcal{G}_\infty^{p\times p}$. With the help of formulas

$$\lambda^2 \int_0^\infty e^{i\lambda t} dt = i\lambda, \quad \lambda^2 \int_0^\infty t e^{i\lambda t} dt = -1 \tag{8.9}$$

and

$$\lambda^2 \int_0^\infty \left\{ e^{-i\mu t} - 1 + \frac{i\mu t}{1+\mu^2} \right\} e^{i\lambda t} dt = \frac{\mu^2}{i} \left\{ \frac{1}{\mu - \lambda} - \frac{\mu}{1+\mu^2} \right\}, \tag{8.10}$$

which are valid for $\mu \in \mathbb{R}$ and $\lambda \in \mathbb{C}_+$, it is readily checked that if $g(t)$ is given by (8.3) for every point $t \in \mathbb{R}$, then

$$\lambda^2 \int_0^\infty e^{i\lambda t} g(t)dt = -i\lambda\beta + i\alpha + \frac{1}{\pi i} \int_{-\infty}^\infty \left\{ \frac{1}{\mu - \lambda} - \frac{\mu}{1+\mu^2} \right\} d\sigma(\mu). \tag{8.11}$$

In view of the representation (3.3), which defines a one to one correspondence between $\{\alpha, \beta, \sigma(\mu)\}$ and mvf's $c(\lambda)$ in the Carathéodory class $\mathcal{C}^{p\times p}$, where α and β are constant $p \times p$ matrices with $\alpha = \alpha^*$ and $\beta \geq 0$, and $\sigma(\mu)$

is a $p \times p$ nondecreasing mvf which is subject to the constraint (8.4), the mapping

$$g \longrightarrow \lambda^2 \int_0^\infty e^{i\lambda t} g(t)dt \quad (\lambda \in \mathbb{C}_+) \tag{8.12}$$

is one to one from

$$\mathcal{G}_\infty^{p\times p}(0) = \{g \in \mathcal{G}_\infty^{p\times p} : \ g(0) \leq 0\} \quad \text{onto} \quad \mathcal{C}^{p\times p}.$$

The identification (8.8) will follow from a more general result that will be established below in Theorem 8.7. □

Given $g^\circ \in \mathcal{G}_a^{p\times p}(0), 0 < a < \infty$, The HEP$(g^\circ; a)$ is said to be

(i) **determinate** if the problem has only one solution.

(ii) **indeterminate** if the problem has more than one solution.

(iii) **completely indeterminate** if for every nonzero vector $\xi \in \mathbb{C}^p$, there exist at least two mvf's $g_1, g_2 \in \mathcal{G}(g^\circ; a)$ such that $\{g_1(t) - g_2(t)\}\xi \not\equiv 0$ on $\mathbb{R}$.

(iv) **strictly completely indeterminate** if there exists at least one $g \in \mathcal{G}(g^\circ; a)$ such that $c_g \in \mathring{\mathcal{C}}^{p\times p}$.

We remark that the identification (8.8) exhibits the fact that a strictly completely indeterminate problem is automatically completely indeterminate.

A two sided connection between completely indeterminate HEP's and the class of mvf's $A \in \mathcal{E} \cap \mathcal{U}(J_p)$ such that $\{e_{a_3}I_p, e_{a_4}I_p\} \in ap_{II}(A)$ for some choice of $a_3 \geq 0$ and $a_4 \geq 0$ will be presented in Theorems 8.20 and 8.21. These theorems will follow from more general results that will be presented in the next section.

The identification (8.8) will be generalized to tangential and bitangential Krein helical extension problems by replacing the special choices $b_3(\lambda) = e_a(\lambda)I_p, b_4(\lambda) = I_p$, with an arbitrary normalized pair $b_3(\lambda), b_4(\lambda)$ of entire inner $p \times p$ mvf's. Correspondingly, the set $\mathcal{G}(g^\circ; a)$ on the right hand side of (8.8) is replaced by the set $\mathcal{G}(g^\circ; \mathcal{F}_\ell, \mathcal{F}_r\}$, of solutions of the following bitangential Krein helical extension problem that we shall call GHEP, an acronym for **generalized helical extension problem**:

GHEP$(g^\circ; \mathcal{F}_\ell, \mathcal{F}_r)$: *Given a mvf $g^\circ \in \mathcal{G}_\infty^{p\times p}(0)$ and a pair of sets $\mathcal{F}_\ell \subseteq L_2^p([0, \alpha_\ell])$ and $\mathcal{F}_r \subseteq L_2^p([0, \alpha_r])$, at least one of which contains a nonzero* *element, describe the set* $\mathcal{G}(g^\circ;\mathcal{F}_\ell,\mathcal{F}_r\}$ *of mvf's* $g\in\mathcal{G}_\infty^{p\times p}(0)$ *that meet the following three conditions for every choice of* $h^\ell\in\mathcal{F}_\ell$ *and* $h^r\in\mathcal{F}_r$*:*

$$\int_t^{\alpha_\ell} h^\ell(u)^*\{g(u-t)-g^\circ(u-t)\}du \;=\; 0 \quad \text{for } 0\le t\le \alpha_\ell. \tag{8.13}$$

$$\int_t^{\alpha_r} \{g(v-t)-g^\circ(v-t)\}h^r(v)dv \;=\; 0 \quad \text{for } 0\le t\le \alpha_r. \tag{8.14}$$

$$\int_t^{\alpha_r}\left[\int_0^{\alpha_\ell} h^\ell(u)^*\{g(u+v-t)-g^\circ(u+v-t)\}du\right]h^r(v)dv=0 \quad \text{for } 0\le t\le\alpha_r. \tag{8.15}$$

The three sets of conditions (8.13), (8.14) and (8.15) are equivalent to the three sets (8.13), (8.14) and

$$\int_t^{\alpha_\ell} h^\ell(u)^*\left[\int_0^{\alpha_r}\{g(u+v-t)-g^\circ(u+v-t)\}h^r(v)dv\right]du=0 \quad \text{for } 0\le t\le\alpha_\ell \tag{8.16}$$

(see Lemma 8.12, below). If $\mathcal{F}_r=\{0\}$ or $\mathcal{F}_r=\emptyset$, then only constraint (8.13) is in effect and the GHEP is a left tangential extension problem.

If $\mathcal{F}_\ell=\{0\}$ or $\mathcal{F}_\ell=\emptyset$, then only constraint (8.14) is in effect and the GHEP is a right tangential extension problem.

Theorem 8.4 *Let* $g^\circ\in\mathcal{G}_\infty^{p\times p}(0)$ *and let* $\mathcal{F}_\ell\subseteq L_2^p([0,\alpha_\ell])$ *and* $\mathcal{F}_r\subseteq L_2^p([0,\alpha_r])$*, where* $\alpha_\ell+\alpha_r>0$*. Then there exists a pair* $\{b_3,b_4\}$ *of normalized entire inner* $p\times p$ *mvf's such that*

$$\mathcal{C}(b_3,b_4;c^\circ)=\mathcal{G}(g^\circ;\mathcal{F}_\ell,\mathcal{F}_r) \quad \textit{with} \quad c^\circ=c_{g^\circ}. \tag{8.17}$$

Moreover, $b_3(\lambda)$ *and* $b_4(\lambda)$ *may be chosen so that*

$$\tau(b_3)\le\alpha_\ell \quad \textit{and} \quad \tau(b_4)\le\alpha_r, \tag{8.18}$$

with $b_4(\lambda)\equiv I_p$ *for left tangential problems and* $b_3(\lambda)\equiv I_p$ *for right tangential problems.*

Conversely, if a mvf $c^\circ\in\mathcal{C}^{p\times p}$ *and a pair* $\{b_3,b_4\}$ *of normalized entire inner* $p\times p$ *mvf's is given, then a pair of sets* $\mathcal{F}_\ell\subseteq L_2^p([0,\alpha_\ell])$ *and* $\mathcal{F}_r\subseteq L_2^p([0,\alpha_r])$ *exists such that the identification (8.17) is in force and equality holds in both of the inequalities in (8.18).*

This theorem will follow from a number of other results on the connection between the $\mathrm{GCIP}(b_3, b_4; c^\circ)$ and the $\mathrm{GHEP}(g^\circ; \mathcal{F}_\ell, \mathcal{F}_r)$ that will be obtained below. We begin with a few preliminary results of a general nature. The notation

$$f_\epsilon(\lambda) = f(\lambda + i\epsilon)$$

will be used in the proof Lemma 8.6 and Theorem 8.7.

Lemma 8.5 *Let $b \in \mathcal{E} \cap H_\infty^{p\times p}$ and suppose that $b(0) = I_p$ and $b(\lambda)$ is of exponential type a. Then there exists a unique $p \times p$ mvf $h_b(t)$ such that*

$$b(\lambda) = I_p + i\lambda \int_0^a e^{i\lambda t} h_b(t)dt \quad \text{with} \quad h_b \in L_2^{p\times p}([0,a]) \tag{8.19}$$

and

$$\int_{a-\epsilon}^a \|h_b(t)\|dt > 0 \quad \text{for every } \epsilon \text{ in the interval } \quad 0 < \epsilon < a. \tag{8.20}$$

Proof The representation (8.19) is immediate from the Paley-Wiener theorem, since

$$\widehat{h_b}(\lambda) = \frac{b(\lambda) - I_p}{\lambda}$$

is an entire $p \times p$ mvf of exponential type a which belongs to the space $H_2^{p\times p}$. □

Lemma 8.6 *Let c and c° belong to the Carathéodory class $\mathcal{C}^{p\times p}$, let $b \in \mathcal{E} \cap \mathcal{S}_{in}^{p\times p}$ and suppose that*

$$b_\varepsilon^{-1}(c_\varepsilon - c_\varepsilon^\circ) \in \mathcal{N}_+^{p\times p}$$

for every $\varepsilon > 0$. Then

$$b^{-1}(c - c^\circ) \in \mathcal{N}_+^{p\times p}.$$

Proof Let

$$s = T_{\mathfrak{V}}[c] = (I_p - c)(I_p + c)^{-1} \quad \text{and} \quad s^\circ = T_{\mathfrak{V}}[c^\circ].$$

Then s, s° and $(I_p + s^\circ)/2$ all belong to $\mathcal{S}^{p\times p}$ and

$$\frac{(I_p + s^\circ)}{2} b = d\varphi, \tag{8.21}$$

where $d \in \mathcal{S}_{in}^{p\times p}$ and $\varphi \in \mathcal{S}_{out}^{p\times p}$. In fact, since $(I_p + s^\circ)/2$ is outer,

$$\det\ d(\lambda) = \gamma_1 \det\ b(\lambda) = \gamma_2 e^{i\delta\lambda}$$

for some unimodular constants γ_1 and γ_2 and $\delta \geq 0$, it follows that $d(\lambda)$ is also entire, by Lemma 3.97. Now

$$\begin{aligned} b^{-1}(c - c^\circ) &= b^{-1}\{(I_p - s)(I_p + s)^{-1} - (I_p + s^\circ)^{-1}(I_p - s^\circ)\} \\ &= 2b^{-1}(I_p + s^\circ)^{-1}(s^\circ - s)(I_p + s)^{-1}, \end{aligned}$$

which in view of (8.21) yields the formula

$$b^{-1}(c - c^\circ) = \varphi^{-1} d^{-1}(s^\circ - s)(I_p + s)^{-1}. \tag{8.22}$$

Therefore, by assumption,

$$b_\varepsilon^{-1}(c_\varepsilon - c_\varepsilon^\circ) = \varphi_\varepsilon^{-1} d_\varepsilon^{-1}(s_\varepsilon^\circ - s_\varepsilon)(I_p + s_\varepsilon)^{-1}$$

belongs to $\mathcal{N}_+^{p\times p}$ for $\varepsilon > 0$ and hence, since φ_ε and $(I_p + s_\varepsilon)$ are both outer mvf's (in $H_\infty^{p\times p}$),

$$d_\varepsilon^{-1}(s_\varepsilon^\circ - s_\varepsilon) \in \mathcal{N}_+^{p\times p}$$

for every $\varepsilon > 0$. Moreover, since

$$d^{-1}(\lambda) = e^{-i\delta\lambda} D(\lambda)$$

for some mvf $D \in \mathcal{E} \cap \mathcal{S}_{in}^{p\times p}$, it follows that

$$\|d_\varepsilon^{-1}(\mu)\{s_\varepsilon^\circ(\mu) - s_\varepsilon(\mu)\}\| \leq 2e^{\delta\varepsilon}$$

for $\varepsilon > 0$ and every point $\mu \in \mathbb{R}$. Therefore,

$$d_\varepsilon^{-1}(s_\varepsilon^\circ - s_\varepsilon) \in L_\infty^{p\times p}(\mathbb{R}) \cap \mathcal{N}_+^{p\times p} = H_\infty^{p\times p}.$$

Thus,

$$\|d_\varepsilon^{-1}(\lambda)\{s_\varepsilon^\circ(\lambda) - s_\varepsilon(\lambda)\}\| \leq 2e^{\delta\varepsilon}$$

for every point $\lambda \in \mathbb{C}_+$, and hence, upon letting $\varepsilon \downarrow 0$, we conclude that

$$\|d^{-1}(\lambda)\{s^\circ(\lambda) - s(\lambda)\}\| \leq 2,$$

i.e.,

$$d^{-1}\{s^\circ - s\} \in H_\infty^{p\times p}.$$

The desired conclusion is now immediate from (8.22) since both φ and $(I_p + s)/2$ belong to $\mathcal{S}_{out}^{p\times p}$. $\square$

Theorem 8.7 *Let b_3 and b_4 be normalized entire inner $p \times p$ mvf's of exponential types a_3 and a_4 respectively, and let the mvf's $h_{b_3}(t)$ and $h_{b_4}(t)$ be defined by the corresponding integral representation formulas (8.19). Then the identification (8.17) holds with*

$$\mathcal{F}_\ell = \{h_{b_3}(t)\xi : \xi \in \mathbb{C}^p\} \quad and \quad \mathcal{F}_r = \{h_{b_4}(t)^*\eta : \eta \in \mathbb{C}^p\}$$

for every choice of $g^\circ \in \mathcal{G}_\infty^{p\times p}(0)$.

Proof The proof is divided into three parts.

1. Let $b_3(\lambda) = b(\lambda)$, $a_3 = a > 0$ and $b_4(\lambda) \equiv I_p$. Then $\mathcal{F}_r = \{0\}$ and hence, in order to verify (8.17), it suffices to show that

$$c \in \mathcal{C}^{p\times p} \quad \text{meets the condition} \quad b^{-1}(c - c^\circ) \in \mathcal{N}_+^{p\times p} \tag{8.23}$$

if and only if $c = c_g$, where $g \in \mathcal{G}_\infty^{p\times p}(0)$ satisfies the condition

$$\int_t^a h_b(u)^*\{g(u-t) - g^\circ(u-t)\}du = 0 \quad \text{for } 0 \le t \le a. \tag{8.24}$$

Suppose first that (8.23) holds and let

$$q(\lambda) = \int_{-a}^0 e^{i\lambda t} h_b(-t)^* dt \quad \text{and} \quad r(\lambda) = \int_0^\infty e^{i\lambda t}\{g(t) - g^\circ(t)\}dt,$$

for short. Then,

$$b(\lambda)^{-1}\{c_g(\lambda) - c_{g^\circ}(\lambda)\} = \{I_p - i\lambda q(\lambda)\}(c_g(\lambda) - c_{g^\circ}(\lambda)) = \{I_p - i\lambda q(\lambda)\}\lambda^2 r(\lambda)$$

and, since $\mathcal{C}^{p\times p} \subset \mathcal{N}_+^{p\times p}$ and $\mathcal{N}_+^{p\times p}$ is closed under addition and multiplication, it is readily seen that

$$\lambda^3 q(\lambda) r(\lambda) \in \mathcal{N}_+^{p\times p}$$

and hence, since $\lambda^{-1} I_p \in \mathcal{N}_+^{p\times p}$, that

$$q(\lambda) r(\lambda) \in \mathcal{N}_+^{p\times p}.$$

Therefore $q_\varepsilon r_\varepsilon \in \mathcal{N}_+^{p\times p}$ for every $\varepsilon > 0$. Moreover, since

$$q_\varepsilon \in L_2^{p\times p}(\mathbb{R}) \cap L_\infty^{p\times p}(\mathbb{R})$$

and

$$r_\varepsilon \in H_2^{p\times p} \cap H_\infty^{p\times p},$$

it follows that the product

$$q_\varepsilon r_\varepsilon \in L_2^{p\times p}(\mathbb{R}) \cap L_\infty^{p\times p}(\mathbb{R}) \cap \mathcal{N}_+^{p\times p}.$$

Thus, by the Smirnov maximum principle for mvf's,

$$q_\varepsilon r_\varepsilon \in H_2^{p\times p} \cap H_\infty^{p\times p}$$

for every $\varepsilon > 0$. Consequently, the inverse Fourier transform

$$(q_\varepsilon r_\varepsilon)^\vee(t) = \int_{-a}^{t} q_\varepsilon^\vee(s) r_\varepsilon^\vee(t-s)ds$$

of $q_\varepsilon r_\varepsilon$ must vanish for $-a \le t \le 0$. But this is the same as to say that

$$\int_{-a}^{t} h_b(-s)^* e^{-\varepsilon s}\{g(t-s) - g^\circ(t-s)\} e^{-\varepsilon(t-s)} ds = 0 \tag{8.25}$$

for $-a \le t \le 0$, which leads easily to (8.24).

Now suppose conversely that (8.24) holds. Then clearly (8.25) holds for every $\varepsilon > 0$ and hence

$$q_\varepsilon r_\varepsilon \in H_2^{p\times p} \cap H_\infty^{p\times p} \subset \mathcal{N}_+^{p\times p}$$

for every $\varepsilon > 0$. Moreover, in view of (8.5), we also have

$$r_\varepsilon \in H_2^{p\times p} \cap H_\infty^{p\times p} \subset \mathcal{N}_+^{p\times p}$$

for every $\varepsilon > 0$. Thus,

$$b(\lambda + i\varepsilon)^{-1}\{c_g(\lambda + i\varepsilon) - c_{g^\circ}(\lambda + i\varepsilon)\} = \{I_p - i(\lambda + i\varepsilon)q_\varepsilon(\lambda)\}(\lambda + i\varepsilon)^2 r_\varepsilon(\lambda)$$

belongs to $\mathcal{N}_+^{p\times p}$ (as a function of λ) for every $\varepsilon > 0$. The desired conclusion now follows by letting $\varepsilon \downarrow 0$ and invoking Lemma 8.6.

2. Let $b_4(\lambda) = b(\lambda)$, $a_4 = a > 0$ and $b_3(\lambda) \equiv I_p$. In this setting $\mathcal{F}_\ell = \{0\}$ and hence, in order to verify (8.17), it suffices to show that

$$c \in \mathcal{C}^{p\times p} \quad \text{meets the condition} \quad (c - c^\circ)b^{-1} \in \mathcal{N}_+^{p\times p} \tag{8.26}$$

if and only if $c = c_g$, where $g \in \mathcal{G}_\infty^{p\times p}(0)$ satisfies the condition

$$\int_t^a \{g(u-t) - g^\circ(u-t)\} h_b(u)^* du = 0 \quad \text{for } 0 \le t \le a. \tag{8.27}$$

This equivalence may be verified in much the same way that the equivalence of (8.23) to (8.24) was justified. Alternatively, it may be reduced to the preceding setting by considering the mvf's $b^\sim(\lambda)$, $c^\sim(\lambda)$ and $(c^\circ)^\sim(\lambda)$ in

(8.23) instead of $b(\lambda)$, $c(\lambda)$ and $c^\circ(\lambda)$ and $h_b(t)^*$, $g(t)^*$ and $g^\circ(t)^*$ in (8.24) instead of $h_b(t)$, $g(t)$ and $g^\circ(t)$.

3. The general setting. If

$$b_3^{-1}(c_g - c_{g^\circ})b_4^{-1} \in \mathcal{N}_+^{p\times p} \tag{8.28}$$

holds, then clearly

$$b_3^{-1}(c_g - c_{g^\circ}) \in \mathcal{N}_+^{p\times p} \quad \text{and} \quad (c_g - c_{g^\circ})b_4^{-1} \in \mathcal{N}_+^{p\times p}, \tag{8.29}$$

Therefore, by the preceding two cases, conditions (8.24) and (8.27) must be met. Furthermore, since $b_3(0) = I_p$, the first of these implies that

$$b_3^{-1}(c_g - c_{g^\circ}) = (c_g - c_{g^\circ}) - i\lambda^3 \int_0^\infty e^{i\lambda t}\left\{\int_0^{a_3} h_{b_3}(u)^* r^\vee(u+t)du\right\}dt,$$

where

$$r^\vee(t) \;=\; g(t) - g^\circ(t) \quad \text{for } t \geq 0.$$

Consequently, condition (8.28) will be met if and only if (8.24) and (8.27) hold and

$$\int_0^\infty e^{i\lambda t}\left\{\int_0^{a_3} h_{b_3}(u)^* r^\vee(u+t)du\right\}dt\; b_4(\lambda)^{-1} \in \mathcal{N}_+^{p\times p}.$$

Much the same sort of analysis as was used before leads to condition (8.15). Moreover, the argument can be run backwards to establish the converse.

Similar considerations lead to the second set of conditions. □

We remark that conditions (8.13)–(8.15) depend at most on the values of $g(t)$ and $g^\circ(t)$ on the interval $-(a_3 + a_4) \leq t \leq a_3 + a_4$.

We shall identify $L_2^p([0,\alpha])$ with the subspace of vvf's in $L_2^p([0,\infty))$ with support in the interval $[0,\alpha]$. With this identification, $L_2^p([0,\alpha])$ is invariant under the action of the semigroup T_τ, $\tau \geq 0$, of backward shift operators in the time domain that are defined by the formula

$$(T_\tau f)(t) \;=\; f(t+\tau) \quad \text{for} \quad f \in L_2^p([0,\infty)) \quad \text{and} \quad \tau \geq 0. \tag{8.30}$$

Then

$$T_\tau f = (\Pi_+ e_{-\tau}\widehat{f})^\vee.$$

Let

$$\widehat{\mathcal{H}} \;=\; \{\widehat{h} : h \in \mathcal{H}\}$$

denote the Fourier transform of a subspace $\mathcal{H}$ of $L_2^p(\mathbb{R})$.

Lemma 8.8 *Let $\mathcal{L}$ be a closed subspace of $L_2^p(\mathbb{R}_+)$. Then $\mathcal{L}$ is invariant with respect to the backwards shift semigroup $T(t)$, $t \geq 0$, in $L_2^p(\mathbb{R}_+)$ if and only if*

$$R_\alpha \widehat{\mathcal{L}} \subseteq \widehat{\mathcal{L}} \text{ for at least one (and hence every) point } \alpha \in \mathbb{C}_+.$$

Proof This follows from Lemma 3.40. □

Lemma 8.9 *Let $\mathcal{H} \subseteq L_2^p([0, \alpha])$ be a closed subspace that is invariant with respect to the semigroup T_τ, $\tau \geq 0$. Then:*

(1) *There exists exactly one mvf $b \in \mathcal{E} \cap \mathcal{S}_{in}^{p\times p}$ with $b(0) = I_p$ such that $\widehat{\mathcal{H}} = \mathcal{H}(b)$.*

Moreover,

(2) *$b(\lambda)$ is of exponential type a with $a \leq \alpha$.*

(3) *The mvf h_b in the representation (8.19) for $b(\lambda)$ belongs to $L_2^p([0, a])$.*

(4) *$\bigvee\{T_\tau h_b \xi : \xi \in \mathbb{C}^p$ and $\tau \geq 0\} = \mathcal{H}$.*

Proof If $\mathcal{H}$ is a closed subspace of $L_2^p([0, \alpha])$ that is invariant under the action of the semigroup T_τ, $\tau \geq 0$, then

$$\widehat{\mathcal{H}} \subseteq H_2^p \ominus e_\alpha H_2^p$$

and $\widehat{\mathcal{H}}$ is invariant under the action of the semigroup of operators

$$\Pi_+ e_{-\tau}|_{H_2^p} \quad \text{for every} \quad \tau \geq 0.$$

Therefore, $H_2^p \ominus \widehat{H}$ is invariant under multiplication by e_t for $t \geq 0$ and hence, by the Beurling-Lax theorem, there exists an essentially unique mvf $b \in \mathcal{S}_{in}^{p\times q}$ such that $\widehat{\mathcal{H}} = \mathcal{H}(b)$. In view of Corollary 3.42, $q = p$, and $b(\lambda)$ is an entire inner $p \times p$ mvf of exponential type $\leq \alpha$, since $\mathcal{H}(b) \subseteq \mathcal{H}(e_\alpha I_p)$ and $R_\alpha b\xi \in \mathcal{H}(b)$ for $\alpha \in \mathbb{C}_+$ and every $\xi \in \mathbb{C}^p$. Moreover, there is only one such mvf $b(\lambda)$ that meets the normalization condition $b(0) = I_p$. This completes the proof of (1) and (2). Assertion (3) then follows from Lemma 8.5.

Next, as

$$\widehat{h_b}(\lambda)\xi = \frac{b(\lambda) - I_p}{i\lambda}\xi = 2\pi k_0^b(\lambda)\xi,$$

belongs to $\mathcal{H}(b)$, it follows that $h_b\xi \in \mathcal{H}$ and hence that $T_\tau h_b\xi \in \mathcal{H}$ for every $\xi \in \mathbb{C}^p$ and every $\tau \geq 0$, since $\mathcal{H}$ is invariant with respect to T_τ for $\tau \geq 0$. Thus, the closed subspace $\mathcal{L}$ that is defined in (4) is a subspace of $\mathcal{H}$. Moreover, in view of the preceding identity, $R_0 b\xi \in \widehat{\mathcal{L}}$ for every $\xi \in \mathbb{C}^p$, and, since $\widehat{\mathcal{L}}$ is invariant with respect to the backward shift operator R_α in $\mathcal{H}(b)$ for every $\alpha \in \mathbb{C}_+$ by Lemma 8.8, it follows that

$$\begin{aligned}(I_p + \omega R_\omega)\widehat{h}_b\xi &= -i(I_p + \omega R_\omega)R_0 b\xi = -i(R_0 + \omega R_\omega R_0)b\xi \\ &= -i(R_0 + R_\omega - R_0)b\xi = -i\frac{b(\lambda) - b(\omega)}{\lambda - \omega}\xi \\ &= -i\frac{b(\lambda)b(\overline{\omega})^* - I_p}{\lambda - \omega}b(\omega)\xi = 2\pi k_{\overline{\omega}}^b(\lambda)b(\omega)\xi;\end{aligned}$$

i.e., $k_\omega^b\xi \in \widehat{\mathcal{L}}$ for every $\omega \in \mathbb{C}$ and $\xi \in \mathbb{C}^p$. Consequently, $\mathcal{H}(b) \subseteq \widehat{\mathcal{L}}$. Thus, $\widehat{\mathcal{L}} = \mathcal{H}(b) = \widehat{\mathcal{H}}$, and $\mathcal{L} = \mathcal{H}$. □

If $\mathcal{F} \subseteq L_2^p([0, \alpha))$, then

$$\bigvee(T_\tau\mathcal{F} : \tau \geq 0)$$

is a closed subspace of $L_2^p([0, \alpha))$ that is invariant under T_τ for every $\tau > 0$. Consequently, there is a unique normalized entire inner $p \times p$ mvf $b(\lambda)$ such that

$$\left(\bigvee\{T_\tau\mathcal{F} : \tau \geq 0\}\right)^{\wedge} = \mathcal{H}(b).$$

Lemma 8.10 *Let $f(t)$ be a continuous $k \times p$ mvf on $[0, a]$, let $h \in L_2^p([0, a])$ and let*

$$\psi(t) = \int_t^a f(v - t)h(v)dv \quad \textit{for } 0 \leq t \leq a. \tag{8.31}$$

Then $\psi \in L_2^k([0, a])$ and

$$\psi(t + \tau) = \int_t^a f(v - t)(T_\tau h)(v)dv \quad \textit{for } 0 \leq t \leq a - \tau \textit{ and } 0 \leq \tau. \tag{8.32}$$

Proof If $a - \tau \leq t \leq a$, then both sides of (8.32) are equal to zero. If $0 \leq t < a - \tau$, then equality (8.32) follows from (8.31) by a change of variables, since $(T_\tau h)(v) = 0$ for $\tau > 0$ and $a - \tau < v \leq a$. □

Lemma 8.11 *Let $f(t)$ be a continuous $p \times k$ mvf on $[0,a]$, let $h \in L_2^p([0,a])$ and let*

$$\varphi(t) = \int_t^a h(u)^* f(u-t)du \quad for \quad 0 \le t \le a.$$

Then $\varphi \in L_2^{1\times k}([0,a])$ and

$$\varphi(t+\tau) = \int_t^a (T_\tau h)(u)^* f(u-t)du.$$

Proof This assertion is equivalent to Lemma 8.10. □

If $0 < \alpha < a < \infty$, then, in the next several lemmas, $L_2^p([0,\alpha])$ is viewed as a subspace of $L_2^p([0,a])$

Lemma 8.12 *Let $f(t)$ be a continuous $p \times p$ mvf on $[0,a]$ and let $h^\ell \in L_2^p([0,\alpha_\ell])$ and $h^r \in L_2^p([0,\alpha_r])$ be such that $0 < \alpha_\ell$, $0 < \alpha_r$, $\alpha_\ell + \alpha_r \le a$ and*

$$\int_t^{\alpha_\ell} h^\ell(u)^* f(u-t)du = 0 \quad for\ 0 \le t \le \alpha_\ell, \tag{8.33}$$

$$\int_t^{\alpha_r} f(v-t)h^r(v)dv = 0 \quad for\ 0 \le t \le \alpha_r. \tag{8.34}$$

Then

$$\int_t^{\alpha_r} \left[\int_0^{\alpha_\ell} h^\ell(u)^* f(u+v-t)du \right] h^r(v)dv = 0 \quad for\ every\ 0 \le t \le \alpha_r \tag{8.35}$$

if and only if

$$\int_t^{\alpha_\ell} \left[\int_0^{\alpha_r} h^\ell(u)^* [f(u+v-t)h^r(v)dv \right] du = 0 \quad for\ every\ 0 \le t \le \alpha_\ell. \tag{8.36}$$

Proof Let the assumption of the lemma be in force and let $\widetilde{f}$ be any continuous extension of f to the interval $0 \le t \le 2a$. Then equations (8.33) and (8.34) are equivalent to the formulas

$$\int_{t-u}^a \widetilde{f}(u+v-t)h^r(v)dv = 0 \quad \text{for } 0 \le u \le t \le a$$

and

$$\int_{t-v}^a h^\ell(u)^* \widetilde{f}(u+v-t)du = 0 \quad \text{for } 0 \le v \le t \le a,$$

respectively, where it is understood that $h^\ell(u) = 0$ if $u \notin [0, \alpha_\ell]$ and $h^r(u) = 0$ if $u \notin [0, \alpha_r]$. Consequently,

$$\begin{aligned}
&\int_t^a \left[\int_0^a h^\ell(u)^* \widetilde{f}(u+v-t)du\right] h^r(v)dv \\
&= \int_0^a h^\ell(u)^* \left[\int_t^a \widetilde{f}(u+v-t)h^r(v)dv\right] du \\
&= \int_0^t h^\ell(u)^* \left[\int_t^a \widetilde{f}(u+v-t)h^r(v)dv\right] du \\
&\quad + \int_t^a h^\ell(u)^* \left[\int_t^a \widetilde{f}(u+v-t)h^r(v)dv\right] du \\
&= -\int_0^t h^\ell(u)^* \left[\int_{t-u}^t \widetilde{f}(u+v-t)h^r(v)dv\right] du \\
&\quad + \int_t^a h^\ell(u)^* \left[\int_t^a \widetilde{f}(u+v-t)h^r(v)dv\right] du \\
&= -\int_0^t \left[\int_{t-v}^t h^\ell(u)^* \widetilde{f}(u+v-t)du\right] h^r(v)dv \\
&\quad + \int_t^a h^\ell(u)^* \left[\int_t^a \widetilde{f}(u+v-t)h^r(v)dv\right] du \\
&= \int_0^t \left[\int_t^a h^\ell(u)^* \widetilde{f}(u+v-t)du\right] h^r(v)dv \\
&\quad + \int_t^a h^\ell(u)^* \left[\int_t^a \widetilde{f}(u+v-t)h^r(v)dv\right] du \\
&= \int_t^a h^\ell(u)^* \left[\int_0^a \widetilde{f}(u+v-t)h^r(v)dv\right] du.
\end{aligned}$$

Thus,

$$\begin{aligned}
&\int_t^a h^\ell(u)^* \left[\int_0^a \widetilde{f}(u+v-t)h^r(v)dv\right] du \\
&\qquad = \int_t^a \left[\int_0^a h^\ell(u)^* \widetilde{f}(u+v-t)du\right] h^r(v)dv \quad \text{for } 0 \le t \le a. \quad (8.37)
\end{aligned}$$

Consequently, (8.36) implies that the left hand side of (8.37) vanishes for $0 \le t \le \alpha_\ell$ and hence for $0 \le t \le a$. Therefore, the right hand side of (8.37) vanishes for $0 \le t \le a$, which serves to justify (8.35). The proof that (8.35) $\Longrightarrow$ (8.36) is similar. □

Lemma 8.13 *Let $f(t)$ be a continuous $k \times p$ mvf on $[0, a]$, let $0 < \alpha_r \le a$ and let $h^r \in L_2^p([0, \alpha_r])$. If (8.34) holds, then*

$$\int_t^a f(v-t)(T_\tau h^r)(v)dv = 0 \quad \text{for } 0 \le t \le a \tag{8.38}$$

and every $\tau \ge 0$.

Proof This follows from Lemma 8.10. □

Lemma 8.14 *Let $f(t)$ be a continuous $p \times k$ mvf on $[0, a]$, let $0 < \alpha_\ell \le a$ and let $h^\ell \in L_2^p([0, \alpha_\ell])$. If (8.33) holds, then*

$$\int_t^a (T_\tau h^\ell(u))^* f(u-t)du = 0 \quad \text{for } 0 \le t \le a \tag{8.39}$$

and every $\tau \ge 0$.

Proof This follows from Lemma 8.11. □

Lemma 8.15 *Let $f(t)$ be a continuous $p \times p$ mvf on $[0, a]$. Let $h^\ell \in L_2^p([0, \alpha_\ell])$, $h^r \in L_2^p([0, \alpha_r])$, $0 < \alpha_\ell + \alpha_r \le a$. If (8.33), (8.34) and (8.35) hold, then (8.38) and (8.39) hold for every $\tau \ge 0$, and*

$$\int_t^{\alpha_\ell} (T_{\tau_1} h^\ell)(u)^* \left[\int_0^{\alpha_r} f(u+v-t)(T_{\tau_2} h^r)(v)dv \right] du = 0, \quad 0 \le t \le \alpha_\ell, \tag{8.40}$$

hold for every $\tau_1 \ge 0$ and $\tau_2 \ge 0$.

Proof If the assumptions of the lemma are in force, then (8.38) and (8.39) hold for $\tau \ge 0$ by Lemmas 8.13 and 8.14. Let

$$f_1(t) = \int_0^{\alpha_r} f(v+t)h^r(v)dv \quad \text{for} \quad 0 \le t \le \alpha_\ell.$$

Then $f_1(t)$ is a continuous $p \times 1$ mvf on $[0, \alpha_\ell]$. In view of Lemma 8.12, (8.35) is equivalent to (8.36) and hence,

$$\int_t^{\alpha_\ell} h^\ell(u)^* f_1(u-t)du = \int_t^{\alpha_\ell} h^\ell(u)^* \left[\int_0^{\alpha_r} f(v+u-t)h^r(v)dv \right] du = 0.$$

Thus, by Lemma 8.14,

$$\int_t^{\alpha_\ell} (T_{\tau_1} h^\ell)(u)^* f_1(u-t)du = 0 \quad \text{for } 0 \le t \le \alpha_\ell \tag{8.41}$$

and every $\tau_1 \ge 0$.

Let $\tau_1 \ge 0$ be fixed. Then

$$\int_t^{\alpha_\ell} (T_{\tau_1} h^\ell)(u)^* f(u-t)du = 0,$$

$$\int_t^{\alpha_r} f(v-t)h^r(v)dv = 0$$

and

$$\int_t^{\alpha_\ell} (T_{\tau_1} h^\ell)(u)^* \left[\int_0^{\alpha_r} f(u+v-t)h^r(v)dv\right] du = 0$$

for $0 \le t \le a$. Thus, by Lemma 8.12,

$$\int_t^{\alpha_r} \left[\int_0^{\alpha_\ell} (T_{\tau_1} h^\ell)^*(u) f(u+v-t)du\right] h^r(v)dv = 0.$$

Now let

$$f_2(t) = \int_0^{\alpha_\ell} (T_{\tau_1} h^\ell)^*(u) f(u+t)du \quad \text{for } 0 \le t \le \alpha_r.$$

Then $f_2(t)$ is a continuous $1 \times p$ mvf on $[0, \alpha_r]$ and

$$\int_t^{\alpha_r} f_2(v-t)h^r(v)dv = 0 \quad \text{for } 0 \le t \le \alpha_r.$$

By Lemma 8.13,

$$\int_t^{\alpha_r} f_2(v-t)(T_{\tau_2} h^r)(v)dv = 0 \quad \text{for} \quad 0 \le t \le \alpha_r.$$

Consequently,

$$\int_t^{\alpha_r} \left[\int_0^{\alpha_\ell} (T_{\tau_1} h^\ell(u)^* f(u+v-t)du\right] (T_{\tau_2} h^r)(v)dv = 0 \text{ for } 0 \le t \le \alpha_r,$$

which is equivalent to (8.40) by Lemma 8.12. □

Lemma 8.16 *Let $f(t)$ be a continuous $p \times p$ mvf on $[0, a]$. Let*

$$\mathcal{F}_r \subseteq L_2^p([0, \alpha_\ell]) \quad \textit{and} \quad \mathcal{F}_\ell \subseteq L_2^p([0, \alpha_r]) \quad \textit{for} \quad 0 < \alpha_\ell + \alpha_r \le a$$

and let

$$\mathcal{H}_\ell = \bigvee_{\tau\geq 0}(T_\tau \mathcal{F}_\ell) \text{ in } L_2^p([0,\alpha_\ell]) \quad \text{and} \quad \mathcal{H}_r = \bigvee_{\tau\geq 0}(T_\tau \mathcal{F}_r) \text{ in } L_2^p([0,\alpha_r]).$$

If (8.33), (8.34) and (8.35) hold for every $h^\ell \in \mathcal{F}_\ell$ and $h^r \in \mathcal{F}_r$, then (8.33), (8.34) and (8.35) hold for every $h^\ell \in \mathcal{H}_\ell$ and $h^r \in \mathcal{H}_r$.

Proof Lemma 8.15 implies that (8.33), (8.34) and (8.35) hold for the vvf's h^ℓ and h^r in the linear span of $\{T_\tau \mathcal{F}_\ell : \tau \geq 0\}$ and the linear span of $\{T_\tau \mathcal{F}_r : \tau \geq 0\}$, respectively. Thus, they are in force for every vvf h^ℓ and h^r in the closure of these linear manifolds, too. □

Lemma 8.17 *Let $\mathcal{F}_\ell \subseteq L_2^p([0,\alpha_\ell])$ and $\mathcal{F}_r \subseteq L_2^p([0,\alpha_r])$ be given and let $b_\ell(\lambda)$ and $b_r(\lambda)$ be the unique pair of normalized entire inner $p \times p$ mvf's such that*

$$\left(\bigvee\{T_\tau \mathcal{F}_\ell : \tau \geq 0\}\right)^\wedge = \mathcal{H}(b_\ell) \quad \text{and} \quad \left(\bigvee\{T_\tau \mathcal{F}_r : \tau \geq 0\}\right)^\wedge = \mathcal{H}(b_r).$$

Let $\mathcal{F}'_\ell$ and $\mathcal{F}'_r$ denote the set of columns of the mvf's $h_{b_\ell}(t)$ and $h_{b_r}(t)^*$, respectively, in the representation formula (8.19). Then for any $g^\circ \in G_\infty^{p\times p}(0)$,

$$\mathcal{G}(g^\circ, \mathcal{F}_\ell, \mathcal{F}_r) = \mathcal{G}(g^\circ, \bigvee_{\tau\geq 0}\{T_\tau \mathcal{F}_\ell\}, \bigvee_{\tau\geq 0}\{T_\tau \mathcal{F}_r\}) = \mathcal{G}(g^\circ, \mathcal{F}'_\ell, \mathcal{F}'_r).$$

Proof Lemma 8.9 implies that $\bigvee\{T_\tau \mathcal{F}'_\ell : \tau \geq 0\} = \bigvee\{T_\tau \mathcal{F}_\ell : \tau \geq 0\}$, and $\bigvee\{T_\tau \mathcal{F}'_r : \tau \geq 0\} = \bigvee\{T_\tau \mathcal{F}_r : \tau \geq 0\}$. Moreover, by Lemmas 8.13-8.15 applied to the mvf's $f = g - g^\circ$, with $g \in \mathcal{G}_\infty^{p\times p}(0)$, we obtain the two identifications

$$\mathcal{G}(g^\circ; \mathcal{F}_\ell, \mathcal{F}_r) = \mathcal{G}(g^\circ, \bigvee_{\tau\geq 0}\{T_\tau \mathcal{F}_\ell\}, \bigvee_{\tau\geq 0}\{T_\tau \mathcal{F}_r\})$$

and

$$\mathcal{G}(g^\circ; \mathcal{F}'_\ell, \mathcal{F}'_r) = \mathcal{G}(g^\circ, \bigvee_{\tau\geq 0}\{T_\tau \mathcal{F}'_\ell\}, \bigvee_{\tau\geq 0}\{T_\tau \mathcal{F}'_r\}) = \mathcal{G}(g^\circ, \bigvee \mathcal{F}_\ell, \bigvee \mathcal{F}_r).$$

□

Proof of Theorem 8.4. Let the mvf $g^\circ \in \mathcal{G}_\infty^{p\times p}(0)$ and the sets $\mathcal{F}_\ell \subseteq L_2^p([0,\alpha_\ell])$ and $\mathcal{F}_r \subseteq L_2^p([0,\alpha_r])$ be given, with $\alpha_\ell + \alpha_r > 0$. Let $\mathcal{F}'_\ell$ and $\mathcal{F}'_r$ be defined by $\mathcal{F}_\ell$ and $\mathcal{F}_r$ as above, as the sets of columns of the mvf's

h_{b_ℓ} and h_{b_r}. Let $b_3(\lambda) = b_\ell(\lambda)$ and $b_4(\lambda) = b_r^\sim(\lambda)$ and let $c^\circ = c_{g^\circ}$. Then $\mathcal{C}(c^\circ; b_3, b_4) = \{c_g : g \in \mathcal{G}(g^\circ; \mathcal{F}'_\ell, \mathcal{F}'_r)\}$ by Theorem 8.7. Consequently,

$$\mathcal{C}(c^\circ; b_3, b_4) = \{c_g : g \in \mathcal{G}(g^\circ; \mathcal{F}_\ell, \mathcal{F}_r)\},$$

since $\mathcal{G}(g^\circ; \mathcal{F}'_\ell, \mathcal{F}'_r) = \mathcal{G}(g^\circ; \mathcal{F}_\ell, \mathcal{F}_r)$ by the previous lemmas.

The converse statement of Theorem 8.4 was established in Theorem 8.7. □

A GHEP is said to be **determinate** if it has only one solution and **indeterminate** otherwise. It is said to be **completely indeterminate** or **strictly completely indeterminate** if the corresponding GCIP is completely indeterminate or strictly completely indeterminate, respectively.

Theorem 8.18 *If a GHEP$(g^\circ; \mathcal{F}_\ell, \mathcal{F}_r)$ is completely indeterminate, then:*

(1) *There exists a mvf $A \in \mathcal{P}(J_p)$ such that*

$$\{g : g \in \mathcal{G}(g^\circ; \mathcal{F}_\ell, \mathcal{F}_r)\} = \mathcal{C}(A). \tag{8.42}$$

(2) *Every mvf $A \in \mathcal{P}(J_p)$ for which the equality (8.42) holds is automatically an entire right regular J-inner mvf and thus may be normalized by the condition $A(0) = I_m$.*

(3) *Every normalized mvf $A \in \mathcal{U}(J_p)$ for which the equality (8.42) holds is uniquely defined by the data of the problem up to a scalar multiplier $e_\alpha(\lambda)$ such that $e_\alpha A \in \mathcal{U}(J_p)$.*

(4) *If $A \in \mathcal{U}(J_p)$ is such that the equality (8.42) holds, then*

$$A \in \mathcal{U}_{rsR}(J_p) \iff \text{GHEP}(g^\circ; \mathcal{F}_\ell, \mathcal{F}_r) \textit{ is strictly completely indeterminate.}$$

Proof Assertion (1) follows from the identification of the solution $GHEP(g^\circ; \mathcal{F}_\ell, \mathcal{F}_r)$ with the solutions of the corresponding $GCIP(c^\circ; b_3, b_4)$, given in Theorem 8.4, and the description of the set $\mathcal{C}(c^\circ; b_3, b_4)$, given in Theorem 7.70. The last theorem also guarantees that $A \in \mathcal{U}_{rR}(J_p)$. Consequently, $A(\lambda)$ is an entire mvf, since $b_3(\lambda)$ and $b_4(\lambda)$ are entire and $\{b_3, b_4\} \in ap_{II}(A)$. Thus, assertion (2) is proved and $A(\lambda)$ may be normalized by the condition $A(0) = I_m$. Then assertion (3) follows from Corollary 4.97.

Assertion (4) follows from formula (8.42) and the definitions of the class $\mathcal{U}_{rsR}(J_p)$ and the class of strictly completely indeterminate GHEP's. □

Theorem 8.19 *Let $A \in \mathcal{E} \cap \mathcal{U}(J_p)$, let $c^\circ \in \mathcal{C}(A)$, $c^\circ = c_{g^\circ}$, $g^\circ \in \mathcal{G}_\infty^{p\times p}(0)$ and let $\{b_3, b_4\} \in ap_{II}(A)$. Then $b_3(\lambda)$ and $b_4(\lambda)$ are entire inner $p \times p$ mvf's that may be assumed to be normalized. Let $\mathcal{F}_\ell$ and $\mathcal{F}_r$ denote the set of columns of the corresponding mvf's $h_{b_3}(t)$ and $h_{b_4}(t)^*$ that arise from the representation formula (8.19). Then the GHEP($g^\circ; \mathcal{F}_\ell, \mathcal{F}_r$) is completely indeterminate and*

$$\mathcal{C}(A) \subseteq \{c_g : g \in \mathcal{G}(g^\circ; \mathcal{F}_\ell, \mathcal{F}_r)\} \tag{8.43}$$

with equality if and only if $A \in \mathcal{U}_{rR}(J_p)$.

Proof The stated assertions follow from Theorems 8.7 and 7.69. □

The preceding two theorems yield the following results for the completely indeterminate classical Krein helical extension problem.

Theorem 8.20 *If the Krein helical extension problem HEP(g°; a) is completely indeterminate, then*

(1) *There exists a mvf $A \in \mathcal{P}(J_p)$ such that*

$$\mathcal{C}(A) = \{c_g : g \in \mathcal{G}(g^\circ; a)\}. \tag{8.44}$$

(2) *If $A \in \mathcal{P}(J_p)$ is such that the equality (8.44) holds, then $A(\lambda)$ is automatically an entire right regular J_p-inner mvf. Moreover,*

$$\{e_{a_3} I_p, e_{a_4} I_p\} \in ap_{II}(A), \quad \text{with } a_3 \geq 0,\ a_4 \geq 0 \text{ and } a_3 + a_4 = a.$$

(3) *If $a_3 \geq 0$, $a_4 \geq 0$ and $a_3 + a_4 = a$, then there exists exactly one normalized mvf $A^\circ \in \mathcal{E} \cap \mathcal{U}(J_p)$ such that (8.44) holds and $\{e_{a_3}, e_{a_4}\} \in ap_{II}(A^\circ)$. Moreover, the set of all normalized $A \in \mathcal{E} \cap \mathcal{U}(J_p)$ such that (8.44) holds is described by the formula*

$$A(\lambda) = e^{i\lambda\alpha} A^\circ(\lambda), \quad -a_3 \leq \alpha \leq a_4.$$

(4) *If $A \in \mathcal{P}(J_p)$ is such that the equality (8.44) holds, then*

$$A \in \mathcal{U}_{rsR}(J_p) \iff \text{the HEP}(g^\circ, a) \text{ is strictly completely indeterminate.}$$

Theorem 8.21 *Let $A \in \mathcal{E} \cap \mathcal{U}(J_p)$ and suppose that $\{e_{a_3} I_p, e_{a_4} I_p\} \in ap_{II}(A)$ for some choice of $a_3 \geq 0$ and $a_4 \geq 0$, with $a = a_3 + a_4 > 0$. Let $c^\circ \in \mathcal{C}(A)$, $c^\circ = c_{\widetilde{g}^\circ}$, where $\widetilde{g}^\circ \in \mathcal{G}_\infty^{p\times p}(0)$ and let g° denote the restriction of the mvf*

$\widetilde{g}^\circ$ to the interval $[-a, a]$. Then the HEP(g°; a) is completely indeterminate and

$$\mathcal{C}(A) \subseteq \{c_g : g \in HEP(g^\circ; a)\}. \tag{8.45}$$

Equality holds in the last inclusion if and only if $A \in \mathcal{U}_{rR}(J_p)$.

Proof The assertion follows from Theorem 8.19. □

The mvf's $A(\lambda)$ considered in formula (8.44) are called A-resolvent matrices of the HEP(g°; a) and the mvf's $B(\lambda) = A(\lambda)\mathfrak{V}$ are called B-resolvent matrices of this problem.

8.2 Positive definite extension problems

The representation (8.7) of a mvf $c \in \mathcal{C}^{p\times p}$ may be rewritten as the Fourier transform

$$c(\lambda) = \int_0^\infty e^{i\lambda t} f(t)dt \tag{8.46}$$

of a matrix valued distribution $f(t)$ of at most second order with a kernel $f(t-s)$ that is positive on $[0, \infty) \times [0, \infty)$ in the sense that:

$$f(t)^* = f(-t) \text{ and } \int_0^\infty \varphi(t)^* \left\{ \int_0^\infty f(t-s)\varphi(s)ds \right\} dt \geq 0 \tag{8.47}$$

for every infinitely differentiable vvf $\varphi(t)$ with support in a finite closed subinterval of $(0, \infty)$. Thus, the classical Krein helical extension problem may be considered as an extension problem for a positive definite $p \times p$ matrix valued distribution $f(t)$ of at most second order that is specified on a finite interval $[-a, a]$ to the interval $(-\infty, \infty)$. In this case, the convolution operator is also understood in the sense of distributions. The corresponding GHEP may also be reformulated in terms of a generalized extension problem for positive definite $p \times p$ matrix valued distributions. Krein solved the extension problem for classical positive definite functions from the interval $[-a, a]$ onto the interval $(-\infty, \infty)$ before he solved the HEP.

A $p \times p$ mvf $f(t)$ is said to belong to the class $\mathcal{P}_a^{p\times p}$ of positive definite mvf's on the finite closed interval $[-a, a]$ if:

(1) $f(t)$ is continuous on the interval $[-a, a]$.

(2) $f(t)^* = f(-t)$ for $t \in [-a, a]$.

(3) The kernel $k(t, s) = f(t-s)$ is positive on $[0, a] \times [0, a]$.

The condition in (3) means that

$$\sum_{i,j=1}^{n} \xi_i^* f(t_i - t_j)\xi_j \geq 0$$

for every finite collection of points $t_1, \ldots, t_n$ from $[0, a]$ and vectors $\xi_1, \ldots, \xi_n$ from $\mathbb{C}^p$. Under conditions (1) and (2), (3) holds if and only if the inequality in (8.47) holds for every vvf $\varphi \in L_2^p([0, a])$. The symbol $\mathcal{P}_\infty^{p\times p}$ denotes the class of $p \times p$ mvf's $f(t)$ which meet the same three conditions except with $(-\infty, \infty)$ in place of $[-a, a]$ in (1) and (2) and $[0, \infty)$ in place of $[0, a]$ in (3).

In [Kr5] Krein considered the following extension problem as an application of the theory of selfadjoint extensions of entire symmetric operators:

PEP$(f; a)$**:** *Given* $f \in \mathcal{P}_a^{p\times p}$*, find the set of all* $\widetilde{f} \in \mathcal{P}_\infty^{p\times p}$ *such that* $\widetilde{f}(t) = f(t)$ *for every point* t *in the interval* $-a \leq t \leq a$*.*

The symbol $\mathcal{P}(f; a)$ will designate the set of solutions to this problem.

It turns out that this extension problem is easily resolved with the help of the following characterization of $\mathcal{P}_a^{p\times p}$:

Theorem 8.22 *A* $p \times p$ *mvf* $f(t)$ *belongs to the class* $\mathcal{P}_a^{p\times p}$ *if and only if it admits an integral representation of the form*

$$f(t) = \frac{1}{\pi}\int_{-\infty}^{\infty} e^{-it\mu} d\sigma(\mu), \qquad |t| < a, \tag{8.48}$$

where $\sigma(\mu)$ *is a nondecreasing bounded* $p \times p$ *mvf on* $\mathbb{R}$*.*

Proof A proof for $p = 1$ is furnished in Theorem 2.1 on p. 127 of [GoGo97]. The proof for $p > 1$ is similar. □

Theorem 8.22 implies that if $f \in \mathcal{P}_a^{p\times p}$, then the set of $\widetilde{f} \in \mathcal{P}(f; a)$ is completely described by the formula

$$\widetilde{f}(t) = \frac{1}{\pi}\int_{-\infty}^{\infty} e^{-it\mu} d\sigma(\mu), \qquad -\infty < t < \infty, \tag{8.49}$$

where $\sigma(\mu)$ in (8.49) is the same as in (8.48). Thus, $\mathcal{P}(f; a) \neq \emptyset$ if $f \in \mathcal{P}_a^{p\times p}$, and the basic issue is to describe all bounded nondecreasing mvf's $\sigma(\mu)$ for which (8.48) holds.

The formula

$$c(\lambda) = \frac{1}{\pi i}\int_{-\infty}^{\infty} \frac{d\sigma(\mu)}{\mu - \lambda} \tag{8.50}$$

defines a one to one correspondence between the set of nondecreasing bounded $p \times p$ mvf's $\sigma(\mu)$ on $\mathbb{R}$ that are subject to the normalizations $\sigma(\mu - 0) = \sigma(\mu)$ and $\sigma(-\infty) = 0$ and the class $\mathcal{C}_0^{p\times p}$ introduced in Section 3.3. Thus, it follows from (8.50) that

$$c(\lambda) = \frac{1}{\pi}\int_{-\infty}^{\infty}\left\{\int_0^{\infty} e^{i(\lambda-\mu)t}dt\right\}d\sigma(\mu)$$

for $\lambda \in \mathbb{C}_+$, and hence, by (8.49), upon interchanging orders of integration, that

$$c(\lambda) = \int_0^{\infty} e^{i\lambda t} f(t)dt \quad \text{for } \lambda \in \mathbb{C}_+. \tag{8.51}$$

Thus, the description of the set of solutions to the PEP$(f;a)$ is equivalent via formula (8.51) to the description of the corresponding functions $c \in \mathcal{C}_0^{p\times p}$.

It follows easily from the representation formulas (8.48) and (8.3) that

$$\mathcal{P}_a^{p\times p} \subseteq \mathcal{G}_a^{p\times p}.$$

8.3 Connections between the positive definite and the helical extension problems

Let $c \in \mathcal{C}_0^{p\times p}$. Then, since $\mathcal{C}_0^{p\times p} \subset \mathcal{C}^{p\times p}$, it follows from Theorem 8.3 and the discussion in subsection 8.2 that $c(\lambda)$ admits two representations:

$$c(\lambda) = \int_0^{\infty} e^{i\lambda t} f(t)dt = \lambda^2 \int_0^{\infty} e^{i\lambda t} g(t)dt \quad (\text{for } \lambda \in \mathbb{C}_+),$$

where $f \in \mathcal{P}_\infty^{p\times p}$ and $g \in \mathcal{G}_\infty^{p\times p}(0)$. Since f is continuous and bounded on $\mathbb{R}$, the formula

$$\begin{aligned}\int_0^{\infty} e^{i\lambda t} f(t)dt &= \int_0^{\infty} e^{i\lambda t}\left\{\frac{d^2}{dt^2}\int_0^t (t-u)f(u)du\right\}dt \\ &= -\lambda^2 \int_0^{\infty} e^{i\lambda t}\left\{\int_0^t (t-u)f(u)du\right\}dt\end{aligned}$$

is readily verified (for $\lambda \in \mathbb{C}_+$) by integration by parts and serves to prove that

$$g(t) = -\int_0^t (t-u)f(u)du. \tag{8.52}$$

on the interval $(0,\infty)$.

Let C_a^2 denote the class of mvf's $g(t)$ for which the second derivative $g''(t)$ exists and is continuous on $[-a, a]$ if $a < \infty$ and on $(-\infty, \infty)$ if $a = \infty$, and let

$$Q \cap \mathcal{G}_a^{p\times p} = \{g \in \mathcal{G}_a^{p\times p} : g \in C_a^2,\ g(0) = 0 \text{ and } g'(0) = 0\}$$

for $0 < a \leq \infty$.

Theorem 8.23 *If $g \in Q \cap \mathcal{G}_a^{p\times p}$, then $f = -g''$ belongs to $\mathcal{P}_a^{p\times p}$. Conversely, if $f \in \mathcal{P}_a^{p\times p}$, then*

$$g(t) = -\int_0^t (t-u)f(u)du \tag{8.53}$$

belongs to $\mathcal{G}_a^{p\times p} \cap Q$ and $f = -g''$.

Proof It suffices to focus on the case $0 < a < \infty$. If $g \in Q \cap \mathcal{G}_a^{p\times p}$, then g'' is continuous on $[-a, a]$ and the kernel

$$g(t-s) - g(t) - g(-s)$$

is positive. Therefore,

$$\int_0^a \varphi'(t)^* \left\{ \int_0^a [g(t-s) - g(t) - g(-s)]\varphi'(s)ds \right\} dt \ \geq\ 0$$

for every choice of $p \times 1$ mvf φ with a continuous derivative on $[0, a]$ and $\varphi(0) = \varphi(a) = 0$. Integrating by parts twice, once with respect to s and once with respect to t, we obtain the inequality

$$-\int_0^a \varphi(t)^* \left\{ \int_0^a g''(t-s)\varphi(s)ds \right\} dt \ \geq\ 0.$$

Thus, the kernel $-g''(t-s)$ is positive and $f = -g''$ belongs to $\mathcal{P}_a^{p\times p}$, as claimed.

To verify the converse, let $f \in \mathcal{P}_a^{p\times p}$ and let $g(t)$ be defined by formula (8.53). Then it follows readily from the representation formula (8.48) that

$$g(t) = \frac{1}{\pi}\int_{-\infty}^{\infty} \frac{e^{-it\mu} - 1 + i\mu t}{\mu^2} d\sigma(\mu)$$

and hence that

$$g(t-s) - g(t) - g(-s) \ =\ \frac{1}{\pi}\int_{-\infty}^{\infty} \left\{ \frac{e^{-it\mu} - 1}{\mu} \right\} \left\{ \frac{e^{is\mu} - 1}{\mu} \right\} d\sigma(\mu)$$

is a positive kernel. □

Corollary 8.24 *A mvf $g \in Q \cap \mathcal{G}_a^{p\times p}$ if and only if it admits a representation of the form (8.3) with*

$$\beta = 0, \quad \sigma \text{ bounded and } \alpha = \frac{1}{\pi}\int_{-\infty}^{\infty} \frac{\mu}{1+\mu^2} d\sigma(\mu). \tag{8.54}$$

In particular, if $g \in \mathcal{G}_\infty^{p\times p}(0)$, then

$$c_g \in \mathcal{C}_0^{p\times p} \Longleftrightarrow g \in Q \cap \mathcal{G}_\infty^{p\times p}. \tag{8.55}$$

Theorem 8.25 *If $g \in Q \cap \mathcal{G}_a^{p\times p}$, then every solution $\widetilde{g}$ of the HEP$(g;a)$ belongs to $Q \cap \mathcal{G}_\infty^{p\times p}$.*

Proof The proof is broken into steps.

1. *There exists at least one solution $\widetilde{g}$ of the HEP$(g;a)$ which belongs to $Q \cap \mathcal{G}_\infty^{p\times p}$.*

Let $g \in Q \cap \mathcal{G}_a^{p\times p}$. Then, by Theorem 8.23, $f = -g''$ belongs to $\mathcal{P}_a^{p\times p}$. Let $\widetilde{f} \in \mathcal{P}(f;a)$ and set

$$\widetilde{g}(t) = -\int_0^t (t-s)\widetilde{f}(s)ds.$$

Then, since $\widetilde{g}(t) = g(t)$ for $t \in (-a,a)$, $\widetilde{g} \in Q \cap \mathcal{G}(g;a)$. This establishes the claim of this step.

2. *Every solution $\widetilde{g}$ of the HEP$(g;a)$ belongs to $Q \cap \mathcal{G}_\infty^{p\times p}$.*

The proof this step will be given in Section 8.5. □

There is also an important connection between $\mathcal{G}_\infty^{p\times p}$ and $\{f \in \mathcal{P}_\infty^{p\times p} : f(0) > 0\}$ that will be discussed in the next section.

8.4 Resolvent matrices for positive definite extension problems

A PEP$(f : a)$ is said to be

(1) **determinate** if the problem has only one solution.

(2) **indeterminate** if the problem has more than one solution.

(3) **completely indeterminate** if for every nonzero vector $\xi \in \mathbb{C}^p$ there exist at least two solutions $\widetilde{f}_1(t)$ and $\widetilde{f}_2(t)$ of the problem such that $\widetilde{f}_1(t)\xi \not\equiv \widetilde{f}_2(t)\xi$ on $\mathbb{R}$.

In this section we shall characterize the class of resolvent matrices for completely indeterminate positive extension problems. We begin by stating four lemmas (without proof), which serve to establish a connection between the solutions of a completely indeterminate PEP and a completely indeterminate HEP. The lemmas are formulated in terms of the following notation, where $\beta \in \mathbb{C}^{p\times p}$ and $\beta > 0$:

$$A_\beta(\lambda) = \begin{bmatrix} I_p & 0 \\ -i\lambda\beta & I_p \end{bmatrix}. \tag{8.56}$$

$$\mathcal{C}_0(\beta) = \{c \in \mathcal{C}_0^{p\times p} : \nu\Re c(i\nu) < \beta^{-1} \quad \text{for every} \quad \nu > 0 \quad \text{and} \quad \lim_{\nu\uparrow\infty} c(i\nu) = 0\} \tag{8.57}$$

and for $f \in \mathcal{P}_a^{p\times p}$, let

$$\mathcal{C}_0(f;a) = \left\{\int_0^\infty e^{i\lambda t}\widetilde{f}(t)dt : \widetilde{f} \in \mathcal{P}(f;a), \quad \lambda \in \mathbb{C}_+\right\}. \tag{8.58}$$

Lemma 8.26 *If $\beta > 0$ is a constant $p\times p$ matrix, then $A_\beta \in \mathcal{E}\cap\mathcal{U}(J_p)$ and*

$$T_{A_\beta}[\mathcal{C}^{p\times p}] = \mathcal{C}_0(\beta).$$

Lemma 8.27 *Let $f \in \mathcal{P}_a^{p\times p}$ and suppose that the PEP$(f;a)$ is completely indeterminate. Then*

$$f(0) > 0 \quad \textit{and} \quad \mathcal{C}_0(f;a) \subseteq \mathcal{C}_0(\beta) \quad \textit{for } \beta = f(0)^{-1}.$$

Lemma 8.28 *Let $f \in \mathcal{P}_a^{p\times p}$ and suppose that the PEP$(f;a)$ is completely indeterminate. Let $c \in \mathcal{C}_0(f;a)$ and let $\beta = f(0)^{-1}$. Then:*

(1) *$c = T_{A_\beta}[c_{\widetilde{g}}]$ for some $\widetilde{g} \in \mathcal{G}_\infty^{p\times p}(0)$.*

(2) *If $g(t) = \widetilde{g}(t)$ for $|t| \le a$, then $g \in \mathcal{G}_a^{p\times p}(0)$, the HEP$(g;a)$ is completely indeterminate and*

$$\mathcal{C}_0(f;a) = T_{A_\beta}[\mathcal{C}(g^\circ;a)], \quad \textit{where} \quad \mathcal{C}(g^\circ;a) = \{c_g : g \in \mathcal{G}_a^{p\times p}\}. \tag{8.59}$$

Lemma 8.29 *Let $g \in \mathcal{G}_a^{p\times p}(0)$ and suppose that the HEP$(g;a)$ is completely indeterminate. Let $\beta > 0$ be a constant $p\times p$ matrix and let $c = T_{A_\beta}[c_{\widetilde{g}}]$ for any choice of $\widetilde{g} \in \mathcal{G}(g;a)$. Then:*

(1) $c(\lambda) = \int_0^\infty e^{i\lambda t}\widetilde{f}(t)dt$ *for* $\lambda \in \mathbb{C}_+$ *and some choice of* $\widetilde{f} \in \mathcal{P}_\infty^{p\times p}$.

(2) *The mvf* $f(t) = \widetilde{f}(t)$ *for* $|t| \le a$ *belongs to* $\mathcal{P}_a^{p\times p}$ *and the HEP*$(f;a)$ *is completely indeterminate.*

(3) *The identity (8.59) holds.*

On the basis of this connection between $\mathcal{C}_0(f;a)$ and $\mathcal{C}(g;a)$ we can now obtain the following results on the resolvent matrices of PEP's from Theorem 8.20.

Theorem 8.30 *Let* $f \in \mathcal{P}_a^{p\times p}$ *and suppose that the PEP*$(f;a)$ *is completely indeterminate (which implies that* $f(0) > 0$*) and let* $\beta = f(0)^{-1}$. *Then there exists a unique mvf* $B \in \mathcal{E} \cap \mathcal{U}(j_p, J_p)$ *such that*

(1) $\mathcal{C}_0(f;a) = T_B[\mathcal{S}^{p\times p} \cap \mathcal{D}(T_B)]$.

(2) $\tau_B^+ = 0$ *and* $\tau_{B^{-1}}^+ = a$.

(3) $B(0) = \mathfrak{V}$.

Moreover, $\det B(\lambda) = (-1)^p e^{ipa\lambda}$ *and:*

(4) *The mvf* $A(\lambda) = B(\lambda)\mathfrak{V}$ *belongs to the class* $\mathcal{U}_{rR}(J_p)$.

(5) $$A_\beta^{-1} A \in \mathcal{E} \cap \mathcal{U}_{rR}(J_p).$$

Theorem 8.31 *Let the setting of Theorem 8.30 be in force and let* $\widetilde{B} \in \mathcal{P}(j_p, J_p)$ *be a second resolvent matrix for the completely indeterminate PEP*$(f;a)$ *under consideration such that*

$$T_{\widetilde{B}}[\mathcal{S}^{p\times p} \cap \mathcal{D}(T_{\widetilde{B}})] = \mathcal{C}_0(f;a). \tag{8.60}$$

Then

$$\widetilde{B}(\lambda) = e^{-ia_1\lambda} B(\lambda)\mathfrak{A}, \tag{8.61}$$

where $B(\lambda)$ *is the unique mvf discussed in Theorem 8.30,* $\mathfrak{A}$ *is a constant* j_p*-unitary matrix and* $0 \le a_1 \le a$. *Moreover,* $\widetilde{B} \in \mathcal{U}_{rR}(j_p, J_p)$.
Conversely, the equality (8.60) holds for every mvf $\widetilde{B}(\lambda)$ *of the form (8.61).*

Theorem 8.32 *Let* $\widetilde{B} \in \mathcal{E} \cap \mathcal{U}(j_p, J_p)$ *and let* $\{e_{a_1} I_p,\ e_{a_2} I_p\} \in ap(\widetilde{B})$, *where* $a_1 \ge 0$, $a_2 \ge 0$ *and* $a = a_1 + a_2 > 0$. *Suppose further that* $c^\circ \in T_{\widetilde{B}}[\mathcal{S}^{p\times p}] \ne \emptyset$ *and let* $c^\circ \in T_{\widetilde{B}}[\mathcal{S}^{p\times p}] \cap \mathcal{C}_0^{p\times p}$ *and* $\widetilde{f}^\circ \in \mathcal{P}_\infty^{p\times p}$ *corresponds to* c° *via formula*

(8.51) and let $f(t) = \widetilde{f}^\circ(t)$ *for* $|t| \le a$. *Then* $f \in \mathcal{P}_a^{p\times p}$ *and the following conclusions hold:*

(1) *The PEP* $(f;a)$ *is completely indeterminate.*

(2) $T_{\widetilde{B}}[\mathcal{S}^{p\times p}] \subseteq \mathcal{C}_0(f;a)$.

(3) *Equality holds in (2) if and only if* $\widetilde{B}\mathfrak{V} \in \mathcal{U}_{rR}(J_p)$.

(4) $\widetilde{B}(\lambda) = A_\beta(\lambda)B(\lambda)$, *where* $\beta = f(0)^{-1}$ *and* $B \in \mathcal{E} \cap \mathcal{U}(j_p, J_p)$.

Theorem 8.33 *A mvf* $\widetilde{B} \in \mathcal{P}(j_p, J_p)$ *is a B-resolvent matrix for a completely indeterminate PEP*$(f;a)$ *for some* $f \in \mathcal{P}_a^{p\times p}$ *(i.e.,* $T_{\widetilde{B}}[\mathcal{S}^{p\times p}] = \mathcal{C}_0(f;a)$*) if and only if* $\widetilde{B}(\lambda)$ *has the following properties:*

(1) $\widetilde{B}\mathfrak{V} \in \mathcal{E} \cap \mathcal{U}_{rR}(J_p)$.

(2) $\{e_{a_3}I_p, e_{a_4}I_p\} \in ap(\widetilde{B})$ *for some choice of* $a_3 \ge 0$ *and* $a_4 \ge 0$, *i.e., the blocks* b_{21} *and* b_{22} *satisfy the growth conditions*

$$a_3 = \tau^+_{b_{21}^\#} = \frac{1}{p}\delta^+_{b_{21}^\#} \quad \textit{and} \quad a_4 = \tau^+_{b_{22}} = \frac{1}{p}\delta^+_{b_{22}}.$$

(3) $\tau^+_{b_{21}^\#} + \tau^+_{b_{22}} = a$.

(4) $A_\beta^{-1}\widetilde{B} \in \mathcal{U}(j_p, J_p)$ *for some constant* $p \times p$ *matrix* $\beta > 0$.

8.5 Tangential and bitangential positive extension problems

The formula

$$\widehat{f}_+(\lambda) = \int_0^\infty e^{i\lambda t} f(t)dt, \quad \text{for } \lambda \in \mathbb{C}_+, \tag{8.62}$$

defines a one to one mapping from $\mathcal{P}_\infty^{p\times p}$ onto $\mathcal{C}_0^{p\times p}$. Moreover, since $\mathcal{P}_\infty^{p\times p} \subset \mathcal{G}_\infty^{p\times p}$, it follows from Theorem 8.3 that if f and f° belong to $\mathcal{P}_\infty^{p\times p}$, then

$$f(t) = f^\circ(t) \text{ for } |t| \le a \Longleftrightarrow e_a^{-1}\{\widehat{f}_+ - \widehat{f}_+^\circ\} \in \mathcal{N}_+^{p\times p}. \tag{8.63}$$

This equivalence serves to identify the solutions of the PEP$(f;a)$ with the set $\mathcal{C}(e_aI_p, I_p; \widehat{f}_+^\circ) \cap \mathcal{C}_0^{p\times p}$. The next lemma enables us to drop the extra intersection with $\mathcal{C}_0^{p\times p}$ and thus reduces the investigation of the PEP$(f;a)$ to a GCIP.

Lemma 8.34 *Let $c^\circ \in \mathcal{C}_0^{p\times p}$ and let $a > 0$. Then*

$$\mathcal{C}(e_a I_p, I_p; c^\circ) \subseteq \mathcal{C}_0^{p\times p}.$$

Proof Let $c \in \mathcal{C}(e_a I_p, I_p; c^\circ)$. Then, since

$$e_a^{-1}(c - c^\circ) = h \in \mathcal{N}_+^{p\times p},$$

it follows that

$$c(i\nu) = c^\circ(i\nu) + e^{-\nu a} h(i\nu)$$

for every choice of $\nu > 0$. But now as

$$h = g^{-1} f,$$

for some choice of $f \in S^{p\times p}$ and $g \in S_{out}^{1\times 1}$, it is readily checked that for any $\varepsilon > 0$, there exists an R such that

$$|h_{ij}(i\nu)| \le \exp\left\{-\frac{\nu}{\pi}\int_{-\infty}^{\infty} \frac{\ln|g(\mu)|}{\mu^2+\nu^2} d\mu\right\} \le \varepsilon$$

for $\nu \ge R$. Therefore, since trace $\{\nu \Re c(i\nu)\}$ is a monotone increasing function of ν on $(0, \infty)$,

$$\text{trace}\,\{\nu \Re c(i\nu)\} = \text{trace}\,\{\nu \Re c^\circ(i\nu)\} + \text{trace}\,\{\nu e^{-\nu} \Re h(iv)\}$$

is bounded on $(0, \infty)$. Moreover,

$$\lim_{\nu\uparrow\infty} c(i\nu) = \lim_{\nu\uparrow\infty} c^\circ(i\nu) + \lim_{\nu\uparrow\infty}\{e^{-\nu a} h(i\nu)\} = 0,$$

since $c^\circ(i\nu) \to 0$ as $\nu \to \infty$ and $h(i\nu)$ is bounded for $\nu \ge R$. Thus, $c \in \mathcal{C}_0^{p\times p}$, as claimed. □

We are now able to complete the proof of Theorem 8.25:

Proof of Step 2 of Theorem 8.25. Let $\widetilde{g}^\circ$ be a solution of the HEP$(g; a)$ which belongs to $Q \cap \mathcal{G}_\infty^{p\times p}$ and let $\widetilde{g} \in \mathcal{G}(g; a)$ be any solution of this extension problem. Then, by (8.55) and Theorem 8.3, respectively, $c_{\widetilde{g}^\circ} \in C_0^{p\times p}$ and

$$e_a^{-1}(c_{\widetilde{g}} - c_{\widetilde{g}^\circ}) \in \mathcal{N}_+^{p\times p}.$$

Therefore, by Lemma 8.34, $c_{\widetilde{g}} \in \mathcal{C}_0^{p\times p}$, which implies in turn that $\widetilde{g} \in Q \cap \mathcal{G}_\infty^{p\times p}$ by another application of (8.55). □

The next step is to consider tangential and bitangential versions of the PEP. Thus, for example, a left tangential version of the PEP may be based on a mvf $b \in \mathcal{E} \cap \mathcal{S}_{in}^{p\times p}$ and a mvf $f \in \mathcal{P}_a^{p\times p}$, where $a = \tau_b$. The objective is to find all $\widetilde{f} \in \mathcal{P}_\infty^{p\times p}$ such that

$$\int_t^a h_b(u)^* \widetilde{f}(u-t)du = \int_t^a h_b(u)^* f(u-t)du \tag{8.64}$$

for $0 \leq t \leq a$, where h_b is obtained from the representation (8.19) for b. There are two other formulations which are analogous to the formulation of the left tangential HEP when $\mathcal{F}_r = \{0\}$ or $\mathcal{F}_r = \emptyset$. Right tangential and bitangential versions of the PEP can be expressed in similar ways too. This problem can also be formulated in the *frequency domain* as in (8.65) below, with $b_3 = b$ and $b_4 = I_p$.

The preceding discussion leads naturally to the problem of finding the set.

$$\mathcal{C}_0(b_3, b_4; c^\circ) = \{c \in \mathcal{C}_0^{p\times p} : b_3^{-1}(c - c^\circ)b_4^{-1} \in \mathcal{N}_+^{p\times p}\} \tag{8.65}$$

for some given $b_3 \in \mathcal{E} \cap \mathcal{S}_{in}^{p\times p}$, $b_4 \in \mathcal{E} \cap \mathcal{S}_{in}^{p\times p}$ and $c^\circ \in \mathcal{C}_0^{p\times p}$. It is clear that

$$\mathcal{C}_0(b_3, b_4; c^\circ) \subseteq \mathcal{C}(b_3, b_4; c^\circ),$$

but (unlike the classical case in which $b_3 = e_a I_p$ and $b_4 = I_p$) equality does not hold in general. Thus this problem cannot be considered as a special case of the GCIP unless extra assumptions are imposed. One such convenient assumption is that either b_3 or b_4 have a nonconstant scalar inner divisor.

Lemma 8.35 *Let $c^\circ \in \mathcal{C}_0^{p\times p}$, $b_3 \in \mathcal{E} \cap \mathcal{S}_{in}^{p\times p}$, $b_4 \in \mathcal{E} \cap \mathcal{S}_{in}^{p\times p}$ and assume further that either b_3 or b_4 has a nonconstant scalar inner divisor. Then*

$$\mathcal{C}_0(b_3, b_4; c^\circ) = \mathcal{C}(b_3, b_4; c^\circ).$$

Proof Suppose for the sake of definiteness that $\beta(\lambda)$ is a nonconstant scalar inner divisor of $b_3(\lambda)$. Then $\beta(\lambda) = e_\tau(\lambda)$ for some $\tau > 0$ and

$$\mathcal{C}(b_3, b_4; c^\circ) \subseteq \mathcal{C}(e_\tau I_p, I_p; c^\circ).$$

However, by Lemma 8.34,

$$\mathcal{C}(e_\tau I_p, I_p; c^\circ) \subseteq \mathcal{C}_0(e_\tau I_p, I_p; c^\circ).$$

Therefore,

$$\mathcal{C}(b_3, b_4; c^\circ) \subseteq \mathcal{C}_0(b_3, b_4; c^\circ),$$

which is to say that equality prevails. □

We remark that if $b \in \mathcal{E} \cap \mathcal{S}_{in}^{p\times p}$ is expressed in the form (8.19), then

$$e_\tau^{-1} b \in \mathcal{S}_{in}^{p\times p} \Longleftrightarrow h_b(t) = b(0) \text{ for } t \in [0, \tau]. \tag{8.66}$$

8.6 The Krein accelerant extension problem

Let $\gamma \in \mathbb{C}^{p\times p}$ and let $\mathcal{A}_a^{p\times p}(\gamma)$ (for $0 < a \leq \infty$) denote the class of mvf's $h \in L_1^{p\times p}((-a, a))$ for which the kernel $f(t-s)$ based on the generalized function

$$f(t) = \delta(t)\Re\gamma + h(t) \tag{8.67}$$

is positive on $[0, a) \times [0, a)$ in the sense that

$$\int_0^a \varphi(t)^* \left\{ (\Re\gamma)\varphi(t) + \int_0^a h(t-s)\varphi(s)ds \right\} dt \geq 0 \tag{8.68}$$

for every $\varphi \in L_2^p((0, a))$. This condition forces $h(t)$ to be **Hermitian**, i.e., $h(t) = h(-t)^*$ for almost all points $t \in (-a, a)$. Following Krein, the function $h(t)$ in the representation (8.67) will be called the **accelerant** and the following extension problem will be called the **accelerant extension problem** (AEP):

AEP$(\gamma, h; a)$: *Given* $h \in \mathcal{A}_a^{p\times p}(\gamma)$, $0 < a < \infty$, *find* $\widetilde{h} \in \mathcal{A}_\infty^{p\times p}(\gamma)$ *such that* $\widetilde{h}(t) = h(t)$ *for a.e.* $t \in (-a, a)$.

The symbol $\mathcal{A}(\gamma, h; a)$ will be used to denote the set of solutions to this problem.

The condition (8.68) for $a = \infty$ can be reexpressed in terms of the Fourier transform $\widehat{h}$ of h.

Lemma 8.36 *Let* $h \in L_1^{p\times p}(\mathbb{R})$. *Then the following three conditions are equivalent:*

(1) $\int_0^\infty \varphi(t)^* \left\{ (\Re\gamma)\varphi(t) + \int_0^\infty h(t-s)\varphi(s)ds \right\} dt \geq 0$ *for every* $\varphi \in L_2^p(\mathbb{R}_+)$.

(2) $\int_{-\infty}^\infty \widehat{\varphi}(\mu)^* \{\Re\gamma + \widehat{h}(\mu)\} \widehat{\varphi}(\mu) d\mu \geq 0$ *for every* $\widehat{\varphi} \in L_2^p(\mathbb{R})$.

(3) $\Re\gamma + \widehat{h}(\mu) \geq 0$ *for every* $\mu \in \mathbb{R}$.

Moreover, the equivalence between these three statements continues to be *valid if the inequality ≥ 0 is replaced by $\geq \delta\|\varphi\|_{st}^2$ in (1), by $\geq \delta\|\widehat{\varphi}\|_{st}^2$ in (2) and $\geq \delta I_p$ in (3) for some $\delta > 0$.*

Proof The Plancherel formula implies that the inequality (1) holds for every $\varphi \in L_2^p((0,\infty))$ if and only if the inequality in (2) holds for every $\widehat{\varphi} \in H_2^p$. But, since the integral in (2) does not change if $\widehat{\varphi}(\mu)$ is replaced by $e^{-ia\mu}\widehat{\varphi}(\mu)$, it is readily checked that the inequality in (2) holds for every $\widehat{\varphi} \in H_2^p$ if and only if it holds for every $\widehat{\varphi} \in L_2^p$. Thus (1) is equivalent to (2). Finally, (2) is equivalent to (3) by a standard argument. □

Condition (3) of Lemma 8.36 implies that if $h \in \mathcal{A}_\infty^{p\times p}(\gamma)$, then

$$c(\lambda) = \gamma + 2\int_0^\infty e^{i\lambda t} h(t)dt \quad (\lambda \in \mathbb{C}_+) \tag{8.69}$$

belongs to $\mathcal{C}^{p\times p}$, because

$$\begin{aligned}\frac{c(\mu+i\nu)+c(\mu+i\nu)^*}{2} &= \Re\gamma + \int_{-\infty}^\infty e^{i\mu t}e^{-\nu|t|}h(t)dt \\ &= \Re\gamma + \frac{\nu}{\pi}\int_{-\infty}^\infty \frac{\widehat{h}(u)}{(u-\mu)^2+\nu^2}du \\ &= \frac{\nu}{\pi}\int_{-\infty}^\infty \frac{\Re\gamma + \widehat{h}(u)}{(u-\mu)^2+\nu^2}du\end{aligned}$$

for $\nu > 0$. Moreover, since this last formula can be reexpressed as

$$(\Re c)(\lambda) = (\Re c_1)(\lambda),$$

where

$$c_1(\lambda) = \frac{1}{\pi i}\int_{-\infty}^\infty \left\{\frac{1}{u-\lambda} - \frac{u}{1+u^2}\right\}\{\Re\gamma + \widehat{h}(u)\}du \tag{8.70}$$

for $\lambda \in \mathbb{C}_+$, it follows that

$$c(\lambda) = i\alpha + c_1(\lambda) \tag{8.71}$$

for a constant Hermitian matrix α. Thus, as $c_1(i) = c_1(i)^*$, we see that

$$i\alpha = \frac{c(i)-c(i)^*}{2} = \frac{1}{2}(\gamma-\gamma^*) + \int_0^\infty e^{-t}\{h(t)-h(t)^*\}dt. \tag{8.72}$$

Assertion (3) of Lemma 8.36 implies that $\Re\gamma \geq 0$ if $\mathcal{A}_a^{p\times p}(\gamma) \neq \emptyset$. If $\Re\gamma > 0$, then it is possible to renormalize the data so that $\gamma = I_p$. However, even in this case, there exist mvf's $h \in \mathcal{A}_a^{p\times p}(I_p)$ for which $\mathrm{AEP}(I_p; h; a)$ is not solvable; see Section 4 of [KrMA86] for an example. Because of this difficulty, we shall focus on the subclass

$$\mathring{\mathcal{A}}_a^{p\times p} = \left\{ h \in \mathcal{A}_a^{p\times p}(I_p) : \int_0^a \varphi(t)^* \left\{ \varphi(t) + \int_0^a h(t-s)\varphi(s)ds \right\} dt > 0 \right.$$
$$\text{for every nonzero } \varphi \in L_2^p(0,a)\} \quad \text{for } 0 < a \leq \infty. \tag{8.73}$$

$\mathring{\mathbf{A}}$EP(h; a): *Given $h \in \mathring{\mathcal{A}}_a^{p\times p}$, find $\widetilde{h} \in \mathring{\mathcal{A}}_\infty^{p\times p}$ such that $\widetilde{h}(t) = h(t)$ for almost all points $t \in (-a, a)$.*

The symbol $\mathring{\mathcal{A}}(h; a)$ will be used to designate the set of solutions to this problem.

If $h \in \mathring{\mathcal{A}}_\infty^{p\times p}$, then the mvf $c(\lambda)$ which is defined by formula (8.69) with $\gamma = I_p$ satisfies the following properties:

(1) c is continuous on $\mathbb{R}$.

(2) $(\Re c)(\mu) = I_p + \widehat{h}(\mu) > 0$ for every point $\mu \in \mathbb{R}$.

(3) $(\Re c)(\mu) \to I_p$ as $|\mu| \to \infty$.

(4) $c \in H_\infty^{p\times p}$.

Thus, $c \in \mathring{\mathcal{C}}^{p\times p}$. This fact will play an important role in the sequel.

8.7 Connections between the accelerant and helical extension problems

The classes $\mathcal{A}_a^{p\times p}(\gamma)$ and $\mathcal{G}_a^{p\times p}$ are connected:

Theorem 8.37 *Let $g \in \mathcal{G}_a^{p\times p}$ enjoy the following properties:*

(1) *g is locally absolutely continuous on $(-a, a)$ and $g(0) = 0$.*

(2) *g' is absolutely continuous on $(-a, 0) \cup (0, a)$, $g'(0+) = -\gamma$ and $g'(0-) = \gamma^*$.*

(3) *$g'' \in L_1^{p\times p}(-a, a)$.*

Then $h = -g''/2$ belongs to $\mathcal{A}_a^{p\times p}(\gamma)$. Conversely, if $h \in \mathcal{A}_a^{p\times p}(\gamma)$, then

$$g(t) = \begin{cases} -t\gamma - 2\int_0^t (t-s)h(s)ds & for \quad 0 \leq t < a \\ t\gamma^* + 2\int_t^0 (t-s)h(-s)^*ds & for \quad -a < t \leq 0 \end{cases} \tag{8.74}$$

belongs to $\mathcal{G}_a^{p\times p}$ and enjoys the properties (1)–(3) and $g(t) = g(-t)^*$.

Proof See [KrMA86]. □

If $g \in \mathcal{G}_a^{p\times p}$ admits a representation of the form (8.74) on some interval $(-a, a)$, then h is termed the **accelerant** of g on the interval $(-a, a)$.

Corollary 8.38 *Let* $0 < a < \infty$ *and suppose that* $g \in \mathcal{G}_a^{p\times p}$ *has an accelerant* $h \in \mathcal{A}_a^{p\times p}(\gamma)$*. Then*

$$\mathcal{A}(\gamma, h; a) = \{-\widetilde{g}'' : \widetilde{g} \in \mathcal{G}(g; a) \quad \text{and has an accelerant on } \mathbb{R}\}.$$

Formulas (8.69)–(8.72) imply that if $g \in \mathcal{G}_\infty^{p\times p}$ admits an accelerant $h \in \mathcal{A}_\infty^{p\times p}(\gamma)$, then $c_g \in \mathcal{C}_a^{p\times p}$ and

$$\Re c_g(\mu) = \Re\gamma + \widehat{h}(\mu) \quad \text{for every } \mu \in \mathbb{R}. \tag{8.75}$$

Conversely, if $c_g \in \mathcal{C}_a^{p\times p}$ and (8.75) holds for some $\gamma \in \mathbb{C}^{p\times p}$ and $h \in L_1^{p\times p}(\mathbb{R})$, then g has accelerant $h \in \mathcal{A}_\infty^{p\times p}(\gamma)$.

8.8 Conditions for the HEP$(g; a)$ to be strictly completely indeterminate

Theorem 8.39 *Let* $g \in \mathcal{G}_a^{p\times p}(0)$ *and* $a < \infty$ *and suppose that the HEP$(g; a)$ is strictly completely indeterminate. Then:*

(1) $g(t) = -\int_0^t v(s)ds$ *for some* $v \in L_2^p([0, a])$ *(i.e.,* $g(t)$ *is absolutely continuous in the interval* $[0, a]$*,* $g(0) = 0$*,* $g'(t) = -v(t)$ *a.e. in* $[0, a]$ *and* $v \in L_2^p([0, a])$*).*

(2) *The* $p\times 1$ *mvf* $\int_0^t v(t-s)\varphi(s)ds$ *is absolutely continuous on the interval* $[0, a]$ *and its derivative*

$$(X^\vee \varphi)(t) = \frac{d}{dt}\int_0^t v(t-s)\varphi(s)ds \tag{8.76}$$

belongs to $L_2^p([0, a])$ *for every choice of* $\varphi \in L_2^p([0, a])$*.*

(3) *The operator* $X^\vee$ *from* $L_2^p([0, a])$ *into itself which is defined by formula (8.76) is bounded.*

Moreover, upon defining

$$v(t) = -v(-t)^*, \quad \text{for almost all points } t \in [-a, 0],$$

we also have the following additional conclusions:

(4) *The $p\times 1$ mvf $\int_t^a v(t-s)\varphi(s)ds$ is absolutely continuous on the interval $[0,a]$ and the adjoint $(X^\vee)^*$ of $X^\vee$ in $L_2^p([0,a])$ is given by the formula*

$$(X^\vee)^*\varphi)(t) = \frac{d}{dt}\int_t^a v(t-s)\varphi(s)ds \tag{8.77}$$

for every $\varphi \in L_2^p([0,a])$.

(5) *There exists a $\delta > 0$ such that the operator $Y^\vee = X^\vee + (X^\vee)^*$, which is given by the formula*

$$(Y^\vee\varphi)(t) = \frac{d}{dt}\int_0^a v(t-s)\varphi(s)ds \tag{8.78}$$

on $L_2^p([0,a])$, is bounded from below by δI, i.e.,

$$\int_0^a \varphi(t)^*\left\{\frac{d}{dt}\int_0^a v(t-s)\varphi(s)ds\right\}dt \ge \delta\int_0^a \varphi(t)^*\varphi(t)dt \tag{8.79}$$

for every $\varphi \in L_2^p([0,a])$.

Proof Let $\widetilde{g}^\circ \in \mathcal{G}(g;a)$ and let

$$c^\circ(\lambda) = \lambda^2\int_0^\infty e^{i\lambda t}\widetilde{g}^\circ(t)dt \quad \text{for } \lambda \in \mathbb{C}_+.$$

Then, under the given assumptions, the GCIP$(e_aI_p, I_p; c^\circ)$ is strictly completely indeterminate. The rest of the proof is long and will be divided into steps.

1. *There exists a mvf $c \in \mathcal{C}(e_aI_p, I_p; c^\circ) \cap \mathring{\mathcal{C}}^{p\times p}$ that is meromorphic in $\mathbb{C}$ with $0 \in \mathfrak{h}_c$ and $c(0) = I_p$.*

Let $A \in \mathcal{E} \cap \mathcal{U}_{rsR}(J_p)$ be such that $\mathcal{C}(A) = \mathcal{C}(e_aI_p, I_p; c^\circ)$, $A(0) = I_m$ and $\{e_aI_p, I_p\} \in ap_{II}(A)$. A mvf $A(\lambda)$ with these properties exists by Theorem 8.20. Let $W = \mathfrak{V}A\mathfrak{V}$ and let $s^\circ = T_{\mathfrak{V}}[c^\circ]$. Then $T_W[\mathcal{S}^{p\times q}] = \mathcal{S}(e_aI_p, I_p; s^\circ)$ and $T_W[\mathcal{S}^{p\times q}] \cap \mathring{\mathcal{S}}^{p\times p} \neq \emptyset$, by Lemma 7.68. Thus, there exists a number ρ, $0 < \rho < 1$, and an mvf $s_\rho \in T_W[\mathcal{S}^{p\times q}]$ such that $\|s_\rho\|_\infty < \rho$. Then

$$\rho\mathcal{S}(e_aI_p, I_p; \rho^{-1}s_\rho) \subseteq T_W[\mathcal{S}^{p\times q}]$$

and $\|\rho^{-1}s_\rho\|_\infty < 1$. By Theorem 7.48 there exists a mvf $W_\rho \in \mathcal{U}_{rsR}(j_{pq})$ such that $T_{W_\rho}[\mathcal{S}^{p\times q}] = \mathcal{S}(e_aI_p, I_p; \rho^{-1}s_\rho)$ and $\{e_aI_p, I_p\} \in ap(W_\rho)$. Since $W_\rho \in \mathcal{U}_{rR}(j_{pq})$ and $\{e_aI_p, I_p\} \in ap(W)$, $W_\rho \in \mathcal{E}^{m\times m}$, by Theorem 4.54. Consequently, W_ρ may be normalized by the condition $W_\rho(0) = I_m$. Let

$s_\rho^\circ \in T_{W_\rho}[0]$. Then $s_\rho^\circ = \mathcal{S}(e_a I_p, I_p; \rho^{-1} s_\rho)$ and s_ρ° is a meromorphic mvf in $\mathbb{C}$ with $0 \in \mathfrak{h}_{s_\rho^\circ}$ and $s_\rho^\circ(0) = 0$. Consequently, the mvf $\rho s_\rho^\circ \in T_W[\mathcal{S}^{p\times q}]$ is meromorphic in $\mathbb{C}$ with $\|\rho s_\rho^\circ\|_\infty \le \rho < 1$. Moreover, $0 \in \mathfrak{h}_{\rho s_\rho^\circ}$ and $(\rho s_\rho^\circ)(0) = 0$.

Let $c = T_{\mathfrak{V}}[\rho s_\rho^\circ]$. Then $c \in \mathcal{C}(A) \cap \mathring{\mathcal{C}}^{p\times p}$ is meromorphic in $\mathbb{C}$ with $0 \in \mathfrak{h}_c$ and $c(0) = I_p$.

2. *The mvf c that was constructed above satisfies the condition*

$$R_0 c \in H_\infty^{p\times p} \cap H_2^{p\times p} \tag{8.80}$$

from which follows that:

(a) $c(\lambda) = I_p - i\lambda \int_0^\infty e^{i\lambda s} u(s) ds$ *for* $\lambda \in \mathbb{C}_+$ *and some* $u \in L_2^{p\times p}$.

(b) $c(\lambda) = c_{\widetilde{g}}(\lambda)$, *where* $\widetilde{g} \in \mathcal{G}_\infty(0)$ *and*

$$\widetilde{g}(t) = -tI_p - \int_0^t u(s) ds, \quad t \ge 0. \tag{8.81}$$

(c) *The mvf's* $\widehat{u}(\lambda)$ *and* $\widehat{u}_1(\lambda) = -i\lambda \widehat{u}(\lambda)$ *belong to* $H_\infty^{p\times p}$.

Since $c \in \mathring{\mathcal{C}}^{p\times p}$, $0 \in \mathfrak{h}_c$ and $c(0) = I_p$, the inclusion (8.80) follows by elementary estimates; and suffices to justify (a) and (c). Then, since

$$\begin{aligned} -\widehat{u}_1(\lambda) &= i\lambda \int_0^\infty e^{i\lambda t} u(t) dt = i\lambda \int_0^\infty e^{i\lambda t} \left\{ \frac{d}{dt} \int_0^t u(s) ds \right\} dt \\ &= \lambda^2 \int_0^\infty e^{i\lambda t} \left\{ \int_0^t u(s) ds \right\} dt \text{ for } \lambda \in \mathbb{C}_+ \end{aligned}$$

and

$$c(\lambda) = I_p + \widehat{u}_1(\lambda) = -\lambda^2 \int_0^\infty e^{i\lambda t} I_p dt + \widehat{u}_1(\lambda),$$

(b) holds.

3. *If* $u(t)$ *is defined as in Step 2, then assertions (1)–(5) of the theorem are valid for*

$$v(t) = \begin{cases} I_p + u(t) & \text{for almost all points } t \in (0, a) \\ -v(-t)^* & \text{for almost all points } t \in (-a, 0). \end{cases} \tag{8.82}$$

The mvf $\widetilde{g} \in \mathcal{G}_\infty(0)$ that was constructed in (8.81) belongs to $\mathcal{G}(g;a)$, i.e., $\widetilde{g}(t) = g(t)$ for $t \in [0,a]$; and formulas (8.81) and (8.82) serve to justify (1). Moreover, since $\widehat{u}, \widehat{u}_1 \in H_\infty^{p\times p}$, the connection between the multiplication operators $M_{\widehat{u}}|_{H_2^p}$ and $M_{\widehat{u}_1}|_{H_2^p}$ with the corresponding convolution operators in the space $L_2^p(\mathbb{R}_+)$ justifies assertions (2)–(4) of the theorem and formula (8.78) for $Y^\vee = X^\vee + (X^\vee)^*$. Then (8.79) follows from the lower bound $2\Re c(\mu) \geq \delta I_p$, $\mu \in \mathbb{R}$ for the mvf $c(\lambda)$ considered above , since

$$\begin{aligned}\langle Y^\vee \varphi^\vee, \varphi^\vee\rangle_{st} &= 2\Re\langle P_{\mathcal{H}(e_a I_p)} M_c \varphi, \varphi\rangle_{st} = 2\langle P_{\mathcal{H}(e_a I_p)} (\Re c)\varphi, \varphi\rangle_{st} \\ &\geq 2\delta\langle \varphi, \varphi\rangle_{st} = 2\delta\langle \varphi^\vee, \varphi^\vee\rangle_{st}\end{aligned}$$

for every $\varphi^\vee \in L_2^p([0,a])$. □

Let Π_a denote the orthogonal projection of H_2^p onto $\mathcal{H}(e_a I_p)$.

Lemma 8.40 *If $f \in H_\infty^{p\times p}$, then there exists a locally summable $p \times p$ mvf v on $\mathbb{R}_+$ such that*

$$f(\lambda) = (i\lambda)^2 \int_0^\infty e^{i\lambda t}\{\int_0^t v(s)ds\}dt \quad \textit{for } \lambda \in \mathbb{C}_+. \tag{8.83}$$

Moreover, if $v(t) = -v(-t)^*$ for almost all points $t \in \mathbb{R}_+$, then for every $a > 0$

$$\int_0^t v(t-s)\varphi(s)ds \quad \textit{and} \quad \int_t^a v(t-s)\varphi(s)ds$$

are absolutely continuous for every $\varphi \in L_2^p([0,a])$ and

$$(\Pi_a f\widehat{\varphi})(\lambda) = \int_0^a e^{i\lambda t}\{\frac{d}{dt}\int_0^t v(t-s)\varphi(s)ds\}dt \tag{8.84}$$

and

$$(\Pi_a f^*\widehat{\varphi})(\lambda) = \int_0^a e^{i\lambda t}\{\frac{d}{dt}\int_t^a v(t-s)\varphi(s)ds\}dt \tag{8.85}$$

for every choice of $\varphi \in L_2^p([0,a])$.

Proof If $f \in H_\infty^{p\times p}$ and $r > \|f\|_\infty$, then $s^\circ = f/r$ belongs to $\mathring{\mathcal{S}}^{p\times p}$ and

$$c^\circ = \begin{bmatrix} I_p & 2s^\circ \\ 0 & I_p \end{bmatrix} \quad \text{belongs to } \mathring{\mathcal{C}}^{p\times p}.$$

Therefore, the GCIP$(e_a I_p, I_p; c^\circ)$ is strictly completely indeterminate. Thus, the asserted results follow from the proof of Theorem 8.39. □

Theorem 8.41 *Let $v \in L_2^p([-a,a])$, $0 < a < \infty$, be such that $v(-t) = v(t)^*$ for almost all points $t \in (-a,a)$ and properties (2)–(5) in Theorem 8.39 are in force (property (4) follows from (2) and (3)). Then the mvf*

$$g(t) = -\int_0^t v(s)ds, \quad -a \le t \le a,$$

belongs to $\mathcal{G}_a^{p\times p}(0)$ and the HEP$(g;a)$ is strictly completely indeterminate.

Proof The bounded operator $X^\vee$ defined on $L_2^p([0,a])$ by formula (8.76) satisfies the condition

$$(X^\vee)^* T_\tau = T_\tau (X^\vee)^*, \quad \tau \ge 0,$$

where T_τ, $\tau \ge 0$, is the semigroup of backward shifts on $L_2^p([0,a])$:

$$(T_\tau \varphi)(t) = \begin{cases} \varphi(t+\tau) & \text{for } 0 \le t < a-\tau \\ 0 & \text{for } a-\tau < t < a. \end{cases}$$

The bounded operator X in $\mathcal{H}(e_a I_p)$ corresponding to $X^\vee$ that is defined by the formula

$$(X\widehat{\varphi}) = (X^\vee \varphi)^\wedge,$$

satisfies the conditions $2\Re X \ge \delta I$ and

$$X^* \Pi_a e_{-\tau} \widehat{\varphi} = \Pi_a e_{-\tau} X^* \widehat{\varphi}, \quad \widehat{\varphi} \in \mathcal{H}(e_a I_p).$$

Consequently,

$$X \Pi_a e_\tau \widehat{\varphi} = \Pi_a e_\tau X \widehat{\varphi}, \quad \tau \ge 0. \tag{8.86}$$

Let

$$\widetilde{X} = X\Pi_a|_{H_2^p} \quad \text{and} \quad \widetilde{T}_\tau^* = e_\tau|_{H_2^p}.$$

Since $\Pi_a e_\tau e_a H_2^p = 0$ for $\tau \ge 0$, the relation (8.86) is equivalent to the relation

$$\widetilde{X}^* \widetilde{T}_\tau = \widetilde{T}_\tau \widetilde{X}^*,$$

where $\widetilde{T}_\tau = \Pi_+ e_{-\tau}|_{H_2^p}$. Moreover,

$$\widetilde{X}|_{\mathcal{H}(e_a I_p)} + (\widetilde{X}|_{\mathcal{H}(e_a I_p)})^* = X + X^* \ge \delta I.$$

Thus, by Theorems 7.75 and 7.76 with $b(\lambda) = e_a(\lambda) I_p$, $\mathring{\mathcal{C}}^{p\times p}(e_a I_p, \widetilde{X}) \ne \emptyset$ and

$$\mathring{\mathcal{C}}(e_a I_p, \widetilde{X}) = \mathcal{C}(A) \cap \mathring{\mathcal{C}}^{p\times p}$$

for some mvf $A \in \mathcal{U}_{rsR}(J_p)$ with $\{e_a I_p, I_p\} \in ap_{II}(A)$. Then, as $A \in \mathcal{E}^{m\times m}$ may be normalized by setting $A(0) = I_m$, the assertion of Step 2 in the previous theorem is valid. The mvf $c(\lambda) \in \mathcal{C}(A) \cap \mathring{\mathcal{C}}^{p\times p}$ such that $0 \in \mathfrak{h}_c$ and $c(0) = I_p$ has the representation (8.80) and $c = c_{\widetilde{g}}$, where $\widetilde{g} \in \mathcal{G}_\infty^{p\times p}(0)$ has the representation (8.81). Furthermore, since $c \in \mathring{\mathcal{C}}(e_a I_p, \widetilde{X})$,

$$\widetilde{X} = \Pi_a M_c|_{H_2^p}, \quad \text{i.e.,} \ X\Pi_a|_{H_2^p} = \Pi_a M_c|_{H_2^p}.$$

Thus, upon expressing the operator X in the terms of the mvf $v(t)$, the last equality implies that

$$\widetilde{g}(t) = g(t), \quad \text{for} \quad 0 \le t \le a,$$

where $g(t) = -\int_0^t v(s)ds$. Therefore, $g \in \mathcal{G}_\infty(0)$ and the HEP$(g; a)$ is strictly completely indeterminate. □

8.9 Formulas for the normalized resolvent matrix for the HEP

In this section we shall apply the general formulas of Section 6 for B-resolvent matrices for the GCIP$(b_3, I_p; c)$ in the strictly completely indeterminate case to the special setting which corresponds to the HEP$(g; a)$. Thus, in particular, we shall choose $b_3 = e_a I_p$ and shall define the time domain versions of the operators X and Y that intervene in the formulas for $B(\lambda)$ directly in terms of $g(t)$ via formulas (8.76) and (8.78).

Theorem 8.42 *Let the HEP (g;a) be strictly completely indeterminate and let*

$$g(t) = -\int_0^t v(s)ds, \quad 0 \le t \le a, \ a < \infty, \tag{8.87}$$

where $v \in L_2^p([0, a])$ and is extended to the full interval $[-a, a]$ by the recipe $v(-t) = -v(t)^*$ for almost all points $t \in [-a, a]$. Then:

(1) *The mvf's $\int_0^t v(t-s)\varphi(s)ds$ and $\int_t^a v(t-s)\varphi(s)ds$ are absolutely continuous for every choice of $\varphi \in L_2^{p\times p}([0, a])$.*

(2) *The system of equations*

$$\frac{d}{dt}\int_0^a v(t-s)\varphi_{22}(s)ds = I_p, \quad 0 \le t \le a, \tag{8.88}$$

$$\frac{d}{dt}\int_0^a v(t-s)\varphi_{21}(s)ds = -v(t), \quad 0 \le t \le a, \tag{8.89}$$

for φ_{22} and φ_{21} is uniquely solvable in $L_2^{p\times p}([0, a])$.

(3) *The mvf's*

$$\varphi_{1j}(t) = -\frac{d}{dt}\int_t^a v(t-s)\varphi_{2j}(s)ds, \quad j=1,2, \quad 0 \le t \le a, \tag{8.90}$$

belong to $L_2^{p\times p}([0,a])$.

(4) *There is exactly one B-resolvent matrix* $B(\lambda)$ *of the* HEP$(g;a)$ *in the class* $\mathcal{N}_+ \cap \mathcal{U}(j_p, J_p)$ *with* $B(0) = \mathfrak{V}$*; it is given by the formula*

$$B(\lambda) = \mathfrak{V} - \frac{i\lambda}{\sqrt{2}}\int_0^a e^{i\lambda t}\begin{bmatrix}\varphi_{11}(t)-\varphi_{12}(t) & \varphi_{11}(t)+\varphi_{12}(t)\\ \varphi_{21}(t)-\varphi_{22}(t) & \varphi_{21}(t)+\varphi_{22}(t)\end{bmatrix}dt. \tag{8.91}$$

Proof If the HEP$(g^\circ;a)$ is strictly completely indeterminate, then, by Theorem 8.20, there exists exactly one mvf $A(\lambda)$ that meets the constraints

(a) $A \in \mathcal{E} \cap \mathcal{U}_{rsR}(J_p)$,

(b) $\{e_a I_p, I_p\} \in ap_{II}(A)$,

(c) $A(0) = I_m$,

such that

$$\{c_{\widetilde{g}} : \widetilde{g} \in \mathcal{G}(g;a)\} = \mathcal{C}(A).$$

Moreover, by Theorem 8.3, there exists a mvf $c^\circ = c_{\widetilde{g}^\circ}$ in $\mathcal{C}(A)$ such that $c^\circ \in \mathring{\mathcal{C}}^{p\times p}$ and

$$\mathcal{C}(e_a I_p, I_p; c^\circ) = \mathcal{C}(A).$$

Thus, the formulas for the resolvent matrix of a strictly completely indeterminate GCIP that were obtained in Section 7.9 are applicable to the GCIP$(b, I_p; c^\circ)$ with $b = e_a I_p$:

$$A(\lambda) = I_m + 2\pi i\lambda K_0(\lambda)J_p, \tag{8.92}$$

where

$$K_0(\lambda) = \left(\begin{bmatrix}-X^*\\ I\end{bmatrix}Y^{-1}\begin{bmatrix}-Xk_0 & k_0\end{bmatrix}\right)(\lambda), \tag{8.93}$$

$$k_0(\lambda) = \frac{I_p - e_a(\lambda)I_p}{-2\pi i\lambda} \quad \text{is the RK of the RKHS } \mathcal{H}(e_a I_p), \tag{8.94}$$

$$X = P_{\mathcal{H}(e_a I_p)} M_{c^\circ}|_{\mathcal{H}(e_a I_p)} \quad \text{and} \quad Y = X + X^*. \tag{8.95}$$

Thus,

$$K_0(\lambda) = \frac{1}{2\pi}\begin{bmatrix} \widehat{\varphi}_{11}(\lambda) & \widehat{\varphi}_{12}(\lambda) \\ \widehat{\varphi}_{21}(\lambda) & \widehat{\varphi}_{22}(\lambda) \end{bmatrix} = \frac{1}{2\pi}\int_0^a e^{i\lambda t}\begin{bmatrix} \varphi_{11}(t) & \varphi_{12}(t) \\ \varphi_{21}(t) & \varphi_{22}(t) \end{bmatrix} dt,$$

where the $\widehat{\varphi}_{ij}$ are $p \times p$ mvf's with columns in $\mathcal{H}(e_a I_p)$ that are obtained by first solving the equations

$$Y\widehat{\varphi}_{21} = -2\pi X k_0 \quad \text{and} \quad Y\widehat{\varphi}_{22} = 2\pi k_0 \tag{8.96}$$

(column by column) for $\widehat{\varphi}_{21}$ and $\widehat{\varphi}_{22}$ and then setting

$$\widehat{\varphi}_{11} = -X^*\widehat{\varphi}_{21} \quad \text{and} \quad \widehat{\varphi}_{12} = -X^*\widehat{\varphi}_{22}. \tag{8.97}$$

The equations in (8.96) are uniquely solvable because Y is strictly positive. With the help of Lemma 8.40 it is readily seen that (8.96) and (8.97) are equivalent to (8.88) and (8.89).

Finally the formula for $B(\lambda)$ follows by first substituting the formula for the kernel $K_0(\lambda)$ into (8.92) to obtain

$$A(\lambda) = I_m - i\lambda \begin{bmatrix} \widehat{\varphi}_{12}(\lambda) & \widehat{\varphi}_{11}(\lambda) \\ \widehat{\varphi}_{22}(\lambda) & \widehat{\varphi}_{21}(\lambda) \end{bmatrix},$$

and then multiplying on the right by $\mathfrak{V}$. □

Corollary 8.43 *In the setting of Theorem 8.42,*

$$B(\lambda) = \mathfrak{V} + \frac{i\lambda}{\sqrt{2}}\int_0^a e^{i\lambda t}\begin{bmatrix} -((X^\vee)^*\varphi_1)(t) & -((X^\vee)^*\varphi_2)(t) \\ \varphi_1(t) & \varphi_2(t) \end{bmatrix} dt, \tag{8.98}$$

where φ_1 and φ_2 are solutions of the equations

$$\frac{d}{dt}\int_0^a v(t-s)\varphi_1(s)ds = v(t) + I_p \tag{8.99}$$

and

$$\frac{d}{dt}\int_0^a v(t-s)\varphi_2(s)ds = v(t) - I_p, \tag{8.100}$$

respectively, in the space $L_2^{p\times p}([0,a])$.

8.10 B-resolvent matrices for the accelerant case

The B-resolvent matrix $B(\lambda)$ of the HEP$(g;a)$ is uniquely defined by the conditions

$$\text{(i) } B \in \mathcal{U}(j_p, J_p), \quad \text{(ii) } \{e_a I_p, I_p\} \in ap(B) \text{ and (iii) } B(0) = \mathfrak{V}.$$

Let $g \in \mathcal{G}_a^{p\times p}(0)$ have an accelerant $h \in \mathcal{A}_a^{p\times p}(\gamma)$. Then

$$g(t) = -\int_0^t v(s)ds \quad \text{for } -a \le t \le a, \tag{8.101}$$

where

$$v(t) = \begin{cases} \gamma + 2\int_0^t h(u)du & \text{for } 0 < t \le a \\ -\gamma^* - 2\int_t^0 h(u)du & \text{for } -a \le t < 0 \end{cases} \quad \text{and} \quad h(-u) = h(u)^*. \tag{8.102}$$

In this case Theorem 8.39 is applicable and formulas (8.76)–(8.78) may be rewritten in terms of the accelerant h:

$$\begin{aligned}(X^\vee\varphi)(t) &= \frac{d}{dt}\int_0^t \left\{\gamma + 2\int_0^{t-s} h(u)du\right\}\varphi(s)ds \\ &= \gamma\varphi(t) + 2\int_0^t h(t-s)\varphi(s)ds,\end{aligned} \tag{8.103}$$

$$\begin{aligned}((X^\vee)^*\varphi)(t) &= -\frac{d}{dt}\int_t^a \left\{\gamma^* + 2\int_0^{s-t} h(u)^*du\right\}\varphi(s)ds \\ &= \gamma^*\varphi(t) + 2\int_t^a h(s-t)^*\varphi(s)ds \\ &= \gamma^*\varphi(t) + 2\int_t^a h(t-s)\varphi(s)ds\end{aligned} \tag{8.104}$$

and

$$(Y^\vee\varphi)(t) = (\gamma+\gamma^*)\varphi(t) + 2\int_0^a h(t-s)\varphi(s)ds, \tag{8.105}$$

where $Y^\vee \ge 0$, since $h \in \mathcal{A}_a^{p\times p}(\gamma)$.

Lemma 8.44 *If $h \in \mathcal{A}_a^{p\times p}(I_p)$ and $Y^\vee$ is defined by formula (8.105) with $\gamma = I_p$, then*

$$\langle Y^\vee\varphi, \varphi\rangle_{st} > 0 \quad \textit{for every nonzero } \varphi \in L_2^p([0,a]) \tag{8.106}$$

if and only if there exists a $\delta > 0$ *such that*

$$\langle Y^\vee \varphi, \varphi \rangle_{st} \geq \delta \langle \varphi, \varphi \rangle_{st} \quad \textit{for every } \varphi \in L_2^p([0,a]). \tag{8.107}$$

Proof If (8.106) is in force, then the integral operator $Y^\vee$ maps $L_2^p([0,a])$ injectively into itself, and thus onto itself, by the Fredholm alternative. Therefore, by a theorem of Banach, $Y^\vee$ has a bounded inverse, i.e., (8.106) implies (8.107). The converse is obvious. □

Theorem 8.45 *If* $g \in \mathcal{G}_a^{p\times p}(0)$ *has an accelerant* $h \in \mathcal{A}_a^{p\times p}(I_p)$, *then the following three conditions are equivalent:*

(1) *The HEP*$(g;a)$ *is completely indeterminate.*

(2) *The HEP*$(g;a)$ *is strictly completely indeterminate.*

(3) $h \in \mathring{\mathcal{A}}^{p\times p}$.

Proof If (1) is in force and $\widetilde{g^\circ} \in \mathcal{G}(g;a)$, then the GSIP$(e_a I_p, I_p; s^\circ)$ based on $s^\circ = T_{\mathfrak{V}}[c_{\widetilde{g^\circ}}]$ is completely indeterminate. Therefore, the NP(Γ) with $\Gamma = \Gamma(e_{-a}s^\circ)$ is completely indeterminate and hence $I - \Gamma^*\Gamma > 0$ by Remark 7.9. But this implies that $Y^\vee > 0$ and hence that (1) $\Longrightarrow$ (2). The converse implication follows from Theorem 7.69. Finally the equivalence of (2) and (3) follows from Lemma 8.44 and Theorem 8.39. □

Theorem 8.46 *If* $g \in \mathcal{G}_a^{p\times p}(0)$ *has an accelerant* $h \in \mathring{\mathcal{A}}_a^{p\times p}$, *then:*

(1) *The HEP*$(g;a)$ *is strictly completely indeterminate.*

(2) *The B-resolvent matrix* $B(\lambda)$ *of the HEP*$(g;a)$ *can be expressed in terms of* h:

$$B(\lambda) = \mathfrak{V} + \frac{i\lambda}{\sqrt{2}} \int_0^a e^{i\lambda t} \mathcal{B}(t)dt, \tag{8.108}$$

where

$$\mathcal{B}(t) = \begin{bmatrix} -((X^\vee)^*\varphi_1)(t) & -((X^\vee)^*\varphi_2)(t) \\ \varphi_1(t) & \varphi_2(t) \end{bmatrix} \tag{8.109}$$

is absolutely continuous on $[0,a]$, φ_1 *and* φ_2 *are solutions of the equations*

$$\varphi_1(t) + \int_0^a h(t-s)\varphi_1(s)ds = I_p + \int_0^t h(s)ds \tag{8.110}$$

and

$$\varphi_2(t) + \int_0^a h(t-s)\varphi_2(s)ds = \int_0^t h(s)ds \tag{8.111}$$

in $L_2^{p\times p}([0,a])$, *respectively,* $h(-s) = h(s)^*$, *and* $(X^\vee)^*$ *is defined by formula (8.104) with* $\gamma = I_p$.

Proof (1) and (2) follow from Theorem 8.45 and Corollary 8.43, respectively. □

Since the operator

$$K : \varphi \in L_2^p([0,a]) \longrightarrow \int_0^a h(t-s)\varphi(s)ds$$

is compact in $L_2^p([0,a])$ when $h \in L_1^{p\times p}([0,a])$, Lemma 7.1 in [AAK71b] guarantees that it has the same nonzero spectrum in $L_2^p([0,a])$ as in the Banach space $\mathfrak{B}$ of absolutely continuous $p \times 1$ vvf's on $[0,a]$ with norm

$$\|\varphi\|_{\mathfrak{B}} = \max_{t\in[0,a]} \|\varphi(t)\| + \int_0^a \|\varphi'(t)\|dt.$$

Therefore, since $Y^\vee \geq \delta I$ in $L_2^p([0,a])$, it also has a bounded inverse in $\mathfrak{B}$. Thus, as the right hand sides of equations (8.110) and (8.111) are absolutely continuous on $[0,a]$, $\varphi_1(t)$ and $\varphi_2(t)$ are also absolutely continuous on $[0,a]$, as is $\mathcal{B}(t)$, due to formulas (8.104) and (8.109). Consequently,

$$B(\lambda) = \mathfrak{V} + \frac{1}{\sqrt{2}}(e^{i\lambda a}\mathcal{B}(a) - \mathcal{B}(0)) - \frac{1}{\sqrt{2}}\int_0^a e^{i\lambda t}\mathcal{B}'(t)dt, \tag{8.112}$$

where $\mathcal{B}' \in L_1^{m\times m}([0,a])$ and, in view of formulas (8.104) and (8.109)–(8.111),

$$\mathcal{B}(0) = \begin{bmatrix} \varphi_1(0) - 2I_p & \varphi_2(0) \\ \varphi_1(0) & \varphi_2(0) \end{bmatrix} \quad \text{and} \quad \mathcal{B}(a) = \begin{bmatrix} -\varphi_1(a) & -\varphi_2(a) \\ \varphi_1(a) & \varphi_2(a) \end{bmatrix}. \tag{8.113}$$

Lemma 8.47 *The mvf*

$$\mathfrak{A}(\lambda) = \begin{bmatrix} e^{-i\lambda a}I_p & 0 \\ 0 & I_p \end{bmatrix} \mathfrak{V}B(\lambda) \tag{8.114}$$

belongs to the class $\mathfrak{M}_r(j_p) \cap \mathcal{W}^{m\times m}(V)$, *where*

$$V = \begin{bmatrix} v_{11} & v_{12} \\ v_{21} & v_{22} \end{bmatrix} = \begin{bmatrix} \varphi_1(a) & \varphi_2(a) \\ I_p - \varphi_1(0) & I_p - \varphi_2(0) \end{bmatrix} \quad \textit{and} \quad V \in \mathcal{U}_{const}(j_p). \tag{8.115}$$

Proof Theorem 7.26 guarantees that $\mathfrak{A} \in \mathfrak{M}_r(j_{pq})$. Moreover, if

$$F(\mu) = \begin{bmatrix} e^{-i\mu a} I_p & 0 \\ 0 & I_p \end{bmatrix} \mathfrak{V} \left\{ \mathfrak{V} + \frac{1}{\sqrt{2}} (e^{i\mu a} \mathcal{B}(a) - \mathcal{B}(0)) \right\},$$

then, by a straight forward calculation based on the formulas in (8.113), $F'(\mu) = 0$. Therefore, $F(\mu) = F(0) = V$, as specified in (8.115). Thus,

$$\mathfrak{A}(\mu) = V - \frac{1}{\sqrt{2}} \int_0^a \begin{bmatrix} e^{-i\lambda(a-t)} I_p & 0 \\ 0 & e^{i\lambda t} I_p \end{bmatrix} \mathfrak{V} \mathcal{B}'(t) dt \tag{8.116}$$

which clearly belongs to $\mathcal{W}^{m \times m}(V)$, since $\mathcal{B}'(t) \in L_1^{m \times m}([0,a)]$. Moreover, $V \in \mathcal{U}_{const}(j_p)$, since $\mathfrak{A}(\mu)$ is j_p-unitary and $\lim_{\mu \uparrow \infty} \mathfrak{A}(\mu) = V$, by the Riemann-Lebesgue lemma. □

The mvf

$$B_1(\lambda) = B(\lambda) V^{-1} \tag{8.117}$$

is also a resolvent matrix for the HEP$(g; a)$ under consideration, since $V \in \mathcal{U}_{const}(j_p)$. Moreover, $B_1(\lambda)$ is uniquely defined by conditions (i), (ii) and a normalization at infinity instead of at zero:

(iii′) The mvf

$$\mathfrak{A}_1(\mu) = \begin{bmatrix} e^{-ia\mu} I_p & 0 \\ 0 & I_p \end{bmatrix} \mathfrak{V} B_1(\mu) \tag{8.118}$$

tends to I_m as $\mu \in \mathbb{R}$ tends to ∞.

Theorem 8.48 *If $g \in \mathcal{G}_a^{p \times p}(0)$ has an accelerant $h \in \mathring{\mathcal{A}}_a^{p \times p}$, then there exists a unique B-resolvent matrix $B_1(\lambda)$ of the HEP$(g; a)$ that has the properties (i), (ii) and (iii′):*

$$B_1(\lambda) = \mathfrak{V} \begin{bmatrix} e^{i\lambda a} I_p & 0 \\ 0 & I_p \end{bmatrix} + \frac{1}{\sqrt{2}} \int_0^a e^{i\lambda t} \mathcal{B}_1(t) dt, \tag{8.119}$$

where

$$\mathcal{B}_1(t) = \begin{bmatrix} (X^\vee e)(t) & -((X^\vee)^* f)(t) \\ e(t) & f(t) \end{bmatrix} \quad \textit{belongs to } L_1([0,a]) \tag{8.120}$$

and $e(t)$ and $f(t)$ are solutions of the equations

$$e(t) + \int_0^a h(t-s)e(s)ds = -h(t-a), \tag{8.121}$$

$$f(t) + \int_0^a h(t-s)f(s)ds = -h(t), \tag{8.122}$$

in $L_1^{p\times p}([0,a])$, respectively. Moreover,

$$\mathfrak{A}_1 \in \mathfrak{M}_r(j_{pq}) \cap \mathcal{W}^{m\times m}(I_m). \tag{8.123}$$

Proof The domain and range of the operators $X^\vee$, $(X^\vee)^*$ and $Y^\vee$ that are defined by formulas (8.103)–(8.105), respectively, is now extended from $L_2^p([0,a])$ to $L_1^p([0,a])$ without change of notation. In view of (8.112), (8.109) and identification $F(\mu) = V$ in the proof of Lemma 8.47,

$$\mathcal{B}_1(t) = -\mathcal{B}'(t)V^{-1} = \begin{bmatrix} ((X^\vee)^*\varphi_1)'(t) & ((X^\vee)^*\varphi_2)'(t) \\ -\varphi_1'(t) & -\varphi_2'(t) \end{bmatrix} V^{-1}.$$

Let

$$[e(t) \quad f(t)] = -[\varphi_1'(t) \quad \varphi_2'(t)]V^{-1}.$$

Then equations (8.110) and (8.111) guarantee that

$$\varphi_j'(t) + \int_0^a h(t-s)\varphi_j'(s)ds = h(t-a)\varphi_j(a) + h(t)(I_p - \varphi_j(0))$$

for $j = 1,2$ a.e. on $\mathbb{R}$ and hence that e and f are solutions of (8.121) and (8.122), respectively. Next, upon differentiating the formula

$$((X^\vee)^*\varphi_j)(t) = \varphi_j(t) + 2\int_t^a h(t-s)\varphi_j(s)ds,$$

a short calculation yields the formula

$$((X^\vee)^*\varphi_j)'(t) = ((X^\vee)^*\varphi_j')(t) - 2h(t-a)\varphi_j(a) \quad \text{a.e. on } [0,a].$$

Therefore, since

$$-2h(t-a)[\varphi_1(a) \quad \varphi_2(a)] = -2h(t-a)[I_p \quad 0]V,$$

the top block row of $\mathcal{B}_1(t)$ is equal to

$$\begin{aligned} &[-((X^\vee)^*e)(t) - 2h(t-a) \quad -((X^\vee)^*f)(t)] \\ &\qquad = [(X^\vee e)(t) \quad -((X^\vee)^*f)(t)] \quad \text{a.e. on } [0,a]. \end{aligned} \tag{8.124}$$

Finally, (8.123) is immediate from Lemma 8.47. □

Theorem 8.49 *If $g \in \mathcal{G}_a^{p\times p}(0)$ has an accelerant $h \in \mathring{\mathcal{A}}_a^{p\times p}$, and if $B(\lambda)$ and $B_1(\lambda)$ are the B-resolvent matrices for the HEP$(g; a)$ considered in Theorems 8.46 and 8.48, respectively, then*

$$\mathcal{G}(g; a) = \{\widetilde{g} : c_{\widetilde{g}} \in T_{B_1}[\mathcal{S}^{p\times p}]\} \tag{8.125}$$

and the formula

$$I_p + (\widetilde{h})^\wedge = T_{B_1}[\varepsilon] \tag{8.126}$$

defines a one to one correspondence between the set of mvf's $\varepsilon \in \mathring{\mathcal{S}}^{p\times p} \cap \mathcal{W}_+^{p\times p}(0)$ and the set $\mathring{\mathcal{A}}(h; a)$ of solutions $\widetilde{h}$ of the $\mathring{A}$EP$(h; a)$, i.e.,

$$T_{B_1}[\mathring{\mathcal{S}}^{p\times p} \cap \mathcal{W}_+^{p\times p}(0)] = T_{B_1}[\mathring{\mathcal{S}}^{p\times p}] \cap \mathcal{W}_+^{p\times p}(I_p)$$

and

$$\mathring{\mathcal{A}}(h; a) = \{\widetilde{h} : I_p + \int_0^\infty e^{i\lambda t}\widetilde{h}(t)dt \in T_{B_1}[\mathring{\mathcal{S}}^{p\times p} \cap \mathcal{W}_+^{p\times p}(0)]\}. \tag{8.127}$$

Moreover, formula (8.127) may be rewritten as

$$\mathring{\mathcal{A}}(h; a) = \left\{\widetilde{h} : I_p + \int_0^\infty e^{i\lambda t}\widetilde{h}(t)dt \in T_B[\mathring{\mathcal{S}}^{p\times p} \cap \mathcal{W}_+^{p\times p}(\gamma_h)]\right\}, \tag{8.128}$$

where $\gamma_h \in \mathring{\mathcal{S}}_{const}^{p\times p}$ and is given by the formula

$$\gamma_h = -\{I_p - \varphi_1(0)^*\}\{I_p - \varphi_2(0)^*\}^{-1}. \tag{8.129}$$

Proof In view of formula (8.119), the entries w_{ij} in the four block decomposition of the j_p-inner mvf $W = \mathfrak{V}B_1$ are entire mvf's,

$$e_{-a}w_{11} \in \mathcal{W}_-^{p\times p}(I_p), \quad w_{12} \in \mathcal{W}_+^{p\times p}(0), \quad w_{21} \in \mathcal{W}_+^{p\times p}(0),$$
$$w_{22} \in \mathcal{W}_+^{p\times p}(I_p) \tag{8.130}$$

and, as follows from Theorem 3.34, the entries s_{ij} in the four block decomposition of $S = PG(W)$ meet the constraints

$$e_a s_{11} = e_a(w_{11}^\#)^{-1} \quad \text{and} \quad s_{22} = w_{22}^{-1} \quad \text{belong to } \mathcal{W}_+^{p\times p}(I_p), \tag{8.131}$$

whereas

$$s_{12} \in \mathring{\mathcal{S}}^{p\times p} \cap \mathcal{W}_+^{p\times p}(0) \quad \text{and} \quad s_{21} \in \mathring{\mathcal{S}}^{p\times p} \cap \mathcal{W}_+^{p\times p}(0). \tag{8.132}$$

Thus, if $\varepsilon \in \mathring{\mathcal{S}}^{p\times p} \cap \mathcal{W}_+^{p\times p}(0)$ and $s = T_W[\varepsilon]$, then

$$s - s_{12} = s_{11}\varepsilon(I_p - s_{21}\varepsilon)^{-1}s_{22} \quad \text{belongs to } e_a\mathcal{W}_+^{p\times p}(0) \tag{8.133}$$

and

$$I_p - s^*s = s_{22}^*\{I_p - s_{21}\varepsilon\}^{-*}\{I_p - \varepsilon^*\varepsilon\}\{I_p - s_{21}\varepsilon\}^{-1}s_{22} \quad \text{on } \mathbb{R}, \tag{8.134}$$

which implies that $s \in \mathring{\mathcal{S}}^{p\times p}$. Thus,

$$T_W[\mathring{\mathcal{S}}^{p\times p} \cap \mathcal{W}_+^{p\times p}(0)] \subseteq \mathring{\mathcal{S}}^{p\times p} \cap \mathcal{W}_+^{p\times p}(0).$$

Suppose next that $\varepsilon \in \mathcal{S}^{p\times p}$ and that $s = T_W[\varepsilon]$ is in $\mathring{\mathcal{S}}^{p\times p} \cap \mathcal{W}_+^{p\times p}(0)$. Then $s - s_{12} \in \mathcal{W}_+^{p\times p}(0)$ and, as the formula in (8.133) is still valid, $s - s_{12} \in e_aH_\infty^{p\times p}$. Therefore, since $e_aH_\infty^{p\times p} \cap \mathcal{W}_+^{p\times p}(0) = e_a\mathcal{W}_+^{p\times p}(0)$, another application of the formula in (8.133) implies that

$$\varepsilon(I_p - s_{21}\varepsilon)^{-1} = w_{11}^{\#}(s - s_{12})w_{22} \quad \text{belongs to } \mathcal{W}_+^{p\times p}(0).$$

Therefore,

$$(I_p - s_{21}\varepsilon)^{-1} = I_p + s_{21}\varepsilon(I_p - s_{21}\varepsilon)^{-1} \quad \text{belongs to } \mathcal{W}_+^{p\times p}(I_p)$$

and, by Theorem 3.34, $I_p - s_{21}\varepsilon \; in \mathcal{W}_+^{p\times p}(I_p)$. Thus,

$$\varepsilon = \varepsilon(I_p - s_{21}\varepsilon)^{-1}(I_p - s_{21}\varepsilon) \quad \text{belongs to } \mathcal{W}_+^{p\times p}(0)$$

and, in view of (8.134), $\varepsilon \in \mathring{\mathcal{S}}^{p\times p}$. This completes the proof that

$$\text{if } \varepsilon \in \mathcal{S}^{p\times p} \text{ and } s = T_W[\varepsilon], \quad \text{then}$$
$$s \in \mathring{\mathcal{S}}^{p\times p} \cap \mathcal{W}_+^{p\times p}(0) \Longleftrightarrow \varepsilon \in \mathring{\mathcal{S}}^{p\times p} \cap \mathcal{W}_+^{p\times p}(0). \tag{8.135}$$

Next, in view of Theorem 3.34,

$$T_{\mathfrak{V}}[\mathring{\mathcal{S}}^{p\times p} \cap \mathcal{W}_+^{p\times p}(0)] = \mathring{\mathcal{C}}^{p\times p} \cap \mathcal{W}_+^{p\times p}(I_p)$$

and hence, $T_{B_1}[\varepsilon] = T_{\mathfrak{V}}[T_W[\varepsilon]]$,

$$T_{B_1}[\mathring{\mathcal{S}}^{p\times p} \cap \mathcal{W}_+^{p\times p}(0)] = \mathring{\mathcal{C}}^{p\times p} \cap \mathcal{W}_+^{p\times p}(I_p) = T_{B_1}[\mathring{\mathcal{S}}^{p\times p}] \cap \mathcal{W}_+^{p\times p}(I_p).$$

Formula (8.127) then follows from the results in Section 8.7.

Finally, formula (8.128) follows from (8.127) and the fact that $B_1(\lambda) = B(\lambda)V^{-1}$, where $V \in \mathcal{U}_{const}(j_p)$ is specified in terms of $\varphi_1(t)$ and $\varphi_2(t)$ evaluated at the end points of the interval $[0, a]$ by formula (8.115): If $\varepsilon \in \mathring{\mathcal{S}}^{p\times p} \cap \mathcal{W}_+^{p\times p}(0)$, then, since $V^{-1} = j_pV^*j_p$ and

$$\det\{-v_{12}^*\varepsilon(\lambda) + v_{22}^*\} \neq 0 \quad \text{for } \lambda \in \overline{\mathbb{C}_+} \cup \{\infty\},$$

$$T_{V^{-1}} : \varepsilon \in \mathring{\mathcal{S}}^{p\times p} \cap \mathcal{W}_+^{p\times p}(0) \longrightarrow$$
$$(v_{11}^*\varepsilon - v_{21}^*)(-v_{12}^*\varepsilon + v_{22}^*)^{-1} \in \mathring{\mathcal{S}}^{p\times p} \cap \mathcal{W}_+^{p\times p}(-v_{21}^*(v_{22}^*)^{-1}).$$

The proof is completed by invoking the formulas $T_{B_1}[\varepsilon] = T_B[T_{V^{-1}}[\varepsilon]]$ and (8.115). □

Theorem 8.50 *Let $B_1(\lambda)$ be any $m \times m$ mvf that satisfies properties (i) and (ii) and is such that*

$$\mathfrak{A}_1(\mu) = \begin{bmatrix} e^{-ia\mu}I_p & 0 \\ 0 & I_p \end{bmatrix} \mathfrak{V}B_1(\mu) \quad \text{belongs to } \mathcal{W}^{m\times m}(I_m).$$

Then

$$T_{B_1}[0] = I_p + (\widetilde{h})^\wedge, \quad \text{where } \widetilde{h} \in \mathring{\mathcal{A}}_\infty^{p\times p}$$

and B_1 is a B-resolvent matrix of the HEP$(g; a)$ based on the mvf $g \in \mathcal{G}_a^{p\times p}(0)$ with accelerant h equal to the restriction of $\widetilde{h}$ to the interval $[0, a]$.

Proof Under the given assumptions, the mvf $A = B_1\mathfrak{V}$ belongs to $\mathcal{E} \cap \mathcal{U}_{rsR}(J_p)$, since $B_1 \in H_\infty^{m\times m}$ and $(e_aI_p, I_p) \in ap_{II}(A)$, thanks to Corollary 4.55. Thus, in view of Theorem 8.21, B_1 is a B-resolvent matrix of the HEP$(g; a)$ in the statement of the theorem. The rest follows easily from the proof of Theorem 8.49, since the blocks w_{ij} of $W = \mathfrak{V}B_1$ and s_{ij} of $S = PG(W)$ have the properties stated in (8.130)–(8.132). □

8.11 Bibliographical notes

Most of the material in this chapter is adapted from [ArD98]. Each positive kernel of the form (8.1) can be interpreted as the metric function

$$k(t, s) = E[\{x(t + r) - x(r)\}\{x(s + r) - x(r)\}^*]$$

of a p-dimensional stochastic process $x(t)$ with weakly stationary increments; see [Kr44b] and [GoGo97]. Theorem 8.1 was established by M. G. Krein [Kr3] for the case $p = 1$. It is easily extended to the matrix case. The integral in (8.3) also arises in the study of characteristic functions of infinitely divisible random variables; see e.g., [GnKo68]. The statement of Theorem 8.3 is adapted from pages 217–218 of N. I. Akhiezer [Ak65].

M. G. Krein [Kr44a] exploited the connection (8.7) between the classes $\mathcal{C}^{p\times p}$ and $\mathcal{G}_\infty^{p\times p}(0)$ extensively to solve the HEP. He also discovered a reformulation of the HEP in the class $\mathcal{C}^{p\times p}$ that is equivalent to the identification

(8.8). Krein obtained his results on the HEP as an application of his theory of entire symmetric operators and their selfadjoint extensions; see [Kr44b] and [GoGo97]. The description of the set $\mathcal{G}(g^\circ; a)$ was obtained by other methods in this chapter.

The matrix version of Theorem 8.22 is easily deduced from the better known scalar version; see e.g., [Kr49] and Chapter 3 of [GoGo97].

The problem of extending positive definite functions was first investigated in the scalar case $p = 1$ by M. G. Krein [Kr40] in the late thirties and subsequently by other methods (based on viewing this extension problem as a continuous analogue of the trigonometric moment problem and iterating a sequence of such moment problems) by his graduate student, A. P. Artemenko [Art84]. This problem was also studied by [KoPo82] using a fundamental matrix inequality that was developed by Potapov in his study of J contractive functions; see also [MiPo81], [Kat84a], [Kat84b], [Kat85a], [Kat85b] and [Kat86]. This class of problems has interesting applications in the theory of stochastic processes. In particular, every function $f \in \mathcal{P}_\infty^{1\times 1}$ with $f(0) = 1$ can be interpreted as the characteristic function $f(t) = E[e^{itX}]$ of a random variable X. The first example of two different characteristic functions which agree on a symmetric interval about zero was proposed by B. V. Gnedenko in 1937 [Gn37]. Another and perhaps more significant connection with the theory of stochastic processes rests on the interpretation of $f \in \mathcal{P}_\infty^{p\times p}$ as the correlation matrix

$$f(t - s) = E[x(t)x(s)^*]$$

of a p-dimensional weakly stationary stochastic process $x(t)$ with zero mean and finite variance.

The first four lemmas in Section 8.4 are adapted from [Ar93]. The connection between the solutions of a completely indeterminate PEP and a completely indeterminate HEP was exploited in [Ar93] to obtain a characterization of resolvent matrices for PEP's from corresponding results for HEP's. A similar strategy was used by Krein and Langer in [KrL85], but in the opposite direction.

The proofs of Lemmas 8.26-8.29 may be found in [Ar93] and in [ArD98].

If $a = \infty$, the representation (8.48) is known as the Bochner–Khinchin formula. For $a < \infty$ the theorem was established by M. G. Krein.

To the best of our knowledge, Krein [Kr54], [Kr55], [Kr56] was the first to observe the deep connections between the solutions of the accelerant extension problem and the solutions of inverse spectral problems for differential systems with potential and for the Schrödinger equation. This line of research was continued in assorted forms in a number of papers: see e.g., [KrMA68], [KrMA70], [KrMA84], [DI84], [KrL85], [KrMA86], [Dy90], and was discussed in greater detail in [ArD05c] and [ArD07]; see also [Den06].

The $\mathring{\mathrm{A}}\mathrm{EP}(h;a)$ was considered in the general context of band extensions in [DG80]. A resolvent matrix B_1 for the $\mathring{\mathrm{A}}\mathrm{EP}(h;a)$ was obtained in [KrMA86], and then by other methods in [Dy90] and [Dy94b]. More information on the accelerant extension problem, its resolvent matrices and connections with the Krein method for solving inverse problems for Dirac and Krein systems is supplied in [ArD05c] and [ArD07]. This method is applicable to the study of a number of bitangential direct and inverse problems (monodromy, scattering, impedance, spectral) for integral and differential systems using the results of the previous chapters; see [ArD05b] and [ArD07b] for a survey and the discussion in Chapter 1 of this monograph.

9

Darlington representations and related inverse problems for J-inner mvf's.

As noted earlier,

$$\mathcal{U}(J) \subset \Pi^{m\times m}$$

for every $m \times m$ signature matrix J. Consequently, every $p \times q$ submatrix of a J-inner mvf belongs to the class $\Pi^{p\times q}$. Conversely, in Section 9.5 it will be shown that every mvf $w \in \Pi^{p\times q}$ is the block w_{12} of a j_{pq}-inner mvf W and the set of all such W will be described. A better known result, related to the inverse scattering problem, will be presented in Section 9.2, where we will prove that every mvf $s \in \Pi \cap \mathcal{S}^{p\times q}$ is the block s_{12} of an inner $n \times n$ mvf S for some n. Moreover, the set of all such mvf's S with minimal possible size n for a given mvf s will be described. The Potapov-Ginzburg transform $W = PG(S)$ for such a mvf S is well defined if and only if

$$I_q - s(\mu)^* s(\mu) > 0 \quad \text{a.e on } \mathbb{R}. \tag{9.1}$$

If $s \in \Pi \cap \mathcal{S}^{p\times q}$ satisfies condition (9.1), then $n = m = p + q$,

$$s = T_W[0_{p\times q}] \quad \text{and} \quad W \in \mathcal{U}(j_{pq}); \tag{9.2}$$

if not, then it admits a representation of the form

$$s = T_W[\varepsilon] \quad \text{for some } W \in \mathcal{U}(j_{pq}) \text{ and } \varepsilon \in \mathcal{S}^{p\times q}_{const}. \tag{9.3}$$

Analogous representations hold for the class $\Pi \cap \mathcal{C}^{p\times p}$. In section 9.3 it will be shown that a mvf c admits a representation of the form

$$c = T_A[I_p] \text{ with } A \in \mathcal{U}(J_p) \tag{9.4}$$

if and only if $c \in \Pi \cap \mathcal{C}^{p\times p}$ and

$$\Delta(\mu) = (\Re c)(\mu) > 0 \quad \text{a.e. on } \mathbb{R}. \tag{9.5}$$

Moreover, the set of all mvf's A that are related to the given c by (9.4) will be described. If the given $c \in \Pi \cap \mathcal{C}^{p\times p}$ does not satisfy the condition (9.5), then

$$c = T_A[\tau], \quad \text{for some } A \in \mathcal{U}(J_p) \text{ and } \tau \in \mathcal{C}_{const}^{p\times p}. \tag{9.6}$$

is valid too. Another representation of mvf's $c \in \Pi \cap \mathcal{C}^{p\times p}$ will be given in Section 9.7.

The inverse impedance (resp., inverse spectral) problem, of describing the set of all mvf's $A \in \mathcal{U}(J_p)$ that are related to a given mvf c (resp., Δ) by formulas (9.4) (resp., (9.4) and (9.5)), is connected with the inverse problem of describing the set of all mvf's $A \in \mathcal{U}(J_p)$ with given bottom $p \times 2p$ block row $[a_{21} \quad a_{22}]$, or equivalently with given mvf's

$$E_- = a_{22} - a_{21} \quad \text{and} \quad E_+ = a_{22} + a_{21}, \tag{9.7}$$

which is considered in Section 9.4. The mvf's E_- and E_+ are solutions of the factorization problems

$$\Delta(\mu) = E_-(\mu)^{-*}E_-(\mu)^{-1} \quad \text{a.e. on } \mathbb{R} \text{ with } \rho_i^{-1}(E_-^\#)^{-1} \in H_2^{p\times p}, \tag{9.8}$$

and

$$\Delta(\mu) = E_+(\mu)^{-*}E_+(\mu)^{-1} \quad \text{a.e. on } \mathbb{R} \text{ with } \rho_i^{-1}E_+^{-1} \in H_2^{p\times p}. \tag{9.9}$$

9.1 $\mathcal{D}$-representations of mvf's $s \in \Pi \cap \mathcal{S}^{p\times q}$

Theorem 9.1 *A mvf $s \in \mathcal{S}^{p\times q}$ is a block of some inner $n \times n$ mvf if and only if $s \in \Pi \cap \mathcal{S}_{in}^{p\times q}$.*

Proof Since $\mathcal{S}_{in}^{n\times n} \subset \Pi^{n\times n}$, one direction of the theorem is obvious. It remains to prove that if $s \in \Pi \cap \mathcal{S}^{p\times q}$, then it is a block of some mvf $S \in \mathcal{S}^{n\times n}$. Four possible cases will be considered.

1. $s \in S_{in}^{p\times p}$: Take $S = s$.

2. $s \in \Pi \cap \mathcal{S}_{in}^{p\times q}$ with $q < p$: Let

$$g(\lambda) = I_p - s(\lambda)s^\#(\lambda) \quad \text{and} \quad r_g = \text{rank } g. \tag{9.10}$$

Then $g \in \Pi^{p\times p}$ and $I_p \geq g(\mu) \geq 0$ a.e. on $\mathbb{R}$. By Theorem 3.110, applied to the mvf $g(-\lambda)$, the factorization problem

$$g(\mu) = \psi(\mu)\psi(\mu)^* \quad \text{a.e. on } \mathbb{R} \text{ with } \psi^\sim \in \mathcal{N}_{out}^{r_g\times p} \tag{9.11}$$

has an essentially unique solution ψ. Moreover, $\psi \in \Pi^{p\times r_g}$ and, by the Smirnov maximum principle, $\psi^\sim \in \mathcal{S}_{out}^{r_g\times p}$. Since $s(\mu)^*s(\mu) = I_q$ a.e. on $\mathbb{R}$, $g(\mu)$ is an orthogonal projection of rank $r_g = p-q$ a.e. on $\mathbb{R}$. Thus, the $p\times p$ mvf

$$S(\lambda) = [\psi(\lambda) \quad s(\lambda)]$$

belongs to $\mathcal{S}_{in}^{p\times p}$, since $S \in H_\infty^{p\times p}$ and

$$S(\mu)S(\mu)^* = \psi(\mu)\psi(\mu)^* + s(\mu)s(\mu)^* = I_p \quad \text{a.e. on } \mathbb{R}.$$

This exhibits $s(\lambda)$ is a block of $S \in \mathcal{S}_{in}^{n\times n}$ with $n = p$. It is clear that $n = p$ is the minimal possible n for a mvf $S \in \mathcal{S}_{in}^{n\times n}$ with block s. The set of all mvf's S with these properties is described by the formula

$$S(\lambda) = [\psi(\lambda)b_1(\lambda) \quad s(\lambda)], \text{ where } b_1 \text{ is any mvf in } \mathcal{S}_{in}^{r_g\times r_g} \tag{9.12}$$

and ψ is a solution of the factorization problem (9.11).

3. $s \in \Pi \cap \mathcal{S}_{*in}^{p\times q}$ **with** $p < q$: Since $s \in \Pi \cap \mathcal{S}_{*in}^{p\times q} \iff s^\sim \in \Pi \cap \mathcal{S}_{in}^{q\times p}$, this case can be reduced to the previous case, i.e., if

$$f(\lambda) = I_q - s^\#(\lambda)s(\lambda) \quad \text{and} \quad r_f = \text{rank } f, \tag{9.13}$$

then there exists an essentially unique solution φ of the factorization problem

$$\varphi(\mu)^*\varphi(\mu) = f(\mu) \quad \text{a.e. on } \mathbb{R}, \quad \text{where } \varphi \in \mathcal{N}_{out}^{r_f\times q}. \tag{9.14}$$

The mvf $\varphi \in \Pi \cap \mathcal{S}_{out}^{r_f\times q}$; and the minimal possible size of a mvf $S \in \mathcal{S}_{in}^{n\times n}$ with s as a block is $n = q$. The set of all such mvf's S is described by the formula

$$S(\lambda) = \begin{bmatrix} s(\lambda) \\ b_2(\lambda)\varphi(\lambda) \end{bmatrix}, \quad \text{where } b_2 \text{ is any mvf in } \mathcal{S}_{in}^{r_f\times r_f} \tag{9.15}$$

and φ is a solution of the factorization problem (9.14).

4. $s \in \Pi \cap \mathcal{S}^{p\times q}$ **and** $s \notin \mathcal{S}_{in}^{p\times q}, s \notin \mathcal{S}_{*in}^{p\times q}$: As above, consider a solution φ of the factorization problem (9.14), where f is defined by formula (9.13).

Then

$$S_2(\lambda) = \begin{bmatrix} s(\lambda) \\ b_2(\lambda)\varphi(\lambda) \end{bmatrix} \in \Pi \cap \mathcal{S}_{in}^{(p+r_f)\times q} \quad \text{if } b_2 \in \mathcal{S}_{in}^{r_f\times r_f}$$

and $p + r_f > q$, since $s \notin \mathcal{S}_{*in}^{p\times q}$ by assumption. Therefore, by the analysis of Case 2, there exists a mvf

$$S(\lambda) = [S_1(\lambda)\ S_2(\lambda)] \in \mathcal{S}_{in}^{(p+r_f)\times(p+r_f)}.$$

Thus, $s(\lambda)$ is the 12 block of an $n \times n$ inner mvf $S(\lambda)$ with $n = p + r_f$, which is the minimal possible size. □

Remark 9.2 *In the proof of the preceding theorem, descriptions of the set of mvf's $S \in \mathcal{S}_{in}^{n\times n}$ with minimal possible n such that s is a block of S are furnished in Cases 1-3. In these three cases $n = \max\{p, q\}$. In Case 4, the embedding*

$$S = \begin{bmatrix} s_{11} & s \\ s_{21} & s_{22} \end{bmatrix} \in \mathcal{S}_{in}^{n\times n} \quad \textit{for some } n$$

yields the equalities

$$s_{22}(\mu)^* s_{22}(\mu) = I_q - s(\mu)^* s(\mu) = f(\mu) \quad \textit{a.e. on } \mathbb{R},$$

which imply that $n - p \ge \operatorname{rank} s_{22} = r_f$. Therefore, $n = p + r_f = q + r_g$ is the minimal possible n.

An $n \times n$ inner mvf $S(\lambda)$ with a block equal to $s(\lambda)$ and n **as small as possible** is called a **scattering $\mathcal{D}$-representation** of $s(\lambda)$.

Theorem 9.3 *Let $s \in \Pi \cap \mathcal{S}^{p\times q}$ and suppose that $s \notin \mathcal{S}_{in}^{p\times q}$ and $s \notin \mathcal{S}_{*in}^{p\times q}$. Let ψ and φ be the essentially unique solutions of the factorization problems (9.11) and (9.14), based on the mvf's $g = I_p - ss^\#$ and $f = I_q - s^\# s$, respectively. Then the set of all mvf's $S \in \mathcal{S}_{in}^{n\times n}$ of minimal size n that contain s as a 12 block is given by the formula*

$$S(\lambda) = \begin{bmatrix} I_p & 0 \\ 0 & b_2(\lambda) \end{bmatrix} \begin{bmatrix} \psi(\lambda) & s(\lambda) \\ h(\lambda) & \varphi(\lambda) \end{bmatrix} \begin{bmatrix} b_1(\lambda) & 0 \\ 0 & I_q \end{bmatrix}, \tag{9.16}$$

where $h \in \Pi^{r_g\times r_f}$ is the unique solution of two equivalent equalities

$$h(\lambda)\psi^\#(\lambda) = -\varphi(\lambda)s^\#(\lambda) \tag{9.17}$$

or

$$\varphi^{\#}(\lambda)h(\lambda) = -s^{\#}(\lambda)\psi(\lambda) \tag{9.18}$$

and the pair $\{b_2, b_1\}$ *is a denominator of* h.

Proof The existence of a mvf $S \in \mathcal{S}_{in}^{n\times n}$ with $s_{12} = s$ and $n = p + r_f$ was established in Theorem 9.1. The fact that $p + r_f$ is the minimal possible choice of n was discussed in Remark 9.2. Let

$$S(\lambda) = \begin{bmatrix} s_{11}(\lambda) & s(\lambda) \\ s_{21}(\lambda) & s_{22}(\lambda) \end{bmatrix} \in \mathcal{S}_{in}^{n\times n}$$

with $n = p + r_f$ be any such $\mathcal{D}$-representation of $s(\lambda)$. Then

$$s_{11}(\mu)s_{11}(\mu)^* = g(\mu) \quad \text{a.e. on } \mathbb{R}, \qquad s_{11} \in \mathcal{S}^{p\times r_g}, \tag{9.19}$$

$$s_{22}(\mu)^* s_{22}(\mu) = f(\mu) \quad \text{a.e. on } \mathbb{R} \quad \text{and} \quad s_{22} \in \mathcal{S}^{r_f\times q}. \tag{9.20}$$

Consequently, in view of Theorem 3.74, $s_{11} = \psi b_1$ and $s_{22} = b_2\varphi$ for some choice of $b_1 \in \mathcal{S}_{in}^{r_g\times r_g}$ and $b_2 \in \mathcal{S}_{in}^{r_f\times r_f}$, where ψ and φ are solutions of the factorization problems (9.11) and (9.14) based on the mvf's $g = I_p - ss^{\#}$ and $f = I_q - s^{\#}s$, respectively. Moreover, since $S \in \mathcal{S}_{in}^{n\times n}$,

$$s_{21}(\lambda)s_{11}^{\#}(\lambda) + s_{22}(\lambda)s^{\#}(\lambda) = 0 \tag{9.21}$$

and

$$s_{22}^{\#}(\lambda)s_{21}(\lambda) + s^{\#}(\lambda)s_{11}(\lambda) = 0, \tag{9.22}$$

the mvf

$$h(\lambda) = b_2(\lambda)^{-1}s_{21}(\lambda)\, b_1(\lambda)^{-1} \tag{9.23}$$

is a solution of equations (9.17) and (9.18) that belongs to $\Pi^{r_f\times r_g}$, and the pair $\{b_2, b_1\}$ is a denominator of h.

The next step is to check that a mvf $h \in L_\infty^{r_f\times r_g}$ is a solution of the equation

$$h(\mu)\,\psi(\mu)^* = -\varphi(\mu)\, s(\mu)^* \quad \text{a.e on } \mathbb{R} \tag{9.24}$$

if and only if it is a solution of the equation

$$\varphi(\mu)^* h(\mu) = -s(\mu)^*\psi(\mu) \quad \text{a.e. on } \mathbb{R}. \tag{9.25}$$

Indeed, if $h \in L_\infty^{r_f\times r_g}$ is a solution of equation (9.24), then

$$\begin{aligned}\varphi(\mu)^*h(\mu)\psi(\mu)^* &= -\varphi(\mu)^*\varphi(\mu)s(\mu)^* = -(I_q - s(\mu)^*s(\mu))s(\mu)^* \\ &= -s(\mu)^*(I_p - s(\mu)s(\mu)^*) = -s(\mu)^*\psi(\mu)\psi(\mu)^*\end{aligned}$$

i.e.,

$$\{\varphi(\mu)^*h(\mu) + s(\mu)^*\psi(\mu)\}\psi(\mu)^* = 0 \quad \text{a.e. on } \mathbb{R}.$$

But this in turn implies that

$$\varphi(\mu)^*h(\mu) + s(\mu)^*\psi(\mu) = 0 \quad \text{a.e. on } \mathbb{R},$$

since $\overline{\psi^* L_2^p} = L_2^{r_g}$, because $\psi^\sim \in \mathcal{S}_{out}^{r_g\times p}$. Thus, h is a solution of equation (9.25). Conversely, if $h \in L_\infty^{r_f\times r_g}$ is a solution of equation (9.25), then

$$\varphi(\mu)^*\{h(\mu)\psi(\mu)^* + \varphi(\mu)s(\mu)^*\} = 0 \quad \text{a.e on } \mathbb{R}$$

and hence, as $\varphi \in \mathcal{S}_{out}^{r_f\times q}$, h is also a solution of equation (9.24). Moreover, since $\overline{\psi^* L_2^p} = L_2^{r_g}$, equation (9.24) has only one solution $h \in L_\infty^{r_f\times r_g}$.

The preceding analysis shows that equation (9.24) has exactly one solution $h \in L_\infty^{r_f\times r_g}$ which is also the one and only solution of equation (9.25). Moreover, formula (9.23) implies that this solution $h(\mu)$ is the nontangential limit of a mvf $h \in \Pi^{r_f\times r_g}$. Consequently, both equations (9.17) and (9.18) have exactly one and the same solution $h \in \Pi^{r_f\times r_g}$. It is also shown that any mvf $S \in \mathcal{S}_{in}^{n\times n}$ with block $s_{12} = s$ and $n = p + r_f$ may be obtained by formula (9.16), where $\{b_2, b_1\}$ is a denominator of the mvf h.

Conversely, if S is defined by formula (9.16), where $\{b_2, b_1\}$ is any denominator of the mvf h, then $S \in H_\infty^{n\times n}$, $s_{12} = s$ and $n = p + r_f$. It remains only to show that $S(\mu)^*S(\mu) = I_n$ a.e. on $\mathbb{R}$. In view of (9.25) and (9.14), it suffices to check that

$$\psi(\mu)^*\psi(\mu) + h(\mu)^*h(\mu) = I_{r_g} \quad \text{a.e. on } \mathbb{R}.$$

However, since $\overline{\psi^* L_2^p} = L_2^{r_g}$, this is equivalent to the equality

$$\{\psi(\mu)^*\psi(\mu) + h(\mu)^*h(\mu) - I_{r_g}\}\,\psi(\mu)^* = 0 \quad \text{a.e. on } \mathbb{R},$$

which holds, because, in view of (10.13) and (10.19),

$$\begin{aligned}\{\psi^*\psi + h^*h - I_{r_g}\}\psi^* &= \psi^*\{I_p - s\, s^*\} + h^*\{-\varphi\, s^*\} - \psi^* \\ &= -\{\psi^*s + h^*\varphi\}s^* = 0.\end{aligned}$$

□

Remark 9.4 *In [Ar79] it is shown that equations (9.24) and (9.25) each have exactly one and the same solution $h \in L_\infty^{r_f \times r_g}$ for any mvf $s \in \mathcal{S}^{p\times q}$ for which the two factorization problems (9.11) and (9.14) are solvable, even if $s \notin \Pi^{p\times q}$. If both of these factorization problems are solvable for a mvf $s \in \mathcal{S}^{p\times q}$ that does not belong to $\Pi^{p\times q}$, then s can not be represented as a block of a mvf $S \in \mathcal{S}_{in}^{n\times n}$ for some n. However, it can be represented as a block of a bi-inner operator valued function $S(\lambda)$ with values that are linear contractive operators that act from the Hilbert space $U \oplus \mathbb{C}^q$ into the Hilbert space $\mathbb{C}^p \oplus Y$ for some infinite dimensional spaces U and Y; see [ArSt??]. The scalar function*

$$s(\lambda) = 1/(\lambda + i)^{1/2}$$

serves as an example.

A scattering $\mathcal{D}$-representation $S \in \mathcal{S}_{in}^{n\times n}$ of a mvf $s \in \Pi \cap \mathcal{S}^{p\times q}$ such that $s \notin \mathcal{S}_{in}^{p\times q}$ and $s \notin \mathcal{S}_{*in}^{p\times q}$ is said to be **minimal** if the implications

$$\begin{bmatrix} I_p & 0 \\ 0 & u(\lambda) \end{bmatrix}^{-1} S \in \mathcal{S}_{in}^{n\times n} \Longrightarrow u(\lambda) \text{ is a constant unitary matrix}$$

and

$$S \begin{bmatrix} v(\lambda) & 0 \\ 0 & I_q \end{bmatrix}^{-1} \in \mathcal{S}_{in}^{n\times n} \Longrightarrow v(\lambda) \text{ is a constant unitary matrix}$$

hold for every $u \in \mathcal{S}_{in}^{q\times q}$ and $v \in \mathcal{S}_{in}^{p\times p}$.

Theorem 9.5 *Let $s \in \Pi \cap \mathcal{S}^{p\times q}$ and suppose that $s \notin \mathcal{S}_{in}^{p\times q}$ and $s \notin \mathcal{S}_{*in}^{p\times q}$. Then a $\mathcal{D}$-representation S of the mvf s that is obtained by formula (9.16) is minimal if and only if the denominator $\{b_2, b_1\}$ of the mvf h considered in this formula is a minimal denominator of h.*

Proof The theorem is an immediate consequence of the definitions of minimal $\mathcal{D}$-representations and minimal denominators. □

A scattering $\mathcal{D}$-representation $S \in \mathcal{S}_{in}^{p\times p}$ of a mvf $s \in \Pi \cap \mathcal{S}_{in}^{p\times q}$ with $p > q$ is said to be **minimal** if the implication

$$S(\lambda) \begin{bmatrix} v(\lambda) & 0 \\ 0 & I_q \end{bmatrix}^{-1} \in \mathcal{S}_{in}^{p\times p} \Longrightarrow v(\lambda) \text{ is a constant unitary matrix}$$

holds for every $v \in \mathcal{S}_{in}^{p\times p}$. It is easy to see that a scattering $\mathcal{D}$ representation $S \in \mathcal{S}_{in}^{p\times p}$ of a mvf s that is obtained from formula (9.12) is minimal if and only if the mvf b_1 considered in this formula is a constant $r_g \times r_g$ unitary matrix.

A scattering $\mathcal{D}$-representation $S \in \mathcal{S}_{in}^{q\times q}$ of a mvf $s \in \Pi \cap \mathcal{S}_{*in}^{p\times q}$ with $q > p$ is said to be minimal if

$$\begin{bmatrix} I_p & 0 \\ 0 & u(\lambda) \end{bmatrix}^{-1} S(\lambda) \in \mathcal{S}_{in}^{q\times q} \Longrightarrow u(\lambda) \text{ is a constant unitary matrix}$$

It is easy to see that a scattering $\mathcal{D}$-representation $S \in \mathcal{S}_{in}^{q\times q}$ of a mvf s that is obtained from formula (9.15) is minimal if and only if the mvf b_2 considered in this formula is a constant $r_f \times r_f$ unitary matrix.

Let $s \in \Pi \cap \mathcal{S}^{p\times q}$ and assume that $s \notin \mathcal{S}_{in}^{p\times q}$ and $s \notin \mathcal{S}_{*in}^{p\times q}$. A scattering $\mathcal{D}$-representation S of mvf s is called **optimal** if $s_{22} \in \mathcal{S}_{out}^{r_f\times q}$ and **$*$-optimal** if $s_{11}^{\sim} \in \mathcal{S}_{out}^{r_g\times p}$.

The optimal (resp., $*$-optimal) scattering $\mathcal{D}$-representation S of a mvf s may be obtained from formula (9.16) with $b_2 = I_{r_f}$ (resp., $b_1 = I_{r_g}$).

Theorem 9.6 *Let $s \in \Pi \cap \mathcal{S}^{p\times q}$ and suppose that $s \notin \mathcal{S}_{in}^{p\times q}$ and $s \notin \mathcal{S}_{*in}^{p\times q}$. Then:*

(1) *There exists a minimal optimal scattering $\mathcal{D}$-representation $S_\circ$ of s which may be obtained from formula (9.16) by setting b_2 equal to a constant $r_f \times r_f$ unitary matrix and setting b_1 equal to a minimal right denominator of the mvf h.*

(2) *There exists a minimal $*$-optimal scattering $\mathcal{D}$-representation $S_\bullet$ of s which may be obtained from formula (9.16) by setting b_1 equal to a constant $r_g \times r_g$ unitary matrix and setting b_2 equal to a minimal left denominator of the mvf h.*

(3) *The two $\mathcal{D}$-representations $S_\circ$ and $S_\bullet$ are uniquely defined by s up to a constant unitary left multiplier diag $\{I_p, v\}$ and a constant unitary right multiplier diag $\{u, I_q\}$, respectively.*

Proof The proof is immediate from the definitions. □

Theorem 9.7 *Let $s \in \Pi \cap \mathcal{S}^{p\times q}$ and let $S \in \mathcal{S}_{in}^{n\times n}$ be a minimal scattering $\mathcal{D}$-representation of s. Then v is a minimal scalar denominator of $s^\#$ if and only if v is a minimal scalar denominator of $S^\#$.*

Proof It is clear that if s is a block of a mvf $S \in \mathcal{S}_{in}^{n\times n}$, then

$$S \in \Pi^{n\times n}(1, v) \Longrightarrow s \in \Pi^{p\times q}(1, v)$$

It remains to show that the converse implication is in force if n is minimal possible size and the $\mathcal{D}$-representation S of s is minimal. There are four cases to consider, just as in the proof of Theorem 9.1.

1. $s \in \mathcal{S}_{in}^{p\times p}$: $S = s$ is the only $\mathcal{D}$-representation of s and the assertion is self-evident.

2. $s \in \Pi \cap \mathcal{S}_{in}^{p\times q}$ with $p > q$: All minimal $\mathcal{D}$-representations S of s are given by formula (9.12) with $b_1(\lambda)$ equal to an $r_g \times r_g$ unitary constant matrix. Thus, if $s \in \Pi^{p\times q}(1, v)$, then $g \in \Pi^{p\times p}(v, v)$ and by Theorem 3.110 a solution ψ of the factorization problem (9.11) belongs to the class $\Pi^{p\times r_g}(1, v)$. Therefore, $s_{11} = \psi b_1 \in \Pi^{p\times r_g}(1, v)$ and, consequently, $S \in \Pi^{n\times n}(1, v)$.

3. $s \in \Pi \cap \mathcal{S}_{*in}^{p\times q}$ with $q > p$: All minimal $\mathcal{D}$-representations S of s are given by formula (9.15) with $b_2(\lambda)$ equal to an $r_f \times r_f$ unitary constant matrix. Consequently, if $s \in \Pi^{p\times q}(1, v)$, then $f \in \Pi^{q\times q}(v, v)$ and, by Theorem 3.110, a solution φ of the factorization problem (9.14) belongs to the class $\Pi^{r_f\times q}(1, v)$. Therefore, $s_{22} = b_2\varphi \in \Pi^{r_f\times q}(1, v)$ and consequently $S \in \Pi^{q\times q}(1, v)$.

4. $s \in \Pi \cap \mathcal{S}^{p\times q}$, $s \notin \mathcal{S}_{in}^{p\times q}$ and $s \notin \mathcal{S}_{*in}^{p\times q}$: Let $S_\circ \in \mathcal{S}_{in}^{m\times m}$ be a minimal optimal scattering $\mathcal{D}$-representation of s with block decomposition

$$S_\circ = \begin{bmatrix} s_{11}^\circ & s \\ s_{21}^\circ & s_{22}^\circ \end{bmatrix}.$$

Then $s_{22}^\circ \in \mathcal{S}_{out}^{r_f\times q}$ and, since it is a solution of the factorization problem (9.14) and $f \in \Pi^{q\times q}(v, v)$, Theorem 3.110 guarantees that $s_{22}^\circ \in \Pi^{r_f\times q}(1, v)$. Thus, the mvf

$$S_2 = \begin{bmatrix} s \\ s_{22}^\circ \end{bmatrix} \quad \text{belongs to} \quad \mathcal{S}_{in}^{m\times q} \cap \Pi^{m\times q}(1, v),$$

and $S_2 \notin \mathcal{S}_{*in}^{m\times q}$. Moreover,

$$S_\circ = [S_1 \quad S_2] \quad \text{with} \quad S_1 = \begin{bmatrix} s_{11}^\circ \\ s_{21}^\circ \end{bmatrix}$$

is a minimal scattering $\mathcal{D}$-representation of S_2. Thus, by Case 2, $S_\circ \in \Pi^{m\times m}(1, v)$, and hence $s_{11}^\circ \in \Pi^{p\times r_g}(1, v)$. Therefore, since $S_\circ$ is a minimal optimal scattering $\mathcal{D}$-representation of s, it may be obtained from formula (9.16) with b_1 equal to a minimal right denominator d_R of the mvf h and b_2 a constant $r_f \times r_f$ unitary matrix. Thus, $s_{11}^\circ = \psi d_R$, where ψ is a solution of the factorization problem (9.11). Consequently, $\psi d_R \in \Pi^{p\times r_g}(1, v)$.

Now let S be a minimal scattering $\mathcal{D}$-representation of s. The mvf

$$S = \begin{bmatrix} s_{11} & s \\ s_{21} & s_{22} \end{bmatrix}$$

may be obtained from the formula (9.16) with some minimal denominator $\{b_2, b_1\}$ of the mvf h. Then b_1 is a minimal right denominator of b_2h and d_R is a right denominator of b_2h. Consequently, $b_1^{-1}d_R \in \mathcal{S}_{in}^{r_g\times r_g}$, which together with the already established fact that $\psi d_R \in \Pi^{p\times r_g}(1, v)$ implies that $\psi b_1 \in \Pi^{p\times r_g}(1, v)$, i.e., $s_{11} \in \Pi^{p\times r_g}(1, v)$. Moreover, since S is a minimal scattering $\mathcal{D}$-representation of the mvf $[s_{11} \quad s]$ and $[s_{11} \quad s] \in \Pi \cap \mathcal{S}_{*in}^{p\times m}$, Case 3 implies that $S \in \Pi^{m\times m}(1, v)$. □

Lemma 9.8 *Let $s \in \Pi \cap \mathcal{S}^{p\times q}$, let $S \in \mathcal{S}_{in}^{n\times n}$ be a mvf with $s_{12} = s$ and suppose that the ranks r_g and r_f of the mvf's g and f defined by formulas (9.10) and (9.13), respectively, are both positive. Then the following conditions are equivalent:*

(1) *The mvf S is a $\mathcal{D}$-representation of s, i.e., $n = p + r_f$.*

(2) $\overline{s_{11}^* L_2^p} = L_2^{r_g}$.

(3) $\overline{s_{22} L_2^q} = L_2^{r_f}$.

Proof Let $S \in \mathcal{S}_{in}^{n\times n}$ with $s_{12} = s$ and diagonal blocks s_{11} and s_{22}. Then $n \geq p + r_f = q + r_g$ and, by Theorem 3.73, the mvf's $s_{11} \in \mathcal{S}^{p\times(n-q)}$ and $s_{22} \in \mathcal{S}^{(n-p)\times q}$ admit factorizations of the form

$$s_{11} \;=\; \psi b_1 \quad \text{where } \psi^\sim \in \mathcal{S}_{out}^{r_g\times p} \text{ and } b_1 \in \mathcal{S}_{*in}^{r_g\times(n-q)}$$

and

$$s_{22} \;=\; b_2\varphi, \quad \text{where } \varphi \in \mathcal{S}_{out}^{r_f\times q} \text{ and } b_2 \in \mathcal{S}_{in}^{(n-p)\times r_f},$$

since ψ and s_{11} are both solutions of the factorization problem (9.11), whereas φ and s_{22} are both solutions of the factorization problem (9.14).

Suppose now that (1) holds, i.e., that $n = p + r_f = q + r_g$. Then $b_1 \in \mathcal{S}_{in}^{r_g \times r_g}$ and $b_2 \in \mathcal{S}_{in}^{r_f \times r_f}$ and hence

$$\overline{s_{11}^* L_2^p} = \overline{b_1^* \psi^* L_2^p} = b_1^* (\overline{\psi^* L_2^p}) = b_1^* L_2^{r_g} = L_2^{r_g}$$

and

$$\overline{s_{22} L_2^q} = \overline{b_2 \varphi L_2^q} = b_2 (\overline{\varphi L_2^q}) = b_2 L_2^{r_f} = L_2^{r_f}.$$

Thus, (1) $\Longrightarrow$ (2) and (1) $\Longrightarrow$ (3).

Suppose next that $n > p + r_f = q + r_g$. Then $b_1 \in \mathcal{S}_{*in}^{r_g \times (n-q)}$, but the vvf $h_1 = \rho_i^{-1}(I_{n-q} - b_1^* b_1)\xi$ has positive norm in L_2^{n-q} for some $\xi \in \mathbb{C}^{n-q}$. Moreover,

$$s_{11}(\mu) h_1(\mu) = \psi(\mu) b_1(\mu) \{I_{n-q} - b_1(\mu)^* b_1(\mu)\} \frac{\xi}{\rho_i} = 0 \quad \text{a.e. on } \mathbb{R},$$

i.e., h_1 is orthogonal to $\overline{s_{11}^* L_2^p}$. Thus (2) fails and consequently, (2) $\Longrightarrow$ (1). Much the same sort of argument shows that (3) $\Longrightarrow$ (1). □

Remark 9.9 *If $S \in \mathcal{S}_{in}^{n \times n}$ and $s_{12} = s$, then the following implications hold:*

(1) *If $r_g > 0$ and $r_f = 0$, then $n = p \Longleftrightarrow \overline{s_{11}^* L_2^p} = L_2^{r_g}$.*

(2) *If $r_g = 0$ and $r_f > 0$, then $n = q \Longleftrightarrow \overline{s_{22} L_2^q} = L_2^{r_f}$.*

If $S \in \mathcal{S}_{in}^{n \times n}$ is a rational $\mathcal{D}$-scattering realization of $s = s_{12}$, then $S(1/\lambda) = S_{\mathfrak{S}}(\lambda)$ is the scattering matrix of a simple (and hence minimal, since $B = iC^* D$) conservative scattering system $\mathfrak{S} = (A, B, C.D; \mathbb{C}^m, \mathbb{C}^n, \mathbb{C}^n)$ with $m = \deg S$ and

$$S_{\mathfrak{S}}(\lambda) = D + C(\lambda I - A)^{-1} B \quad \text{for } \lambda \in \rho(A).$$

If $r_f > 0$ and $r_g > 0$, then B, C and D have block decompositions

$$B = [* \quad B_1], \quad C = \begin{bmatrix} C_1 \\ * \end{bmatrix} \quad \text{and} \quad D = \begin{bmatrix} * & D_1 \\ * & * \end{bmatrix}$$

with respect to the decompositions

$$\mathbb{C}^n = \mathbb{C}^{r_f} \oplus \mathbb{C}^q \quad \text{and} \quad \mathbb{C}^n = \mathbb{C}^p \oplus \mathbb{C}^{r_g}$$

of the input and output spaces of $\mathfrak{S}$ corresponding to the four block decomposition of S. The system $\mathfrak{S}_1 = (A, B_1, C_1, D_1; \mathbb{C}^m, \mathbb{C}^q, \mathbb{C}^p)$ is a passive scattering system with scattering matrix

$$S_{\mathfrak{S}_1}(\lambda) = D_1 + C_1(\lambda I - A_1)^{-1}B_1 = s(\lambda).$$

This is a minimal passive scattering realization of $s(\lambda)$ if and only if $S(\lambda)$ is a minimal scattering $\mathcal{D}$ representation of $s(\lambda)$. Moreover, the system $\mathfrak{S}$ will be a minimal optimal (resp., $*$-optimal) passive scattering system if and only if $S(\lambda)$ is a minimal optimal (resp., $*$-optimal) scattering $\mathcal{D}$-representation of $s(\lambda)$.

9.2 Chain scattering $\mathcal{D}$-representations

If $S \in \mathcal{S}_{in}^{n\times n}$ with $n = p + r_f (= q + r_g)$ is a mvf with $s_{12} = s$, then:

$$r_g = p \Longleftrightarrow r_f = q \Longleftrightarrow s_{11}(\lambda) \not\equiv 0 \Longleftrightarrow \det s_{22}(\lambda) \not\equiv 0. \qquad (9.26)$$

The Potapov-Ginzburg transform $W = PG(S)$ of the mvf S is well defined if and only if the four equivalent conditions in (9.26) are satisfied. In this case the scattering $\mathcal{D}$-representation of s as the 12 block of S is equivalent to the representation

$$s = T_W[0_{p\times q}], \quad \text{where } W \in \mathcal{U}(j_{pq}). \qquad (9.27)$$

The analysis that was carried out in the previous section yields the following results.

Theorem 9.10 *Let $s \in \mathcal{S}^{p\times q}$. There exists at least one mvf W such that (9.27) holds if and only if $s \in \Pi^{p\times q}$ and*

$$g(\mu) = I_p - s(\mu)s(\mu)^* > 0 \quad \text{and} \quad f(\mu) = I_q - s(\mu)^*s(\mu) > 0 \quad \text{a.e. on } \mathbb{R}, \qquad (9.28)$$

i.e., if and only if $r_g = p$ and $r_f = q$. If these conditions are satisfied, then the solution ψ (resp., φ) of the factorization problem (9.11) (resp., 9.14) belongs to $\Pi \cap \mathcal{S}_{out}^{p\times p}$ (resp., $\Pi \cap \mathcal{S}_{out}^{q\times q}$), the mvf h considered in (9.17) may be defined by the formula

$$h(\lambda) = -\varphi(\lambda)s^\#(\lambda)\psi^\#(\lambda)^{-1} = -(\varphi^\#)^{-1}(\lambda)s^\#(\lambda)\psi(\lambda) \qquad (9.29)$$

and the set of all mvf's $W \in \mathcal{U}(j_{pq})$ such that (9.27) holds is described by the formula

$$W(\lambda) = W^\circ(\lambda)\begin{bmatrix} b_1(\lambda) & 0 \\ 0 & b_2(\lambda)^{-1} \end{bmatrix}, \qquad (9.30)$$

where

$$W^\circ = \begin{bmatrix} \mathfrak{d}_- & \mathfrak{c}_+ \\ \mathfrak{c}_- & \mathfrak{d}_+ \end{bmatrix} = \begin{bmatrix} (\psi^\#)^{-1} & s\varphi^{-1} \\ s^\#(\psi^\#)^{-1} & \varphi^{-1} \end{bmatrix} \tag{9.31}$$

belongs to $\mathfrak{M}_l(j_{pq})$ *and* $\{b_2, b_1\}$ *is any denominator of the mvf* h.

Proof The theorem follows from Theorem 9.3 and formula (4.40) for the Potapov-Ginzburg transform. □

A representation of a mvf s in the form (9.27) is called a **chain scattering $\mathcal{D}$-representation** of the mvf s. Thus, Theorem 9.10 gives a criterion for such a representation to exist and a description of the set of all chain scattering $\mathcal{D}$-representations of the given s.

Corollary 9.11 *If* $W \in \mathcal{U}(j_{pq})$, *then* $T_W[0] = 0$ *if and only if it can be expressed in the form*

$$W(\lambda) = \begin{bmatrix} u(\lambda) & 0 \\ 0 & v(\lambda)^{-1} \end{bmatrix}, \quad \textit{where } u \in \mathcal{S}_{in}^{p\times p} \textit{ and } v \in \mathcal{S}_{in}^{q\times q}. \tag{9.32}$$

A chain scattering $\mathcal{D}$-representation (9.27) of a mvf s is called **minimal** if the implication

$$W(\widetilde{W})^{-1} \in \mathcal{U}(j_{pq}) \implies \widetilde{W} \in \mathcal{U}_{const}(j_{pq}) \tag{9.33}$$

for every mvf $\widetilde{W} \in \mathcal{U}(j_{pq})$ such that $T_{\widetilde{W}}[0] = 0$. In this case W is also called minimal.

Theorem 9.12 *A chain scattering $\mathcal{D}$-representation (9.27) of s is minimal if and only if the scattering $\mathcal{D}$-representation $S = PG(W)$ of s is minimal, i.e., if and only if the pair $\{b_2, b_1\}$ considered in formula (9.30) is a minimal denominator of the mvf h defined by formula (9.29).*

Proof This follows easily from Theorem 9.10. □

A chain scattering $\mathcal{D}$-representation (9.27) and the mvf W in it is called **optimal** if $W \in \mathcal{N}_+^{m\times m}$ and **$*$-optimal** if $W^\# \in \mathcal{N}_+^{m\times m}$. The mvf's that are considered in minimal optimal and minimal $*$-optimal representations will be denoted $W_\circ$ and $W_\bullet$, respectively. They are the PG transforms of the mvf's $S_\circ$ and $S_\bullet$, respectively, that were introduced in Theorem 9.6.

Theorem 9.13 *Let* $s \in \Pi \cap \mathcal{S}^{p\times q}$ *and assume that* $\det\{I_q - s^\# s\} \not\equiv 0$. *Let* $W \in \mathcal{U}(j_{pq})$ *be a minimal chain scattering $\mathcal{D}$-representations (9.27) of s and* *let v be a minimal scalar denominator of the mvf h that is defined by formula (9.29). Then:*

(1) *v is a scalar denominator of both W and $W^\#$, i.e.,*

$$W \in \Pi^{m\times m}(v,v). \tag{9.34}$$

(2) *v is a minimal scalar denominator of $W_\bullet$ and $W_\circ^\#$, where $W_\bullet$ is a minimal $*$-optimal and $W_\circ$ is a minimal optimal chain scattering $\mathcal{D}$-representation of s.*

Proof If $\{b_2, b_1\}$ is a minimal denominator of the mvf h defined by formula (9.29), then, in view of Theorem 9.12, formulas (9.30)-(9.31) define a minimal mvf W. Let $g = I_p - ss^\#$, $f = I_q - s^\# s$, and let

$$h_1 = s^\# g^{-1} = f^{-1} s^\#. \tag{9.35}$$

Then

$$h = -\varphi h_1 \psi, \tag{9.36}$$

where φ and ψ are solutions of the problems (9.14) and (9.11), respectively. Since $\varphi \in \mathcal{N}_{out}^{q\times q}$ and $\psi \in \mathcal{N}_{out}^{p\times p}$, the functions h and h_1 have the same minimal scalar denominator v. Since

$$I_p + sh_1 = g^{-1} \text{ and } I_q + h_1 s = f^{-1},$$

v is a scalar denominator of the mvf's g^{-1} and f^{-1}. Moreover, since b_1 is a minimal right denominator of the mvf $b_2 h$ and b_2 is a minimal left denominator of hb_1, the function v is a scalar denominator of the mvf's b_1^{-1} and b_2^{-1}.

Let w_{ij} be the blocks of the mvf W. Then

$$\begin{aligned} vw_{11} &= v\psi^{-\#}b_1 = (vg^{-1})\psi b_1 \in \mathcal{N}_+^{p\times p}, \\ vw_{12} &= s\varphi^{-1}(vb_2^{-1}) \in \mathcal{N}_+^{p\times q}, \\ vw_{21} &= -vs^\#\psi^{-\#}b_1 = -\varphi^{-1}(vh)b_1 \in \mathcal{N}_+^{q\times p}, \\ vw_{22} &= v\varphi^{-1}b_2^{-1} = -\varphi^{-1}(vb_2^{-1}) \in \mathcal{N}_+^{q\times q}, \end{aligned}$$

i.e., $W \in \mathcal{N}^{m\times m}(v)$. Furthermore,

$$\begin{aligned} vw_{11}^{\#} &= (vb_1^{-1})\psi^{-1} \in \mathcal{N}_+^{p\times p}, \\ vw_{12}^{\#} &= vb_2\varphi^{-\#}s^{\#} = b_2\varphi(vh_1) \in \mathcal{N}_+^{q\times p}, \\ vw_{21}^{\#} &= -(vb_1^{-1})\psi^{-1}s \in \mathcal{N}_+^{p\times q}, \\ vw_{22}^{\#} &= b_2(v\varphi^{-\#}) = b_2\varphi(vf^{-1}) \in \mathcal{N}_+^{q\times q}, \end{aligned}$$

i.e., $W^{\#} \in \mathcal{N}^{m\times m}(v)$. Thus, $W \in \Pi^{m\times m}(v, v)$.

If $W = W_\bullet$, then b_1 is constant and, consequently, v is a minimal scalar denominator of w_{21}, since $vw_{21} = \varphi^{-1}(vh)b_1$ and $\varphi \in \mathcal{N}_{out}^{q\times q}$. Therefore, the denominator v for $W_\bullet$ is minimal. Similarly, if $W = W_\circ$, then b_2 is constant and v is a minimal scalar denominator of $w_{12}^{\#}$, since $vw_{12}^{\#} = b_2\varphi(vh_1)$ and $\varphi \in \mathcal{N}_{out}^{q\times q}$. Thus the denominator v of $W_\circ^{\#}$ is minimal. □

Theorem 9.14 *A mvf $s \in \mathcal{S}^{p\times q}$ admits a representation of the form (9.3) with some $\varepsilon \in \mathcal{S}_{const}^{p\times q}$ if and only if $s \in \Pi^{p\times q}$. If this condition is satisfied, then, in the considered representation,*

$$\operatorname{rank}(I_p - \varepsilon\varepsilon^*) = \operatorname{rank}(I_p - s\,s^*) \quad \textit{and} \quad \operatorname{rank}(I_q - \varepsilon^*\varepsilon) = \operatorname{rank}(I_q - s^*s).$$

Proof See [Ar73]. □

9.3 $\mathcal{D}$-representations in the class $\mathcal{C}^{p\times p}$

Analogues of the representations (9.27) and (9.3) with $\varepsilon \in \mathcal{S}_{\text{const}}^{p\times q}$ exist for mvf's $c \in \Pi \cap \mathcal{C}^{p\times p}$:

Theorem 9.15 *A mvf $c \in \mathcal{C}^{p\times p}$ admits a representation of the form*

$$c = T_A[\tau] \quad \textit{with } A \in \mathcal{U}(J_p) \textit{ and } \tau \in \mathcal{C}_{const}^{p\times p} \tag{9.37}$$

if and only if $c \in \Pi^{p\times p}$. Moreover, for any such representation,

$$\operatorname{rank} \Re\tau \;=\; \operatorname{rank} \Re c.$$

Proof Since

$$c \in \mathcal{C}^{p\times p} \Longrightarrow T_{\mathfrak{V}}[c] \in \mathcal{S}^{p\times p} \quad \text{and} \quad A \in \mathcal{U}(J_p) \Longrightarrow \mathfrak{V}A\mathfrak{V} \in \mathcal{U}(j_p), \tag{9.38}$$

the assertion follows from Theorem 9.14 with $s = T_{\mathfrak{V}}[c]$ and $W = \mathfrak{V}A\mathfrak{V}$ because in the representation (9.3) ε may be chosen in $\mathcal{D}(T_{\mathfrak{V}})$. □

Theorem 9.16 *A mvf c admits a representation of the form*

$$c = T_A[I_p] \text{ with } A \in \mathcal{U}(J_p) \tag{9.39}$$

if and only if

$$c \in \Pi \cap \mathcal{C}^{p\times p} \quad \text{and} \quad \operatorname{rank} \Re c = p. \tag{9.40}$$

If this condition is satisfied, then the set of all mvf's $A \in \mathcal{U}(J_p)$ such that $T_A[I_p] = c$ is described by the formula

$$A(\lambda) = B^\circ(\lambda) \begin{bmatrix} b_1(\lambda) & 0 \\ 0 & b_2(\lambda)^{-1} \end{bmatrix} \mathfrak{V}, \tag{9.41}$$

where

$$B^\circ(\lambda) = \frac{1}{\sqrt{2}} \begin{bmatrix} -c^\#(\lambda)x_2^\#(\lambda) & c(\lambda)x_1(\lambda) \\ x_2^\#(\lambda) & x_1(\lambda) \end{bmatrix} \tag{9.42}$$

and

(1) *$x_1(\lambda)$ and $x_2(\lambda)$ are essentially unique solutions of the factorization problems*

$$x_1(\mu)x_1(\mu)^* = \{(\Re c)(\mu)\}^{-1} \quad \text{a.e. on } \mathbb{R} \text{ and } x_1 \in \mathcal{N}_{out}^{p\times p}, \tag{9.43}$$

$$x_2(\mu)^*x_2(\mu) = \{(\Re c)(\mu)\}^{-1} \quad \text{a.e. on } \mathbb{R} \text{ and } x_2 \in \mathcal{N}_{out}^{p\times p}. \tag{9.44}$$

(2) *$x_1 \in \Pi^{p\times p}$ and $x_2 \in \Pi^{p\times p}$.*

(3) *$\{b_2, b_1)\}$ is a denominator of the mvf*

$$\theta(\lambda) = x_1^\#(\lambda)x_2(\lambda)^{-1} = x_1(\lambda)^{-1}x_2^\#(\lambda). \tag{9.45}$$

Proof The theorem follows from Theorem 9.10 and formula (9.37), since $(I_p + s) \in \mathcal{N}_{out}^{p\times p}$ and $\{b_2, b_1\} \in \text{den } h \iff \{b_2, b_1\} \in \text{den } \theta$, because the mvf's h and θ, considered in (9.29) and in (9.45) are connected by the relation

$$\theta(\lambda) - h(\lambda) = x_1^{-1}(\lambda)\{I_p + s(\lambda)\}x_2^{-1}(\lambda),$$

where the right hand side belongs to $\mathcal{N}_{out}^{p\times p}$ by Lemma 3.54. □

Remark 9.17 *If $c(\lambda) \equiv I_p$, then $x_1(\lambda) = x_2(\lambda) \equiv I_p$, $B^\circ(\lambda) \equiv \mathfrak{V}$ and $\theta(\lambda) \equiv I_p$. Consequently every pair $\{b_2, b_1\}$ of mvf's $b_1, b_2 \in \mathcal{S}_{in}^{p\times p}$ is a* *denominator of the mvf* $\theta = I_p$ *and the formula*

$$A^\circ(\lambda) = \mathfrak{V} \begin{bmatrix} b_1 & 0 \\ 0 & b_2^{-1} \end{bmatrix} \mathfrak{V}, \quad \textit{where } b_1, b_2 \in \mathcal{S}_{in}^{p\times p}, \tag{9.46}$$

gives a description of the set of mvf's A° *such that*

$$A^\circ \in \mathcal{U}(J_p) \quad \textit{and} \quad I_p = T_{A^\circ}[I_p].$$

A $\mathcal{D}$-representation (9.39) of c is said to be:

(a) **minimal** if the implication

$$A(\lambda)A^\circ(\lambda)^{-1} \in \mathcal{U}(J_p) \Longrightarrow A^\circ \in \mathcal{U}_{const}(J_p)$$

holds for every mvf A° of the form (9.46). In this case, A is also said to be minimal.

(b) **optimal** if $A \in \mathcal{N}_+^{m\times m}$; in this case, A is also said to be optimal.

(c) $*$**-optimal** if $A^\# \in \mathcal{N}_+^{m\times m}$; in this case, A is also said to be $*$-optimal.

If $c \in \Pi \cap \mathcal{C}^{p\times p}$ and $\Re c(\mu) > 0$ a.e. on $\mathbb{R}$, then there exist minimal optimal and minimal $*$-optimal mvf's $A_\circ$ and $A_\bullet$ for c. They are connected with minimal optimal and minimal $*$-optimal mvf's $W_\circ$ and $W_\bullet$ for the mvf $s = T_{\mathfrak{V}}[c]$ by the relations

$$A_\circ(\lambda) = \mathfrak{V}\, W_\circ(\lambda)\mathfrak{V} \quad \text{and} \quad A_\bullet(\lambda) = \mathfrak{V}\, W_\bullet(\lambda)\mathfrak{V}. \tag{9.47}$$

Thus, $A_\circ(\lambda)$ may be obtained from the general formula (9.41) with $b_2(\lambda) \equiv I_p$ and $b_1(\lambda) = d_R(\lambda)$, where $d_R(\lambda)$ is a minimal right denominator of the mvf θ, and $A_\bullet$ may be obtained from formula (9.41) with $b_1(\lambda) \equiv I_p$ and $b_2(\lambda) = d_L(\lambda)$, where $d_L(\lambda)$ is a minimal left denominator of the mvf $\theta(\lambda)$.

9.4 Inverse problems for de Branges matrices and spectral functions

In this section two inverse problems for J_p-inner mvf's will be studied.

The inverse de Branges matrix problem: *Given a* $p \times 2p$ *mvf* $\mathfrak{E}(\lambda) = [E_-(\lambda) \quad E_+(\lambda)]$, *describe the set of mvf's* $A(\lambda)$ *such that*

$$\mathfrak{E}(\lambda) = \sqrt{2}[0 \quad I_p]A(\lambda)\mathfrak{V} \quad \textit{and} \quad A \in \mathcal{U}(J_p). \tag{9.48}$$

In view of formula (5.131) this problem can be reformulated as follows: *Given two $p \times p$ mvf's $a_{21}(\lambda)$ and $a_{22}(\lambda)$, describe the set of all mvf's $A \in \mathcal{U}(J_p)$ such that $[a_{21}\ a_{22}]$ is the bottom block row of A, i.e.,*

$$[a_{21}(\lambda) \quad a_{22}(\lambda)] = [0 \quad I_p]A(\lambda). \tag{9.49}$$

The inverse spectral problem: *Given a $p \times p$ mvf $\Delta(\mu)$ on $\mathbb{R}$, describe the set of all mvf's $A \in \mathcal{U}(J_p)$ such that*

$$\Delta(\mu) = \Re c(\mu) \quad a.e. \text{ on } \mathbb{R} \quad when \quad c = T_A[I_p], \tag{9.50}$$

i.e., $\Delta(\mu)$ is the spectral density of the component c_a in the decomposition (3.14) of the mvf $c \in \mathcal{C}^{p\times p}$ considered in (9.50).

The inverse de Branges matrix problem is connected with the inverse spectral problem because if formulas (9.50) and (9.48) hold for the same mvf $A \in \mathcal{U}(J_p)$, then

$$\Delta(\mu) = E_+(\mu)^{-*}E_+(\mu)^{-1} = E_-(\mu)^{-*}E_-(\mu)^{-1} \quad \text{a.e. on } \mathbb{R}. \tag{9.51}$$

The solutions of these inverse problems will be obtained on the basis of the solution of the inverse impedance problem considered in Theorem 9.16.

Theorem 9.18 *Let $\mathfrak{E} = [E_- \quad E_+]$ be a given $p \times 2p$ mvf. Then there exists a mvf $A \in \mathcal{U}(J_p)$ such that (9.48) holds if and only if $\mathfrak{E}$ is a regular de Branges matrix. Moreover, if $\mathfrak{E}$ is a regular de Branges matrix, then:*

(1) *There exists a mvf $c \in \mathcal{C}^{p\times p}$ with*

$$\Re c(\mu) = E_+(\mu)^{-*}E_+(\mu)^{-1} \quad a.e. \text{ on } \mathbb{R} \tag{9.52}$$

and every such c belongs to $\Pi \cap \mathcal{C}^{p\times p}$.

(2) *The set of all mvf's $A \in \mathcal{U}(J_p)$ such that (9.48) holds is described by formula (5.132).*

(3) *The parameter c in formula (5.132) can be recovered from A by the formula $c = T_A[I_p]$.*

Proof If $A \in \mathcal{U}(J_p)$, then $\mathfrak{E} = \sqrt{2}[0 \quad I_p]A\mathfrak{V}$ is a regular de Branges matrix by Theorem 5.71. Moreover, the mvf $c = T_A[I_p]$ belongs to the class $\Pi\cap\mathcal{C}^{p\times p}$ and formula (5.132) holds, thanks to (9.41), (9.42) and (9.48).

Conversely, if $\mathfrak{E} = [E_- \quad E_+]$ is a regular de Branges matrix, then $E_+^{-*}E_+^{-1} \in \widetilde{L}_1^{p\times p}$. Therefore the mvf c_a defined by formula (5.134) satisfies the condition

(9.52). Moreover, every mvf $c \in \mathcal{C}^{p\times p}$ for which (9.52) holds may be written as $c = c_s + c_a$, where c_s is any mvf in $\mathcal{C}_{sing}^{p\times p}$ and c_a is defined by formula (5.134. The mvf

$$c_a(\mu) = -c_a(\mu)^* + 2E_+(\mu)^{-*}E_+(\mu)^{-1} \quad \text{a.e. on } \mathbb{R}$$

is the nontangential boundary value of the mvf $-c_a^{\#}(\lambda)+2E_+(\lambda)^{-\#}E_+(\lambda)^{-1}$, which belongs to the class of meromorphic mvf's f in $\mathbb{C}_-$ such that $f^{\#} \in \mathcal{N}^{p\times p}$. Therefore, $c_a \in \Pi^{p\times p}$; and, since the inclusion $c_s \in \Pi^{p\times p}$ is always true, it follows that $c \in \Pi^{p\times p}$.

The final step is to obtain the description (5.132) from Theorem 9.16. Since $\det E_+(\lambda) \not\equiv 0$ and $\det E_-^{\#}(\lambda) \not\equiv 0$ and the mvf's E_+^{-1} and $E_-^{-\#}$ belong to $\mathcal{N}_+^{p\times p}$, they admit factorizations of the form

$$E_+^{-1} = b_2\varphi \quad \text{and} \quad E_-^{-\#} = \psi b_1,$$

where

$$b_1 \in \mathcal{S}_{in}^{p\times p}, \quad b_2 \in \mathcal{S}_{in}^{p\times p}, \quad \varphi \in \mathcal{N}_{out}^{p\times p} \quad \text{and} \quad \psi \in \mathcal{N}_{out}^{p\times p}.$$

Moreover, $x_1 = \varphi^{-1}$ and $x_2 = \psi^{-1}$ are solutions of the factorization problems (9.43) and (9.44), respectively, and $\{b_2, b_1\}$ is a denominator of the mvf θ defined in (9.45), since

$$\theta = x_1^{-1}x_2^{\#} = \varphi\psi^{-\#} = b_1^{-1}E_+^{-1}E_-b_2^{-1} = b_1^{-1}\chi b_2^{-1},$$

and $\chi \in \mathcal{S}_{in}^{p\times p}$. Thus, Theorem 9.16 is applicable to this choice of b_1, b_2 and c and formulas (9.41) and (9.42) are equivalent to (5.132). □

The solution of the inverse spectral problem is given in the following theorem.

Theorem 9.19 *The inverse spectral problem for a given $p\times p$ mvf $\Delta(\mu)$ on $\mathbb{R}$ admits a solution $A(\lambda)$ if and only if (1) $\Delta \in \widetilde{L}_1^{n\times n}$; (2) $\Delta(\mu) > 0$ a.e. on $\mathbb{R}$; and (3) $\Delta(\mu)$ is the nontangential boundary value of a mvf $\Delta \in \Pi^{p\times p}$. Moreover, if these three conditions are satisfied, then formulas (5.133) and (5.134) (with $\Delta(\mu)$ in place of $E_+^{-*}(\mu)E_+^{-1}(\mu)$) define a mvf $c \in \Pi \cap \mathcal{C}^{p\times p}$ with $\Re c(\mu) = \Delta(\mu) > 0$ and the set of solutions $A(\lambda)$ of the inverse spectral problem for $\Delta(\mu)$ coincides with the set of solutions of the inverse impedance problem for c described in Theorem 9.16.*

Proof In the proof of Theorem 9.18 it is shown that if Δ is obtained from a mvf $A \in \mathcal{U}(J_p)$ by (9.50), then Δ has the three listed properties. Conversely, if the mvf $\Delta(\mu)$ has these three properties, then formulas (5.133) and (5.134) (with $\Delta(\mu)$ in place of $E_+^{-*}(\mu)E_+^{-1}(\mu)$) define a mvf $c \in \Pi \cap \mathcal{C}^{p\times p}$ with $\Delta(\mu) = \Re c(\mu)$ a.e. on $\mathbb{R}$. □

9.5 The inverse 12 block problem for j_{pq}-inner mvf's

In this section we consider a third inverse problem:

IP3 *Given a $p \times q$ mvf w, describe the set of all mvf's W such that*

$$W \in \mathcal{U}(j_{pq}) \quad \textit{and} \quad W = \begin{bmatrix} w_{11} & w \\ w_{21} & w_{22} \end{bmatrix}. \tag{9.53}$$

The next theorem provides a solution of this problem.

Theorem 9.20 *Let $w(\lambda)$ be a $p \times q$ mvf that is meromorphic on $\mathbb{C}_+$. Then:*

(1) *There exists a mvf W such that (9.53) holds if and only if $w \in \Pi^{p\times q}$.*

(2) *If $w \in \Pi^{p\times q}$, then the set of mvf's W such that (9.53) holds is described by the formula*

$$W = \begin{bmatrix} I_p & 0 \\ 0 & b_2^{-1} \end{bmatrix} \begin{bmatrix} \varphi_1^{\#} & w \\ \varphi_2 w^{\#} \varphi_1^{-1} & \varphi_2 \end{bmatrix} \begin{bmatrix} b_1 & 0 \\ 0 & I_q \end{bmatrix}, \tag{9.54}$$

where φ_1 and φ_2 are solutions of the factorization problems

$$\varphi_1^{\#}(\lambda)\varphi_1(\lambda) = I_p + w(\lambda)w^{\#}(\lambda) \quad \textit{with } \varphi_1 \in \mathcal{N}_{out}^{p\times p}, \tag{9.55}$$

$$\varphi_2^{\#}(\lambda)\varphi_2(\lambda) = I_q + w^{\#}(\lambda)w(\lambda) \quad \textit{with } \varphi_2 \in \mathcal{N}_{out}^{q\times q}, \tag{9.56}$$

respectively, and b_1 and b_2 are right denominators of the mvf's $w^{\#}\varphi_1^{-1}$ and $w\varphi_2^{-1}$, respectively.

Proof The details are left to the reader since the proof is similar to the proof of Theorem 9.3. □

A solution W of IP3 is said to be **minimal** if $v \in \mathcal{S}_{in}^{q\times q}$, $u \in \mathcal{S}_{in}^{p\times p}$ and

$$\begin{bmatrix} I_p & 0 \\ 0 & v^{-1} \end{bmatrix}^{-1} W \begin{bmatrix} u & 0 \\ 0 & I_q \end{bmatrix}^{-1} \in \mathcal{U}(j_{pq})$$

implies that both $v(\lambda)$ and $u(\lambda)$ are constant unitary matrices.

Theorem 9.21 *Let $w \in \Pi^{p\times q}$. Then there exists a unique minimal solution W of IP3, up to left and right constant unitary multipliers of the form*

$$\begin{bmatrix} I_p & 0 \\ 0 & v^* \end{bmatrix} \quad \text{and} \quad \begin{bmatrix} u & 0 \\ 0 & I_q \end{bmatrix},$$

respectively. A minimal solution W of IP3 may be obtained from formula (9.54), where b_1 and b_2 are minimal right denominators of the mvf's $w^{\#}\varphi_1^{-1}$ and $w\varphi_2^{-1}$, respectively.

Proof The proof is similar to the proof of Theorem 9.5. The details are left to the reader. □

9.6 Inverse problems in special classes

This section focuses on solutions of the inverse problems considered above in one or more of the following special classes:

(1) rational mvf's.

(2) entire mvf's.

(3) real mvf's f in the sense that $\overline{f(-\overline{\lambda})} = f(\lambda)$.

(4) symmetric mvf's f in the sense that $f(\lambda)^\tau = f(\lambda)$ and symplectic mvf's in the sense that $f^\tau(\lambda)\mathcal{J}_p f(\lambda) = \mathcal{J}_p$.

These classes are important in a variety of applications in physics and engineering.

The inverse scattering problem

Let $s \in \Pi \cap S^{p\times q}$, let f, g, r_f and r_g be defined by formulas (9.13) and (9.10) and assume that $r_f > 0$ and $r_g > 0$. This corresponds to Case 4 in the proof of Theorem 9.1. Cases 2 and 3 are simpler and may be analyzed in the same way as Case 4; Case 1 is not of interest, because then $S = s$.

Let φ and ψ be the unique solutions of the factorization problems (9.14) and (9.11) that are normalized by the conditions

$$\varphi(i\nu)R_f \begin{bmatrix} I_{r_f} \\ 0 \end{bmatrix} > 0 \quad \text{and} \quad [I_{r_g} \quad 0]L_g\psi(i\nu) > 0 \quad \text{for some } \nu > 0, \tag{9.57}$$

where R_f and L_g are appropriately chosen permutation matrices that depend upon f and g, but are independent of ν. Let h be the unique mvf that is defined by the relation (9.17).

In the next theorem, the given mvf s will be considered as the block s_{11} of the mvf $S \in \mathcal{S}_{in}^{m\times m}$, instead of as the block s_{12}. This identification is convenient if s is symmetric. Then, in view of Theorem 9.3, the set of all mvf's $S \in \mathcal{S}_{in}^{n\times n}$ of minimal size with $s = s_{11}$ is given by the formula

$$S = \begin{bmatrix} I_p & 0 \\ 0 & b_2 \end{bmatrix} S^\circ \begin{bmatrix} I_q & 0 \\ 0 & b_1 \end{bmatrix}, \tag{9.58}$$

where

$$S^\circ = \begin{bmatrix} s & \psi \\ \varphi & h \end{bmatrix}. \tag{9.59}$$

Theorem 9.22 *Let $s \in \Pi \cap \mathcal{S}^{p\times q}$ and assume that $r_f > 0$ and $r_g > 0$. Then the set of all scattering $\mathcal{D}$-representations $S \in \mathcal{S}_{in}^{n\times n}$ with $s_{11} = s$ is described by formula (9.58), where S° is uniquely specified by formula (9.59) and $\{b_2, b_1\}$ is an arbitrary denominator of the mvf h.*
Suppose further that s is subject to one or more of the extra constraints that are presented in the first column of Table 8.1. Then the mvf's S°, S, b_1 and b_2 that are considered in formulas (9.58) and (9.59) satisfy the conditions that are displayed in the remaining columns of the table.

s	$\overset{\circ}{S}$	minimal S	S	$\{b_2\ b_1\}$
real	real	$S_\circ$ and $S_\bullet$ may be chosen real	real $\iff$	b_1 and b_2 are real
symmetric	is chosen symmetric	-	symmetric $\iff$	$b_2 = b_1^\tau$
rational	rational	rational	rational $\iff$	b_1 and b_2 are rational
entire of exp type		entire of exp type		

Table 9.1.

Thus, for example, if $s \in \mathcal{S}^{p\times q}$ is a real rational mvf such that $r_f > 0$ and $r_g > 0$, then the mvf S° defined by formula (9.59) with normalized ψ and φ is also a real rational mvf and the scattering $\mathcal{D}$-representation S with $s_{11} = s$ defined by formula (9.58) is a real rational mvf if and only *if the denominator $\{b_2, b_1\}$ of the mvf h is chosen to be real and rational. Moreover, every minimal scattering $\mathcal{D}$-representation S of s is rational. The minimal optimal and minimal $*$-optimal $\mathcal{D}$ representations $S_\circ$ and $S_\bullet$ may be chosen real. If s is also symmetric, then we may choose $L_g = R_f^\tau$ in the normalization condition (9.57). Then S° will be symmetric too and then S is real symmetric and rational if and only if $b_2^\tau = b_1$ and b_1 is real and rational.*

Proof For the sake of simplicity, the proof will be organized in tabular form, that serves to justify Table 8.1.

s	f, g	ψ, φ	h	$\mathring{S}$
real	both real	both real	real	real
symmetric	$f^\tau(\lambda) = g(\lambda)$	$\psi^\tau = \varphi$	symmetric	symmetric
rational	both rational	both rational	rational	rational
entire of exp type	both entire of exp type	both entire of exp type		

Table 9.2.

The statements in the second column of Table 8.2 follow easily from the constraints imposed on s in the first column and the formulas for f and g. Then the uniqueness of the normalized solutions of the factorization problems guarantee the validity of the first two entries in the third column, since R_f and L_g are both real and may be chosen with $L_g = R_f^\tau$ when $s = s^\tau$. The last two entries follow from Theorem 3.70.

Next, the uniqueness of the solution h of equation (9.17) serves to justify the first two entries in the fourth column. The third entry in this column also follows from (9.17), because $s^\#$, φ, and $\psi^\#$ are rational mvf's and the $r_g \times p$ mvf $\psi^\#$ has a rational right inverse, since rank $\psi^\# = r_g$. The entries in the fifth column of Table 8.2 follow from the entries in columns one, three and four. This yields the second column of Table 8.1, and hence by formula (9.58), justifies the equivalences between the fourth and fifth columns of Table 8.1. The last conclusion depends upon the formulas

$$s_{12} = \psi b_1 \quad \text{and} \quad s_{21} = b_2\varphi.$$

If $S^\tau = S$, then $\psi^\tau = \varphi$ and hence, by the essential uniqueness of the inner-outer factorization of s_{21}, it follows that

$$s_{12}^\tau = s_{21} \Longleftrightarrow b_1^\tau = b_2.$$

If s is rational or real, then since the $p \times r_g$ mvf ψ has rank r_g, it has a left inverse that is real if ψ is real and is rational if ψ is rational. Therefore s_{12} is real or rational if and only if b_1 is real or rational, respectively. Similar conclusions hold for s_{21} and b_2.

The verification of the entries in the third column of Table 8.1 follows from Theorem 9.7, which guarantees that if S is a minimal $\mathcal{D}$-representation of a mvf $s \in \mathcal{S}^{p\times q} \cap \Pi^{p\times q}(1, v)$, then $S \in \Pi^{n\times n}(1, v)$. If s is a rational mvf, then a minimal scalar denominator v of $s^{\#}$ is a rational inner function, i.e., v is a finite Blaschke product and hence S is rational mvf, since $vS^{-1} \in S_{in}^{n\times n}$. If the mvf s is entire of exponential type a , then $v = e_a$ and, consequently, $e_a S^{-1} \in S_{in}^{n\times n}$. Therefore, by Lemma 3.98, S is an entire mvf. Finally, the verification of the first entry in the third column of Table 8.1 is obtained from formula (9.58) with $b_1 = d_R$ and $b_2 = I_{r_f}$ for $S_\circ$ and $b_1 = I_{r_g}$ and $b_2 = d_L$ for $S_\bullet$, where d_R and d_L are minimal right and left denominators of the mvf h. Since S° is real in this setting, it suffices to show that d_R and d_L may both be chosen real when h is real. The mvf's $d_R(\lambda)$ and $d_L(\lambda)$ may be uniquely specified by imposing the normalization $d_R(i\nu) > 0$ and $d_L(i\nu) > 0$ for some $\nu > 0$. Then, if h is real, $\overline{d_R(-\overline{\lambda})}$ and $d_R(\lambda)$ are both minimal right denominators of h that meet the same normalization condition. Therefore $\overline{d_R(-\overline{\lambda})} = d_R(\lambda)$, i.e., d_R is real. Similar considerations serve to establish the existence of a real minimal left denominator of h. □

The inverse chain scattering problem

The next theorem focuses on special properties of the mvf W in the chain scattering $\mathcal{D}$-representation (9.27) when the given matrix s is subject to one or more of the extra constraints that are under consideration in this section.

Theorem 9.23 *Let $s \in \Pi \cap \mathcal{S}^{p\times q}$, assume that $r_g = p$ and $r_f = q$ and let the set of mvf's W such that*

$$W \in \mathcal{U}(j_{pq}) \quad and \quad s = T_W[0]$$

be described by formula (9.30) , where W° is uniquely specified by formula (9.31) and the mvf's b_2, b_1 are defined by s as in Theorem 9.10.
Suppose further that s is subject to one or more of the extra constraints that are presented in the first column of Table 8.3. Then the mvf's W°, W, b_1

and b_2 that are considered in formulas (9.30)–(9.31) satisfy the conditions that are displayed in the remaining columns of the table.

s	W°	minimal W	W	$\{b_2\ b_1\}$
real	real	$W_\circ$ and $W_\bullet$ may be chosen real	real $\iff$	b_1 and b_2 are real
symmetric	is chosen symplectic	-	symplectic $\iff$	$b_2 = b_1^\tau$
rational	rational	rational	rational $\iff$	b_1 and b_2 are rational

Table 9.3.

Thus, for example, if $s \in \mathcal{S}^{p\times p}$ is a real rational mvf such that $r_f > 0$ and $r_g > 0$, then the mvf W° defined by formula (9.31) with normalized ψ and φ is also a real rational mvf and the mvf W in the chain scattering $\mathcal{D}$-representation $s = T_W[0]$ is a real rational mvf if and only if the denominator $\{b_2,\ b_1\}$ of the mvf h is chosen so that b_1 and b_2 are real and rational. Moreover, every minimal W in the chain scattering $\mathcal{D}$-representation of s is rational. The minimal optimal and minimal $*$-optimal mvf's $W_\circ$ and $W_\bullet$ may be chosen real. If s is also symmetric, then W° may be chosen symplectic and then W is real symplectic and rational if and only if $b_2^\tau = b_1$ and b_1 is real and rational.

Proof Let S be a scattering $\mathcal{D}$-representation of s with block $s_{11} = s$ as in Theorem 9.22 and let

$$\widetilde{S} = S\begin{bmatrix} 0 & I_q \\ I_p & 0 \end{bmatrix} = \begin{bmatrix} \widetilde{s}_{11} & s \\ \widetilde{s}_{21} & \widetilde{s}_{22} \end{bmatrix}. \tag{9.60}$$

The PG transform $W = PG(\widetilde{S})$ defines a one to one correspondence between the mvf's S and the mvf's W in the chain scattering $\mathcal{D}$-representations (9.27). Moreover,

$$\begin{aligned} W \text{ is real} &\iff S \text{ is real} \\ W \text{ is symplectic} &\iff S \text{ is symmetric} \\ W \text{ is rational} &\iff S \text{ is rational} \\ W \text{ is minimal} &\iff S \text{ is minimal} \\ W = W_\circ &\iff S = S_\circ \quad \text{and} \quad W = W_\bullet \iff S = S_\bullet. \end{aligned}$$

Thus, as W° and S° are related in the same way as W and S, the theorem follows from Theorem 9.22. □

Remark 9.24 *The third column of Table 8.1 guarantees the existence of a real S when s is real and a rational S when s is rational. If s is symmetric, then a symmetric S may be obtained from formula (5.132) with $b_1 = d_R$ and $b_2 = d_R^T$, where d_R is any minimal right denominator of h.*

Theorem 9.25 *Let $s \in \mathcal{S}^{p\times q}$ and let $r_g = p$ and $r_f = q$. Then the mvf s admits the representation*

$$s = T_W[0] \text{ with some } W \in \mathcal{E} \cap \mathcal{U}(j_{pq}) \tag{9.61}$$

if and only if the following conditions are satisfied:

(1) *The mvf s is meromorphic in $\mathbb{C}$.*

(2) *The mvf*

$$f(\lambda) = I_q - s^\#(\lambda)s(\lambda) \tag{9.62}$$

has the properties

$$\det f(\lambda) \not\equiv 0, \quad \ln \det f \in \widetilde{L}_1. \tag{9.63}$$

(3) *The mvf*

$$h_1(\lambda) = f(\lambda)^{-1}s^\#(\lambda) \tag{9.64}$$

is entire of exponential type.

Moreover, if these conditions are satisfied, then:

(i) *The set of all mvf's W such that (9.61) holds is described by formula (9.30) with entire inner mvf's b_1 and b_2.*

(ii) *The mvf W° that is defined by formula(9.31) is entire.*

(iii) *Every minimal W in the chain scattering D-representation s (9.27) is entire.*

(iv) *If, in addition to (1)–(3), a mvf s is also (a) real or (b) symmetric or (c) real and symmetric, then the mvf W in the representation (9.27) of s may be chosen to be (a) real and entire, (b) symplectic and entire, (c) real, symplectic and entire, respectively, by appropriate choices of b_1 and b_2.*

Proof Let s admit the representation (9.61). Then (1) is obvious and (2) holds, since

$$f(\lambda) = w_{22}^{\#}(\lambda)^{-1} w_{22}(\lambda)^{-1}.$$

Moreover, (3) holds, since

$$h_1(\lambda) = w_{22}(\lambda) w_{12}^{\#}(\lambda)$$

and the mvf W is assumed to be entire of exponential type.

Conversely, if properties (1)–(3) are in force, then the mvf's f^{-1} and g^{-1} are both entire of exponential type, since

$$I_q + h_1(\lambda)s(\lambda) = f(\lambda)^{-1} \quad \text{and} \quad f^{\#}(\lambda) = f(\lambda),$$

$$I_p + s(\lambda)h_1(\lambda) = g(\lambda)^{-1} \quad \text{and} \quad g^{\#}(\lambda) = g(\lambda)$$

and h_1 is an entire mvf of exponential type.(Since s is meromorphic in $\mathbb{C}$ and bounded in $\mathbb{C}_+$, it can not have poles on $\mathbb{R}$.) Moreover, since $f^{-1} \in \widetilde{L}_1^{q\times q}$, the bound

$$\int_{-\infty}^{\infty} \frac{\ln^+ \|h_1(\mu)\|}{1+\mu^2} d\mu \le \int_{-\infty}^{\infty} \frac{\ln \|f(\mu)^{-1}\|}{1+\mu^2} d\mu \le \int_{-\infty}^{\infty} \frac{\ln \det f(\mu)^{-1}}{1+\mu^2} d\mu$$

implies that h_1 satisfies the Cartwright condition. Therefore, by Theorem 3.108, $h_1 \in \mathcal{E} \cap \Pi^{q\times p}$, $f^{-1} \in \mathcal{E} \cap \Pi^{q\times q}$ and $g^{-1} \in \mathcal{E} \cap \Pi^{p\times p}$.

By Theorem 3.111, the solutions φ_1 and φ_2 of the factorization problems

$$f^{-1}(\lambda) = \varphi_1(\lambda)\varphi_1^{\#}(\lambda) \quad \text{with} \quad \varphi_1 \in \mathcal{N}_{out}^{q\times q} \tag{9.65}$$

and

$$g^{-1}(\lambda) = \varphi_2^{\#}(\lambda)\varphi_2(\lambda) \quad \text{with} \quad \varphi_2 \in \mathcal{N}_{out}^{p\times p} \tag{9.66}$$

are entire mvf's of exponential type that satisfy the Cartwright condition, i.e., $\varphi_1 \in \mathcal{E} \cap \Pi^{q\times q}$ and $\varphi_2 \in \mathcal{E} \cap \Pi^{p\times p}$. Moreover, the mvf's φ_1 and φ_2 are uniquely defined by s and the normalization conditions $\varphi_1(0) > 0$ and $\varphi_2(0) > 0$. Thus, the mvf W° defined by formula (9.31) may be written in terms of the mvf's φ_1 and φ_2 and s:

$$W^\circ = \begin{bmatrix} \varphi_2^{\#} & s\varphi_1 \\ s^{\#}\varphi_2^{\#} & \varphi_1 \end{bmatrix}. \tag{9.67}$$

The mvf's $s\varphi_1$ and $s^\# \varphi_2^\#$ are holomorphic and have exponential growth in $\overline{\mathbb{C}_+}$ and $\overline{\mathbb{C}_-}$, respectively. Moreover, since

$$s\varphi_1 = h_1^\# \varphi^\# \quad \text{and} \quad s^\# \varphi_2^\# = h_1 \psi,$$

the mvf's $s\varphi_1$ and $s^\# \varphi_2^\#$ are holomorphic and have exponential growth also in $\mathbb{C}_-$ and $\mathbb{C}_+$, respectively. Consequently, the mvf's $s\varphi_1$ and $s^\# \varphi_2^\#$ are both entire of exponential type. Thus, the mvf W°, defined in (9.67) is entire of exponential type. Moreover, $W^\circ \in \mathcal{E} \cap \Pi^{m\times m}$.

The mvf's h and h_1, that are defined by formulas (9.29) and (9.64) respectively, are connected by the relation

$$h = -\varphi h_1 \psi, \tag{9.68}$$

since

$$h = -\varphi s^\# \psi^{\#-1} = -\varphi s^\# g^{-1} \psi = -\varphi f^{-1} s^\# \psi = -\varphi h_1 \psi.$$

Since h_1 is an entire mvf, its minimal scalar denominator is entire, i.e., it is e_a, $a \geq 0$. Formula (9.68) implies that e_a is a minimal scalar denominator of the mvf h too, since φ and ψ are outer. Let $\{b_2, b_1\}$ be a minimal denominator of the mvf h. Then b_2 is a minimal left denominator of the mvf hb_1. Since e_a is a scalar denominator of hb_1, it is a scalar denominator of b_2^{-1}. Therefore, b_2 is an entire mvf, by Lemma 3.98. In the same way, it can be shown that the mvf b_1 is entire. Thus, all minimal W are entire, since W° is entire and b_1 and b_2 are both entire, since $\{b_2, b_1\}$ is a minimal denominator of h. Moreover, formula (9.30) implies that the mvf W in a chain scattering $\mathcal{D}$-representation (9.27) is entire if and only if the mvf's b_1 and b_2, that are considered in formula (9.27) are both entire. The rest of the theorem follows from Theorem 9.22. □

The inverse impedance problem.

The next two theorems on the inverse impedance problem are analogues of the preceding two theorems.

Theorem 9.26 *Let $c \in \Pi \cap \mathcal{C}^{p\times p}$, assume that*

$$\det \{c(\lambda) + c^\#(\lambda)\} \not\equiv 0,$$

and let the set of mvf's A such that

$$A \in \mathcal{U}(j_{pq}) \quad \textit{and} \quad c = T_A[I_p]$$

be described by formulas (9.41) and (9.42). Suppose further that c is subject to one or more of the extra constraints that are presented in the first column of Table 8.4. Then the mvf's B°, A, b_1 and b_2 that are considered in formulas (9.41)–(9.45) satisfy the conditions that are displayed in the remaining columns of the table.

c	$\overset{\circ}{B}$	minimal A	A	$\{b_2\ b_1\}$
real	real	$A_\circ$ and $A_\bullet$ may be chosen real	real	$\Longleftrightarrow$ b_1 and b_2 are real
symmetric	is chosen antisymplectic	-	symplectic	$\Longleftrightarrow$ $b_2 = b_1^\tau$
rational	rational	rational	rational	$\Longleftrightarrow$ b_1 and b_2 are rational

Table 9.4.

Thus, for example, if $c \in \mathcal{C}^{p\times p}$ is a real rational mvf, then the mvf B° defined by formula (9.42) with normalized x_1 and x_2 is also a real rational mvf and the mvf A in the $\mathcal{D}$-representation of c defined by formula (9.41) is a real rational mvf if and only if the denominator $\{b_2,\ b_1\}$ of the mvf h is chosen to be real and rational. Moreover, every minimal A in the $\mathcal{D}$-representation of c is rational. The minimal optimal and minimal $*$-optimal mvf's $A_\circ$ and $A_\bullet$ may be chosen real. If c is also symmetric, then B° may be chosen antisymplectic and then A is real symplectic and rational if and only if $b_2^\tau = b_1$ and b_1 is real and rational.

Proof The theorem follows from Theorem 9.23 applied to the mvf $s = T_{\mathfrak{V}}[c]$, since the mvf's A and B° are connected by the relations

$$A = \mathfrak{V}\, W\, \mathfrak{V} \quad \text{and} \quad B^\circ = \mathfrak{V} W^\circ$$

and the matrix $\mathfrak{V}$ is real and antisymplectic: $\mathfrak{V}^\tau \mathcal{J}_p \mathfrak{V} = -\mathcal{J}_p$. The mvf's b_1 and b_2 in formulas (9.41) and (9.30) are the same. □

Theorem 9.27 *A mvf $c \in \mathcal{C}^{p\times p}$ has a $\mathcal{D}$-representation (9.39) with an entire mvf A if and only if the following conditions are satisfied:*

(1) *The mvf c is meromorphic in $\mathbb{C}$ and has no poles on $\mathbb{R}$.*

(2) $\det[c(\lambda) + c^\#(\lambda)] \not\equiv 0$ *and* $\ln\det \Re c \in \widetilde{L}_1^{p\times p}$.

(3) *The mvf*

$$h_2(\lambda) = \left[\frac{c(\lambda) + c^\#(\lambda)}{2}\right]^{-1} \tag{9.69}$$

is entire of exponential type.

If these conditions are in force, then, in formula (9.41), the mvf $A^\circ \in \mathcal{E} \cap \Pi^{m\times m}$ and every minimal A in the $\mathcal{D}$-representations (9.39) is entire. Moreover, if, in addition, c is (a) real, (b) symmetric, or (c) real and symmetric, then the mvf A° may be (and will be) chosen (a) real, (b) symplectic, or (c) both real and symplectic, respectively, and after such a choice of A°:

(i) *A is a real entire mvf $\Longleftrightarrow$ b_1 and b_2 are both real entire mvf's.*

(ii) *A is a symplectic entire mvf $\Longleftrightarrow$ $b_2 = b_1^\tau$ and b_1 is an entire mvf.*

(iii) *A is a real symplectic entire mvf $\Longleftrightarrow$ $b_2 = b_1^\tau$ and b_1 is a real entire mvf.*

Proof The theorem follows from Theorem 9.25 applied to the mvf $s = T_{\mathfrak{V}}[c]$, from the connections between the mvf's B° and W°, and A and W, mentioned in the previous theorem and the connections

$$\begin{aligned} h(\lambda) &= -\varphi(\lambda)h_1(\lambda)\psi(\lambda) \\ &= -v_1^* x_1(\lambda)^{-1}\{I_p + s(\lambda)\}x_2(\lambda)^{-1}v_2^* + v_1^* x_1(\lambda)^{-1}h_2(\lambda)x_2(\lambda)^{-1}v_2^*, \end{aligned}$$

between the mvf's h, h_1 and h_2 defined by formulas (9.29), (9.64) and (9.69). In the preceding equality, v_1 and v_2 are $p\times p$ unitary matrices such that

$$\{I_p + s(\lambda)\}\varphi(\lambda)^{-1} = x_2(\lambda)v_1 \quad \text{and} \quad \psi(\lambda)^{-1}\{I_p + s(\lambda)\} = v_2 x_2(\lambda).$$

□

9.7 $J_{p,r}$-inner SI-dilations of $c \in \Pi \cap \mathcal{C}^{p\times p}$

There is another representation of a mvf $c \in \Pi \cap \mathcal{C}^{p\times p}$, as the 22 block of a $J_{p,r}$-inner mvf Θ in $\mathbb{C}_+$, where

$$J_{p,r} = \begin{bmatrix} I_r & 0 \\ 0 & J_p \end{bmatrix} \quad \text{if } r>0 \quad \text{and} \quad J_{p,0} = J_p \text{ if } r = 0, \tag{9.70}$$

and Θ has the block decomposition

$$\Theta = \begin{bmatrix} \theta_{11} & \theta_{12} & 0 \\ \theta_{21} & \theta_{22} & I_p \\ 0 & I_p & 0 \end{bmatrix} \tag{9.71}$$

that is conformal with $J_{p,r}$.

It is easy to verify the equivalence

$$c \in \mathcal{C}_{sing}^{p\times p} \Longleftrightarrow \Theta = \begin{bmatrix} c & I_p \\ I_p & 0 \end{bmatrix} \text{ belongs to } \mathcal{U}(J_p). \tag{9.72}$$

Moreover, if $c \in \Pi \cap \mathcal{C}^{p\times p}$, then

$$f_c = c + c^{\#} \text{ belongs to } \Pi^{p\times p} \tag{9.73}$$

and, by Lemma 3.109, the number

$$r_c = \text{rank } f_c(\lambda) \tag{9.74}$$

is constant on $\mathfrak{h}_{f_c}$, except possibly at a set of isolated points (where rank $f_c(\lambda) < r_c$), and rank $f(\mu) = r_c$ a.e. on $\mathbb{R}$. Clearly,

$$r_c = 0 \Longleftrightarrow c \in \mathcal{C}_{sing}^{p\times p}.$$

A $J_{p,r}$-inner mvf of the form (9.71) with $\theta_{22} = c$ will be called a $J_{p,r}$**-inner SI (scattering-impedance)-dilation** of the mvf c. We shall show:

(1) If c has a $J_{p,r}$-inner SI-dilation, then $c \in \Pi \cap \mathcal{C}^{p\times p}$ and $r \geq r_c$.

(2) If $c \in \Pi \cap \mathcal{C}^{p\times p}$, then there exists a $J_{p,r}$-inner SI-dilation of c with $r = r_c$.

Let $c \in \Pi \cap \mathcal{C}^{p\times p}$, let $r = r_c$ and assume that $r_c > 0$. Then, by Theorem 3.110, the factorization problems

$$\varphi(\mu)^*\varphi(\mu) = f_c(\mu) \quad \text{a.e. on } \mathbb{R} \tag{9.75}$$

and

$$\psi(\mu)\psi(\mu)^* = f_c(\mu) \quad \text{a.e. on } \mathbb{R} \tag{9.76}$$

admit an essentially unique pair of solutions $\varphi = \varphi_c$ and $\psi = \psi_c$ such that φ_c and ψ_c^τ are outer mvf's that belong to $\mathcal{N}_{out}^{r\times p}$. Since $f_c \in \widetilde{L}_1^{p\times p}$, Theorem 3.110 also guarantees that

$$\rho_i^{-1}\varphi_c \in \Pi \cap H_2^{r\times p} \quad \text{and} \quad \rho_i^{-1}\psi_c \in \Pi \cap H_2^{p\times r}.$$

Lemma 9.28 *Let $c \in \Pi \cap \mathcal{C}^{p\times p}$, let $r = r_c$ and suppose that $r_c > 0$ and let φ_c and ψ_c be solutions of the factorization problems (9.75) and (9.76) such that φ_c and ψ_c^τ belong to $\mathcal{N}_{out}^{r\times p}$. Then there exists a unique mvf $s_c \in \Pi^{r\times r}$ such that*

$$s_c\psi_c^{\#} = \varphi_c. \tag{9.77}$$

Moreover, s_c has unitary nontangential boundary values $s_c(\mu)$ a.e. on $\mathbb{R}$.

Proof If $r = p$, then it is readily seen that

$$s_c = \varphi_c(\psi_c^{\#})^{-1} \tag{9.78}$$

meets the stated requirements.

If $r < p$, then there exists an $r \times r$ submatrix of f_c with rank equal to r a.e. on $\mathbb{R}$. Suppose first that the upper left hand $r \times r$ block of f_c is such and consider the corresponding block decompositions

$$\varphi_c = [\varphi_1 \quad \varphi_2] \text{ and } \psi_c^{\tau} = [\psi_1^{\tau} \quad \psi_2^{\tau}]$$

with blocks φ_1 and ψ_1 of size $r \times r$. Then the formulas

$$\begin{bmatrix} \varphi_1^{\#} \\ \varphi_2^{\#} \end{bmatrix} [\varphi_1 \quad \varphi_2] = \begin{bmatrix} \psi_1 \\ \psi_2 \end{bmatrix} [\psi_1^{\#} \quad \psi_2^{\#}]$$

imply that

$$\varphi_1(\psi_1^{\#})^{-1} = (\varphi_1^{\#})^{-1}\psi_1 \quad \text{and} \quad \varphi_1^{\#}\varphi_2 = \psi_1\psi_2^{\#}$$

and hence that

$$\varphi_1(\psi_1^{\#})^{-1}\psi_2^{\#} = (\varphi_1^{\#})^{-1}\psi_1\psi_2^{\#} = (\varphi_1^{\#})^{-1}\varphi_1^{\#}\varphi_2 = \varphi_2.$$

Thus, the mvf

$$s_c = \varphi_1(\psi_1^{\#})^{-1} \tag{9.79}$$

meets the stated requirements.

If the upper left-hand $r \times r$ block of f_c is degenerate, then there exists a $p \times p$ permutation matrix K such that the rank of the upper left-hand $r \times r$ block of $K^{\tau} f_c K$ is equal to r. Then the preceding argument may be repeated for the block decompositions

$$\varphi_c K = [\varphi_1 \quad \varphi_2] \quad \text{and} \quad \psi_c^{\tau} K = [\psi_1^{\tau} \quad \psi_2^{\tau}].$$

The equality (9.77) defines s_c uniquely, because the specified mvf's φ_c and $\psi_c^{\#}$ are both outer. □

Lemma 9.29 *Let $c \in \Pi \cap \mathcal{C}^{p\times p}$, let $r_c > 0$ and let*

$$\Theta_c = \begin{bmatrix} s_c & \varphi_c & 0 \\ \psi_c & c & I_p \\ 0 & I_p & 0 \end{bmatrix} \tag{9.80}$$

be defined in terms of the mvf's φ_c, ψ_c and s_c that were introduced in Lemma 9.28. Then

$$\Theta_c \in \Pi^{(r+2p)\times(r+2p)} \quad \textit{and} \quad \Theta_c^{\#} J_{p,r}\Theta_c = J_{p,r} = \Theta_c J_{p,r}\Theta_c^{\#}. \tag{9.81}$$

Proof The inclusion in (9.81) is self-evident, since all the blocks of Θ belong to $\Pi^{k\times\ell}$ for appropriate choices of k and ℓ. The first identity in (9.81) is justified by a straightforward calculation; the second is immediate from the first. □

Theorem 9.30 *If a $p \times p$ mvf c admits a $J_{p,r}$-inner SI-dilation, then $c \in \Pi \cap \mathcal{C}^{p\times p}$ and $r \geq r_c$. Conversely, if $c \in \Pi \cap \mathcal{C}^{p\times p}$ and $r_c > 0$, then c admits infinitely many $J_{p,r}$-inner SI-dilations Θ with $r = r_c$ and all of them may be obtained by the formula*

$$\Theta = \begin{bmatrix} b_1 & 0 & 0 \\ 0 & I_p & 0 \\ 0 & 0 & I_p \end{bmatrix} \Theta_c \begin{bmatrix} b_2 & 0 & 0 \\ 0 & I_p & 0 \\ 0 & 0 & I_p \end{bmatrix}, \tag{9.82}$$

where Θ_c is defined by formula (9.80) and $\{b_1, b_2\}$ is a denominator of s_c, i.e.,

$$b_1 \in \mathcal{S}_{in}^{r\times r}, \quad b_2 \in \mathcal{S}_{in}^{r\times r} \quad \textit{and} \quad b_1 s_c b_2 \in \mathcal{S}_{in}^{r\times r}. \tag{9.83}$$

Proof If Θ is a $J_{p,r}$-inner SI-dilation of c, then $c = \Theta_{22}$ belongs to $\Pi^{p\times p}$, since $\Theta \in \mathcal{U}(J_{p,r})$ and $\mathcal{U}(J_{p,r}) \subset \Pi^{(r+2p)\times(r+2p)}$. Moreover, as

$$\Theta(\lambda)^* J_{p,r}\Theta(\lambda) \leq J_{p,r} \quad \text{on } \mathfrak{h}_\Theta^+$$

with equality a.e. on $\mathbb{R}$, it is readily seen that

$$\theta_{11}(\lambda)^*\theta_{11}(\lambda) \leq I_r \quad \text{and} \quad 2(\Re c)(\lambda) \geq \theta_{12}(\lambda)^*\theta_{12}(\lambda) \text{ on } \mathfrak{h}_\Theta^+ \tag{9.84}$$

with equality a.e. on $\mathbb{R}$ and

$$\theta_{21}^{\#}(\lambda) = \theta_{11}^{\#}(\lambda)\theta_{12}(\lambda). \tag{9.85}$$

The bounds (9.84) guarantee that θ_{11}, θ_{12} and $c = \theta_{22}$ have holomorphic extensions to all of $\mathbb{C}_+$ and that the extensions

$$\theta_{11} \in \mathcal{S}_{in}^{r\times r} \quad \text{and} \quad c \in \mathcal{C}^{p\times p}.$$

Thus, $c \in \Pi \cap \mathcal{C}^{p\times p}$, and the supplementary identity

$$\theta_{12}^{\#}\theta_{12} = f_c \tag{9.86}$$

implies that $r \geq r_c$. The second inequality in (9.84) implies that the mvf $f(\lambda) = \rho_i(\lambda)^{-1}\theta_{12}(\lambda)$ is subject to the bound

$$f(a+ib)^* f(a+ib) \leq \frac{(\Re c)(a+ib)}{2\pi^2\{a^2+(b+1)^2\}}$$

for every choice of $a+ib$ with $a \in \mathbb{R}$ and $b > 0$. Thus, the inequality (3.19) implies that

$$\int_{-\infty}^{\infty} f(a+ib)^* f(a+ib)da \leq \frac{1}{\pi}(\Re c)(i)$$

for every $b > 0$, and hence that

$$\rho_i^{-1}\theta_{12} \in H_2^{r\times p}. \tag{9.87}$$

In view of (9.86) and (9.87),

$$\theta_{12} = b_1\varphi_c, \quad \text{where } b_1 \in \mathcal{S}_{in}^{r\times r}. \tag{9.88}$$

Similar considerations based on the fact that

$$\Theta(\lambda)J_{p,r}\Theta(\lambda)^* - J_{p,r} \leq 0 \text{ in } \mathfrak{h}_\Theta^+$$

with equality a.e. on $\mathbb{R}$ lead to the inequality

$$2(\Re c)(\lambda) \geq \theta_{21}(\lambda)\theta_{21}(\lambda)^* \quad \text{on } \mathfrak{h}_\Theta^+$$

with equality a.e. on $\mathbb{R}$. Consequently, θ_{21} has a holomorphic extension to all of $\mathbb{C}_+$, and this extension, which we continue to denote by θ_{21}, has the following properties

$$\rho_i^{-1}\theta_{21} \in H_2^{p\times r} \quad \text{and} \quad \theta_{21}\theta_{21}^\# = f_c. \tag{9.89}$$

Thus,

$$\theta_{21} = \psi_c b_2, \quad \text{where } b_2 \in \mathcal{S}_{in}^{r\times r}, \tag{9.90}$$

and hence, upon substituting (9.88) and (9.91) into (9.85), it follows that

$$b_2^\#\psi_c^\# = \theta_{11}^\# b_1\varphi_c, \quad \text{or} \quad (\text{as } \theta_{11}^\#\theta_{11} = I_r) \quad b_1^\#\theta_{11}b_2^\#\psi_c^\# = \varphi_c.$$

Consequently, by Lemma 9.28, $b_1^\#\theta_{11}b_2^\# = s_c$, i.e.,

$$b_1 s_c b_2 = \theta_{11}. \tag{9.91}$$

Therefore, since $\theta_{11} \in \mathcal{S}_{in}^{r\times r}$, $\{b_1, b_2\}$ is a denominator of s_c; and, the system of equalities (9.88), (9.90), (9.91) and $\theta_{22} = c$ is equivalent to the equality (9.82).

Now let $c \in \Pi \cap \mathcal{C}^{p\times p}$ and $r_c > 0$, let $\{b_1, b_2\}$ be any denominator of the mvf s_c and let Θ_c be defined by formula (9.80). Then, since $\Theta_c \in \Pi^{(r+2p)\times(r+2p)}$ and is $J_{p,r}$ unitary a.e. on $\mathbb{R}$ with $r = r_c$, by Lemma 9.29, it follows that the mvf Θ that is defined by formula (9.82) has the same properties. Therefore, the PG transform

$$S = (P\Theta + Q)(P + Q\Theta)^{-1}, \quad \text{with } P = \frac{I_{r+2p} + J_{p,r}}{2} = I_{r+2p} - Q,$$

belongs to $\Pi^{(r+2p)\times(r+2p)}$ and is unitary a.e. on $\mathbb{R}$. Thus, in order to complete the proof that Θ is a $J_{p,r}$-inner SI-dilation of the mvf c, it remains only to check that $\Theta \in \mathcal{U}(J_{p,r})$, or equivalently, that $S = PG(\Theta)$ belongs to the class $\mathcal{S}_{in}^{(r+2p)\times(r+2p)}$. The verification makes use of the fact that $b_1 s_c b_2 \in \mathcal{S}_{in}^{r\times r}$, since $\{b_1, b_2\}$ is a denominator of s_c, and

$$(I_p + \frac{1}{2}c)^{-1} = I_p + T_{\mathfrak{V}}[\frac{c}{2}] \quad \text{belongs to } H_\infty^{p\times p}, \text{ since } T_{\mathfrak{V}}[\frac{c}{2}] \in \mathcal{S}^{p\times p}.$$

By direct calculation,

$$S = \begin{bmatrix} s_{11} & s_{12} & 0 \\ 0 & 0 & I_p \\ s_{31} & s_{32} & 0 \end{bmatrix},$$

where:

$$\begin{aligned} s_{32} &= \left(I_p - \tfrac{1}{2}c\right)\left(I_p + \tfrac{1}{2}c\right)^{-1} \text{ belongs to } \mathcal{S}^{p\times p}, \text{ since } \tfrac{1}{2}c \in \mathcal{C}^{p\times p}. \\ s_{12} &= b_1\varphi_c\left(I_p + \tfrac{1}{2}c\right)^{-1} \text{ belongs to } \mathcal{N}_+^{r\times p}. \\ s_{31} &= -\left(I_p + \tfrac{1}{2}c\right)^{-1}\psi_c b_2 \text{ belongs to } \mathcal{N}_+^{p\times r}. \\ s_{11} &= b_1 s_c b_2 - \tfrac{1}{2}b_1\psi_c\left(I_p + \tfrac{1}{2}c\right)^{-1}\psi_c b_2 \text{ belongs to } \mathcal{N}_+^{r\times r}. \end{aligned}$$

Consequently, $S \in \mathcal{N}_+^{(r+2p)\times(r+2p)}$ and, by the Smirnov maximum principle, $S \in \mathcal{S}_{in}^{r\times r}$. Thus $\Theta \in \mathcal{U}(J_{p,r})$.

If $\widetilde{\Theta}$ is a $J_{p,r}$-inner SI-dilation of c with $r = r_c$ and if $u \in \mathcal{S}_{in}^{r\times r}$ and $v \in \mathcal{S}_{in}^{r\times r}$ are two inner $m \times m$ mvf's, then

$$\Theta = \begin{bmatrix} u & 0 & 0 \\ 0 & I_p & 0 \\ 0 & 0 & I_p \end{bmatrix} \widetilde{\Theta} \begin{bmatrix} v & 0 & 0 \\ 0 & I_p & 0 \\ 0 & 0 & I_p \end{bmatrix} \tag{9.92}$$

is also a $J_{p,r}$-inner SI-dilation of c with $r = r_c$. A $J_{p,r}$-inner SI-dilation Θ of c with $r = r_c$ is said to be **minimal** if in every factorization of the form (9.92) the mvf's u and v are constant unitary $r \times r$ matrices. □

Theorem 9.31 *If a $J_{p,r}$-inner SI-dilation Θ of c with $r = r_c$ is expressed in the form (9.82) then it is minimal if and only if $\{b_1, b_2\}$ is a minimal denominator of the mvf s_c.*

Proof This follows easily from the definitions and formula (9.82). □

Theorem 9.32 *Let $c \in \mathcal{C}^{p\times p}$ with $r_c > 0$ be a rational mvf. Then every minimal $J_{p,r}$-inner SI-dilation Θ of c with $r = r_c$ is a rational mvf. Moreover, the mvf's Θ and c have the same McMillan degree.*

Proof The solutions φ_c and ψ_c of the factorization problems (9.75) and (9.76) are rational, by Theorem 3.110. Thus, the mvf s_c, that is defined by formula (9.77) is rational, since it is given explicitly by formula (9.78) if $r_c = p$, and by formula (9.79) if $0 < r_c < p$. If $\{b_1, b_2\}$ is a minimal denominator of a rational mvf, s_c, then b_1 and b_2 are rational mvf's by assertion (1) of Theorem 3.70. Thus a minimal Θ is a rational mvf, since all the mvf's in formulas (9.82) and (9.80) are rational. The proof that the McMillan degrees of Θ and c coincide may be based on Kalman minimal realization theory for systems and will not be presented here. □

Theorem 9.33 *If $c \in \mathcal{C}^{p\times p}$ is an entire mvf of exponential type, then*

(1) $c \in \Pi \cap \mathcal{C}^{p\times p}$.

(2) *Every minimal $J_{p,r}$-inner SI-dilation of c with minimal left or right denominators of s_c is an entire mvf.*

Proof Let $c \in \mathcal{E} \cap \mathcal{C}^{p\times p}$. Then $c \in \Pi^{p\times p}$ if and only if c is an entire mvf of exponential type, since $\mathcal{C}^{p\times p} \subset \mathcal{N}^{p\times p}$ and, consequently, Theorem 3.108 is applicable to the entries of the mvf c. Let $c \in \Pi \cap \mathcal{C}^{p\times p}$ be an entire mvf. Then $r_c > 0$, except in the case when $c(\lambda) = i\alpha - i\beta\lambda$, $\alpha^* = \alpha \in \mathbb{C}^{p\times p}$ and $\beta \geq 0$, $\beta \in \mathbb{C}^{p\times p}$. Let $r_c > 0$. Then f_c is an entire mvf and, by Theorem 3.111, the mvf's φ_c and ψ_c are entire. Let $r = r_c$ and let b be a minimal right denominator of $\varphi^\#$. Since $\varphi_c^\# \in \Pi^{p\times m}$ is entire, b is entire: $b \in \mathcal{E} \cap \mathcal{S}_{in}^{m\times m}$, by Theorem 3.70, and hence $\psi = \varphi_c^\# b \in \mathcal{E} \cap \mathcal{N}_+^{p\times m}$. Moreover, $\psi = \theta_{21}$ is a solution of the factorization problem (9.76). Consequently,

$$\psi = \psi_c b_2 \quad \text{for some } b_2 \in \mathcal{S}_{in}^{m\times m}.$$

Therefore,

$$\varphi_c = b b_2^\# \psi_c^\#$$

and hence, $bb_2^\# = s_c$, since s_c is the unique solution of equation (9.77). Since $b = s_c b_2 \in \mathcal{S}_{in}^{m\times m}$, b_2 is a right denominator of s_c. Moreover, if d_R is a minimal right denominator of s_c, then $s_c d_R \in \mathcal{S}_{in}^{r\times r}$, $u = d_R^{-1} b_2 \in \mathcal{S}_{in}^{r\times n}$ and $bu^{-1} = s_c b_2 u^{-1} = s_c d_R$ belongs to $\mathcal{N}_+^{r\times r}$. Thus, $bu^{-1} \in \mathcal{S}_{in}^{r\times r}$ and

$$\varphi_c^\# b u^{-1} = \psi_c b_2 u^{-1} = \psi_c d_R \in \mathcal{N}_+^{p\times r}.$$

Therefore, since b is a minimal right denominator of $\varphi_c^\#$, u must be constant. Thus, b_2 is a minimal right denominator of s_c. By Theorem 9.31, the mvf Θ with blocks

$$\theta_{11} = s_c b_2, \quad \theta_{12} = \psi_c, \quad \theta_{21} = \psi_c b_2 \quad \text{and} \quad \theta_{22} = c$$

is a minimal $J_{p,r}$-inner SI-dilation of c. But

$$s_c b_2 = b \in \mathcal{E}^{r\times r}, \quad \varphi_c \in \mathcal{E}^{r\times r}, \quad \psi_c b_2 = \varphi_c^\# b \in \mathcal{E}^{p\times r} \quad \text{and} \quad c \in \mathcal{E}^{p\times p}.$$

Thus, Θ is an entire mvf.

In the same way it may be checked that if d is a minimal left denominator of $\psi_c^\#$ and $\varphi = d\psi_c^\#$, then $\varphi = b_1\varphi_c$, where b_1 is a minimal left denominator of s_c, and that a mvf Θ of the form (9.71) with

$$\theta_{11} = b_1 s_c, \quad \theta_{12} = b_1 \varphi_c, \quad \theta_{21} = \psi_c \quad \text{and} \quad \theta_{22} = c$$

is an entire minimal $J_{p,r}$-inner SI-dilation of c. □

9.8 Rational approximation of mvf's of the class $\Pi \cap H_\infty^{p\times q}$

In this section an application of $\mathcal{D}$-representation to the solution of a rational approximation problems for mvf's of the class $\Pi \cap H_\infty^{p\times q}$ will be presented.

A mvf $f \in \Pi \cap H_\infty^{p\times q}$ is called **quasimeromorphic** if a minimal scalar denominator v of $f^\#$ is a Blaschke product, i.e., if

$$f = v g_-, \quad \text{where} \quad g_-^\# \in \Pi \cap H_\infty^{p\times q} \tag{9.93}$$

and

$$v(\lambda) = \prod_{k=1}^{\overline{n}} \gamma_k[(\lambda - \lambda_k)/(\lambda - \overline{\lambda_k})]^{m_k}, \tag{9.94}$$

where the $\overline{\lambda}_k$ are poles of f of orders m_k, and the γ_k are constants with $|\gamma_k| = 1$ that are chosen so that the Blaschke product converges if $\overline{n} = \infty$.

If $\lambda_k \neq 0$ for all $k \geq 1$ and $\lambda = 0$ is not a limit point of the sequence $\{\lambda_k\}$, i.e., if $0 \in \mathfrak{h}_f$, then γ_k may be chosen such that

$$v(\lambda) = \prod_{k=1}^{\overline{n}} [(1 - \lambda/\lambda_k)/(1 - \lambda/\overline{\lambda_k})]^{m_k}, \tag{9.95}$$

The Blaschke condition

$$\sum_{k=1}^{\overline{n}} \frac{m_k \Im \lambda_k}{|\lambda_k|^2 + 1} < \infty, \quad \text{where} \quad \lambda_k \neq \lambda_j \quad \text{if} \quad k \neq j, \tag{9.96}$$

is necessary to insure that there exists a quasimeromorphic mvf $f \in \Pi \cap H_\infty^{p\times q}$ with poles $\overline{\lambda}_k$ of orders $m_k, k \geq 1$.

Theorem 9.34 *Let the sequence $\{\overline{\lambda}_k, m_k\}, k \geq 1$, of complex numbers $\overline{\lambda}_k \in \mathbb{C}_-$ and integers $m_k, k \geq 1$, satisfy the Blaschke condition (9.96). Let $f \in L_\infty^{p\times q}$ be such that*

$$f(\mu) = \lim_{n\to\infty} f_n(\mu) \quad \text{a.e. on } \mathbb{R} \tag{9.97}$$

where $\{f_n\}$, $n \geq 1$, is a sequence of rational $p \times q$ mvf's with properties:

(a) *The poles of f_n are contained in the set $\{\overline{\lambda}_k\}, k \geq 1$, and the order of each pole $\overline{\lambda}_j$ of f_n is at most m_j.*

(b) *The sequence of mvf's f_n is uniformly bounded in $\mathbb{C}_+$, i.e.,*

$$\|f_n\|_\infty \leq c, \quad n \geq 1, \quad \text{for some } c > 0. \tag{9.98}$$

Then:

(1) *The mvf f is the nontangential limit of a quasimeromorphic mvf that will also be denoted f.*

(2) *The poles of the quasimeromorphic mvf f are contained in the set $\{\overline{\lambda}_k\}$, $k \geq 1$, and the order of every pole $\overline{\lambda}_j$ of f is at most m_j, i.e., the Blaschke product $v(\lambda)$, defined in (9.94) is a scalar denominator of the mvf $f^\#$.*

(3) *The mvf f is the limit of the sequence f_n in the complex plane, i.e.,*

$$f(\lambda) = \lim_{n\to\infty} f_n(\lambda) \quad \text{for } \lambda \in \mathbb{C} \setminus \{\overline{\lambda}_n\}_1^\infty. \tag{9.99}$$

Moreover, f is the uniform limit of the sequence f_n on every compact set $K \subset \mathbb{C}_+ \cup (\mathbb{C}_- \setminus \{\overline{\lambda}_n\}^\infty)_1$.

Proof Let $f \in L_\infty^{p\times q}$ and assume that the sequences $\{\overline{\lambda_k}, m_k\}$, $k \geq 1$, and $\{f_n\}$, $n \geq 1$, satisfy the conditions of the theorem. Let

$$F(\mu) = \frac{f(\mu)}{(\mu+i)^2} \quad \text{and} \quad F_n(\lambda) = \frac{f_n(\lambda)}{(\lambda+i)^2}.$$

Then $F \in L_1^{p\times q} \cap L_2^{p\times q}$, $F_n \in H_1^{p\times q} \cap H_\infty^{p\times q}$ and $F(\mu) = \lim_{n\to\infty} F_n(\mu)$ a.e. on $\mathbb{R}$. Moreover, since

$$F_n^\vee(t) = \frac{1}{2\pi} \int_{-\infty}^{\infty} e^{-i\mu t} F_n(\mu) d\mu = 0 \quad \text{for every } t < 0$$

and

$$\|F_n(\mu)\| \leq \frac{c}{\mu^2+1},$$

Lebesgue's dominated convergence theorem implies that

$$F^\vee(t) = \frac{1}{2\pi} \int_{-\infty}^{\infty} e^{-i\mu t} F(\mu) d\mu = 0 \quad \text{for every } t < 0.$$

Consequently,

$$F(\mu) = \int_0^\infty e^{i\mu t} F^\vee(t) dt, \quad \text{with } F^\vee \in L_2^{p\times q}(\mathbb{R}_+);$$

i.e., F is the nontangential limit of a mvf from $H_2^{p\times q}$, that will also be denoted by F. Thus, $f(\mu)$ is the nontangential limit of the mvf $f(\lambda) = (\lambda+i)^2 F(\lambda)$, that belongs to $\mathcal{N}_+^{p\times q}$. Since $f \in L_\infty^{p\times q}$, the Smirnov maximum principle guarantees that $f \in H_\infty^{p\times q}$.

Similar arguments applied to the mvf $g(\mu) = v(\mu)f(\mu)^*$ and the sequence $g_n(\lambda) = v(\lambda)f_n^\#(\lambda)$ imply that $g(\mu)$ is the nontangential limit of a mvf in $H_\infty^{q\times p}$ that will also be denoted g. Thus, $f \in H_\infty^{p\times q}$ and

$$f(\mu) = v(\mu)g^*(\mu) \quad \text{a.e. on } \mathbb{R}, \text{ where } g \in H_\infty^{q\times p}.$$

This means that $f \in \Pi \cap H_\infty^{p\times q}$ is a quasimeromorphic function and v is a scalar denominator of $f^\#$.

It remains to prove assertion (3) of the theorem. The Cauchy formula

$$\frac{f(\lambda)-f_n(\lambda)}{(\lambda+i)^2} = \frac{1}{2\pi i}\int_{-\infty}^{\infty}\frac{f(\mu)-f_n(\mu)}{(\mu+i)^2(\mu-\lambda)}d\mu$$

yields the bounds

$$\begin{aligned}&\|f(\lambda)-f_n(\lambda)\|^2\\ &\le\ |\lambda+i|^4\frac{1}{(2\pi)^2}\int_{-\infty}^{\infty}\frac{d\mu}{|\mu-\lambda|^2}\int_{-\infty}^{\infty}\frac{\|f(\mu)-f_n(\mu)\|^2}{(\mu^2+1)^2}d\mu\\ &=\ \frac{|\lambda+i|^4}{4\pi\Im\lambda}\int_{-\infty}^{\infty}\frac{\|f(\mu)-f_n(\mu)\|^2}{(\mu^2+1)^2}d\mu \quad \text{for } \lambda\in\mathbb{C}_+.\end{aligned}$$

Thus, as the Lebesgue dominated convergence theorem guarantees that the integral on the right hand side of the last inequality tends to zero, it follows that $f(\lambda)$ is the uniform limit of the sequence $f_n(\lambda)$ on every compact set $K_1\subset\mathbb{C}_+$. By the same arguments $g(\lambda)$ is the uniform limit of the sequence $g_n(\lambda)$ on every compact set $K_2\subset\mathbb{C}_+$. Since

$$f(\lambda)=v(\lambda)g^{\#}(\lambda) \quad \text{and} \quad f_n(\lambda)=v(\lambda)g_n^{\#}(\lambda),$$

the mvf $f(\lambda)$ is the uniform limit of the sequence $f_n(\lambda)$ on every compact set $K_3\subset\mathbb{C}_-\backslash\{\overline{\lambda}_k\}_1^\infty$. Thus, $f(\lambda)$ is the uniform limit of the sequence $\{f_n(\lambda)\}_1^\infty$ on every compact set $K\subset\mathbb{C}_+\cup(\mathbb{C}_-\setminus\{\overline{\lambda}_k\}_1^\infty)$. □

Theorem 9.35 *Let $f\in\Pi\cap H_\infty^{p\times q}$, $f\not\equiv 0$, be a quasimeromorphic function. and let $v(\lambda)$ be the minimal scalar denominator of $f^{\#}$ with poles $\overline{\lambda_k}$ in $\mathbb{C}_-$ of orders m_k, $k\ge 1$, that is defined by formula (9.94). Then there exists a sequence of rational mvf's f_n such that:*

(1) *The poles of each mvf f_n are contained in the sequence $\{\overline{\lambda_k}\}_{k=1}^{\overline{m}}$ and the order of each pole $\overline{\lambda_j}$ of f_n is at most m_j, i.e., v is a scalar denominator of $f_n^{\#}$.*

(2) *$f(\lambda)=\lim_{n\uparrow\infty}f_n(\lambda)$ for $\lambda\in\mathbb{C}\setminus\{\overline{\lambda_k}\}_1^{\overline{m}}$, $\overline{m}\le\infty$. Moreover, the convergence is uniform on every compact set $K\subset\mathbb{C}\setminus\{\overline{\lambda_k}\}_{k=1}^{\overline{m}}$ and*

$$f(\mu)=\lim_{n\uparrow\infty}f_n(\mu) \quad \text{a.e. on } \mathbb{R}.$$

(3) $\|f_n\|_\infty\le\|f\|_\infty$

Proof Let the mvf f satisfy the conditions of the theorem and let $s = f/\|f\|_\infty$. Then $s \in \Pi \cap \mathcal{S}^{p\times q}$, $\|s\|_\infty = 1$ and a minimal scalar denominator v of $s^\#$ is a Blaschke product. Let $S \in \mathcal{S}_{in}^{n\times n}$ be a minimal scattering $\mathcal{D}$-representation of the mvf s with $s_{12} = s$. Then, by Theorem 9.7, v is a minimal scalar denominator of the mvf $S^\#$. Therefore, by Theorem 3.88, S is a BP product. If $S(\lambda)$ is rational, then $s(\lambda)$ is rational and there is nothing to prove. Suppose therefore that

$$S(\lambda) = \overset{\infty}{\underset{k=1}{\overset{\curvearrowright}{\prod}}} B_k(\lambda),$$

is an infinite product of elementary BP factors and let

$$\mathcal{S}_m(\lambda) = \overset{m}{\underset{k=1}{\overset{\curvearrowright}{\prod}}} B_k(\lambda)$$

be the partial products. Then

$$S(\lambda) = \lim_{m\to\infty} \mathcal{S}_m(\lambda)$$

and the convergence is uniform on every compact set $K \subset \mathbb{C} \setminus \{\overline{\lambda_k}\}_1^\infty$, by Theorem 4.9. Since $v(\lambda)$ is a scalar denominator of mvf $S^\#$ and $\mathcal{S}_m^{-1}\mathcal{S} \in \mathcal{S}_{in}^{n\times n}$, v is a scalar denominator of every mvf $\mathcal{S}_m^\#$, $m \geq 1$, i.e., the set of poles of every mvf $\mathcal{S}_m$ is contained in $\{\overline{\lambda}_k\}$ and $\overline{\lambda}_j$ is a pole of $\mathcal{S}_m$ of order at most m_j. The 12 blocks $s_m(\lambda)$ of the mvf's $\mathcal{S}_m(\lambda)$, $m \geq 1$, are rational mvf's such that:

(1) The mvf $s(\lambda)$ is uniform limit of the sequence $s_m(\lambda)$ of every compact set $K \subset \mathbb{C} \setminus \{\overline{\lambda_k}\}_1^\infty$.

(2) The poles of s_m are contained in the set $\{\overline{\lambda}_k\}_1^\infty$ and a pole $\overline{\lambda}_j$ of s_m has order at most m_j.

(3) $\|s_m\|_\infty \leq 1 \quad (= \|s\|_\infty)$.

Moreover, there exists a subsequence $s_{m_k}(\lambda)$ such that

$$s(\mu) = \lim_{k\to\infty} s_{m_k}(\mu) \quad \text{a.e. on } \mathbb{R},$$

see Theorem 4.60. Then the subsequence $s_{m_k}(\lambda)$ has also the property

$$s(\mu) = \lim_{k\to\infty} s_{m_k}(\mu) \quad \text{a.e. on } \mathbb{R},$$

and the sequence of rational mvf's $f_k(\lambda) = \|f\|_\infty s_{m_k}(\lambda)$ have all the properties (1)–(3) that are formulated in the theorem. □

Lemma 9.36 *Let $f \in H_\infty^{p\times q}$. Then $f \in \Pi \cap H_\infty^{p\times q}$ if and only if there exists a sequence f_n of quasimeromorphic mvf's $f_n \in H_\infty^{p\times q}$ such that:*

(1) *The distinct poles $\lambda_1^{(n)}, \ldots, \lambda_{k_n}^{(n)}$ of f_n satisfy the condition*

$$\sup_n \sum_{j=1}^{k_n} \frac{m_j^{(n)} |\Im \lambda_j^{(n)}|}{1 + |\lambda_j^{(n)}|^2} < \infty, \quad \textit{where} \quad m_j^{(n)} = \textit{ the order of } \lambda_j^{(n)}.$$

(2) $f(\lambda) = \lim_{n\to\infty} f_n(\lambda)$ *uniformly on every compact set and the convergence is uniform on every compact*

$$K \subset \mathbb{C} \setminus \bigcup_{n=1}^{\infty} \{\lambda_j^{(n)}\}_{j=1}^{k_n}.$$

(3) $f(\mu) = \lim_{n\to\infty} f_n(\mu)$ *a.e. on* $\mathbb{R}$.

(4) $\|f_n\|_\infty \leq \|f\|_\infty$.

Proof One direction of the theorem follows from a corresponding result of Tumarkin [Tum66] on scalar functions of the class $H_p \cap \Pi$: if f_n is a sequence of quasimeromorphic mvf's from $H_\infty^{p\times q}$ that satisfy the conditions (1)–(4), then $f \in \Pi \cap H_\infty^{p\times q}$.

To prove the converse, it is enough to verify the assertion for $f \in \Pi \cap H_\infty^{p\times q}$ with $\|f\|_\infty = 1$. Let $F \in \mathcal{S}_{in}^{m\times m}$ be a scattering $\mathcal{D}$-representation of f, which exists by Theorem 9.1, i.e., let f be the block f_{12} of mvf F. A generalization of a theorem of Frostman to mvf's by Ginzburg [Gi97] implies that the mvf's

$$F_a = (F - aI_m)(I_m - \bar{a}F)^{-1} \quad \text{for } a \in \mathbb{D}$$

are BP products for almost all parameters $a \in \mathbb{D}$ with respect to Lebesgue measure. Then there exists a sequence $a_n \in \mathbb{D}$ such that $a_n \to 0$ as $n \to \infty$ and F_{a_n} are BP products that satisfy conditions (1)–(4) with F in place of f. Then the sequence f_n of the 12 blocks of F_{a_n} is a sequence of quasimeromorphic mvf's from $H_\infty^{p\times q}$ that satisfy the conditions (1)–(4). □

Theorem 9.37 *Let $f \in H_\infty^{p\times q}$. Then $f \in \Pi \cap H_\infty^{p\times q}$ if and only if there exists a sequence f_n of rational mvf's from $H_\infty^{p\times q}$ that satisfy the conditions (1)–(4) of Lemma 9.36.*

Proof This follows from Theorems 9.34, 9.35 and Lemma 9.36. □

9.9 Bibliographical notes

The representations of the mvf's $s \in \Pi \cap \mathcal{S}^{p\times q}$ and $c \in \Pi \cap \mathcal{C}^{p\times p}$ that were discussed above are called Darlington representations, even though the first results of this nature seem to have been obtained by V. Belevich for rational mvf's. More precisely, he showed that every rational *real* mvf $s \in \mathcal{S}^{p\times p}$ may be considered as a block of a rational *real* mvf $S \in \mathcal{S}_{in}^{n\times n}$ for some n, see the monograph [Bel68] and the references cited therein. This scattering formalism permitted Belevich to generalize the Darlington realization [Dar39] of a *real* scalar rational function $c \in \mathcal{C}$ with $(\Re c)(\mu) > 0$ on $\mathbb{R}$ (except for at most finitely many points) as the impedance of a passive linear one-port circuit with only one resistor to the more general setting of multiports. Belevich's result was generalized to the nonrational case by P. Dewilde [De71] and [De76]. The Darlington representations (9.4) and (9.6) for rational *real* mvf's $c \in \mathcal{C}^{p\times p}$ were obtained independently of Belevich by V. P. Potapov (see e.g., [EfPo73]), E. Ya. Melamud [Me72] and [Me79] by other methods that were based on the theory of J-contractive mvf's. Subsequently, influenced by his participation in Potapov's seminar, D. Z. Arov obtained Darlington representations for mvf's $s \in \Pi \cap \mathcal{S}^{p\times q}$ and $c \in \Pi \cap \mathcal{C}^{p\times p}$ in [Ar71] and [Ar73].

Darlington chain scattering representations with entire mvf's W were studied in connection with inverse problems for canonical differential systems with dissipative boundary conditions in [Ar71] and [Ar75].

Scattering $\mathcal{D}$-representations for operator valued functions $s \in \mathcal{S}(U, Y)$ were obtained in [DoH73], [Ar71], [Ar74a] and [Ar79a]. Inverse scattering, impedance and spectral problems for inner mvf's $S(\lambda)$ and J-inner mvf's $W(\lambda)$ and $A(\lambda)$ were also studied in [De71], [De76], [DeD81a], [DeD81b], [DeD84] and [ADD89].

The following inverse chain scattering problem was studied in Chapter 8 of [Dy89b]: given $s \in \mathcal{S}^{p\times q}$ describe the set of all mvf's $W \in \mathcal{U}(j_{pq})$ such that

$$s = T_W[\varepsilon] \text{ for some } \varepsilon \in \mathcal{S}^{p\times q}; \tag{9.100}$$

see also [BoD06] and, for another variant, [AlD84]. Connections with the Carathéodory-Julia theorem are discussed in [Sar94]; extensions of this theorem may be found in [BoK06] and a number of other papers by the same authors. A solution W of the inverse chain scattering problem with $\varepsilon \in \mathring{\mathcal{S}}^{p\times q}_{const}$ in (9.100) exists if and only if the conditions of Theorem 9.10 are satisfied. Moreover, in this case,

$$\varepsilon = T_{W_\varepsilon^\circ}[0_{p\times q}], \text{ where } W_\varepsilon^\circ = \begin{bmatrix} (I-\varepsilon\varepsilon^*)^{-1/2} & \varepsilon(I-\varepsilon^*\varepsilon)^{-1/2} \\ \varepsilon^*(I-\varepsilon\varepsilon^*)^{-1/2} & (I-\varepsilon^*\varepsilon)^{-1/2} \end{bmatrix}$$

and

$$s = T_{WW_\varepsilon^\circ}[0_{p\times q}]. \tag{9.101}$$

Representations of the form (9.100) with $W \in \mathcal{E}\cap\mathcal{U}(j_{pq})$ and rational mvf's $\varepsilon \in \mathcal{S}^{p\times q}$ were considered in [Ar01].

The $J_{p,r}$-inner SI-dilations in (9.71) is adapted from the papers [ArR07a] and [ArR07b], which were motivated by the realizations of stationary stochastic processes discussed in [LiPi82], [LiPa84] and [LiPi85]. The four block mvf $[\theta_{ij}]$, $i,j = 1,2$ that carries the information in (9.71) was also considered independently in [BEO00] in their study of acoustic wave filters. Subsequently, rational symmetric $\mathcal{D}$-representations S with minimal McMillan degree were studied in [BEGO07].

The equivalence of conditions (2) and (3) in Lemma 9.8 was observed by D. Pik [Pik03] in his study of the time varying case.

The inverse problems discussed in Section 9.4 were first considered by L. de Branges in the special case that the mvf's

$$\mathfrak{E}(\lambda), \quad \Delta(\lambda)^{-1} = E_+(\lambda)E_+^\#(\lambda) \quad \text{and} \quad A(\lambda)$$

are entire and $p = 1$ in connection with his study of inverse problems for canonical integral and differential systems; see the monograph [Br68a] and the references cited therein; some generalizations to matrix and operator valued functions are presented in [Br68b]. The problem of reconstructing an entire $\mathcal{J}_p$-inner mvf (resp., 2×2 $\mathcal{J}_1$-inner mvf that is meromorphic in the full complex plane $\mathbb{C}$) when either one or two of its four blocks are given was studied in [KrOv82] (resp., [GoM97]).

A complete proof of Theorem 9.20 may be found in Theorem 1.2 of [Ar97]. A number of related block completion problems are considered in [AFK93], [AFK94a] and [AFK94b].

The results in Section 9.8 are adapted from [Ar78]. Connections between the classes $\Pi\cap H_r$ and $\Pi\cap H_r^{p\times q}$ for $1\le r<\infty$ and rational approximation were studied by G. Ts. Tumarkin [Tum66] and V. E. Katsnelson [Kat93] and [Kat94], respectively.

Scattering $\mathcal{D}$-representations were used in [Ar74a] to study simple conservative scattering systems with losses by the Darlington method. This is equivalent to studying simple contractive and maximal dissipative operators by their characteristic functions. Passive scattering systems with extra properties: minimal, optimal, minimal and optimal, etc, were studied by the same methods in [Ar79a]; see [Ar95b] for a review. An application of scattering and chain scattering $\mathcal{D}$-representations to the passive system realization of rational mvf's $s \in \mathcal{S}^{p\times q}$ may be found in Sections 4–6 of [ArD02b]. Connections of $J_{p,r}$-inner SI-dilations with passive system theory were considered in [ArR07b]. There exist time varying generalizations of scattering $\mathcal{D}$-representations; see, e.g., [De99] and the references cited therein, and [Pik03].

10

More criteria for strong regularity

In this section a number of criteria for a mvf to belong to the class of right (left) strongly regular J-inner mvf's for a fixed $m \times m$ signature matrix $J \neq \pm I_m$ and for the specific signature matrices j_{pq}, J_p and $\mathcal{J}_p$ will be obtained and formulated in the terms of the matrix Muckenhoupt (A_2) condition that was introduced by S. Treil and A. Volberg and used by them to characterize those weighted $L_2^n(\Delta)$ spaces (with matrix valued weights $\Delta(\mu) \geq 0$ a.e. on $\mathbb{R}$) for which the Hilbert transform is bounded; see [TrV97]. Enroute to these criteria we shall establish analogous criteria for a right (resp., left) γ-generating matrix to belong to the class of right (resp., left) strongly regular γ-generating matrices.

10.1 The matrix Muckenhoupt (A_2) condition

Let

$$A_I(\Delta) = \frac{1}{|I|} \int_I \Delta(\mu) d\mu \tag{10.1}$$

denote the average of a mvf $\Delta \in L_{1,loc}^{n \times n}$ over a finite interval $I \subset \mathbb{R}$ of length $|I|$. If $\Delta(\mu) > 0$ a.e. on $\mathbb{R}$ and $\Delta^{\pm 1} \in L_{1,loc}^{n \times n}$, then $\Delta(\mu)$ is said to satisfy the **matrix Muckenhoupt (A_2) condition** if

$$\sup_I \|\{A_I(\Delta)\}^{1/2}\{A_I(\Delta^{-1})\}^{1/2}\| < \infty. \tag{10.2}$$

Lemma 10.1 *If an $n \times n$ mvf $\Delta(\mu)$ that is positive definite a.e. on $\mathbb{R}$ satisfies the matrix Muckenhoupt (A_2) condition (10.2), then*

$$\Delta \in \widetilde{L}_1^{n \times n} \quad \text{and} \quad \Delta^{-1} \in \widetilde{L}_1^{n \times n}. \tag{10.3}$$

Proof See Lemma 2.1 in [TrV97]. □

Lemma 10.2 *Let* $\Delta^{\pm 1} \in \widetilde{L}_{1,loc}^{n\times n}$ *and* $\Delta(\mu) > 0$ *a.e. on* $\mathbb{R}$. *Then condition (10.2) is equivalent to the condition*

$$\sup_I \det \{A_I(\Delta) \cdot A_I(\Delta^{-1})\} < \infty. \tag{10.4}$$

Proof In [TrV97] it is shown that the matrix

$$X = \{A_I(\Delta)\}^{1/2}\{A_I(\Delta^{-1})\}^{1/2} \tag{10.5}$$

is expansive, i.e., $X^* X \geq I_n$. Therefore, in view of Lemma 2.68, it is subject to the inequalities

$$\|X\| \leq |\det X| \leq \|X\|^n, \tag{10.6}$$

which lead easily to the equivalence of the two conditions (10.2) and (10.4). □

Condition (10.4) will be referred to as the **determinantal Muckenhoupt condition**.

Let $\Delta \in \widetilde{L}_1^{n\times n}$ be positive semidefinite a.e. on $\mathbb{R}$ and let $L_2^n(\Delta)$ denote the Hilbert space of $n \times 1$ measurable vvf's f on $\mathbb{R}$ with norm

$$\|f\|_\Delta = \|\Delta^{1/2} f\|_{st}.$$

Then the subspaces

$$\mathcal{D}_+(\Delta) = \bigvee_{t\geq 0} R_0 e_t \mathbb{C}^n \quad \text{and} \quad \mathcal{D}_-(\Delta) = \bigvee_{t\leq 0} R_0 e_t \mathbb{C}^n \quad \text{in } L_2^n(\Delta) \tag{10.7}$$

are well defined and the inclusions $e_t\mathcal{D}_+(\Delta) \subseteq \mathcal{D}_+(\Delta)$, $e_{-t}\mathcal{D}_-(\Delta) \subseteq \mathcal{D}_-(\Delta)$ hold for every $t \geq 0$.

Theorem 10.3 *Let* Δ *be a measurable* $n \times n$ *mvf on* $\mathbb{R}$ *that is positive definite a.e. on* $\mathbb{R}$. *Then* Δ *meets the matrix Muckenhoupt* (A_2) *condition if and only if*

$$\Delta \in \widetilde{L}_1^{n\times n} \quad \textit{and} \quad \|P_{\mathcal{D}_-(\Delta)}|_{\mathcal{D}_+(\Delta)}\| < 1. \tag{10.8}$$

Proof See Lemmas 2.1 and 2.2 in [TrV]. □

The second condition in (10.8) states that the angle between the **past** $\mathcal{D}_-(\Delta)$ and the **future** $\mathcal{D}_+(\Delta)$ in the Hilbert space $L_2^n(\Delta)$ is strictly positive.

Lemma 10.4 *Let Δ be an $n \times n$ mvf that is positive definite a.e. on $\mathbb{R}$ and suppose that*

$$\Delta \in \widetilde{L}_1^{n\times n} \quad \textit{and} \quad \ln\det \Delta \in \widetilde{L}_1. \tag{10.9}$$

Then:

(1) *The factorization problems*

$$\Delta(\mu) = \psi_-^*(\mu)\, \psi_-(\mu) \quad \textit{a.e. on } \mathbb{R}, \quad \textit{with} \quad \psi_-^\# \in \mathcal{N}_{out}^{n\times n}, \tag{10.10}$$

and

$$\Delta(\mu) = \psi_+^*(\mu)\psi_+(\mu) \quad \textit{a.e. on } \mathbb{R}, \quad \textit{with} \quad \psi_+ \in \mathcal{N}_{out}^{n\times n}, \tag{10.11}$$

have essentially unique solutions $\psi_-^\#$ and ψ_+.

(2) *The equalities*

$$\bigvee_{t\le 0} R_0 e_t \psi_- \mathbb{C}^n = K_2^n \quad \textit{and} \quad \bigvee_{t\ge 0} R_0 e_t \psi_+ \mathbb{C}^n = H_2^n \tag{10.12}$$

hold, where the closed linear spans are considered in L_2^n.

(3) *The mvf*

$$g(\mu) = \psi_-(\mu)\, \psi_+^{-1}(\mu) \tag{10.13}$$

is unitary a.e. on $\mathbb{R}$.

(4) *The norm of the Hankel operator*

$$\Gamma(g) = \Pi_- M_g|_{H_2^n} \tag{10.14}$$

is equal to the norm of restricted projection, considered in Theorem 10.3, i.e.,

$$\|\Gamma(g)\| = \|P_{\mathcal{D}_-(\Delta)}|_{\mathcal{D}_+(\Delta)}\|. \tag{10.15}$$

Proof Assertions (1) and (2) follow from Theorems 3.78 and 3.79 applied to $\Delta(\mu)$ and $\Delta(-\mu)$. Assertion (3) follows from (10.10) and (10.11). To verify (4), we first observe that

$$f_+ \in \mathcal{D}_+(\Delta) \quad \text{if and only if} \quad \psi_+ f_+ \in H_2^n,$$

whereas

$$f_- \in \mathcal{D}_-(\Delta) \quad \text{if and only if} \quad \psi_- f_- \in K_2^n.$$

Thus, as

$$\langle f_+, f_- \rangle_\Delta = \langle \Delta f_+, f_- \rangle_{st},$$

we see that

$$\|P_{\mathcal{D}_-} f_+\|_\Delta^2 = \sup\{|\langle \Delta f_+, f_- \rangle_{st}| : f_- \in \mathcal{D}_-(\Delta) \quad \text{and} \quad \|f_-\|_\Delta = 1\},$$

while

$$\begin{aligned} \langle \Delta f_+, f_- \rangle_{st} &= \langle \psi_-^* \psi_- \psi_+^{-1} \psi_+ f_+, f_- \rangle_{st} \\ &= \langle g \psi_+ f_+, \psi_- f_- \rangle_{st} = \langle g h_+, h_- \rangle_{st}, \end{aligned}$$

where $h_+ = \psi_+ f_+$ belongs to H_2^n, $h_- = \psi_- f_-$ belongs to K_2^n and

$$\|h_\pm\|_{st} = \|f_\pm\|_\Delta.$$

Thus,

$$\|P_{\mathcal{D}_-} f_+\|_\Delta^2 = \sup\{|\langle g h_+, h_- \rangle_{st}| : h_- \in K_2^n \text{ and } \|h_-\|_{st} = 1\}$$

and hence the desired result (10.15) now follows by standard arguments. □

10.2 Criteria of strong regularity for γ-generating matrices

The first criterion that $\mathfrak{A} \in \mathfrak{M}_{rsR}(j_{pq})$ (resp., $\mathfrak{A} \in \mathfrak{M}_{\ell sR}(j_{pq})$) is easily obtained from the previous analysis.

Theorem 10.5 *If $\mathfrak{A} \in \mathfrak{M}_r(j_{pq})$ and Δ_r is defined by the formula in (7.29), then*

$\mathfrak{A} \in \mathfrak{M}_{rsR}(j_{pq})$

$\iff$ *Δ_r satisfies the matrix Muckenhoupt (A_2) condition*

$\iff$ *Δ_r satisfies the determinantal Muckenhoupt condition.*

If $\mathfrak{A} \in \mathfrak{M}_\ell(j_{pq})$ and Δ_ℓ is defined by the formula in (7.29), then

$\mathfrak{A} \in \mathfrak{M}_{\ell sR}(j_{pq})$

$\iff$ *Δ_ℓ satisfies the matrix Muckenhoupt (A_2)condition*

$\iff$ *Δ_ℓ satisfies the determinantal Muckenhoupt condition.*

Proof Let $\mathfrak{A} \in \mathfrak{M}_r(j_{pq})$ and let $\widetilde{\mathfrak{A}}$ be defined by formulas (7.86) and (7.87). Then $\widetilde{\mathfrak{A}} \in \mathfrak{M}_r(j_m)$ and, by Lemma 7.36, $\mathfrak{A} \in \mathfrak{M}_{rsR}(j_{pq}) \Longleftrightarrow \widetilde{\mathfrak{A}} \in \mathfrak{M}_{rsR}(j_m)$. Let

$$\widetilde{g} = T_{\widetilde{\mathfrak{A}}}[I_m].$$

Then, in view of Theorems 7.38 and 7.22,

$$\mathfrak{A} \in \mathfrak{M}_{rsR}(j_{pq}) \Longleftrightarrow \mathfrak{A} \in \mathfrak{M}_{rR}(j_{pq}) \quad \text{and} \quad \|\Gamma(\widetilde{g})\| < 1. \tag{10.16}$$

Suppose now that Δ_r satisfies the matrix Muckenhoupt (A_2) condition. Then $\Delta_r^{-1} \in \widetilde{L}_1^{m\times m}$ and hence by Lemma 7.39, $\mathfrak{A} \in \mathfrak{M}_{rR}(j_{pq})$. Moreover, by Theorem 10.3 and Lemma 10.4, $\|\Gamma(\widetilde{g})\| < 1$. Consequently, $\mathfrak{A} \in \mathfrak{M}_{rsR}(j_{pq})$. Conversely, if $\mathfrak{A} \in \mathfrak{M}_{rsR}(j_{pq})$, then $\mathfrak{A} \in \mathfrak{M}_{rR}(j_{pq})$ and $\|\Gamma(\widetilde{g})\| < 1$ by Theorem 7.22. Therefore, by Theorem 10.3 and Lemma 10.4, Δ_r must satisfy the matrix Muckenhoupt (A_2) condition. □

Theorem 10.6 *Let $\mathfrak{A} \in \mathfrak{M}_r(j_{pq})$ and let*

$$s_{21}^{\circ} = \begin{cases} s_{21} & \text{if } \; q = p \\ \begin{bmatrix} 0_{k\times p} \\ s_{21} \end{bmatrix} & \text{if } \; p > q \\ \begin{bmatrix} 0_{q\times k} & s_{21} \end{bmatrix} & \text{if } \; p < q, \end{cases} \tag{10.17}$$

where $k = |p - q|$ and $s_{21} = -\mathfrak{a}_{22}^{-1}\mathfrak{a}_{21}$. Let $p_\circ = \max\{p, q\}$, let ε be a unitary $p_\circ \times p_\circ$ matrix and define the mvf Δ_ε by the formula

$$\Delta_\varepsilon(\mu) = \{I_{p_\circ} - s_{21}^{\circ}(\mu)\varepsilon\}^*\{I_{p_\circ} - s_{21}^{\circ}(\mu)s_{21}^{\circ}(\mu)^*\}^{-1}\{I_{p_\circ} - s_{21}^{\circ}(\mu)\varepsilon\}. \tag{10.18}$$

Then the following conditions are equivalent:

(1) $\mathfrak{A} \in \mathfrak{M}_{rsR}(j_{pq})$.

(2) *At least one (and hence every one) of the five conditions in Lemma 7.13 is satisfied and $\Delta_\varepsilon(\mu)$ satisfies the matrix Muckenhoupt (A_2) condition for at least one unitary $p_\circ \times p_\circ$ matrix ε.*

(3) *The two mvf's $\Delta_\varepsilon(\mu)$ and $\Delta_{-\varepsilon}(\mu)$ satisfy the matrix Muckenhoupt (A_2) condition for at least one unitary $p_\circ \times p_\circ$ matrix ε.*

(4) *The mvf's $\Delta_\varepsilon(\mu)$ satisfy the matrix Muckenhoupt (A_2) condition for every unitary $p_\circ \times p_\circ$ matrix ε.*

Proof Let $\mathfrak{A} \in \mathfrak{M}_r(j_{pq})$ and let the mvf $\mathfrak{A}^\circ$ be defined as in Lemma 7.34 if $p \neq q$ and let $\mathfrak{A}^\circ = \mathfrak{A}$ if $p = q$. Then $\mathfrak{A}^\circ \in \mathfrak{M}_r(j_{p_\circ})$ and

$$\mathfrak{A} \in \mathfrak{M}_{rsR}(j_{pq}) \Longleftrightarrow \mathfrak{A}^\circ \in \mathfrak{M}_{rsR}(j_{p_\circ}),$$

by Lemma 7.34. Moreover, $\mathfrak{A} \in \widetilde{L}_2^{m\times m} \Longleftrightarrow \mathfrak{A}^\circ \in \widetilde{L}_2^{2p_\circ\times 2p_\circ}$. Thus, without loss of generality we may assume that $q = p$. Then $s_{21}^\circ = s_{21}$ and $p_\circ = p$. Moreover, it is enough to verify the equivalences (1) $\Longleftrightarrow$ (2) $\Longleftrightarrow$ (3) for the specific unitary matrix $\varepsilon = I_p$, since

$$\mathfrak{A} \in \mathfrak{M}_r(j_p) \Longleftrightarrow \mathfrak{A}\begin{bmatrix} \varepsilon & 0 \\ 0 & I_p \end{bmatrix} \in \mathfrak{M}_r(j_p) \tag{10.19}$$

and

$$\mathfrak{A} \in \mathfrak{M}_{rsR}(j_p) \Longleftrightarrow \mathfrak{A}\begin{bmatrix} \varepsilon & 0 \\ 0 & I_p \end{bmatrix} \in \mathfrak{M}_{rsR}(j_p) \tag{10.20}$$

for any $p \times p$ unitary matrix ε. Let

$$\begin{aligned} \Delta_{I_p}(\mu) &= \{I_p - s_{21}(\mu)^*\}\{I_p - s_{21}(\mu)s_{21}(\mu)^*\}^{-1}\{I_p - s_{21}(\mu)\} \\ &= \{I_p - s_{21}(\mu)\}\{I_p - s_{21}(\mu)^*s_{21}(\mu)\}^{-1}\{I_p - s_{21}(\mu)^*\} \end{aligned} \tag{10.21}$$

and set

$$g(\mu) = T_{\mathfrak{A}}[I_p]. \tag{10.22}$$

Then

$$\Delta_{I_p}(\mu) = \psi_-(\mu)^*\psi_-(\mu) = \psi_+(\mu)^*\psi_+(\mu) \tag{10.23}$$

and

$$g(\mu) = \psi_-(\mu)\psi_+(\mu)^{-1}, \tag{10.24}$$

where, in terms of the blocks in the decomposition (7.32),

$$\psi_- = \mathfrak{a}_- + \mathfrak{b}_- \quad \text{and} \quad \psi_+ = \mathfrak{b}_+ + \mathfrak{a}_+. \tag{10.25}$$

By Lemma 7.21,

$$\rho_i^{-1}(\psi_-^\#)^{-1} \in H_2^{p\times p} \quad \text{and} \quad \rho_i^{-1}\psi_+^{-1} \in H_2^{p\times p}.$$

Suppose now that (1) holds. Then $\|\Gamma(g)\| < 1$ and, by Lemma 7.15, $\mathfrak{A} \in \widetilde{L}_2^{m\times m}$. Moreover, in view of Lemma 10.4 and Theorem 10.3, $\Delta_{I_p}(\mu)$ satisfies the matrix Muckenhoupt (A_2) condition. Thus (1) $\Longrightarrow$ (2).

Suppose next that (2) holds. Then $\mathfrak{A} \in \widetilde{L}_2^{m\times m}$ and hence

(a) $\mathfrak{A} \in \mathfrak{M}_{rR}(j_p)$, by Lemma 7.13.

(b) $\Delta_{I_p} \in \widetilde{L}_1^{p\times p}$, by formulas (10.23) and (10.25).

(c) $\|\Gamma(g)\| < 1$, by Lemma 10.4 and Theorem 10.3.

Thus, in view of Theorem 7.22, $\mathfrak{A} \in \mathfrak{M}_{rsR}(j_p)$, i.e., (2) $\Longrightarrow$ (1).

The equivalence (1) $\Longleftrightarrow$ (2) guarantees that (1) $\Longrightarrow$ (3) and that (2) $\Longrightarrow$ (3).

Suppose now that (3) holds for $\varepsilon = I_p$. Then, by Lemma 10.1,

$$\Delta_{I_p} \in \widetilde{L}_1^{p\times p} \quad \text{and} \quad \Delta_{I_p}^{-1} \in \widetilde{L}_1^{p\times p}.$$

Therefore,

$$(\mathfrak{a}_- \pm \mathfrak{b}_-) \in \widetilde{L}_2^{p\times p} \quad \text{and} \quad (\mathfrak{b}_+ \pm \mathfrak{a}_+) \in \widetilde{L}_2^{p\times p}.$$

Thus, $\mathfrak{A} \in \widetilde{L}_2^{m\times m}$, i.e., (3) $\Longrightarrow$ (2) and hence the proof of the equivalences (1) $\Longleftrightarrow$ (2) $\Longleftrightarrow$ (3) is complete.

Finally, the implication (4) $\Longrightarrow$ (3) is obvious and the implication (3) $\Longrightarrow$ (4) follows from the equivalence (10.20). □

10.3 Strong regularity for J-inner mvf's

Let J be an $m \times m$ signature matrix. Then $J \neq \pm I_m \Longrightarrow J = V^* j_{pq} V$ for some $m \times m$ unitary matrix V and, if correspondingly,

$$W(\lambda) = VU(\lambda)V^*,$$

then:

$$\begin{aligned}
U \in \mathcal{U}(J) \quad &\Longleftrightarrow \quad W \in \mathcal{U}(j_{pq}) \\
U \in \mathcal{U}_{rR}(J) \quad &\Longleftrightarrow \quad W \in \mathcal{U}_{rR}(j_{pq}), \quad U \in \mathcal{U}_{\ell R}(J) \Longleftrightarrow W \in \mathcal{U}_{\ell R}(j_{pq}) \\
U \in \mathcal{U}_{rsR}(J) \quad &\Longleftrightarrow \quad W \in \mathcal{U}_{rsR}(j_{pq}) \quad \text{and} \quad U \in \mathcal{U}_{\ell sR}(J) \Longleftrightarrow W \in \mathcal{U}_{\ell sR}(j_{pq}).
\end{aligned}$$

In view of Theorem 7.27, the criteria for right (resp., left) strong regularity for mvf's in $\mathcal{U}(j_{pq})$ may be obtained from the corresponding criteria for the right (resp., left) strong regularity of γ-generating matrices that were discussed in the previous section and the parametrization formula based on formula (7.57), i.e.,

$$W = \begin{bmatrix} b_1 & 0 \\ 0 & b_2^{-1} \end{bmatrix} \mathfrak{A} \quad \text{with} \quad \mathfrak{A} = \begin{bmatrix} \mathfrak{a}_- & \mathfrak{b}_- \\ \mathfrak{b}_+ & \mathfrak{a}_+ \end{bmatrix}, \tag{10.26}$$

where $\{b_1, b_2\} \in ap(W)$ and $\mathfrak{A} \in \mathfrak{M}_r(j_{pq})$.

Let

$$s_{21} = -w_{22}^{-1} w_{21}, \quad s_{12} = w_{12} w_{22}^{-1}, \tag{10.27}$$

$$\Delta_r(\mu) = \begin{bmatrix} I_p & -s_{21}(\mu)^* \\ -s_{21}(\mu) & I_q \end{bmatrix} \quad \text{and} \quad \Delta_\ell(\mu) = \begin{bmatrix} I_p & s_{12}(\mu) \\ s_{12}(\mu)^* & I_q \end{bmatrix}.$$

Theorem 10.7 *Let $W \in \mathcal{U}(j_{pq})$. Then:*

(1) *$W \in \mathcal{U}_{rsR}(j_{pq}) \Longleftrightarrow \Delta_r$ satisfies the matrix (or, equivalently, the determinantal) Muckenhoupt (A_2) condition.*

(2) *$W \in \mathcal{U}_{\ell sR}(j_{pq}) \Longleftrightarrow \Delta_\ell$ satisfies the matrix (or, equivalently, the determinantal) Muckenhoupt (A_2) condition.*

Proof The first assertion follows from Theorems 10.5 and 7.27, and the observation that

$$w_{22}(\lambda)^{-1}\, w_{21}(\lambda) = \mathfrak{a}_+(\lambda)^{-1}\, \mathfrak{b}_+(\lambda).$$

The second assertion then follows from (1) and the fact that

$$W \in \mathcal{U}_{\ell sR}(j_{pq}) \Longleftrightarrow W^\sim \in \mathcal{U}_{rsR}(j_{pq}),$$

since

$$W^\sim(\lambda) = \begin{bmatrix} w_{11}^\sim(\lambda) & w_{21}^\sim(\lambda) \\ w_{12}^\sim(\lambda) & w_{22}^\sim(\lambda) \end{bmatrix}$$

implies that

$$-w_{22}^\sim(\lambda)^{-1} w_{12}^\sim(\lambda) = -(w_{12}(\lambda) w_{22}(\lambda)^{-1})^\sim = -s_{12}^\sim(\lambda). \qquad \square$$

Theorem 10.8 *Let $W \in \mathcal{U}(j_{pq})$ and let the mvf $\Delta_\varepsilon(\mu)$ be defined by the $p \times q$ mvf $s_{21} = -w_{22}^{-1} w_{21}$ and the $p_\circ \times p_\circ$ unitary matrix ε by formulas (10.17) and (10.18). Then the following statements are equivalent:*

(1) *$W \in \mathcal{U}_{rsR}(j_{pq})$.*

(2) *$W \in \widetilde{L}_2^{m\times m}$ and the mvf $\Delta_\varepsilon(\mu)$ satisfies the matrix Muckenhoupt (A_2) condition for at least one (and hence every) $p_\circ \times p_\circ$ unitary matrix ε.*

(3) *The two mvf's $\Delta_\varepsilon(\mu)$ and $\Delta_{-\varepsilon}(\mu)$ satisfy the matrix Muckenhoupt (A_2) condition for at least one $p_\circ \times p_\circ$ unitary matrix ε.*

(4) *The mvf's $\Delta_\varepsilon(\mu)$ satisfy the matrix Muckenhoupt (A_2) condition for every $p_\circ \times p_\circ$ unitary matrix ε.*

Proof The theorem follows from Theorem 7.27 and the corresponding criteria for right strongly γ-generating matrices given in Theorem 10.6. □

Analogous criteria hold for left strongly regular j_{pq}-inner matrices in which the mvf $s_{12}^{\sim}(\lambda) = w_{22}^{\sim}(\lambda)^{-1} w_{12}^{\sim}(\lambda)$ is considered in place of the mvf $s_{21}(\lambda)$.

Theorem 10.9 *Let $A \in \mathcal{U}(J_p)$, let*

$$c_\ell(\lambda) = T_A[I_p], \quad c_r^{\sim}(\lambda) = T_{A^\sim}[I_p] \tag{10.28}$$

and let

$$\delta_r(\mu) = \Re c_r(\mu), \ \delta_\ell(\mu) = \Re c_\ell(\mu) \ \text{a.e. on } \mathbb{R}. \tag{10.29}$$

Then:

(1) *The following conditions are equivalent:*

(a) $A \in \mathcal{U}_{rsR}(J_p)$.

(b) $A \in \widetilde{L}_2^{m\times m}$ *and $\delta_r(\mu)$ satisfies the matrix Muckenhoupt (A_2) condition.*

(c) $c_r \in \mathcal{C}_a^{p\times p}$ *and its spectral density $\delta_r(\mu)$ satisfies the matrix Muckenhoupt (A_2) condition.*

(2) *The following conditions are equivalent:*

(a) $A \in \mathcal{U}_{\ell sR}(J_p)$.

(b) $A \in \widetilde{L}_2^{m\times m}$ *and $\delta_\ell(\mu)$ satisfies the matrix Muckenhoupt (A_2) condition.*

(c) $c_\ell \in \mathcal{C}_a^{p\times p}$ *and its spectral density $\delta_\ell(\mu)$ satisfies the matrix Muckenhoupt (A_2) condition.*

Proof Let $A \in \mathcal{U}(J_p)$ and $W = \mathfrak{V}A\mathfrak{V}$. Then

$$A \in \mathcal{U}_{rsR}(J_p) \Longleftrightarrow W \in \mathcal{U}_{rsR}(j_p). \tag{10.30}$$

Let $s_{12} = w_{12}w_{22}^{-1}$, $s_{21} = -w_{22}^{-1}w_{21}$,

$$c_\ell(\lambda) = T_{\mathfrak{V}}[s_{12}] \quad \text{and} \quad c_r(\lambda) = T_{\mathfrak{V}}[-s_{21}]. \tag{10.31}$$

The equivalences (a) $\Longleftrightarrow$ (b) in (1) and (2) are obtained from Theorems

10.6 and 10.8, taking into account the equivalence (10.30) and the formulas (10.31).

Next, let $\mathfrak{A} \in \mathfrak{M}_r(j_p)$ be the mvf considered in the factorization formula (7.57) for the mvf W and let $g = T_{\mathfrak{A}}[I_p]$. Then in view of Theorem 7.27,

$$A \in \mathcal{U}_{rsR}(J_p) \Longleftrightarrow \mathfrak{A} \in \mathfrak{M}_{rsR}(j_p)$$

and hence, by Theorem 7.33,

$$A \in \mathcal{U}_{rsR}(J_p) \Longleftrightarrow \text{ index } g = 0, \quad c_r \in \mathcal{C}_a^{p\times p} \quad \text{and} \quad \|\Gamma(g)\| < 1. \quad (10.32)$$

Thus, the implication (1a) $\Longrightarrow$ (1c) follows from the already established implication (1a) $\Longrightarrow$ (1b).

Conversely, if (1c) is in force, then Theorem 10.3 and Lemmas 10.1 and 10.4 applied to the mvf $\delta_r(\mu)$ imply that $\|\Gamma(g)\| < 1$ and index $g = 0$ Thus in view of (10.32), (1a) holds and hence the equivalences in (1) are established.

Finally, (2) follows from (1), since $A \in \mathcal{U}_{\ell sR}(J_p) \Longleftrightarrow A^{\sim} \in \mathcal{U}_{rsR}(J_p)$. □

Remark 10.10 *If* $A \in \mathcal{U}(J_p)$ *and* $[E_- \quad E_+] = \sqrt{2}A\mathfrak{V}$, *then*

$$\delta_\ell(\mu) = E_+(\mu)^{-*}E_+(\mu)^{-1} = E_-(\mu)^{-*}E_-(\mu)^{-1} \quad a.e. \text{ on } \mathbb{R}. \quad (10.33)$$

Analogously, if $[E_-^\bullet \quad E_+^\bullet] = \sqrt{2}A^{\sim}\mathfrak{V}$, *then*

$$\delta_r(\mu) = E_-^\bullet(\mu)^{-*}E_-^\bullet(\mu)^{-1} = E_+^\bullet(\mu)^{-*}E_+^\bullet(\mu)^{-1} \quad a.e. \text{ on } \mathbb{R}. \quad (10.34)$$

Remark 10.11 *If* $U \in \mathcal{U}(\mathcal{J}_p)$ *and*

$$A(\lambda) = \begin{bmatrix} -iI_p & 0 \\ 0 & I_p \end{bmatrix} U(\lambda) \begin{bmatrix} iI_p & 0 \\ 0 & I_p \end{bmatrix},$$

then

$$U \in \mathcal{U}_{rsR}(\mathcal{J}_p) \Longleftrightarrow A \in \mathcal{U}_{rsR}(J_p)$$

and

$$U \in \mathcal{U}_{\ell sR}(\mathcal{J}_p) \Longleftrightarrow A \in \mathcal{U}_{\ell sR}(J_p).$$

Moreover,

$$T_A[I_p] = -iT_U[iI_p] \quad and \quad T_{A^\sim}[I_p] = -iT_{U^\sim}[iI_p]. \quad (10.35)$$

Theorem 10.12 *Let $U \in \mathcal{U}(J)$, $J \neq \pm I_m$, and let the complementary $m\times m$ orthoprojection matrices P and Q be defined by (2.1). Let*

$$G_r(\mu) = P + U(\mu)^*QU(\mu) \quad \text{and} \quad G_\ell(\mu) = P + U(\mu)QU(\mu)^*. \tag{10.36}$$

Then:

(1) *$U(\lambda) \in \mathcal{U}_{rsR}(J)$ if and only if the mvf $G_r(\mu)$ satisfies the matrix (or, equivalently, the determinantal) Muckenhoupt (A_2) condition.*

(2) *$U(\lambda) \in \mathcal{U}_{\ell sR}(J)$ if and only if the mvf $G_\ell(\mu)$ satisfies the matrix (or, equivalently, the determinantal) Muckenhoupt (A_2) condition.*

Proof The theorem follows from Theorem 10.7, applied to the mvf $W(\lambda) = VU(\lambda)V^*$, where V is an $m \times m$ unitary matrix such that $J = V^*j_{pq}V$, and the formulas

$$G_r(\mu)^{-1} = V^* \begin{bmatrix} I_p & s_{21}(\mu)^* \\ s_{21}(\mu) & I_q \end{bmatrix} V \tag{10.37}$$

and

$$G_\ell(\mu)^{-1} = V \begin{bmatrix} I_p & -s_{12}(\mu) \\ -s_{12}(\mu)^* & I_q \end{bmatrix} V^*, \tag{10.38}$$

which are obtained by direct calculation. Thus, for example, in view of (10.36),

$$\begin{aligned} VG_r(\mu)V^* &= \begin{bmatrix} I_p & 0 \\ 0 & 0 \end{bmatrix} + \begin{bmatrix} 0 & w_{21}(\mu)^* \\ 0 & w_{22}(\mu)^* \end{bmatrix} \begin{bmatrix} 0 & 0 \\ w_{21}(\mu) & w_{22}(\mu) \end{bmatrix} \\ &= \begin{bmatrix} I_p + w_{21}(\mu)^*w_{21}(\mu) & w_{21}(\mu)^*w_{22}(\mu) \\ w_{22}(\mu)^*w_{21}(\mu) & w_{22}(\mu)^*w_{22}(\mu) \end{bmatrix} \\ &= \begin{bmatrix} I_p & -s_{21}(\mu)^* \\ 0 & I_q \end{bmatrix} \begin{bmatrix} I_p & 0 \\ 0 & w_{22}(\mu)^*w_{22}(\mu) \end{bmatrix} \begin{bmatrix} I_p & 0 \\ -s_{21}(\mu) & I_q \end{bmatrix} \end{aligned}$$

for almost all points $\mu \in \mathbb{R}$. Therefore, $G_r(\mu)$ is invertible and (10.37) holds. Similar considerations lead to (10.38). □

10.4 A mvf $U \in \mathcal{E} \cap \mathcal{U}_{rsR}(J_1)$ that is not in $L_\infty^{2\times 2}$

In this section, a one parameter family of 2×2 mvf's $A_t(\lambda)$ such that $A_t \in \mathcal{E} \cap \mathcal{U}_{\ell sR}(J_1)$ and $A_t \notin L_\infty^{2\times 2}$ and hence $A_t^\sim \in \mathcal{E} \cap \mathcal{U}_{rsR}(J)$ and $A_t^\sim \notin L_\infty^{2\times 2}$ will be presented. This example serves to show that the inclusions $\mathcal{U}(J) \cap L_\infty^{m\times m} \subseteq \mathcal{U}_{rsR}(J)$ and $\mathcal{U}(J) \cap L_\infty^{m\times m} \subseteq \mathcal{U}_{\ell sR}(J)$ are proper.

Let

$$S_t(\lambda) = \sum_{n=0}^{\infty} (-1)^n \frac{\lambda^{2n+1}}{\Gamma(1+t+2n)} \quad \text{for } t > -1, \tag{10.39}$$

where $\Gamma(\lambda)$ denotes the Gamma function.

In [Dz84] the integral representation formula

$$S_t(\lambda) = \begin{cases} \dfrac{\lambda}{\Gamma(t)} \displaystyle\int_0^1 (1-x)^{t-1} \cos \lambda x dx & \text{for} \quad 0 < t \leq 1 \\ \dfrac{1}{\Gamma(t-1)} \displaystyle\int_0^1 (1-x)^{t-2} \sin \lambda x dx & \text{for} \quad 1 < t < 2, \end{cases} \tag{10.40}$$

is used to show that $S_t(\lambda)$ is an entire function of exponential type and that for $0 < t < 2$, $S_t(\lambda)$ has real simple roots and that for $\mu > 0$,

$$S_t(\mu) = \mu^{1-t} \cos\left(\mu - \frac{\pi}{2} t\right) + O(\mu^{-1}) \quad \text{as} \quad \mu \to +\infty. \tag{10.41}$$

The supplementary asymptotic formula

$$S_t'(\mu) = \begin{cases} \mu^{1-t} \sin(\mu - \frac{\pi}{2} t) + O(1) & \text{if} \quad 0 < t < 1 \\ -\mu^{1-t} \sin(\mu - \frac{\pi}{2} t) + O(1) & \text{if} \quad 1 < t < 2, \end{cases} \tag{10.42}$$

for the derivative $S_t'(\lambda)$ with respect to λ of $S_t(\lambda)$ as $\mu \uparrow \infty$ was obtained in [ArD01b], pp. 294–295.

Let

$$f_t(\lambda) = S_t(\lambda) + i S_t'(\lambda) \quad \text{and} \quad \Delta_t(\lambda) = f_t(\lambda) f_t^{\#}(\lambda).$$

Then

$$\Delta_t(\mu) = |f_t(\mu)|^2 = |S_t(\mu)|^2 + |S_t'(\mu)|^2 \quad \text{for} \quad \mu \in \mathbb{R}$$

and, in view of formulas (10.40)–(10.42),

$$f_1(\lambda) = i e^{-i\lambda}, \quad \Delta_1(\lambda) \equiv 1$$

and

$$\lim_{\mu \uparrow \infty} \frac{\Delta_t(\mu)}{|\mu|^{2-2t}} = 1 \quad \text{for } 0 < t < 2. \tag{10.43}$$

Clearly, $\Delta_t(\lambda)$ is an entire function of exponential type. Moreover, if $1/2 < t < 3/2$, then:

(1) $\Delta_t(\mu)$ satisfies the Muckenhoupt (A_2) condition.

(2) $\Delta_t^{\pm 1} \in \widetilde{L}_1$.

(3) Δ_t is of Cartwright class.

(4) $\Delta_t \in \Pi$.

Items (1)–(3) follow from (10.43); (4) then follows from Theorem 3.108.

Thus, the factorization problems

$$\Delta_t(\mu) = |\varphi_t(\mu)|^2, \quad \varphi_t \in \mathcal{N}_{out}, \tag{10.44}$$

have unique solutions that meet the normalization conditions $\varphi_t(0) > 0$. By Theorem 3.111, they are entire functions of exponential type. Let

$$c_t(\lambda) = \frac{1}{\pi i}\int_{-\infty}^{\infty}\left\{\frac{1}{\mu-\lambda} - \frac{\mu}{1+\mu^2}\right\}\Delta_t^{-1}(\mu)du, \quad \alpha_t = \tau(\varphi_t),$$

$$E_+^t(\lambda) = \varphi_t(\lambda), \quad E_-^t(\lambda) = e^{i\alpha_t\lambda}\varphi_t^{\#}(\lambda)$$

and

$$A_t(\lambda) = \frac{1}{\sqrt{2}}\begin{bmatrix} -c_t^{\#}(\lambda)E_-^t(\lambda) & c_t(\lambda)E_+^t(\lambda) \\ E_-^t(\lambda) & E_+^t(\lambda)\end{bmatrix}\mathfrak{V}.$$

Then, in view of Theorems 10.9 and 9.15, $c_t \in \Pi$ and

$$A_t \in \mathcal{E} \cap \mathcal{U}_{\ell sR}(J_1) \quad \text{if} \quad 1/2 < t < 3/2.$$

However,

$$A_t \notin L_\infty^{2\times 2} \quad \text{if} \quad 1/2 < t < 3/2 \text{ and } t \neq 1,$$

because

$$A_t \in L_\infty^{2\times 2} \Longrightarrow \Delta_t \in L_\infty \text{ and } \Delta_t^{-1} \in L_\infty$$

whereas, the asymptotic formula (10.43) implies that

$$\Delta_t(\mu) \longrightarrow \infty \quad \text{as } \mu \longrightarrow \infty \quad \text{if} \quad 1/2 < t < 1$$

and

$$\Delta_t(\mu)^{-1} \longrightarrow \infty \quad \text{as } \mu \longrightarrow \infty \quad \text{if} \quad 1 < t < 3/2.$$

10.5 Right and left strongly regular de Branges matrices

Let $\mathfrak{E}$ be a regular de Branges matrix that meets the condition (5.129) and let $A_{\mathfrak{E}}$ denote the characteristic mvf of the simple LB J_p-node $\Sigma_{\mathfrak{E}} = (R_0, C_{\mathfrak{E}}; \mathcal{B}(\mathfrak{E}), \mathbb{C}^m)$ that was defined in Theorem 6.15. Then $A_{\mathfrak{E}}$ is the unique perfect mvf in $\mathcal{U}_0(J_p)$ that is related to $\mathfrak{E}$ by (5.130). By Theorem 6.15, the unitary operator U_2^* from $\mathcal{B}(\mathfrak{E})$ onto $\mathcal{H}(A_{\mathfrak{E}})$ that is given by formula (6.22) is a unitary similarity operator from the node $\Sigma_{\mathfrak{E}}$ onto the simple LB J_p-node $\mathring{\Sigma} = (R_0, \mathring{C}; \mathcal{H}(A_{\mathfrak{E}}), \mathbb{C}^m)$. Moreover, the characteristic mvf of the dual simple LB J_p-node $(\Sigma_{\mathfrak{E}})_* = (-R_0^*, C_{\mathfrak{E}}; \mathcal{B}(\mathfrak{E}), \mathbb{C}^m)$ is equal to $A_{\mathfrak{E}}^{\sim}(\lambda) = A_{\mathfrak{E}}(-\overline{\lambda})^*$; and $(\Sigma_{\mathfrak{E}})_*$ is unitarily similar to the simple LB J_p-node $\mathring{\Sigma}_* = (-R_0^*, \mathring{C}; \mathcal{H}(A_{\mathfrak{E}}^{\sim}), \mathbb{C}^m)$.

In view of Theorem 6.6, the unitary similarity operators from $\Sigma_{\mathfrak{E}}$ to $\mathring{\Sigma}$ and $(\Sigma_{\mathfrak{E}})_*$ to $\mathring{\Sigma}_*$ are given by the formulas

$$(\mathcal{F}_r g)(\lambda) = \frac{1}{\sqrt{2\pi}} C_{\mathfrak{E}}(I - \lambda R_0)^{-1} g \quad \text{for } g \in \mathcal{B}(\mathfrak{E}) \tag{10.45}$$

and

$$(\mathcal{F}_\ell g)(\lambda) = \frac{1}{\sqrt{2\pi}} C_{\mathfrak{E}}(I + \lambda R_0^*)^{-1} g \quad \text{for } g \in \mathcal{B}(\mathfrak{E}), \tag{10.46}$$

respectively. Moreover, $\mathcal{F}_r = U_2^*$,

$$(\mathcal{F}_\ell g)(\lambda) = A_{\mathfrak{E}}^{\sim}(\lambda)(\mathcal{F}_r g)(-\lambda) \tag{10.47}$$

and explicit expressions for U_2^* and $C_{\mathfrak{E}}$ are given in Theorem 6.15.

Since $\mathcal{F}_r g \in \mathcal{H}(A_{\mathfrak{E}})$ and $\mathcal{F}_\ell g \in \mathcal{H}(A_{\mathfrak{E}}^{\sim})$ when $g \in \mathcal{B}(\mathfrak{E})$ and $\mathcal{H}(U) \subset \Pi^m$ when $U \in \mathcal{U}(J)$, both of these vvf's have nontangential boundary values a.e. on $\mathbb{R}$.

A regular de Branges matrix $\mathfrak{E}(\lambda)$ that meets the constraint (5.129) will be called **right strongly regular** (resp., **left strongly regular**) if $\mathcal{F}_r g \in L_2^m$ for every $g \in \mathcal{B}(\mathfrak{E})$ (resp., $\mathcal{F}_\ell g \in L_2^m(\mathbb{R})$) for every $g \in \mathcal{B}(\mathfrak{E})$; the corresponding de Branges space $\mathcal{B}(\mathfrak{E})$ will be called right or left strongly regular accordingly.

Recall that if $\mathfrak{E} = [E_- \quad E_+]$ is a regular de Branges matrix that satisfies the condition (5.129), then, by Theorem 5.69, $(R_0 E_\pm)u \in \mathcal{B}(\mathfrak{E})$ for every $u \in \mathbb{C}^p$; and $\mathfrak{E}$ is right (resp., left) strongly regular if and only if $A_{\mathfrak{E}} \in \mathcal{U}_{rsR}(J_p)$ (resp., $A_{\mathfrak{E}}^{\sim} \in \mathcal{U}_{rsR}(J_p)$). Moreover, $A_{\mathfrak{E}}^{\sim} \in \mathcal{U}_{rsR}(J_p) \iff A_{\mathfrak{E}} \in \mathcal{U}_{\ell sR}(J_p)$.

Theorem 10.13 *Let $\mathfrak{E}(\lambda)$ be a regular de Branges matrix that satisfies the condition (5.129). Then the following statements are equivalent:*

(1) *$\mathfrak{E}(\lambda)$ is right strongly regular.*

(2) *$\{\mathcal{F}_r g : g \in \mathcal{B}(\mathfrak{E})\}$ is a closed subspace of $L_2^m(\mathbb{R})$.*

(3) *There exist a pair of positive constants γ_1 and γ_2 such that*

$$\gamma_1 \|\mathcal{F}_r g\|_{st} \leq \|g\|_{\mathcal{B}(\mathfrak{E})} \leq \gamma_2 \|\mathcal{F}_r g\|_{st}.$$

(4) *At least one of the functions*

$$\langle g, (I - \lambda R_0^*)^{-1} G_+ u\rangle_{\mathcal{B}(\mathfrak{E})}, \quad \langle g, (I - \lambda R_0^*)^{-1} G_- u\rangle_{\mathcal{B}(\mathfrak{E})}$$

has nontangential boundary limits that belong to $L_2(\mathbb{R})$ for every choice of $g \in \mathcal{B}(\mathfrak{E})$ and $u \in \mathbb{C}^p$.

Proof The equivalence of the conditions (1), (2) and (3) follows from Theorem 5.92. It remains only to verify the equivalence of (1) and (4). To this end, let $W(\lambda) = \mathfrak{V} A_{\mathfrak{E}} \mathfrak{V}$ and let $f \in \mathcal{H}(A_{\mathfrak{E}})$. Then, since $\mathfrak{V}\mathcal{H}(A_{\mathfrak{E}}) = \mathcal{H}(W)$, Corollary 5.56 applied to $\mathfrak{V} f$ guarantees that

$$f = \begin{bmatrix} f_1 \\ f_2 \end{bmatrix} \in L_2^m \iff f_1 - f_2 \in L_2^p \iff f_1 + f_2 \in L_2^p.$$

The equivalence (1) $\iff$ (4) follows from the last equivalence applied to $f = \mathcal{F}_r g$ and the formulas (10.45) and

$$C = \begin{bmatrix} C_1 \\ C_2 \end{bmatrix}, \tag{10.48}$$

where

$$C_1 : g \in \mathcal{B}(\mathfrak{E}) \longrightarrow \frac{\sqrt{\pi}}{2\pi i} \int_{-\infty}^{\infty} (G_+(\mu) + G_-(\mu))^* \Delta_{\mathfrak{E}}(\mu) g(\mu) d\mu \tag{10.49}$$

and

$$C_2 : g \in \mathcal{B}(\mathfrak{E}) \longrightarrow \frac{\sqrt{\pi}}{2\pi i} \int_{-\infty}^{\infty} (G_+(\mu) - G_-(\mu))^* \Delta_{\mathfrak{E}}(\mu) g(\mu) d\mu = \sqrt{\pi} g(0). \tag{10.50}$$

□

Remark 10.14 *If $\mathfrak{E}(\lambda)$ is a right strongly regular de Branges matrix and the pair $\{b_3, b_4\}$ of $p \times p$ inner mvf's is obtained from the factorizations in (5.195), then $\{b_3, b_4\} \in ap_{II}(A_{\mathfrak{E}})$ and, in view of Theorem 5.98,*

$$\mathcal{B}(\mathfrak{E}) = \mathcal{H}(b_3) \oplus \mathcal{H}_*(b_4), \tag{10.51}$$

as linear topological spaces with scalar product

$$\langle g, h \rangle_{\mathcal{B}(\mathfrak{E})} = \int_{-\infty}^{\infty} h(\mu)^* E_+(\mu)^{-*} E_+(\mu)^{-1} g(\mu) d\mu. \tag{10.52}$$

We remark that if $\mathfrak{E}(\lambda)$ is a regular de Branges matrix and (5.129) is in force, then the generalized transform $\mathcal{F}_\ell$ is a unitary operator from $\mathcal{B}(\mathfrak{E})$ onto $\mathcal{H}(A_{\mathfrak{E}}^{\sim})$ and

$$\mathfrak{E}(\lambda) \text{ is left strongly regular} \Longleftrightarrow A_{\mathfrak{E}} \in \mathcal{U}_{\ell sR}(J_p) \Longleftrightarrow A_{\mathfrak{E}}^{\sim} \in \mathcal{U}_{rsR}(J_p).$$

Thus, an application of Theorem 5.92 to $A_{\mathfrak{E}}^{\sim}$ yields the following result:

Theorem 10.15 *Let $\mathfrak{E}(\lambda)$ be a regular de Branges matrix that satisfies the conditions (5.129). Then the following statements are equivalent:*

(1) *$\mathfrak{E}(\lambda)$ is left strongly regular.*

(2) *$\{\mathcal{F}_\ell g : g \in \mathcal{B}(\mathfrak{E})\}$ is a closed subspace of $L_2^m(\mathbb{R})$.*

(3) *There exist a pair of positive constants γ_1 and γ_2 such that*

$$\gamma_1 \|\mathcal{F}_\ell g\|_{st} \le \|g\|_{\mathcal{B}(\mathfrak{E})} \le \gamma_2 \|\mathcal{F}_\ell g\|_{st}.$$

(4) *At least one of the functions*

$$\langle g, (I - \lambda R_0)^{-1} G_+ u \rangle_{\mathcal{B}(\mathfrak{E})}, \quad \langle g, (I - \lambda R_0)^{-1} G_- u \rangle_{\mathcal{B}(\mathfrak{E})}$$

has nontangential boundary limits that belong to $L_2(\mathbb{R})$ for every choice of $g \in \mathcal{B}(\mathfrak{E})$ and $u \in \mathbb{C}^p$.

Theorem 10.16 *Let $\mathfrak{E}(\lambda)$ be a regular de Branges matrix. Then $\mathfrak{E}(\lambda)$ is left strongly regular if and only if the density $\Delta_{\mathfrak{E}}(\mu)$ satisfies the matrix Muckenhoupt (A_2) condition.*

Proof This is immediate from Theorem 10.9, applied to the perfect mvf $A_{\mathfrak{E}}$. □

10.6 LB J-nodes with characteristic mvf's in $\mathcal{U}_{rsR}(J)$

Theorem 10.17 *Let $\Sigma = (A, C; X, \mathbb{C}^m)$ be a simple LB J_p-node with characteristic mvf $U_\Sigma \in \mathcal{U}_{rsR}(J_p)$. Let*

$$\{b_3, b_4\} \in ap_{II}(U_\Sigma) \quad \text{and} \quad b_2(0) = b_4(0) = I_p \tag{10.53}$$

and let $\Sigma_+ = (A_+, C_+; X_+, \mathbb{C}^p)$ be a simple LB I-node with characteristic mvf $b_3(\lambda)$ and $\Sigma_- = (A_-, C_-; X_-, \mathbb{C}^p)$ be a simple LB $(-I)$-node with characteristic mvf $b_4(\lambda)^{-1}$. Then there exists an invertible operator

$$R \in \mathcal{L}(X, X_+ \oplus X_-) \quad \text{with} \quad R^{-1} \in \mathcal{L}(X_+ \oplus X_-, X) \tag{10.54}$$

such that

$$A = R^{-1} \begin{bmatrix} A_+ & 0 \\ 0 & A_- \end{bmatrix} R, \quad \text{and} \quad [0_{p\times p} \quad I_p]C = [C_+ \quad C_-]R. \tag{10.55}$$

Moreover,

$$X_- = \{0\} \Longleftrightarrow b_4(\lambda) \equiv I_p \Longleftrightarrow U_\Sigma \in \mathcal{N}_+^{m\times m}$$

and

$$X_+ = \{0\} \Longleftrightarrow b_3(\lambda) \equiv I_p \Longleftrightarrow U_\Sigma \in \mathcal{N}_-^{m\times m}.$$

Proof The given simple LB nodes Σ, Σ_+, Σ_- can be replaced by their functional models. In these models, A, A_-, A_+, are backwards shifts in the spaces $X = \mathcal{H}(U_\Sigma)$, $X_+ = \mathcal{H}(b_3)$ and $X_- = \mathcal{H}_*(b_4)$, respectively, and the operators C, C_+, C_- map vvf's g from their respective domains X, X_+, X_- into $\sqrt{2\pi}g(0)$. Let L_A denote the operator from $\mathcal{H}(b_3) \oplus \mathcal{H}_*(b_4)$ onto $\mathcal{H}(A)$ that is defined by formula (5.205). Then, in the given setting, Theorem 5.102 is applicable to the mvf $A(\lambda) = U_\Sigma(\lambda)$ and the operator $R = (L_A)^{-1}$ satisfies the stated assertions. □

Theorem 10.18 *Let $\Sigma = (A, C; X, \mathbb{C}^m)$ be an LB j_{pq}-node with characteristic mvf $U_\Sigma \in \mathcal{U}(j_{pq})$. Let*

$$\{b_1, b_2\} \in ap(U_\Sigma), \quad b_1(0) = I_p, \quad b_2(0) = I_q, \tag{10.56}$$

and let $\Sigma_+ = (A_+, C_+; X_+, \mathbb{C}^p)$ be an LB I-node with characteristic mvf $b_1(\lambda)$ and let $\Sigma_- = (A_-, C_-; X_-, \mathbb{C}^q)$ be an LB $(-I)$-node with characteristic mvf $b_2(\lambda)^{-1}$.

(1) *If the nodes* Σ, Σ_+ *and* Σ_- *are simple and* $U_\Sigma \in \mathcal{U}_{rsR}(j_{pq})$, *then there exists an invertible operator*

$$R \in \mathcal{L}(X, X_+ \oplus X_-) \quad \text{with} \quad R^{-1} \in \mathcal{L}(X_+ \oplus X_-, X), \tag{10.57}$$

such that

$$A = R^{-1} \begin{bmatrix} A_+ & 0 \\ 0 & A_- \end{bmatrix} R, \quad C_\pm = P_\pm C R^{-1}|_{X_\pm}, \tag{10.58}$$

where $P_+ = P = (I_m + j_{pq})/2$ *and* $P_- = Q = (I_m - j_{pq})/2$. *Moreover,*

$$X_- = \{0\} \Longleftrightarrow b_2(\lambda) \equiv I_q \Longleftrightarrow U_\Sigma \in \mathcal{N}_+^{m\times m}$$

and

$$X_+ = \{0\} \Longleftrightarrow b_1(\lambda) \equiv I_p \Longleftrightarrow U_\Sigma \in \mathcal{N}_-^{m\times m}.$$

(2) *Conversely, if (10.58) and (10.57) hold for the nodes* Σ, Σ_+ *and* Σ_-, *then* $U_\Sigma \in \mathcal{U}_{rsR}(j_{pq})$.

Proof A proof of the first statement may be based on the description of the space $\mathcal{H}(U_\Sigma)$ for $U_\Sigma \in \mathcal{U}_{rsR}(j_{pq})$ that is furnished in the proof of Theorem 5.81, in much the same way that Theorem 10.17 was verified.

The verification of the second assertion rests on the fact that the transforms T_{Σ_+} and T_{Σ_-} based on the nodes Σ_+ and Σ_- map X_+ and X_- into L_2^p and L_2^q, respectively, and the fact that if

$$f = \begin{bmatrix} g \\ h \end{bmatrix} \in \mathcal{H}(U_\Sigma) \quad \text{and} \quad U_\Sigma \in \mathcal{U}(j_{pq}),$$

then

$$f \in L_2^m \Longleftrightarrow g \in L_2^p \Longleftrightarrow h \in L_2^q. \qquad \square$$

An LB J-node $\Sigma = (A, C; X, Y)$ is called a **Volterra node** if its main operator A is a Volterra operator (i.e., if A is a compact operator and its spectrum $\sigma(A) = \{0\}$) and C is compact). A simple LB J-node $\Sigma = (A, C; X, \mathbb{C}^m)$ is a Volterra node if and only if $U_\Sigma \in \mathcal{E} \cap \mathcal{U}_0(J)$; see e.g., [Bro72].

If $U_\Sigma \in \mathcal{E} \cap \mathcal{U}_0(J)$, i.e., if the simple node Σ is a Volterra node, then the $b_j(\lambda)$ are entire inner functions and consequently the nodes Σ_+ and Σ_- considered in the preceding two theorems are Volterra nodes. If $m = 2$, then $b_1(\lambda) = b_3(\lambda) = e^{i\lambda\tau_3}$ and $b_2(\lambda) = b_4(\lambda) = e^{i\lambda\tau_4}$, where $\tau_3 = \tau_U^- \geq 0$, $\tau_4 = \tau_U^+ \geq 0$ and $U = U_\Sigma$. The functional models based on the backwards shift acting in $\mathcal{H}(b_3)$ and $\mathcal{H}_*(b_4)$ are of course still applicable. However, since $\mathcal{H}(b_3) = L_2([0,\tau_3])^\wedge$ and $\mathcal{H}_*(b_4) = L_2([-\tau_4,0])^\wedge$, the identities

$$h(\lambda) = \int_0^{\tau_3} e^{i\lambda a} h^\vee(a)da \Longrightarrow (R_0 h)(\lambda) = i\int_0^{\tau_3} e^{i\lambda b}\int_b^{\tau_3} h^\vee(a)dadb$$

and

$$h(\lambda) = \int_{-\tau_4}^0 e^{i\lambda a} h^\vee(a)da \Longrightarrow (R_0 h)(\lambda) = -i\int_{-\tau_4}^0 e^{i\lambda b}\int_{-\tau_4}^b h^\vee(a)dadb$$

lead to well known functional models based on integration operators acting in the indicated subspaces of L_2. In particular, the functional model of a simple Volterra LB I-node with characteristic function $e^{i\lambda\tau_3}$ may be chosen as $\Sigma_+ = (A_+, C_+; X_+, \mathbb{C})$, where $X_+ = L_2([0,\tau_3])$,

$$(A_+u)(t) = i\int_t^{\tau_3} u(a)da \quad \text{and} \quad C_+u = \int_0^{\tau_3} u(a)da \quad \text{for} \quad u \in X_+.$$

In much the same way, the functional model of a simple Volterra LB $(-I)$-node with characteristic function $e^{-i\lambda\tau_4}$ may be chosen equal to the node $\Sigma_- = (A_-, C_-; X_-, \mathbb{C})$, where $X_- = L_2([-\tau_4, 0]))$,

$$(A_-u)(t) = i\int_{-\tau_4}^{t} u(a)da \quad \text{and} \quad C_-u = \int_{-\tau_4}^{0} u(a)da \quad \text{for } u \in X_-.$$

Correspondingly, the operator R considered in the preceding two theorems acts from X onto $L_2([0,\tau_3]) \oplus L_2([-\tau_4,0])$.

Analogues of the preceding two theorems for the case $U_\Sigma \in \mathcal{U}_{\ell sR}(J)$ may be obtained by applying the preceding results to $(U_\Sigma)^\sim$ and recalling that

$$U_\Sigma \in \mathcal{U}_{\ell sR}(J) \Longleftrightarrow (U_\Sigma)^\sim \in \mathcal{U}_{rsR}(J).$$

10.7 Bibliographical notes

Most of the material in this chapter is adopted from [ArD01b] and [ArD03b]. The properties of left strongly regular spaces $\mathcal{B}(\mathfrak{E})$ and of operators in these spaces related to the backwards shift R_0 were studied by other methods in the case that $E_+(\lambda)$ and $E_-(\lambda)$ are scalar meromorphic functions in $\mathbb{C}$ by G. M. Gubreev [Gu00b], as an application of his theory of regular quasiexponentials. In particular, he noted the connection between the class $\mathcal{U}_{\ell sR}(Jp)$ and the class of left strongly regular de Branges spaces $\mathcal{B}(\mathfrak{E})$ when $p = 1$

and $\mathfrak{E}(\lambda)$ is entire. Some of his results may be obtained from the analysis in the last two sections.

The example presented in Section 10.4 is adapted from [ArD01b].

The function $S_t(\lambda)$ defined in formula (10.39) has been investigated extensively by M. M. Dzhrbashyan [Dz84] in his study of interpolation problems for entire functions of finite order and finite type that belong to L_2 with some weight $\omega(\mu)$.

Theorem 10.18 was announced in [Ar00a]; a proof was furnished in [ArD04a].

The evolution semigroup $T(t) = e^{itA}$, $t \geq 0$, based on the main operator A of a simple generalized LB I-node $\Sigma = (A, B, N : X, U, Y)$ is bistable in the sense that $T(t) \to 0$ and $T(t)^* \to 0$ strongly as $t \uparrow \infty$ if and only if the characteristic function $U_\Sigma(\lambda)$ has unitary nontangential boundary values a.e. on $\mathbb{R}$ (i.e., U_Σ is biinner). see [ArNu96] and [St05]. Moreover, if $\Sigma = (A, C; X, \mathbb{C}^m)$ is a simple LB J-node and its characteristic mvf $U_\Sigma \in \mathcal{U}_{rsR}(J)$, then Theorem 10.18 implies that:

(1) $T(t)$ is bistable if $U_\Sigma \in \mathcal{N}_+^{m\times m}$.

(2) $T(-t)$ is bistable if $U_\Sigma^\# \in \mathcal{N}_+^{m\times m}$.

(3) If $U_\Sigma \notin \mathcal{N}_+^{m\times m}$ and $U_\Sigma^\# \notin \mathcal{N}_+^{m\times m}$, then $X = X_+ \dot{+} X_-$, $T(t)X_+ \subseteq X_+$, $T(t)X_- = X_-$ for $t \geq 0$ and the semigroups $T_+(t) = T(t)|_{X_+}$ and $T_-(t) = T(-t)|_{X_-}$ for $t \geq 0$ are bistable, i.e., $T(t)$ is a group of operators with a two sided dichotomy property.

Analogues of Theorems 10.17 and 10.18 on LB J-nodes and J-unitary nodes with strongly regular J-inner characteristic mvf's and the properties of the corresponding evolution semigroups were obtained by Z. D. Arova in her PhD thesis [Ara02]; see also [Ara00a], [Ara00b], [Ara01] and [Ara05]. Her characterizations of the class of simple operator nodes Σ with characteristic mvf $U_\Sigma \in \mathcal{U}_{rsR}(J)$ used somewhat different nodes than were exhibited here. Thus, for example, in place of the relations (10.58), she used the relations

$$A = R^{-1} \begin{bmatrix} A_+ & 0 \\ 0 & A_- \end{bmatrix} R, \quad C_\pm = CR^{-1}|_{X_\pm}, \tag{10.59}$$

where $\Sigma_+ = (A_+, C_+; X_+, \mathbb{C}^m)$ and $\Sigma_- = (A_-, C_-; X_-, \mathbb{C}^m)$ are simple LB I-nodes and simple LB $(-I)$-nodes, respectively.

Ball and Raney [BaR07] introduced the class of L_2-regular mvf's (on the circle), which is larger than the class $\mathcal{U}_{rsR}(J)$, and studied the two sided dichotomy property of linear discrete time invariant system realizations of L_2-regular mvf's. They also gave another characterization of right strong regularity (on the circle) in terms of the class of $L_2^{m\times m}$ regular mvf's and presented applications to problems of interpolation and realization.

11
Formulas for entropy functionals and their extremal values

In this chapter entropy formulas for a number of the extension problems considered in Chapters 7 and 8 will be obtained in a uniform way that exploits the description of the set of solutions to these problems in terms of linear fractional transformations based on gamma-generating matrices and J-inner mvf's. Formulas will be obtained for the extremal value of the entropy over the set of solutions to each of the extension problems under consideration in the completely indeterminate case. In this treatment, the entropy of a contractive mvf is defined to be nonnegative. Consequently, the extremal value will be a minimum. If the sign in the definition of entropy is reversed (as is also common in the literature), then the extremal value will be a maximum.

11.1 Definitions of the entropy functionals

The entropy functional $\mathcal{I}(f;\omega)$ is defined on the set of mvf's $f \in L_\infty^{p\times q}$ with $\|f\|_\infty \leq 1$ by the formula

$$\mathcal{I}(f;\omega) = -\frac{\Im\omega}{2\pi}\int_{-\infty}^{\infty} \frac{\ln\det\{I_q - f(\mu)^* f(\mu)\}}{|\mu-\omega|^2} d\mu \quad \text{for } \omega \in \mathbb{C}_+. \tag{11.1}$$

It is clear that:

(1) $0 \leq \mathcal{I}(f;\omega) \leq \infty$.

(2) $\mathcal{I}(f;\omega) = 0$ for at least one (and hence every) $\omega \in \mathbb{C}_+ \Longleftrightarrow f = 0_{p\times q}$.

(3) $\mathcal{I}(f;\omega) < \infty$ for at least one (and hence every) $\omega \in \mathbb{C}_+ \Longleftrightarrow \mathcal{I}(f;i) < \infty$.

(4) $\mathcal{I}(f;i) < \infty \Longrightarrow f(\mu)^* f(\mu) < I_q$ a.e. on $\mathbb{R}$.

If $f \in S^{p\times q}$, then:

$$\mathcal{I}(f;i) < \infty \Longrightarrow f(\lambda)^* f(\lambda) < I_q \quad \text{for every } \lambda \in \mathbb{C}_+.$$

The entropy functional $h(c;\omega)$ is defined on mvf's $c \in \mathcal{C}^{p\times p}$ by the formula

$$h(c;\omega) = -\frac{\Im\omega}{2\pi}\int_{-\infty}^{\infty}\frac{\ln\det\Re c(\mu)}{|\mu-\omega|^2}d\mu \quad \text{for } \omega \in \mathbb{C}_+. \tag{11.2}$$

If $c \in \mathcal{C}^{p\times p}$ and $s = T_{\mathfrak{V}}[c]$, then $s \in \mathcal{S}^{p\times p}$ and

$$\Re c = (I+s)^{-*}(I_p - s^*s)(I_p+s)^{-1} \tag{11.3}$$

Therefore,

$$h(c;\omega) = \mathcal{I}(s;\omega) + \ln|\det\{I_p + s(\omega)\}|, \tag{11.4}$$

since

$$\ln|\det\{I_p+s(\omega)\}| = \frac{\Im\omega}{\pi}\int_{-\infty}^{\infty}\frac{\ln|\det\{I_p+s(\mu)\}|}{|\mu-\omega|^2}d\mu, \tag{11.5}$$

because $\det\{I_p + s(\lambda)\}$ is an outer function in H_∞ by Lemma 3.54. Consequently,

(1) $-\infty < h(c;\omega) \leq \infty$.

(2) $h(c;\omega) < \infty$ for at least one (and hence every) $\omega \in \mathbb{C}_+ \Longleftrightarrow h(c;i) < \infty$.

11.2 Motivation for the terminology entropy functional

If $x(t)$ is a complex p-dimensional continuous time stationary Gaussian process with zero mean, then it admits a representation of the form

$$x(t) = \int_{-\infty}^{\infty} e^{it\mu}dy(\mu) \quad \text{for } t \in \mathbb{R},$$

where $y(t)$ is a p-dimensional process with independent increments. If $x(t)$ is full rank and regular in the sense of Kolmogorov, then the covariance

$$E\{x(t)x(s)^*\} = \int_{-\infty}^{\infty} e^{i(t-s)\mu}f(\mu)d\mu,$$

where $f \in L_1^{p\times p}$, $f(\mu) \geq 0$ a.e. on $\mathbb{R}$ and $\ln\det f \in \widetilde{L_1}$. Let

$$x_\omega(n) = \int_{-\infty}^{\infty} b_\omega(\mu)^n dy(\mu) \quad \text{for } n = 0, \pm 1, \dots,$$

where $b_\omega(\mu) = (\mu - \omega)/(\mu - \overline{\omega})$ and $\omega \in \mathbb{C}_+$. Then $x_\omega(n)$ is a stationary Gaussian sequence with zero mean and correlation matrix

$$E\{x_\omega(n)x_\omega(k)^*\} = \int_{-\infty}^{\infty} b_\omega(\mu)^{n-k} f(\mu)d\mu = \int_{-\pi}^{\pi} e^{i(n-k)\theta} g(\theta)d\theta,$$

where

$$\theta = \theta(\mu) = \frac{\omega - \overline{\omega}}{i} \int_0^\mu \frac{1}{|s-\omega|^2} ds \quad \text{and} \quad g(\theta) = f(\mu).$$

By a formula due to Pinsker [Pi54], the differential entropy $h(x_\omega)$ of the discrete time process $\{x_\omega(n)\}$ is

$$h(x_\omega) = k_p + \frac{1}{4\pi} \int_{-\pi}^{\pi} \ln\det g(\theta)d\theta = k_p - h(c;\omega), \tag{11.6}$$

where $c \in \mathcal{C}_a^{p\times p}$, $f(\mu) = \Re c(\mu)$ a.e. on $\mathbb{R}$ and $k_p = (p/2)\ln(4\pi^2 e)$.

If $f \in L_\infty^{p\times q}$ and $\|f\|_\infty \le 1$, then the mvf

$$\begin{bmatrix} I_p & f(\mu) \\ f(\mu)^* & I_q \end{bmatrix}$$

may be viewed as the spectral density of a generalized m-dimensional stationary Gaussian process $x(t) = \mathrm{col}(u(t), v(t))$, where $u(t)$ and $v(t)$ are white noise processes with spectral densitities $f_{uu}(\mu) = I_p$ and $f_{vv}(\mu) = I_q$ and cross spectral density $f_{uv}(\mu) = f(\mu)$.

By another formula of Pinsker that is presented in Theorem 10.4.1 of [Pi64], $\mathcal{I}(f;\omega) = I(u_\omega, v_\omega)$, the amount of information per unit time in the component $u_\omega(n)$ about the component $v_\omega(n)$ when $x_\omega(n) = \mathrm{col}(u_\omega(n), v_\omega(n))$.

11.3 Entropy of completely indeterminate interpolation problems

The functionals

$$\mathcal{I}(f) = \mathcal{I}(f;i) = -\frac{1}{2\pi} \int_{-\infty}^{\infty} \frac{\ln\det\{I_q - f(\mu)^* f(\mu)\}}{1+\mu^2} d\mu \tag{11.7}$$

and

$$h(c) = h(c;i) = -\frac{1}{2\pi} \int_{-\infty}^{\infty} \frac{\ln\det \Re c(\mu)}{1+\mu^2} d\mu \tag{11.8}$$

are of particular interest.

Theorem 11.1 *Let Γ be a Hankel operator with $\|\Gamma\| \leq 1$, acting from H_2^q into K_2^p. The $NP(\Gamma)$ is completely indeterminate if and only if*

$$\mathcal{I}(f) < \infty \quad \textit{for some } f \in \mathcal{N}(\Gamma). \tag{11.9}$$

Proof Let NP(Γ) be completely indeterminate. Then

$$\mathcal{N}(\Gamma) = T_{\mathfrak{A}}[S^{p\times q}] \quad \text{for some } \mathfrak{A} \in \mathfrak{M}_r(j_{pq})$$

by Theorem 7.20. Let

$$f_0 = T_{\mathfrak{A}}[0].$$

Then $f_0 \in \mathcal{N}(\Gamma)$ and

$$I_q - f_0(\mu)^* f_0(\mu) = \mathfrak{a}_+(\mu)^{-*}\mathfrak{a}_+(\mu)^{-1} \quad \text{a.e. on } \mathbb{R}, \tag{11.10}$$

where $\mathfrak{a}_+$ is the 22 block of $\mathfrak{A}$ and

$$\mathfrak{a}_+^{-1} \in \mathcal{S}_{out}^{q\times q},$$

by definition of the class $\mathfrak{M}_r(j_{pq})$. Therefore,

$$\mathcal{I}(f_0;\omega) = -\frac{\Im\omega}{\pi}\int_{-\infty}^{\infty} \frac{\ln|\det \mathfrak{a}_+(\mu)|}{|\mu-\omega|^2} d\mu = -\ln|\det \mathfrak{a}_+(\omega)| < \infty, \tag{11.11}$$

since $\det \mathfrak{a}_+(\lambda)$ is an outer function.

Conversely, if $\mathcal{I}(f) < \infty$ for some mvf $f \in \mathcal{N}(\Gamma)$, then

$$\int_{-\infty}^{\infty} \frac{\ln(1-\|f(\mu)\|)}{1+\mu^2} d\mu > -\infty, \tag{11.12}$$

since

$$(1-\|A\|^2)^q \leq \det(I_q - A^*A) \leq 1 - \|A\|^2$$

for every contractive $p \times q$ matrix A. Thus, the function

$$\varphi(\lambda) = \exp\left\{\frac{1}{\pi i}\int_{-\infty}^{\infty}\left\{\frac{1}{\mu-\lambda} - \frac{\mu}{1+\mu^2}\right\}\ln(1-\|f(\mu)\|)d\mu\right\}$$

belongs to the class S_{out} and

$$1 - \|f(\mu)\| = |\varphi(\mu)| \quad \text{a.e. on } \mathbb{R}.$$

Moreover, if $\xi \in \mathbb{C}^p$ and $\eta \in \mathbb{C}^q$ are unit vectors and

$$f_1(\mu) = f(\mu) + \varphi(\mu)\xi\eta^*,$$

then $f_1 \in L_\infty^{p\times q}$, $f_1 - f \in H_\infty^{p\times q}$ and

$$\|f_1(\mu)\| \le \|f(\mu)\| + |\varphi(\mu)| = 1 \quad \text{a.e. on } \mathbb{R}.$$

Consequently, $f_1 \in \mathcal{N}(\Gamma)$, and

$$(f_1(\mu) - f(\mu))\eta = \varphi(\mu)\xi \Longrightarrow \|(f_1 - f)\eta\|_\infty > 0;$$

i.e., the NP(Γ) is completely indeterminate. □

Theorem 11.2 *Let $b_1 \in \mathcal{S}_{in}^{p\times p}$, $b_2 \in \mathcal{S}_{in}^{q\times q}$ and $s^\circ \in \mathcal{S}^{p\times q}$. The GSIP($b_1, b_2; s^\circ$) is completely indeterminate if and only if*

$$\mathcal{I}(s) < \infty \quad \textit{for some} \quad s \in \mathcal{S}(b_1, b_2; s^\circ).$$

Proof Let $b_1 \in \mathcal{S}_{in}^{p\times p}$, $b_2 \in \mathcal{S}_{in}^{q\times q}$, $s^\circ \in \mathcal{S}^{p\times q}$, $f^\circ(\mu) = b_1(\mu)^* s^\circ(\mu) b_2(\mu)^*$ and $\Gamma = \Gamma(f^\circ)$. Then, since

$$s \in \mathcal{S}(b_1, b_2; s^\circ) \Longleftrightarrow b_1^*(\mu)s(\mu)b_2^*(\mu) \in \mathcal{N}(\Gamma),$$

the GSIP($b_1, b_2; s^\circ$) is completely indeterminate if and only if the NP(Γ) is completely indeterminate. Moreover, if $f = b_1^* s b_2^*$ then

$$\mathcal{I}(s) = \mathcal{I}(f).$$

Thus, the theorem follows from Theorem 11.1. □

Theorem 11.3 *Let $b_3 \in \mathcal{S}_{in}^{p\times p}$, $b_4 \in \mathcal{S}_{in}^{p\times p}$ and $c^\circ \in \mathcal{C}^{p\times p}$. The GCIP($b_3, b_4; c^\circ$) is completely indeterminate if and only if*

$$h(c) < \infty \quad \textit{for some} \quad c \in \mathcal{C}(b_3, b_4; c^\circ).$$

Proof This follows from Lemma 7.68 and Theorem 11.2. □

11.4 Formulas for entropy functionals and their minimum

Let

$$\mathfrak{A}(\mu) = \begin{bmatrix} u_{11}(\mu) & u_{12}(\mu) \\ u_{21}(\mu) & u_{22}(\mu) \end{bmatrix}$$

be a measurable $m \times m$ mvf on $\mathbb{R}$ with blocks u_{11} and u_{22} of sizes $p \times p$ and $q \times q$, respectively, such that $\mathfrak{A}(\mu)$ has j_{pq}-unitary values a.e. on $\mathbb{R}$. Then

$$u_{22}(\mu)u_{22}(\mu)^* - u_{21}(\mu)u_{21}(\mu)^* = I_q \quad \text{a.e. on } \mathbb{R}, \tag{11.13}$$

and hence det $u_{22}(\mu) \neq 0$ a.e. on $\mathbb{R}$,

$$u_{22}^{-1} \in L_\infty^{q\times q}, \quad \text{and} \quad \|u_{22}^{-1}\|_\infty \le 1. \tag{11.14}$$

Thus, the mvf

$$s_{21}(\mu) = -u_{22}(\mu)^{-1}u_{21}(\mu) \tag{11.15}$$

enjoys the properties

$$s_{21} \in L_\infty^{q\times p} \quad \text{and} \quad \|s_{21}\|_\infty \le 1.$$

Lemma 11.4 *Let $\mathfrak{A}(\mu)$ be a measurable $m\times m$ mvf on $\mathbb{R}$ that is j_{pq}-unitary valued a.e. on $\mathbb{R}$ such that $u_{21}(\mu)$ and $u_{22}(\mu)$ are the boundary values of mvf's $u_{21}(\lambda)$ and $u_{22}(\lambda)$ that are holomorphic in $\mathbb{C}_+$ and*

$$s_{21} = -u_{22}^{-1}u_{21} \in \mathcal{S}^{q\times p} \quad \text{and} \quad u_{22}^{-1} \in \mathcal{S}_{out}^{q\times q} \tag{11.16}$$

and let

$$f_\varepsilon = T_{\mathfrak{A}}[\varepsilon], \quad \varepsilon \in \mathcal{S}^{p\times q}. \tag{11.17}$$

Then

$$\mathcal{I}(s_{21};\omega) = \mathcal{I}(s_{12};\omega) = \ln|\det u_{22}(\omega)| \tag{11.18}$$

and:

(1) *The entropy functional is given by the formula*

$$\mathcal{I}(f_\varepsilon;\omega) = \mathcal{I}(s_{21};\omega) + \mathcal{I}(\varepsilon;\omega) + \ln|\det\{I_q - s_{21}(\omega)\varepsilon(\omega)\}| \tag{11.19}$$

for every point $\omega \in \mathbb{C}_+$. Formula (11.19) can also be expressed as the algebraic sum of three entropy functionals:

$$\mathcal{I}(f_\varepsilon;\omega) = \mathcal{I}(s_{21};\omega) - \mathcal{I}(k_\omega;\omega) + \mathcal{I}(\varepsilon_\omega;\omega), \tag{11.20}$$

where

$$k_\omega(\lambda) \equiv s_{21}(\omega)^*, \tag{11.21}$$

$$\varepsilon_\omega(\lambda) = T_{\mathfrak{A}_\omega}[\varepsilon], \quad \varepsilon \in \mathcal{S}^{p\times q} \tag{11.22}$$

and $\mathfrak{A}_\omega$ is the j_{pq}-unitary matrix that is defined by the formula

$$\mathfrak{A}_\omega = \begin{bmatrix} (I_p - k_\omega k_\omega^*)^{-1/2} & -(I_p - k_\omega k_\omega^*)^{-1/2}k_\omega \\ -(I_q - k_\omega^* k_\omega)^{-1/2}k_\omega^* & (I_q - k_\omega^* k_\omega)^{-1/2} \end{bmatrix}. \tag{11.23}$$

(2) *The entropy functional is subject to the bound*

$$\mathcal{I}(f_\varepsilon;\omega) \geq \mathcal{I}(s_{21};\omega) - \mathcal{I}(k_\omega;\omega) \tag{11.24}$$

for every $\varepsilon \in \mathcal{S}^{p\times q}$*, with equality if and only if*

$$\varepsilon(\lambda) \equiv s_{21}(\omega)^*.$$

(3) *The lower bound in (11.24) may be written as*

$$\mathcal{I}(s_{21};\omega) - \mathcal{I}(k_\omega;\omega) = \frac{1}{2}\ln\det[u_{22}(\omega)u_{22}(\omega)^* - u_{21}(\omega)u_{21}(\omega)^*]. \tag{11.25}$$

Proof Under the given assumptions,

$$I_q - s_{21}(\mu)s_{21}(\mu)^* = u_{22}(\mu)^{-1}u_{22}(\mu)^{-*} \quad \text{a.e. on } \mathbb{R} \tag{11.26}$$

and

$$I_q - s_{12}(\mu)^*s_{12}(\mu) = u_{22}(\mu)^{-*}u_{22}(\mu)^{-1} \quad \text{a.e. on } \mathbb{R}. \tag{11.27}$$

The equalities in (11.18) follow from (11.26), (11.27) and the fact that, in view of Theorem 3.50, $(\det u_{22})^{-1} \in \mathcal{S}_{out}$.

Since the mvf $\mathfrak{A}(\mu)$ is assumed to be j_{pq}-unitary a.e. on $\mathbb{R}$, formulas (11.15), and (11.17) imply that

$$I_q - f_\varepsilon^* f_\varepsilon = u_{22}^{-*}\{I_q - s_{21}\varepsilon\}^{-*}\{I_q - \varepsilon^*\varepsilon\}\{I_q - s_{21}\varepsilon\}^{-1}u_{22}^{-1} \tag{11.28}$$

a.e. on $\mathbb{R}$. The identity

$$\ln|\det[I_q - s_{21}(\omega)\varepsilon(\omega)]| = \frac{\Im\omega}{\pi}\int_{-\infty}^{\infty}\frac{\ln|\det\{I_q - s_{21}(\mu)\varepsilon(\mu)\}|}{|\mu-\omega|^2}d\mu \tag{11.29}$$

holds, since $\det\{I_q - s_{21}(\lambda)\varepsilon(\lambda)\}$ is an outer function in H^∞, by Lemma 3.54 and Theorem 3.50. Formula (11.19) now follows easily from (11.18) and (11.29).

Next, upon applying formula (11.22) to the mvf $\varepsilon_\omega(\lambda) = T_{\mathfrak{A}_\omega}[\varepsilon]$, it follows that

$$\mathcal{I}(\varepsilon_\omega;\omega) = \mathcal{I}(k_\omega^*;\omega) + \mathcal{I}(\varepsilon;\omega) + \ln|\det\{I_q - s_{21}(\omega)\varepsilon(\omega)\}|. \tag{11.30}$$

Formula (11.20) follows from (11.19) and (11.30), since

$$\det\{I_p - k_\omega k_\omega^*\} = \det\{I_q - k_\omega^* k_\omega\}.$$

Finally, (2) follows from formula (11.20), since $\mathcal{I}(s_{21};\omega)$ and $\mathcal{I}(k_\omega;\omega)$ do not depend upon the choice of the mvf $\varepsilon \in \mathcal{S}^{p\times q}$ and

$$\mathcal{I}(\varepsilon_\omega;\omega) \geq 0 \quad \text{with equality if and only if} \quad \varepsilon_\omega = 0\,;$$

i.e., if and only if $\varepsilon(\lambda) = s_{21}(\omega)^*$. □

Corollary 11.5 *Let $[\mathfrak{b}_+ \quad \mathfrak{a}_+]$ be the bottom block row of a mvf $\mathfrak{A} \in \mathfrak{M}_r(j_{pq})$, $f_\varepsilon = T_{\mathfrak{A}}[\varepsilon]$ and $s_{21} = \mathfrak{a}_+^{-1}\mathfrak{b}_+$. Then the assertions of Lemma 11.4 are in force. Moreover,*

$$\mathcal{I}(f_\varepsilon,\omega) \geq \frac{1}{2}\ln\det[\mathfrak{a}_+(\omega)\mathfrak{a}_+(\omega)^* - \mathfrak{b}_+(\omega)\mathfrak{b}_+(\omega)^*] \tag{11.31}$$

for every $\varepsilon \in \mathcal{S}^{p\times q}$ with equality if and only if

$$\varepsilon(\lambda) \equiv s_{21}(\omega)^*.$$

Proof The mvf $\mathfrak{A} \in \mathfrak{M}_r(j_{pq})$ satisfies the conditions of Lemma 11.4. Moreover,

$$\mathcal{I}(s_{21};\omega) = \mathcal{I}(f_0;\omega) = \ln|\det\mathfrak{a}_+(\omega)| \tag{11.32}$$

by formula (11.11). Thus, (11.31) follows from the formula (11.24). □

Corollary 11.6 *Let Γ be a Hankel operator acting from H_2^q into K_2^p and let $NP(\Gamma)$ be completely indeterminate. Then*

$$\mathcal{N}(\Gamma) = T_{\mathfrak{A}}[\mathcal{S}^{p\times q}] \quad \textit{for some} \quad \mathfrak{A} \in \mathfrak{M}_r(j_{pq}) \tag{11.33}$$

and the conclusions of Corollary 11.5 are in force.

Proof This follows from Theorem 7.20 and Corollary 11.5. □

Theorem 11.7 *Let $W \in \mathcal{U}(j_{pq})$, let $\{b_1, b_2\} \in ap(W)$ and let*

$$\delta_W(\omega) = b_2(\omega)[w_{22}(\omega)w_{22}(\omega)^* - w_{21}(\omega)w_{21}(\omega)^*]b_2(\omega)^*. \tag{11.34}$$

Then

$$\mathcal{I}(s;\omega) \geq \frac{1}{2}\ln\det\delta_W(\omega) \quad \textit{for every} \quad s \in T_W[\mathcal{S}^{p\times q}] \tag{11.35}$$

with equality if and only if

$$s(\lambda) = T_W[\varepsilon] \quad \textit{with} \quad \varepsilon(\lambda) \equiv s_{21}(\omega)^*. \tag{11.36}$$

Proof Under the given conditions, the mvf

$$\mathfrak{A}(\mu) = \begin{bmatrix} b_1(\mu)^* & 0 \\ 0 & b_2(\mu) \end{bmatrix} W(\mu)$$

belongs to $\mathfrak{M}_r(j_{pq})$ and, if $f_\varepsilon = T_{\mathfrak{A}}[\varepsilon]$, then $s_\varepsilon = T_W[\varepsilon] = b_1 f_\varepsilon b_2$. Therefore,

$$\mathcal{I}(s_\varepsilon; \omega) = \mathcal{I}(f_\varepsilon; \omega)$$

and hence the theorem follows from Corollary 11.5. □

Corollary 11.8 *Let* $b_1 \in \mathcal{S}_{in}^{p\times p}$, $b_2 \in \mathcal{S}_{in}^{q\times q}$ *and* $s^\circ \in \mathcal{S}^{p\times q}$, *and let the* GSIP$(b_1, b_2; s^\circ)$ *be completely indeterminate. Then* $\mathcal{S}(b_1, b_2; s^\circ) = T_W[\mathcal{S}^{p\times q}]$ *for some* $W \in \mathcal{U}(j_{pq})$ *with* $\{b_1, b_2\} \in ap(W)$ *and the conclusions of Theorem 11.7 are in force.*

Proof This follows from Theorems 7.48 and 11.7. □

Lemma 11.9 *Let* $A \in \mathcal{U}(J_p)$, $\{b_3, b_4\} \in ap_{II}(A)$, $B(\lambda) = A(\lambda)\mathfrak{V}$,

$$\begin{bmatrix} E_-(\lambda) & E_+(\lambda) \end{bmatrix} = \sqrt{2}\begin{bmatrix} I_p & 0 \end{bmatrix} B(\lambda), \quad \chi(\lambda) = E_+(\lambda)^{-1}E_-(\lambda),$$

let

$$\delta_{\mathfrak{E}}(\omega) = b_4(\omega)\{E_+(\omega)E_+(\omega)^* - E_-(\omega)E_-(\omega)^*\}b_4(\omega)^* \tag{11.37}$$

and suppose that $\delta_{\mathfrak{E}}(\omega) > 0$ *for some* $\omega \in \mathbb{C}_+$. *Then:*

(1) $\mathcal{C}(A) = T_B[\mathcal{S}^{p\times p}]$, *i.e.,* $\mathcal{S}^{p\times p} \subseteq \mathcal{D}(T_B)$.

(2) *If* $c_\varepsilon = T_B[\varepsilon]$ *for some* $\varepsilon \in \mathcal{S}^{p\times p}$, *then*

$$h(c_\varepsilon; \omega) = \ln|\det b_4(\omega)E_+(\omega)| + \mathcal{I}(\varepsilon; \omega) + \ln|\det\{I_p - \chi(\omega)\varepsilon(\omega)\}| \tag{11.38}$$

and

$$h(c_\varepsilon; \omega) \geq \frac{1}{2}\ln\det\delta_{\mathfrak{E}}(\omega), \tag{11.39}$$

with equality if and only if $\varepsilon(\lambda) \equiv \chi(\omega)^*$.

Proof In view of condition (11.37), Lemma 4.70 guarantees that $\mathcal{S}^{p\times p} \subset \mathcal{D}(T_B)$. The condition $B^*(\mu)J_pB(\mu) = j_p$ a.e. on $\mathbb{R}$ implies that

$$\Re c_\varepsilon = \Re T_B[\varepsilon] = E_+^{-*}\{I_p + \chi\varepsilon\}^{-*}\{I_p - \varepsilon^*\varepsilon\}\{I_p + \chi\varepsilon\}^{-1}E_+^{-1} \tag{11.40}$$

a.e. on $\mathbb{R}$. Thus, as the mvf's $I_p + \chi\varepsilon$ and $\varphi = b_4E_+$ belong to the class $\mathcal{N}_{out}^{p\times p}$, it follows that

$$h(c_\varepsilon;\omega) = \ln|\det\varphi(\omega)| + \ln|\det[I_p + \chi(\omega)\varepsilon(\omega)]| + \mathcal{I}(\varepsilon;\omega).$$

Thus, by formula (11.30) with $s_{21}(\omega) = -\chi(\omega)$,

$$h(c_\varepsilon;\omega) = \ln|\det\varphi(\omega)| + \mathcal{I}(\varepsilon_\omega;\omega) - \mathcal{I}(k_\omega;\omega),$$

where $k_\omega(\lambda) \equiv \chi(\omega)^*$ and the mvf $\varepsilon_\omega \in \mathcal{S}^{p\times p}$ is obtained from formulas (11.22) and (11.23). Consequently, $h(c_\varepsilon;\omega)$ has minimum value when ε varies over $\mathcal{S}^{p\times p}$ if and only if $\varepsilon_\omega(\lambda) \equiv 0$, i.e., if and only if $\varepsilon(\lambda) \equiv \chi(\omega)^* = k_\omega(\lambda)$, and, if $\varepsilon(\lambda) \equiv \chi(\omega)^*$, then

$$\begin{aligned}
h(c_\varepsilon) &= \ln|\det\varphi(\omega)| - \mathcal{I}(k_\omega;\omega)\\
&= \ln|\det\varphi(\omega)| + \frac{1}{2}\ln\det[\{_p - \chi(\omega)\chi(\omega)^*\}\\
&= \ln|\det\varphi(\omega)| - \ln|\det E_+(\omega)|\\
&\quad + \frac{1}{2}\ln\det\{E_+(\omega)E_+(\omega)^* - E_-(\omega)E_-(\omega)^*\}\\
&= \ln|\det b_4(\omega)| + \frac{1}{2}\ln\det\{E_+(\omega)E_+(\omega)^* - E_-(\omega)E_-(\omega)^*\}\\
&= \frac{1}{2}\ln\det\delta_{\mathfrak{E}}(\omega).
\end{aligned}$$

□

Theorem 11.10 *Let $b_3 \in \mathcal{S}_{in}^{p\times p}$, $b_4 \in \mathcal{S}_{in}^{p\times p}$, $c^\circ \in \mathcal{C}^{p\times p}$ and let the GCIP$(b_3, b_4; c^\circ)$ be completely indeterminate. Let*

$$\mathcal{C}(b_3, b_4; c^\circ) = \mathcal{C}(A), \text{ where } A \in \mathcal{U}(J_p), \quad \{b_3, b_4\} \in ap_{II}(A)$$

and assume that $\det\delta_{\mathfrak{E}}(\omega) > 0$ and that $B = A\mathfrak{V}$. Then the inequality (11.39) holds for $c_\varepsilon = T_B[\varepsilon]$ for every $\varepsilon \in \mathcal{S}^{p\times p}$, with equality if and only if $\varepsilon(\lambda) = \chi(\omega)^*$ and $\chi = b_{22}^{-1}b_{21}$.

Proof The assertions of the theorem follow from Lemma 11.9 and Theorem 7.70. □

11.5 A matricial generalization of the entropy inequality

Theorem 11.11 *If $f \in L_\infty^{p\times q}$ and $\|f\|_\infty \le 1$, then $\mathcal{I}(f) < \infty$ if and only if there exists a mvf $\varphi_f(\lambda) \in \mathcal{S}_{out}^{q\times q}$ such that*

$$\varphi_f(\mu)^*\varphi_f(\mu) = I_q - f(\mu)^* f(\mu) \quad a.e.\ on\ \mathbb{R}.$$

Moreover, this mvf φ_f is uniquely defined up to a constant unitary left factor.

Proof This follows by applying Theorem 3.78 to the mvf $g = (I_q - f^* f)|\rho_i|^{-2}$ and then invoking the Smirnov maximum principle. □

If the conditions of Theorem 11.11 are met, then the entropy functionals $\mathcal{I}(f;\omega)$ may be calculated in terms of the mvf φ_f:

$$\mathcal{I}(f;\omega) = -\ln|\det \varphi_f(\omega)|. \tag{11.41}$$

Thus, for example, formula (11.20) may be rewritten as

$$\ln|\det \varphi_{f_\varepsilon}(\omega)| = \ln|\det \varphi_{s_{21}}(\omega)| + \ln|\det \varphi_{\varepsilon_\omega}(\omega)| - \ln|\det \varphi_{k_\varepsilon}(\omega)|.$$

The following matricial generalization of assertion (2) of Lemma 11.4 exists if, in addition to (11.16), the block $u_{22}(\mu)$ of the mvf $\mathfrak{A}(\mu)$ satisfies the extra condition

$$u_{22}(\mu) = \beta(\mu)\varphi(\mu)^{-1} \quad \text{a.e. on } \mathbb{R}, \tag{11.42}$$

where $\beta(\mu)$ is a scalar function with $|\beta(\mu)| = 1$ a.e. on $\mathbb{R}$ and $\varphi(\mu)$ is the nontangential limit a.e. on $\mathbb{R}$ of a mvf $\varphi \in \mathcal{S}_{out}^{q\times q}$.

Under this assumption we consider

$$\Delta_1(\omega) = \varphi(\omega)^*[I_q - s_{21}(\omega)s_{21}(\omega)^*]^{-1}\varphi(\omega). \tag{11.43}$$

Theorem 11.12 *Let $\mathfrak{A}(\mu)$ be a measurable mvf that is j_{pq}-unitary a.e. on $\mathbb{R}$ such that condition (11.16) is in force and the block u_{22} admits a factorization of the form (11.42), and let $\Delta_1(\omega)$ be defined by (11.43). Then for each point $\omega \in \mathbb{C}_+$ the mvf $f_\varepsilon = T_{\mathfrak{A}}[\varepsilon]$ is subject to the inequality*

$$\varphi_{f_\varepsilon}(\omega)^*\varphi_{f_\varepsilon}(\omega) \le \Delta_1(\omega), \tag{11.44}$$

for every $\varepsilon \in \mathcal{S}^{p\times q}$ with $\mathcal{I}(\varepsilon) < \infty$. Moreover, equality holds if and only if $\varepsilon(\lambda) \equiv s_{21}(\omega)^*$.

Proof Under the given conditions, the equality (11.28) may be rewritten for $\varepsilon \in S^{p\times q}$ with $\mathcal{I}(\varepsilon) < \infty$ as

$$\varphi_{f_\varepsilon}(\mu)^*\varphi_{f_\varepsilon}(\mu) = \varphi(\mu)^*\{I_q - s_{21}(\mu)\varepsilon(\mu)\}^{-*}\varphi_\varepsilon(\mu)^*\varphi_\varepsilon(\mu)\{I_q - s_{21}(\mu)\varepsilon(\mu)\}^{-1}\varphi(\mu).$$

Therefore, since

$$\varphi_\varepsilon\{I_q - s_{21}\varepsilon\}^{-1}\varphi \in S_{out}^{q\times q},$$

it follows that

$$\varphi_{f_\varepsilon}(\lambda) = v_\varepsilon\varphi_\varepsilon(\lambda)\{I_q - s_{21}(\lambda)\varepsilon(\lambda)\}^{-1}\varphi(\lambda), \quad \lambda \in \mathbb{C}_+,$$

for some constant unitary $q \times q$ matrix v_ε. Thus,

$$\varphi_{f_\varepsilon}(\omega)^*\varphi_{f_\varepsilon}(\omega) = \varphi(\omega)^*\{I_q - s_{21}(\omega)\varepsilon(\omega)\}^{-*}\varphi_\varepsilon(\omega)^*\varphi_\varepsilon(\omega)\{I_q - s_{21}(\omega)\varepsilon(\omega)\}^{-1}\varphi(\omega). \quad (11.45)$$

If this equality is applied to the mvf $\varepsilon_\omega(\mu) = T_{\mathfrak{A}_\omega}[\varepsilon]$ with $\mathfrak{A}_\omega$ as in (11.23) and $\mathfrak{k}_\omega = (I_q - k_\omega^* k_\omega)^{1/2}$, then

$$\varphi_{\varepsilon_\omega}(\omega)^*\varphi_{\varepsilon_\omega}(\omega) = \mathfrak{k}_\omega\{I_q - s_{21}(\omega)\varepsilon(\omega)\}^{-*}\varphi_\varepsilon(\omega)^*\varphi_\varepsilon(\omega)\{I - s_{21}(\omega)\varepsilon(\omega)\}^{-1}\mathfrak{k}_\omega.$$

Thus, the equality (11.45) may be rewritten as

$$\varphi_{f_\varepsilon}(\omega)^*\varphi_{f_\varepsilon}(\omega) = \varphi(\omega)^*\mathfrak{k}_\omega^{-1}\varphi_{\varepsilon_\omega}(\omega)^*\varphi_{\varepsilon_\omega}(\omega)\mathfrak{k}_\omega^{-1}\varphi(\omega).$$

Therefore, since

$$\varphi_{\varepsilon_\omega}(\omega)^*\varphi_{\varepsilon_\omega}(\omega) \le I_q$$

with equality if and only if $\varepsilon_\omega(\lambda) \equiv 0$,

$$\varphi_{f_\varepsilon}(\omega)^*\varphi_{f_\varepsilon}(\omega) \le \varphi(\omega)^*\{I_q - s_{21}(\omega)s_{21}(\omega)^*\}^{-1}\varphi(\omega)$$

with equality if and only if $\varepsilon_\omega(\lambda) \equiv 0_{p\times q}$, i.e., if and only if $\varepsilon(\lambda) \equiv s_{21}(\omega)^*$. □

Theorem 11.13 *If the $NP(\Gamma)$ is completely indeterminate and $\mathcal{N}(\Gamma) = T_{\mathfrak{A}}[\mathcal{S}^{p\times q}]$ for some mvf $\mathfrak{A} \in \mathfrak{M}_{rR}(j_{pq})$ with bottom block row $\begin{bmatrix}\mathfrak{b}_+ & \mathfrak{a}_+\end{bmatrix}$, then*

$$\varphi_f(\omega)^*\varphi_f(\omega) \le \{\mathfrak{a}_+(\omega)\mathfrak{a}_+(\omega)^* - \mathfrak{b}_+(\omega)\mathfrak{b}_+(\omega)^*\}^{-1} \quad (11.46)$$

for each point $\omega \in \mathbb{C}_+$ *and every* $f \in \mathcal{N}(\Gamma)$ *with* $\mathcal{I}(f) < \infty$. *Moreover, equality holds in (11.46) if and only if*

$$f = T_{\mathfrak{A}}[s_{21}(\omega)^*], \quad \textit{where} \quad s_{21} = -\mathfrak{a}_+^{-1}\mathfrak{b}_+. \tag{11.47}$$

Proof Theorem 11.12 is applicable with $\beta = 1$ and $\varphi = \mathfrak{a}_+^{-1}$ in (11.42). □

Corollary 11.14 *Let the GSIP* $(b_1, I_q; s^\circ)$ *be completely indeterminate and let*

$$\mathcal{S}(b_1, I_q; s^\circ) = T_W[S^{p\times q}],$$

where

$$W \in \mathcal{U}(J_{pq}) \textit{ and } \{b_1, I_q\} \in ap(W).$$

Then

$$\varphi_s(\omega)^*\varphi_s(\omega) \leq \{w_{22}(\omega)w_{22}(\omega)^* - w_{21}(\omega)w_{21}(\omega)^*\}^{-1}. \tag{11.48}$$

for each point $\omega \in \mathbb{C}_+$ *and every* $s \in \mathcal{S}(b_1, I_q; s^\circ)$ *with* $\mathcal{I}(s) < \infty$. *Moreover, equality holds in (11.48) if and only if* $s = T_W[\varepsilon]$ *with* $\varepsilon(\lambda) \equiv s_{21}(\omega)^*$, *where* $s_{21} = -w_{22}^{-1}w_{21}$.

Proof This follows from Theorem 11.12. □

If $c \in \mathcal{C}^{p\times p}$ and $h(c) < \infty$, then, by Theorem 3.78, there exists a mvf $\psi_c \in \mathcal{N}_{out}^{p\times p}$ such that

$$\psi_c(\mu)^*\psi_c(\mu) = \Re c(\mu) \quad \text{a.e. on } \mathbb{R};$$

ψ_c is defined by c up to a constant unitary left factor.

Theorem 11.15 *Let* $A \in \mathcal{U}(J_p)$, $B(\lambda) = A(\lambda)\mathfrak{V}$,

$$\mathfrak{E}(\lambda) = [E_-(\lambda) \quad E_+(\lambda] = \sqrt{2}[0 \quad I_p]B(\lambda)$$

and assume that $\mathfrak{E} \in \mathcal{N}_+^{p\times 2p}$ *and the RK* $K_\omega^{\mathfrak{E}}(\omega) > 0$ *for some point* $\omega \in \mathbb{C}_+$ *(and hence for every* $\omega \in \mathbb{C}_+$*). Then*

$$\psi_c(\omega)^*\psi_c(\omega) \leq (E_+(\omega)E_+(\omega)^* - E_-(\omega)E_-(\omega)^*)^{-1} \tag{11.49}$$

for every $c \in T_B[S^{p\times p}]$ *with* $h(c) < \infty$. *Moreover, equality prevails in (11.49) if and only if* $c(\lambda) = T_B[\chi(\omega)^*]$, *where* $\chi = b_{22}^{-1}b_{21}$.

Proof Under the given assumptions, $c = T_B[\varepsilon]$, and $h(c) < \infty$. (11.40) may be rewritten for mvf's c with $h(c) < \infty$ as

$$\begin{aligned}&\psi_c^*(\mu)\psi_c(\mu)\\ &\quad = E_+(\mu)^{-*}\{I_p + \chi(\mu)\varepsilon(\mu)\}^{-*}\varphi_\varepsilon(\mu)^*\varphi_\varepsilon(\mu)\{I_p + \chi(\mu)\varepsilon(\mu)\}^{-1}E_+(\mu)\end{aligned}$$

a.e. on $\mathbb{R}$. Consequently,

$$\psi_c(\omega)^*\psi_c(\omega) = E_+(\omega)^{-*}\mathfrak{k}_\omega\varphi_{\varepsilon_\omega}(\omega)^*\varphi_{\varepsilon_\omega}(\omega)\mathfrak{k}_\omega E_+(\omega)^{-1},$$

where

$$\varphi_{\varepsilon_\omega}(\lambda) = \varphi_\varepsilon(\lambda)\{I_p + \chi(\omega)\varepsilon(\lambda)\}^{-1}, \quad \mathfrak{k}_\omega = (I_p - k_\omega^* k_\omega)^{-1/2},$$

$\varepsilon_\omega(\lambda) = T_{\mathfrak{A}_\omega}[\varepsilon]$ and $\mathfrak{A}_\omega$ is defined by formula (11.23) but with $k_\omega(\lambda) \equiv \chi(\omega)^*$. Thus,

$$\psi_c(\omega)^*\psi_c(\omega) \le E_+(\omega)^{-*}(I - k_\omega^* k_\omega)^{-1}E_+(\omega)^{-1}$$

with equality if and only if $\varepsilon_\omega(\lambda) \equiv 0$, i.e., if and only if $\varepsilon(\lambda) \equiv \chi(\lambda)^*$. □

Corollary 11.16 *If the GCIP($b_3, I_p; c^\circ$) is completely indeterminate and*

$$\mathcal{C}(b_3, I_p; c^\circ) = T_B[S^{p\times p}] \tag{11.50}$$

for some mvf $B \in \mathcal{U}(j_p, J_p)$ that belongs to $\mathcal{N}_+^{p\times p}$ and if $\mathfrak{E} = \sqrt{2}[0 \quad I_p]B$ and $K_\omega^{\mathfrak{E}}(\omega) > 0$ for at least one point $\omega \in \mathfrak{h}_{\mathfrak{E}}^+$, then the conclusions of Theorem 11.15 are in force for every $c \in \mathcal{C}(b_3, I_p; c^\circ)$ with $h(c) < \infty$.

Proof Under the given assumptions, Theorem 11.15 is applicable and yields the stated assertions. □

11.6 Bibliographical notes

Chover [Ch61] is possibly the first paper to consider the relevance of entropy integrals to extension problems. However, there seemed to be little interest in the subject before Burg considered maximum entropy interpolants in the setting of the discrete covariance extension problem, first for the scalar case [Bu67], and then several years later in his PhD thesis [Bu75] for the matrix case. A linear fractional description of the set of spectral densitities for a Feller-Krein string was used to find the density with maximal (minimal) entropy in [DMc76]. Maximum entropy extensions were identified with band extensions in a number of different settings by Dym and Gohberg; see, e.g.,

[DG79], [DG80], [DG81], [DG86] and [DG88]. This theme was extended to the setting of C^* algebras in [GKW91].

Most of the results in this chapter were adapted from the papers [ArK81] and [ArK83]. Minimum entropy solutions for scalar and tangential Nevanlinna-Pick problems were obtained independently in the same way in [DeD81a] and [DeD81b]. For additional discussion, references and examples see e.g., Chapter 11 of [Dy89b], Section 7 of [DI84], [Dy89c] and [Lan87]. Connections of entropy evaluations with matrix balls are discussed in Chapter 11 of [Dy89b]. Explicit formulas for the maximum entropy and the semiradii in terms of reproducing kernels are furnished in Theorems 7.4 and 8.3 of [DI84].

Applications of extremal entropy solutions to control theory may be found e.g., in [MuGl90], [PI97], [Fe98] and the references cited therein; applications to rational interpolation with degree constraints are discussed in [BGL01]. Connections of a class of matrix extremal problems with maximal entropy integrals and tangential (resp., bitangential) interpolation problems are considered in [DG95] (resp., [Dy96]). Entropy integrals for interpolation problems with singular Pick matrices are considered in [BoD98]. Maximum entropy problems in the setting of upper triangular operators are treated in [DF97] and a number of the references cited therein.

Bibliography

[Ad73a] V. M. Adamjan, Nondegenerate unitary couplings of semiunitary operators. (Russian) Funktsional. Anal. i Prilozhen. **7** (1973), no. 4, 1–16.

[Ad73b] V. M. Adamjan, The Theory of Couplings of Semi-unitary Operators. (Russian) Doctoral Thesis, Odessa State University, Odessa, 1973.

[AdAr66] V. M. Adamjan and D. Z. Arov, Unitary couplings of semi-unitary operators. (Russian) Mat. Issled. **1** (1966), no. 2, 3–64.

[AAK68] V. M. Adamjan, D. Z. Arov and M. G. Krein, Infinite Hankel matrices and generalized problems of Carathodory-Fejr and I. Schur problems. (Russian) Funktsional. Anal. i Prilozhen. **2** (1968), no. 4, 1–17.

[AAK71a] V. M. Adamjan, D. Z. Arov and M. G. Krein, Infinite Hankel block matrices and related problems of extension. (Russian) Izv. Akad. Nauk Armjan. SSR Ser. Mat. **6** (1971), no. 2-3, 87–112.

[AAK71b] V. M. Adamjan, D. Z. Arov and M. G. Krein, Analytic properties of the Schmidt pairs of a Hankel operator and the generalized Schur-Takagi problem. (Russian) Mat. Sb. (N.S.) **86** (128) (1971), 34–75.

[Ak65] N. I. Akhiezer, *The Classical Moment Problem and Some Related Questions in Analysis.* Translated by N. Kemmer Hafner Publishing Co., New York 1965.

[Aku56] E. J. Akutowicz, A qualitative characterization of Blaschke products in a half-plane. Amer. J. Math. **7** 8 (1956), 677–684.

[Al01] D. Alpay, *The Schur Algorithm, Reproducing Kernel Spaces and System Theory.* SMF/AMS TEXTS and Monographs, Amer. Math. Soc., Providence, RI, 2001.

[AlD84] D. Alpay and H. Dym, Hilbert spaces of analytic functions, inverse scattering and operator models, I., Integ. Equat. Oper. Th. **7** (1984), 589–741.

[AlD85] D. Alpay and H. Dym, Hilbert spaces of analytic functions, inverse scattering and operator models, II. Integ. Equat. Oper. Th. **8** (1985), 145–180.

[AlD86] D. Alpay and H. Dym, On applications of reproducing kernel spaces to the Schur algorithm and rational J unitary factorization, in *I. Schur Methods in Operator Theory and Signal Processing* (I. Gohberg, ed.), Oper. Theory Adv. Appl., **18**, Birkhäuser, Basel, 1986, pp. 89–159.

[AlD93] D. Alpay and H. Dym, On a new class of structured reproducing kernel spaces. J. Funct. Anal. **111** (1993), 1–28.

[ADD89] D. Alpay, P. Dewilde and H. Dym, On the existence and construction of solutions to the partial lossless inverse scattering problem with applications to estimation theory. IEEE Trans. Inform. Theory **35** (1989), no. 6, 1184–1205.

[ADRS97] D. Alpay, A. Dijksma, J. Rovnyak and H. de Snoo, *Schur Functions, Operator Colligations, and Reproducing Kernel Pontryagin Spaces.* Oper. Theory Adv. Appl., **96** Birkhäuser Verlag, Basel, 1997.

[An90] T. Ando, *de Branges Spaces and Analytic Operator Functions.* Division of Applied Mathematics, Research Institute of Applied Electricity, Hokkaido University, Sapporo, 1990.

[An04] T. Ando, Löwner inequality of indefinite type. Linear Algebra Appl. **385** (2004) 73–80.

[Arn50] N. Aronszajn, Theory of reproducing kernels. Trans. Amer. Math. Soc., **68** (1950), 337–404.

[Arc94] R. Arocena, Unitary colligations and parametrization formulas. Ukran. Mat. Zh. **46** (1994), no. 3, 147–154; translation in Ukrainian Math. J. **46** (1994), no. 3, 151–158.

[Ar71] D. Z. Arov, Darlington's method in the study of dissipative systems. (Russian) Dokl. Akad. Nauk SSSR 201 (1971), no. 3, 559–562; translation in Soviet Physics Dokl. **16** (1971), 954–956.

[Ar73] D. Z. Arov, Realization of matrix-valued functions according to Darlington. (Russian) Izv. Akad. Nauk SSSR Ser. Mat. **37** (1973), 1299–1331.

[Ar74a] D. Z. Arov, Unitary couplings with losses (a theory of scattering with losses). (Russian) Funktsional. Anal. i Prilozhen. **8** (1974), no. 4, 5–22.

[Ar74b] D. Z. Arov, Scattering theory with dissipation of energy. (Russian) Dokl. Akad. Nauk SSSR **216** (1974) 713–716.

[Ar75] D. Z. Arov, Realization of a canonical system with a dissipative boundary condition at one end of the segment in terms of the coefficient of dynamical compliance. (Russian) Sibirsk. Mat. **16** (1975), no. 3, 440–463, 643.

[Ar78] D. Z. Arov, An approximation characteristic of functions of the class $B\Pi$. (Russian) Funktsional. Anal. i Prilozhen. **12** (1978), no. 2, 70–71.

[Ar79a] D. Z. Arov, Stable dissipative linear stationary dynamical scattering systems. (Russian) J. Operator Theory **2** (1979), no. 1, 95–126; translation with appendices by the author and J. Rovnyak in Oper. Theory Adv. Appl., **134**, *Interpolation Theory, Systems theory and Related Topics* (D. Alpay, I. Gohberg and V. Vinnikov, eds.), (Tel Aviv/Rehovot, 1999), Birkhäuser, Basel, pp. 99–136.

[Ar79b] D. Z. Arov, Boundary values of a convergent sequence of meromorphic matrix-valued functions. (Russian) Mat. Zametki **25** (1979), no. 3, 335–339, 475.

[Ar79c] D. Z. Arov, Passive linear steady-state dynamical systems. (Russian) Sibirsk. Mat. Zh. **20** (1979), no. 2, 211–228, 457.

[Ar84a] D. Z. Arov, Functions of class Π. (Russian) Investigations on linear operators and the theory of functions, XIII. Zap. Nauchn. Sem. Leningrad. Otdel. Mat. Inst. Steklov. (LOMI) **135** (1984), 5–30.

[Ar84b] D. Z. Arov, Three problems about J-inner matrix functions, in: Springer Lecture Notes in Mathematics, **1043** (1984), pp. 164–168.

[Ar88] D. Z. Arov, Regular and singular J-inner matrix functions and corresponding extrapolation problems. (Russian) Funktsional. Anal. i Prilozhen. **22** (1988), no. 1, 57–59; translation in Funct. Anal. Appl. **22** (1988), no. 1, 46–48.

[Ar89] D. Z. Arov, γ-generating matrices, j-inner matrix-functions and related extrapolation problems. Teor. Funktsii Funktsional. Anal. i Prilozhen, I, **51** (1989), 61–67; II, **52** (1989), 103–109; translation in J. Soviet Math. I, **52** (1990), 3487–3491; III, **52** (1990), 3421–3425.

[Ar90] D. Z. Arov, Regular J-inner matrix-functions and related continuation problems, in *Linear Operators in Function Spaces* (H. Helson, B. Sz.-Nagy, F.-H. Vasilescu and Gr. Arsene, eds.), (Timişoara, 1988) Oper. Theory Adv. Appl., **43**, Birkhäuser, Basel, 1990, pp. 63–87.

[Ar93] D. Z. Arov, The generalized bitangent Carathodory-Nevanlinna-Pick problem and (j, J_0)-inner matrix functions. (Russian) Izv. Ross. Akad. Nauk Ser. Mat. **57** (1993), no. 1, 3–32; translation in Russian Acad. Sci. Izv. Math. **42** (1994), no. 1, 1–26.

[Ar95a] D. Z. Arov, γ-generating matrices, j-inner matrix-functions and related extrapolation problems, IV. Mathematical Physics, Analysis, Geometry **2** (1995), 3–14.

[Ar95b] D. Z. Arov, A survey on passive networks and scattering systems which are lossless or have minimal Losses. AEU-International Journal of Electronics and Communications. **49** (1995) 252–265.

[Ar97] D. Z. Arov, On monotone families of J-contractive matrix functions. (Russian) Algebra i Analiz **9** (1997), no. 6, 3–37; translation in St. Petersburg Math. J. **9** (1998), no. 6, 1025–1051.

[Ar00a] D. Z. Arov, Conservative linear time-invariant lossless systems with strongly regular J-inner transfer functions, in Proceedings, MTNS-2000, Perpignon, France, 2000.

[Ar00b] D. Z. Arov, Conditions for the similarity of all minimal passive scattering systems with a given scattering matrix. (Russian) Funktsional. Anal. i Prilozhen. **34** (2000), no. 4, 71–74; translation in Funct. Anal. Appl. **34** (2000), no. 4, 293–295.

[Ar01] D. Z. Arov, The scattering matrix and impedance of a canonical differential system with a dissipative boundary condition in which the coefficient is a rational matrix function of the spectral parameter. (Russian) Algebra i Analiz **13** (2001), no. 4, 26–53; translation in St. Petersburg Math. J. **13** (2002), no. 4, 527–547.

[ArD97] D. Z. Arov and H. Dym, J-inner matrix functions, interpolation and inverse problems for canonical systems, I: Foundations, Integ. Equat. Oper. Th. **29** (1997), 373–454.

[ArD98] D. Z. Arov and H. Dym, On the three Krein extension problems and some generalizations. Integ. Equat. Oper. Th. **31** (1998), 1–91.

[ArD00a] D. Z. Arov and H. Dym, J-inner matrix functions, interpolation and inverse problems for canonical systems, II: The inverse monodromy problem. Integ. Equat. Oper. Th. **36** (2000), 11–70.

[ArD00b] D. Z. Arov and H. Dym, J-inner matrix functions, interpolation and inverse problems for canonical systems, III: More on the inverse monodromy problem. Integ. Equat. Oper. Th. **36** (2000), 127–181.

[ArD01a] D. Z. Arov and H. Dym, Some remarks on the inverse monodromy problem for 2×2 canonical differential systems, in *Operator Theory and Analysis* (H. Bart, I. Gohberg and A. C. M. Ran, eds.), (Amsterdam, 1997), Oper. Theory Adv. Appl., **122**, Birkhäuser, Basel, 2001, pp. 53–87.

[ArD01b] D. Z. Arov and H. Dym, Matricial Nehari problems J-inner matrix functions and the Muckenhoupt condition. J. Funct. Anal. **181** (2001), 227–299.

[ArD02a] D. Z. Arov and H. Dym, J-inner matrix functions, interpolation and inverse problems for canonical systems, IV: Direct and inverse bitangential input scattering problem. Integ. Equat. Oper. Th. **43** (2002), 1–67.

[ArD02b] D. Z. Arov and H. Dym, J-inner matrix functions, interpolation and inverse problems for canonical systems, V: The inverse input scattering problem for Wiener class and rational $p \times q$ input scattering matrices. Integ. Equat. Oper. Th. **43** (2002), 68–129.

[ArD03a] D. Z. Arov and H. Dym, The bitangential inverse input impendance problem for canonical systems, I: Weyl-Titchmash classification, and existence and uniqueness theorems, Integ. Equat. Oper. Th. **47** (2003), 3–49.

[ArD03b] D. Z. Arov and H.Dym, Criteria for the strongly regularity of J-inner functions and γ-generating functions. J. Math. Anal. Appl. **280** (2003), 387–399.

[ArD04a] D. Z. Arov and H. Dym, Strongly regular J-inner matrix functions and related problems, in: *Current Trends in Operator Theory and its Applications* (J.A. Ball, J.W. Helton, M. Klaus and L. Rodman, eds), Oper. Theor. Adv. Appl., **149**, Birkhäuser, Basel, 2004, pp. 79–106.

[ArD04b] D. Z. Arov and H. Dym, The bitangential inverse spectral problem for canonical systems, J. Funct. Anal. **214** (2004), 312–385.

[ArD05a] D. Z. Arov and H. Dym, The bitangential inverse input impedance problem for canonical systems, II.: Formulas and examples, Integ. Equat. Oper. Th. **51** (2005), 155–213.

[ArD05b] D. Z. Arov and H. Dym, Strongly regular J-inner matrix-vaued functions and inverse problems for canonical systems, in: *Recent Advances in Operator Theory and its Applications* (M.A. Kaashoek, S. Seatzu and C. van der Mee, eds), Oper. Theor. Adv. Appl., **160**, Birkhäuser, Basel, 2005, pp. 101–160.

[ArD05c] D. Z. Arov and H. Dym, Direct and inverse problems for differential systems connected with Dirac systems and related factorization problems, Indiana J. **54** (2005), 1769–1815.

[ArD07] D. Z. Arov and H. Dym, Direct and inverse asymptotic scattering problems for Dirac-Krein systems. (Russian) Funktsional. Anal. i Prilozhen. 41 (2007), no. 3, 17–33.

[ArD07b] D. Z. Arov and H. Dym, Bitangential direct and inverse problems for systems of differential equations, in *Probability, Geometry and Integrable Systems* (M. Pinsky and B. Birnir, eds.) MSRI Publications, **55**, Cambridge University Press, Cambridge, 2007, pp. 1–28.

[AFK93] D. Z. Arov, B. Fritzsche and B. Kirstein, On block completion problems for various subclasses of j_{pq}-inner functions, in *Challenges of a generalized system theory* (Amsterdam, 1992), Konink. Nederl. Akad. Wetensch. Verh. Afd. Natuurk. Eerste Reeks, **40**, North-Holland, Amsterdam, 1993, pp. 179–194.

[AFK94a] D. Z. Arov, B. Fritzsche and B. Kirstein, On block completion problems for j_{qq}-J_q-inner functions. I. The case of a given block column. Integ. Equat. Oper. Th. **18** (1994), no. 1, 1–29.

[AFK94b] D. Z. Arov, B. Fritzsche and B. Kirstein, On block completion problems for j_{qq}-J_q-inner functions. II. The case of a given $q \times q$ block. Integ. Equat. Oper. Th. **18** (1994), no. 3, 245–260.

[AFK95] D. Z. Arov, B. Fritzsche and B. Kirstein, On some aspects of V. E. Katsnelson's investigations on interrelations between left and right Blaschke-Potapov products, in *Operator Theory and Boundary Eigenvalue Problems* (Vienna, 1993) (I. Gohberg and H. Langer, eds.) Oper. Theory Adv. Appl., **80**, Birkhäuser, Basel, 1995, pp. 21–41.

[AFK98] D. Z. Arov, B. Fritzsche and B. Kirstein, A function-theoretic approach to a parametrization of the set of solutions of a completely indeterminate matricial Nehari problem. Integ. Equat. Oper. Th. **30** (1998), no. 1, 1–66.

[AFK00] D. Z. Arov, B. Fritzsche and B. Kirstein, On a parametrization formula for the solution set of a completely indeterminate generalized matricial Carathodory-Fejr problem. Math. Nachr. **219** (2000), 5–43.

[ArG83] D. Z. Arov and L. Z. Grossman, Scattering matrices in the theory of extensions of isometric operators. (Russian) Dokl. Akad. Nauk SSSR **270** (1983), no. 1, 17–20.

[ArG94] D. Z. Arov and L. Z. Grossman, Scattering matrices in the theory of unitary extension of isometric operators. Math. Nachr. **157** (1992), 105–123.

[AKP05] D. Z. Arov, M. A. Kaaashoek and D. R. Pik, The Kalman-Yakubovich-Popov inequality for discrete time systems of infinite dimension. J. Operator Theory **55** (2006), no. 2, 393–438.

[ArK81] D. Z. Arov and M. G. Krein, The problem of finding the minimum entropy in indeterminate problems of continuation. (Russian) Funktsional. Anal. i Prilozhen. **15** (1981), no. 2, 61–64.

[ArK83] D. Z. Arov and M. G. Krein, Calculation of entropy functionals and their minima in indeterminate continuation problems. (Russian) Acta Sci. Math. (Szeged) **45** (1983), no. 1-16-4, 33–50.

[ArNu96] D. Z. Arov and M. A. Nudelman, Passive linear stationary dynamical scattering systems with continuous time. Integ. Equat. Oper. Th. **24** (1996), no. 1, 1–45.

[ArNu00] D. Z. Arov and M. A. Nudelman, A criterion for the unitary similarity of minimal passive systems of scattering with a given transfer function. (Russian) Ukran. Mat. Zh. **52** (2000), no. 2, 147–156; translation in Ukrainian Math. J. **52** (2000), no. 2, 161–172.

[ArNu02] D. Z. Arov and M. A. Nudelman, Conditions for the similarity of all minimal passive realizations of a given transfer function (scattering and resistance matrices). (Russian) Mat. Sb. **193** (2002), no. 6, 3–24; translation in Sb. Math. **193** (2002), no. 5-6, 791–810.

[ArR07a] D. Z. Arov and N. A. Rozhenko, $J_{p,m}$-inner dilations of matrix functions of the Carathéodory class that have pseudo-extension. (Russian) Algebra i Analiz **19** (2007), no. 3, 76–105.

[ArR07b] D. Z. Arov and N. A. Rozhenko, Passive impedance systems with losses of scattering channels. (Russian) Ukrain. Mat. Zh. **59** (2007), no. 5, 618–649.

[ArSi76] D. Z. Arov and L. A. Simakova, The boundary values of a convergent sequence of J-contractive matrix-valued functions. (Russian) Mat. Zametki **19** (1976), no. 4, 491–500.

[ArSt05] D. Z. Arov and O. J. Staffans, The infinite-dimensional continuous time Kalman-Yakubovich-Popov inequality. *The extended field of operator theory* (M. Dritschel, ed.) Oper. Theory Adv. Appl., **171**, Birkhäuser, Basel, 2007, pp. 37–72,

[ArSt07] D. Z. Arov and O. J. Staffans, State/signal linear time-invariant systems theory, Part IV: Affine Representations of discrete time systems. Complex Anal. Oper. Th. **1** (2007) 457–521.

[ArSt??] D. Z. Arov and O. J. Staffans, Bi-inner dilations and bi-stable passive scattering realizations of Schur class operator-valued functions. Integ. Equat. Oper. Th., in press.

[Ara97] Z. D. Arova, The functional model of J-unitary node with a given J-inner characteristic matrix function, Integ. Equat. Oper. Th. **28** (1997), 1–16.

[Ara00a] Z. D. Arova, J-unitary nodes with strongly regular J-inner characteristic matrix functions. Methods of Func. Analysis and Topology, **6** (2000), no. 3, 9–23.

[Ara00b] Z. D. Arova, On J-unitary regular J-inner characteristic functions in the Hardy class $H_2^{n\times n}$, in *Operator Theoretical Methods,* 12th International Conference on Operator Theory, Timisoara (Romania), 2000, pp. 29–38.

[Ara01] Z. D. Arova, On Livsic-Brodskii nodes with strongly regular J-inner characteristic matrix functions in the Hardy class, in *Recent Advances in Operator Theory and Related Topics,* **127**, Birkhäuser Verlag, Basel, 2001, pp. 83–97.

[Ara02] Z. D. Arova, *Conservative Linear Time Invariant Systems with Infinite Dimensional State Spaces and J-inner Transfer Matrix Functions*, PhD Thesis, Free University of Amsterdam, Submitted 2002.

[Ara05] Z. D. Arova, On H_2 strongly regular operator pairs and stability of semigroups of operators, Visnik Khark. Nac. Univ. **55** (2005), no. 711, 127–131.

[Art84] A. P. Artemenko, Hermitian-positive functions and positive functionals. I. (Russian) Teor. Funktsii Funktsional. Anal. i Prilozhen. **41** (1984) 3–16; Hermitian-positive functions and positive functionals. II. (Russian) Teor. Funktsii Funktsional. Anal. i Prilozhen. **42** (1984) 3–21.

[Ba75] J. A. Ball, Models for noncontractions, J. Math. Anal. Appl. **52** (1975) 235–254.

[BaC91] J. A. Ball and N. Cohen, de Branges-Rovnyak operator models and systems theory: a survey, in *Topics in Matrix and Operator Theory* (H. Bart, I. Gohberg and M. A. Kaashoek, eds.), Oper. Theory Adv. Appl., **50**, Birkhäuser, Basel, 1991, pp. 93–136.

[BGR91] J. A. Ball, I. Gohberg and L. Rodman, *Interpolation of rational matrix functions.* Oper. Theor. Adv. Appl. **45** Birkhäuser, Basel, 1990.

[BaR07] J. A. Ball and M. W. Raney, Discrete-time dichotomous well-posed linear systems and generalized Schur-Nevanlinna-Pick interpolation. Complex Anal. Oper. Theory **1** (2007), no. 1, 1–54.

[BEO00] L. Baratchart, A. Gombani, M. Olivi, Parameter determination for surface acoustic wave filters, Proceedings IEEE Conference on Decision and Control, Sydney, Australia, December 2000.

[BEGO07] L. Baratchart, P. Enqvist, A. Gombani and M. Olivi, Minimal symmetric Darlington synthesis. Math. Control Signals Systems **19** (2007), no. 4, 283–311.

[Bel68] V. Belevich, *Classical Network Theory.* Holden Day, San Francisco, 1968.

[Be48] A. Beurling, On two problems concerning linear transformations in Hilbert space. Acta Math. **81** (1949), 239–255.

[BoD98] V. Bolotnikov and H. Dym, On degenerate interpolation, entropy and extremal problems for matrix Schur functions. Integ. Equat. Oper. Th. **32** (1998), no. 4, 367–435.

[BoD06] V. Bolotnikov and H. Dym, On boundary interpolation for matrix valued Schur functions. Mem. Amer. Math. Soc., **181** (2006), no. 856.

[BoK06] V. Bolotnokov and A. Kheifets, A higher order analogue of the Carathéodory-Julia theorem. J. Funct. Anal. **237** (2006), no. 1, 350–371.

[Br63] L. de Branges, Some Hilbert spaces of analytic functions I. Trans. Amer. Math. Soc. **106** (1963), 445–668.

[Br65] L. de Branges, Some Hilbert spaces of analytic functions II. J. Math. Anal. Appl. **11** (1965), 44–72.

[Br68a] L. de Branges, *Hilbert Spaces of Entire Functions.* Prentice-Hall, Englewood Cliffs, 1968.

[Br68b] L. de Branges, The expansion theorem for Hilbert spaces of entire functions, in *Entire Functions and Related Parts of Analysis* Amer. Math. Soc., Providence, 1968, pp.

[BrR66] L. de Branges and J. Rovnyak, Canonical models in quantum scattering theory, in *Perturbation Theory and its Applications in Quantum Mechanics* (C. Wilcox, ed.) Wiley, New York, 1966, pp. 295–392.

[Bro72] M. S. Brodskii, *Triangular and Jordan Representations of Linear Operators.* Transl. Math Monographs, **32** Amer. Math. Soc. Providence, R.I., 1972.

[BulM98] A. Bultheel and K. Müller, On several aspects of J-inner functions in Schur analysis. Bull. Belg. Math. Soc. Simon Stevin **5** (1998), no. 5, 603–648.

[Bu67] J. P. Burg, Maximum entropy spectral analysis, Proceedings of the 37th meeting of the Society of Exploration Geophysicists, Oklahoma City, Oklahoma, 1967; reprinted in *Modern Spectrum Analysis* (ed., D. G. Childers) IEEE Press, New York, 1978, pp. 34–39.

[Bu75] J. P. Burg, *Maximum Entropy Spectral Analysis* PhD Dissertation, Dept. of Geophysics, Stanford University, Stanford, California, 1975.

[BGL01] C. J. Byrnes, T. T. Georgiou and A. Lindquist, A generalized entropy criterion for Nevanlinna-Pick interpolation with degree constraint. IEEE Trans. Automat. Control **46** (2001), no. 6, 822–839.

[Ch61] J. Chover, On normalized entropy and the extensions of a positive definite function. J. Math. Mech. **10**(1961), 927–945.

[Car67] T. Carleman, *L'intégrale de Fourier et Questions qui s'y Rattachent*, Almquist & Wiksells, Uppsala, 1967.

[Dar39] S. Darlington, Synthesis of reactance 4-poles which produce prescribed insertion loss characteristics. J. Math. Phys. **18** (1939), 257–355.

[Den06] Sergey A. Denisov, Continuous analogs of polynomials orthogonal on the unit circle and Krein systems. IMRS Int. Math. Res. Surv. 2006, Art. ID 54517.

[Dev61] A. Devinatz, The factorization of operator valued functions. Ann. of Math. **73** (1961), no. 2, 458–495.

[De71] P. Dewilde, *Roomy Scattering Matrix Synthesis*, Tech. Rep., Dept. of Mathematics, University of California at Berkeley, Berkeley, 1971.

[De76] P. Dewilde, Input-output description of roomy systems. SIAM J. Control Optimization **14** (1976), no. 4, 712–736.

[DvdV98] P. Dewilde and A. van der Veen, *Time-varying Systems and Computations.* Kluwer, Boston, MA, 1998.

[De99] P. Dewilde, Generalized Darlington synthesis. Darlington memorial issue. IEEE Trans. Circuits Systems I Fund. Theory Appl. **46** (1999), no. 1, 41–58.

[DBN71] P. Dewilde, V. Belevich, R. N. Newcomb, On the problem of degree reduction of a scattering matrix by factorization. J. Franklin Inst. **291** (1971), 387–401.

[DeD81a] P. Dewilde and H. Dym, Schur recursions, error formulas, and convergence of rational estimators for stationary stochastic sequences. IEEE Trans. Inform. Theory **27** (1981), no. 4, 446–461.

[DeD81b] P. Dewilde and H. Dym, Lossless chain scattering matrices and optimum linear prediction: the vector case. Internat. J. Circuit Theory Appl. **9** (1981), no. 2, 135–175.

[DeD84] P. Dewilde and H. Dym, Lossless inverse scattering, digital filters, and estimation theory. IEEE Trans. Inform. Theory **30** (1984), no. 4, 644–662.

[DF79] J. D. Dollard and C. N. Friedman, *Product integration with applications to differential equations.* With a foreword by Felix E. Browder. With an appendix by P. R. Masani. Encyclopedia of Mathematics and its Applications, **10**, Addison-Wesley Publishing Co., Reading, Mass., 1979.

[DoH73] R. G. Douglas and J. W. Helton, Inner dilations of analytic matrix functions and Darlington synthesis. Acta Sci. Math. (Szeged) **34** (1973), 61–67.

[DSS70] R. G. Douglas, H. Shapiro and A. Shields, Cyclic vectors and invariant subspaces for the backward shift operator. Ann. Inst. Fourier (Grenoble) **20** (1970) fasc. 1, 37–76.

[DFK92] V. K. Dubovoj, B. Fritsche and B. Kirstein, *Matricial Version of the Classical Schur Problem*, B. G. Teubner, Stuttgart, 1992.

[Du70] P. Duren, *Theory of H^p Spaces*, Academic Press, New York, 1970.

[Dy70] H. Dym, An introduction to de Branges spaces of entire functions with applications to differential equations of the Sturm-Liouville type. Adv. Math. **5** (1970), 395–471.

[Dy74] H. Dym, An extremal problem in the theory of Hardy functions. Israel J. Math. **18** (1974), 391–399.

[Dy89a] H. Dym, On reproducing kernel spaces, J unitary matrix functions, interpolation and displacement rank, in *The Gohberg anniversary collection* (H. Dym, S. Goldberg, M. A. Kaashoek and P. Lancaster, eds.), Vol. II (Calgary, AB, 1988), Oper. Theory Adv. Appl., **41**, Birkhäuser, Basel, 1989, pp. 173–239.

[Dy89b] H. Dym, *J-contractive matrix functions, reproducing Kernel Hilbert spaces and interpolation,* CBMS Regional Conference series, **71**, AMS, Providence, RI, 1989.

[Dy89c] H. Dym, On Hermitian block Hankel matrices, matrix polynomials, the Hamburger moment problem, interpolation and maximum entropy. Integ. Equat. Oper. Th. **12** (1989), no. 6, 757–812.

[Dy90] H. Dym, On reproducing kernels and the continuous covariance extension problem, in *Analysis and Partial Differential Equations: A Collection of Papers Dedicated to Mischa Cotlar* (C. Sadosky, ed.), Marcel Dekker, New York, 1990, pp. 427–482.

[Dy94a] H. Dym, Shifts, realizations and interpolation, redux. in *Nonselfadjoint operators and related topics* (A. Feintuch and I. Gohberg, eds.), (Beer Sheva, 1992), Oper. Theory Adv. Appl., **73**, Birkhuser, Basel, 1994, pp. 182–243.

[Dy94b] H. Dym, On the zeros of some continuous analogues of matrix orthogonal polynomials and a related extension problem with negative squares. Comm. Pure Appl. Math. **47** (1994), 207–256.

[Dy96] H. Dym, More on maximum entropy interpolants and maximum determinant completions of associated Pick matrices. Integ. Equat. Oper. Th. **24** (1996), no. 2, 188–229.

[Dy98] H. Dym, A basic interpolation problem, in *Holomorphic Spaces* (eds. S. Axler, J. E. McCarthy and D. Sarason), Cambridge U. Press, Cambridge 1998, pp. 381–423.

[Dy01a] H. Dym, On Riccati equations and reproducing kernel spaces, in *Recent advances in operator theory* (A. Dijksma, M. A. Kaashoek and A. C. M. Ran, eds.), (Groningen, 1998), Oper. Theory Adv. Appl., **124**, Birkhäuser, Basel, 2001, pp. 189–215.

[Dy01b] H. Dym, Reproducing kernels and Riccati equations. Mathematical theory of networks and systems (Perpignan, 2000). Int. J. Appl. Math. Comput. Sci. **11** (2001), no. 1, 35–53.

[Dy03a] H. Dym, Riccati equations and bitangential interpolation problems with singular Pick matrices, in *Fast algorithms for structured matrices: theory and applications* (South Hadley, MA, 2001), 361–391, Contemp. Math., **323**, Amer. Math. Soc., Providence, RI, 2003, pp. 361–391.

[Dy03b] H. Dym, Linear fractional transformations, Riccati equations and bitangential interpolation, revisited, in *Reproducing kernel spaces and applications* (D. Alpay, ed.), Oper. Theory Adv. Appl., **143**, Birkhäuser, Basel, 2003, pp. 171–212.

[Dy07] H. Dym, *Linear Algebra in Action.* Graduate Studies in Mathematics, **78**, Amer. Math. Soc., Providence, RI, 2007.

[DF97] H. Dym and B. Freydin, Bitangential interpolation for triangular operators when the Pick operator is strictly positive. in *Topics in interpolation theory* (Leipzig, 1994) (H. Dym, B. Frizsche, V. Katsnelson and B. Kirstein, eds.) Oper. Theory Adv. Appl., **95**, Birkhäuser, Basel, 1997, pp. 143–164.

[DG79] H. Dym and I. Gohberg, Extensions of matrix valued functions with rational polynomial inverses. Integ. Equat. Oper. Th. **2** (1979), no. 4, 503–528.

[DG80] H. Dym and I. Gohberg, On an extension problem, generalized Fourier analysis, and an entropy formula. Integ. Equat. Oper. Th. **3** (1980), no. 2, 143–215; addendum ibid, **3** (1980), no. 4, 603.

[DG81] H. Dym and I. Gohberg, Extensions of band matrices with band inverses. Lin. Alg. Appl. **36** (1981), 1–24.

[DG86] H. Dym and I. Gohberg, A maximum entropy principle for contractive interpolants. J. Funct. Anal. **65** (1986), no. 1, 83–125.

[DG88] H. Dym and I. Gohberg, A new class of contractive interpolants and maximum entropy principles, in *Topics in operator theory and interpolation* (I. Gohberg, ed.), Oper. Theory Adv. Appl., **29**, Birkhäuser, Basel, 1988, pp. 117–150.

[DG95] H. Dym and I. Gohberg, On maximum entropy interpolants and maximum determinant completions of associated Pick matrices. Integ. Equat. Oper. Th. **23** (1995), no. 1, 61–88.

[DI84] H. Dym and A. Iacob, Positive definite extensions, canonical equations and inverse problems, in *Topics in Operator Theory, Systems and Networks* (H. Dym and I. Gohberg, eds.), Oper. Theory Adv. Appl. **12**, Birkhäuser, Basel, 1984, pp. 141–240.

[DK78] H. Dym and N. Kravitsky, On the inverse spectral problem for the string equation. Integ. Equat. Oper. Th. **1/2** (1978), 270–277.

[DMc72] H. Dym and H.P. McKean, *Fourier Series and Integrals*, Academic Press, New York, 1972.

[DMc76] H. Dym and H.P. McKean, *Gaussian Processes, Function Theory, and the Inverse Spectral Problem*, Academic Press, New York, 1976; reprinted by Dover, New York, 2008.

[Dz84] M. M. Dzhrbashyan, Interpolation and spectral expansions associated with differential operators of fractional order. (Russian) Izv. Akad. Nauk Armyan. SSR Ser. Mat. **19** (1984), no. 2, 81–181.

[EfPo73] A. V. Efimov and V. P. Potapov, J-expanding matrix-valued functions, and their role in the analytic theory of electrical circuits. (Russian) Uspekhi Mat. Nauk **28** (1973), no. 1(169), 65–130.

[Fe98] A. Feintuch, *Robust Control Theory in Hilbert Space*, Springer, New York, 1998.

[FW71] P. A. Fillmore and J. P. Williams, On operator ranges. Advances in Math. **7** (1971), 254–281.

[Fi37] P. Finsler, Über das Vorkommen definiterund semidefiniterFormen in Scharen quadratischer Formen, Commentarii Math. Helvetici, **9** (1937), 188-192.

[FoFr90] C. P. Foias and A. E. Frazho, *The Commutant Lifting Approach to Interpolation Problems.* Oper. Theory Adv. Appl. **44**. Birkhäuser Verlag, Basel, 1990.

[FFGK98] C. P. Foias, A. E. Frazho and I. Gohberg and M. A. Kaashoek, *Metric Constrained Interpolation, Commutant Lifting and Systems.* Oper. Theory Adv. Appl. **100**, Birkhäuser Verlag, Basel, 1998.

[Fr87] B. A. Francis, *A course in H_∞ Control Theory.* Lecture Notes in Control and Information Sciences, **88**. Springer-Verlag, Berlin, 1987.

[Fu81] P. A. Fuhrmann, *Linear Systems and Operators in Hilbert Space.* McGraw-Hill International Book Co., New York, 1981.

[Ga81] J. B. Garnett, *Bounded Analytic Functions*, Academic Press, New York, 1981.

[GRS64] I. Gelfand, D. Raikov and G. Shilov, *Commutative normed rings.* Translated from the Russian, with a supplementary chapter. Chelsea Publishing Co., New York 1964.

[Gi57] Yu. P. Ginzburg, On J-contractive operator functions. (Russian) Dokl. Akad. Nauk SSSR (N.S.) **117** (1957) 171–173.

[Gi67] Yu. P. Ginzburg, Multiplicative representations and minorants of bounded analytic operator functions. (Russian) Funktsional. Anal. i Prilozhen. **1** 1967 no. 3, 9–23.

[GiZe90] Yu. P. Ginzburg and L. M. Zemskov, Multiplicative representations of operator-functions of bounded type. (Russian) Teor. Funktsii Funktsional. Anal. i Prilozhen. **53** (1990), 108–119; translation in J. Soviet Math. **58** (1992), no. 6, 569–576.

[GiSh94] Yu. P. Ginzburg and L. V. Shevchuk, On the Potapov theory of multiplicative representations, in *Matrix and Operator Valued Functions* (I. Gohberg and L. A. Sakhnovich eds.), Oper. Theory Adv. Appl. **72**, Birkhäuser, Basel, 1994, pp. 28–47.

[Gi97] Yu. P. Ginzburg, Analogues of a theorem of Frostman on linear fractional transformations of inner functions and the typical spectral structure of analytic families of weak contractions, in *Operator Theory, System Theory and Related Topics* (Beer-Sheva/Rehovot, 1997), Oper. Theory Adv. Appl., **123**, Birkhäuser, Basel, 2001, pp. 323–336.

[GLK05] I. Gohberg, P. Lancaster and L. Rodman, *Indefinite Linear Algebra and Applications.* Birkhäuser Verlag, Basel, 2005

[Gn37] B. V. Gnedenko, On characteristic functions, Mat. Bul. Moscov Univ., A. **1** (1937), 17–18.

[GnKo68] B. V. Gnedenko and A. N. Kolmogorov, *Limit distributions for sums of independent random variables* . Translated from the Russian, annotated, and revised by K. L. Chung. With appendices by J. L. Doob and P. L. Hsu. Revised edition Addison-Wesley Publishing Co., Reading, Mass.-London-Don Mills., Ont. 1968

[GKW91] I. Gohberg, M. A. Kaashoek and H. J. Woerdeman, A maximum entropy principle in the general framework of the band method. J. Funct. Anal. **95** (1991), no. 2, 231–254.

[GoM97] L. Golinskii and I Mikhailova, Hilbert spaces of entire functions as a J theory subject [Preprint No. 28–80, Inst. Low Temp. Phys. Engg., Kharkov, 1980]. Edited by V. P. Potapov. (Russian), translation in Oper. Theory Adv. Appl., **95**, *Topics in Interpolation Theory* (Leipzig, 1994) (H. Dym, B. Frizsche, V. Katsnelson and B. Kirstein, eds.) Birkhäuser, Basel, 1997, pp. 205–251.

[GoGo97] M. L. Gorbachuk and V. I. Gorbachuk, *M. G. Krein's lectures on entire operators.* Oper. Theory Adv. Appl., **97**, Birkhäuser Verlag, Basel, 1997.

[Gu00a] G. M. Gubreev, The structure of model Volterra operators, biorthogonal expansions, and interpolation in regular de Branges spaces. (Russian) Funktsional. Anal. i Prilozhen. **35** (2001), no. 2, 74–78; translation in Funct. Anal. Appl. **35** (2001), no. 2, 142–145.

[Gu00b] G. M. Gubreev, Spectral theory of regular and regular B-representative vector functions (projection method: 20 years after), Algebra and Anal. **12**, (2000), no. 6, 1–97.

[GuT03] G. M. Gubreev and A. A. Tarasenko, Representability of the de Branges matrix as the Blaschke-Potapov product and the completeness of some families of functions. (Russian) Mat. Zametki **73** (2003), no. 6, 841–847; translation in Math. Notes **73** (2003), no. 5–6, 796–801.

[GuT06] G. M. Gubreev and A. A. Tarasenko, Unconditional bases of de Branges spaces constructed from values of reproducing kernels. (Russian) Funktsional. Anal. i Prilozhen. **40** (2006), no. 1, 71–75; translation in Funct. Anal. Appl. **40** (2006), no. 1, 58–61.

[Ha61] P. R. Halmos, Shifts on Hilbert spaces. J. Reine Angew. Math. **208** 1961 102–112.

[He64] H. Helson, *Lectures on Invariant Subspaces.* Academic Press, New York, 1964.

[Hel74] J. W. Helton, Discrete time systems, operator models, and scattering theory. J. Functional Analysis **16** (1974) 15–38.

[KaKr74] I. S. Kac and M. G. Krein, R-functions-analytic functions mapping the upper halfplane into itself. Amer. Math. Soc. Transl. (2), **103** (1974), 1–18.

[Kai74] T. Kailath, A view of three decades of linear filtering theory. IEEE Trans. Information Theory IT-20 (1974) 146–181.

[Kal63a] R. E. Kalman, Mathematical description of linear dynamical systems. J. SIAM Control Ser. A 1 (1963) 152–192.

[Kal63b] R. E. Kalman, Lyapunov functions for the problem of Luré in automatic control. Proc. Nat. Acad. Sci. U.S.A. **49** (1963) 201–205.

[Kal65] R. E. Kalman, Irreducible realizations and the degree of a rational matrix. J. Soc. Indust. Appl. Math. **13** (1965) 520–544.

[Kat84a] V. E. Katsnelson, Regularization of the fundamental matrix inequality of the problem on the decomposition of a positive-definite kernel into elementary products. (Russian) Dokl. Akad. Nauk Ukrain. SSR Ser. A (1984), no. 3, 6–8.

[Kat84b] V. E. Katsnelson, Fundamental matrix inequality of the problem of decomposition of a positive definite kernel into elementary products. (Russian) Dokl. Akad. Nauk Ukrain. SSR Ser. A (1984), no. 2, 10–12.

[Kat85a] V. E. Katsnelson, Integral representation of Hermitian positive kernels of mixed type and the generalized Nehari problem. I. (Russian) Teor. Funktsii Funktsional. Anal. i Prilozhen. **43** (1985), 54–70; translation in J. Soviet Math. **48** (1990), no. 2, 162–176

[Kat85b] V. E. Katsnelson, *Methods of J-theory in Continuous Interpolation Problems of Analysis.* Part I. Translated from the Russian and with a foreword by T. Ando. Hokkaido University, Sapporo, 1985.

[Kat86] V. E. Katsnelson, Extremal and factorization properties of the radii of a problem on the representation of a matrix Hermitian positive function. I. (Russian) Math. Physics, Funct. Anal, (1986) 80–94, 146, Naukova Dumka, Kiev.

[Kat89] V. E. Katsnelson, A left Blaschke-Potapov product is not necessarily a right Blaschke-Potapov product. (Russian) Dokl. Akad. Nauk Ukrain. SSR Ser. A (1989), no. 10, 15–17, 86.

[Kat90] V. E. Katsnelson, Left and right Blaschke-Potapov products and Arov-singular matrix-valued functions. Integ. Equat. Oper. Th. **13** (1990), no. 6, 836–848.

[Kat93] V. Katsnelson, Weight spaces of pseodocontinuable functions and approximations by rational functions with prescribed poles. Z. Anal. Anwend. **12** (1993), 27–67.

[Kat94] V. Katsnelson, Description of a class of functions which admit an approximation by rational functions with preassigned poles, in *Matrix and Operator Valued Functions* (I. Gohberg and L. A. Sakhnovich eds.), Oper. Theory Adv. Appl. **72**, Birkhäuser, Basel, 1994, pp. 87–132.

[Kat95] V. E. Katsnelson, On transformations of Potapov's fundamental matrix inequality, in *Topics in interpolation theory* (H. Dym, B. Frizsche, V. Katsnelson and B. Kirstein, eds.) (Leipzig, 1994), Oper. Theory Adv. Appl. **95**, Birkhuser, Basel, 1997, pp. 253–281.

[KKY87] V. E. Katsnelson, A. Ya. Kheifets and P. M. Yuditskii, An abstract i nterpolation problem and the extension theory of isometric operators. (Russian) Operators in function spaces and problems in function theory (Russian), (1987) 83–96, 146, Naukova Dumka, Kiev; translation in *Topics in interpolation theory* (Lcipzig, 1994) (H. Dym, B. Frizsche, V. Katsnelson and B. Kirstein, eds.), Oper. Theory Adv. Appl. **95**, Birkhäuser, Basel, 1997, pp. 283–288.

[KK95] V. Katsnelson and B. Kirstein, On the theory of matrix-valued functions belonging to the Smirnov class, in *Topics in interpolation theory* (Leipzig, 1994)(H. Dym, B. Frizsche, V. Katsnelson and B. Kirstein, eds.), Oper. Theory Adv. Appl. **95**, Birkhäuser, Basel, 1997, pp. 299–350.

[Kh90] A. Kheifets, Generalized bitangential Schur-Nevanlinna-Pick problem and the related Parseval equality. (Russian), Teor. Funktsii Funktsional Anal. i Prilozhen. **54** (1990), 89–96.; translation in J. Sov. Math. **58** (1992), 358–364.

[Kh95] A. Kheifets, On regularization of C-generating pairs, J. Funct. Anal. **130** (1995), 310–333.

[Kh96] A. Kheifets, Hamburger moment problem: Parseval equality and A-singularity, J. Funct. Anal. **141** (1996), 374–420.

[Kh00] A. Kheifets, Parametrization of solutions of the Nehari problem and nonorthogonal dynamics, in *Operator Theory and Interpolation* (H. Bercovici and C. Foias, eds.) Oper. Theory Adv. Appl. **115**, Birkhäuser, Basel, 2000, pp. 213–233.

[Ko80] A. N. Kochubei, On extensions and characteristic functions of symmetric operators. (Russian) Izv. Akad. Nauk Armenian SSSR **15** (1980), no. 3, 219–232.

[Kov83] I. V. Kovalishina, Analytic theory of a class of interpolation problems. (Russian) Izv. Akad. Nauk SSSR Ser. Mat. **47** (1983), no. 3, 455–497.

[KoPo74] I. V. Kovalishina and V. P. Potapov, An indefinite metric in the Nevanlinna-Pick problem. (Russian) Akad. Nauk Armjan. SSR Dokl. **59** (1974), no. 1, 17–22.

[KoPo82] I. V. Kovalishiva and V. P. Potapov, *Integral representation of Hermitian positive functions.* Translated from the Russian by T. Ando, Hokkaido University, Sapporo, 1982.

[Kr40] M. G. Krein, Sur le problème du prolongement des fonctions hermitiennes positives et continues. (French) C. R. (Dokl.) Acad. Sci. URSS (N.S.) **26** (1940) 17–22.

[Kr44a] M. G. Krein, On the logarithm of an infinitely decomposible Hermite-positive function. C. R. (Dokl.) Acad. Sci. URSS (N.S.) **45** (1944) 91–94.

[Kr44b] M. G. Krein, On the problem of continuation of helical arcs in Hilbert space. C. R. (Dokl.) Acad. Sci. URSS (N.S.) **45** (1944) 139–142.

[Kr47a] M. G. Krein, On the theory of entire functions of exponential type, Izvestiya Akad. Nauk SSSR, Ser. Matem. **11** (1947) 309–326.

[Kr47b] M. G. Krein, Compact linear operators on functional spaces with two norms. Translated from the Ukranian Sb. Trudov Inst. Mat. Akad. Nauk Ukranian SSR (1947), 104–129; translation in Integ, Equat. Oper. Th. **30** (1998), no. 2, 140–162.

[Kr49] M. G. Krein, The fundamental propositions of the theory of representations of Hermitian operators with deficiency index (m, m). (Russian) Ukrain. Mat. Zh. **1** (1949), no. 2, 3–66.

[Kr51] M. G. Krein, On the theory of entire matrix functions of exponential type. (Russian) Ukrain. Mat. Zh. **3** (1951) 164–173.

[Kr54] M. G. Krein, On a method of effective solution of an inverse boundary problem. (Russian) Dokl. Akad. Nauk SSSR (N.S.) **94** (1954) 987–990.

[Kr55] M. G. Krein, Continuous analogues of propositions on polynomials orthogonal on the unit circle. (Russian) Dokl. Akad. Nauk SSSR (N.S.) **105** (1955) 637–640.

[Kr56] M. G. Krein, On the theory of accelerants and S-matrices of canonical differential systems. (Russian) Dokl. Akad. Nauk SSSR (N.S.) **111** (1956) 1167–1170.

[KrL85] M. G. Krein and Heinz Langer, On some continuation problems which are closely related to the theory of operators in spaces Π_κ. IV. Continuous analogues of orthogonal polynomials on the unit circle with respect to an indefinite weight and related continuation problems for some classes of functions. J. Oper. Th. **13** (1985), no. 2, 299–417.

[KrMA68] M.G Krein and F.E. Melik-Adamyan, On the theory of S-matrices of canonical differential systems with summable potential. Dokl. Akad. Nauk Armyan SSR. **46** (1968), no. 1, 150–155.

[KrMA70] M.G Krein and F.E. Melik-Adamyan, Some applications of theorems on the factorization of a unitary matrix, Funktsional. Anal. i Prilozhen. **4** (1970), no. 4, 73–75.

[KrMA84] M.G Krein and F.E. Melik-Adamyan, Integral Hankel operators and related continuation problems. Izv. Akad. Nauk Armyan SSR, Ser. Mat. **19** (1984), 311–332.

[KrMA86] M. G. Krein and F. E. Melik-Adamyan, Matrix-continuous analogues of the Schur and the Carathéodory-Toeplitz problem. (Russian) Izv. Akad. Nauk Armyan. SSR Ser. Mat. **21** (1986), no. 2, 107–141, 207.

[KrOv82] M. G. Krein and I. E. Ovcharenko, On the theory of inverse problems for the canonical differential equation. (Russian) Dokl. Akad. Nauk Ukrain. SSR Ser. A 1982, no. 2, 14–18; translation in *On the Theory of Inverse Problems for the Canonical Differential Equation, Matrix and Operator Valued Functions* (I. Gohberg and L. A. Sakhnovich, eds.), **72** Oper. Theory Adv. Appl., Birkhäuser, Basel, 1994, pp. 162–170.

[KrS70] M. G. Krein and S. N. Saakjan, The resolvent matrix of a Hermitian operator and the characteristic functions connected with it. (Russian) Funktsional. Anal. i Prilozhen **4** 1970, no. 3, 103–104.

[KrS96a] M. G. Krein and Ju. L. Shmuljan, On the plus operators in a space with indefinite metric. (Russian), Mat. Issled. **1** (1996), no. l, 131–161.

[KrS96b] M. G. Krein and Ju. L. Shmuljan, $\mathfrak{J}$-polar representations of plus-operators. (Russian) Mat. Issled. **1** (1966), no. 2, 172–210.

[KrS97] M. G. Krein and Ju. L. Shmuljan, Fractional linear transformations with operator coefficients. (Russian) Mat. Issled. **2** (1967), no. 3, 64–96.

[Ku96] A. Kuzhel, *Characteristic functions and models of nonselfadjoint operators.* Mathematics and its Applications, **349**. Kluwer , Dordrecht, 1996.

[Lan87] H. Landau, Maximum entropy and the moment problem. Bull. Amer. Math. Soc. (N.S.) **16** (1987), no. 1, 47–77.

[La59] P. D. Lax, Translation invariant subspaces, Acta Math. **101** (1959) 163–178.

[LiYa6] A. Lihtarnikov and V. A. Yakubovich, A frequency theorem for equations of evolution type. Sibirsk. Mat. Z. **17** (1976), no. 5, 1069–1085, 1198, translation in Sib. Math. J. **17** (1976) 790–803(1977).

[LiPa84] A. Lindquist and M. Pavon, On the structure of state-space models for discrete-time stochastic vector processes. IEEE Trans. Automat. Control **29** (1984), no. 5, 418–432.

[LiPi82] A. Lindquist and G. Picci, On a condition for minimality of Markovian splitting subspaces. Systems Control Lett. **1** (1981/82), no. 4, 264–269.

[LiPi85] A. Lindquist and G. Picci, Realization theory for multivariate stationary Gaussian processes. SIAM J. Control Optim. **23** (1985), no. 6, 809–857.

[Liv54] M. S. Livsic, On the spectral resolution of linear non-selfadjoint operators, Mat. Sbornik N.S. **34** (1954) 145–199; translation in Amer. Math. Soc. Transl. (2) **5** (1957) 67–114.

[Liv73] M. S. Livsic, *Operators, Oscillations, Waves. Open Systems.* Transl. Math. Monographs, **34**, Amer. Math. Soc., Providence, RI, 1973.

[Liv97] M. S. Livsic, The Blaschke-Potapov factorization theorem and the theory of non-selfadjoint operators, in *Topics in Interpolation Theory* (Leipzig, 1994) (H. Dym, B, Fritzsche, V. Katsnelson and B. Kirstein, eds.) Oper. Theory Adv. Appl., **95**, Birkhäuser, Basel, 1997, pp. 391–396.

[Me72] E. Ya. Melamud, A certain generalization of Darlington's theorem. (Russian) Izv. Akad. Nauk Armjan. SSR Ser. Mat. **7** (1972), no. 3, 183–195, 226.

[Me79] E. Ya. Melamud, Realization of positive matrix-valued functions according to Darlington. Polynomial realization in the case of degeneration. (Russian) Izv. Akad. Nauk Armyan. SSR Ser. Mat. **14** (1979), no. 4, 237–250, 314.

[MiPo81] I. V. Mikhailova and V. P. Potapov, A criterion for Hermitian positivity. (Russian) Dokl. Akad. Nauk Ukrain. SSR Ser. A (1981), no. 9, 22–27, 95.

[Mo93] Maria D. Morán, Unitary extensions of a system of commuting isometric operators, in *Operator Extensions, Interpolation of Functions and Related Topics* (A. Gheondea, D. Timotin, and F.-H. Vascilescu, eds.), (Timişoara, 1992), Oper. Theory Adv. Appl. **61**, Birkhäuser, Basel, 1993, pp. 163–169.

[MuGl90] D. Mustafa and K. Glover, *Minimum Entropy H_∞ Control.* Lecture Notes in Control and Information Sciences, **146**. Springer-Verlag, Berlin, 1990.

[Na76] S. N. Naboko, Absolutely continuous spectrum of a nondissipative operator, and a functional model. I, (Russian) in it Investigations on Linear Operators and the Theory of Functions, VII. Zap. Naučn. Sem. Leningrad. Otdel Mat. Inst. Steklov. (LOMI) **65** (1976), 90–102, 204–205.

[Na78] S. N. Naboko, Absolutely continuous spectrum of a nondissipative operator, and a functional model. II. (Russian) Investigations on linear operators and the theory of functions, VIII. Zap. Nauchn. Sem. Leningrad. Otdel. Mat. Inst. Steklov. (LOMI) **73** (1977) 118–135, (1978) 232–233.

[Ne57] Z. Nehari, On bounded bilinear forms. Ann. of Math. (2) **65** (1957) 153–162.

[Ni02] N. Nikolskii, *Operators, Functions, and Systems: an Easy Reading, I, II* Translated from the French by Andreas Hartmann and revised by the author. Math. Surveys and Monographs, **92**, **93**, Amer. Math. Soc., Providence, RI, 2002.

[Nu77] A. A. Nudelman, A new problem of the type of the moment problem. (Russian) Dokl. Akad. Nauk SSSR **233** (1977), no. 5, 792–795.

[Nu81] A. A. Nudelman, A generalization of classical interpolation problems. (Russian) Dokl. Akad. Nauk SSSR **256** (1981), no. 4, 790–793.

[Or76] S. A. Orlov, Nested matrix discs that depend analytically on a parameter, and theorems on the invariance of the ranks of the radii of the limit matrix discs. (Russian) Izv. Akad. Nauk SSSR Ser. Mat. **40** (1976), no. 3, 593–644, 710.

[PaW34] R. E. A. C. Paley and N. Wiener, *Fourier transforms in the complex domain.* Amer. Math. Society Colloq. Publications, **19**. Amer. Math. Soc., Providence, RI, 1934.

[Pa70] L. A. Page, Bounded and compact vectorial Hankel operators. Trans. Amer. Math. Soc. **150** (1970) 529–539.

[Pe03] V. V. Peller, *Hankel Operators and their Applications.* Springer-Verlag, New York, 2003.

[PI97] M. A. Peters and P. A. Iglesias, *Minimum Entropy Control for Time-varying Systems.* Systems & Control: Foundations & Applications. Birkhäser, Boston, 1997.

[Pik03] D. Pik, Time-variant Darlington synthesis and induced realizations, in Infinite-dimensional systems theory and operator theory (Perpignan, 2000). Int. J. Appl. Math. Comput. Sci. **11** (2001), no. 6, 1331–1360.

[Pi54] M. S. Pinsker, The quantity of information about a Gaussian random stationary process, contained in a second process connected with it in a stationary manner. (Russian) Dokl. Akad. Nauk SSSR (N.S.) **99** (1954) 213–216.

[Pi58] M. S. Pinsker, The extrapolation of random vector process and the quantity of information contained in a stationary random vector process relative to another one stationarily connected with it. Dokl. Akad. Nauk. SSSR. **121** (1958), no. 1, 49–51.

[Pi64] M. S. Pinsker, *Information and Information Stability of Random Variables and Processes.* Translated and edited by Amiel Feinstein. Holden-Day, San Francisco, 1964.

[Pop61] V. M. Popov, Absolute stability of nonlinear systems of automatic control. Avtomat. i Telemeh. **22** (1961) 961–979 (Russian), translation in Automat. Remote Control **22** (1961) 857–875.

[Pop73] V. M. Popov, Hyperstability of control systems. Translated from the Romanian by Radu Georgescu. Die Grundlehren der mathematischen Wissenschaften, **204**, Editura Academiei, Bucharest; Springer-Verlag, Berlin-New York, 1973.

[Po60] V. P. Potapov, The multiplicative structure of J-contractive matrix functions. Amer. Math. Soc. Transl. (2) **15** (1960) 131–243.

[Po88a] V. P. Potapov, A theorem on the modulus. I. Fundamental concepts. The modulus. (Russian) Teor. Funktsii Funktsional. Anal. i Prilozhen. **38** (1982), 91–101, 129. Amer. Math. Soc. Transl. (2) **138** (1988), 55–65.

[Po88b] V. P. Potapov, A theorem on the modulus. II. Amer. Math. Soc. Transl. (2) **138** (1988), 67–77.

[Re60] R. Redheffer, On a certain linear fractional transformation. J. Math. and Phys. **39** (1960) 269–286.

[Re62] R. Redheffer, On the relation of transmission-line theory to scattering and transfer. J. Math. and Phys. **41** (1962) 1–41.

[Re02] C. Remling, Schrödinger operators and de Branges spaces. J. Funct. Anal. **196** (2002), 323–394.

[RSzN55] F. Riesz and B. Sz.-Nagy, *Functional Analysis*, Ungar, New York, 1955; reprinted by Dover, New York, New York, 1990.

[RR85] M. Rosenblum and J. Rovnyak, *Hardy Classes and Operator Theory*, Oxford University Press, New York, 1985; reprinted by Dover, New York, 1997.

[RR94] M. Rosenblum and J. Rovnyak, *Topics in Hardy Classes and Univalent Functions*, Birkhäuser, Basel, 1994.

[RS02] W. T. Ross and H. S. Shapiro, *Generalized Analytic Continuation.* University Lecture Series, **25**. Amer. Math. Soc., Providence, RI, 2002.

[Rov68] J. Rovnyak, Characterizations of spaces $\mathbf{K}(M)$, unpublished manuscript, 1968; http://people.virginia.edu/ jlr5m/

[Roz58] Yu. A. Rozanov, Spectral theory of multi-dimensional stationary processes with discrete time, Uspekhi Mat. Nauk (N.S.) **13** (1958); translation in *Selected Translations in Mathematical Statistics and Probability*, **1**, Amer. Math. Soc. Providence RI, 1961, pp. 253–306.

[Roz60] Yu. A. Rozanov, Spectral properties of multivariate stationary processes and boundary properties of analytic matrices. (Russian) Teor. Verojatnost. i Primenen. **5** (1960) 399–414.

[Roz67] Yu. A. Rozanov, *Stationary Random Processes.* Translated from the Russian by A. Feinstein, Holden-Day, San Francisco, 1967

[Ru74] W. Rudin, *Real and complex analysis.* Second edition. McGraw-Hill, New York, 1974.

[Sai97] S. Saitoh, *Integral Transforms, Reproducing kKrnels and their Applications.* Pitman Research Notes in Mathematics Series, **369**. Longman, Harlow, 1997.

[Sak92] A. L. Sakhnovich, Spectral functions of a canonical system of order $2n$. Math. USSR Sbornik. **71** (1992), no. 2, 355–369.

[Sak93] L. A. Sakhnovich, The method of operator identities and problems in analysis. (Russian) Algebra i Analiz **5** (1993), no. 1, 3–80; translation in St. Petersburg Math. J. **5** (1994), no. 1, 1–69.

[Sak97] L. A. Sakhnovich, *Interpolation Theory and its Applications.* Mathematics and its Applications, **428**, Kluwer Academic Publishers, Dordrecht, 1997.

[Sak99] L. A. Sakhnovich, *Spectral Theory of Canonical Differential Systems. Method of operator identities.* Translated from the Russian manuscript by E. Melnichenko. Oper. Theory Adv. Appl. **107**, Birkhäuser Verlag, Basel, 1999.

[Sar67] D. Sarason, Generalized interpolation in H^∞. Trans. Amer. Math. Soc. **127** (1967) 179–203.

[Sar89] D. Sarason, Exposed points in H_1, in *The Gohberg anniversary collection* (H. Dym, S. Goldberg, M. A. Kaashoek and P. Lancaster, eds.), vol. II (Calgary, AB, 1988), Oper. Theory Adv. Appl., **41**, Birkhäuser, Basel, 1989, pp. 485–496.

[Sar94] D. Sarason, *Sub-Hardy Hilbert spaces in the unit disk*, Wiley, New York, 1994.

[Sch64] L. Schwartz, Sous-espaces hilbertiens d'espaces vectoriels topologiques et noyaux associs (noyaux reproduisants). (French) J. Analyse Math. **13** (1964) 115–256.

[Shm68] Ju. L. Shmuljan, Operator balls. Teor. Funktsii. Funktsional. Anal. i Prilozhen. **6** (1968), 68–81; translation in Integ. Equat. Oper. Th. **13** (1990), no. 6, 864–882.

[Sht60] A. V. Shtraus, Characteristic functions of linear operators. (Russian) Izv. Akad. Nauk SSSR. Ser. Mat. **24** (1960), no. 1, 43–74.

[Shv70] Ya. S. Shvartsman, Invariant subspaces of dissipative operators and the divisor of their characteristic functions. (Russian) Funktsional. Anal. i Prilozhen. **4** (1970), no. 4, 85–86.

[Si74] L. A. Simakova, Plus-matrix-valued functions of bounded characteristic. (Russian) Mat. Issled. **9** (1974), no. 2(32), 149–171, 252–253.

[Si75] L. A. Simakova, On meromorphic plus-matrix functions. Mat. Issled. **10** (1975), no. 1, 287–292.

[Si03] L. A. Simakova, On real and "symplectic" meromorphic plus-matrix functions and corresponding linear-fractional transformations. (Russian) Mat. Fiz. Anal. Geom. **10** (2003), no. 4, 557–568.

[Smi28] V. E. Smirnov, Sur la théorie des polynomes orthogonaux à une variable complexe, Zh. Leningr. Phis. Mat. **11** (1928), no. 1, 155–179.

[Smi32] V. E. Smirnov, Sur les formules de Cauchy et Green et quelques problèmes qui s'y rattachent, Izv. AN SSSR, Phis. Mat. **3** (1932), 338–372.

[St05] O. Staffans, *Well-posed Linear Systems.* Encyclopedia of Mathematics and its Applications, **103**. Cambridge University Press, Cambridge, 2005.

[St81] A. Stray, A formula by V. M. Adamjan, D. Z. Arov and M. G. Krein. Proc. Amer. Math. Soc. **83** (1981), no. 2, 337–340.

[SzNK56] Béla Sz.-Nagy and Adam Koranyi, Relations d'un problme de Nevanlinna et Pick avec la thorie des oprateurs de l'espace hilbertien. (French) Acta Math. Acad. Sci. Hungar. **7** (1956) 295–303.

[SzNF70] Béla Sz.-Nagy and Ciprian Foias, *Harmonic Analysis of Operators on Hilbert Space*, North Holland, Amsterdam, 1970.

[Tik02] A. S. Tikhonov, Inner-outer factorization of J-contractive-valued functions. in *Operator Theory and Related Topics*, Vol. II (M. A. Kaashoek, H. Langer and G. Popov, eds.), (Odessa, 1997), Oper. Theory Adv. Appl., **118**, Birkhäuser, Basel, 2000, pp. 405–415.

[Ti60] E. C. Titchmarsh, *The Theory of Functions* (Second Edition), Oxford University Press, London, 1960.

[TrV97] S. Treil and A. Volberg, Wavelets and the angle between past and future, J. Funct. Anal. **143** (1997) 269–308.

[TsS77] E. R. Tsekanovskii and Yu. L. Shmulyan, The theory of bi-extensions of operators on rigged Hilbert spaces. Unbounded operator colligations and characteristic functions, Russian Math. Surveys, **32** (1977), no. 5, 73–131.

[Tum66] G. Ts. Tumarkin, Description of a class of functions admitting an approximation by functions with preassigned poles. Izv. Akad. Nauk Armjan SSR. Ser Mat. **1** (1966), no 2, 85–109.

[Ve91a] V. F. Veselov, Regular factorizations of a characteristic function, and invariant subspaces of a nonselfadjoint operator. (Russian) in *Differential Equations. Spectral theory. Wave propagation* (Russian), Probl. Mat. Fiz. **13**, Leningrad. Univ., Leningrad, 1991, pp. 87–107, 306.

[Ve91b] V. F. Veselov, Regular factorizations of a characteristic function, and singular operator functions. (Russian) Funktsional. Anal. i Prilozhen. **25** (1991), no. 2, 58–60; translation in Funct. Anal. Appl. **25** (1991), no. 2, 129–131.

[WM57] N. Wiener and P. Masani, The prediction theory of multivariate stochastic processes. I. The regularity condition. Acta Math. **98** (1957) 111–150.

[WM58] N. Wiener and P. Masani, The prediction theory of multivariate stochastic processes. II. The linear predictor. Acta Math. **99** (1958) 93–137.

[Wi72a] J. C. Willems, Dissipative dynamical systems. I. General theory. Arch. Rational Mech. Anal. **45** (1972) 321–351.

[Wi72b] J. C. Willems, Dissipative dynamical systems. II. Linear systems with quadratic supply rates. Arch. Rational Mech. Anal. **45** (1972) 352–393.

[Yak62] V. A. Yakubovich, The solution of some matrix inequalities encountered in automatic control theory, Dokl. Akad. Nauk SSSR **143** (1962) 1304–1307.

[Yak74] V. A. Yakubovich, The frequency theorem for the case in which the state space and the control space are Hilbert spaces and its application in certain problems in the synthesis of optimal control. I. Sibirsk. Mat Z. **15** (1974) 639–668, 703, translation in Sib. Math. J. **15** (1974) 457–476 (1975).

[Yak75] V. A. Yakubovich, The frequency theorem for the case in which the state space and the control space are Hilbert spaces and its application in certain problems in the synthesis of optimal control. II. Sibirsk. Mat Z. **16** (1975), no. 5, 1081–1102, 1132, translation in Sib. Math. J. **16** (1975) 828–845 (1976).

[Za41] V. N. Zasukhin, On the theory of multidimensional stationary random processes. C. R. (Doklady) Acad. Sci. URSS (N.S.) **33** (1941) 435–437.

[Zol03] V. A. Zolotarev, *Analytical Methods of Spectral Representations of Nonselfadjoint and Nonunitary Operators* (Russian) Kharkov, Kharkov National University, 2003

Notation index

Subject index